Contemporary Cryptology
The Science of Information Integrity

Edited by
Gustavus J. Simmons
Sandia National Laboratories

IEEE
PRESS

The Institute of Electrical and Electronics Engineers, Inc., New York

IEEE PRESS
445 Hoes Lane, PO Box 1331
Piscataway, NJ 08855-1331

©1992 by the Institute of Electrical and Electronics Engineers, Inc.
345 East 47th Street, New York, NY 10017-2394

Printed in the United States of America
10 9 8 7 6 5 4 3 2 1

ISBN   0-87942-277-7
IEEE Order Number: PC0271-7

Library of Congress Cataloging-in-Publication Data
Contemporary cryptology : the science of information integrity /
edited by Gustavus J. Simmons.
    p.   cm.
Includes bibliographical references and index.
ISBN 0-87942-277-7
1. Computer security. 2. Telecommunication systems—Security
measures. 3. Cryptography. I. Simmons, Gustavus J.
QA76.9.A25C6678   1992
005.8'2—dc20
                                        91–19684
                                        CIP

Contents

SECTION 1   CRYPTOGRAPHY   41

SECTION 3   PROTOCOLS   421

SECTION 4   CRYPTANALYSIS   499

SECTION 5   APPLICATIONS   559

Contemporary Cryptology
A Foreword

Cryptology (from the Greek kryptós, "hidden," and lógos, "word") has come to be understood to be the science of secure (often interpreted to mean secret) communications. Although secrecy is certainly an important element to the security or integrity of information, it is only one element, as demonstrated by the contributing authors of this book. Information integrity is also concerned with questions of authenticity, authority, concurrence, timeliness, etc., as well as with all the problems normally addressed by documentary records. The intent of the editor and of the authors in putting this book together was to treat the subject of information integrity as comprehensively as possible—with special emphasis on those questions of information integrity whose resolution is primarily cryptographic in nature.

As the most casual reader of the technical literature, or even of the popular press, must be aware, an enormous amount of public activity in the field of cryptology has occurred during the past decade and a half. This has been marked by the appearance of several fundamental new ideas such as two-key (also public key or asymmetric) cryptography, provably secure protocols whose security is derived from mathematical problems of classifiable complexity, interactive proof systems and zero-knowledge protocols, etc., and, of course, by the widespread recognition of an urgent need for means to provide for the integrity of information in all phases of our information-intensive society [1]. This perceived need is the driving force responsible for much of the public activity.

The conduct of commerce, affairs of state, military actions, and personal affairs all depend on the parties to a transaction having confidence in there being means of accomplishing such functions as privacy, proof of identity, authority, ownership, license, signature, witnessing or notarization, date of action, certification of origination and/or receipt, etc. As a result, an elaborate, and legally accepted, collection of procedural and physical protocols have evolved that specify how to create records (information in documentary form) in such a way that later disputes as to who is liable, or of the nature of that liability, or of when a liability was incurred, etc., can be arbitrated by a third party (typically in a court of law). The essential point is that existing precedent depends on information having a physical existence in the form of a document which may have been signed, witnessed, notarized, recorded, dated, etc.

The "proof" process, if it must be invoked, depends almost entirely on the physical instrument(s) as the means for establishing the integrity of the recorded information. In an information-intensive society however, in which the possession, control, transfer, or access to real assets is frequently based on incorporeal information—that is, information whose existence is not essentially linked to any physical record, and in which a license (to use, modify, copy, etc., valuable or sensitive information) is similarly determined, it is essential that means be found to carry out all of the functions associated with establishing the integrity of information mentioned above based only on the internal evidence present in the information itself, since this is the only thing available. Table 1 lists several of the more common information integrity functions; a complete list would be much longer. All of these functions are mentioned in one or more of the chapters that make up this book. Some of them—such as authentication, digital signatures and shared capability—even have full chapters devoted to them.

TABLE 1 A PARTIAL LIST OF COMMON INFORMATION
INTEGRITY FUNCTIONS

- Identification
- Authorization
- License and/or certification
- Signature
- Witnessing (notarization)
- Concurrence
- Liability
- Receipts
- Certification of origination and/or receipt
- Endorsement
- Access (egress)
- Validation
- Time of occurrence
- Authenticity—software and/or files
- Vote
- Ownership
- Registration
- Approval/disapproval
- Privacy (secrecy)

For example, there are many applications that need or even require a digital signature for digital information that would serve all the purposes now served by a handwritten signature to a document. There is no single technical means of solution to these problems, and, as a matter of fact, it remains an open question as to whether some of them even have feasible or legally acceptable solutions. There is a common element, however, to the solution to many of them, and that is cryptography or more precisely, crypto-like transformations on the information whose integrity is to be insured. These are the technical means that make it possible for one or more parties who know a private piece (or pieces) of information to carry out an operation on the information which (probably) cannot be duplicated by someone not "in the know." The advantage or knowledge gained by being able to do this varies from application to application. In some cases it may be as simple as being granted access or entry to an automated teller machine (ATM) or to a remote computer or data bank; in others it may be the ability to conceal information or to recover hidden (encrypted) information, or it may be as complex as being able to "prove" to impartial third parties the culpability of a treaty signatory who has violated the terms of a treaty.

Simply put, information integrity is about how to prevent cheating, or failing that, to detect cheating in information-based systems wherein the information itself has no meaningful physical existence. Because there are so many different objectives for cheating where information is concerned, the subject of information integrity, and hence for the application of cryptographic principles, is consequently very broad. For example, the cheater may wish to impersonate some other participant in the system, or to eavesdrop on communications between other participants, or to intercept and modify information being communicated between other users of the system. The cheater may be an insider who either wishes to disavow communications that he actually originated or to claim to have received messages that were not sent. He may wish to enlarge his license to gain access to information that he has some level of authorized access for, or to subvert the system to alter (without authorization) the access license of others. The point is that since information can be enormously valuable or critical *so can its misuse.* Consequently, information integrity is concerned with devising means for either preventing or detecting all forms of cheating that depend on tampering with the information in information-based systems, where the means depend only on the information itself for their realization as distinguished from other noninformation-dependent means such as documentary records, physical security, etc.

Unless the reader has wrestled with real-world problems of protecting critical information from would-be cheaters, he is probably unaware of the gamut of reasons for cheating in information-based systems. Table 2 lists some of the more obvious reasons for cheating, each of which has arisen in one or more real-world situations. Not all these reasons have cryptographic solutions, but many do, and of these, most are discussed in one or more of the chapters in this book.

As mentioned earlier, the solution to problems of this type depends on the availability of operations (or transformations) on the information that is feasible for one or more participants in an information-based protocol to carry out because they know some private piece(s) of additional information, but which are (probably) impossible to do without knowing the private information. We will adopt this viewpoint to introduce the papers that make up *Contemporary Cryptology.*

In classical cryptography (secret key cryptography in the terminology used by Massey in his chapter of this book, "Contemporary Cryptology: An Introduction," or single-key cryptography in the terminology used by Brickell, Diffie, Moore, Odlyzko, and Simmons) there is only a single piece of private and necessarily secret information—the key—known to and used by the originator to encrypt information into a cipher and also known to and used by the intended recipient to decrypt the cipher. It is this operation of encryption and/or decryption that is assumed to (probably) be impossible to carry out without a knowledge of the secret key.

TABLE 2 REASONS FOR CHEATING

 1. Gain unauthorized access to information, i.e., violate secrecy or privacy.
 2. Impersonate another user either to shift responsibility, i.e., liability, or else to use his license for the purpose of:
    a. originating fraudulent information,
    b. modifying legitimate information,
    c. utilizing fraudulent identity to gain unauthorized access,
    d. fraudulently authorizing transactions or endorsing them.
 3. Disavow responsibility or liability for information the cheater did originate.
 4. Claim to have received from some other user information that the cheater created, i.e., fraudulent attribution of responsibility or liability.
 5. Claim to have sent to a receiver (at a specified time) information that was not sent (or was sent at a different time).
 6. Either disavow receipt of information that was in fact received, or claim a false time of receipt.
 7. Enlarge his legitimate license (for access, origination, distribution, etc.).
 8. Modify (without authority to do so) the license of others (fraudulently enroll others, restrict or enlarge existing licenses, etc.).
 9. Conceal the presence of some information (a covert communication) in other information (the overt communication).
10. Insert himself into a communications link between other users as an active (undetected) relay point.
11. Learn who accesses which information (sources, files, etc.) and when the accesses are made (even if the information itself remains concealed), i.e., a generalization of traffic analysis from communications channels to data bases, software, etc.
12. Impeach an information integrity protocol by revealing information the cheater is supposed to (by the terms of the protocol) keep secret.
13. Pervert the function of software, typically by adding a convert function.
14. Cause others to violate a protocol by means of introducing incorrect information.
15. Undermine confidence in a protocol by causing apparent failures in the system.
16. Prevent communication among other users, in particular surreptitious interference to cause authentic communications to be rejected as unauthentic.

In public key cryptography, there are two pieces of information, at least one of which is computationally infeasible to recover from a knowledge of the other. One is the private piece of information (key) used by the originator to encrypt the information whose integrity is to be secured and the other is the private information (key) used by a recipient to decrypt the resulting ciphers. Depending on the application, both of these pieces of information need not be kept secret.

If it is computationally infeasible to recover the decryption key from the encryption key, then the encryption key need not be kept secret in order to insure the secrecy of the encrypted information using it. It must, however, be protected against substitution and/or modification, otherwise the transmitter could be deceived into encrypting information using a bogus encryption key for which the matching decryption key is known to an opponent (cheater). The decryption key must, of course, be kept secret and be physically secured against substitution and/or modification to insure the secrecy of the information concealed in the ciphers. This is the *secrecy channel*.

Conversely, if it is computationally infeasible to recover the encryption key from the decryption key, then the decryption key need not be kept secret. In this case, if a cipher, when decrypted, contains authenticating information (previously agreed on by the authorized transmitter or originator of the information and the intended recipients), then it was in all probability generated by the purported originator. This is the *authen-*

tication channel. The separation of these two functions by virtue of the separation of the two pieces of information needed to carry out the two complementary operations of encryption and decryption is the essential concept involved in public key cryptography, whose genesis is recounted by its inventor, Whitfield Diffie, in the chapter, "The First Ten Years of Public Key Cryptography."

One might at first think that this is the end of the process—that is, that having separated encryption and decryption and having put a computationally infeasible-to-overcome barrier between the pieces of information needed to carry them out, that nothing more is possible. To see this is not the case, one needs only to examine the list of reasons for cheating tabulated in Table 2. For example, if the party that is supposed to physically protect and keep secret the encryption key for an authentication channel, either deliberately or inadvertently allows it to be compromised (an example of deception #12 in Table 2), it then becomes impossible for an arbiter to establish who originated a cipher, even though the cipher contains the expected authenticating information. This example also illustrates the essential difference between actual signatures and digital signatures but more importantly it illustrates the first step in a natural "taxonomy of trust" in information integrity schemes described in detail in the chapter by Simmons, "A Survey of Information Authentication."

For commercial and private applications, probably the most important single information integrity function is a means to create digital signatures. As pointed out earlier, digital signatures differ in a critical respect from handwritten signatures because the author of a handwritten signature cannot transfer the ability to utter his signature to another party—no matter how great the desire to do so—while all that needs to be done to transfer the ability to utter the digital signature is to share the private piece of information used to generate it. Signature protocols can be devised to deal with this problem, reducing the likelihood of an attempted deception either being successful or else going undetected. Mitchell, Piper, and Wild provide a comprehensive treatment of the technical aspects of this topic in their chapter, "Digital Signatures." Because the applications for signatures (handwritten and digital) have to do with liability, concurrence, ownership, records, etc., all of which have legal implications, there is an evolving area of law concerned with the legal status and acceptability of digital signatures. A deliberate decision was made to limit the discussion here to the technical questions associated with creating digital signatures; however the reader should be aware that there are equally important, nontechnical issues.

In single-key cryptography, the transmitter and receiver have no choice but to trust each other unconditionally since either is capable of doing anything the other can. In the case of two-key cryptography only one specified participant (which can be either the transmitter or the receiver) must be assumed to be unconditionally trustworthy. The other participant is unable to carry out (some) actions that the other can, which means that the participant does not have to be trusted to not impersonate the other party insofar as those actions are concerned because he is not capable of doing so. But there are many applications in which no participant is *a priori* unconditionally trustworthy. It may, however, be reasonable to assume that some (unknown) elements in the system are trustworthy. Applications of this sort are discussed in the chapter "How to Insure That Data Acquired to Verify Treaty Compliance Are Trustworthy" by Simmons. As shown in that chapter, in order to prevent a unilateral action by one of the participants making it impossible to logically arbitrate disputes between mutually deceitful and distrusting parties, it becomes necessary for the operations on the information to depend on three or more separate (but related) private pieces of information, all of which are necessary to correctly carry out the operations. The underlying idea is simple: to separate functional capability by separating the additional information needed to carry out the operations on the information, and then to give these separate pieces of information privately to the various participants in the protocol; in some cases to enable them to work cooperatively to carry out an operation, and in other cases to individually verify the authenticity of operations carried out by other participants. The logical extension of this notion of requiring the participation of three parties in order to carry out operations on information is to require the concurrence of specified—but arbitrary—subsets of the participants in order to do so. These concepts are discussed in detail in the chapter "An Introduction to Shared Secret and/or Shared Control Schemes and Their Application" by Simmons.

Cryptographic systems are commonly classified into block and stream ciphers—a rather artificial classification based on the size of the objects to which the cryptographic transformation is applied. If the objects are single symbols (normally an alphabetic or numeric character) the system is called a stream cipher, while if the object is made up of several symbols, the system is said to be a block cipher. In the second case, blocks could be considered to be symbols from a larger symbol alphabet, however the distinction between stream and block ciphers is a useful device when considering such problems as error propagation, synchronization, and especially of the achievable communication data rates and delays.

The search for, and often the catastrophic consequences of a failure to find, secure cryptoalgorithms for military and diplomatic applications, although conducted in great secrecy at the time, is well known up to a period shortly after World War II [2]. The development of algorithms for public use, however, has been carried out in full public view. The origins and subsequent development of what is certainly the best known, and arguably the most widely used, single-key cryptoalgorithm in history, the Data Encryption Standard (DES), are recounted by two of the principals in its adoption as a federal standard: Branstad and Smid, in their chapter "The Data Encryption Standard: Past and Future." An unusual aspect of this algorithm is that from its inception every detail of the DES operation has been public knowledge, an attribute common to almost all of the algorithms that have been the subject matter of the recent activity in cryptology.

Stream ciphers are of great practical importance, especially in applications where high data rates are required (secure video for example), and in which minimal communication delay is important. A comprehensive treatment of this subject is given by Rueppel in "Stream Ciphers." It should be pointed out that most fielded single-key secure communications technology is based on stream ciphers.

Diffie, in his chapter, "The First Ten Years of Public Key Cryptography," describes in detail the several attempts to devise secure two-key cryptoalgorithms and the gradual evolution of a variety of protocols based on them. A comprehensive treatment of this cornerstone of contemporary cryptology is given by Nechvatal in the chapter "Public Key Cryptography." Brickell and Odlyzko describe the efforts to disprove (or prove) the security of these schemes. Their chapter, "Cryptanalysis: A Survey of Recent Results," is the first compilation and cohesive presentation of the exciting sequence of cryptographic proposals and cryptanalytic breaks that have characterized public cryptology in the past decade, by two of the main contributors to those cryptanalytic successes. Rather than being discouraged by the cryptanalytic successes described there, one should be encouraged by the emergence of algorithms and protocols whose security can be shown to be as "good" as some hard mathematical problem is difficult to solve, which is a new development in the science of cryptology. It is only the intense scrutiny and combined efforts of an active public research community that has brought this about. The bottom line is that after a decade and a half of effort, there are available acceptably secure single-key and two-key cryptoalgorithms in a variety of VLSI implementations whose operation is well understood and widely known. van Oorschot, in his chapter, "A Comparison of Practical Public Key Cryptosystems Based on Integer Factorization and Discrete Logarithms" gives a very thorough comparison of the relative merits of the principle contenders for two-key cryptoalgorithms—both from the algorithmic standpoint and from the efficiency of their best VLSI implementations to date. These comparisons (of apples and oranges to be sure) should be invaluable to a system designer faced with a choice among several algorithms, an even larger number of implementations, and of competing security, speed and protocol requirements.

As Massey points out in his chapter, "Contemporary Cryptography: An Introduction," even after suitable crypto-like operations have been devised, there still remain substantial cryptographic problems to be solved. How do the participants get the private pieces of information they need to perform their functions in the protocol and how can they be guaranteed of the integrity of what they receive? In an oversimplified form, this is the key distribution problem that was one of the stimuli for the discovery of public key cryptography (see the description by Diffie of the reasoning process that led to this discovery). The underlying problem, though, is broader than simple-key distribution and is concerned with the entire question of how a participant in an information-based protocol can trust his part of the protocol and hence the soundness of the protocol itself, even though he cannot trust any of the other participants or the communications channel (data bank, software, etc.) from which the information is acquired. In its simplest form, this may reduce to how a user can be confident that his personal identification number (PIN) cannot be learned by someone at a financial institution and used to (undetectably) impersonate the user, or it may be as complex as how a participant can trust a nondeterministic, interactive, protocol between himself and a collection of other participants in which individual responses are complex functions of all of the prior responses, some of which are random, and in which the user must assume the other participants will collude to deceive or defraud him.

Even if one has a secure crypto-like operation or algorithm, and a trustworthy (that is, secure) means of distributing the private pieces of information to the participants, there is yet another way in which an information-based system can fail. These are protocol failures, discussed in the pioneering paper (and reprinted as a chapter in this book) "Protocol Failures in Cryptosystems" by Moore. Obviously, if the way private information is distributed in a protocol allows a compromise of information that should be kept secret, the cryptoalgorithm is broken, or if a collection of insiders can pool their private pieces of information to recover information that is supposed to be kept secret from them, then ordinary cryptanalysis may be possible. These sorts of failures, although potentially devastating to the integrity of a protocol, are not surprising, nor is their prevention particularly interesting. The cases of interest are those in which the intended function of the overall system or protocol can be defeated, even though the underlying cryptoalgorithm remains secure against cryptanalysis—that is, the system failure does not come about as a result of breaking the cryptoalgorithm. In a sense, protocol failures are a result of cryptanalysis at the system level instead of at the algorithm level. As Moore makes clear through several examples, this type of failure occurs by exploiting information in unexpected ways. It is important, therefore, for understanding how cheating can occur in information-based systems, and hence, for understanding how to prevent cheating, to realize that information can be passed from one part of a protocol to another by a variety of channels other than the intended overt one.

One of the reasons the popular press has been so attracted by developments in contemporary cryptology is that many of the problems appear to be impossible to solve—making their solutions seem paradoxical. For example, problems such as how to make a single cipher mean different things to different people or how to conceal information in a cipher so that even someone who knows the cryptographic key used to produce the cipher will be unable to detect the presence of the concealed information have been solved. Other examples of seemingly impossible problems that also have been solved are how to authenticate a message even though nothing about the message can be kept secret from the very persons who wish to create fraudulent messages that would be accepted as authentic, or how to communicate securely despite the fact that none of the parties to the communication can be trusted. Perhaps the most paradoxical result of all is how one participant can prove to another that he knows a particular piece of information without revealing the information itself, and indeed without revealing anything about it that would aid someone else in pretending to know it. These protocols, which have formed the basis for a number of schemes for proof of identity, are introduced and discussed by Feigenbaum in her chapter "Overview of Interactive Proof Systems and Zero-Knowledge." Even after the concept is explained, the results still seem paradoxical. Interactive proof systems and zero-knowledge protocols are prototypes illustrating the impact of theoretical computer science on contemporary cryptology.

Nonspecialists are surrounded by transparent instances of information integrity schemes of the sort described here. They regularly identify themselves to ATMs, share access control to their safety deposit boxes with the institution, rely on the integrity of credit card numbers containing a low-level of security self-authenticating capability, etc. Less transparent examples are code-controlled scramblers on cable and/or satellite TV broadcasts, security for telephones (ranging from simple—and not very secure—analog schemes to the STU III NSA certified secure telephone units) etc. Almost everyone has daily contact with some of these information-integrity schemes; however, there is a new area of information technology that promises to eventually replace the ubiquitous plastic credit cards: smart cards. Smart cards that draw on several information-integrity technologies (cryptography, proof of identity, authentication, etc.) are described in the chapter, "Smart Card: A Standardized Security Device Dedicated to Public Cryptology" by three of the prime movers in their development: Guillou, Quisquater, and Ugon. This application will put a sophisticated information-integrity device in the wallet or purse of practically every person in the industrialized world, and will therefore probably be the most extensive application ever made of cryptographic schemes.

Finally, we note that our initial motivation in putting this book together and our concluding observation are the same; namely, that given the social, commercial, and personal importance of being able to protect information against all forms of information-based cheating, and given the apparent essential dependence of solutions to this class of problems on crypto-like transformations, it is desirable that computer scientists, communications engineers, systems designers, and others who may need to provide for the integrity of information and, of course, the ultimate end users who must depend on the integrity of information, be acquainted with the essential concepts and principles of cryptography. The authors and the editor wish to thank the IEEE PRESS for their support in the publication of *Contemporary Cryptology* to satisfy this need.

REFERENCES

[1] G. J. Simmons, "Cryptology," in *Encyclopedia Britannica,* 16th Edition. Chicago, IL: Encyclopedia Britannica Inc., pp. 913–924B, 1986.
[2] D. Kahn, *The Codebreakers.* New York: Macmillan, 1967 (abridged edition, New York: New American Library, 1974).

Contemporary Cryptology
An Introduction

JAMES L. MASSEY
Institute for Signal and Information Processing
Swiss Federal Institute of Technology
Zürich, Switzerland

An appraisal is given of the current status, both technical and nontechnical, of cryptologic research. The principal concepts of both secret-key and public key cryptography are described. Shannon's theory of secrecy and Simmons's theory of authenticity are reviewed for the insight that they give into practical cryptographic systems. Public key concepts are illustrated through consideration of the Diffie–Hellman public key-distribution system and the Rivest–Shamir–Adleman public key cryptosystem. The subtleties of cryptographic protocols are shown through consideration of some specific such protocols.

1 PRELIMINARIES

1.1 Introduction

That cryptology is a "hot" research area hardly needs saying. The exploits of cryptographic researchers are reported today not only in an increasing number of scholarly journals and popular scientific magazines, but also in the public press. One hears of conflicts between cryptologic researchers and government security agencies, insinuations of built-in "trapdoors" in commonly used ciphers, claims about new ciphers that would take millions of years to break and counterclaims that no cipher is secure—all the stuff of high drama. To ferret out the truth in such controversies, one needs a basic understanding of cryptology, of its goals and methods, and of its capabilities and limitations. The aim of this chapter is to provide a brief, self-contained introduction to cryptology that may help the reader to reach such a basic understanding of the subject, and that may give him or her additional insight into the more specialized papers on cryptology that form the rest of this book.

The present chapter is an updated, expanded, and slightly revised version of our earlier paper [45], large sections of which appear virtually unchanged herein. The reader who is familiar with this earlier paper may wish to concentrate his or her attention on the new material that appears in this one. Reference numbers [45] and onward denote references added to the earlier report and their appearance will flag such a reader's attention to the substantially new segments of this chapter.

Only scant attention will be given in this chapter to the long and rich history of cryptology. For an excellent short history, the reader is referred to that given in a splendid earlier survey of cryptology [1] or that in an unusually penetrating encyclopedia article [2]. But Kahn's voluminous history, *The Codebreakers* [3], is indispensable to anyone who wishes to dig deeply into cryptologic history. The abridged paperback edition [4] of Kahn's book can be especially recommended as it packs as much suspense as the best spy fiction has to offer, but will also satisfy the historical curiosity of most readers.

1.2 Cryptologic Nomenclature and Assumptions

The word *cryptology* stems from Greek roots meaning ''hidden'' and ''word,'' and is the umbrella term used to describe the entire field of secret communications. For instance, the 8-year-old scientific society formed by researchers in this field is appropriately called the International Association for Cryptologic Research.

Cryptology splits rather cleanly into two subdivisions: cryptography and cryptanalysis. The cryptographer seeks to find methods to ensure the secrecy and/or authenticity of messages. The cryptanalyst seeks to undo the former's work by breaking a cipher or by forging coded signals that will be accepted as authentic. The original message on which the cryptographer plies his art is called the plaintext message, or simply the *plaintext;* the product of his labors is called the ciphertext message, or just the *ciphertext* or, most often, the *cryptogram*. The cryptographer always employs a *secret key* to control the enciphering process. Often (but not always) the secret key is delivered by some secure means (e.g., in an attaché case handcuffed to the wrists of a courier) to the person (or machine) to whom he expects later to send a cryptogram formed using that key.

The almost universal assumption of cryptography is that the enemy cryptanalyst has full access to the cryptogram. Almost as universally, the cryptographer adopts the precept, first enunciated by the Dutchman A. Kerckhoff (1835–1903), that the security of the cipher must reside entirely in the secret key. Equivalently, *Kerckhoff's assumption* is that the entire mechanism of encipherment, except for the value of the secret key, is known to the enemy cryptanalyst. If the cryptographer makes only these two assumptions, then he is designing the system for security against a *ciphertext-only attack* by the enemy cryptanalyst. If the cryptographer further assumes that the enemy cryptanalyst will have acquired (''by hook or by crook'') some plaintext-cryptogram pairs formed with the actual secret key, then he is designing against a *known-plaintext attack*. The cryptographer may even wish to assume that the enemy cryptanalyst can submit any plaintext message of his own and receive in return the correct cryptogram for the actual secret key (a *chosen-plaintext attack*), or to assume that the enemy cryptanalyst can submit purported ''cryptosystems'' and receive in return the unintelligible garble to which they (usually) decrypt under the actual key (a *chosen-ciphertext attack*), or to assume both of these possibilities (a *chosen-text attack*). Most cipher systems in use today are intended by their designers to be secure against at least a chosen-plaintext attack, even if it is hoped that the enemy cryptanalyst will never have the opportunity to mount more than a ciphertext-only attack.

1.3 The Need for Cryptology

Cryptography has been used for millenia to safeguard military and diplomatic communications. Indeed, the obvious need for cryptography in the government sector led to the rather general acceptance, until quite recently, of cryptography as a prerogative of government. Most governments today exercise some control of cryptographic apparatus if not of cryptographic research. The United States, for instance, applies the same export/import controls to cryptographic devices as to military weapons. But the dawning of the Information Age revealed an urgent need for cryptography in the private sector. Today vast amounts of sensitive information such as health and legal records, financial transactions, credit ratings, and the like are routinely exchanged between computers via public communication facilities. Society turns to the cryptographer for help in ensuring the privacy and authenticity of such sensitive information.

While the need for cryptography in both the government and private sectors is generally accepted, the need for cryptanalysis is less well acknowledged. "Gentlemen do not read each other's mail," was the response of U.S. Secretary of State H. L. Stimson in 1929 on learning that the U.S. State Department's "Black Chamber" was routinely breaking the coded diplomatic cables of many countries. Stimson forthwith abolished the Black Chamber, although as secretary of war in 1940 he relented in his distaste of cryptanalysis enough to condone the breaking of Japanese ciphers [4, p. 178]. In today's less innocent world, cryptanalysis is generally regarded as a proper and prudent activity in the government sector, but as akin to keyhole-peeping or industrial espionage in the private sector. However, even in the private sector, cryptanalysis can play a valuable and ethical role. The "friendly cryptanalyst" can expose the unsuspected weaknesses of ciphers so that they can be taken out of service or their designs remedied. A paradigm is Shamir's recent breaking of the Merkle–Hellman trapdoor-knapsack public key cryptosystem [5]. By publishing his ingenious cryptanalysis [6] of this clever and very practical cipher, Shamir forestalled its likely adoption in practice with subsequent exposure to the attacks of cryptanalysts seeking rewards more tangible than scientific recognition. Shamir's reward was the 1986 IEEE W. R. G. Baker Award.

In the preceding discussion, we abided by the long-accepted attribution of the dogmatic pronouncement, "Gentlemen do not read each other's mail," to H. L. Stimson in 1929. Kruh [46] has recently given a convincing historical argument suggesting that these famous words may in fact have been uttered by Stimson first in 1946, rather than 1929, during his interviews with McGeorge Bundy, who was then preparing Stimson's authorized biography [47]. Kruh [46, p. 80] concludes, "It thus seems highly likely that Stimson's 1946 remark accurately described his motivation for closing the Cipher Bureau in 1929. But whether he also said it then remains unknown."

1.4 Secret and Open Cryptologic Research

If one regards cryptology as the prerogative of government, one accepts that most cryptologic research is conducted behind closed doors. Without doubt, the number of workers engaged today in such secret research in cryptology far exceeds that of those engaged in open research in cryptology. For only about 15 years has there in fact been widespread open research in cryptology. There have been, and will continue to be, conflicts between these two research communities. Open research is a common quest for knowledge that depends for its vitality on the open exchange of ideas via conference presentations and publications in scholarly journals. But can a government agency, charged with the responsibility of breaking the ciphers of other nations, countenance publication of a cipher that it could not break? Can a researcher in good conscience publish such a cipher that might undermine the effectiveness of his own government's code-breakers? One might argue that publication of a provably secure cipher would force all governments to behave like Stimson's "gentlemen," but one must be aware that open research in cryptology is fraught with political and ethical considerations of a severity much greater than in most scientific fields. The wonder is not that some conflicts have occurred between government agencies and open researchers in cryptology, but rather that these conflicts (at least those of which we are aware) have been so few and so mild.

One can even argue that the greatest threat to the present vigorous open cryptologic research activity in the United States stems not from the intransigence of government but rather from its largesse. A recent U.S. government policy will require governmental agencies to rely on cryptographic devices at whose heart are tamper-proof modules incorporating secret algorithms devised by the National Security Agency (NSA) and loaded with master keys distributed by NSA [7]. Moreover, NSA will make these modules available to certified manufacturers for use in private-sector cryptography, and will presumably also supply the master keys for these applications. If, as appears likely, these systems find widespread acceptance in the American private sector, it will weaken the practical incentive for further basic open research in cryptography in the United States. The main practical application for such research will be restricted to international systems where the NSA technology will not be available.

1.5 Epochs in Cryptology

The entire period from antiquity until 1949 can justly be regarded as the *era of prescientific cryptology;* which is not to say that the cryptologic history of these times is devoid of interest today, but rather that cryptology was then plied almost exclusively as an art rather than as a science. Julius Caesar wrote to Cicero and his other friends in Rome more than 2000 years ago, employing a cipher in which each letter in the plaintext was replaced by the third (cyclically) later letter in the Latin alphabet [4, p. 77]. Thus, the plaintext CAESAR would yield the ciphertext FDHVDU. Today, we would express Caesar's cipher as

$$y = x \oplus z \tag{1}$$

where x is the plaintext letter ($A = 0, B = 1, \ldots, Z = 25$), z is the secret key (which Julius Caesar always chose as 3—Caesar Augustus chose 4), y is the ciphertext letter, and $\oplus$ here denotes addition modulo 26 (so that $23 \oplus 3 = 0$, $23 \oplus 4 = 1$, etc.). There is no historical evidence to suggest that Brutus broke Caesar's cipher, but a schoolchild today, who knew a little Latin and who had read the elementary cryptanalysis described in Edgar Allen Poe's masterful short story, "The Gold Bug," would have no difficulty succeeding in a ciphertext-only attack on a few sentences of ciphertext. In fact, for the next almost two thousand years after Caesar, the cryptanalysts generally had a clear upper hand over the cryptographers. Then, in 1926, G. S. Vernam, an engineer with the American Telephone and Telegraph Company, published a remarkable cipher to be used with the binary Baudot code [8]. Vernam's cipher is similar to Caesar's in that it is described by Eq. (1), except that now x, y, and z take values in the binary alphabet $\{0, 1\}$ and $\oplus$ denotes addition modulo 2 ($0 \oplus 0 = 0$, $0 \oplus 1 = 1$, $1 \oplus 1 = 0$). The new idea advanced by Vernam was to *use the key only one time*, that is, to encipher each bit of plaintext with a new randomly chosen bit of key. This necessitates the secure transfer of as much secret key as one will later have plaintext to encipher, but it yields a truly unbreakable cipher as we shall see below. Vernam indeed believed that his cipher was unbreakable and was aware that it would not be so if the randomly chosen key bits were to be reused later, but he offered no proofs of these facts. Moreover, he cited in [8] field tests that had confirmed the unbreakability of his cipher, something no amount of field testing could in fact confirm. Our reason for calling the period up to 1949 the prescientific era of cryptology is that cryptologists then generally proceeded by intuition and "beliefs," which they could not buttress by proofs. It was not until the outbreak of World War II, for instance, that the English cryptologic community recognized that mathematicians might have a contribution to make to cryptology [9, p. 148] and enlisted among others, A. Turing, in their service.

The publication in 1949 by C. E. Shannon of the paper, "Communication Theory of Secrecy Systems" [10], ushered in the era of scientific secret key cryptology. Shannon, educated both as an electrical engineer and mathematician, provided a theory of secrecy systems almost as comprehensive as the theory of communications that he had published the year before [11]. Indeed, he built his 1949 paper on the foundation of the 1948 one, which had established the new discipline of information theory. Shannon not only proved the unbreakability of the random Vernam cipher, but also established sharp bounds on the required amount of secret key that must be transferred securely to the intended receiver when any perfect cipher is used.

For reasons that will become clear in the sequel, Shannon's 1949 work did not lead to the same explosion of research in cryptology that his 1948 report had triggered in information theory. The real explosion came with the publication in 1976 by W. Diffie and M. E. Hellman of their work, "New Directions in Cryptography" [12]. Diffie and Hellman showed for the first time that secret communications was possible without any transfer of a secret key between sender and receiver, thus establishing the turbulent *epoch of public key cryptography* that continues unabated today. R. C. Merkle, who had submitted his paper about the same time as Diffie and Hellman but to another journal, independently introduced some of the essential ideas of public key cryptography. Unfortunately, the long delay in publishing his work [13] has often deprived him of due scientific credit. A detailed first hand account by W. Diffie of this formative period for public cryptography is given in Chapter 3, "The First Ten Years of Public Key Cryptography," in this volume.

1.6 Plan of This Chapter

In the next section, we review briefly the theory of secret key cryptography, essentially following Shannon's original approach and making Shannon's important distinction between theoretic and practical security. We also indicate the directions of some contemporary research in secret key cryptography. Section 3 gives a short exposition of public key cryptography, together with a description of some of the most important public key systems thus far advanced. In Section 4 we touch on the delicate subject of crypto-graphic protocols, and show how cryptographic techniques can be used to accomplish nonstandard, but very useful, tasks.

2 SECRET KEY CRYPTOGRAPHY

2.1 Model and Notation

By a secret key cryptosystem, we mean a system that corresponds to the block diagram of Fig. 1. The essential feature of such a system is the "secure channel" by which the

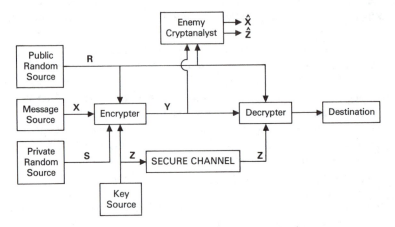

Figure 1   Model of a secret key cryptosystem.

secret key, $Z = [Z_1, Z_2, \ldots, Z_K]$, after generation by the *key source*, is delivered to the intended receiver, protected from the prying eyes of the enemy cryptanalyst. To emphasize that the same secret key is used by both the encrypter and decrypter, secret key cryptosystems have also been called *one-key cryptosystems* and *symmetric crypto-systems*. The K digits of the key are letters in some finite alphabet that we will often choose to be the binary alphabet $\{0, 1\}$. The *message source* generates the plaintext, $X = [X_1, X_2, \ldots, X_M]$. The private random source (whose purpose will soon be evi-dent) generates the *private randomizer*, $S = [S_1, S_2, \ldots, S_J]$, and the public random source (whose purpose will be seen later) generates the *public randomizer*, $R = [R_1, R_2, \ldots, R_T]$. The encrypter forms the cryptogram, $Y = [Y_1, Y_2, \ldots, Y_N]$, as a function of X, R, S, and Z. We write this encrypting transformation as

$$Y = E_{ZRS}(X) \tag{2}$$

to underscore the fact that it is useful to think of the cryptogram Y as a function of the plaintext X with the particular function being specified by the values of the secret key Z and of the randomizing sequences R and S. As Fig. 1 implies, the decrypter must be able to invert this transformation without knowledge of the *private* randomizing sequence **S**. That is,

$$X = D_{ZR}(Y) \tag{3}$$

which expresses the fact that the plaintext X must be a function of the cryptogram Y where the particular function is determined only by the secret key Z and the public randomizer R. The enemy cryptanalyst observes the cryptogram Y and the public randomizer R but nothing else. The enemy cryptanalyst then forms an estimate $\hat{X}$ of the plaintext X and/or an estimate $\hat{Z}$ of the secret key Z. The enemy cryptanalyst, in accordance with Kerckhoff's precept, is assumed to know all details of the encrypter and decrypter, but of course to have no knowledge of X, S, and, in particular, of Z.

Our Fig. 1 differs from the "schematic of a general secrecy system" that appears as Fig. 1 in Shannon's 1949 paper [10] only in that we have included a private and a public randomizer in our model.

Private randomization is an old cryptographic trick. In English text the letter e appears much more frequently than any other letter. If English text is first converted into text in some larger alphabet by replacing e each time with a randomly chosen letter from the large "e-group" of letters in the larger alphabet, and similarly replacing other frequently chosen English letters with random choices of a letter from appropriately sized groups in the larger alphabet, one obtains a new text in which all letters of the larger alphabet have (approximately) the same frequency. Enciphering of this randomized text frustrates a single-letter frequency analysis by the enemy cryptanalyst. But, after deciphering the randomized text, the legitimate receiver can remove the randomization merely by replacing each letter in the e-group of the larger alphabet by the letter e, and so on—such a reader does not need to be told in advance which random substitutions would be made. Such randomized ciphers are known as "multiple-substitution ciphers" and also as "homophonic ciphers." The great mathematician, Gauss, deceived himself into believing that, by using homophonic substitution, he had devised an unbreakable cipher [2]; but, without question, private randomization is a useful cryptographic tool. We will see later that the newer cryptographic trick of using a public randomizer can be even more powerful in enhancing the security of a cryptographic system. For these reasons and because their inclusion scarcely complicates Shannon's theory of secrecy, we have included both types of randomizers in our Fig. 1.

It is important to recognize that X, Z, R, and S are *random quantities*. The statistics of the plaintext X are of course determined by the message source, but the statistics of the secret key Z and of the randomizing sequences R and S are under the control of the cryptographer. As Fig. 1 suggests, we shall always assume that the random quantities X, Z, R, and S are statistically independent.

2.2 Theoretical and Practical Security

Shannon considered two very different notions of security for cryptographic systems. He first considered the question of *theoretical security*, by which he meant, "How secure is a system against cryptanalysis when the enemy has unlimited time and manpower available for the analysis of intercepted cryptograms?" [10, p. 658]. Shannon's theory of theoretical security, which we shall review next, casts much light into cryptography, but leads to the pessimistic conclusion that the amount of secret key needed to build a theoretically secure cipher will be impractically large for most applications. Thus, Shannon also treated the question of *practical security*, by which he meant: Is the system secure against a cryptanalyst who has a certain limited amount of time and computational power available for the analysis of intercepted cryptograms? Public key systems, to be discussed in Section 3, are intended to provide practical security—they cannot provide theoretical security.

2.3 Perfect Secrecy

The first assumption in Shannon's theory of theoretical security is that the secret key will be used only one time, or equivalently that the M digits of the plaintext X form the total of messages that will be enciphered before the secret key Z and the randomizers R and S are changed. Because the enemy cryptanalyst observes only the cryptogram Y and the public randomizer R, it is appropriate, following Shannon [10], to define *perfect secrecy* to mean that the plaintext X is statistically independent of the pair Y and R, that is, that

$$P_{X \mid YR}(x \mid y, r) = P_X(x)$$

holds for all x, y, and r. This is the same as saying that the enemy cryptanalyst can do no better estimating X with knowledge of Y and R than could be done in the absence of this knowledge, no matter how much time and computing power the enemy cryptanalyst has at his disposal. Having made the right mathematical formulation of the problem, it was then child's play for Shannon to show that perfect secrecy systems exist.

Consider the case of a nonrandomized cipher in which the plaintext, ciphertext, and key digits all takes values in the L-ary alphabet $\{0, 1, \ldots, L - 1\}$, and in which the length K of the key and length N of the cryptogram coincide with the length M of the plaintext, that is, $K = N = M$. Suppose that the key is chosen to be *completely random*, that is, $P(Z = z) = L^{-M}$ for all L^M possible values z of the secret key, and that the enciphering transformation is

$$Y_i = X_i \oplus Z_i, \qquad i = 1, 2, \ldots, M \qquad (4)$$

where $\oplus$ denotes addition mod L. Because for each possible choice x_i and y_i of X_i and Y_i, respectively, there is a unique z_i such that $Z_i = z_i$ satisfies Eq. (4), it follows that $P(Y = y \mid X = x) = L^{-M}$ for every possible particular y and x, no matter what the statistics of X may be. Thus X and Y are statistically independent, and hence this *modulo-L Vernam system* (to use Shannon's terminology) provides perfect secrecy. The modulo-L Vernam system is better known under the name, the *one-time pad*, from its use shortly before, during, and after World War II by spies of several nationalities who were given a pad of paper containing the randomly chosen secret key and told that it could be used for only one encipherment. There appears to have been a general belief in cryptographic circles that this cipher was unbreakable, but Shannon seems to have been the first to publish a proof of this theoretical unbreakability.

It is worth noting here that the one-time pad offers perfect secrecy no matter what the statistics of the plaintext X may be. In fact, we will show shortly that it also uses the least possible amount of secret key for any cipher that provides perfect secrecy independent of the statistics of the plaintext—this is a most desirable attribute; one would not usually wish the security of the cipher system to depend on the statistical nature of the message source. But the fact that the one-time pad requires one digit of secret key for each digit of plaintext makes it impractical in all but the few cryptographic applications, such as encrypting the Moscow–Washington hotline, where the need for secrecy is paramount and the amount of plaintext is quite limited.

We have learned recently from a reliable source that the Washington–Moscow hotline is no longer encrypted with a one-time pad, but that in its stead a conventional secret-key cipher that requires much less key is used. This change is apparently the result of increased confidence within the closed cryptographic community in the security of the secret key ciphers at their disposal.

2.4 Key Requirements for Perfect Secrecy

To go further in the study of theoretical security, we need to make use of some properties of "uncertainty" (or "entropy"), the fundamental quantity in Shannon's information theory [11]. *Uncertainty* is always defined as the mathematical expectation of the negative logarithm of a corresponding probability distribution. For instance, $H(X \mid Y)$ (which should be read as "the uncertainty about X given knowledge of Y") is the expectation of the negative logarithm of $P_{X \mid Y}(X \mid Y)$, that is,

$$H(X \mid Y) = \sum_{xy \, \varepsilon \, \text{supp} \, (P_{XY})} P_{XY}(x,y) \, (-log \, P_{X \mid Y}(x \mid y))$$

where supp (P_{XY}) denotes the set of all x, y such that $P_{XY}(x, y) \neq 0$. (The reason that in information theory one takes an expectation by summing only over the *support* of the joint probability distribution of the random variables involved is that this permits one to deal with the expectation of functions such as $-log \, P_{X \mid Y}(x \mid y)$ that can take on the values $-\infty$ or $+\infty$.) Uncertainties obey intuitively pleasing rules, such as $H(X, Y) = H(X) + H(Y \mid X)$, which we will use in our discussion of theoretical secrecy without further justification—the reader is referred to [11] or to the introductory chapters of any standard textbook on information theory for proofs of the validity of these "obvious" manipulations of uncertainties.

Equations (2) and (3) can be written equivalently in terms of uncertainties as

$$H(Y \mid X,Z,R,S) = 0 \tag{5}$$

and

$$H(X \mid Y,R,Z) = 0 \tag{6}$$

respectively, because, for instance, $H(Y \mid X,Z,R,S)$ is zero if and only if X,Z,R, and S together uniquely determine Y. Shannon's definition of perfect secrecy can then be written as

$$H(X \mid Y,R) = H(X) \tag{7}$$

since this equality holds if and only if X is statistically independent of the pair Y and R.

For any secret key cryptosystem, one has

$$H(X \mid Y,R) \leq H(X,Z \mid Y,R)$$
$$= H(Z \mid Y,R) + H(X \mid Y,R,Z)$$

$$= H(Z \mid Y, R)$$
$$\leq H(Z) \tag{8}$$

where we have made use of Eq. (6) and of the fact that the removal of given knowledge can only increase uncertainty. If the system gives perfect secrecy, it follows from Eqs. (7) and (8) that

$$H(Z) \geq H(X) \tag{9}$$

Inequality [Eq. (9)] is *Shannon's fundamental bound for perfect secrecy; the uncertainty of the secret key must be at least as great as the uncertainty of the plaintext that it is concealing.* If the K digits in the key are chosen from an alphabet of size L_z, then

$$H(Z) \leq \log (L_z^K) = K \log L_z \tag{10}$$

with equality if and only if the key is completely random. Similarly,

$$H(X) \leq M \log L_x \tag{11}$$

(where L_x is the size of the plaintext alphabet) with equality if and only if the plaintext is completely random. Thus, if $L_x = L_z$ (as in the one-time pad) and if the plaintext is completely random, Shannon's bound [Eq. (9)] for perfect secrecy yields, with the aid of Eq. (10) and of equality in Eq. (11),

$$K \geq M \tag{12}$$

That is, the key must be at least as long as the plaintext, a lower bound that holds with equality for the one-time pad.

2.5 Breaking an Imperfect Cipher

Shannon also considered the question of when the enemy cryptanalyst would be able in theory to break an imperfect cipher. To this end, he introduced the *key equivocation function*

$$f(n) = H(Z \mid Y_1, Y_2, \ldots, Y_n) \tag{13}$$

which measures the uncertainty that the enemy cryptanalyst has about the key given that he has examined the first n digits of the cryptogram. Shannon then defined the *unicity distance u* as the smallest n such that $f(n) \approx 0$. Given u digits of the ciphertext and not before, there will be essentially only one value of the secret key consistent with $Y_1, Y_2, \ldots, Y_n$, so it is precisely at this point that the enemy cryptanalyst with unlimited time and computing power could deduce the secret key and thus break the cipher. Shannon showed for a certain well-defined "random cipher" that

$$u \approx \frac{H(Z)}{r \log L_y} \tag{14}$$

where

$$r = 1 - \frac{H(X)}{N \log L_y} \tag{15}$$

is the *percentage redundancy* of the message information contained in the N digit cryptogram, whose letters are from an alphabet of size L_y. When $N = M$ and $L_x = L_y$ (as is true in most cryptosystems), r is just the percentage redundancy of the plaintext itself, which is about $\frac{3}{4}$ for typical English text. When $L_x = L_z$ and the key is chosen completely at random to maximize the unicity distance, Eq. (14) gives

$$u \approx \frac{K}{r} \qquad (16)$$

Thus, a cryptosystem with $L_x = L_y = L_z$ used to encipher typical English text can be broken after only about $N = \frac{4}{3}K$ ciphertext digits are received. For instance, a secret key of 56 bits (8 American Standard Code for Information Interchange [ASCII] 7-bit symbols) can be found in principle from examination of only about 11 ASCII 7-bit symbols of ciphertext.

Although Shannon's derivation of Eq. (14) assumes a particular kind of "random" cipher, he remarked "that the random cipher analysis can be used to estimate equivocation characteristics and the unicity distance for the ordinary types of ciphers" [10, p. 698]. Wherever it has been possible to test this assertion of Shannon's, it has been found to be true. Shannon's approximation [Eq. (14)] is routinely used today to estimate the unicity distance of "ordinary" secret key ciphers.

The reader may well be worrying about the validity of Eqs. (14) and (16) when $r = 0$, as it would in the case when $N = M$, $L_x = L_y$, and the message source emitted completely random plaintext so that $H(X) = M \log L_x = N \log L_y$. The answer is somewhat surprising: The enemy cryptanalyst can never break the system ($u = \infty$ is indeed the correct unicity distance!), even if $K \ll M$ so that Eq. (12) tells us that the system does not give perfect secrecy. The resolution of this paradox is that perfect secrecy demands that Y provide no information at all about X, whereas breaking the system demands that Y determines X essentially uniquely, that is, that Y must provide the maximum possible information about X. If the secret key Z were also chosen completely at random in the cipher for the completely random message source described above, there would always be L_z^K different plaintext-key pairs consistent with any possible cryptogram y, and all would be equally likely alternatives to the hapless cryptanalyst. This suggests, as Shannon was quick to note, that *data compression is a useful cryptographic tool*. An ideal data compression algorithm transforms a message source into the completely random (or "nonredundant") source that we have just been considering. Unfortunately, no one has yet devised a data compression scheme for realistic sources that is both ideal and practical (nor is anyone ever likely to do so), but even a nonideal scheme can be used to decrease r significantly, and thus to increase the unicity distance u significantly. Experience had long ago taught cryptographers that redundancy removal was a useful trick. In the days when messages were hand-processed, cryptographers would often delete from the plaintext many letters and blanks that could be recognized as missing and be replaced by the legitimate receiver. THSISASIMPLFORMOFDATACOMPRESION.

Shannon's derivation of Eq. (14) assumed a cryptographic system without the two randomizers that we have included in our Fig. 1. When a private randomizer S is included in the system, then $H(X)$ in Eq. (15) must be replaced by the joint uncertainty $H(X,S)$ for Eq. (14) still to hold. This suggests that randomization can also be used to reduce the redundancy r in the cryptogram. This, too, old-time cryptographers had learned from experience. They frequently inserted extra symbols into the plaintext, often an X, to hide the real statistics of the message. THXISISAXNEXAMXPLE.

Homophonic substitution, which was discussed in Section 2.1, is also a method for using a private randomizer to reduce the redundancy r in the cryptogram. Günther [48] quite recently suggested an ingenious variant of homophonic substitution in which the substitutes for a single plaintext letter are binary strings of varying length. Günther showed that it is possible to make the redundancy of the ciphertext *exactly* zero while at the same time making only a modest expansion in the number of binary digits needed to represent the plaintext. Jendahl, Kuhn, and Massey [49] modified Günther's scheme to achieve the minimum possible expansion of the plaintext and showed that, on the average, less than 4 bits of a completely random binary private randomizer suffice to determine the homophonic string for replacing each plaintext letter (whether or not the plaintext alphabet is also binary). What keeps both of these schemes from achieving zero redundancy in practice (and hence from yielding unbreakable practical ciphers) is that both schemes require complete and exact knowledge of the plaintext statistics, something that is never available for real information sources. However, both schemes can make use of available partial knowledge of the plaintext statistics (such as knowledge of the statistics of single letters, pairs of letters, and triplets of letters) to reduce greatly the redundancy r of the cryptogram and hence to increase greatly the unicity distance of an "ordinary" cipher.

2.6 Authenticity and Deception

We have mentioned several times that cryptography seeks to ensure the secrecy and/or authenticity of messages. But it is in fact quite a recent realization that secrecy and authenticity are independent attributes. If one receives a cryptogram that decrypts under the actual secret key to a sensible message, cannot one be sure that this cryptogram was sent by one's friend who is the only other person privy to this secret key? The answer, as we shall see, in general is: No! The systematic study of authenticity is the work of G. J. Simmons [14], who has developed a theory of authenticity that in many respects is analogous to Shannon's theory of secrecy.

To treat the theoretical security of authenticity systems as formulated by Simmons, we must give the enemy cryptanalyst more freedom than he is allowed in the model of Fig. 1. Figure 2 shows the necessary modification to Fig. 1. The enemy cryptanalyst is now the one who originates the "fraudulent" cryptogram $\tilde{Y}$ that goes to the decrypter. The line from the decrypter to the destination is shown dotted in Fig. 2 to suggest that the decrypter might recognize $\tilde{Y}$ as fraudulent and thus not be deceived into passing a fraudulent plaintext $\tilde{X}$ to the destination. The authentic cryptogram Y is shown on a dotted input line to the enemy cryptanalyst in Fig. 2 to suggest that the latter may have to form his fraudulent cryptogram $\tilde{Y}$ without ever seeing the authentic cryptogram itself.

As did Shannon, Simmons assumes that the secret key Z will be used only one time, that is, to form only one authentic cryptogram Y. But Simmons recognized that even in this case, three quite different attacks need to be distinguished. First, the enemy may of necessity form a fraudulent cryptogram $\tilde{Y}$ without knowledge of the authentic cryptogram Y [the *impersonation attack*], indeed Y might not yet exist. The impersonation attack is said to succeed if the decrypter accepts $\tilde{Y}$ as a valid cryptogram—even if it should turn out later that $\tilde{Y}$ coincides with the valid cryptogram Y. The *probability of successful impersonation*, P_I, is defined as the enemy's probability of success when he

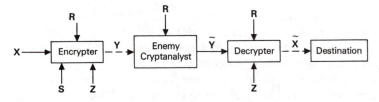

Figure 2 Modifications to Fig. 1 for consideration of authenticity attacks.

or she employs an optimum impersonation strategy. Second, the enemy cryptanalyst may be able to intercept the authentic cryptogram Y and replace it with his fraudulent cryptogram $\tilde{Y}$ where $\tilde{Y} \neq Y$ [the *substitution attack*]. The substitution attack succeeds if the decrypter accepts $\tilde{Y}$ as a valid cryptogram, and the *probability of successful substitution, P_S,* is defined as the probability of success when the enemy employs an optimum substitution strategy. And third, the enemy may be able to choose freely between an impersonation attack and a substitution attack [the *deception attack*]; the *probability of successful deception, P_d,* is then defined as the probability of success for an optimum deception strategy.

It may appear obvious that $P_d = \max(P_I, P_S)$. Simmons, however, used a game-theoretic authentication model, which was appropriate for the treaty-compliance-and-verification problems that he was considering and in which the cryptographer has the freedom to choose the key statistics to foil the type of attack that the enemy cryptanalyst may choose. In this case, one can only assert that $P_d \geq \max(P_I, P_S)$, since the best choice of key statistics for foiling a deception attack can differ from that for foiling an impersonation attack or for foiling a substitution attack. Our adoption of Kerckhoff's assumption (see Section 2, above), however, forces us to assume that the key statistics are fixed once and for all by the cryptographer, independently of the attack used by the enemy cryptanalyst. In this case, which we assume hereafter, it is indeed true that $P_d = \max(P_I, P_S)$.

The theory of authenticity is in many ways more subtle than the corresponding theory of secrecy. In particular, it is not at all obvious how ''perfect authenticity'' should be defined. Let $\#\{Y\}$ denote the number of cryptograms y such that $P(Y = y) \neq 0$, and let $\#\{X\}$ and $\#\{Z\}$ be similarly defined as the number of plaintexts and cryptograms, respectively, with nonzero probability. It follows from Eq. (3) that, for every z, there must be at least $\#\{X\}$ different cryptograms y such that $P(Y = y \mid Z = z) \neq 0$. Hence, if the enemy cryptanalyst in an impersonation attack selects Y completely at random from the $\#\{Y\}$ cryptograms with nonzero probability, his probability of success will be at least $\#\{X\}/\#\{Y\}$. Thus, P_I, the probability of success in an optimal impersonation attack satisfies

$$P_I \geq \#\{X\}/\#\{Y\} \tag{17}$$

This equation shows that good protection against an impersonation attack demands that $\#\{Y\}$ be much greater than $\#\{X\}$, and shows that *complete protection* (that is, $P_I = 0$) *is impossible*. We note further that Eq. (17) can hold with equality only when there are exactly $\#\{X\}$ valid cryptograms y for each key z, which means that a randomized cipher cannot achieve equality in Eq. (17).

Because complete protection against deception is impossible, the only recourse is to define ''perfect authenticity'' to mean as much protection against deception as is possible given the size of the space of valid cryptograms (even if this means that we must call a system "perfect" for which $\#\{Y\} = \#\{X\}$ and hence $P_d = 1$). This is what Simmons has done, but we must develop the theory a little further before introducing his precise definition of "perfect authenticity."

Let the *authentication function*, $\phi(y, z)$ be defined to be 1 if y is a valid cryptogram for the secret key z and to be 0, otherwise. Note that if $Z = z$ the decrypter will accept $\tilde{Y} = y$ as a valid cryptogram just when $\phi(y, z) = 1$. The probability that a particular y is a valid cryptogram can be written

$$P(y \text{ valid}) = \sum_y \phi(y,z)P_Z(z) \tag{18}$$

which is just the total probability of the keys z for which y is a valid cryptogram. The best impersonation attack is for the enemy cryptanalyst to choose $\tilde{Y} = y$ for that y that maximizes $P(y \text{ valid})$. Thus

$$P_I = \max_y P(y \text{ valid}) \tag{19}$$

In [14], Simmons derived the following fundamental lower bound on the probability of successful impersonation:

$$P_I \geq 2^{-I(Y;\,Z)} \tag{20}$$

which reveals the quite surprising fact that P_I can be made small *only if the cryptogram gives away much information about the secret key*—at least in principle, exploiting this information is another matter. One of the minor original contributions of our earlier paper [45] was a shortened proof of the bound [Eq. (20)] that allowed one to identify the necessary and sufficient conditions for equality. This simplification motivated Sgarro [50] to provide a still simpler proof of Eq. (2) based on properties of "informational divergence." Johannesson and Sgarro then observed that the bound [Eq. (20)] could be strengthened and, in their paper [51] thereon, included an even simpler proof of Eq. (20) that was suggested to them by Körner and is based on the "logsum inequality" [52, p. 48]. This led in turn to our finding yet a new proof of Eq. (20) that we now present.

It is immediately apparent from Eq. (19) that

$$P_I \geq \sum_y P_Y(y)P(y \text{ valid}) \tag{21}$$

with equality if and only if $P(y \text{ valid})$ is constant for all y. Substituting Eq. (18) into Eq. (21) gives

$$P(y \text{ valid}) \geq \sum_{yz} P_Y(y)P_Z(z)\phi(y,z) \tag{22}$$

But the pair y and z is in supp P_{YZ} precisely when $\phi(y,z) = 1$ and $P_Z(z) \neq 0$. Thus, this last inequality can be written equivalently in terms of an expectation as

$$P(y \text{ valid}) \geq E\left[\frac{P_Y(y)P_Z(z)}{P_{YZ}(y,z)}\right]$$

as follows from the discussion of expectations in Section 2.4, above. This inequality is of course equivalent to

$$\log P(y \text{ valid}) \geq \log E \left[\frac{P_Y(y)P_Z(z)}{P_{YZ}(y,z)} \right] \tag{23}$$

Because the logarithm is a strictly concave function, Jensen's well-known inequality [15, pp. 151–152] can be applied to give

$$\log E \left[\frac{P_Y(y)P_Z(z)}{P_{YZ}(y,z)} \right] \geq E \left[\log \frac{P_Y(y)P_Z(z)}{P_{YZ}(y, z)} \right] \tag{24}$$

with equality if and only if $(P_Y(y)P_Z(z))/P_{YZ}(y,z)$ is constant for all pairs y and z in supp P_{YZ}. The final step in the derivation of Eq. (20) is to note that

$$E \left[\log \frac{P_Y(y)P_Z(z)}{P_{YZ}(y,z)} \right] = H(YZ) - H(Y) - H(Z) = -I(Y; Z) \tag{25}$$

Combining Eqs. (23)–(25) gives

$$\log P_I \geq -I(Y; Z) \tag{26}$$

which is equivalent to Eq. (20). The necessary and sufficient conditions for equality in Eq. (20) are seen to be that both (i) $P(y \text{ valid})$ is constant for all y (or, equivalently, that *every impersonation strategy is optimum*) and (ii) that $(P_Y(y)P_Z(z))/P_{YZ}(y,z)$ is constant for all pairs y and z in supp P_{YZ}.

Johannesson and Sgarro [51] obtained a first strengthening of Simmons's bound [Eq. (20)] by noting that although P_I does not depend on the statistics of the plaintext X (as can be seen from Eqs. (18) and (19)), the mutual information $I(Y; Z)$ generally does depend on P_X. Thus

$$P_I \geq 2^{-\text{inf1 } I(Y; Z)} \tag{27}$$

where inf1 here denotes the *infimum* (or "minimum") of $I(Y; Z)$ over all choices of P_X that leaves the authentication function $\phi(y,z)$ unchanged. They further strengthened this bound by noting that nothing in the derivation of Eq. (20) demands that the plaintext X and the key Z be statistically independent (although they always are in our model and in practice) and hence that

$$P_I \geq 2^{-\text{inf2 } I(Y; Z)} \tag{28}$$

where inf2 denotes the infimum of $I(Y; Z)$ over all choices of *conditional* probability distributions for the plaintext X given the key Z.

Because for our Kerckhoffian assumption we have that

$$P_d = \max(P_I, P_S) \tag{29}$$

it follows that Eq. (20) gives also *Simmons's lower bound on the probability of successful deception*, namely,

$$P_d \geq 2^{-I(Y; Z)} \tag{30}$$

where conditions (i) and (ii) above are necessary, but no longer sufficient, conditions for equality.

Simmons [14] has defined *perfect authenticity* to mean that equality holds in Eq. (30). Even with perfect authenticity, however, it must be remembered that the probability of deception P_d will be small only when $I(Y; Z)$ is large; that is, only when the cryptogram provides the enemy cryptanalyst with much information about the key! The information that Y gives about Z is a measure of how much of the secret key is used to provide authenticity.

[It might seem more appropriate to define "perfect authenticity" to mean that equality holds when the stronger bounds inf1 $I(Y; Z)$ or inf2 $I(Y; Z)$ are used on the right of Eq. (30). However, it seems to us better to abide by Simmons's use of $I(Y; Z)$ and then to consider the case when inf1 $I(Y; Z)$ or inf2 $I(Y; Z)$ is less than $I(Y; Z)$ as indicating that the authenticity system is "wasting" part of the information $I(Y; Z)$ that the cryptogram Y betrays about the key Z and thus does not deserve the appellation "perfect."]

The theory of the theoretical security of authenticity systems is less well developed than is that of secrecy systems. In particular, it is not known in general under what conditions systems offering perfect authenticity exist, although constructions of particular such systems have been given. Thus, we will content ourselves here with giving a series of simple examples that illuminate the main ideas of authentication theory and show the relation between authentication and secrecy.

In the following examples, the plaintext is always a single binary digit X, the cryptogram $Y = [Y_1, Y_2]$ a binary sequence of length 2, the key $Z = [Z_1, \ldots, Z_K]$ is a completely random binary sequence so that $P(Z = z) = 2^{-K}$ for all z, and all logarithms are taken to the base 2 so that $H(Z) = K$ bits.

Example 1. Consider the encipherment scheme with a key of length $K = 1$ described by the following table.

z \ x	0	1
0	00	10
1	01	11

The meaning is that, for instance, $Y = [1, 0]$ when $X = 1$ and $Z = 0$. The enciphering rule is simply $Y = [X, Z]$, that is, the key is appended as a "signature" to the plaintext to form the cryptogram. Thus, this system provides *no secrecy* at all. Moreover, $H(Z \mid Y) = 0$ so that $I(Y; Z) = 1$ bit, and the bound [Eq. (28)] becomes $P_I \geq \frac{1}{2}$. But if $P_X(0) = \frac{1}{2}$, then $P(y \text{ valid}) = \frac{1}{2}$ for all y so that in fact $P_I = \frac{1}{2}$, which is as small as possible. But upon observing $Y = y$, the enemy cryptanalyst always knows the other valid cryptogram so that he can always succeed in a substitution attack. Hence $P_S = 1 = P_d > 2^{-I(Y; Z)} = \frac{1}{2}$, that is, the authenticity is not perfect.

Example 2. Consider the randomized encipherment system in which the private randomizer S is a binary random variable with $P(S = 0) = \frac{1}{2}$.

z	s \ x	0	1
0	0	00	10
0	1	01	11
1	0	00	11
1	1	01	10

Note that $Y_1 = X$ so that again there is *no secrecy*. Given $Y = y$ for any y, the two possible values of Z are equally likely so that $H(Z \mid Y) = 1$, and thus $I(Y; Z) = 0$. It follows then from Eq. (28) that this system must have $P_I = 1 = P_d = 2^{-I(Y;Z)}$ and thus trivially provides perfect authenticity. But on observing, say, $Y = [0, 0]$, the enemy cryptanalyst is faced with two equally likely alternatives, $[1, 0]$ and $[1, 1]$, for the other valid cryptogram, only one of which will be accepted by the receiver, who knows Z, as authentic. Thus $P_S = \frac{1}{2}$. This example shows that a randomized cipher can satisfy Eq. (30) with equality, and also that $-I(Y; Z)$ is *not* in general a lower bound on $\log P_S$.

Examples 1 and 2 show that *the substitution attack can be stronger than the impersonation attack*, and *vice versa*.

Example 3. Consider the same system as in Example 2 except that z and s are now the two digits z_1 and z_2, respectively, of the secret key, and hence both are known to the legitimate receiver. There is still *no secrecy* because $Y_1 = X$. Given $Y = y$ for any y, there are still two equally likely possibilities for Z so that $H(Z|Y) = 1$ and hence $I(Y; Z) = 1$ bit. But $P(y$ valid$) = \frac{1}{2}$ for all four cryptograms y and thus $P_I = \frac{1}{2}$. Moreover, given that he observes $Y = y$, the enemy cryptanalyst is faced with two equally likely choices for the other valid cryptogram so that $P_S = \frac{1}{2}$. Thus $P_d = \frac{1}{2} = 2^{-I(Y;Z)}$ and hence this system offers (nontrivial) *perfect authenticity*, no matter what the statistics of the plaintext X may be.

Example 4. Consider the following encipherment system.

z_1 z_2 \\ x	0	1
0 0	00	11
0 1	01	10
1 0	10	01
1 1	11	00

Because $P(Y = y \mid X = x) = \frac{1}{4}$ for all x and y, the system provides *perfect secrecy*. By the now familiar arguments, $I(Y; Z) = H(X)$ and $P_I = \frac{1}{2}$, the corresponding best possible protection against impersonation when $H(X) = 1$, that is, when $P(X = 0) = \frac{1}{2}$. But, on observing $Y = y$, the enemy can always succeed in a substitution attack by choosing Y to be the complement of y. Thus $P_S = 1 = P_d$ and hence this system provides *no protection against deception* by substitution.

Example 5. Consider the following encipherment system.

z_1 z_2 \\ x	0	1
0 0	00	10
0 1	01	00
1 0	11	01
1 1	10	11

This cipher provides *perfect secrecy* and has $I(Y; Z) = 1$ bit. Moreover, $P(y$ valid$) = \frac{1}{2}$ for all y so that $P_I = \frac{1}{2}$. Upon observing that $Y = y$, say $y = [0, 0]$, the enemy cryptanalyst is faced with the two alternatives $[1,0]$ and $[0,1]$ for the other valid cryptogram with the probabilities $P(X = 0)$ and $P(X = 1)$, respectively. Thus, $P_S \geq \frac{1}{2}$ with equality if and only if $P(X = 0) = \frac{1}{2}$. It follows that $P_d = P_S \geq 2^{-I(Y;Z)} = \frac{1}{2}$ with equality if and only if $P(X = 0) = \frac{1}{2}$. Thus, if $P(X = 0) = \frac{1}{2}$, this cipher also provides *perfect authenticity*.

Examples 3, 4, and 5 illustrate the fact that *secrecy and authenticity are independent attributes of a cryptographic system*—a lesson that is too often forgotten in practice.

2.7 Practical Security

In Section 2.5, we noted the possibility for a cipher system with a limited key [i.e., with $K \ll H(X)$] to have an infinite unicity distance and hence to be theoretically "unbreakable." Shannon called such ciphers *ideal*, but noted that their design poses virtually insurmountable practical problems [10, p. 700]. Most practical ciphers must depend for their security not on the theoretical impossibility of their being broken, but on the practical difficulty of such breaking. Indeed, Shannon postulated that every cipher has a *work characteristic* $W(n)$ which can be defined as the average amount of work (measured in some convenient units such as hours of computing time on a CRAY 2) required to find the key when given n digits of the ciphertext. Shannon was thinking here of a ciphertext-only attack, but a similar definition can be made for any form of cryptanalytic attack. The quantity of greatest interest is the limit of $W(n)$ as n approaches infinity, which we shall denote by $W(\infty)$ and which can be considered the average work needed to "break the cipher." Implicit in the definition of $W(n)$ is that the *best possible cryptanalytic algorithm* is employed to break the cipher. Thus to compute or underbound $W(n)$ for a given cipher, we are faced with the extremely difficult task of finding the best possible way to break that cipher, or at least of proving lower bounds on the work required in the best possible attack. There is no practical cipher known today (at least to researchers outside the secret research community) for which even an interesting lower bound on $W(\infty)$ is known. Such practical ciphers are generally evaluated in terms of what one might call the *historical work characteristic*, $W_h(n)$, which can be defined as the average amount of work to find the key from n digits of ciphertext when one uses the best of *known attacks* on the cipher. When one reads about a "cipher that requires millions of years to break," one can be sure that the writer is talking about $W_h(\infty)$. When calculated by a cryptographer who is fully acquainted with the techniques of cryptanalysis, $W_h(\infty)$ can be a trustworthy measure of the real security of the cipher, particularly if the cryptographer includes a judicious "margin of error" in his calculations. But there always lurks the danger that $W(\infty) \ll W_h(\infty)$, and hence that an enemy cryptanalyst might devise a new and totally unexpected attack that will, when it is ultimately revealed, greatly reduce $W_h(\infty)$—the history of cryptography is rife with examples!

2.8 Diffusion and Confusion

Shannon suggested two general principles, which he called diffusion and confusion [10, p. 708], to guide the design of practical ciphers. By *diffusion*, he meant the spreading out of the influence of a single plaintext digit over many ciphertext digits so as to hide the statistical structure of the plaintext. An extension of this idea is to spread the influence of a single key digit over many digits of ciphertext so as to frustrate a piecemeal attack on the key. By *confusion*, Shannon meant the use of enciphering transformations that complicate the determination of how the statistics of the ciphertext depend on the statistics of the plaintext. But a cipher should not only be difficult to break, it must also be easy to encipher and decipher when one knows the secret key. Thus, a very common approach to creating diffusion and confusion is to use a *product cipher,* that is, a cipher that can be implemented as a succession of simple ciphers, each of which adds its modest share to the overall large amount of diffusion and confusion.

Product ciphers most often employ both transposition ciphers and substitution ciphers as the component simple ciphers. A *transposition cipher* merely permutes the letters in the plaintext, the particular permutation being determined by the secret key. For instance, a transposition cipher acting on six-letter blocks of Latin letters might cause CAESAR to encipher to AESRAC. The single-letter statistics of the ciphertext are the same as for the plaintext, but the higher-order statistics of the plaintext are altered in a confusing way. A *substitution cipher* merely replaces each plaintext letter with another letter from the same alphabet, the particular substitution rule being determined by the secret key. The single-letter statistics of the ciphertext are the same as for the plaintext. The Caesar cipher discussed in Section 1.5 is a simple substitution cipher with only 26 possible values of the secret key. But if the substitution is made on a very large alphabet so that it is not likely that any plaintext letter will occur more than once in the lifetime of the secret key, then the statistics of the plaintext are of little use to the enemy cryptanalyst and a substitution cipher becomes quite attractive. To achieve this condition, the cryptographer can choose the "single letters" on which the substitution is applied to be groups of several letters from the original plaintext alphabet. For instance, a substitution upon pairs of Latin letters, in which CA was replaced by WK, ES by LB, and AR by UT, would result in CAESAR being enciphered to WKLBUT. If this ciphertext was then further enciphered by the above-considered transposition cipher, the resulting ciphertext would be KLBTUW. Such interleaving of simple transpositions and substitutions, when performed many times, can yield a very strong cipher, that is, one with very good diffusion and confusion.

2.9 The Data Encryption Standard

Perhaps the best example of a cipher designed in accordance with Shannon's diffusion and confusion principles is the Data Encryption Standard (DES). In the DES, the plaintext X, the cryptogram Y, and the key Z are binary sequences with lengths $M = 64$, $N = 64$, and $K = 56$, respectively. All 2^{64} possible values of X are, in general, allowed. Because $M = N = 64$, DES is in fact a substitution cipher, albeit on a very large alphabet of $2^{64} \approx 10^{19}$ "letters"! In its so-called *electronic code book mode,* successive 64-bit "blocks" of plaintext are enciphered using the same secret key, but otherwise independently. Any cipher used in this manner is called a *block cipher.*

The DES is a product cipher that employs 16 "rounds" of successive encipherment, each round consisting of rather simple transpositions and substitutions of 4-bit groups. Only 48 key bits are used to control each round, but these are selected in a random-appearing way for successive rounds from the full 56-bit key. We shall not pursue further details of the DES here; a good short description of the DES algorithm appears in [1] and the complete description is readily available [16]. It suffices here to note that it appears hopeless to give a useful description of how a single plaintext bit (or a single key bit) affects the ciphertext (good diffusion!), or of how the statistics of the plaintext affect those of the ciphertext (good confusion!)

The DES algorithm was submitted by the IBM Corporation in 1974 in response to the second of two public invitations by the U.S. National Bureau of Standards (NBS) for designers to submit algorithms that might be used as a standard for data encryption by government and private entities. One design requirement was that the algorithm could be made public without compromising its security—a requirement that Kerckhoff would have admired! The IBM design was a modification of the company's older Lucifer cipher that used a 128-bit key. The original design submitted by IBM permitted all $16 \times 48 = 768$ bits of key used in the 16 rounds to be selected independently. A U.S. Senate Select Committee ascertained in 1977 that the NSA was instrumental in reducing the DES secret key to 56 bits that are each used many times, although this had previously been denied by IBM and NBS [17]. NSA also classified the *design principles* that IBM had used to select the particular substitutions that are used within the DES algorithm. But the entire algorithm in full detail was published by NBS in 1977 as a U.S. Federal Information Processing Standard [16], to become effective in July of that year.

Almost from the beginning, the DES was embroiled in controversy. W. Diffie and M. E. Hellman, cryptologic researchers at Stanford University, led a chorus of skepticism over the security of the DES that focused on the smallness of the secret key. With $2^{56} \approx 10^{17}$ possible keys, a *brute-force attack* or "exhaustive cryptanalysis" (in which the cryptanalyst tries one key after another until the cryptogram deciphers to sensible plaintext) on the DES was beyond feasibility, but only barely so. Diffie and Hellman published the conceptual blueprint for a highly parallel special-purpose computer that, by their reckoning, would cost about 20 million dollars and would break DES cryptograms by essentially brute force in about 12 hours [18]; Hellman later proposed a variant machine, that, by his reckoning, would cost only four million dollars and, after a year of initial computation, would break 100 cryptograms in parallel each day [19]. Countercritics have attacked both of these proposals as wildly optimistic. But the hornet's nest of public adverse criticism of DES did lead the NBS to hold workshops of experts in 1976 and 1977 to "answer the criticisms" [17] and did give rise to the Senate hearing mentioned above. The general consensus of the workshops seems to have been that DES would be safe from a Diffie–Hellman-style attack for only about ten years, but that the 56-bit key provided no margin of safety [17]. Almost fifteen years have now passed, and the DES appears to have justified the faith of its defenders. Despite intensive scrutiny of the DES algorithm by cryptologic researchers, no one has yet publicly revealed any weakness of DES that could be exploited in an attack that would be significantly better than exhaustive cryptanalysis. The general consensus of cryptologic researchers today is that DES is an extremely good cipher with an unfortunately small key. But it should not be forgotten that the effective size of the secret key can be increased by using multiple DES encryptions with different keys, that is, by making a product cipher with DES used for the component ciphers. At least three encryptions should be used to foil the "meet-in-the-middle attack" proposed by Merkle and Hellman [20].

2.10 Stream Ciphers

In a block cipher, a plaintext block identical to a previous such block would give rise to the identical ciphertext block as well. This is avoided in the so-called *stream ciphers* in which the enciphering transformation on a plaintext "unit" changes from unit to unit.

For instance, in the *cipher-block chaining* (CBC) mode proposed for the DES algorithm [16], the current 64-bit plaintext block is added bit-by-bit modulo 2 to the previous 64 bit ciphertext block to produce the 64-bit block that is then enciphered with the DES algorithm to produce the current ciphertext block. CBC converts a block cipher into a stream cipher with the advantage that tampering with ciphertext blocks is more readily detected, that is, impersonation or substitution attacks become much more difficult. But cryptographers generally reserve the term *stream cipher* for use only in the case when the plaintext "units" are very small, say a single Latin letter or a single bit.

The most popular stream ciphers today are what can be called *binary additive stream ciphers*. In such a cipher, the K bit secret key Z, is used only to control a *running-key generator* (RKG) that emits a binary sequence, $Z'_1, Z'_2, \ldots, Z'_N$, called the *running key*, where in general $N \gg K$. The ciphertext digits are then formed from the binary plaintext digits by simple modulo 2 addition in the manner

$$Y_n = X_n \oplus Z'_n, \qquad n = 1, 2, \ldots N \tag{31}$$

Because modulo 2 addition and subtraction coincide, Eq. (31) implies

$$X_n = Y_n \oplus Z'_n, \qquad n = 1, 2, \ldots N \tag{32}$$

which shows that encryption and decryption can be performed by identical devices. A single plaintext bit affects only a single ciphertext bit, which is the worst possible diffusion; but each secret key bit can influence many ciphertext bits so the key diffusion can be good.

There is an obvious similarity between the binary additive stream cipher and a binary one-time pad. In fact, if $Z_n = Z'_n$ (that is, if the secret key is used as the running key), then the additive stream cipher is identical to the one-time pad. This similarity undoubtedly accounts in part for the widespread faith in additive stream ciphers that one encounters in many cryptographers and in many users of ciphers. But, of course, in practical stream ciphers, the ciphertext length N greatly exceeds the secret key length K. The best that one can then hope to do is to build an RKG whose output sequence cannot be distinguished by a resource-limited cryptanalyst from a completely random binary sequence. The trick is to build the RKG in such a way that, on observing $Z'_1, Z'_2, \ldots, Z'_n$, the resource-limited cryptanalyst can do no better than to guess Z'_{n+1} at random. If this can be done, one has a cipher that is secure against even a chosen-plaintext attack (by which one would mean that the enemy cryptanalyst could freely select, say the first n bits of the plaintext sequence).

Stream ciphers have the advantage over block ciphers in that analytic measures of their quality are more easily formulated. For instance, stream cipher designers are greatly concerned with the *linear complexity* or "linear span" of the running-key sequence, which is defined as the length L of the shortest linear-feedback shift-register (LFSR) that could produce the sequence. Figure 3 shows a typical LFSR of length 6.

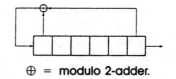

Figure 3 A "typical" linear-feedback shift-register.

$\oplus$ = modulo 2-adder.

The reason for this concern is that there is a simple algorithm that would quickly find this shortest LFSR after examining only the first $2L$ bits of the running key [21]. Thus, large linear complexity of the running-key sequence is a necessary (but far from sufficient) condition for the practical security of an additive stream cipher. (An up-to-date treatment of linear complexity in connection with stream ciphers may be found in the book by Rueppel [44].) The RKG of an additive stream cipher is often built by the nonlinear combining of the output sequences of several LFSRs, as such combining can create a sequence with large linear complexity. There arises then the danger that individual LFSR sequences will be correlated with the running-key sequence so that the enemy cryptanalyst can attack the cipher piecemeal. Siegenthaler [22] has shown recently that the "correlation-immunity" of nonlinear combining functions can be precisely quantified and that the designer has to make an explicit tradeoff between correlation-immunity and linear complexity. There are many other known analytic approaches to stream cipher design. Taken together, they still leave one far from the point where one could say that the true work characteristic of a practical stream cipher is known, but they tend to give many cryptographers and users (perhaps misleadingly) greater trust in the historical work characteristics computed for stream ciphers than in those computed for block ciphers.

2.11 Provably Secure Ciphers?

When dealing with the practical security of ciphers, "It is difficult to define the pertinent ideas involved with sufficient precision to obtain results in the form of mathematical theorems," as Shannon said nearly 40 years ago [10, p. 702] in an eloquent understatement that needs no alteration today. It is an open question whether it is even possible to compute the true work characteristic $W(n)$ or its asymptotic value $W(\infty)$. A slender ray of hope lies in a totally impractical cipher proposed by this writer and I. Ingemarsson [23]. This cipher is a randomized stream cipher with a secret key of K bits. One can prove that $W(\infty) \approx 2^{K/2}$ where the unit of computation is a binary test, that is, a test with two outcomes. The "catch" is that the legitimate receiver must wait (during which time he does no testing or other computational work) until about 2^K bits have arrived before deciphering is begun. One can easily guarantee that the enemy cryptanalyst will need thousands of years to break the cipher, if one is willing to wait millions of years to read the plaintext! Such a cipher would be tolerable perhaps only to Rip van Winkle, the lazy and sleep-prone hero of Washington Irving's delightful short story, after whom both the story and the cipher have been named. Randomization, which was the feature that allowed the calculation of $W(\infty)$ for the impractical Rip van Winkle cipher, may turn out to be useful in developing a practical provably secure cipher, if in fact this can be done at all.

The previous words, which appeared in our earlier paper [45], have taken on a prophetic ring. At Eurocrypt'90, Maurer [53] presented a new cipher that exploits a very large public randomizer $\mathbf{R}$, that is provably secure, and that is at least arguably on the verge of being practical. Perhaps the most interesting facet of Maurer's work was his introduction of a new information-theoretic approach to cryptography that allows one to overcome the "bottleneck" of Shannon's inequality [Eq. (9)] for perfect secrecy. Maurer's trick was to introduce a *security event, S,* with the property that the cipher provides *perfect secrecy given that the event S occurs* [and even if $H(X) \gg H(Z)$]—but "all bets are off" when S does not occur! For his "strongly randomized" cipher, Maurer showed that the probability that S does *not* occur will be negligibly small unless the enemy cryptanalyst examines a substantial fraction of all the bits in the very large public randomizer R. The legitimate sender and receiver need examine only the very small portion of the public randomizer that is specified by the short secret key Z. The conclusion from Maurer's work is the (in retrospect obvious) fact that Shannon's bound [Eq. (9)] governs the needed key size only when one demands that his cipher provide *perfect secrecy with probability 1*.

3 PUBLIC KEY CRYPTOGRAPHY

3.1 One-Way Functions

That the publication of Shannon's 1949 paper [10] resulted in no discernible upsurge in open cryptologic research is due to several factors. First, the theory of theoretical security of secrecy systems that it provided was virtually complete in itself, and showed conclusively that theoretically secure secrecy systems demand the secure transfer of far more secret key than is generally practicable. Moreover, the insights that Shannon provided into the practical security of secrecy systems tended to reinforce accepted cryptographic approaches rather than to suggest new ones. But Shannon's observation that "The problem of good cipher design is essentially one of finding difficult problems, subject to certain other conditions. . . . We may construct our cipher in such a way that breaking it is equivalent to (or requires at some point in the process) the solution of some problems known to be laborious" [10, p. 704] took root in the fertile imaginations of the Stanford cryptologic researchers, W. Diffie and M. E. Hellman. The fruit was their 1976 paper, "New Directions in Cryptography," [12] that stunned the cryptologic world with the startling news that *practically secure secrecy systems can be built that require no secure transfer of any secret key whatsoever*.

The crucial contribution of the Diffie–Hellman paper lies in two unusually subtle definitions, that of a "one-way function," which was borrowed from work by R. M. Needham on secure computer login techniques, and that of a "trapdoor one-way function," which was totally new. A *one-way function* is defined as a function f such that for every x in the domain of f, $f(x)$ is easy to compute; but for virtually all y in the range of f, it is computationally infeasible to find an x such that $y = f(x)$. The first thing to note is that this is not a precise mathematical definition. What do "easy," "virtually all" (which we have substituted for Diffie and Hellman's "almost all," as the latter can have a precise mathematical meaning that was not intended in the definition), and "computationally infeasible" mean precisely? Yet the definition is sufficiently precise that one has no doubt as to what Diffie and Hellman essentially meant by a one-way function, and one has the feeling that it could be made completely precise in a particular context. It is less clear how such a function could be of use cryptographically—to build a cipher that not even the legitimate receiver could decipher seems the obvious (and worthless) application! A *trapdoor one-way* function is defined as a family of invertible functions f_z, indexed by z, such that, given z, it is easy to find algorithms E_z and D_z that easily compute $f_z(x)$ and $f_z^{-1}(y)$ for all x and y in the domain and range, respectively, of f_z; but for virtually all z and for virtually all y in the range of f_z, it is computationally infeasible to compute $f_z^{-1}(y)$ even when one knows

E_z. Again, this is only a semimathematical definition, but this time the cryptologic utility is nakedly apparent.

3.2 Public Key Distribution

As a likely candidate for a one-way function, Diffie and Hellman [12] suggested the *discrete exponential function*

$$f(x) = \alpha^x \quad (\text{mod } p) \tag{33}$$

where x is an integer between 1 and $p - 1$ inclusive, where, as indicated, the arithmetic is done modulo p, a very large prime number, and where $\alpha(1 \leq \alpha < p)$ is an integer such that $\alpha, \alpha^2, \ldots, \alpha^{p-1}$ are, in some order, equal to $1, 2, \ldots, p - 1$. For example, with $p = 7$, one could take $\alpha = 3$ since $\alpha = 3$, $\alpha^2 = 2$, $\alpha^3 = 6$, $\alpha^4 = 4$, $\alpha^5 = 5$, and $\alpha^6 = 1$. (In algebraic terminology, such an α is called a *primitive element* of the finite field $GF(p)$, and such α's are known always to exist.) If $y = \alpha^x$, then it is natural to write

$$x = \log_\alpha (y) \tag{34}$$

so that inverting $f(x)$ is the problem of calculating *discrete logarithms*. Even for very large p, say $p \approx 2^{1000}$, it is quite easy to calculate $f(x)$ by the trick of square-and-multiply. For instance, to compute $\alpha^{53} = \alpha^{32+16+4+1}$, one would first form α^2, $\alpha^4 = (\alpha^2)^2$, $\alpha^8 = (\alpha^4)^2$, $\alpha^{16} = (\alpha^8)^2$, and $\alpha^{32} = (\alpha^{16})^2$, which requires five multiplications. Then one would multiply α^{32}, α^{16}, α^4, and α together, which takes three more multiplications for a total of eight multiplications (mod p). Even with $p \approx 2^{1000}$, calculation of $f(x)$ for any integer x, $1 \leq x < p$, would take less than 2000 multiplications (mod p).

     If the discrete exponential function is indeed one-way, then for virtually all integers y, $1 \leq y < p$, it must be computationally infeasible to compute $\log_x y$. It was soon realized by Hellman and Pohlig that it was not enough that p be large, $p - 1$ must also have a large prime factor (ideally, $p - 1$ would be twice another prime) if the discrete logarithm is indeed to be hard to compute [24]. With this proviso, the best of known algorithms for computing the discrete logarithm require roughly $\sqrt{p}$ multiplies (mod p), compared to only about $2 \log_2 p$ multiplies for discrete exponentiation. If the discrete logarithm is truly this hard to compute, then the discrete exponential with the proviso on $p - 1$ is indeed a one-way function. But as of this writing *there is no proof that the discrete exponential, or any other function for that matter, is truly one-way.*

     Diffie and Hellman suggested an astoundingly simple way in which the discrete exponential could be used to create secret keys between pairs of users in a network using only public messages. All users are presumed to know α and p. Each user, say user i, randomly selects an integer X_i between 1 and $p - 1$ that is kept as his *private secret*. The user then computes

$$Y_i = \alpha^{X_i} \quad (\text{mod } p) \tag{35}$$

Rather than keeping Y_i secret, the user places Y_i in a *certified public directory* accessible to all users. If users i and j later wish to communicate secretly, user i fetches Y_j from the directory, then uses the private secret X_i to form

$$Z_{ij} = (Y_j)^{X_i} = (\alpha^{X_j})^{X_i} = \alpha^{X_i X_j} \quad (\text{mod } p) \tag{36}$$

In a similar manner, user j forms Z_{ji}. But $Z_{ij} = Z_{ji}$ so that users i and j can now use Z_{ij} as the secret key in a conventional cryptosystem. If an enemy could solve the discrete logarithm problem, he could take Y_i and Y_j from the directory, solve for $X_i = log_\alpha Y_i$, and then form Z_{ij} in the same manner as did user i—there seems to be no other way for an enemy to find Z_{ij} (but there is no proof of this). The scheme just described is the Diffie–Hellman *public key-distribution system*. Although it is the oldest proposal for eliminating the transfer of secret keys in cryptography, it is still generally considered today to be one of the most secure and most practical public key schemes.

It should not be overlooked that the Diffie–Hellman public key-distribution scheme (and indeed every public key technique) *eliminates the need for a secure channel to pass along secrets, but does not eliminate the need for authentication*. The custodian of the public directory must be certain that it is indeed user i who puts the (nonsecret) Y_i into the directory, and user i must be certain that Y_i was actually sent to him by the custodian of the public directory. But it must not be forgotten that in secret key cryptography, (see Fig. 1) the receiver must not only be sure that the key Z was kept secret en route to him, but also that the key Z was actually sent by the legitimate sender. *Public key methods* remove one of these two problems; they *do not create a new authentication problem*, but rather make the old authentication problem more apparent.

3.3 The Rivest–Shamir–Adleman Public Key Cryptosystem

Having defined a trapdoor one-way function, it was an easy step for Diffie and Hellman to propose the structure of a *public key cryptosystem* for a network of many users. Each user, say user i, randomly chooses a value Z_i of the index and keeps Z_i as his *private secret*. The user next forms the algorithm E_{Z_i} which is then *published* in the certified public directory. Each user also forms the algorithm D_{Z_i} that is *kept secret* for each user's own use. If user j wishes to send a secret message X to user i, he fetches E_{Z_i} from the directory. User j then uses this algorithm to compute the cryptogram $Y = f_{Z_i}(X)$ that is then sent to user i. User i uses his private algorithm D_{Z_i} to compute $f_{Z_i}^{-1}(Y) = X$. If f_z is truly a trapdoor one-way function, this cryptosystem provides unassailable practical security.

When, for every index z, the domain and range of f_z coincide, Diffie and Hellman noted that a trapdoor one-way function can be used to create *digital signatures*. If user i wishes to send a *nonsecret* message X (to any or all users in the system) that he wishes to "sign" in a way that the recipient will recognize him unmistakably as the author, he merely uses his private algorithm to form $Y = f_{Z_i}^{-1}(X)$ and transmits Y. Every user can fetch the public algorithm E_{Z_i} and then compute $f_{Z_i}(Y) = X$; but no one except user i could have known how to write an intelligible message X in the form $Y = f_{Z_i}^{-1}(X)$, since no one except user i can compute $f_{Z_i}^{-1}$. Of course, user i could also send a signed secret message to user j by encrypting Y in user j's public key E_{Z_j}, rather than sending Y in the clear (he might first need to break Y into smaller pieces if Y is "too large to fit" into the domain of f_{Z_j}).

It was not at all clear to Diffie and Hellman in 1976 whether trapdoor one-way functions existed, and they did not hazard a conjectured such function in their paper. It was left to R. L. Rivest, A. Shamir, and L. Adleman (RSA) of the Massachusetts Institute of Technology (MIT) to make the first proposal of a possible trapdoor one-way function in their remarkable 1978 work, "A Method for Obtaining Digital Signatures and Public-Key Cryptosystems" [25]—it is interesting to note that authentication received higher billing than secrecy in their title. The RSA trapdoor one-way function is the essence of simplicity, but to describe it we need a few ideas from elementary number theory.

Let $GCD(i, n)$ denote the greatest common divisor of the integers i and n (not both 0). For example, $GCD(12, 18) = 6$. The *Euler totient function* $\phi(n)$, where n is a positive integer, is defined as the number of positive integers i less than n such that $GCD(i, n) = 1$ (except that $\phi(1)$ is defined to be 1). For instance, $\phi(6) = 2$ since for $1 \leq i < 6$ only $i = 1$ and $i = 5$ give $GCD(i, 6) = 1$. One sees immediately that for a prime p, $\phi(p) = p - 1$; and just a little thought more shows that if p and q are distinct primes, then

$$\phi(pq) = (p - 1)(q - 1) \tag{37}$$

For instance, $\phi(6) = \phi(2 \times 3) = 1 \times 2 = 2$. A celebrated theorem of Euler (1707–1783) states that for any positive integers x and n with $x < n$

$$x^{\phi(n)} = 1 \pmod{n} \tag{38}$$

provided that $GCD(x, n) = 1$. For example

$$5^2 = 1 \pmod 6$$

The last fact from number theory that we need is that if e and m satisfy $0 < e < m$ and $GCD(m, e) = 1$, then there is a unique d such that $0 < d < m$ and

$$de = 1 \pmod m \tag{39}$$

moreover d can be found in the process of using Euclid's "extended" algorithm for computing $GCD(m, e)$, (see [26, p. 14]).

The *RSA trapdoor one-way function* is just the discrete exponentiation

$$f_z(x) = x^e \pmod n \tag{40}$$

where x is a nonnegative integer less than $n = pq$ and where the "trapdoor" $z = \{p, q, e\}$; here p and q are distinct very large primes such that $\phi(n) = (p - 1)(q - 1)$ has a very large prime factor, and e is a positive integer less than $\phi(n)$ such that $GCD(e, \phi(n)) = 1$. The easy-to-find algorithm E_z to compute f_z easily is exponentiation by square-and-multiply; *publishing this algorithm amounts just to publishing n and e.* The inverse function is

$$f_z^{-1} = y^d \pmod n \tag{41}$$

where d is the unique positive integer less than n such that

$$de = 1 \pmod{\phi(n)}. \tag{42}$$

The easy-to-find (when one knows z) algorithm D_z to compute f_z^{-1} is also exponentiation by square-and-multiply; the decrypting exponent d is found using Euclid's algorithm for computing $GCD(e, \phi(n))$.

Note that the domain and range of the RSA trapdoor one-way function coincide, both are the set of integers from 0 to $m - 1$ inclusive. This means that the RSA function can be used to form digital signatures in the manner suggested by Diffie and Hellman. This digital signature capability is one of the most important and useful features of the RSA function.

That Eq. (41) really gives the inverse function for Eq. (40) can be seen as follows. Equation (42) is equivalent to the statement (in ordinary integer arithmetic) that

$$de = \phi(n)Q + 1 \tag{43}$$

for some integer Q. From Eqs. (40) and (43), we obtain

$$\begin{aligned}
(x^e)^d &= x^{\phi(n)Q+1} &&(\text{mod } n) \\
&= (x^{\phi(n)})^Q x &&(\text{mod } n) \\
&= x &&(\text{mod } n)
\end{aligned} \tag{44}$$

where at the last step we used Euler's theorem [Eq. (38)]. [The wary reader will have noted that Euler's theorem requires $\text{GCD}(x, n) = 1$; but in fact Eq. (44) holds for all nonnegative integers x less than n in the special case when n is the product of two distinct primes.] Equation (44) shows that raising a number to the d power (mod n) is indeed the inverse of raising a number to the e power (mod n). It remains to show why RSA believed (as do most cryptographers today) that it is computationally impossible to invert this function f_z when one knows only n and e, and also how it is possible easily to choose *randomly* the two distinct and very large primes p and q, as must be done for an enemy to be unable to guess p and q.

The enemy knows only n and e. But if he can factor $n = pq$, then he knows the entire trapdoor $z = \{p, q, e\}$, and hence can decrypt just as readily as the legitimate receiver. The security of the RSA public key cryptosystem depends on the assumption that *any way of inverting f_z* is equivalent *to factoring $n = pq$*, that is, given any way to invert f_z, one could with at most a little more computational work go on to factor n. In their paper [25], RSA show that this is true for the most likely ways that one might try to factor n, but the assumption has never been proved. But is the attack by factoring n computationally infeasible? The answer is yes if one chooses p and q on the order of 100 decimal digits each (as RSA suggested thirteen years ago) *and* if there is no revolutionary breakthrough in factoring algorithms. As Rivest [27] recently pointed out, all of the best factoring algorithms today have running times upper-bounded by the same peculiar-looking function which, for numbers to be factored between 50 and 200 decimal digits, increases by a factor of 10 for every additional 15 digits (roughly) in the number. Today it takes about 1 day on a supercomputer to factor a number of about 80 decimal digits. It would take 10^8 times that long to factor a 200 digit number $n = pq$, roughly half a million years! One of the by-products of the RSA paper has been a revival of interest in factoring, but this accelerated research effort has produced no revolutionary breakthrough. Proponents of the RSA public key cryptosystem believe that it never will. An interesting fact is that the best algorithms today for solving the (mod p) discrete logarithm problem [28] and the best algorithms for factoring n [29] require a computational effort that grows asymptotically in the same manner with p and n, respectively. Thus the RSA trapdoor function [Eq. (49)] and the Diffie–Hellman function [Eq. (33)] have, as of today, about the same claim to be called "one-way." For given $n \approx p$, however, the Diffie–Hellman function appears more difficult to invert.

It remains to consider how one can randomly choose the very large primes, p and q, required for RSA. A theorem of Tchebychef, (see [30, pp. 9–10]), states that the fraction of positive integers less than any large integer m that are primes is close to

$(\ln m)^{-1}$. For instance, the fraction of integers less than 10^{100} that are primes is about $(\ln 10^{100})^{-1} \approx \frac{1}{230}$. Because 90 percent of these integers lie between 10^{99} and 10^{100}, the fraction of primes in this range is also about $\frac{1}{230}$. Thus, if one chooses an integer between 10^{99} and 10^{100} completely at random, the chances that one chooses a prime are about $\frac{1}{230}$. One easily doubles the odds to $\frac{1}{115}$ if one is sensible enough to choose only odd integers. One needs then only about 115 such choices on the average before one has chosen a prime. But how does one recognize a prime? It is a curious fact that one can rather easily test quite reliably whether an integer is a prime or not, even if one cannot factor that integer after one discovers that it is not a prime. Such primality tests rely on a *theorem of Fermat* (1601–1665) that asserts that for any positive integer b less than a prime p

$$b^{p-1} = 1 \quad (\text{mod } p) \tag{45}$$

For instance, $2^4 = 1 \pmod 5$. [The reader may have noticed that Eq. (45) is a special case of Eq. (38), but should remember that Fermat lived a century before Euler!] If one has an integer r that one wishes to test for primeness, one can choose any positive integer b less than r and check whether

$$b^{r-1} \overset{?}{=} 1 \quad (\text{mod } r) \tag{46}$$

If the answer is no, one has the absolute assurance of Fermat that r is not a prime. If the answer is yes, one can begin to suspect that r is a prime, and one then christens r a *pseudoprime* to the base b. If r is not a prime, it turns out that it can be a pseudoprime for less (actually much less) than about half of the possible bases b. Thus if r is very large, and one independently chooses t bases b completely at random, the probability is less than about 2^{-t} that r will pass Fermat's test [Eq. (46)] for all these bases if r is not truly a prime. If we take, say $t = 100$, then we can be virtually certain that r is a prime if it passes t indepedent Fermat tests. Such "probabilistic tests for primeness" were introduced by Solovay and Strassen, and have been further refined by Rabin [31]. Such tests are today being used to check randomly chosen odd integers for primeness until one has found the two distinct large primes one needs for the RSA trapdoor one-way function, or, more precisely, until one is sufficiently sure that he has found two such primes.

    The technique just described leads to the formation of large randomly chosen "probable primes" and is the technique currently in widest use for finding the large primes needed with RSA. There is an alternative approach, however, that leads to *sure* primes that are "probably randomly chosen." It is not hard to "grow" large primes with a probabilistic algorithm; the trick is to make the primes appear to be chosen according to a probability distribution that is as uniform as possible over some interval. Maurer [54] has recently given such an algorithm that is very fast (its running time is about the same as for $t = 4$ Fermat tests) and plausibly gives an almost uniform distribution for the selected primes. It would not be surprising should this or similar algorithms eventually replace prime-testing as the method of choice for finding the large primes needed in the RSA public key cryptosystem or in the Diffie–Hellman public key-distribution system.

    There are very large-scale integration (VLSI) chips today that can implement the RSA encrypting and decrypting function at a data rate of a few kilobits per second. (These same chips can also be used to implement Fermat's test, and thus to find the needed 100 decimal digit primes, p and q). Rivest [27] has given convincing arguments that significantly higher data rates will never be achieved. For many cryptographic applications, these data rates are too low. In such cases, the RSA public key cryptosystem may still desirably be used to distribute the secret keys that will then be used in high-speed secret key ciphers, such as DES or certain stream ciphers. And the RSA algorithm may still desirably be used for authentication in its "digital signature" mode.

Before closing this section on the RSA system, we should mention that Rabin [32] has developed a variant of the RSA public key system for which he *proved* that being able to find the plaintext X from the cryptogram Y is *equivalent to factoring* $n = pq$. The system is somewhat more complicated than basic RSA, but Williams [33] refined the variant so that the extra complication is quite tolerable. This might seem to be the ultimate "RSA system," but paradoxically the breaking-is-provably-equivalent-to-factoring versions of RSA have a new weakness that was pointed out by Rivest. The proof of their equivalence to factoring is *constructive,* that is, one shows that if one could solve $Y = X^e$ (mod n) for X in these systems [which differ from RSA in that now $GCD(e, \phi(n)) \neq 1$], then one could easily go on to factor X. But this means that these systems *succumb to a chosen-ciphertext attack* in which an enemy randomly chooses X', computes $Y = (X')^e$ and then submits Y to the decrypter, who returns a solution X of $Y = X^e$ [where the fact that $GCD(e, \phi(n)) \neq 1$ results in the situation that the solution is not unique so that $X \neq X'$ is possible]. The chances are $\frac{1}{2}$ that the returned X together with X' will give the enemy the information necessary to factor $n = pq$ and thus to break the system. In a public key environment, such a chosen-ciphertext attack becomes a distinct possibility. The net result is that most cryptographers prefer to use the original RSA public key cryptosystem, and to pray for the day when a *nonconstructive* proof is given that breaking it is equivalent to factoring.

This is perhaps the appropriate point to mention that a *public key cryptosystem, if it is secure at all, is secure against a chosen-plaintext attack.* For the enemy cryptanalyst is always welcome to fetch the algorithm E_z from the public directory and then to compute the cryptograms, $y = f_y(x)$, for as many plaintexts x as he pleases. This shows that a trapdoor one-way function must necessarily be much more difficult to invert than the encrypting function of a conventional secret key cipher that is also secure against a chosen-plaintext attack. In the latter case, the enemy can still (by assumption) obtain the cryptograms y, for whatever plaintexts x, he pleases. But the enemy no longer has the luxury of watching the encryption algorithm execute its encryptions, because the secret key is an ingredient of the algorithm.

3.4 Some Remarks on Public Key Cryptography

The Diffie–Hellman one-way function and the RSA trapdoor one-way function suffice to illustrate the main ideas of public key cryptography, which is why we have given them rather much attention. But a myriad of other such functions have been proposed. Some have almost immediately been exposed as insecure, others appear promising. But no one has yet produced a proof that any function is a one-way function or a trapdoor one-way function. Even the security of the Rabin variant of RSA rests on the unproved (but very plausible) assumption that factoring large integers is computationally infeasible.

There has been some hope that the new, but rapidly evolving, theory of computational complexity, particularly Cook and Karp's theory of nondeterministic-polynomial (NP) completeness, (see [34]), will lead to provably one-way functions or provably trapdoor one-way functions. This hope was first expressed by Diffie and Hellman [12], but has thus far led mainly to failures such as the spectacular failure of the Merkle–Hellman trapdoor-knapsack public key cryptosystem. Part of the difficulty has been that NP-completeness is a worst-case phenomenon, not a "virtually all cases" phenomenon as one requires in public key cryptography. For instance, Even, Lempel, and Yacobi have constructed an amusing example of a public key cryptosystem whose breaking is equivalent to solving an "NP-hard" problem, but which can virtually always be broken [35]. [A problem is NP-hard if its solution is at least as difficult as the solution of an NP-complete problem.] But the greater difficulty has been to formulate a trapdoor one-way function whose inversion would require the solution of an NP-complete problem; this has not yet been accomplished. For instance, the inversion of the Merkle–Hellman trapdoor-knapsack one-way function is actually an easy problem disguised to resemble an NP-hard problem; Shamir broke this public key cipher, not by solving the NP-hard problem, but by stripping off the disguise.

We are grateful to J. Denés for calling our attention to the fact that the notion of "one-wayness" is much older than we had suspected. W. S. Jevons, in his book [55] first published in 1873, wrote:

> There are many cases in which we can easily and infallibly do a certain thing but may have much trouble in undoing it. . . . Given any two numbers, we may by a simple and infallible process obtain their product, but when a large number is given it is quite another matter to determine its factors. Can the reader say what two numbers multiplied together will produce the number 8 616 460 799? I think it is unlikely that anyone but myself will ever know; for they are *two large prime numbers* (emphasis added).

Thirty years later, Lehmer [56] announced the "two large prime numbers" to be 89 681 and 96 079, but added "I think that the number has been resolved before, but I do not know by whom." Such anecdotes as that just recounted here serve to feed the suspicions of those who innately mistrust public key cryptography and who will continue to do so until a provably secure public key cipher is produced. But, as we have stressed above, the security of all known practical secret key ciphers also rests on conjectures. Neither the secret key advocate nor the public key advocate is in a good position to hurl stones at the other.

4 CRYPTOGRAPHIC PROTOCOLS

4.1 What Is a Protocol?

It is difficult to give a definition of "protocol" that is both precise and general enough to encompass most things to which people apply this label in cryptography and elsewhere. Roughly speaking, we might say that a *protocol* is a multiparty algorithm, that is, a specified sequence of actions by which two or more parties cooperatively accomplish some task. Sending a secret message from one user to another in a large network by means of a public key cryptosystem, for instance, can be considered a protocol, based on a trapdoor one-way function, by means of which the users of the system and the custodian of the public directory cooperate to ensure the privacy of messages sent from one user to another.

4.2 A Key-Distribution Protocol

Many cryptographers, particularly those skeptical of public key ideas, consider the *key management problem* (that is, the problem of securely distributing and changing secret keys) to be the main practical problem in cryptography. For example, if there are S users in the system, one will need $S(S - 1)/2$ different secret keys if one is to have a dedicated secret key for every possible pair of users—an unwelcome prospect in a large system. It is unlikely that any user will ever wish to send secret messages to more than a few other users, but in advance one usually does not know who will later want to talk secretly to whom. A popular solution to this problem is the following key-distribution protocol that requires the advance distribution of only S secret keys, but still permits any pair of users to communicate secretly; there is a needed new entity, however, the *trusted key distribution center* (TKDC).

Key-Distribution Protocol

1. The TKDC securely delivers a randomly chosen secret key Z_i to user i in the system, for $i = 1, 2, . . . , S$.

2. When user i wishes to communicate secretly to user j, he sends the TKDC a request (which can be in the clear) over the public network for a secret key to be used for this communication.

3. The TKDC randomly chooses a new secret key Z_{ij} which it treats as part of the plaintext. The other part of the plaintext is a "header" in which user i and user j are identified. The TKDC encrypts this plaintext in both key Z_i and key Z_j with whatever secret key cipher is installed in the system, then sends the first cryptogram to user i and the second to user j over the public network.

4. Users i and j decrypt the cryptograms they have just received and thereby obtain the secret key to be used for encrypting further messages between these two users.

This protocol sounds innocent enough, but its security against a ciphertext-only attack requires more than ciphertext-only security of the system's secret key cipher. Why? Because in step (3) we see that an enemy cryptanalyst will have access to two cryptograms in different keys for the *same* plaintext. This can be helpful to the cryptanalyst, although it does not give him as much information as he could get in a chosen-plaintext attack on the individual ciphers. Thus, security of the system's cipher against a chosen-plaintext attack will make this protocol also secure against chosen-plaintext attacks. The point to be made here is that when one embeds a cipher into a protocol, *one must be very careful to ensure that whatever security is assumed for the cipher is not compromised by the protocol.*

4.3 Shamir's Three-Pass Protocol

One of the most interesting cryptographic protocols, due to A. Shamir in unpublished work, shows that secrecy can be obtained with no advance distribution of either secret keys or public keys. The protocol assumes two users connected by a link (such as a seamless optical fiber or a trustworthy but curious postman) that guarantees that the enemy cannot insert, or tamper with, messages but allows the enemy to read all messages sent over the link. The users are assumed to have a secret key cipher system whose encrypting function $E_z(\cdot)$ has the *commutative property,* that, for all plaintexts, x, and all keys, z_1 and z_2,

$$E_{z_2}(E_{z_1}(x)) = E_{z_1}(E_{z_2}(x)) \tag{47}$$

that is, the result of a double encryption is the same whether one uses first the key z_1 and then the key z_2 or vice versa. There are many such ciphers, for example, the one-time pad [Eq. (4)] fits the bill because $(x \oplus z_1) \oplus z_2 = (x \oplus z_2) \oplus z_1$, where the addition is bit-by-bit mod 2.

Shamir's Three-Pass Protocol

1. Users A and B randomly choose their own private secret keys, Z_A and Z_B, respectively.

2. When user A wishes to send a secret message X to user B, user A encrypts X with his own key Z_A and sends the resulting cryptogram $Y_1 = E_{Z_A}(X)$ on the open-but-tamperproof link to user B.

3. User B, upon receipt of Y_1, treats Y_1 as plaintext and encrypts Y_1 with his own key Z_B and sends the resulting cryptogram $Y_2 = E_{Z_B}(Y_1) = E_{Z_B}(E_{Z_A}(X))$ on the open-but-tamperproof link to user A.

4. User A, upon receipt of Y_2, decrypts Y_2 with his own key Z_A. Because of the commutative property [Eq. (47)], this removes the former encryption by Z_A and results in $Y_3 = E_{Z_B}(X)$. User A then sends Y_3 over the open-but-tamper-proof link to user B.

5. User B, upon receipt of Y_3, decrypts Y_3 with his own key Z_B to obtain X, the message that A has now successfully sent to him secretly.

What secret key cipher shall we use in this protocol? Why not the one-time pad, a cipher that gives perfect secrecy? If we use the one-time pad, the three cryptograms become

$$Y_1 = X \oplus Z_A$$
$$Y_2 = X \oplus Z_A \oplus Z_B$$
$$Y_3 = X \oplus Z_B \tag{48}$$

The enemy cryptanalyst sees all three cryptograms, and hence can form

$$Y_1 \oplus Y_2 \oplus Y_3 = X$$

where we have used the fact that two identical quantities sum to O mod 2. Thus, the three-pass protocol is completely insecure when we use the one-time pad for the embedded cipher! The reason for this is, as Eq. (48) shows, that the effect of the protocol is that each of the two ciphers get used "$1\frac{1}{2}$ times," rather than only once as is required for the security of the "one-time" pad.

Is there a cipher that can be used in the Shamir three-pass protocol and still retain its security? There seems to be. Let p be any large prime for which $p - 1$ has a large prime factor (to make the discrete logarithm problem in mod p arithmetic computationally infeasible to solve). Randomly choose a positive integer e less than $p - 1$ such that $GCD(e, p - 1) = 1$, and let d be the unique positive integer less than $p - 1$ such that

$$de = 1 \quad (\text{mod } p - 1) \tag{49}$$

Let $Z = (d, e)$ be the secret key and take the encrypting and decrypting functions to be

$$y = E_z(x) = x^e \quad (\text{mod } p)$$
$$x = D_z(y) = y^d \quad (\text{mod } p) \tag{50}$$

where x and y are positive integers less than p. [The fact that $y^d = x^{de} = x$ (mod p) is an easy consequence of Fermat's theorem [Eq. (45)] and the fact that Eq. (49) implies

$de = Q(p - 1) + 1$ for some integer Q.] That this cipher has the commutative property in Eq. (47) follows from Eq. (50) because

$$(x^{e_1})^{e_2} = x^{e_1 e_2} = (x^{e_2})^{e_1} \pmod{p}$$

When this cipher is used in the three-pass protocol, the three cryptograms become

$$y_1 = x^{e_A} \pmod{p}$$
$$y_2 = x^{e_A e_B} \pmod{p}$$
$$y_3 = x^{e_B} \pmod{p} \tag{51}$$

If one can solve the discrete logarithm problem, one can obtain

$$\log_\alpha y_1 = e_A \log_\alpha x \pmod{p - 1} \tag{52a}$$
$$\log_\alpha y_2 = e_A e_B \log_\alpha x \pmod{p - 1} \tag{52b}$$

where α is any chosen primitive element for arithmetic mod p, and where we have used the fact that the arithmetic of discrete logarithms is mod $(p - 1)$ arithmetic—this follows from Fermat's theorem [Eq. (45)] that gives $\alpha^{p-1} = 1 = \alpha^0$. We can now use Euclid's extended GCD algorithm (see Section 3.3) to find the positive integer b less than $p - 1$ such that

$$b \log_\alpha y_1 = 1 \pmod{p - 1}$$

which from Eq. (52a) further implies

$$b e_A \log_\alpha x = 1 \pmod{p - 1} \tag{53}$$

Multiplying Eq. (52b) by b on both sides, then using Eq. (53), we obtain

$$b \log_\alpha y_2 = e_B \pmod{p - 1} \tag{54}$$

Thus, an enemy who can solve the discrete logarithm problem for modulo-p arithmetic can find e_B, hence also d_B, and thus read the message x just as well as user B. There seems to be no way for the enemy to find x without equivalently solving the discrete logarithm problem, but (like so many other things in public key cryptography) this has never been proved. This particular cipher for the three-pass protocol was proposed by Shamir (and independently but later by J. Omura, who was aware of Shamir's three-pass protocol, but unaware of his proposed cipher for the protocol).

4.4 Conclusion

There are many protocols that have been proposed recently by cryptologic researchers. One of the most amusing is the Shamir–Rivest–Adleman protocol for "mental poker," a protocol that manages to allow an honest game of poker to be played with no cards [36]. Such frivolous-sounding protocols have a serious cryptographic purpose; however, in this case one could take the purpose to be a protocol for assuring the authenticity of randomly chosen numbers. Similarly, Chaum [37] has proposed an interesting protocol by which parties making transactions through a bank can do so without the bank ever knowing who is paying what to whom that also suggests a cryptographic application in key distribution. Protocol formulation has recently gained new momentum and has become one of the most active areas of current cryptologic research, as well as one of the most difficult, particularly when one seeks particular cryptographic functions to imbed in the protocol without compromise of their security. The RSA trapdoor one-way function is far and away the most frequently used function for this purpose.

We have not mentioned many of the important contributions to cryptology made in the past 10 years. It has *not* been our purpose to *survey* research in cryptology, but rather to sketch the intellectual outlines of the subject. Readers wishing to keep abreast of current research in cryptology will find the Proceedings of the CRYPTO conference (held annually in Santa Barbara, California since 1981) and of the EUROCRYPT conference (held annually since 1982) to be invaluable. There are also several journals either completely dedicated to the field, such as the *Journal of Cryptology* or *Cryptologia,* or else with special emphasis on the subject, such as *Designs, Codes and Cryptography* or the *IEEE Transactions on Information Theory.* In addition several recent general textbooks [38]–[42], [58], [59] on cryptology will give the reader an orderly development of the subject. Two recent texts [43], [57] give a broad treatment of the number-theoretic concepts on which much of present-day public key cryptology depends. The book by Rueppel [44] is a good source of information about stream ciphers.

REFERENCES

[1] W. Diffie and M. E. Hellman, "Privacy and authentication: An introduction to cryptography," *Proc. IEEE,* vol. 67, pp. 397–427, March 1979.

[2] G. J. Simmons, "Cryptology," in *Encyclopaedia Britannica,* ed. 16. Chicago: Encyclopaedia Britannica Inc., 1986, pp. 913–924B.

[3] D. Kahn, *The Codebreakers, The Story of Secret Writing.* New York: Macmillan, 1967.

[4] ———— , *The Codebreakers, The Story of Secret Writing,* abridged ed. New York: New American Library, 1973.

[5] R. C. Merkle and M. E. Hellman, "Hiding information and signatures in trapdoor knapsacks," *IEEE Trans. Informat. Theory,* vol. IT-24, pp. 525–530, Sept. 1978.

[6] A. Shamir, "A polynomial-time algorithm for breaking the basic Merkle-Hellman cryptosystem," *IEEE Trans. Informat. Theory,* vol. IT-30, pp. 699–704, Sept. 1984.

[7] D. B. Newman, Jr., and R. L. Pickholtz, "Cryptography in the private sector," *IEEE Commun. Mag.,* vol. 24, pp. 7–10, Aug. 1986.

[8] G. S. Vernam, "Cipher printing telegraph systems for secret wire and radio telegraphic communications," *J. Am. Inst. Elec. Eng.,* vol. 55, pp. 109–115, 1926.

[9] A. Hodges, *Alan Turing, The Enigma.* New York: Simon and Schuster, 1983.

[10] C. E. Shannon, "Communication theory of secrecy systems," *Bell Syst. Tech. J.,* vol. 28, pp. 656–715, Oct. 1949.

[11] ———— , "A mathematical theory of communication," *Bell Syst. Tech. J.,* vol. 27, pp. 379–423, 623–656, July and Oct. 1948.

[12] W. Diffie and M. E. Hellman, "New directions in cryptography," *IEEE Trans. Informat. Theory,* vol. IT-22, pp. 644–654, Nov. 1976.

[13] R. C. Merkle, "Secure communication over insecure channels," *Comm. ACM,* vol. 21, pp. 294–299, Apr. 1978.

[14] G. J. Simmons, "Authentication theory/coding theory," in *Advances in Cryptology, Proceedings of CRYPTO 84,* G. R. Blakley and D. Chaum, Eds. Lecture Notes in Computer Science, No. 196. Berlin: Springer-Verlag, 1985, pp. 411–431.

[15] W. Feller, *An Introduction to Probability Theory and Its Applications,* vol. 2. New York: Wiley, 1966.

[16] "Data encryption standard," Federal Information Processing Standard PUB 46, National Tech. Info. Service, Springfield, VA, 1977.

[17] R. Morris, "The data encryption standard—retrospective and prospects," *IEEE Commun. Mag.,* vol. 16, pp. 11–14, Nov. 1978.

[18] W. Diffie and M. E. Hellman, "Exhaustive cryptanalysis of the NBS data encryption standard," *Computer,* vol. 10, pp. 74–84, June 1977.

[19] M. Hellman, "A cryptanalytic time-memory trade-off," *IEEE Trans. Informat. Theory,* vol. IT-26, pp. 401–406, July 1980.

[20] R. C. Merkle and M. E. Hellman, "On the security of multiple encryption," *Comm. ACM,* vol. 24, pp. 465–467, July 1981.

[21] J. L. Massey, "Shift-register synthesis and BCH decoding," *IEEE Trans. Informat. Theory,* vol. IT-15, pp. 122–127, Jan. 1969.

[22] T. Siegenthaler, "Correlation-immunity of nonlinear combining functions for cryptographic applications," *IEEE Trans. Informat. Theory,* vol. IT-30, pp. 776–780, Sept. 1984.

[23] J. L. Massey and I. Ingemarsson, "The Rip van Winkle cipher—A simple and provably computationally secure cipher with a finite key," in *IEEE Int. Symp. on Informat. Theory,* (Brighton, England) (abstr.), p. 146, June 24–28, 1985.

[24] S. C. Pohlig and M. E. Hellman, "An improved algorithm for computing logarithms in GF(p) and its cryptographic significance," *IEEE Trans. Informat. Theory,* vol. IT-24, pp. 106–110, Jan. 1978.

[25] R. L. Rivest, A. Shamir, and L. Adleman, "A method for obtaining digital signatures and public-key cryptosystems," *Comm. ACM,* vol. 21, pp 120–126, Feb. 1978.

[26] D. E. Knuth, *The Art of Computer Programming.* vol. 1. *Fundamental Algorithms.* Reading, MA: Addison-Wesley, 1973.

[27] R. L. Rivest, "RSA chips (past/present/future)," presented at Eurocrypt 84, Paris, Apr. 9–11, 1984.

[28] A. M. Odlyzko, "On the complexity of computing discrete logarithms and factoring integers," in *Open Problems in Communication and Computation,* B. Gopinath and T. Loven, Eds. New York: Springer, pp. 113–116.

[29] C. Pomerance, "Analysis and comparison of some integer factoring algorithms," in *Computational Number Theory,* H. W. Lenstra, Jr., and R. Tijdeman, Eds. Amsterdam, The Netherlands: Mathematics Centre Tract, 1982.

[30] G. H. Hardy and E. M. Wright, *An Introduction to the Theory of Numbers,* ed. 4. London: Oxford, 1960.

[31] M. O. Rabin, "Probabilistic algorithm for primality testing," *J. Number Theory,* vol. 12, pp. 128–138, 1980.

[32] —— , "Digital signatures and public-key functions as intractable as factorization," Tech. Rep. LCS/TR212, Massachusetts Institute of Technology Laboratory for Computer Science, Cambridge, MA, 1979.

[33] H. C. Williams, "An M^3 public-key encryption scheme," in *Advances in Cryptology, Proceedings of CRYPTO 85*, H. C. Williams, Ed. Lecture Notes in Computer Science, No. 218. Berlin: Springer-Verlag, 1985, pp. 358–368.

[34] M. R. Garey and D. S. Johnson, *Computers and Intractability, A Guide to the Theory of NP-Completeness*, New York: W. H. Freeman, 1979.

[35] A. Lempel, "Cryptology in transition," *Computing Surv.*, vol. 11, pp. 285–303, Dec. 1979.

[36] A. Shamir, R. L. Rivest, and L. Adleman, "Mental poker," in *Mathematical Gardener*, D. E. Klarner, Ed. New York: Wadsworth, 1981, pp. 37–43.

[37] D. Chaum, "Security without identification: Transaction systems to make big brother obsolete," *Comm. ACM*, vol. 28, pp. 1030–1044, Oct. 1985.

[38] H. Beker and F. Piper, *Cipher Systems, The Protection of Communications*. London: Northwood Books, 1982.

[39] D. W. Davies and W. L. Price, *Security for Computer Networks*. New York: Wiley, 1984.

[40] D. E. R. Denning, *Cryptography and Data Security*. Reading, MA: Addison-Wesley, 1982.

[41] A. C. Konheim, *Cryptography, A Primer*. New York: Wiley, 1981.

[42] C. H. Meyer and S. M. Matyas, *Cryptography: A New Dimension in Computer Data Security*. New York: Wiley, 1982.

[43] E. Kranakis, *Primality and Cryptography*. New York: Wiley, 1986.

[44] R. Rueppel, *Analysis and Design of Stream Ciphers*. New York: Springer, 1986.

[45] J. L. Massey, "An introduction to contemporary cryptology," *Proc. IEEE*, vol. 76, pp. 533–549, May 1988.

[46] L. Kruh, "Stimson, the black chamber, and the gentlemen's mail quote," *Cryptologia*, vol. 12, pp. 65–89, Apr. 1988.

[47] H. L. Stimson and McG. Bundy, *On Active Service in Peace and War*. New York: Harper & Bros., 1947.

[48] C. G. Günther, "A universal algorithm for homophonic coding," in *Advances in Cryptology—Eurocrypt'88*, C. G. Gunther, Ed. Lecture Notes in Computer Science, No. 330. Berlin: Springer-Verlag, 1988.

[49] H. N. Jendahl, Y. J. B. Kuhn, and J. L. Massey, "An information-theoretic treatment of homophonic substitution," in *Advances in Cryptology—Eurocrypt'89*, J. -J. Quisquater and J. Vandewalle, Eds. Lecture Notes in Computer Science, No. 434. Berlin: Springer-Verlag, 1990, pp. 382–394.

[50] A. Sgarro, "Informational divergence bounds for authentication codes," in *Advances in Cryptology—Eurocrypt'89*, J. -J. Quisquater and J. Vandewalle, Eds. Lecture Notes in Computer Science No.434. Berlin: Springer-Verlag, 1990, pp. 93–101.

[51] R. Johannesson and A. Sgarro, "Strengthening Simmons' bound on impersonation," *IEEE Trans. Informat. Theory*, vol. 37, no. 4, pp. 1182–1185, July 1991.

[52] I. Csiszár and J. Körner, *Information Theory: Coding Theorems for Discrete Memoryless Systems*. New York: Academic Press, 1981.

[53] U. M. Maurer, "A provably-secure strongly-randomized cipher," in *Advances in Cryptology—Eurocrypt'90*, I. Dåmgard, Ed. Lecture Notes in Computer Science. New York and Heidelberg: Springer-Verlag, no. 473, pp. 361–373, 1991.

[54] U. M. Maurer, "Fast Generation of secure RSA-moduli with almost maximum diversity," in *Advances in Cryptology—Eurocrypt'89*, J. -J. Quisquater and J.

Vandewalle, Eds. Lecture Notes in Computer Science, No. 434. Berlin: Springer-Verlag, 1990, pp. 636–647.

[55] W. S. Jevons, *The Principles of Science* (1st ed. 1873, 2nd ed. 1883). New York: Dover, 1958.

[56] D. H. Lehmer, "A theorem in the theory of numbers," *Bull. Amer. Math. Soc.*, Vol.13, no.2, pp. 501–502, July 1907.

[57] N. Koblitz, *A Course in Number Theory and Cryptography*. New York: Springer, 1987.

[58] H. C. A. van Tilborg, *An Introduction to Cryptology*. Norwell, MA: Kluwer Academic, 1988.

[59] J. Seberry and J. Pieprzyk, *Cryptography: An Introduction to Computer Security*. Englewood Cliffs, NJ: Prentice Hall, 1988.

SECTION 1

Cryptography

CHAPTER 1

The Data Encryption Standard
*Past and Future**

Miles E. Smid and Dennis K. Branstad
National Institute of Standards and Technology

1. The Birth of the DES

2. The DES Controversy

3. Acceptance by Government and Commercial Sectors

4. Applications

5. New Algorithms

6. DES: The Next Decade

7. Conclusions

*First appeared in *Proceedings of the IEEE,* vol. 76, no. 5, pp. 550–559, May 1988. U.S. government work not protected by U.S. copyright.

The Data Encryption Standard (DES) is the first, and to the present date, only, publicly available cryptographic algorithm that has been endorsed by the U.S. government. This chapter deals with the past and future of the DES. It discusses the forces leading to the development of the standard during the early 1970s, the controversy regarding the proposed standard during the mid-1970s, the growing acceptance and use of the standard in the 1980s, and some recent developments that could affect the future of the standard.

1 THE BIRTH OF THE DES

1.1 The Development of Security Standards

In 1972, the National Bureau of Standards (NBS), a part of the U.S. Department of Commerce, initiated a program to develop standards for the protection of computer data. The Institute for Computer Sciences and Technology (ICST), one of the major operating units of the National Bureau of Standards, had been recently established in response to a 1965 federal law known as the Brooks Act (PL89-306) that required new standards for improving utilization of computers by the federal government. Computer security had been identified by an ICST study as one of the high-priority areas requiring standards if computers were to be effectively used. A set of guidelines and standards were defined by the ICST that were to be developed as resources became available in computer security. The guidelines were to include areas such as physical security, risk management, contingency planning, and security auditing. Guidelines were adequate in areas not requiring interoperability among various computers. Standards were required in areas such as encryption, personal authentication, access control, secure data storage, and transmission because they could affect interoperability.

Standards come in different "flavors": basic, interoperability, interface, and implementation.

1. *Basic standards* (also called "standards of good practice") are used to specify generic functions (services, methods, results) required to achieve a certain set of common goals. Examples include standards for purity of chemicals, contents of food products, and in the computer field, structured programming practices.

2. *Interoperability standards* specify functions and formats so that data transmitted from one computer can be properly acted on when received by another computer. The implementation (hardware, firmware, software) or structure (integrated, isolated, interfaced layers) need not be specified in interoperability standards, since there is no intent of replacing one implementation or structure within a system with another.

3. *Interface standards* specify not only the function and format of data crossing the interface, but also include physical, electrical, and logical specifications sufficient to replace one implementation (device, program, component) on either side of the interface with another.

4. *Implementation standards* not only specify the interfaces, functions, and formats, but also the structure and the method of implementation. These may be necessary to assure that secondary characteristics such as speed, reliability, physical security, etc. also meet certain needs. Such standards are often used to permit component replacement in an overall system.

Each of the above types of standards was considered for the specification of the DES. A basic standard did not achieve telecommunications interoperability if different algorithms were selected by the communicating parties. Although an interface standard was desirable in some applications (e.g., data encryption on a RS-232C interface device) it would not be applicable in other applications (e.g., secure mail systems). An implementation standard was rejected because it would restrict vendors from using new technologies. Therefore, the DES was developed as an interoperability standard, requiring complete specification of basic function and format yet remaining independent of physical implementation.

1.2 Public Perception of Cryptography

Cryptography is a word that has been derived from the Greek words for "secret writing." It generally implies that information that is secret or sensitive may be converted from an intelligible form to an unintelligible form. The intelligible form of information or data is called plaintext and the unintelligible form is called ciphertext. The process of converting from plaintext to ciphertext is called encryption and the reverse process is called decryption. Most cryptographic algorithms make use of a secret value called the key. Encryption and decryption are easy when the key is known, but decryption should be virtually impossible without the use of the correct key. The process of attempting to find a shortcut method, not envisioned by the designer, for decrypting the ciphertext when the key is unknown is called *cryptanalysis.*

In the early 1970s, there was little public understanding of cryptography. Most people knew that the military and intelligence organizations used special codes or code equipment to communicate, but few understood the science of cryptography. The International Business Machines Corp. (IBM) initiated a research program in cryptography because of the perceived need to protect electronic information during transmission between terminals and computers and between computers (especially where the transmissions were to authorize the transfer or dispensing of money). Several small companies in the United States made cryptographic equipment for sale, much of it overseas. Several major companies made cryptographic equipment under contract to the U.S. government, but most such equipment was itself classified.

There was an interest in the mathematics of cryptography at several universities, including Stanford and MIT. Cryptographic algorithms were frequently based on mathematics or statistics and hence were often of interest to mathematicians. Making and breaking cryptographic algorithms was considered an intellectual challenge. However, there was only a limited market for expertise in cryptography outside the military and intelligence circles.

The NBS project in computer security identified a number of areas requiring research and the development of standards. A cryptographic algorithm that could be used in a broad spectrum of applications by many different users to protect computer data during transmission and storage was identified as a needed standard. A standard cryptographic algorithm was considered necessary so that only one algorithm needed to be implemented and maintained, and so that interoperability could be easily achieved. This led to the initiation of the NBS project in data encryption and the first solicitation for candidate algorithms.

1.3 The NBS–NSA–IBM Roles

The National Bureau of Standards initiated development of the DES when it published in the *Federal Register* of May 15, 1973, a solicitation for encryption algorithms for computer data protection. Responses to this solicitation demonstrated that there was an interest in developing such a standard, but that little technology in encryption was publicly available. NBS requested assistance from the National Security Agency (NSA) in evaluating encryption algorithms if any were received or in providing an encryption algorithm if none were received.

IBM had initiated a research project in the late 1960s in computer cryptography. The research activity, led by Dr. Horst Feistel, resulted in a system called LUCIFER [1]. In the early 1970s, Dr. W. Tuchman became leader of a development team in cryptographic systems at IBM. This development activity resulted in several publications, patents, cryptographic algorithms, and products. One of the algorithms was to become the Data Encryption Standard.

IBM submitted its cryptographic algorithm to NBS in response to a second solicitation in the *Federal Register* of August 27, 1974. NBS requested that the NSA evaluate the algorithm against an informal set of requirements and simultaneously requested that IBM consider granting nonexclusive, royalty-free licenses to make, use, and sell apparatus that implemented the algorithm. A great deal of discussion was conducted by NBS with both organizations in response to these requests.

On March 17, 1975, nearly 2 years following the first solicitation, NBS published two notices in the *Federal Register*. First, the proposed ''Encryption Algorithm for

Computer Data Protection'' was published in its entirety. NBS stated that it satisfied the primary technical requirements for the algorithm of a DES. It also notified readers to be aware that certain U.S. and foreign patents contain claims that may cover implementation and use of this algorithm and that cryptographic devices and technical data relating to them may come under the export control. The second notice contained a statement by IBM that it would grant the requested nonexclusive, royalty-free licenses provided that the Department of Commerce established the DES by September 1, 1976.

On August 1, 1975, NBS published in the *Federal Register* the fourth notice of a proposed Federal Information Processing Data Encryption Standard. Comments were requested from federal agencies and the public regarding the proposed standard. On October 22, 1975, Dr. M. Hellman sent his criticism of the proposed standard. His letter began, ''Whit Diffie and I have become concerned that the proposed data encryption standard, while probably secure against commercial assault, may be extremely vulnerable to attack by an intelligence organization.'' He then outlined a ''brute force'' attack on the proposed algorithm, using a special-purpose ''parallel computer using one million chips to try one million keys each'' per second. He estimated the financial requirements to build such a machine to be twenty million dollars [2].

Because of the concern for adequate protection to be provided by the DES, NBS continued to evaluate the algorithm, the requirements for security in the private and public sectors, and the alternatives to issuing the standard. Finally, NBS recommended that the standard be issued and it was published on January 15, 1977. The standard included provisions for a review by NBS every 5 years.

2 THE DES CONTROVERSY

2.1 How Long Is Long Enough?

The DES security controversy forced consideration of basic security questions about how good is good enough and how long is long enough. Every practical security system must be evaluated with respect to security, costs (initial, operational, maintenance), and user ''friendliness.'' These factors were studied in great depth during the evaluation of the proposed standard.

The effective key length of the DES is 56 binary digits (bits) and the straightforward ''work factor'' of the algorithm is 2^{56} (i.e., the number of keys that would have to be tried is 2^{56} or approximately 7.6×10^{16}). Hellman and Diffie argued that, in certain situations, a symmetric characteristic of the algorithm would cut this number in half and that on the average, only half of these would have to be tried to find the correct key. They also noted that increasing the key length by 8 bits would ''appear to outstrip even the intelligence agencies' budgets'' but that ''decreasing the key size by 8 bits would decrease the cost, . . . making the system vulnerable to attack by almost any reasonable sized organization.'' It was thus argued that the length of the key was critical to the maximum security provided by the proposed standard.

2.2 S-Boxes and Trapdoors

The second criticism of the proposed standard was that of the fundamental design of the algorithm which is based on a set of eight fixed substitution tables, or S-boxes, that are used in the encryption and decryption processes. It was argued that, since the design criteria of the tables were not publicly available, the entries could have been selected in such a manner as to hide a "trapdoor." The argument was that the people or organizations who selected the tables might be able to cryptanalyze the algorithm while everyone else could not.

2.3 Resolution

NBS, NSA, and IBM were the principals in the development of the Data Encryption Standard as noted above. Since NBS had initiated the development of the DES, NBS was responsible for assuring that the proposed standard met all of the requirements, and that it was acceptable to many potential users with a large number of applications. NBS continued to assess the requirements for the standard, analyze the security concerns regarding the proposed standard, and evaluate the costs and benefits of modifying or replacing the proposed standard. The principals involved in developing the proposed standard decided, after 2 years of evaluation, to rely on a public peer review process in order to make a final decision. Two workshops were organized by NBS; one on the mathematics of the algorithm to analyze the "trapdoor" concern [3], and one on the economic trade-offs of modifying the algorithm to increase its key length [4]. The designers, evaluators, implementors, vendors, and potential users of the algorithm, along with the vocal critics of the proposed standard, were invited to both workshops. A number of mathematicians were also invited to the mathematics workshop.

The workshops were extremely lively. The critics were given an opportunity to state their concerns to the audience. The designers stated that some of the design criteria were classified, but outlined many of the criteria used in the design. The evaluators stated the results of their evaluations. The implementors stated they needed a standard in order to justify implementation costs, and the users stated they wanted a resolution of the issue so that they could obtain effective cryptographic protection of their data.

The decision to publish the proposed standard without modification was made immediately following the workshop. There were no "trapdoors" identified in the algorithm. The potential users and vendors of the algorithm agreed that while the key could have been longer at little additional cost, it was considered adequate for their needs for 10–15 years. There was also concern that any change in the key length would make implementations of the algorithm unexportable to all potential markets. It was therefore recommended that the standard be reviewed every few years to evaluate its continued adequacy for meeting all of its intended applications and meeting all of its requirements. This recommendation has been fulfilled by NBS in 1983 and again in 1988.

3 ACCEPTANCE BY GOVERNMENT AND COMMERCIAL SECTORS

3.1 No Attack Demonstrated

Despite the controversy over the security of the Data Encryption Standard, it is the most widely accepted, publicly available, cryptoalgorithm today. And with the exception of the Rivest–Shamir–Adleman (RSA) public key algorithm, no other algorithm is even a significant contender. The DES has been accepted for two main reasons.

First, despite all the claims of discovered or imagined flaws, no one has demonstrated a fundamental weakness of the DES algorithm. In fact, the only seriously proposed attacks involve exhaustively testing keys until the correct key is found. This method is precisely what designers of cryptoalgorithms hope their adversaries will be forced to attempt. If the number of possible keys is sufficiently large to dissuade the attacker from attempting exhaustively testing keys, and no easier attack on the algorithm can be found, then the designer of the algorithm has succeeded in providing adequate security. Today, most security applications can be subverted for much less than the tens of millions of dollars required to break the DES.

Second, the DES has been accepted because of its endorsement by the federal government. No other publicly available algorithm has ever been endorsed by the U.S. government. Federal agencies are required to use DES for the protection of unclassified data, but the private sector has adopted DES as well because government endorsement implies an approved degree of security. Thus, the DES has become the most widely accepted mechanism for the cryptographic protection of unclassified data.

3.2 DES Validations

Since publishing the Data Encryption Standard, NBS has validated 45 (as of May 7, 1991) hardware and firmware implementations. Approximately three implementations are validated each year. The list of companies with validated chips is quite varied. It contains very small companies as well as many of the large U.S. electronics corporations. The implementations range from firmware programmable read-only memories (PROMs), which implement only the basic DES algorithm, to electronic chips that provide several different modes of operation running at speeds up to 45 million bits per second. The motivations of the companies vary as well. Some sell their implementations to other companies that embody the devices into cryptographic equipments; some of the companies embody the DES devices into equipment that they sell directly; and still others use their devices for their own internal security purposes with no intentions of offering security products for sale. Hardware implementations of the DES are widely available in the United States at prices under $100; DES encryption boards that can encrypt stored and transmitted data in a personal computer are available for under $1000; and stand-alone encryption units may be purchased for under $3000. No other public encryption algorithm can claim such availability.

The Data Encryption Standard requires that the DES algorithm be implemented in hardware (or firmware) for federal applications, but many individuals and corporations have programmed it in software. The number of software implementations is unknown. Reported maximum encryption speeds vary from 100,000 bit/sec on a VAX 780 to 20,000 bit/sec on a personal computer. In many applications, however, low cost is more important than maximum speed. Some vendors offer assembled versions of the DES free of charge, and NBS has provided Fortran and C language DES source listings for testing purposes. The cost of a software implementation depends mostly on the supporting software that is desired along with the algorithm.

3.3 DES Standards-Making Organizations

The widespread acceptance of the Data Encryption Standard is evident from the organizations that have produced DES-based standards. The belief that future communications and data storage systems will require cryptographic protection, and the additional belief that standards are necessary to establish common levels of security and interoperability, led five standards-making organizations to participate in the development of DES-based cryptographic standards. These organizations produce standards in many diverse fields, including security.

1. *The American Bankers Association (ABA):* The ABA develops voluntary standards related to financial matters for their own members. DES cryptography has had applications in both retail and wholesale banking. Generally speaking, retail banking involves transactions between private individuals and a financial institution, while wholesale banking involves transactions among financial institutions and corporate customers. Automatic teller machines and point-of-sale terminals identify customers by means of personal identification numbers (PINs) submitted by the customers at the time of the transaction. The DES is widely used to protect these numbers from disclosure and the information contained in the transactions from alteration. Wholesale electronic fund transfers of 2 million dollars are quite common. U.S. banks collectively transfer more than 400 billion dollars daily. The Clearing House Interbank Payments System (CHIPS) which processes 560,000 messages a week with a total dollar value of 1.5 trillion dollars, uses the DES to protect the messages from unauthorized modification.

The ABA has published a standard recommending the use of the DES whenever encryption is needed to protect sensitive financial data [5]. It has also published a standard for the management of cryptographic keys [6].

2. *The American National Standards Institute (ANSI):* The American National Standards Institute produces voluntary standards in many technical areas. Two committees within ANSI have been involved in developing DES-based cryptographic standards: Accredited Standards Committee (ASC) X3 deals with information processing systems and Accredited Standards Committee (ASC) X9 is responsible for financial services. The Computer and Business Equipment Manufacturers Association (CBEMA) is the secretariat for ASC X3 and the American Bankers Association is the secretariat for ASC X9. ASC X3 standards are published and copyrighted by ANSI while ASC X9 standards are published and copyrighted by the ABA.

Under each committee are subcommittees and working groups. The X3T1 (Data Encryption) subcommittee has standardized the DES as the Data Encryption Algorithm (ANSI X3.92) [7] and produced a Data Encryption Algorithm Modes of Operation Standard (ANSI X3.106) [8]. In the field of network security, X3T1 produced a standard for Information Systems—Data Link Encryption (ANSI X3.105) [9] which makes use of the Data Encryption Algorithm. X3T1 has developed draft standards for encryption at the Transport and Presentation layers of networks which conform to the Open Systems Interconnection Reference Model [10]. The further development of these standards is now taking place in the International Organization for Standardization.

The X9A3 (Financial Institution Retail Security) working group developed DES-based standards for the management and security of PINs (ANSI X9.8) [11], and for the authentication of retail financial messages (ANSI X9.19) [12]. The PIN standard and the use of DES for PIN encryption has been in use for several years. The working group is now developing a key management standard which will provide for the secure distribution of cryptographic keys to the various terminals and host computers used in retail networks (ANSI X9.24) [13].

The X9E9 (Financial Institution Wholesale Security) working group developed DES-based standards for message authentication (ANSI X9.9) [14] and key management (ANSI X9.17) [15]. ANSI X9.17 and its international counterpart are currently the only standards that fully specify automated key distribution protocols. X9E9 is currently in the process of developing DES-based standards for encryption (ANSI X9.23) [16] and for secure personal and node authentication [17].

3. *The General Services Administration (GSA):* The GSA is responsible for the promulgation of federal procurement regulations. Prior to the passage of the Computer Security Act of 1987 [18], GSA was responsible for the development of federal telecommunications standards. GSA had delegated the responsibility for producing and coordinating telecommunications standards to the National Communications System (NCS). However, under the Computer Security Act of 1987, NBS has recently been given the responsibility for computer and related telecommunications standards.

NCS produced three DES-based standards: "Telecommunications: Interoperability and Security Requirements for Use of the Data Encryption Standard in the Physical and Data Link Layers of Data Communications" (Federal Standard 1026) [19], "Telecommunications: General Security Requirements for Equipment Using the Data Encryption Standard" (Federal Standard 1027) [20], and "Interoperability and Security Requirements for Use of the Data Encryption Standard with CCITT Group 3 Facsimile Equipment" (Federal Standard 1028) [21]. Federal Standard 1027 is the only public standard for securely implementing a cryptoalgorithm in electronic equipment. Until January 1, 1988, the National Security Agency endorsed products as conforming to the standard.

4. *The International Organization for Standardization (ISO):* ISO has become increasingly involved in telecommunications security standards. In 1986 ISO voted to approve the DES as an international standard called the DEA-1. However, the approval of the DEA-1 led to a rethinking of the role that ISO should play in the standardization of cryptography. A resolution was passed that ISO should not standardize any cryptoalgorithms, and the ISO Council approved a proposal that the DEA-1 should not progress to publication. As an alternative some ISO members believe that ISO should maintain a public registry of cryptoalgorithms. At a minimum, the registry would contain an agreed on name for each algorithm, thereby providing an international referencing capability.

ISO/TC-68/SC-2/WG-2 (International Wholesale Financial Standards) has produced a message authentication standard [22] and key management [23] standard. Both standards, which permit the use of the DES as well as other cryptoalgorithms, are highly compatible with the corresponding ANSI wholesale authentication and key management standards.

Currently, several ISO groups are involved in developing standards that use cryptography as a mechanism for network security. The standards will provide for data confidentiality, data integrity, peer entity authentication, access control, key distribution, and digital signatures. It is expected that these standards will be compatible with a variety of cryptoalgorithms and applicable to open systems conforming to the Open Systems Interconnection (OSI) standards.

5. *The National Bureau of Standards (NBS):* Under the provisions of Public Law 89-306 and the Computer Security Act of 1987, the Secretary of Commerce is authorized to establish uniform federal automatic data processing standards. Within the

Department of Commerce, standards for computer security (and the protection of unclassified automatic data processing [ADP] data by various means, including the *application* of cryptography) are the responsibility of the Institute for Computer Sciences and Technology (ICST) of the National Bureau of Standards. The Computer Security Act of 1987 affirms and enhances NBS's responsibility for computer security standards and guidance.

NBS has published the Data Encryption Standard (Federal Information Processing Standard [FIPS] 46) [24], Guidelines for Implementing and Using the DES (FIPS 74) [25], DES Modes of Operation (FIPS 81) [26], and Computer Data Authentication (FIPS 113) [27]. These standards have been used as the basis of standards by other standards-making organizations. Additionally, NBS hosts the Workshop for OSI Implementors and chairs its Special Interest Group on Security. This group is selecting which security options in the OSI architecture will be initially implemented.

3.4 Validation and Certification

While cryptographic standards are most useful in defining accepted security methods, often there are no means for determining whether a particular product or implementation does, in fact, conform to a given standard. To satisfy a need for such means, the Department of Treasury, the National Security Agency, and the National Bureau of Standards have developed interrelated validation programs for certain cryptographic systems.

When the Data Encryption Standard was published, NBS felt that it must establish a program for validating hardware implementations. A set of tests were devised so that any device passing all tests was very likely to correctly implement the standard. The success of the program has been previously discussed in this chapter.

Federal Standard 1027 placed additional requirements on equipments beyond the basic DES algorithm. The DES had to be securely embodied into an enclosure with physical access controls including locks and alarms, and the equipment had to be frequently tested for proper operation so that failures would not cause the compromise of sensitive data. The National Security Agency has endorsed at least 32 vendor equipments as properly implementing FS 1027.

In 1984, the U.S. Department of Treasury wrote a policy directive requiring that the Department's electronic funds transfer (EFT) messages be properly authenticated in all new systems immediately and in all systems by 1988 [28]. This policy was affirmed by Treasury Secretary James Baker III on October 2, 1986 [29]. The Treasury also decided to certify vendor authentication devices and wrote the criteria that such devices must meet [30]. Such equipments must implement the DES and conform to FS 1027. NBS and the NSA have assisted Treasury with its certification program.

As a part of this cooperative effort, NBS agreed to develop a validation system which would test conformance of systems to the FIPS 113 and ANSI X9.9 authentication standards. The tests are automated so that a product vendor can call a remote bulletin board system at NBS and validate the product over the telephone. To date, 29 remote validations, including two transatlantic validations, have been performed (as of May 7, 1991). A subsequent security examination is required for Treasury certification, but passing the NBS validation gives the vendor a strong indication that the product functions in accordance with commercial and federal standards. NBS is now developing a key management validation program which will test vendor products for conformance

to the DES-based ANSI wholesale key management standards (ANSI X9.17). The Department of Treasury will use the results of the NBS validation program when certifying the key management capabilities of products intended for Treasury applications.

Since the Data Encryption Standard is a federal standard, the federal government
has established validation and certification programs to ensure product conformance.
No other publicly available algorithm has been validated to this extent.

3.5 Increased Public Knowledge of Cryptography

After the publication of the Data Encryption Standard in 1977 it quickly became clear
that there was much more to the implementation of a secure cryptographic system than
a high-quality cryptographic algorithm. It can be argued that the development of a secure cryptoalgorithm is an essential tool, but only one building block, of a secure data
system. The above mentioned organizations have developed data security standards for
security applications. Their goal was to achieve a common level of security and interoperability. While this goal was not always attained, great strides have been achieved as
a result of their efforts.

The efforts of the standards-making organizations have also served a purpose far
beyond the actual standards that were developed. Standardization, validation, and certification programs greatly increased the public's interest in cryptography and raised the
level of confidence that it could be a cost-effective solution to practical security problems. There is still much to decide about the best use of cryptography, but there is now
no doubt that it will be used far beyond its original military applications.

4 APPLICATIONS

The DES is a basic building block for data protection. The algorithm provides the user
with a set of functions each of which transforms a 64-bit input to a 64-bit output. The
user selects which one of over 70 quadrillion transformation functions is to be used by
selecting a particular 56-bit key. Anyone knowing the key can calculate both the function and its inverse, but without the key it is infeasible to determine which function was
used, even when several inputs and outputs are provided. Since an independent set of 70
quadrillion functions would be impossible to support, the DES provides a simple means
of simulating the family of functions.

4.1 General Applications

The basic DES algorithm can be used for both data encryption and data authentication.

1. *Data Encryption:* It is easy to see how the DES may be used to encrypt a
64-bit plaintext input to a 64-bit ciphertext output, but data are seldom limited to 64
bits. In order to use DES in a variety of cryptographic applications, four modes of
operation were developed: electronic codebook (ECB); cipher feedback (CFB); cipher
block chaining (CBC); and output feedback (OFB) [26] (Figs. 1–4). Each mode has its
advantages and disadvantages. ECB is excellent for encrypting keys; CFB is typically
used for encrypting individual characters; and OFB is often used for encrypting satellite
communications. Both CBC and CFB can be used to authenticate data. These modes of

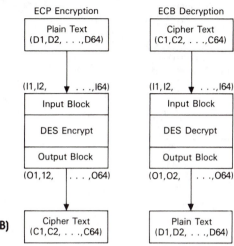

Figure 1 Electronic codebook (ECB) mode.

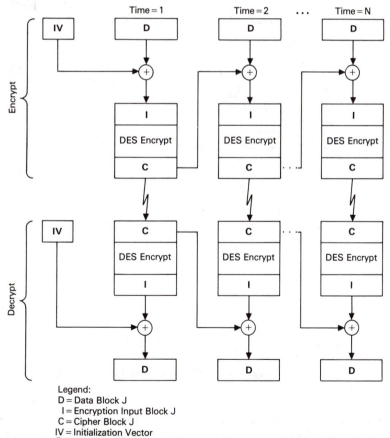

Legend:
D = Data Block J
I = Encryption Input Block J
C = Cipher Block J
IV = Initialization Vector
⊕ = Exclusive-OR

Figure 2 Cipher block chaining (CBC) mode.

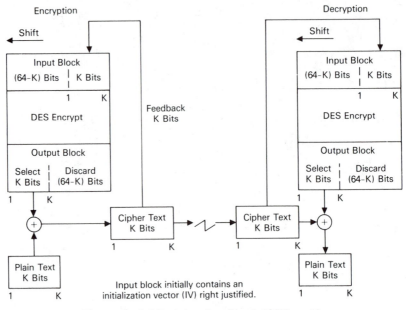

Figure 3 k-bit cipher feedback (CFB) mode.

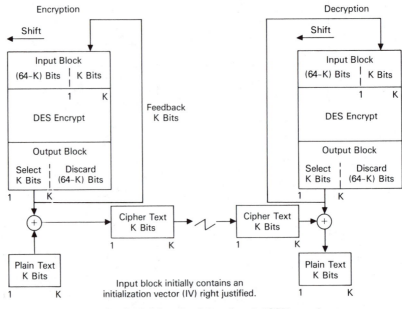

Figure 4 k-bit output feedback (OFB) mode.

operation permit the use of DES for interactive terminal to host encryption, crypto-graphic key encryption for automated key management applications, file encryption, mail encryption, satellite data encryption, and other applications. In fact, it is extremely difficult, if not impossible, to find a cryptographic application where the DES cannot be applied.

2. *Data Authentication:* Originally the Data Encryption Standard was intended for the encryption and decryption of computer data. However, its application has been extended to data authentication as well. In automated data processing systems it is often not possible for humans to scan data to determine if the data have been modified. Examination may be too time consuming for the vast quantities of data involved in modern data processing, or the data may have insufficient redundancy for error detection. Even if human scanning were possible, the data could have been modified in such a manner that it would be very difficult for the human to detect the modification. For example, "do" may have been changed to "do not" or "$1900" may have been changed to "$9100". Without additional information the human scanner could easily accept the altered data as authentic. These threats may still exist even when data encryption is used. It is therefore desirable to have an automated means of detecting both intentional and unintentional modifications to data. Ordinary error detecting codes are not adequate because, if the algorithm for generating the code is known, an adversary can generate the correct code after modifying the data. Intentional modification is undetectable with such codes. However, DES can be used to produce a cryptographic checksum that can protect against both accidental and intentional, but unauthorized, data modification. NBS Standard for Computer Data Authentication (FIPS 113) [27] describes the process. Essentially the data are encrypted using either the cipher feedback or the cipher block chaining mode which yields a final cipher block that is a function of all the plaintext bits. The plaintext message may then be transmitted with the computed final cipher block used as the cryptographic checksum.

3. *Data Encryption and Authentication:* The same data may be protected by both encryption and authentication. The data are protected from disclosure by encryption and modification is detected by authentication. The authentication algorithm may be applied to either the plaintext or the cipher. In most financial applications where both encryption and authentication are implemented, authentication is applied to the plaintext.

4.2 Specific Applications

1. *Data Storage and Mail Systems:* Encryption and authentication may be used to protect data stored in computers. Many computer systems encrypt passwords in a one-way fashion for storage in the computer memory. When a user signs on the computer and enters the password, it is encrypted and compared with the stored value. If the two encryptions are equal the user is permitted access to the computer; otherwise access is denied. The encrypted password is often created by using DES; setting the key equal to the password and the plaintext equal to the user's identity. A Fortran program for implementing this function is given in the NBS Standard for Password Usage (FIPS 112) [31].

The DES can also be used to encrypt computer files for storage. NBS Special Publication 500-54 [32] describes a key notarization system which may be integrated into computer systems to protect files from undetected modification and disclosure, and to provide a digital signature capability using the DES. Users have the capability of exercising a set of commands for key management as well as for data encryption and authentication functions. The facilities perform notarization which, on encryption, seals a key or password with the identities of the transmitter and intended receiver. Thus, in order to decrypt a message, the receiver must be authenticated and must supply the correct identity of the transmitter. This notarization technique is used in ANSI standard X9.17 to protect against key substitutions which could lead to the compromise of sensitive data.

The key notarization system that incorporates the DES may also be used in conjunction with a mail system to provide for secure mail. A cryptographic header that contains the information necessary to decrypt and authenticate a mail file is automatically appended to the file that is transmitted to the receiver. The receiver may then decrypt and authenticate the file in a near transparent manner.

2. *Electronic Funds Transfers (Retail and Wholesale):* Perhaps the most significant use of the DES is for the protection of retail and wholesale electronic funds transfer messages. The retail and wholesale financial communities have developed standards for the authentication of EFT messages (ANSI X9.9 and ANSI X9.19), and these efforts have led to encryption (ANSI X9.23 Draft) and key management (ANSI X9.17 and ANSI X9.24 Draft) standards. DES is used in automatic teller machines, point of sale terminals, workstations, and host computers. The data that it protects range from a $50 charge to a multi-million-dollar transfer. The flexibility of the basic DES algorithm permits its use in a wide variety of EFT applications. The standards that have been developed for U.S. EFT applications are now being developed into international standards in the ISO community. Therefore, these authentication, encryption, and key management techniques will be used worldwide.

The U.S. government is responsible for transferring billions of dollars daily. In order that these transfers be secure, the Department of Treasury initiated its (previously cited) policy on the authentication of EFT messages. The Federal Reserve Bank is cooperating with the Treasury to insure that this policy is successful. One system, which the Treasury is considering, makes use of hand-held tokens that contain DES keys that are generated for a particular individual. The token is used to supply a key that authenticates an EFT message containing the individual's identity. This authenticated message, containing the individual's identity, is the electronic substitute for a signed paper document.

3. *Electronic Business Data Interchange:* Large corporations are now in the process of automating their business transactions to reduce costs and increase efficiency. Business transactions will be accomplished via electronic means rather than by traditional paper-based systems, and ANSI Accredited Standards Committee X12 (Electronic Business Data Interchange) is now in the process of developing the formats that will be used for these communications. Electronic transmissions among buyer, seller, and banker will have to be protected from modification and eavesdropping. In most cases cryptography provides the only effective mechanism for providing such protection.

Electronic business data interchange will incorporate several DES-based standards [33–34]. ANSI X9.9 will provide protection against unauthorized modification and replay; the methods of draft ANSI Standard X9.23 will prevent unauthorized disclosure; and the secure generation, distribution, and storage of DES keys will be accomplished using the techniques specified in ANSI Standard X9.17. Currently General Motors and seven associated banks are using the method specified in these standards to protect their business transactions.

5 NEW ALGORITHMS

5.1 Forces for New Algorithms

From its initial specification, the Data Encryption Standard was intended to be a publicly known algorithm. Previously, most cryptographic algorithms fell into one of three categories: outdated algorithms developed during the Second World War, proprietary algorithms known only to the vendors who designed them, and classified government algorithms. Therefore, commercial and nonclassified government users did not have confidence that the algorithms available to them offered a reasonable level of security. NBS developed the DES to provide a high-quality, modern cryptoalgorithm that could be used to protect unclassified sensitive data.

In addition, the DES was intended to be widely available. DES has been published, dissected, and analyzed in the open literature. It can be built and used without a clearance or license (in the United States). It can be implemented in hardware, firmware, or software by anyone from a large corporation to a private individual.

Making a cryptographic algorithm publicly known has its disadvantages as well. Even though the DES is designed to be secure as long as the secret key is kept secret, algorithms that are kept secret can make the attacker's task more difficult since the algorithm often has to be deduced before the algorithm can be broken. Also, if a known algorithm becomes popular and is widely used, as is the case with the DES, it becomes a more attractive target for the attacker. Since the potential payoff is greater, the attacker may be willing to put forth an increased effort in breaking the algorithm.

On the other hand, one should not put too much value into the secrecy of the algorithm. First of all, poorly designed secret algorithms can often be deduced by the attacker. Consider, for example, the recent article in which five secret algorithms were easily recovered and broken [35]. Second, algorithms that are themselves secret are usually compromised (i.e., disclosed) sooner or later. For this reason, governments design their classified algorithms assuming the details of the design have been, or will be, compromised.

Since the DES has been publicly known for more than 10 years and since it is becoming very widely used, the National Security Agency (NSA) has decided to develop new algorithms. These algorithms will provide the cryptosecurity for the program discussed in the following section.

5.2 CCEP: The New Way of Doing Business

In 1984, the NSA initiated the Commercial COMSEC Endorsement Program (CCEP) which was intended by NSA to provide cryptographic algorithms that would eventually replace the DES [36]. NSA has stated that in 1988 it would no longer endorse equipments as complying with Federal Standard 1027, and that CCEP would provide government-endorsed cryptographic equipments [37]. Two types of cryptographic equipment are intended by NSA to be produced: type 1 and type 2. Type 1 equipment would protect classified data while type 2 equipment is intended by NSA to replace DES for the protection of unclassified data. The CCEP differs from the Federal Standard 1027 endorsement program in three respects.

1. The cryptoalgorithms would be designed only by NSA.

2. The cryptoalgorithms would not be made public. A protective coating will be used on electronic chip implementations to prevent reverse engineering.

3. The manufacturers of CCEP products and NSA would follow a seven-step process leading to product production: initial contact; program decision (approval); memorandum of understanding and transfer of technology by NSA; memorandum of agreement and product specification; program execution and product development and evaluation; endorsement; and production.

NSA's intent of the CCEP is that less expensive and technologically more sophisticated products will be produced as a result of an increased market base (both government and commercial) and the technical guidance provided by the NSA.

5.3 Unresolved Issues

The CCEP program still has several unresolved issues.

1. Since vendors permitted to enter the program must meet certain criteria, competition is restricted. Restricted competition can lead to higher customer costs.

2. Since the CCEP algorithms are secret and their implementation is restricted to vendors participating in the program, software implementations that do not lend themselves to the physical security provided by the protective coating would defeat the secrecy of the algorithm and therefore would not be permitted.

3. Since CCEP algorithms are secret and their implementation by foreign manufacturers will likely be restricted, end-to-end cryptography for many international security applications will be impossible. Future international networks may require cryptographic gateways between countries where the data are translated from the cryptographic protection of one country to the cryptographic protection of the other. In such networks, end users would have to be satisfied that their data remained secure within these gateways.

4. It is not clear whether the user will be able to select the key or if the user will have to use a key provided by NSA.

5. Since sophisticated cryptography and highly secure implementations often result in increased costs, the number of customers is usually reduced which in turn increases the cost of individual equipments.

It is still too early to determine whether the CCEP will be successful in meeting its goals, especially in unclassified government applications and in the commercial sector.

6 DES: THE NEXT DECADE

6.1 Renewing DES for Another 5 Years

On March 6, 1987, NBS published in the *Federal Register* a request for comments on the second Five Year Review of the Data Encryption Standard. Three alternatives were suggested for consideration.

1. Reaffirm the standard for another 5 years. The National Bureau of Standards would continue to validate equipment that implements the standard. The DES would continue to be an approved method for protecting unclassified computer data against unauthorized modification or disclosure.

2. Withdraw the standard. The National Bureau of Standards would no longer continue to support the standard. Organizations could continue to utilize existing equipment that implements the standard, and nongovernment organizations could continue to develop new implementations as desired.

3. Revise the applicability of the standard. The applicability statement of the standard would be changed to specify certain uses, such as using the standard for protecting electronic funds transfers. Proposed technical changes to the algorithm will not be considered during this review.

Thirty-three comments were received; 12 were from federal agencies and the remainder were from the private sector. The federal agency responses were often at the department level, and the private sector responses included comments from industry organizations such as the Computer and Business Equipment Manufacturers Association and the American Bankers Association. Thirty-one comments supported the reaffirmation of the standard for another 5 years. One organization stated that it had no comments but did not oppose reaffirmation, and one organization recommended that the DES be modified to apply only to the protection of financial transactions.

Many of the comments pointed out that the DES is widely available as a commercial product, that it is used extensively by both commercial and government organizations for a variety of applications extending far beyond financial transactions, and that no adequate alternative currently exists. Withdrawal of the standard or the limitation of it to financial transactions would leave many organizations without adequate protection for their information.

NBS reviewed all comments, and made its recommendation to the secretary of commerce. After considering all available information, the secretary of commerce reaffirmed the standard, in its present form, for another 5 years. The standard will be reviewed again beginning on or before January 1992.

Waivers will be considered for devices certified by the National Security Agency as complying with its commercial COMSEC Endorsement Program when such devices offer equivalent cost and performance features as compared to devices conforming with the DES.

6.2 Government Use

The DES is now a basic security mechanism employed by several government organizations. For example, the Department of Energy has more than 30 active networks using DES devices, and the Justice Department is in the process of installing 20,000 DES radio units. It is likely that the DES will continue to provide protection for network communications, stored data, passwords, and access control systems.

6.3 Commercial and Government Financial Applications

Many commercial and certain government applications have already committed to the DES. DES is the basis of the Department of the Treasury's Electronic Funds Transfer program, and the Federal Reserve System uses DES to encrypt connections between depository financial institutions and Federal Reserve banks. In addition, many financial and electronic business data applications already use DES and are unlikely to change for some time.

6.4 Gradual Progression of New Security Devices

In the past, the cryptography industry has not experienced rapid growth. Indications are that the interest and commitment to security by U.S. corporations is increasing and therefore the market for security products will increase as well. It is important that new products be developed that can offer cost, performance, and security advantages. However, it is also important to make use of existing technologies. Since the DES offers a substantial security improvement to the vast majority of government and commercial data security applications, sensitive data should not be left unprotected while waiting for future cryptographic systems.

7 CONCLUSIONS

As we move toward a society where automated information resources are increasingly shared, cryptography will continue to increase in importance as a security mechanism. Electronic networks for banking, shopping, inventory control, benefit and service delivery, information storage and retrieval, distributed processing, and government applications will need improved methods for access control and data security. The DES algorithm has been a successful effort in the early development of security mechanisms. It is the most widely analyzed, tested, and used cryptoalgorithm and it will continue to be for some time yet to come. But perhaps the most important contribution of the DES is that it has led us to other security considerations, beyond the algorithm itself, that must be made in order to have secure computer systems and networks.

REFERENCES

[1] H. Feistel, "Cryptography and computer privacy," *Sci. Amer.,* vol. 228, no. 5, pp. 15–23, May 1973.

[2] W. Diffie, M. Hellman, "Exhaustive cryptanalysis of the NBS data encryption standard," *Computer,* pp. 74–78, June 1977.

[3] D. Branstad, J. Gait, and S. Katzke, "Report on the workshop on cryptography in support of computer security", Sept. 21–22, 1976, NBSIR-771291, Sept. 1977.

[4] "Report of the workshop on estimation of significant advances in computer technology," NBSIR 76-1189, National Bureau of Standards, Dec. 1976.

[5] "Management and use of personal identification numbers," ABA Bank Card Standard, Aids from ABA, Catalog no. 207213, 1979.

[6] "Key management standard," Document 4.3, American Bankers Association, Washington, DC, 1980.

[7] "American national standard for data encryption algorithm (DEA)" ANSI X3.92–1981, American National Standards Institute, New York.

[8] "American national standard for information systems—Data encryption algorithm—Modes of operation," ANSI X3.106–1983, American National Standards Institute, New York.

[9] "American national standard for information systems—Data link encryption," ANSI X3.105–1983, American National Standards Institute, New York.

[10] "Information processing systems—Open systems interconnection—Basic reference model," IS 7498–1984, International Organization for Standardization, Geneva, Switzerland.

[11] "American national standard for personal identification number (PIN) management and security," ANSI X9.8–1982, American Bankers Association, Washington, DC.

[12] "American national standard for retail message authentication," ANSI X9.19–1985, American Bankers Association, Washington, DC.

[13] "Draft proposed American national standard for retail key management," ANSI X9.24–1988, American Bankers Association, Washington, DC.

[14] "American national standard for financial institution message authentication (wholesale)," ANSI X9.9–1986 (Revised), American Bankers Association, Washington, DC.

[15] "American national standard for financial institution key management (wholesale)," ANSI X9.17–1985 (Revised), American Bankers Association, Washington, DC.

[16] "American national standard for financial institution message encryption," ANSI X9.23–1988, American Bankers Association, Washington, DC.

[17] "American national standard for financial institution sign-on authentication for wholesale financial transactions," ANSI X9.26–1990, American Bankers Association, Washington, DC.

[18] Computer Security Act of 1987, PL 100–235.

[19] "Telecommunications: Interoperability and security requirements for use of the data encryption standard in the physical and data link layers of data communications," Federal Standard 1026, General Services Administration, Washington, DC, Jan. 1983.

[20] "Telecommunications: General security requirements for equipment using the data encryption standard," Federal Standard 1027, General Services Administration, Washington, DC, Apr. 1982.

[21] "Interoperability and security requirements for use of the data encryption standard with CCITT group 3 facsimile equipment," Federal Standard 1028, General Services Administration, Washington, DC, Apr. 1985.

[22] "Banking—Requirements for message authentication (wholesale)," DIS 8730, Association for Payment Clearing Services, London, July 1987.

[23] "Banking—Key management (wholesale)," DIS 8732, Association for Payment Clearing Services, London, Dec. 1987.

[24] "Data encryption standard (DES)," National Bureau of Standards (U.S.), Federal Information Processing Standards Publication 46, National Technical Information Service, Springfield, VA, Apr. 1977.

[25] "Guidelines for implementing and using the NBS data encryption standard," National Bureau of Standards (U.S.), Federal Information Processing Standards Publication 74, National Technical Information Service, Springfield, VA, Apr. 1981.

[26] "DES modes of operation," National Bureau of Standards (U.S.), Federal Information Processing Standards Publication 81, National Technical Information Service, Springfield, VA, Dec. 1980.

[27] "Computer data authentication," National Bureau of Standards (U.S.), Federal Information Processing Standards Publication 113, National Technical Information Service, Springfield, VA, May 1985.

[28] "Electronic funds and securities transfer policy," Department of the Treasury Directives Manual, Chapter TD 81, Section 80, Department of the Treasury, Washington, DC, Aug. 16, 1984.

[29] "Electronic funds and securities transfer policy—Message authentication and enhanced security," Department of the Treasury Order number 106-09, Department of the Treasury, Washington, DC, Oct. 2, 1986.

[30] "Criteria and procedures for testing, evaluating, and certifying message authentication devices for federal E.F.T. use," United States Department of the Treasury, May 1, 1985.

[31] "Password usage," National Bureau of Standards (U.S.), Federal Information Processing Standards Publication 112, National Technical Information Service, Springfield, VA, May 1985.

[32] "A key notarization system for computer networks," National Bureau of Standards (U.S.), Special Publication 500-54, National Technical Information Service, Springfield, VA, Oct. 1979.

[33] "American standards committee X12 draft standard for trial use for managing electronic data interchange, cryptographic service, message transaction set (815)," ANSI X12.42–1990, Data Interchange Standards Association, Inc., Alexandria, VA.

[34] "American standards committee X12 draft standard for trial use for managing electronic data interchange, security structures," ANSI X12.58–1990, Data Interchange Standards Association, Inc., Alexandria, VA.

[35] M. Kochanski, "A survey of data insecurity package," *Cryptologia,* pp. 1–15, Jan. 1987.

[36] C. Barker, "An industry perspective of the CCEP," presented at the 2nd Annual AIAA Computer Security Conf. Proceedings, Dec. 1986.

[37] Letter from H. E. Daniels, Jr., NSA Deputy Director for Information Security, to Datapro Research Corporation, Dec. 23, 1985.

CHAPTER 2

Stream Ciphers

Rainer A. Rueppel
R^3 Security Engineering

1 INTRODUCTION

Symmetric cryptosystems can be subdivided into block and stream ciphers. *Block ciphers* operate with a fixed transformation on large blocks of plaintext data; *stream ciphers* operate with a time-varying transformation on individual plaintext digits. Typically, a stream cipher consists of a keystream generator whose pseudo-random output sequence is added modulo 2 to the plaintext bits. A major goal in stream cipher design is to efficiently produce random-looking sequences, that is, sequences that as closely as possible resemble coin-tossing sequences. In general, one considers a sequence random if no patterns can be recognized in it, if no predictions can be made about it, and no simple description of it can be found. But if in fact the keystream can be generated efficiently, there certainly exists such a simple description. Nevertheless, the generator may produce "indistinguishable" sequences if no computations done on the keystream can reveal this simple description. The original key must be transformed in such a complicated way that it is computationally infeasible to recover the key. The level of randomness of a sequence can be defined in terms of the classes of computations which cannot detect statistical irregularities in it. Statistical tests of randomness emulate various simple computations encountered in practice, and check that the statistical properties of the sequence under investigation agree with those predicted if every sequence element was drawn from a uniform probability distribution.

We distingish four principal approaches to the construction of stream cipher systems [102]. These approaches differ in their assumptions about the capabilities and opportunities of the cryptanalyst, in the definition of cryptanalytic success, and in the notion of security.

1. *Information-theoretic approach.* In the information-theoretic approach the cryptanalyst is assumed to have unlimited time and computing power. Cryptanalysis is the process of determining the message (or the particular key) given only the cryptogram and the a priori probabilities of various keys and messages. The secrecy system is considered broken when there is a "unique" solution to the cryptogram: one message with probability essentially unity while all others are practically zero. The assumption of an infinitely powerful adversary implies that the notion of security in the Shannon model is independent of the complexity of the encryption or decryption method. A cipher system is said to be *perfectly secure* if the plaintext and the ciphertext are statistically independent. That is, the cryptanalyst is no better off after observation of the cryptogram than he was before. A cipher system is said to be *ideally secure* if the cryptanalyst cannot find a unique solution for the plaintext, no matter how much ciphertext he is allowed to observe.

An interesting subproblem, called the local randomization problem, arises when the basic information-theoretic model is modified in such a way that the opponent's observations are assumed to be limited to less than a certain number of (not necessarily consecutive) ciphertext digits.

2. *System-theoretic approach.* The objective of the system-theoretic approach is to make sure that each new cryptosystem creates a difficult and previously unknown problem for the cryptanalyst. Breaking such a system is an "unglamorous" problem as compared to the cryptanalysis of a system based on some "famous" problem such as factoring or the discrete log. The designer's goal is to make sure that none of the fundamental cryptanalytic principles (such as substitution, divide and conquer, statistical analysis) are applicable to his system. To prevent cryptanalysis based on these fundamental principles, a set of general design criteria for keystream generators has evolved over time. Examples are period, linear complexity, statistical criteria, confusion, diffusion, and nonlinearity criteria for Boolean functions. Because the criteria are mainly of system-theoretic nature, we refer to this approach as the system-theoretic approach. A (stream) cipher system is designed in such a way that it directly satisfies the set of applicable design criteria. Doubtless this is today the most widely used practical design methodology.

3. *Complexity-theoretic approach.* In the complexity-theoretic approach all computations are parametrized by a security parameter, usually the key length, and an asymptotic analysis is carried out. Only algorithms whose running times can be expressed as polynomials in the size of the input are considered to be computationally feasible. Cryptanalysis is the process of (a) predicting a single digit of the keystream, or (b) distinguishing the keystream sequence from a truly random sequence. The designer's goal is to base his stream cipher system on, or make it equivalent to, some computationally infeasible problem. A keystream generator is defined to be *perfect* if it is (a) unpredictable, or (b) indistinguishable by all polynomial time statistical tests. Unfortunately, the perfect generator is a hypothetical device; it is not known if it exists. The proposed generators are all based on the assumed difficulty of one out of a few "famous" problems such as for instance, discrete logarithm, quadratic residuosity, and inverting the Rivest–Shamir–Adleman (RSA) encryption algorithm.

4. *Randomized stream ciphers.* The designer may, instead of aiming to ensure that the cryptanalytic process involves an infeasible work effort, try to ensure that the cryptanalytic problem has an infeasible size. The objective is to increase the number of bits the cryptanalyst has to examine in the cryptanalytic process while keeping the secret key small. This can done by making use of a large publicly accessible random string in the encryption and decryption process. The key then specifies which parts of the large randomizer are to be used, whereas the opponent, not knowing the secret key, is forced to search through all the random data. The security of randomized stream cipher systems may be expressed by (a lower bound on) the average number of bits the cryptanalyst must examine before his chances of determining the key or the plaintext improve over pure guessing. Different interpretations of such a result are possible. The expected number of bittests is a lower bound on the number of steps that any algorithm breaking the system must perform, and thus leads to a notion of computational security. But the expected number of bittests is also a lower bound on the number of bits the opponent has to observe before his a posteriori probabilities of the various keys improve, and thus leads to a notion of information-theoretic security. These possible interpretations are the reason we have treated these stream cipher systems in a separate section.

In this survey, we will discuss and illustrate these four approaches. Many of the comments apply not only to stream ciphers but also to cryptosystems in general. Desmedt [24] recently looked at the dual problem of classifying cryptosystems according to the methods that have to be used to cryptanalyze them.

Often, linear feedback shift registers (LFSR) and linear congruential generators are recommended as pseudorandom sequence generators. But being linear devices they are usually easily cryptanalyzed no matter how many of the parameters are kept secret. Since we do not consider LFSRs and linear congruential generators as possible keystream generators (although they may be potential building blocks in a stream cipher design), and since there is a wealth of literature on their characteristic properties [43,60,130], we refrain from describing them in this survey.

As history has shown, most cryptanalytic successes were caused by human failure and poor administrative procedures, especially in the key management domain. In addition, a secure cipher design does not necessarily imply a secure implementation, as was pointed out explicitly by Desmedt [24]. It seems obvious that not only does one have to verify that the implemented cipher algorithm performs as specified in the design, but also that the assumptions of the design, such as physical security of the keys, are satisfied. There also remains the larger issue of verifying that no additional functionality in the implementation compromises security. A description of historic designs and failures or implementation issues is beyond the scope of this survey, where we want to concentrate on current research directions in stream cipher design.

First we give a classification of stream cipher systems and their different modes of operation.

1.1 Terminology and Modes of Operation

Automata theory and, in particular, finite-state machine theory are well suited to describe stream cipher systems and their different modes of operation. Let $\mathcal{X}$ denote the plaintext alphabet, $\mathcal{Y}$ the ciphertext alphabet, $\mathcal{Z}$ the keystream alphabet, $\mathcal{S}$ the state space (internal state) of the stream cipher, $\mathcal{K}$ the key space. Let x_i, y_i, z_i, and s_i denote the plaintext digit, ciphertext digit, keystream digit, and the internal state at time i. A key $k \in \mathcal{K}$ is selected according to a probability distribution P_K. Usually the key is drawn according to the uniform probability distribution, but occasionally it may be infeasible to select the key completely randomly.

A general stream encryptor is described by the equations

$$s_{i+1} = F(k, s_i, x_i)$$
$$y_i = f(k, s_i, x_i)$$

where F is the next-state function and f is the output function. Typically

$$y_i = x_i + F(k_i, s_i)$$

a condition that is necessary and sufficient for the stream decryptor to operate without delay [66]. The sequence

$$\{z_i = f(k, s_i) : i \geq 1\}$$

is referred to as the keystream. To provide secure encryption the keystream must be as random as possible.

1.1.1   Synchronous stream ciphers and pseudorandom generators.   Stream ciphers are commonly subdivided into synchronous and self-synchronizing systems. In a synchronous stream cipher the keystream is generated independently of the message stream (and the ciphertext stream). The corresponding device is referred to as keystream generator, running-key generator, or pseudorandom sequence generator. The operation of the keystream generator is governed by the two rules:

$$s_{i+1} = F(k, s_i)$$
$$z_i = f(k, s_i)$$

The initial state s_0 may be a function of the key k, and possibly, some randomization variable. The purpose of a keystream generator is to expand a short random key k into a long pseudorandom string $z^l = z_1, \ldots, z_l$. This can be written compactly as

$$G : \mathcal{K} \to \mathcal{Z}^l$$
$$z^l = G(k)$$

which emphasizes the functional relationship between the key k and the keystream z^l. For a binary key k^n of length n and a binary keystream of length l the (n, l) sequence generator is a function from $\{0, 1\}^n$ to $\{0, 1\}^l$ such as that $z^l = G(k^n)$.

In complexity theory a generator is defined asymptotically as an infinite class $\{G_n : n \geq 1\}$ of $(n, l(n))$ sequence generators, where l is a polynomial function of the index n and the computation time of each sequence generator is upperbounded by a polynomial function of n.

Synchronous stream ciphers can be further classified according to the mode they are operated in :

Counter mode [25]:

$$s_{i+1} = F(s_i)$$
$$z_i = f(k, s_i)$$

The next-state function does not depend on the key but is guaranteed to step through all (or most) of the state space. Examples for such F are ordinary counters and maximum-length LFSRs. The cryptographic strength necessarily resides in the output function f.

Output feedback (internal feedback) mode [15]:

$$s_{i+1} = F(k, s_i)$$
$$z_i = f(s_i)$$

The output function f does not depend on the key. Very often it simply consists of a 1-bit predicate of the current state (for instance, the least significant bit or the parity). For a discussion of using the Data Encryption Standard (DES) in output feedback mode see [31]. Sometimes one sees a variant of this mode where the key k determines only the initial state.

$$s_0 = k$$
$$s_{i+1} = F(s_i)$$
$$z_i = f(s_i)$$

A synchronous stream cipher has no error propagation. The decryption of a distorted ciphertext digit only affects the corresponding plaintext digit. But a synchronous stream cipher (as its name indicates) requires perfect synchronization between sender and receiver. If a digit is gained or lost during transmission the receiver only obtains garbled data after the point of synchronization loss. To regain synchronization would typically require searching over the possible offsets between the sender's and the receiver's clock. In general, to "enter late" into an ongoing ciphertext sequence is only possible for the decrypter if he can determine the correct time instant registered by the encryptor's clock.

1.1.2   Self-synchronizing stream ciphers and scramblers.   The most common mode of a self-synchronizing stream cipher is the *cipher feedback mode:*

$$s_i = F(y_{i-1}, y_{i-2}, \ldots, y_{i-N})$$
$$z_i = f(k, s_i)$$

The state of the stream cipher is determined by the previous N ciphertext symbols. The stream encryptor employs feedback, but the corresponding finite-state machine inverse, the decryptor, uses feedforward. The cryptographic strength resides in the output function. Note that input (the ciphertext symbols) and output (the keystream) of f are known to the cryptanalyst under a known plaintext attack. A self-synchronizing stream decryptor has, as its name indicates, the ability to automatically synchronize itself without knowledge of the encryptor's clock time. On reception of N correct consecutive channel digits its state becomes identical to the encryptor's state and synchronization is regained. A self-synchronizing stream cipher has limited error propagation. A distorted channel digit remains in the internal state of the stream decryptor for N consecutive plaintext digits (as it is shifted through the state). After reception of N consecutive undistorted channel digits the stream decryptor is able to decipher correctly.

A *scrambler* is also defined as a finite-state machine that, regardless of the initial state, converts a periodic input sequence into a periodic output sequence with generally higher, but at least the same, period [103]. Scramblers have a randomizing effect on the data patterns transmitted and thus can reduce the sensitivity of synchronization systems to specific periodic data patterns [127]. Typically scramblers are linear devices whose purpose is to facilitate synchronization. Nonlinear key-dependent scramblers (self-synchronizing stream ciphers) have the potential to combine secrecy with ease of synchronization. Some basic results and a case study are reported in [66].

Synchronous stream ciphers can be equipped with a self-synchronizing feature. This typically involves periodically checking the internal state of the decryptor or reinitializing the states of both the encryptor and the decryptor on some preagreed condition.

2 INFORMATION-THEORETIC APPROACH

In his landmark paper [109] Shannon developed the basic information-theoretic approach to secrecy systems. In the Shannon model the cryptanalyst is assumed to have unlimited time and computing power. He is restricted to a ciphertext-only attack, but as Diffie and Hellman [25] pointed out, a known segment of plaintext can be taken into account as added redundancy since it improves the a priori information of the cryptanalyst about the plaintext. The a posteriori knowledge consists of the (a posteriori) probabilities of the various messages and keys that may have produced the observed ciphertext. Cryptanalysis in the Shannon model is defined as the process of determining the message (or the particular key) given only the cryptogram and the a priori probabilities of various keys and messages. The secrecy system is considered broken when there is a "unique" solution to the cryptogram: one message with probability essentially unity while all others are practically zero. Note that the assumption of an infinitely powerful adversary implies that the notion of security in the Shannon model is independent of the complexity of the encryption or decryption method.

Let $X^n = X_1, \ldots, X_n$ denote an n-bit plaintext message, let $Y^n = Y_1, \ldots, Y_n$ denote the corresponding n-bit ciphertext message, and let K be the key drawn according to the probability distribution P_K. If $H(X^n)$ denotes the uncertainty about the plaintext and $H(X^n \mid Y^n)$ the conditional uncertainty after observation of Y^n, then the mutual information $I(X^n; Y^n) = H(X^n) - H(X^n \mid Y^n)$ between X^n and Y^n is a basic measure of security in the Shannon model. Three cases may occur:

1. $I(X^n; Y^n) = 0$ for all n; then X^n and Y^n are statistically independent, and the cryptosystem is said to be *perfectly secure*. The cryptanalyst cannot do better than guess the message according to the probability distribution of meaningful messages. In other words, the (a posteriori) probability distribution of decrypted messages $D(K, y^n)$ after observation of y^n with K chosen according to P_K is, for all y^n, identical to the (a priori) probability distribution over the message space. In a perfect cryptosystem the basic inequality $H(K) \geq H(X^n)$ holds. Thus, fresh key has to be supplied at a constant rate. Knowing a plaintext segment of a perfect cipher may reveal the subkey used to encrypt that plaintext segment but will in general not allow to sharpen the a posteriori probabilities of the remaining message segments. The one-time pad discussed below is a perfect cipher.

2. $0 < I(X^n; Y^n) < H(X^n)$ for large n; then there remains a residual uncertainty about the plaintext that cannot be resolved (not even with unlimited computing power). This condition implies that the key equivocation $H(K \mid Y^n) > 0$. Cryptosystems with a finite key for which $H(K \mid Y^n) > 0$ when $n \to \infty$ are called *ideally secure*. The cryptanalyst can sharpen the probability distribution of possible messages X^n, but he cannot find a unique solution for X^n. This does not necessarily mean that the system is secure; the residual uncertainty may only affect a small portion of the plaintext. To reach ideal security usually demands perfect data compression, or randomization of the plaintext [47]. Note that ideal security hinges on complete confidentiality of the plaintext.

3. $I(X^n; Y^n) \approx H(X^n)$ for large n; after a certain number of ciphertext bits there remains only one solution for the corresponding plaintext (and the applied key). This is typically the case with practical cryptosystems.

As a theoretical security index for cryptosystems Shannon introduced the *unicity distance* $n_u = \min \{n : H(K \mid Y^n) \approx 0\}$ which is the required amount of ciphertext before the cryptanalyst can, in principle, solve for the key. Once he knows the key he also knows the message. For typical cryptosystems $n_u = H(K)/(1 - h)$ where h denotes the information rate per plaintext bit, and $1 - h$ is the redundancy per plaintext bit. Note that both for perfect and ideal cryptosystems $n_u = \infty$, and hence it became common-place to call perfect and ideal cryptosystems *unconditionally secure* [22,25]. For perfect cryptosystems this is a useful definition, but an ideal cryptosystem can become disastrously insecure in a known plaintext attack. Thus it seems inappropriate to call an ideal cryptosystem unconditionally secure.

One of the most famous cipher systems is doubtless the one-time pad (sometimes called the Vernam cipher) [109,118]. It received its name from the requirement that every part of the key be only used once.

One-time pad:

Input: message: a sequence of message bits x_i, $i = 1, 2, \ldots$

key: a sequence of independent and uniformly distributed bits k_i, $i = 1, 2, \ldots$

Output: ciphertext $y_i = x_i \oplus k_i$, $i = 1, 2, \ldots$, consisting of the bitwise exor of the message stream with the key sequence.

The one-time pad is a stream cipher in the sense that every message bit is transformed independently with a time-varying transformation. There is no keystream generator since the key is directly added to the message. The one-time pad is perfect, $I(X^n; Y^n) = 0$. Knowing part of the plaintext does not violate the statistical independence condition for the remaining segments. The number of key bits has to be larger than the number of message bits. The key generation, distribution, and management problem is enormous, so that it is only used for certain untypical applications such as two users sharing a hotline with high confidentiality requirements.

Equivalent to the unicity distance of a cipher system, one may define the *unicity distance of a keystream generator* as the number of keystream symbols that need to be observed in a known plaintext attack before the key can be uniquely determined. In a known plaintext attack there is no uncertainty about the plaintext, $h = 0$, and consequently $n_{G,u} \approx H(K)$.

2.1 Local Randomization

Since the unicity distance of a keystream generator with a finite key K is about $n_{G,u} \approx H(K)$ bits, there is no security left in the Shannon sense once the opponent has acquired more than $n_{G,u}$ bits of keystream. An interesting subproblem arises when the Shannon model is modified in such a way that the opponent is given direct access to the keystream, but only to a limited number of e (not necessarily consecutive) bits of his choice. The problem of proving security under this additional constraint is called the local randomization problem [73]. Note that e has to be smaller than $H(K)$. Schnorr [105] motivated the local randomization problem by presenting the following construction:

Schnorr's pseudorandom generator $G = \{G_n\}$:

Input: the key (seed) k is a random function $f : I_m \rightarrow I_m$; size of description $m2^m$.

1. Set $y_i^0 = i$ for $i = 0, 1, \ldots, 2^{2m} - 1$.
2. For $j = 0, 1, 2$, do

$$y_i^{j+1} = (R(y_i^j), L(y_i^j) \oplus f(R(y_i^j)))$$

(L and R mean left and right half of argument.)

Output $G_n(k)$: the sequence of y_i^3, $i = 0, 1, 2, \ldots, 2^{2m} - 1$.

Note that the key k has size $n = m2^m$ (the function description) and that $G_n(k)$ has length $2m2^{2m}$. Thus, the generator stretches a seed of length n into roughly n^2 pseudorandom bits. Schnorr's generator is an application of the permutation function generator developed by Luby and Rackoff [63]. This permutation generator consists of an M-round DES-like structure, where in each round i a different (pseudo) random function f_i is applied. It is shown that three rounds suffice to prove perfectness (polynomial-time indistinguishability) of the resulting permutation generator, provided the functions f_i are indistinguishable.

Schnorr used the following asymptotic notion of security borrowed from complexity theory (see Section 3). A generator $G = \{G_n\}$ is provably locally randomized if it passes all (even those with unlimited time-bound) statistical tests $T = \{T_n\}$ that depend on at most $e(n)$ bits of $G_n(k)$. Schnorr claimed that his generator is provably locally randomized with $e(m) = 2^{m/3-(\log m)^2}$. In [101] it was shown that this is not true. In fact, a statistical test was given that efficiently distinguishes Schnorr's generator with only $e(m) = 4m$ bits of $G_n(k)$. This result even holds for an arbitrary number of internal rounds (instead of the proposed 3), or, for an odd number of internal rounds, when the seed is doubled and two functions f, g are used alternately. The flaw in Schnorr's construction was independently discovered by Ohnishi [88], and Zheng, Matsumoto, and Imai [129], who were studying the construction of pseudorandom permutations, as proposed by Luby and Rackoff [63]. Ohnishi discovered that, for an odd number of internal rounds, G can be distinguished for any palindromic arrangement of functions.

If Schnorr's generator is subjected to a system-theoretic analysis the result is disastrous. There exists a linear time attack that recovers the key from a segment of $m2^m + O(m)$ bits of keystream [101]. Note that this segment is only slightly larger than the unicity distance of G_n. This attack reaches the performance limits of any attack, since simply reading the key from memory requires linear time and linear space in the size of the key. This result shows that, if the assumptions of the security model do not coincide with the practical application, security can break down.

Since the basic model is still the Shannon model it seems more natural to define security directly in information-theoretic terms. Maurer and Massey [73] used the following notion of security: an (n, l)-sequence generator G is an (n, l, e)-perfect local randomizer if every subset of e bits of the l output bits is a set of independent and uniformly distributed (i.u.d.) binary random variables (provided the key bits are independent and uniformly distributed random variables). By interpreting an (n, l)-sequence generator as the encoder of a binary block code with 2^n codewords of length l, where the n information bits correspond to the key, they were able to use coding-theoretic arguments and tools. As a consequence, the problem of determining the maximum achievable degree of perfect local randomness of any linear (n, l) sequence generator is equivalent to the problem of determining the maximum achievable minimum distance d of a linear binary $[l, l - n, d]$ code. They showed that the encoder of an extended Reed–Solomon (RS) code over $GF(2^m)$ with e information symbols, codeword length 2^m, and design distance $2^m - e + 1$ is a linear $(me, m2^m, e)$-perfect local randomizer when the symbols are appropriately represented by m binary digits. Note that $e > n/\log l$. The parameters of an extended RS code can be chosen to coincide with the parameters $n = m2^m$ and $l = 2m2^{2m}$ of Schnorr's generator. If the extended RS code is defined over $GF(2^{2m})$ with 2^{m-1} information symbols then the corresponding encoder is a $(m2^m, 2m2^{2m}, 2^{m-1})$-perfect local randomizer. This local randomizer not only achieves true local randomness instead of indistinguishability but also gives a degree of local randomness that is greater than the third power of what Schnorr hoped for. In [73] it was also shown that nonlinear perfect local randomizers based on nonlinear codes exist whose degree of local randomness surpasses the degree of the corresponding best linear local randomizer, but no constructions were given.

3 SYSTEM-THEORETIC APPROACH

The system-theoretic approach to stream cipher design is twofold. One objective is to develop methods and building blocks that have provable properties with respect to certain system-theoretic measures such as period, linear complexity, frequency distribution, and distance to linear structures. A second objective is to study cryptanalytic principles and to develop design rules that render attacks based on these principles impossible. Such fundamental cryptanalytic principles are, for instance, substitution and approximation (preferrably by linear components), divide and conquer (on the key-space), and exploitation of statistical deficiencies (such as intersymbol dependencies). To prevent cryptanalysis based on these fundamental principles a set of general design criteria for keystream generators [2,89,98,109] has evolved over time, for instance,

1. Long period, no repetitions
2. Linear complexity criteria: large linear complexity, linear complexity profile, local linear complexity, etc.
3. Statistical criteria such as ideal k-tuple distributions
4. Confusion: every keystream bit must be a complex transformation of all or most of the key bits
5. Diffusion: Redundancies in substructures must be dissipated into long-range statistics
6. Nonlinearity criteria for Boolean functions like mth-order correlation immunity, distance to linear functions, avalanche criterion, etc.

Any secure keystream generator must satisfy this set of general design criteria. In the system-theoretic approach, the ciphers are designed to directly satisfy the criteria (in general, under the most favorable conditions for the cryptanalyst). Doubtless this is today the most widely used approach to practical security. The DES, although in its basic mode not a stream cipher, provides a good example of a system-theoretic design. With a finite key of only 56 bits (which is much smaller than the necessary keysizes for most other systems inluding public-key systems) the DES has shown considerable resistance to (public) cryptanalysis over the past 15 years.

But a major problem is that the design criteria only partially reflect the general cryptanalytic principles. It may happen that a generator provably satisfies all design criteria and yet is insecure. Or, in other words, the design criteria form a set of necessary but not sufficient conditions for the security of a keystream generator. The advantage of the system-theoretic approach is that each new cryptosystem creates a difficult and previously unknown problem for the cryptanalyst. Breaking such a system is an ''unglamorous'' problem as compared to the cryptanalysis of a system based on some ''famous'' problem such as factoring or the discrete log.

3.1 Transform Techniques

In this section we introduce some basic definitions and describe transform techniques that are useful in the analysis of sequences and cryptographic transformations, especially of Boolean functions.

    3.1.1   **Discrete Fourier transform and linear complexity.**   If α is a primitive N-th root of unity in any field $\mathbb{F}$, then the discrete Fourier transform (DFT) of the ''time-domain'' sequence $a^N = (a_0, a_1, \ldots, a_{N-1})$ with components from F is the ''frequency-domain'' sequence $A^N = (A_0, A_1, \ldots, A_{N-1})$ where

$$A_i = \sum_{j=0}^{N-1} a_j \alpha^{ij} \qquad i = 0, 1, \ldots, (N-1)$$

The inverse DFT relation is

$$a_j = \frac{1}{N^*} \sum_{i=0}^{N-1} A_i \alpha^{-ij} \qquad j = 0, 1, \ldots, (N-1)$$

where $N^* = N \bmod p$ if the characteristic of F is p and $N^* = N$ if the characteristic of F is 0.

We proceed to discuss the close relationship between the DFT and the linear complexity of a sequence, and we follow the treatment given in [104]. The computation of the ith component A_i of the DFT may be viewed as the inner product between a^N and the sequence $(\alpha^{0i}, \alpha^{1i}, \alpha^{2i}, \ldots, \alpha^{(N-1)i})$. Therefore, it is possible to describe the DFT as a matrix transform

$$A^N = a^N \cdot F$$

where

$$F = \begin{pmatrix} 1 & 1 & 1 & \cdots & 1 \\ 1 & \alpha^1 & \alpha^2 & \cdots & \alpha^{N-1} \\ \vdots & & & & \\ 1 & \alpha^{N-1} & \alpha^{2(N-1)} & \cdots & \alpha^{(N-1)(N-1)} \end{pmatrix}$$

Analogously, the inverse DFT is given as

$$a^N = \frac{1}{N*} A^N \cdot F^{-1}$$

where

$$F^{-1} = \begin{pmatrix} 1 & 1 & 1 & \ldots & 1 \\ 1 & \alpha^{-1} & \alpha^{-2} & \ldots & \alpha^{-(N-1)} \\ \vdots & & & & \\ 1 & \alpha^{-(N-1)} & \alpha^{-2(N-1)} & \ldots & \alpha^{-(N-1)(N-1)} \end{pmatrix}$$

Now define the circulant matrix of a^N [104], denoted $M(a^N)$, to be the matrix whose rows consist of the N cyclic left shifts of a^N.

$$M(a^N) = \begin{pmatrix} a_0 & a_1 & \ldots & a_{N-1} \\ a_1 & a_2 & \ldots & a_0 \\ \vdots & & & \\ a_{N-1} & a_0 & \ldots & a_{N-2} \end{pmatrix}$$

Let $L = \Lambda(\bar{a})$ be the linear complexity of the sequence $\bar{a} = a_0, a_1, a_2, \ldots$. Then, by the definition of the linear complexity, L is the smallest integer such that the ith row of $M(a^N)$ can be written as a linear combination of the previous rows. Hence, the rank of $M(a^N)$ is at least L. On the other hand, every row with index $L \le i \le N - 1$ can be written as a linear combination of previous rows (in fact, of the L previous rows). Hence, the rank of $M(a^N)$ is L. This establishes that the linear complexity of a periodic, semi-infinite sequence $\bar{a} = (a^N)^\infty$ is equal to the rank of the circulant $M(a^N)$.

The circulant matrix $M(a^N)$ may be written as

$$M(a^N) = F^{-1} \cdot D_A \cdot F^{-1}$$

where F^{-1} is the inverse DFT matrix, and D_A is the N by N diagonal matrix whose diagonal entries are the elements of A^N. Under the assumption that α is a primitive Nth root of unity the matrices F and F^{-1} have full rank. Hence,

$$\text{rank } (M(a^N)) = \text{rank } (D_A)$$

But the rank of D_A is equal to the number of nonzero elements in A^N, that is, equal to $W_H(A^N)$. This establishes the first part of the following theorem which has been used implicitly by Blahut [7], but was first established by Massey in this explicit form [70]. The second part is proved analogously.

Blahut's theorem: *The linear complexity of the periodic, semi-infinite "time-domain" sequence $\bar{a} = (a^N)^\infty$ is equal to the Hamming weight of the finite-length, "frequency-domain" sequence A^N, where A^N is the DFT of a^N,*

$$\Lambda((a^N)^\infty) = w_H(A^N)$$

Similarly, the linear complexity of the periodic, semi-infinite sequence $(A^N)^\infty$ is equal to the Hamming weight of the finite-length sequence a^N, where a^N is the inverse DFT of A^N.

$$w_H(a^N) = \Lambda((A^N)^\infty)$$

Many linear complexity results about nonlinear combinations of shift register sequences have alternate proofs in the "frequency" domain.

3.1.2 Walsh transform and Boolean functions. The Walsh transform of a real-valued function $f : \mathbb{F}_2^n \to \mathbb{R}$ is defined as

$$F(\omega) = \sum_x f(x)(-1)^{\omega \cdot x}$$

where $\omega \cdot x$ denotes the dot product $\omega_1 x_1 \oplus \omega_2 x_2 \oplus \ldots \omega_n x_n$.
The function f can be recovered by the inverse Walsh transform

$$f(x) = 2^{-n} \sum_\omega F(\omega)(-1)^{\omega \cdot x}$$

Let the real-valued function $f : \mathbb{F}_2^n \to \mathbb{R}$ be specified by the naturally ordered vector

$$[f(x)] = [f(0), f(1), \ldots, f(2^n - 1)]$$

where $f(x) = f(x_1, x_2, \ldots, x_n)$ and $x = x_1 + x_2 2 + \ldots + x_n 2^{n-1}$. Analogously, let the Walsh transform $F : \mathbb{F}_2^n \to \mathbb{R}$ of the function f be given by the naturally ordered vector

$$[F(\omega)] = [F(0), F(1), \ldots, F(2^n - 1)]$$

Then, the Walsh transform may be represented as the matrix transform

$$[F(\omega)] = [f(x)] \cdot H_n$$

where H_n denotes the Hadamard matrix of order n. H_n is defined recursively by

$$H_0 = [1]$$

$$H_n = \begin{bmatrix} 1 & 1 \\ 1 & -1 \end{bmatrix} \otimes H_{n-1}$$

where $\otimes$ denotes the Kronecker product of matrices. Since $H_n^2 = 2^n I_n$, the inverse Walsh transform is given by

$$[f(x)] = 2^{-n}[F(\omega)] \cdot H_n$$

By treating the values of a Boolean function $f : \mathbb{F}_2^n \to \mathbb{F}_2$ as the real numbers 0 and 1, one can define the Walsh transform of such Boolean f. Sometimes it is more convenient to work with the equivalent binary function $\hat{f} : \mathbb{F}_2^n \to \{-1, +1\}$ defined by

$$\hat{f}(x) = (-1)^{f(x)}$$

whose Walsh transform has the form

$$\hat{F}(\omega) = \sum_x (-1)^{f(x) \oplus \omega \cdot x}$$

The relationship between $\hat{F}(\omega)$ and $F(\omega)$ is given by

$$\hat{F}(\omega) = 2^n \delta(\omega) - 2F(\omega)$$

$$F(\omega) = 2^{n-1} \delta(\omega) - \frac{1}{2}\hat{F}(\omega)$$

where $\delta(\omega) = 1$ for $\omega = 0$ and 0 otherwise.

In this survey, whenever we apply the Walsh transform to establish properties of the Boolean function f, we use the equivalent binary function $\hat{f}$ to compute $\hat{F}$.

3.1.3   Algebraic normal form transform.   Let $f : \mathbb{F}_q^N \to \mathbb{F}_q$ be a $\mathbb{F}_q$ switching function. Then f may be written in algebraic normal form (ANF) as

$$f(\mathbf{x}) = \sum_{\mathbf{i} \in \mathbb{Z}_q^N} c_{\mathbf{i}} \mathbf{x}^{\mathbf{i}}$$

where $\mathbf{x} = (x_1, \ldots, x_N) \in \mathbb{F}_q^N$, $\mathbf{i} = (i_1, \ldots, i_N) \in \mathbb{Z}_q^N$, $c_{\mathbf{i}} \in \mathbb{F}_q$, and

$$\mathbf{x}^{\mathbf{i}} = \prod_{n=1}^{N} x_n^{i_n}$$

The order of a product term is defined as $\sum_{n=1}^{N} i_n$. The nonzero constant term is defined to have 0th order. The linear terms have order 1. The order of the function f is defined to be the maximum of the order of its product terms that have a nonzero coefficient.

Now consider the Boolean function: $f : \mathbb{F}_2^n \to \mathbb{F}_2$. Let f be specified by the naturally ordered vector

$$[f(x)] = [f(0), f(1), \ldots, f(2^n - 1)]$$

where $f(x) = f(x_1, x_2, \ldots, x_n)$ and $x = x_1 + x_2 2 + x_3 2^2 + \ldots x_n 2^{n-1}$.

Let the ANF of f be specified by the naturally ordered vector

$$[a_i] = [a_0, a_1, \ldots, a_{2^n - 1}]$$

where $a_i = a(i_1, i_2, \ldots, i_n)$ and $i = i_1 + i_2 2 + i_3 2^2 + \ldots i_n 2^{n-1}$. Then the ANF transform is given by [50,98]

$$[a_i] = [f(x)]A_n$$

and the inverse ANF transform is

$$[f(x)] = [a_i]A_n$$

where A_n is defined recursively by

$$A_0 = [1]$$

$$A_n = \begin{bmatrix} 1 & 1 \\ 0 & 1 \end{bmatrix} \otimes A_{n-1}$$

and $\otimes$ denote the Kronecker product of matrices. Note that A_n is an involution, i.e., that $A_n^2 = I$.

There is a fast way to iteratively evaluate the binary ANF transform [50,98]. Let $[f^1(x)]$ and $[f^2(x)]$ be the first and second half of the vector $[f(x)]$. Then

$$[a_i] = [f(x)] \cdot A_n = [[f^1(x)] \cdot A_{n-1}, ([f^1(x)] + [f^2(x)]) \cdot A_{n-1}]$$

which can be iterated until A_0 is reached.

The general ANF transform for GF(q)-functions has been described in [50].

3.2 Period and Linear Complexity of Sequences

The linear complexity $\Lambda(s^l)$ of the sequence $s^l = s_0, s_1, \ldots, s_{l-1}$ is the length L of the shortest LFSR that can generate s^l when the first L digits of s^l are initially loaded into the register. Equivalently, the linear complexity $\Lambda(s^l)$ is defined to be the smallest nonnegative integer L such that there exists a linear recursion with fixed constants $c_1, c_2, \ldots, c_L$ satisfying

$$s_j + c_1 s_{j-1} + \ldots + c_L s_{j-L} = 0 \qquad L \leq j < l$$

The linear complexity is a very useful concept in the study of stream ciphers. As was pointed out by Massey [71] any sequence that can be generated by a finite-state machine (whether linear or nonlinear) over a finite field also has a finite linear complexity. Moreover, there exists an efficient algorithm, the Berlekamp–Massey algorithm [65], that computes the linear complexity of a given sequence s^l in time lL. This algorithm forms a universal attack for running key generators since it carries the potential of substituting any running key generator by its shortest linear equivalent. Therefore, a necessary condition that any running key generator must meet, is the requirement for a large linear complexity [89,98].

3.2.1   Random sequences.   Define L_i to be the linear complexity of the subsequence $s^i = (s_0, s_1, \ldots, s_{i-1})$. Then the sequence $L_1, L_2, \ldots, L_l$ is called the *linear complexity profile* of s^l. For binary independent and uniformly distributed sequences L_i becomes a random variable. In [95] it is shown that

$$\text{E}(L_i) = \frac{i}{2} + \frac{1}{4} + \frac{(-1)^i}{36} + O(i2^{-i})$$

$$\text{Var}(L_i) = \frac{86}{81} + O(i2^{-i})$$

These results have been generalized by Smeets [114] to sequences over arbitrary finite fields $\mathbb{F}_q$. He showed that the variance of the linear complexity decreases with the size of the sequence alphabet; for large values of q the variance will be approximately $1/q$. Thus, the larger the size of $\mathbb{F}_q$ the closer the complexity profile will follow the line $i/2$.

The close relationship between the linear complexity profile and continued fractions was studied in [21,82,83,121]. Knowing that the linear complexity profile of random sequences will typically stay close to the $i/2$ line, one may reverse the problem and ask how random is a sequence whose linear complexity profile perfectly follows the $i/2$ line, that is, where $L_i = \lfloor (i+1)/2 \rfloor$ for $i \geq 1$. In [95] it was conjectured that the sequence $\tilde{s}$ which contains a 1 at the locations $i = 2^j - 1$ and is zero otherwise possesses such a perfect linear complexity profile. This conjecture was proved in [21] using the Euclidean algorithm. An adapted version of the proof was given in [69] using the

Berlekamp–Massey synthesis algorithm. In [120] a complete characterization of all binary sequences that possess a perfect linear complexity profile was derived. It was shown that a binary sequence $\bar{s}$ has a perfect linear complexity profile if and only if $s_0 = 1$ and $s_{2i} = s_{2i-1} \oplus s_{i-1}$ for $i \geq 1$. A relatively short proof of this characterization can also be found in [83].

Instead of investigating the ensemble of all sequences of a given length i, one may consider how the linear complexity of a randomly chosen and then fixed sequence will behave as i varies. Niederreiter [84] has shown that $\lim_{i \to \infty} L_i/i = 1/2$ for almost all sequences over $\mathbb{F}_q$. He defined a sequence $\bar{s}$ to possess a good linear complexity profile [85] if there exists a constant c (that may depend on the sequence) such that

$$\left| L_i - \frac{1}{2} \right| \leq c \, \text{Log} \, i \qquad \text{for all } i \geq 1$$

where $\text{Log} \, i = \max\{1, \log i\}$, and showed that almost all $\mathbb{F}_q$ sequences possess such a good linear complexity profile.

Piper [89] pointed out that a keystream sequence should have an acceptable linear complexity profile for every starting point. Niederreiter showed [85] also that $\lim_{i \to \infty} L_i(\bar{s}_m)/i = 1/2$ for all shifted versions $\bar{s}_m = s_m, s_{m+1}, \ldots$, $m \geq 0$, of almost all sequences $\bar{s}$ over $\mathbb{F}_q$. Naturally extending the notion of good linear complexity profile, he defined a sequence to possess a uniformly good linear complexity profile if there exists a constant c (that may depend on the sequence but not on the shift factor m) such that

$$\left| \Lambda_i(\bar{s}_m) - \frac{1}{2} \right| \leq c \, \text{Log} \, i \quad \text{for all } m \geq 0 \text{ and } i \geq 1$$

However, a uniformly good linear complexity profile is not a typical property of random sequences, since it was shown in [85] that almost no sequence has such a uniformly good linear complexity profile.

The linear complexity profile quite naturally suggests a statistical test: The linear complexity profile of well-designed running key generators is indistinguishable from the linear complexity profile of random sequences. An experimental study of the practical applicability of the linear complexity profile test was reported in [114].

3.2.2 Periodic sequences. Consider the periodic $\mathbb{F}_q$-sequence $\bar{s} = (s^T)^\infty$ and assume that

$$s_i + c_1 s_{i-1} + \ldots + c_L s_{i-L} = 0 \qquad L \leq i$$

is the linear recursion of least possible order that can generate the sequence $\bar{s}$. The polynomial

$$m_{\bar{s}}(x) = x^L + c_1 x^{L-1} + \ldots + c_L$$

associated with the linear recursion is called the minimal polynomial of $\bar{s}$. $m_{\bar{s}}$ is monic and unique; it divides the characteristic polynomial of any other linear recursion that is satisfied by $\bar{s}$. The sequence $\bar{s}$ may be represented by a linear combination of fundamental solutions of the associated linear recursion, or equivalently, by a linear combination of the roots of the corresponding minimal polynomial, thereby yielding a direct "time-domain solution" (see, e.g., [54,81,131]. Let $m_{\tilde{s}}(x)$ consist of K irreducible factors $m_k(x)$ with multiplicity e_k,

$$m_{\tilde{s}}(x) = \prod_{k=1}^{K} m_k^{e_k}(x)$$

Furthermore, let d_k be the degree and α_k be one of the roots of the kth irreducible factor. Then the ith digit of $\tilde{s}$ is uniquely determined by

$$s_i = \sum_{k=1}^{K} \sum_{j=0}^{d_k-1} \alpha_k^{iq^j} \sum_{l=0}^{e_k-1} A_{k,l}^{q^j} \binom{i+l}{l}$$

with $A_{k,l}^{q^j} \in GF(q^{d_k})$, $A_{k,e_k} \neq 0$ being determined by the initial terms of the sequence, and the binomial coefficients being computed modulo the characteristic p of the field $\mathbb{F}_q$.

If $m_{\tilde{s}}(x)$ is the minimal polynomial of the sequence $\tilde{s}$ then the least positive period T of $\tilde{s}$ is equal to the exponent (also called order or period) of $m_{\tilde{s}}(x)$. If T_k denotes the exponent of $m_k(x)$ or equivalently, the multiplicative order of the root α_k, then [4,60]

$$T = p^e \, \mathrm{lcm}(T_1, \ldots, T_k)$$

where e is the smallest integer with $p^e \geq \max\{e_1, \ldots, e_k\}$. When $\mathbb{F}_q$-sequences are combined then the two operations of termwise addition and termwise multiplication are of principal interest. Consider first the sum of two sequences $\tilde{z} = \tilde{x} + \tilde{y}$. In [45] the following bounds on the linear complexity of the sum sequence are given

$$\Lambda(\tilde{x}) + \Lambda(\tilde{y}) - 2\gcd(T_x, T_y) \leq \Lambda(\tilde{z}) \leq \Lambda(\tilde{x}) + \Lambda(\tilde{y})$$

Equivalently, one may prove that $\Lambda(\tilde{z}) = \Lambda(\tilde{x}) + \Lambda(\tilde{y})$ if $\gcd(T_x, T_y) = 1$ and $x - 1$ divides at most one of the two minimal polynomials $m_{\tilde{x}}$, $m_{\tilde{y}}$ [38]. The period of the sum sequence is implicitly bounded by the following divisibility condition [45,98].

$$\frac{\mathrm{lcm}(T_x, T_y)}{\gcd(T_x, T_y)} \Big| T_z \Big| \, \mathrm{lcm}(T_x, T_y)$$

whose right side is known at least since Selmer's work [106]. As a consequence, $T_z = T_x T_y$ if and only if the two periods are relatively prime [38].

3.2.3 Products of periodic sequences. Now consider the termwise product of two sequences (Herlestam [48] uses the term *Hadamard product* to distinguish the term-wise product of two sequences from the multiplication rule in the ring of formal power series).

$$\tilde{z} = \tilde{x} \wedge \tilde{y}$$
$$z_i = x_i \cdot y_i, \quad i = 0, 1, 2, \ldots$$

It is a basic fact that the linear complexity of a product sequence can never exceed the product of the linear complexities of the sequences being multiplied [48,131]. Consequently, whenever the linear complexity of a product sequence satisfies this upperbound with equality, one says that this product sequence attains maximum linear complexity.

Considerable interest has been paid to the question of what conditions (in particular, what easily verfied conditions) have to be imposed on the minimal polynomials of the sequences such that the resulting product sequence will exhibit maximum linear complexity. At least since Selmer's work [106] it has been known that if the two minimal polynomials are irreducible and have relatively prime degree then the product sequence attains maximum linear complexity. In fact, Selmer provides a reference dated 1881 where this result appears without proof.

Herlestam [48,49] established the following necessary and sufficient condition on the roots of the involved minimal polynomials that would guarantee maximum linear complexity of the product sequence: $\Lambda(\bar{z}) = \Lambda(\bar{x}) \cdot \Lambda(\bar{y})$ if and only if at most one of $m_{\bar{x}}, m_{\bar{y}}$ has multiple roots and the mutual product of their roots are all distinct. Unfortunately, this is not a readily verified condition.

Based on arithmetic properties of the periods T_x and T_y Rueppel and Staffelbach [99] developed the following condition: Let $m_{\bar{x}}$ have no multiple roots and let $m_{\bar{y}}$ be irreducible then the product sequence will attain maximum linear complexity if

$$\text{ord}(q) \bmod t_y = L_x$$

where $\text{ord}(q) \bmod t$ denotes the multiplicative order of q modulo t and t_y is defined as $T_y/\gcd(T_x, T_y)$. If both $m_{\bar{x}}$ and $m_{\bar{y}}$ are irreducible, then either $\text{ord}(q) \bmod t_y = L_x$ or $\text{ord}(q) \bmod t_x = L_y$ satisfies to guarantee maximum linear complexity. Golić [38] subsequently showed that this condition is the most general sufficiency condition expressed only in terms of the periods of the minimal polynomials. He also derived a second sufficiency condition (we refer to the original paper for the very involved formula) which is of some theoretical interest, since one of the two conditions has to be satisfied if the product sequence is to exhibit maximum linear complexity.

If it is easy to find divisors t of $q^L - 1$ that satisfy the condition $\text{ord}(q) \bmod t = L$, it is also easy to achieve maximum linear complexity of the product sequence. One must ensure that the period of the first (irreducible) minimal polynomial contains t as a factor, and that the period of the second (not necessarily irreducible) minimal polynomial is relatively prime to t. Any integer $t > 1$ that divides $q^L - 1$ but is relatively prime to $q^i - 1$, $i < L$, is called a primitive factor of $q^L - 1$. The greatest such factor is denoted F_L. As an immediate consequence of this definition it follows that for any divisor t of F_L, the multiplicative order of q modulo t is L. Therefore maximum linear complexity is attained when t_y contains a primitive factor of $q^{L_y} - 1$. Interestingly enough, for all integers $q > 1$ and $L > 2$, every $q^L - 1$ contains a primitive factor, except for the single case $q = 2$ and $L = 6$. Moreover, the primitive factor F_L can actually be computed from the Lth cyclotomic polynomial. As a practical consequence, any product of N m-sequences will attain maximum linear complexity when the corresponding degrees are distinct and greater than 2 [99].

Finally, a less general but simpler condition in the sense that it only depends on the periods T_x and T_y is given in both [38,45]: the product sequence will attain maximum linear complexity if $\gcd(T_x, T_y) = 1$, (and provided the sequences are nonzero). The same condition also ensures maximum period of the product sequence, see [60, p. 435] or [106].

3.3 Functions of Periodic Sequences

From a practical viewpoint two situations are important: nonlinearly filtering the state of a LFSR (or, more generally, the state of any counting mechanism), and combining the output sequences from several LFSRs. The former type of generator is referred to as a *filter generator* and the latter as a *combination generator*. It is obvious that the filter generator could be considered as a special case of the combination generator, where all the sequences combined satisfy the same recursion. But, since analysis and correlation attacks (see Section 3.4) are somewhat different, it is useful to consider each class separately.

Many of the published generator proposals (see Section 3.7) belong to one of the two classes. First, technology has provided some stimulus, since shift registers and combination components such as flip-flops, multiplexers, random access memories (RAMs), etc., were readily available. Second, the system-theoretic properties of these generators such as period, linear complexity, and statistics can frequently be analyzed by the feedforward nature of the nonlinearities. When considering these two classes of generators one may think alternatively of periodic sequences or LFSRs feeding the non-linearities, since any periodic sequence has a finite and unique linear equivalent. A basic requirement for a generator is that it be unpredictable, which directly implies that the linear complexity of its output sequence should be large enough to prevent any attack based on the Berlekamp–Massey shift register synthesis algorithm [65].

There exists a canonical form for $\mathbb{F}_q$ switching functions, the so-called algebraic normal form [3], which directly reflects the sum of products property (see Section 3.1.3 for the description of the ANF). When a Galois switching function is applied to combine several $\mathbb{F}_q$ sequences, the resulting output sequence is formed by termwise combination of the sequence digits.

3.3.1   Filter generator.   We proceed with the discussion of nonlinear combinations of different phases of one sequence. Consider the binary case $q = 2$.

Filter generator:

Input: parameters: one m-LFSR $< C(D), L >$
        key: output function $f : \mathbb{F}_2^L \to \mathbb{F}_2$
        initial state a_0 of the LFSR.

For $i = 1, 2, \ldots$ do

1. Shift LFSR (state transform F): $a_i = F(a_{i-1})$.
2. Compute $z_i = f(a_i)$

Output: the sequence of z_i, $i = 1, 2, \ldots$

Groth [44] concentrated on the use of second-order products which he applied to the stages of an LFSR with a primitive connection polynomial. No stage was allowed to be used more than once. To obtain reasonable statistics, Groth summed as many of the second-order products as possible. Higher-order nonlinearities were achieved by layering, that is, the sequences generated by summing second-order products were combined in the next layer by more second-order products, and so on. Groth was able to show the expected growth in linear complexity of the generated sequence as a function of the growing order of the nonlinearities. But he neglected to notice that maximum linear complexity is not always achieved. Hence, his results are only upper bounds on the true linear complexity.

It is well known that, for a shift register, or a counter, of period T and a sequence of period dividing T, there exists a function f that maps the counter into the sequence.

Consider a binary filter generator, its general linear equivalent will consist of all LFSRs whose characteristic polynomial divides $x^{2^L-1} - 1$. The $2^L - 1$ sequences that correspond to the $2^L - 1$ state bits of this general linear equivalent are linearly independent. It can also be shown [98] that the $2^L - 1$ sequences that correspond to the $2^L - 1$ coefficients in the ANF of f are linearly independent. This suggests, at least in principle, a solution to the problem of generating functions that produce sequences of guaranteed linear complexity. First, pick the desired linear complexity by turning on state bits in the general linear equivalent, that is, by choosing a linear combination of the $2^L - 1$ basis sequences specified by the general linear equivalent. Then express the resulting sequence in terms of the ANF basis, which directly yields the required function f. But specifying the function f in general needs as much data as the first period of the sequence. Hence, the above procedure is impractical, and a more realistic goal is to find classes of functions with reasonable keyspace that generate large linear complexity when used in a filter generator.

Key [54] has shown that if f has order k then the linear complexity of $\bar{z}$ is bounded from above by

$$\Lambda(\bar{z}) \le L_k = \sum_{j=1}^{k} \binom{L}{j}$$

But in the cryptographic application one is rather interested in generating sequences with guaranteed large mimimum linear complexity. For a subset of the class of bent functions, Kumar and Scholtz [57] were able to derive a lower bound that slightly exceeds

$$\Lambda(\bar{z}) \ge \binom{L/2}{L/4} 2^{L/4}$$

where L is restricted to be a multiple of 4. Bernasconi and Günther [5] and, independently, Rueppel [98] developed slightly different bounds whose combination yields the following lower bound: let $\bar{z}$ be produced by any nonzero linear combination of N consecutive kth-order products ($k < L$) of equidistant phases of the m-sequence $\bar{s}$,

$$\bar{z} = \sum_{j=0}^{N-1} c_j \bar{s}^j \bar{s}^{j+\delta} \ . \ . \ . \ \bar{s}^{j+(k-1)\delta}$$

where $\gcd(\delta, 2^L - 1) = 1$, then the linear complexity of $\bar{z}$ has the lower bound

$$\Lambda(\bar{z}) \ge \binom{L}{k} - (N - 1)$$

This lower bound remains valid when an arbitrary function f' is added to this sum of kth-order products, provided its order k' is smaller than k. As a practical consequence, one has at hand for implementation a large class of functions with guaranteed minimum linear complexity.

In [98] a probabilistic analysis is carried out for the case where a randomly selected kth-order function f is applied to the stages of a m-LFSR of prime length L. It is shown that the fraction P_n of filter generators with nonlinear order k producing sequences of linear complexity L_k is lower bounded by

$$P_n \approx e^{-L_k/L2^L} > e^{-1/L}$$

It follows that the fraction of degenerate sequences approaches zero with increasing L.

3.3.2 Combination generator. Now consider the more general case where N periodic $\mathbb{F}_q$ sequences are combined in a memoryless output function $f : \mathbb{F}_q^N \to \mathbb{F}_q$. Restrict the algebraic normal form of f to powers $i \in \mathbb{Z}_2^N$.

Combination generator:

Input: parameters: N shift registers $< L_j, C_j(D) >$
             nonlinear output function $f : \mathbb{F}_q^N \to \mathbb{F}_q$
             key: N initial states $a_0^{(j)}$ of the N LFSRs.

For $i = 1, 2, \ldots$ do

1. For $j = 1, \ldots, N$ do
    shift $LFSR_j$.
    extract $a_i^{(j)}$.
2. $z_i = f(a_i^{(1)}, \ldots, a_i^{(N)})$

Output: the sequence of z_i, $i = 1, 2, \ldots$

Define the integer function $f^* : \mathbb{Z}^N \to \mathbb{Z}$ corresponding to f by

$$f^*(\mathbf{x}) = \sum_i c_i^* \, \mathbf{x^i}$$

where $c_i^* = 0$ if $c_i = 0$ and $c_i^* = 1$ if $c_i \neq 0$ [99]. From the discussion of sum and product sequences in the previous section it follows that the linear complexity and the period of the keystream produced by the combination generator are upperbounded by [48,99]

$$\Lambda(\bar{z}) \leq f^*(L_1, \ldots, L_N)$$
$$T_z \leq \text{lcm}(T_1, \ldots, T_N)$$

Key [54] showed for the binary case that the upper bound can be reached when m-sequences of relatively prime degree are combined in f.

In [98] a set of conditions is developed each of which guarantees maximum linear complexity. The two that have most practical relevance are:

1. If the N $\mathbb{F}_q$–LFSRs employed in the combination generator are chosen to have primitive connection polynomials $C_j(D)$ and different lengths L_j, then the keystream will exhibit maximum linear complexity $\Lambda(\bar{z}) = f^*(L_1, \ldots, L_N)$.

2. If the N $\mathbb{F}_2$- "LFSRs" employed in the combination generator are chosen to be filter generators of pairwise relatively prime lengths L_j, then the keystream will exhibit maximum linear complexity $\Lambda(\bar{z}) = f^*(\Lambda_1, \ldots, \Lambda_N)$ where Λ_j denotes the linear complexity of the jth filter generator.

Note that with arrangement 2 one can generate sequences with a linear complexity that is exponential in the number of storage cells employed. In [99] a simple generator arrangement is given that produces with 150 memory cells a keystream whose linear complexity is at least 2^{146}.

In [38] it is shown that based on the pairwise coprimality of the periods (see Section 3.2.3) a lower bound on the linear complexity can be derived. Let the N GF(q)-sequences be nonzero and have relatively prime periods, then

$$\Lambda(\tilde{z}) \geq f*(\Lambda_1 - 1, \ldots, \Lambda_N - 1)$$

where Λ_j represents the degree of the jth minimal polynomial.

Now consider the power function $f: \mathbb{F}_q \rightarrow \mathbb{F}_q$, $q = p^m$, defined by $f(x) = x^e$ where e is fixed and lies between $0 \leq e \leq q - 1$. Herlestam [48,49] showed that the linear complexity of $\tilde{z} = f(\tilde{x}) = \tilde{x}^e$ (computed termwise) is upperbounded by

$$\Lambda(\tilde{x}^e) \leq \prod_k \binom{\Lambda(\tilde{x}) + e_k - 1}{e_k}$$

where e_k are the digits of the p-ary expansion $e = e_0 + e_1 p + \ldots + e_{m-1} p^{m-1}$. Brynielsson [12] proved that the upper bound is reached if $\tilde{x}$ is a maximum-length sequence of degree $\Lambda(\tilde{x}) = L$. When the characteristic $p = 2$, the formula becomes particularly easy,

$$\Lambda(\tilde{x}^e) = L^{W_H(e)}$$

where $W_H(e)$ denotes the Hamming weight of e. This result generalizes to the case of an arbitrary polynomial $f(x) = \sum_{i=0}^{q-1} a_i x^i$. In [12] it is shown that for $p = 2$ and a maximum-length sequence $\tilde{x}$ of degree L,

$$\Lambda(f(\tilde{x})) = \sum_{i:a_i \neq 0} L^{W_H(i)}$$

Chan and Games [18] studied a similar problem. Let $\tilde{s}$ be a maximum-length sequence of degree L with digits from $\mathbb{F}_q$, q odd, and let $f: \mathbb{F}_q \rightarrow \mathbb{F}_2$ be an arbitrary mapping (the nonlinear feedforward function). Consider the binary sequence $\tilde{z} = f(\tilde{s})$. Chan and Games showed that the resulting sequences have linear complexity

$$\Lambda((\tilde{z})) = \frac{q^L - 1}{q - 1} \Lambda(\tilde{u})$$

where $\tilde{u}$ denotes the binary sequence obtained by applying the defining mapping f to a listing of the nonzero elements of $\mathbb{F}_q$ according to the powers of some primitive element. Because the sequences can be viewed as being obtained from finite geometries, they are termed binary geometric sequences. In [19] the autocorrelation and the cross-correlation functions of such binary geometric sequences is studied, including the case q even.

Now consider an arbitrary function $f: \mathbb{F}_q^2 \rightarrow \mathbb{F}_q$ whose ANF is a polynominal in two variables:

$$f(x, y) = \sum_{(i,j) \in \mathbb{Z}_q^2} a_{ij} x^i y^j$$

Brynielsson also proved [12] that for $p = 2$, $\bar{x}$ and $\bar{y}$ maximum-length sequences of degree L_1 and L_2, respectively, the linear complexity of $\bar{z} = f(\bar{x}, \bar{y})$ satisfies

$$\Lambda(f(\bar{x}, \bar{y})) = \sum_{(i,j):a_{ij} \neq 0} L_1^{W_H(i)} L_2^{W_H(j)}$$

The filter and the combination generator are special classes of keystream generators. We refer the reader to Section 3.7 for a list of specific generators and the corresponding system-theoretic analysis.

3.4 Correlation Attacks

Correlation attacks have been proposed for two principal keystream generator arrangements, the combination generator consisting of a set of shift registers and a nonlinear output function f [8,77,112,128], and the filter generator consisting of a shift register with nonlinear state filter f, [30,111] (see Section 3.3). Interestingly, a structure identical to the combination generator was used in an interplanetary ranging system that was developed at the Jet Propulsion Laboratory (JPL) under the guidance of S. W. Golomb in the late 1950s [42,116]. The objective then was to design a Boolean combiner for shift register sequences of short pairwise relatively prime periods to produce a sequence whose period was the product of the component periods and which was highly correlated with each component sequence to facilitate the calculation of range. This ranging problem is virtually dual to the cryptographic problem, where one desires to combine a set of shift register sequences in such a way that the resulting sequence has least possible correlation with the component sequences. Blaser and Heinzmann [8] noticed that correlation methods could be used as a means to attack combination generators with "poorly" chosen combiners f (i.e., combiners that leak information about the component sequences). Their work in the cryptographic setting was independent of the much earlier work at the JPL. Siegenthaler [110] then developed the basic correlation attack (described in Section 3.4.1), thereby spurring much research activity. A first objective was to improve the efficiency of the algorithms and to enlarge their applicability. The general importance of correlation-immunity as a design criterion for combiners was recognized.

The applicability of the correlation attacks depends first on the choice of the output function f, and second on the parameters of the LFSRs.

3.4.1   Correlation attacks on combination generators.   Consider a binary combination generator (see Section 3.3) and assume that $f : \mathbb{F}_2^N \to \mathbb{F}_2$ leaks information about the jth shift register sequence $\bar{a}^{(j)}$ into the keystream $\bar{z}$. Modeling the input to f as sequences of independent and uniformly distributed binary random variables, and taking into account the memoryless nature of f, one may compute the probability $p_j = P(f(A_1, \ldots, A_N) = A_j)$; p_j approximates the probability that the keystream digit coincides with a digit from the shift register sequence $\bar{a}^{(j)}$. To isolate the effect of the jth LFSR on the keystream, one can model the rest of the combination generator as a binary symmetric channel (BSC) with error probability $1 - p_j$. Thus, conceptually, one considers the keystream as a distorted version of the shift register sequence $\bar{a}^{(j)}$.

The problem now reduces to finding the correct phase $a_0^{(j)}$ from a finite string of keystream z^n and from the redundancy contained in $\bar{a}^{(j)}$ (the linear relations governing the behavior of $\bar{a}^{(j)}$). Siegenthaler [112] developed and analyzed the (basic) correlation attack (here denoted as algorithm A). It is assumed that the cryptanalyst is in possession of a complete description of the combination generator, except for the key.

Algorithm A: Basic correlation attack on combination generator [112]

Input: description of combination generator $< \{L_i, C_i(D)\}, f >$
segment of keystream z^n of length n

For $j = 1, 2, \ldots, N$ do

1. Compute leakage (correlation) probability $p_j = P(z = a^{(j)})$ from function f assuming that N BSS's form the input to f.
   If $p_j = \frac{1}{2}$ abort attempt on LFSR$_j$ and continue with $j = j + 1$.
2. Correlation phase: for $d = 1$ up to at most $2^{L_j} - 1$ do
   a. Compute cross-correlation function

$$C_{\bar{a}^{(j)}, \bar{z}}(d) = \frac{1}{n} \sum_{i=1}^{n} (-1)^{z_i}(-1)^{a_{i+d}^{(j)}}$$

   b. If $C_{\bar{a}^{(j)}, \bar{z}}(d)$ is greater than some suitable threshold T go to the verification step.
   c. Verify d by correlating an additional keystream segment $z'^{n'}$ with the corresponding segment from $\bar{a}_d^{(j)}$.
   d. If test is successful assume d_j is correct phase and proceed with $j + 1$st LFSR, otherwise proceed with $d = d + 1$.

Ouput: the set of initial states $\{a_0^{(j)}\}$ for those LFSRs that leak information into the keystream.

For the attack to be successful a certain amount of keystream is needed and a suitable threshold T has to be chosen. Define $P_f = P(C_{\bar{a}^{(j)}, \bar{z}}(d) \geq T \mid$ incorrect phase) to be the probability of a false alarm. Define $P_m = P(C_{\bar{a}^{(j)}, \bar{z}}(d) < T \mid$ correct phase) to be the probability of a miss. P_m can be chosen freely, for example, $P_m = 0.01$, according to the risk one is willing to undergo in missing the correct phase. To minimize the number of verification steps in algorithm A choose $P_f = 2^{-L}$. The threshold T and the needed number n of keystream digits then follow from these choices [112]. For example, for a leakage probability $p = 0.75$, a probability $P_m = 0.01$ of missing the correct phase, a shift register length $L = 41$ at most $n = 355$ digits of keystream are needed with a threshold $T = 0.394$. A natural method is to use the maximum likelihood test to minimize $P_m + 2^L P_f$ which is approximately the probability of error [13]. Then, for the given values, the result is a threshold $T = 0.408$ with $P_m = 0.022$ and $P_f = 0.017 \cdot 2^{-41}$. Another possibility is to consider the k most correlated phases. An approximation for the probability that the correct one is among the selected ones can be found in [14]. For the given values, for $n = 300$ digits of keystream one needs to look at the 1000 largest correlation values to have a success probability 0.98 of finding the correct one.

The computational effort to run the algorithm A is an average of $\sum_{j=1}^{N} 2^{L_j}$ (provided f leaks) which compares favorably with the effort $\prod_{j=1}^{N} 2^{L_j}$ required by an exhaustive search. Such attacks that sequentially solve for the individual subkeys residing in the shift registers are also referred to as "divide and conquer" attacks. A well-designed generator must not succumb to divide and conquer attacks. Algorithm A is only applicable when the shift register lengths L_j are small, and when the nonlinear combiner f leaks information about individual input variables. For $L_j \geq 50$ algorithm A becomes infeasible.

Algorithm A attacks the subkey residing in LFSR$_j$ by exhausting all its possible phases. But not all phases are equally likely; for instance, by the very fact that there is correlation between $\bar{a}^{(j)}$ and $\bar{z}$, phases of $\bar{a}^{(j)}$ that have smaller Hamming distance to $\bar{z}$ are more likely than phases with high Hamming distance. One improvement [77] over algorithm A that may be pursued at this point is to try to select those digits in z^n that have high probability of being undistorted copies of the corresponding digits in a^n, and correlate with all phases that are close to those highly reliable digits. The process of estimating which digits may be unaltered is based on the linear recursion satisfied by $\bar{a}^{(j)}$. Let t denote the number of taps of the LFSR that produces $\bar{a}^{(j)}$. Then every sequence digit a_i participates exactly $t + 1$ times in the linear recursion implemented by this LFSR:

$$a_j = c_1 a_{j-1} \oplus c_2 a_{j-2} \oplus \ldots \oplus c_L a_{j-L}$$

namely, at those positions where $c_k \neq 0$, inclusively at $c_0 = 1$. More linear relations involving a_i can be found by allowing polynomial multiples of the feedback polynomial $C(D)$. In particular,

$$(C(D))^{2^t} = C(D^{2^t})$$

a transformation that preserves the number of taps t. This property is important since the feasibility of the correlation attack strongly depends on a small number of taps. Suppose that a total of m linear relations is at disposal. A fixed digit a in the sequence $\bar{a}^{(j)}$ must satisfy all m relations,

$$LR_k = a \oplus A_k = 0 \qquad k = 1, \ldots, m$$

where each A_k is the sum of t different terms of $\bar{a}^{(j)}$. After passing $\bar{a}^{(j)}$ through the BSC with error probability $1 - p$ the linear relations become distorted:

$$LR_k = z \oplus Z_k \qquad k = 1, \ldots, m$$

where now each Z_k is the sum of t different terms of $\bar{z}$ and not necessarily sums to zero with z. For the considered digit z count the number h of linear relations LR_k that are satisfied. Conditioned on the event that h linear relations are satisfied the a priori probability $p = P(z = a)$ can be modified into the a posteriori probability p^*. When $z = a$ then a higher number h of satisfied relations is to be expected than for the case $z \neq a$. Or alternatively, the probability distribution of h given $z = a$ can be distinguished from the probability distribution of h given $z \neq a$ if sufficient keystream is available. Consequently, the basic correlation algorithm can be improved by adding an estimation phase [77] that first determines a set of highly reliable digits with the aim of reducing the work of the computationally expensive correlation phase.

Algorithm B: Estimation/correlation attack on combination generator [77]

Input: description of combination generator $< \{L_j, C_j(D)\}, f >$
           n digits of keystream z^n

For $j = 1$ to N do

1. Compute leakage (correlation) probability $p_j = P(z = a_j)$ from function f assuming that N BSS's form the input to f.
   If $p_j = \frac{1}{2}$ abort attempt on LFSR$_j$ and continue with $j = j + 1$.
2. Estimation phase: find set of highly reliable digits in z^n
   a. Compute expected number m of $t + 1$ digit relations available to test z^n

$$m = (t_j + 1) \log \left(\frac{n}{2L_j} \right)$$

   b. Select $h \leq m$ in such a way that at least L_j digits of z^n are expected to satisfy h or more linear relations.
   c. For any digit of z^n compute a posteriori correlation probability p_j^* given that h out of m relations for z^n are satisfied. Select those linearly independent L_j digits that have highest probability p_j^* as a reference guess I_0 for a_j^n.
3. Correlation phase: detect and correct remaining errors
   a. Compute the corresponding initial state $a_0^{(j)}$ and test its correctness by correlation of the resulting LFSR-sequence $(a^{(j)})^n$ with z^n.
   b. If the correlation indicates that the initial state $a_0^{(j)}$ is incorrect, systematically modify the initial guess by adding correction patterns of Hamming weight 1, 2, . . . to I_0 and return to the verification step until the correct initial state $a_0^{(j)}$ is found.

Output: the set of initial states $a_0^{(j)}$, $j = 1, \ldots, N$ for those LFSRs that leak information to the output of the keystream generator.

The computational effort of this algorithm is dominated by the correlation phase. The more errors were made in the estimation phase, the more initial points a_0 have to be tested in the correlation phase. In the extreme, the algorithm degenerates to an exhaustive search over all initial phases of the LFSR under investigation; but still a slight improvement over algorithm A remains by starting the correlation with the most probable initial states, instead of a random initial state. In general, algorithm B has computational complexity $O(2^{cL})$ with $0 \leq c \leq 1$; c depends on the leakage probability p, the number t of taps, and on the ratio n/L. Algorithm B is successful if a long enough segment of keystream z^n is available, if the number t of taps is small, $t \leq 10$, and if the LFSR substantially leaks into the keystream, $p > 0.6$.

On the other hand, a clear limit to algorithm B exists: when $t > 16$ and $p \leq 0.75$ then no improvement over exhaustive search is gained. As a consequence, algorithm B suggests concrete design rules for combination generators:

1. Choose f such that it is kth-order correlation-immune (see Section 3.5, below).
2. Choose long LFSRs ($L \geq 100$) with connection polynomials that have many taps ($t \geq 10$).

In algorithm B only the digits having high probability of being correct were selected and processed. In addition, one can also make use of digits that have a high probability of being distorted. Another improvement over algorithm A that may be pursued [77,128] is to try to correct the errors in the distorted "codeword" z^n to get back the undistorted "codeword" a^n thereby recovering the information symbols a^L (the initial state); in this second approach computation of the cross-correlation function is no longer necessary. In [128] a method is proposed where those digits in z^n are complemented that satisfy less than a certain number of parity-check equations (linear relations). This threshold is set to about $m/2$ when m is the total number of parity-check equations. After the correction step the procedure is iterated. This approach bears similarity to Gallager's iterated decoding algorithm [32].

In [77] another algorithm is proposed based on the idea of iterating the process of assigning a posteriori probabilities p^* several times before complementation of those digits whose probability of being correct is below a suitable threshold. This threshold is chosen to yield optimal correction effect. Then the whole process is restarted with the original probability p reassigned to every digit. In the end the algorithm should converge to the correct solution a^n. The behavior of these iterative algorithms has not been fully analyzed, one problem being that the correction step introduces statistical dependencies in the remaining error pattern. But in terms of applicability they result in similar design rules as algorithm B. We refer the reader to [77,128] for a full account.

3.4.2 Correlation attacks on filter generators.

Consider a binary filter generator (see Section 3.3). To extend a given keystream segment z^n systematically, the cryptanalyst may try to determine the used key directly, or he may try to determine an equivalent system that produces the same keystream. A specific filter generator could always be simulated by a period $(2^L - 1)$ counter and a suitably chosen output function. Note that this argument implies that any m-LFSR of length L with appropriately chosen output function is able to generate the keystream. Another approach to determining an equivalent system could be the linear equivalent generator for the keystream. In both approaches the cryptanalyst does not make use of his knowledge of the connection polynomial used in the original system. Taking into account the connection polynomial the original system can be simulated by a combination generator with m-LFSRs all possessing the same known primitive connection polynomial and a suitably chosen nonlinear combining function g. Such equivalent systems always exist; when k denotes the number of taps for f then put simply the k phases tapped by f into separate shift registers and set $g = f$ to obtain a trivial equivalent system. Occasionally, it may happen that $m < k$. Consider the periodic cross-correlation function.

$$C_{\tilde{a},\tilde{z}}(d) = \frac{1}{T} \sum_{i=0}^{T-1} (-1)^{z_i}(-1)^{a_{i-d}}$$

between the driving register sequence $\tilde{s}$ and the keystream $\tilde{z}$. In [111] an analytic expression for $C_{\tilde{a},\tilde{z}}$ is derived involving the Walsh transform $\hat{F}$ of $\hat{f}(x) = (-1)^{f(x)}$

$$C_{\tilde{a},\tilde{z}}(d) = 2^{-k}\left(1 + \frac{1}{T}\right) \hat{F}(v(d)) - \frac{\hat{f}(0)}{T}$$

where $v(d)$ denotes the linear combination of the L stage sequences that produces the dth shift $\tilde{s}(d)$ of $\tilde{s}$. Inspection of the above expression shows that a peak in the cross-correlation function can only occur if $|\hat{F}(v(d))|$ is large. But the Walsh transform $\hat{F}(v(d))$

is zero for all linear combinations $v(d)$ that select a state sequence that is not tapped by f. Thus, there are at most $2^k - 1$ peaks in the cross-correlation function. Of these at most k are at positions d_i such that the corresponding sequences $\tilde{s}(d_i)$ are linearly independent. In principle one could directly compute the filter function f by inverting the Walsh transform $\hat{F}$ whose values are known from the cross-correlation function $C_{\tilde{a},\tilde{z}}$. But in practice this is infeasible since it would require to know and computationally handle a full period of keystream. Thus, in practice one must compute the cross-correlation with a short segment of n bits of keystream which introduces a certain amount of noise. Depending on the segment length n, the number of taps k, and the properties of the filter function f it may be that the relevant peaks of the cross-correlation $C_{\tilde{a},\tilde{z}}$ disappear in the noise. The following algorithm suggested in [111] attempts to construct an equivalent generator from a set of peaks observed in the cross-correlation whose positions correspond to linearly independent sequences $\tilde{s}(d_i)$.

Algorithm: Cryptanalyst's representation of filter generator [111]

Input: description $< C(D), L >$

keystream segment z^n of length n

1. Compute cross-correlation function for all d between $0 \le d \le 2^L - 2$.

$$C_{\tilde{a},\tilde{z}}(d) = \frac{1}{n} \sum_{i=0}^{n-1} (-1)^{z_i}(-1)^{a_{i-d}}$$

2. Keep a list with the L strongest values of $C_{\tilde{a},\tilde{z}}(d)$.
3. Select a basis $\tilde{a}(d_1), \ldots, \tilde{a}(d_m)$ of dimension $m \le n$ for the sequences $\tilde{a}(d_1)$, $\tilde{a}(d_2), \ldots$ corresponding to the positions $d_1, d_2, \ldots$ of the observed peaks.
4. Compute g from the m input sequences $\tilde{a}(d_1), \ldots, \tilde{a}(d_m)$ and the corresponding output sequence $\tilde{z}$.

Output: the set of m linearly independent initial states $\{a_0^{(d_i)}\}$ and the function $g : \mathbb{F}_2^m \to \mathbb{F}_2$.

It is an open problem when the above attack will be successful, since for any other m-LFSR of length L there exists a corresponding function f' that generates the same keystream as the original arrangement [98]. Thus, one might think that correlation with any maximum-length shift register of length L would produce another usable equivalent system. However, practice has shown that the method works only for small functions (small number of taps n) with pronounced correlation between output and linear combinations of inputs. Since the above algorithm involves correlation with a significant part of the LFSR period, it is only feasible when $L < 50$ [111].

3.5 Nonlinearity Criteria

The analysis of Boolean functions with respect to their suitability as cryptographic transformations is based on the following model: Let $f, g : \mathbb{F}_2^n \to \mathbb{F}_2$ be Boolean functions and let X be a random variable that takes on values $x \in \mathbb{F}_2^n$ with uniform probability 2^{-n}. We are interested in describing the statistical dependencies of the random variables $Z = f(X)$ and $Y = g(X)$ in terms of their producing functions f and g. The cross-correlation function of f and g is defined as

$$C_{fg}(t) = \sum_x (-1)^{f(x)}(-1)^{g(x \oplus t)}$$

Note that the correlation coefficient $c(f, g)$, as defined in [78], satisfies

$$c(f, g) = 2^{-n}C_{fg}(0)$$

On the other hand,

$$c(f, g) = P(Y = Z) - P(Y \neq Z)$$

Hence, it follows that

$$P(Y = Z) = \frac{1}{2} + \frac{c(f, g)}{2}$$

Suppose g is a (nonconstant) linear function $g(x) = \omega \cdot x = \omega_1 x_1 \oplus \ldots \oplus \omega_n x_n$. Such linear functions will be denoted $L_\omega(x)$. Then

$$c(f, L_\omega) = 2^{-n}\sum_x (-1)^{f(x)}(-1)^{\omega x} = 2^{-n}\hat{F}(\omega)$$

It is easy to see that $Z = f(X)$ and $L_\omega(X)$ are statistically independent if and only if $c(f, L_\omega) = 0$, or equivalently, if and only if $\hat{F}(\omega) = 0$. To prevent divide and conquer correlation attacks on combination generators, Siegenthaler [110] introduced the notion of correlation-immunity. A Boolean function $f : \mathbb{F}_2^n \to \mathbb{F}_2$ is defined to be mth order correlation-immune if $Z = f(X)$ is statistically independent of any subset $X_{i_1}, X_{i_2}, \ldots, X_{i_m}$ of m input variables. Xiao and Massey [124] showed that this condition is equivalent to the condition that $Z = f(X)$ is statistically independent of the sum

$$\omega_1 X_1 \oplus \omega_2 X_2 \oplus \ldots \oplus \omega_n X_n$$

for every choice of $(\omega_1, \ldots, \omega_n) \in \mathbb{F}_2^n$ such that $W_H(\omega) \leq m$. The proof of the above condition was subsequently simplified by Brynielsson [14] using the following argument. Consider the Walsh transform of the conditional probability distribution $p(x \mid z)$ for a subset $X_1, \ldots, X_m$ of m binary variables

$$\sum_{x \in \mathbb{F}_2^m} p(x \mid z)(-1)^{\omega \cdot x} = E[(-1)^{\omega \cdot X} \mid Z = z]$$

$$= E[(-1)^{\omega \cdot X}]$$

$$= \sum_{x \in \mathbb{F}_2^m} p(x)(-1)^{\omega \cdot x}$$

Hence, $p(x \mid z)$ and $p(x)$ are identical since their Walsh transforms are identical. Using the correlation coefficient $c(f, L_\omega)$ it follows that a Boolean function f is mth-order correlation-immune if and only if

$$\hat{F}(\omega) = 0 \qquad (\forall \omega : 1 \leq W_H(\omega) \leq m)$$

This characterization of correlation-immune functions in terms of their Walsh transforms was developed by Xiao and Massey [124].

Siegenthaler [110] showed that there exists a trade-off between the attainable non-linear order k and the attainable level of correlation immunity m. A memoryless mth-order correlation-immune function, for $1 \le m < n$, can attain at most nonlinear order $n - m$. For balanced f the attainable nonlinear order is $n - m - 1$ unless $m = n - 1$. Xiao and Massey slightly extended these results by showing that for any mth-order correlation-immune function all coefficients of $(n - m)$th-order product terms in the algebraic normal form of f must be equal, that is, they either all are present or all vanish. When f is required to be balanced, then they all vanish, as was already noted [110]. Since the ANF transform and the Walsh transform are both linear, their composition is a linear map which directly computes the algebraic normal form of f from the Walsh coefficients. The upper bound on the attainable order of an mth-order correlation-immune function then follows from the vanishing Walsh coefficients and the structure of the Walsh-to-ANF-transform.

Siegenthaler addressed the problem of constructing balanced output mth order correlation immune combining functions with maximum nonlinear order $n - m - 1$. His recursive procedure is based on the fact that two mth-order correlation-immune functions f_1 and f_2 of n variables can be combined into a mth-order correlation-immune function f of $n + 1$ variables,

$$f(x_1, \ldots, x_{n+1}) = x_{n+1} f_1(x_1, \ldots, x_n) \oplus (x_n \oplus 1) f_2(x_1, \ldots, x_n)$$

If the product terms in f_1 and f_2 do not coincide completely, then the nonlinear order of f is equal to the larger of the two orders of f_1 and f_2 plus 1.

In [98] it was shown that the Walsh transform can be used to find the best affine approximation of a given Boolean function f. As can be seen from the correlation coefficient $c(f, L_\omega)$ this problem reduces to finding the largest Walsh transform coefficient $F(\omega)$. If $L_\omega = \omega \cdot X$ correlates positively with $Y = f(X)$ then $\omega \cdot x$ is the best approximation, otherwise $\omega \cdot x \oplus 1$ is the best approximation. There is an interesting connection to coding theory. Finding the best affine approximation to a function f is formally equivalent to maximum likelihood decoding, when the codewords are the 2^{n+1} affine function descriptions, and the information bits are the $n + 1$ coefficients of the affine function. In fact, this code is known as a Reed–Muller code of order 1 [64, p. 373].

The result was also extended [98] to solve the problem of finding the best affine approximation to a set $\mathcal{F} = \{f_1, f_2, \ldots, f_M\}$ of Boolean functions of n input variables. For instance, in a keystream generator part of the key may be devoted to selecting a specific nonlinear combiner from a set of suitable ones. In a first step compute the Walsh transforms $\hat{F}_i(\omega)$ of the M functions. Then sum the Walsh transforms component-wise $\hat{F}_\mathcal{F}(\omega) = \Sigma_{i=1}^{M} \hat{F}_i(\omega)$ (it is assumed that the f_i are selected with equal probability in the application). Find that nonzero ω that maximizes $|\hat{F}_\mathcal{F}(\omega)|$ and denote it by ω_a. If $\hat{F}_\mathcal{F}(\omega_a)$ is positive then $L_{\omega_a}(x) = \omega_a \cdot x$ is the best linear approximation, otherwise $1 \oplus \omega_a \cdot x$ is the best affine approximation.

Application of the above approximation technique to the S-boxes used in the DES showed that (1) linearizing the S-box mappings typically results in success rates of 75% up to 87.5% (a balanced Boolean function of four variables always has an affine approximation with success probability $p \ge \frac{3}{4}$, see below), with slightly better results for the inverse mappings; (2) approximating the four S-box mappings jointly by a single affine transformation may produce high success rates, occasionally even up to 81%.

The objective in the design of correlation-immune functions is to prevent correlation with linear functions L_ω, especially those for which $W_H(\omega)$ is small. Now consider the total correlation of a Boolean function f with the set of all linear functions [78], defined as

$$C^2(f) = \sum_\omega c^2(f, L_\omega) = 2^{-2n} \sum_\omega \hat{F}^2(\omega)$$

$$= 2^{-n} \sum_x f^2(x)$$

$$= 1$$

where we have made use of Parseval's theorem. The total correlation is independent of the function f, in fact it is constant. Hence, there are limits to how much correlation-immunity can be designed into a Boolean function f. The larger the subset of linear functions L_ω for which we require the correlation coefficient to vanish, the stronger will be the correlation to the remaining linear functions. In the extreme, maximum correlation-immunity of order $n - 1$ is only satisfied by the (linear) function $x_1 \oplus x_2 \oplus \ldots \oplus x_n$ (or its complement), thereby introducing maximum correlation to the sum of all variables.

In [97] it was investigated how memory can help to address the problem of designing correlation-immune combiners. It was shown that finite-state combiners can decouple the requirements for correlation immunity and nonlinearity. Combiners of the form

$$z_i = \sum_{j=1}^n x_{j,i} \oplus g(\sigma_{i-1})$$

$$\sigma_i = F(\sigma_{i-1}, \mathbf{x}_{i-1})$$

where σ_i denotes the internal state of the combiner at time i, attain maximum order $n - 1$ of correlation immunity. The functions F and g can be designed to satisfy any suitable nonlinearity criterion, such as maximum nonlinear order, without being restricted by the desired order of correlation immunity. In particular, 1 bit of memory is sufficient to decouple the requirements. It is also shown that the (bitserial) integer addition of n integers inherently defines a maximum-order correlation-immune finite-state combiner. The input sequences can be interpreted as binary representations of integer numbers starting with the least significant bit first. Computing the ith bit of the sum requires all the ith bits of the input sequences and the carry from lower positions of the sum. Keeping the carry information requires some finite memory.

In [79] the total correlation of finite-state combiners with 1 bit of memory is investigated. Since the output digit z_i of the finite-state combiner at time i depends on at most the digits $x_{j,0}, x_{j,1}, \ldots, x_{j,i}$ for $1 \leq j \leq n$, there must be correlation to linear functions of the form

$$\sum_{k=0}^i \sum_{j=1}^n \omega_{j,k} x_{j,k}$$

There are $2^{n(i+1)}$ such functions. Now consider a general combiner with 1 bit of memory with balanced output and next-state functions $f_0, f_s : \mathbb{F}_2^{n+1} \to \mathbb{F}_2$

$$z_i = f_0(x_{1,i}, \ldots, x_{n,i}, \sigma_{i-1})$$

$$\sigma_i = f_s(x_{1,i}, \ldots, x_{n,i}, \sigma_{i-1})$$

Let $\alpha = (\omega_0, \omega_1, \ldots, \omega_{n-1}, 0)$ and $\beta = (\omega_0, \omega_1, \ldots, \omega_{n-1}, 1)$ denote linear combinations that explicitly exclude and include the state σ. Decompose the total correlation $C^2(f_0)$ into

$$C^2(f_0) = \sum_\alpha c^2(f_0, L_\alpha) \sum_\beta c^2(f_0, L_\beta)$$

$$= C_0^2(f_0) + C_1^2(f_0) = 1$$

Equivalently, $C^2(f_s)$ may be decomposed into $C_0^2 (f_s) + C_1^2 (f_s)$. Then it is proved in [79] that the total correlation of the output digit z_i of the 1-bit combiner with the $N = 2^{n(m+1)}$, $1 \leq m \leq i$, linear functions of the form

$$L_{\omega,m} = \sum_{k=i-m}^{i} \sum_{j=1}^{n} \omega_{j,k} x_{j,k}$$

satisfies

$$\sum_{n=1}^{N} c_n^2 = C_0^2(f_0) + C_1^2(f_0)[1 - (C_1^2(f_s))^m]$$

Note that this total correlation converges to 1, except in the singular case $C_0(f_s) = 0$ where the limit is $C_0^2(f_0)$. Hence, in the nonsingular case the total correlation is independent of the functions f_0, f_s employed in the 1-bit combiner. This generalizes the corresponding result for (memoryless) combining functions. When LFSR-sequences are used to drive the 1-bit combiner then nonzero correlation to functions $L_{\omega,m}$ leads to nonzero correlation to sums $\sum_{j=1}^{n} s_j$ of LFSR-sequences $s_j = \sum_{k=i-m}^{i} \omega_{j,k} x_{j,k}$ where s_j is a different phase of the jth LFSR. The best one can do in this situation is to design the 1-bit combiner to be maximum order correlation-immune [98]. This condition corresponds to requiring that for all nonvanishing correlation coefficients there is no correlation to a sum of less than N LFSR-sequences [79]. The summation combiner [97] for any number n of input sequences is an example of such a maximum-order correlation-immune finite-state combiner.

Meier and Staffelbach recently [78] developed a framework for nonlinearity criteria of Boolean functions. They classified nonlinearity criteria, such as nonlinear order, correlation immunity, and best affine approximation, with respect to their suitability for cryptographic design. They argued that a function has to be considered weak whenever it can be turned into a cryptographically weak function by means of simple (e.g., affine) transformations, and that a criterion is only useful if it is invariant under such a group of transformations. The distance to sets of cryptographically weak functions appears to be a fundamentally useful measure to express and analyze nonlinearity criteria.

The *distance* between two binary functions f and g is defined as

$$d(f, g) = |\{x \in \{0, 1\}^n : f(x) \neq g(x)\}|$$

that is, as the Hamming distance between the two function tables. The distance of f to a linear function L_ω is closely related to the correlation coefficient $c(f, L_\omega)$

$$d(f, L_\omega) = 2^{n-1}(1 - c(f, L_\omega))$$

and may equivalently be expressed in terms of the Walsh transform $\hat{F}(\omega)$

$$d(f, L_\omega) = 2^{n-1} - \frac{1}{2} \hat{F}(\omega)$$

The maximum distance to an affine function is 2^{n-1}, since $d(f, \omega \cdot x) = d$ implies $d(f, \omega \cdot x \oplus 1) = 2^n - d$. Consequently, a Boolean function f is mth-order correlation-immune if its distance $d(f, L_\omega) = 2^{n-1}$ for all linear functions $L_\omega(x) = \omega \cdot x$ with $W_H(\omega) \le m$. But as with the total correlation it is impossible to achieve maximum distance to all linear functions. Let the distance $d(f, S)$ of f to a set of functions S be defined as

$$d(f, S) = \min_{g \in S} \{d(f, g)\}$$

that is, the distance between f and the nearest neighbor of f in S. As a consequence, the distance of f to the set $\mathcal{A}$ of affine functions is given as

$$d(f, \mathcal{A}) = 2^{n-1} - \frac{1}{2} \max_\omega \{|\hat{F}(\omega)|\}$$

Hence, the distance $d(f, \mathcal{A})$ may be upperbounded if we can give a lower bound on $\hat{F}^2(\omega)$. Using Parseval's theorem one finds

$$\sum_\omega \hat{F}^2(\omega) = 2^{2n}$$

Therefore, there exists an ω for which $\hat{F}^2(\omega) \ge 2^n$. We conclude that for even n and any Boolean function f

$$d(f, \mathcal{A}) \le 2^{n-1} - 2^{\frac{n}{2}-1}$$

Similarly, it may be shown that for even n and any balanced Boolean function f

$$d(f, \mathcal{A}) \le 2^{n-1} - 2^{\frac{n}{2}-1} - 2$$

It was shown in [78] that $d(f, \mathcal{A})$ satisfies the invariance property, that is, $d(f, \mathcal{A})$ remains invariant under the group of all affine transformations. Another set of potentially weak functions is the set $\mathcal{L}$ of functions with linear structures. A Boolean function f has a *linear structure* [20, 28] at $a \in \mathbb{F}_2^n$ if there exists $ab \in \mathbb{F}_2^n$ such that for all $x \in \mathbb{F}_2^n$

$$f(x) = f(x \oplus a) \quad \text{or} \quad f(x) \ne f(x \oplus a)$$

Affine functions have a linear structure for all $a \ne 0$. But a function with a linear structure is not necessarily affine. Consider as an example the nonlinear function $f(x) = x_1 x_2 \oplus x_2 \oplus x_2 x_3$ which has a linear structure for $a = (1, 0, 1)$. In [78] it was proved that

$$d(f, \mathcal{L}) \le 2^{n-2}$$

and it was shown that $d(f, \mathcal{L})$ also satisfies the invariance property, that is, $d(f, \mathcal{L})$ remains invariant under the group of all affine transformations.

Motivated by the definition of functions with linear structures, Meier and Staffel-bach [78] defined a function to be *perfect nonlinear* (with respect to linear structures) if

$$(\forall a \neq 0) \quad \left|\{x \in \{0, 1\}^n : f(x) = f(x \oplus a)\}\right| = \left|\{x \in \{0, 1\}^n : f(x) \neq f(x \oplus a)\}\right| = \frac{1}{2}2^n$$

that is, if for every nonzero vector a the values $f(x)$ and $f(x \oplus a)$ agree for exactly half the arguments x and disagree for the other half. Equivalently, f is perfect nonlinear if the autocorrelation function $C_f(a)$ satisfies

$$C_f(a) = \sum_x \hat{f}(x \oplus a)\hat{f}(x) = 0 \qquad (\forall a \neq 0)$$

Invoking the Wiener–Khintchine theorem (which states that $C_f(a)$ and the spectrum $\hat{F}^2(\omega)$ are Walsh transform pairs) yields the alternative characterization of perfect non-linear functions in the transform domain

$$\hat{F}^2(\omega) = 2^n$$

Therefore, a ± 1–valued Boolean function is perfect nonlinear if and only if $|\hat{F}(\omega)| = 2^{n/2}$ for all ω. But this is exactly the defining property of *bent* functions as introduced by Rothaus [93] in combinatorial theory. It is shown that the class of perfect nonlinear (bent) functions is optimum with respect to both distances; for an even number of arguments n, the perfect nonlinear (bent) functions simultaneously have maximum distance $2^{n-1} - 2^{(n/2)-1}$ to all affine functions as well as maximum distance 2^{n-2} to all functions with linear structure. Rothaus shows [93] that bent functions only exist for an even number n of arguments, and that their nonlinear order is bounded by $n/2$. He gives explicit constructions for bent functions:

1. Let $n = 2m$ and g be an arbitrary function; then functions of the form

$$f(x_1, \ldots, x_n) = g(x_1, \ldots, x_m) \oplus x_1 x_{m+1} \oplus \ldots \oplus x_m x_n$$

are bent.

2. Let $x = (x_1, \ldots, x_n)$ and let $a(x)$, $b(x)$ and $c(x)$ be bent functions such that $a(x) \oplus b(x) \oplus c(x)$ is also bent. Then the function

$$f(x, x_{n+1}, x_{n+2}) = a(x)b(x) \oplus a(x)c(x) \oplus b(x)c(x)$$
$$\oplus [a(x) \oplus b(x)]x_{n+1} \oplus [a(x) \oplus c(x)]x_{n+2} \oplus x_{n+1}x_{n+2}$$

is also a bent function.

A more general construction for Boolean bent functions was given by Maiorana (see [27,86]). Let $g : \mathbb{F}_2^n \rightarrow \mathbb{F}_2$ be any function and let $\pi : \mathbb{F}_2^n \rightarrow \mathbb{F}_2^n$ be any bijective trans-formation. Then the function $f : \mathbb{F}_2^n \times \mathbb{F}_2^n \rightarrow \mathbb{F}_2$ defined by

$$f(x_1, x_2) = \pi(x_1) \cdot x_2 \oplus g(x_1)$$

is a binary bent function. An alternative characterization of bent functions may be given in terms of combinatorial structures called difference sets [27]. A function $f : \mathbb{F}_2^n \rightarrow \mathbb{F}_2$ is bent if and only if

The set $S_1 = \{x : f(x) = 1\}$ is a difference set in $\mathbb{F}_2^n$ with parameters $(2^n, 2^{n-1} \pm 2^{\frac{n}{2}-1}, 2^{n-2} \pm 2^{\frac{n}{2}-1})$. (That is, there are $2^{n-1} \pm 2^{\frac{n}{2}-1}$ elements in S_1 and each nonzero $x \in S_1$ can be expressed in $2^{n-2} \pm 2^{\frac{n}{2}-1}$ ways as a difference $x = y \oplus z$ of elements $y, z \in S_1$.)

Constructions of difference sets have been known for a long time. In [76] a construction is given which is equivalent to Maiorana's construction. Generalized bent functions from $\mathbb{Z}_q^n$ to $\mathbb{Z}_q$, q a positive integer, are investigated by Kumar, Scholtz, and Welch [58]. Subsequently, Nyberg [86] considered the cryptographic properties of such generalized bent functions. Nyberg extended the notion of perfect nonlinearity to the q-ary case: a function $f : \mathbb{Z}_q^n \rightarrow \mathbb{Z}_q$ is perfect nonlinear if for all nonzero $a \in \mathbb{Z}_q^n$ and $b \in \mathbb{Z}_q$,

$$\left| \{x : f(x) = f(x + a) + b\} \right| = q^{n-1}$$

Clearly a perfect nonlinear function is bent. But a bent function is perfect nonlinear, only if q is prime. It can be shown that bent (perfect nonlinear) functions $f : \mathbb{Z}_p^n \rightarrow \mathbb{Z}_p$ exist for every odd prime p and every positive integer n. Nyberg [86] considers in particular the value distributions of p-ary bent functions, p prime, and their Hamming distances to the set of affine functions. The construction of (Boolean) bent functions has intrigued researchers for quite some time. The number of Boolean bent functions remains an open problem. In [126] a list of known constructions is given. In [91] a construction based on concatenation of rows of the Hadamard matrix H_m is given which was subsequently shown to be equivalent to Maiorana's construction (Nyberg).

Webster and Tavares [122] introduced the notion of *strict avalanche criterion* (SAC) when they studied S-boxes (see also [23]). A Boolean function satisfies the strict avalanche criterion if any output bit changes with probability exactly $\frac{1}{2}$ on complementation of one input bit. That is, if

$$(\forall a : W_H(a) = 1) \qquad P(f(X) = f(X \oplus a)) = \frac{1}{2}$$

Equivalently, the function f satisfies the SAC if the autocorrelation function vanishes at all arguments of Hamming weight 1

$$(\forall a : W_H(a) = 1) \qquad C_f(a) = 0$$

By the Wiener–Khintchine theorem the inverse Walsh transform of $\hat{F}^2(\omega)$ must satisfy

$$(\forall a : W_H(a) = 1) \qquad \sum_\omega \hat{F}^2(\omega)(-1)^{\omega \cdot a} = 0$$

This is exactly Forré's spectral characterization of Boolean functions satisfying the SAC [29]. Forré then extended the basic notion by defining the SAC of order m. A function f satisfies the SAC of order m [29,61] if and only if any partial function obtained from f by keeping m of its inputs constant satisfies the SAC (for any choice of the positions and of the values of the constant bits). In [61] it is proved that there are 2^{n+1} functions of n variables that satisfy the SAC of maximum order $n - 2$.

Perfect nonlinear (bent) functions also satisfy the strict avalanche criterion. In fact, the definition of the SAC corresponds to the definition of perfect nonlinearity restricted to offset vectors a of Hamming weight 1. Thus perfect nonlinearity is a stronger requirement than the SAC. In [91] a construction for all 2^{n+1} functions that satisfy the SAC of order $n - 2$ is given. Noting the similarity in the definitions of perfect nonlinearity and SAC, Preneel et al. [91] subsequently proposed a nonlinearity criterion that encompasses both. They defined a Boolean function $f : \mathbb{F}_2^n \to \mathbb{F}_2$ to satisfy the *propagation criterion of degree k* if

$$(\forall a : 1 \le W_H(a) \le k) \qquad P(f(X) = f(X \oplus a)) = \frac{1}{2}$$

Let $PC(k)$ denote the class of functions satisfying the propagation criterion of degree k. Equivalently, f is in $PC(k)$ if

$$(\forall a : 1 \le W_H(a) \le k) \qquad C_f(a) = 0$$

or, by invoking the Wiener–Khintchine theorem, if

$$\sum_{\omega : 1 \le W_H(\omega) \le k} \hat{F}^2(\omega)(-1)^\omega = 0$$

Note that for Boolean functions SAC corresponds to $PC(1)$ and that perfect nonlinearity (bentness) corresponds to $PC(n)$.

There are some consequences for the design of keystream generators. For a nonlinearly filtered LFSR, the keystream is correlated to at most 2^n phases of the same LFSR with a constant sum $\Sigma c^2 = 1$. Thus, zero correlation to some phases necessarily implies higher correlation to other phases. The best one can do is to uniformly minimize the cross-correlation, which is done by choosing f as close as possible to a perfect nonlinear function. For a memoryless nonlinear combiner of n sequences from N different LFSRs, the keystream is correlated to at most 2^n linear combinations of phases from N shift registers. A divide and conquer attack is possible when there is a correlation with a linear combination of phases from less than N shift registers. Thus, the best one can do [78,97] is to choose f to be maximum-order correlation-immune, such that there is only correlation with sums of LFSR phases involving all N shift registers. The remaining unavoidable correlation can be uniformly minimized. When an application demands balanced functions, Meier and Staffelbach [78] recommend to randomly select a bent function (if there is one) and to complement an arbitrary set of $2^{(n/2)-1}$ ones (zeros) in the function table. The resulting balanced function is called nearly perfect. When other design criteria have to be met in an application, they recommend to systematically generate a random perfect nonlinear function, and then to search for the closest function also satisfying those other criteria.

3.6 Clock-Controlled Shift Registers

One technique that has attracted designers of keystream generators ever since is to irregularly clock certain parts of the generator in order to achieve nonlinear effects. Recently, Gollmann and Chambers published a comprehensive survey on clock-controlled shift registers [41]. We distinguish two classes of clock control techniques applicable to shift registers: forward clock control and feedback clock control.

3.6.1   Forward clock control.   Basic forward clock control refers to the situation where one regularly clocked shift register is used to control the clock of another shift register.

Basic clock-controlled generator:

Input: parameters: a control register $< L_1, C_1(D) >$ with period T_1
a generating register $< L_2, C_2(D) >$ with period T_2
mapping $f : \mathbb{F}_2^{L_1} \to \mathbb{Z}_{T_2}$
key: initial states x_0, y_0 of the two registers
and, possibly, the function f

For $i = 1, 2, \ldots$ do

1. Clock LFSR_1 once and extract $f(\mathbf{x}_i)$.
2. Clock LFSR_2 $f(\mathbf{x}_i)$ times and extract resulting output $y_{\sigma(i)}$, where $\sigma(i)$ denotes the accumulated sum of all clock pulses that have reached LFSR_2 at time i.
3. Assign $z_i = y_{\sigma(i)}$

Output: the sequence of z_i, $i = 1, 2, \ldots$

The operation of this generator is governed by

$$z_i = y_{\sigma(i)} \qquad \sigma(i) = \sum_{k=1}^{i} f(x_k)$$

Depending on the realization of the function f one distinguishes different types of forward clock-controlled generators. If $f(\mathbf{x}_i) = x_i$, that is, if the current output bit of the control register directly is fed to the clock input of the generating register, the generator is called a stop-and-go generator [6,119]. This generator has poor cryptologic and statistical properties. Whenever the keystream bit changes value from time i to time $i + 1$ one knows that the control register has emitted a 1. Moreover, the generator has strong intersymbol dependency since half the time (depending on the number of zeros in the control sequence) the previous keystream symbol is copied to the next position. The stop-and-go generator can be improved by simply adding a 1 to the control sequence, that is, $f(\mathbf{x}_i) = 1 + x_i$, thereby transforming the generator into a step-once-twice generator. For the binary rate multiplier generator [16] f takes on the form

$$f(\mathbf{x}_i) = 1 + \sum_{j=0}^{n-1} x_{i-j} 2^j \qquad n \le L_1$$

An early investigation of the period of clock-controlled generators was done by Tretter [117]. The control register produces a constant number of ones and zeros counted over its period. Consequently, the number of clock pulses that reach the generating register during one period of the control register is constant. Define

$$S = \sigma(T_1) = \sum_{k=1}^{T_1} f(x_k)$$

After $\text{lcm}(S, T_2)$ clock pulses have been applied to the generating register both registers are back in their starting positions. Hence, the generator is periodic with $T_1 (\text{lcm}(S, T_2)/S) = T_1 T_2 / \gcd(S, T_2)$. If $\gcd(S, T_2) = 1$ then it can be shown that the overall period reaches its maximum value $T = T_1 T_2$. For the remaining discussion of forward clock control it is assumed that the condition $\gcd(S, T_2) = 1$ is always satisfied.

An early investigation of the linear complexity achievable with clock controlled generators was done by Nyffeler [87]. An upper bound on the linear complexity of the keystream can be derived as follows. Decimate the keystream by T_1, that is, consider the set of sequences

$$\{z_{j+kT_1}\}_k = \{y_{\sigma(j+kT_1)}\}_k = \{y_{\sigma(j)+kS}\}_k$$

for $1 \leq j \leq T_1$. Each of these T_1 sequences is an S-decimation of the regular output sequence $\bar{y}$ of the generating register. Let $C_s(D)$ denote the connection polynomial that can produce all these S-decimations. Then the polynomial $C_s(D^{T_1})$ can produce the key-stream for all choices of parameters and keys. Hence, the linear complexity of the key-stream is upperbounded by $\Lambda(\bar{z}) \leq L_2 T_1$. For the stop-and-go generator the upper bound is reached if the control register produces an m-sequence, that is, if $T_1 = 2^{L_1} - 1$, and if $\gcd(L_1, L_2) = 1$ [6].

One way of proving that the upper bound is reached is to prove that the generating polynomial of the keystream $C_s(D^{T_1})$ is irreducible. Suppose $C_s(D)$ of degree L is irreducible and has order T. If, for any prime factor p of T_1, holds that p divides T but not $(2^L - 1)/T$ then $C_s(D^{T_1})$ is irreducible. In particular, the degree of $C_s(D^{T_1})$ is LT_1. Versions of this result can be found in [16,87,107,113,119]. For instance, the stop-and-go generator and the step-once-twice generator achieve maximum linear complexity if both the control and the generating register produce the same m-sequence of period $T = 2^L - 1$ [119], since then the keystream has period T^2 and linear complexity TL. For the special case $T_1 = 2^n$ Günther derived a "tight" lower bound [46] $\Lambda(\bar{z}) > \frac{1}{2} L_2 T_1$ which holds provided $C_2(D)$ is irreducible. When the generating register of the step-once-twice generator produces an m-sequence then it can be shown that all l-tuples of length $l \leq (L_2 + 1)/2$ appear in the output sequence with the same frequency as in the original m-sequence $\bar{y}$. For delays bounded by $T_2/2$ the magnitude of the out-of-phase autocorrelation function is bounded by $1/T_2$.

Golić and Zivković [37] consider the following generalization of the basic clock-controlled generator. Let the generating register be a m-LFSR, and let the control register be a pure cycling register of length T_c that is randomly filled with digits $0 \leq d_i \leq 2^{L_2} - 1$. At time i the digit d_i specifies how many steps the generating register has to take. The sequence $\bar{d}$ is called the difference decimation sequence. It is shown that maximum linear complexity $L_2 T_c$ of the output sequence can be reached only if the multiplicative order of 2 modulo $T_2/\gcd(S, T_2)$ is equal to L_2. Furthermore, when the difference decimation sequence is assumed to be chosen at random and uniformly, a lowerbound on the probability that a decimated m-sequence has maximum linear complexity $T_c L_2$ is established. This bound can be made arbitrarily close to 1 for appropriate choices of the parameters T_c and L_2.

To improve the bad statistics of the stop-and-go generator while maintaining maximum speed Günther proposed the alternating step generator [46]. This generator consists essentially of two stop-and-go generators that share the same control register; the corresponding output sequences are then added to produce the keystream.

Alternating step generator:

Input: parameters: a control register $< L_c, C_c(D) >$ with period T_c
a pair of generating registers $< L_1, C_1(D), L_2, C_2(D) >$
key: initial states $\mathbf{x}_0, \mathbf{y}_0^{(1)}, \mathbf{y}_0^{(2)}$ of the three registers

For $i = 1, 2, \ldots$ do

1. Clock LFSR_1 once and produce x_i

2. If $x_i = 1$ then shift LFSR_1 and produce $y_{\sigma(i)}^{(1)}$

    If $x_i = 0$ then shift LFSR_2 and produce $y_{i-\sigma(i)}^{(2)}$

3. Set $z_i = y_{\sigma(i)}^{(1)} \oplus y_{i-\sigma(i)}^{(2)}$

Output: the sequence of z_i, $i = 1, 2, \ldots$

For the analysis in [46] the control register is assumed to produce a de Bruijn sequence of period $T_c = 2^k$. If $C_1(D)$ and $C_2(D)$ are different and irreducible, and if $\gcd(T_1, T_2) = 1$ then the period of the keystream is maximal, $T = 2^k T_1 T_2$, and the linear complexity of the keystream is bounded, $(L_1 + L_2)2^{k-1} < \Lambda(\bar{z}) \le (L_1 + L_2)2^k$. More generally, Günther has also shown that the conditions $\gcd(S, T_1) = \gcd(T_c - S, T_2) = \gcd(T_1, T_2) = 1$ and $T_1, T_2 \ge 2$ are sufficient to guarantee a maximal period $T = T_c T_1 T_2$. Using a result from [46] that states the sum of two sequences $\bar{x} \oplus \bar{x}'$ has linear complexity at least $L + L' - 2\gcd(T, T')$, one obtains the lower bound $\Lambda(\bar{z}) \ge (L_1 + L_2 - 2)T_c$ for the general case.

    If the two generating registers regularly produce m-sequences, then the frequency of all tuples of length $l \le \min\{L_1, L_2\}$ is 2^{-l} up to a deviation of order $O(1/2^{L_1-l}) + O(1/2^{L_2-l})$ that is, these frequencies are close to ideal. A similar result holds for the autocorrelation function when the delays are restricted to values $|\tau| \le T_c - 1$.

    It was pointed out in [46] that the alternating step generator succumbs to a divide and conquer attack with respect to the control register. With a keystream segment of about $4L_c$ a search through all phases of the control register will reveal the correct one. With knowledge of the state of the control register the rest of the key is easily recovered. Thus, assuming that all registers have comparable lengths, the effective keysize of the alternating generator is at most the third root of the number of possible keys.

    In a natural extension, the basic clock-controlled generator may be used to control the clock of a third shift register, its output may then control the clock of a fourth shift register, and so forth. This arrangement is referred to as a cascade generator [17,39,55,117,119].

Cascade generator:

Input: parameters: a basic shift register $< L, C(D) >$ with period T_0
        the number N of stages, each consisting of a basic shift register
        key: initial states $s_0^{(n)}$, $n = 1 \ldots N$

For $i = 1, 2, \ldots$ do

1. Shift stage 1 and produce $y_i^{(1)}$.

2. For $n = 2 \ldots N$ do
    Shift the nth stage $y_i^{(n-1)}$ times and produce

$$y_i^{(n)} = y_i^{(n-1)} \oplus s_{\sigma_{n-1}(i)}^{(n)}$$

$$\sigma_{n-1}(i) = \sum_{k=1}^{i} y_i^{(k-1)}$$

Alternatively, one may shift the nth stage $y_i^{(n-1)} + 1$ times

3. Set $z_i = y_i^{(N)}$

Output: the sequence of z_i, $i = 1, 2, \ldots$

Two arrangements have been analyzed: the m-sequence cascade [41,55,117,119] and the p-cycle cascade [17,39,40]. In the *m*-sequence cascade the basic shift register produces a maximum-length sequence of period $T_0 = 2^L - 1$. For each stage a different primitive polynomial may be chosen. It can be shown that the immediate output sequence of the shift register in the *n*th stage has period T_0^n and linear complexity LT_0^{n-1}. The keystream then is the sum of all the immediate outputs and has period T_0^N and linear complexity $1 + L(T_0 + \ldots + T_0^{N-1})$ that is greater than LT_0^{N-1}.

In the *p*-cycle cascade the basic shift register consists of a pure cycling register of prime length *p*. If one assumes that each register is loaded with a state of even parity then the minimal polynomial of a *p*-cycle register can be at most $C_p(D) = 1 + D + \ldots + D^{p-1}$. Now $C_p(D)$ is irreducible if and only if 2 is a primitive element in GF(*p*). Such primes are referred to as 2-primes [17]. Artin has conjectured [56] that about 37% of the primes are 2-primes. If p^2 does not divide $2^{p-1} - 1$ then the linear complexity of the *p*-cycle cascade is p^N, exactly identical to the period. If there is a register whose state has odd parity it can be shown that the linear complexity can at most decrease by one, that is, $p^N - 1$. In [40] a curious effect is described: Suppose one knows the sequence produced by the basic shift register but not the corresponding stage phases, then it is possible to unravel the correct phases stage by stage. This effect has been termed the *lock-in* effect. For the *p*-cycle cascade the effective key space is reduced from $(2^p - 2)^n$ to $((2^p - 2)/p)^n$ which may be serious when *p* is small.

3.6.2 Feedback clock control.   In [100] the idea of a shift register that controls its own clock is developed. Whenever the output symbol is a zero, *d* clock pulses are applied to the LFSR, and, in case the output symbol is a 1, *k* clock pulses are applied to the LFSR. The effect on the output sequence is that of an irregular decimation; correspondingly, the resulting sequence is termed a [*d, k*]-self-decimated sequence.

[*d, k*]-Self-decimation generator:

Input: parameters: a shift register $< L, C(D) >$ with period T_0
stepping rule $f : \{0, 1\} \to \mathbb{Z}_{T_0}$
key: initial state $\mathbf{y}_0$

For $i = 0, 1, 2, \ldots$ do

1. Extract $z_i = y_{\sigma(i)}$

2. Apply $f(y_{\sigma(i)})$ clock pulses to LFSR and update $\sigma(i)$

$$\sigma(i + 1) = \sigma(i) + f(y_{\sigma(i)})$$

Output: the sequence of z_i, $i = 0, 1, 2, \ldots$

Clearly the [*d, k*]-self-decimated sequence generator is a singular device, that is, the state diagram will in general contain (one or more) cycles and tails. Depending on the initial state there may be a preperiod in the self-decimated sequence. Assume that the shift register produces a maximum-length sequence, that is, $T_0 = 2^L - 1$. Then there are 2^{L-2} states with no predecessor in the state diagram provided only that $d \neq k$. This result immediately establishes a universal upper bound on the period of the generator, $T \leq (3/4)(2^L - 1)$. If d is a unit modulo T_0 then the general self-decimation rule can be separated, $[d, k] = [d][1, k']$, into a constant decimation followed by a $[1, k']$-decimation. It can be shown that the state sequence of the self-decimation generator has period

$$T_L = \left| \frac{2}{3} (2^L - 1) \right|$$

for all self-decimation rules $[d, k] = g[1, 2] \bmod (2^L - 1)$ with $\gcd(g, 2^L - 1) = 1$. A similar result holds for $[d, k] = g[1, 2^{L-1}] \bmod (2^L - 1)$ with $\gcd(g, 2^L - 1) = 1$. The distribution of zeros and ones in a self-decimated sequence is balanced and the same is true for l-tuples of short lengths. Experimental data on the linear complexity of $[1, 2]$-self-decimated sequences is consistent with the lower bound 2^{L-1} but no proof of this bound exists.

It is obvious that the self-decimation generator could not be used directly as a keystream generator since any of its output digits directly reveals a state bit, and, when the stepping rule is known, the position of the next state bit to be copied to the output sequence. A simple improvement suggested by Günther is to use different elements of the shift register sequence for the feedback clock control and the output. If both d and $k \geq 2$ then

$$z_i = y_{\sigma(i)+1} \qquad \sigma(i + 1) = \sigma(i) + f(y_{\sigma(i)})$$

will not reveal the position of the next keystream digit.

Chambers and Gollmann [17] suggested a modification of the self-decimation generator which allowed them to find cases in which the output sequence achieves maximum linear complexity. Let, as before, $z_i = y_{\sigma(i)}$ and clock the shift register regularly unless the condition

$$y_{\sigma(i)}, \ldots, y_{\sigma(i)-(z-1)} = \underline{0}$$

is satisfied. Then step twice. For $L - z$ even, it can be shown that the self-decimation generator has period $T = 2^L - 1 - (2/3)(2^{L-z} - 1)$. An exhaustive search was carried out to find pairs (L, z) that result in 2-prime values for the period T. If T is a 2-prime then $C_p(D) = 1 + D + \ldots + D^{T-1}$ is irreducible and the output sequence has linear complexity T or $T - 1$. Such a modified self-decimation generator could then be used as basic stage in a p-cycle cascade, since for large values of p pure cycling registers become unfeasible.

3.7 Generators

In this section we describe a collection of generators that have been proposed and published in the open literature. Those generators that utilize clock control such as

1. Basic clock-controlled generator
2. Alternating step generator

3. Cascade generator

4. Self-decimation generator

are described separately in Section 3.6. Some of the following generators are specific implementations of the generic arrangements described in Section 3.3.

3.7.1 Geffe's generator. In 1973 [34] Geffe proposed as a basic building block a 3-LFSR keystream generator which employed the nonlinear combining function $f : \mathbb{F}_2^3 \to \mathbb{F}_2$ defined by

$$f(x_1, x_2, x_3) = x_3 \oplus x_1 x_2 \oplus x_2 x_3$$

The lengths L_j and the feedback polynomials $C_j(D)$ are parameters of choice, subject to the constraints that the lengths are pairwise relatively prime and the feedback polynomials are primitive.

Geffe's generator:

Input: parameters: 3 LFSRs $< L_j, C_j(D) >$
key: initial states $\mathbf{a}_{j0}$ of the three LFSRs.

For $i = 1, 2, \ldots$ do

    1. For $j = 1, \ldots, 3$ do
    shift LFSR$_j$.
    extract a_{ji}.

    2. $z_i = a_{3i} \oplus a_{1i} a_{2i} \oplus a_{2i} a_{3i}$

    Output: the sequence of z_i, $i = 1, 2, \ldots$

The system-theoretic analysis shows that the overall period $T = T_1 T_2 T_3$ is maximized. The linear complexity has value $\Lambda(\bar{z}) = L_3 + L_1 L_2 + L_2 L_3$. Although the combination principle yields good statistical results and large linear complexity it is cryptographically weak. The function f leaks information about the states of LFSR 1 and 3 into the keystream, since $P(f(X) = X_1) = P(f(X) = X_3) = \frac{3}{4}$. Later it was noted that the function $f' : \mathbb{F}_2^3 \to \mathbb{F}_2$ defined by

$$f'(x_1, x_2, x_3) = x_1 x_2 \oplus x_1 x_3 \oplus x_2 x_3$$

obtained by a slight modification of f, exhibits equally good distribution properties, but allows one to generate sequences with higher linear complexity. For a discussion of this function, see Section 3.7.4, below.

3.7.2 Pless generator. The proposal of this generator [90] appears to be technology-driven, that is, the availability of the J–K flip-flop circuits seems to have stimulated the construction. In algebraic normal form, a J–K flip-flop implements the function

$$y_j = x_j^{(1)} \oplus y_{j-1}(1 \oplus x_j^{(1)} \oplus x_j^{(2)})$$

where y_j denotes the internal state at time j. Pless noted that knowledge of the current output bit of the J–K flip-flop identifies both the source and the value of the next output bit. In order to remedy this weakness Pless proposed to suppress alternate bits of the output sequence.

The generator consists of eight driving LFSRs and four *J–K* flip-flops each of which act as a nonlinear combiner for a separate pair of LFSRs. The four flip-flop output sequences are decimated by 4 and then interleaved to yield the keystream. The lengths L_j and the feedback polynomials $C_j(D)$ are parameters of choice.

Pless generator:

Input: parameters: 8 LFSRs $< L_j, C_j(D) >$,
           key: initial states $\mathbf{a}_0^{(1)}, \ldots, \mathbf{a}_0^{(8)}$ of the 8 LFSRs.

For $i = 0, 1, 2, \ldots$ do

  1. For $k = 1, \ldots, 4$ do
     a. Shift each LFSR
     b. Compute *k*th *J–K* flip-flop function for corresponding pair of LFSRs

$$y_{4i}^{(k)} = a_{4i}^{(2k-1)} \oplus y_{4i-1}^{(k)}(1 \oplus a_{4i}^{(2k-1)} \oplus a_{4i}^{(2k)})$$

     c. Collect four keystream bits as $z_{4i+k} = y_{4i}^{(k)}$

  Output: the sequence of z_l, $l = 1, 2, \ldots$

When the periods of the eight LFSRs are chosen to be pairwise relatively prime then the overall period of the keystream is their product. Rubin [94] subsequently noticed that a "divide and conquer" strategy can be used to attack each of the four subgenerators independently, and he developed a known-plaintext attack which is successful with as little as 15 characters. Moreover, and more fundamentally, the *J–K* flip-flop as a combination principle is cryptographically weak. The information-theoretic analysis of *f* under the assumption of an unknown and random state bit and independent and uniformly distributed binary input variables yields a nonzero mutual information $I(Z; X_1) = I(Z; X_2) = 0.189$ bits for either of the two input variables. Consequently, this generator succumbs to attacks as described in Section 3.4.

3.7.3 Multiplexer generator.   The proposal of this generator [51,52] appears to be technology-driven, that is, the availability of multiplexer circuits seems to have stimulated the construction. The generator consists of two driving LFSRs and a multiplexer as nonlinear combiner *f*. The multiplexer, controlled by the state of LFSR$_1$, selects at each time instant one stage of LFSR$_2$. The content of this stage then forms the current term of the keystream. The number *h* of control inputs to the multiplexer must satisfy $1 \leq h \leq L_1$ and $2^h \leq L_2$. The lengths L_j, the feedback polynomials $C_j(D)$, and *h* are parameters of choice.

Multiplexer generator:

Input: parameters: 2 LFSRs $< L_j, C_j(D) >$,
         h and control vector $j = (j_0, j_1, \ldots, j_{h-1})$ such that
         $0 \leq j_0 \leq j_1 < \ldots < j_{h-1} \leq L_1$.
         key: initial states $s_0^{(1)}, s_0^{(2)}$ of the 2 LFSRs.

For $i = 1, 2, \ldots$ do

1. Shift LFSR$_1$ and LFSR$_2$
2. Compute the integer

$$a_i = \sum_{k=0}^{h-1} 2^k \mathbf{s}_i^{(1)}(j_k)$$

3. Extract

$$z_i = \mathbf{s}_i^{(2)}(\theta(a_i))$$

where θ is an invertible mapping from $\{0, 1, \ldots, 2^h - 1\}$ to $\{0, 1, \ldots, L_2 - 1\}$

Output: the sequence of z_i, $i = 1, 2, \ldots$

Assume that the two connection polynomials are primitive, and that the lengths are relatively prime, that is, gcd $(L_1, L_2) = 1$. Then it can be proved [51] that the period assumes the maximum value $T = (2^{L_1} - 1)(2^{L_2} - 1)$. The linear complexity can be upperbounded by

$$\Lambda(\bar{z}) \leq L_2 \left(1 + \sum_{i=1}^{h} \binom{L_1}{i}\right)$$

if $2 \leq h < L_1 - 1$ with equality if all h stages are spaced at equal intervals. If $h = L_1 - 1$ or $h = L_1$ then $\Lambda(\bar{z}) = L_2(2^{L_1} - 1)$. It is also shown [52] that the periodic autocorrelation function $C(\tau) = -1/(2^{L_2} - 1)$ for most out-of-phase values of τ, $1 \leq \tau \leq T - 1$.

3.7.4 Threshold generator. This generator was proposed in 1984 [11] as a simple and efficient way of obtaining keystreams with guaranteed large linear complexity while maintaining good statistical properties. The generator consists of a set of M driving LFSRs and a nonlinear combination rule f. The number M of LFSRs, the lengths L_j, and the feedback polynomials $C_j(D)$ are parameters of choice, subject to the constraints that the lengths are pairwise relatively prime and the feedback polynomials are primitive.

Threshold generator:

Input: parameters: M LFSRs $< L_j, C_j(D) >$
            key: initial state $\mathbf{a}_{10}, \ldots, \mathbf{a}_{M0}$ of the M LFSRs.

For $i = 1, 2, \ldots$ do

1. For $j = 1, \ldots, M$ do
   Shift LFSR$_j$.
   Extract a_{ji}.
2. Compute the integer sum of the current output bits and decide

$$z_i = \begin{cases} 1 & \text{if } \left(\sum_{j=1}^{M} a_{ji}\right) > \dfrac{M}{2} \\ 0 & \text{otherwise} \end{cases}$$

Output: the sequence of z_i, $i = 1, 2, \ldots$

The output sequence will only be balanced if the number M of shift registers employed is odd. Consider the case $M = 3$, then the output mapping f can equivalently be written in algebraic normal form:

$$z_i = a_{1i}a_{2i} \oplus a_{1i}a_{3i} \oplus a_{2i}a_{3i}$$

which is the mod 2 sum of all second-order products. The system-theoretic analysis shows that the overall period $T = T_1T_2T_3$ is maximized. The linear complexity has value $\Lambda(\bar{z}) = L_1L_2 + L_1L_3 + L_2L_3$. Although the combination principle yields good statistics and large linear complexity it is cryptographically weak. The information-theoretic analysis of f under the assumption of independent and uniformly distributed binary input variables yields a nonzero mutual information $I(Z; X_j)$ of 0.189 bits for every j. As a consequence, there is positive correlation between the keystream and every input sequence.

3.7.5 Inner product generator.   In [67] it is shown that perfect linear cipher systems exist. The linear encryption transformation has the form

$$y_i = F_k(x_i, \ldots, x_{i-M})$$

$$= x_i + \sum_{j=1}^{M} c_j(i, k)x_{i-j}$$

where the coefficients $c_j(i, k)$ depend on both the time instant i and the key k. If the plaintext is restricted never to contain M consecutive zeros, a perfect linear cipher (where plaintext and ciphertext are statistically independent) can be obtained with two digits of key per plaintext digit. The construction makes use of the M-fold inner product of M consecutive plaintext digits with M consecutive key digits shifted at double speed. This perfect linear cipher suggests that the inner product of two LFSRs that are operated at different speeds may provide an interesting sequence generator. The lengths L_1, L_2, and the speed factors d_1, d_2 are parameters of choice.

Inner product generator:

Input: parameters: two LFSRs $< L_j, C_j(D) >$ and two speed factors d_1, d_2
key: initial states $\mathbf{a}_0^{(1)}$, $\mathbf{a}_0^{(2)}$ of the registers.

For $i = 1, 2, \ldots$ do

1. Shift LFSR$_1$ d_1 times.
2. Shift LFSR$_2$ d_2 times.
3. Compute the inner product of the 2 LFSR states

$$z_i = \sum_{k=1}^{min\{L_1,L_2\}} \mathbf{a}_{1k} \cdot \mathbf{a}_{2k}$$

Output: the sequence of z_i, $i = 1, 2, \ldots$

The system-theoretic analysis shows [67] that the output sequence will have linear complexity $\Lambda(\tilde{z}) = L_1 L_2$ provided $C_1(D)$ and $C_2(D)$ are irreducible, $\gcd(L_1, L_2) = 1$, $\gcd(d_1, T_1) = \gcd(d_2, T_2) = 1$, where T_1 and T_2 denote the periods of the original shift registers, and the shift registers are initially loaded with nonzero contents. The period T of $\tilde{z}$ is lowerbounded by $T_z \geq T_1 T_2/(q - 1)$. For the analysis of the global randomness properties, assume that $C_1(D)$ and $C_2(D)$ are primitive polynomials over $GF(2)$, and that $L_1 > L_2$. Then it can be proved that the number of zeros within a period is $(2^{L_2} - 1)(2^{L_1-1} - 1)$ and the difference between the number of ones and the number of zeros relative to the period length is $1/(2^{L_1} - 1)$. The inner product combination enforces "almost" ideal distribution properties of the output sequence. This generator hence also carries the potential of synthesizing LFSRs with irreducible but nonprimitive characteristic polynomials that possess a state cycle with good global randomness properties. The described generator may be cascaded in a natural way: combine the output sequence $\tilde{z}$ in a second inner product with a third LFSR of length L_3; since $\tilde{z}$ has no identifiable speed factor attached to it, it is sufficient to choose d_3 greater than 1.

3.7.6 Wolfram's cellular automaton generator.

At Crypto'85 Wolfram [123] proposed to use a binary one-dimensional cellular automaton with nonlinear next-state function as sequence generator. A one-dimensional cellular automaton consists of a (possibly infinite) line of sites with values $a_i \in \mathbb{Z}_n$. These values are updated in parallel (synchronously) in discrete time steps according to a fixed rule of the form

$$a'_k = \Phi(a_{k-r}, a_{k-r+1}, \ldots, a_{k+r})$$

where r denotes the span of input arguments to Φ. Any practical implementation of a cellular automaton must contain a finite number of sites N. These are typically arranged in a circular register with periodic boundary conditions, that is, the cell indexes of the arguments of next-state rule Φ are computed mod N. In [123] the next-state update rule

$$a'_k = a_{k-1} \oplus (a_k \vee a_{k+1})$$

is singled out as the most promising among all three-argument rules Φ.

Wolfram's cellular automaton generator:

Input: parameters: number of cells N
                (and possibly the next-state function f)
                key: initial state $\mathbf{a}(0) = (a_0(0), a_1(0), \ldots, a_{N-1}(0))$.

For $i = 1, 2, \ldots$ do

1. Update cellular automaton using the rule

$$a_k(i) = a_{k-1}(i - 1) \oplus (a_k(i - 1) \vee a_{k+1}(i - 1))$$

for $k = 0, 1, \ldots, N - 1$, with the cell indexes being computed mod N.

2. Extract keystream from any single site k

$$z_i = a_k(i)$$

Output: the sequence of z_i, $i = 1, 2, \ldots$

For infinite N it is shown in [123] that all length S spatial sequences and all length T temporal sequences occur with equal probabilities, provided the probability distribution of initial configurations is uniform. Thus, the corresponding entropies are maximal. But the deterministic nature of the above cellular automaton rule implies that only certain space-time patterns of values can occur. In fact, knowing the temporal sequences of two adjacent sites allows the spatial reconstruction of the cellular automaton configuration. Similarly, all the site values in a particular space-time patch are completely determined by the values that appear on its upper, left, and right boundaries. It is shown that the entropy contained in the cellular automaton configuration per time unit is at most 1.2 bits. Hence knowledge of the time sequences of values of about 1.2 sites suffice in principle to determine the values of all other sites.

In addition to studying the global properties of cellular automata evolution, one may also look at the effects caused by local changes in the cellular automaton configuration. The form of the cellular automaton rule implies that a change in one site propagates to the right with 1 bit per time unit. Empirical measurements indicate that the rate of information propagation to the left is about $\frac{1}{4}$ bit per application of the cellular automaton rule [123].

Regarding the finite-size behavior, one may examine the state transition diagram of the N site circular cellular automaton. Typically the state transition diagram consists of a set of cycles, fed by trees representing transients. For the above generator it is shown that the number of states that have two (branch nodes) or zero predecessors (root nodes) tends to zero asymptotically with N. For large N, the state transition diagrams appear to be increasingly dominated by a single cycle. The actual maximal cycle lengths Π_N were computed for $N < 55$ [123]; for the determined range Π_N can be approximated by

$$\log_2 \Pi_N \approx 0.61(N + 1)$$

Note that for a random next-state mapping one would expect cycles of average length $2^{N/2}$. Empirical results indicate that all 2^S possible length S keystreams can be generated from any of the 2^N keys (initial states of the size N cellular automaton) provided $N \geq S + 2$. Finally, a battery of statistical tests was applied to the generator, which all showed no significant deviations from randomness.

3.7.7 $1/p$ generator. In [10] the $1/p$ generator is proposed and analyzed. It is based on the state transformation $F(x) = bx \bmod N$ of a linear congruential generator. In the same paper the quadratic congruential generator whose state transformation is $F(x) = x^2 \bmod N$ is analyzed and compared to the $1/p$ generator. Let $b > 1$ denote a fixed base. The seed for the $1/p$ generator consists of an integer p that is relatively prime to b, and a random $x_0 \in Z_p^*$. Basically, the $1/p$ generator expands the seed x_0/p to the base b, that is, it produces the sequence of b-ary quotient digits when x_0 is divided by p to the base b. In contrast, the linear congruential generator produces the sequence of b-ary remainder digits when x_0 is divided by p to the base b.

$1/p$ generator:

Input: parameters: base $b > 1$
           key: integer p with $\gcd(p, b) = 1$, a random $x_0 \in Z_p^*$.

For $i = 1, 2, \ldots$ do

   1. Compute next-state function

$$x_i = F(x_{i-1}) = bx_{i-1} \bmod p$$

   2. Compute output function

$$z_i = f(x_{i-1}) = bx_{i-1} \operatorname{div} p$$

   Output: the sequence of z_i, $i = 1, 2, \ldots$

The analysis in [10] shows that the period of the keystream reaches its maximum $T = p - 1$ for p prime and b a primitive root mod p. For this subclass of generators interesting statistical properties result. The corresponding keystreams can be proved to be generalized de Bruijn sequences of period $p - 1$. A generalized de Bruijn sequence has the property that every b-ary string of $|p| - 1$ digits appears at least once, and every b-ary string of $|p|$ digits appears at most once in a given period of the sequence ($|p|$ denotes the length of the b-ary expansion of p). These sequences resemble the maximum-length shift register sequences. Hence, it is not surprising that knowledge of the base b and of $k = \lceil \log_b(2p^2) \rceil$ digits $z_{m+1}, \ldots, z_{m+k}$ of the keystream is sufficient to predict the generator in both forward and backward direction. This is done by applying the continued fraction expansion algorithm to the fraction $z_{m+1} \ldots, z_{m+k} / b^k$. The convergent is guaranteed to be x_m/p if $1/b^k \le 1/2p^2$. Since $\gcd(b, p) = 1$ it is also easy to extend the sequence in backward direction by computing $x_{i-1} = b^{-1}x_i \bmod p$.

   3.7.8 Summation generator. In [97] the suitability of integer addition as a combining function is investigated. Let the input sequences be binary expansions of some integers (possibly infinite, if the sequences are assumed to be semi-infinite). The sum function $f : \mathbb{Z}^N \to \mathbb{Z}$ defined by $z = \Sigma_{i=1}^N x_i$ can be computed in bit-serial fashion starting with the least significant bit. The investigation was motivated by the fact that integer addition is highly nonlinear, when considered over $\mathbb{F}_2$, and at the same time provides the maximum order of correlation immunity. The following generator [97] is a simple application:

   Summation generator:

   Input: parameters: N LFSRs $< L_j, C_j(D) >$
             key: initial states of the N LFSRs and carry C_0

For $i = 1, 2, \ldots$ do

   1. Step each shift register once to produce $x_{1i}, x_{2i}, \ldots, x_{Ni}$.
   2. Compute the integer sum

$$S_i = \sum_{k=1}^{N} x_{ki} + C_{i-1}$$

   3. Set

$$z_i = S_i \bmod 2$$

$$C_i = \left\lfloor \frac{S_i}{2} \right\rfloor$$

Output: the sequence z_i, $i = 1, 2, \ldots$

The analysis in [97] shows that the real sum of N periodic sequences with r-ary digits is ultimately periodic with period $T = \Pi T_i$ if the periods T_i of the individual sequences are pairwise relatively prime. When two binary m-sequences of relatively prime degrees are added over the reals then the sum sequence exhibits linear complexity close to its period length, $\Lambda(\bar{z}) \leq (2^{L_1} - 1)(2^{L_2} - 1)$. If the input vectors are independent and uniformly distributed then the keystream will also consist of i.u.d. random variables. The real sum directly provides a $(N - 1)$th-order correlation-immune combination principle, which is the maximum order of immunity possible.

Subsequently, in [79] the correlation analysis of memoryless functions was extended to combiners with 1 bit of memory (see also Section 3.5, above). The summation generator with two input sequences is such a 1-bit combiner and it was shown that there is positive correlation with $x_{1,i} \oplus x_{1,i-1} \oplus x_{2,i}$ (which is in accordance with result that the summation generator is maximum-order correlation-immune when fed with shift register sequences). In [115] the probability distribution of the carry for addition of random integers is analyzed. It is proved that the carry is balanced for even N and biased for odd N and that the bias converges to 0 as N tends to ∞.

3.7.9 Knapsack generator. In [96] the suitability of the knapsack as a nonlinear transformation is investigated. The knapsack problem, also known as a subset sum problem, is an NP-complete problem [33]. A problem instance consists of a finite set of L positive integer-valued weights and a positive integer S. The problem is to decide if there is a subset of the weights that sums exactly to S. For a given set of weights the knapsack may be viewed as a function from the selection vector $\mathbf{x} = x_1, \ldots, x_L$ to the integer S which is the sum of those weights selected by $\mathbf{x}$. The difficulty of the knapsack problem motivated its application as a nonlinear function in a filter generator [96]. Note that unlike the NP-complete problem the weights are kept secret in this application.

Knapsack generator:

Input: parameters: LFSR $< L, C(D) >$, modulus Q
key: L knapsack weights $w_1, \ldots, w_L$ of size N bits each
initial state $\mathbf{x}_0 = x_{10}, \ldots, x_{L0}$ of the LFSR

For $i = 1, 2, \ldots$ do

1. Step LFSR to produce next state
2. Compute knapsack sum

$$S_i = \sum_{k=1}^{L} x_{ki} w_i \mod Q$$

3. Extract some bits of S_i to form Z_i

Output: the sequence Z_i, $i = 1, 2, \ldots$

The knapsack sum sequence is periodic with period $2^L - 1$ which implies that the jth bit sequence $\bar{s}_j$ also is periodic with period $2^L - 1$. The analysis in [98] shows that, if $Q = 2^N$ then the jth bit sequence $\bar{s}_j$ of the knapsack sum sequence $\bar{S}$ has linear complexity

$$\Lambda(\bar{s}_j) \leq \sum_{k=1}^{2^j} \binom{L}{k} \qquad j < \lceil \log L \rceil$$

$$\Lambda(\bar{s}_j) \leq \sum_{k=1}^{L} \binom{L}{k} = 2^L - 1 \qquad j \geq \lceil \log L \rceil$$

which shows that the log L least significant bits of a mod 2^L knapsack are cryptographically weaker than the higher-order bits. The effective keysize for output stage j is at least $L \log L$ bits for all $j \geq \lceil \log L \rceil$.

4 COMPLEXITY-THEORETIC APPROACH

The basis of the complexity-theoretic approach is the notion of computationally accessible information. If two random variables are statistically independent then there is no way to compute information about the second from observation of the first. But even if there is complete statistical dependence (as for ciphertext-plaintext pairs in some public key cryptosystems) it may not be computationally accessible.

In the complexity-theoretic model all computations are parametrized by a security parameter, usually the key length, and an asymptotic analysis is carried out. Only algorithms whose running times can be expressed as polynomials in the size of the input are considered to be computationally feasible. The opponent is assumed to be limited to polynomial time attacks. Cryptanalysis is the process of (1) predicting a single digit of the keystream, or (2) distinguishing the keystream sequence from a truly random sequence. A keystream generator is defined to be *perfect* if it is (1) unpredictable, or (2) indistinguishable by all polynomial-time statistical tests. But such a perfect generator is a hypothetical device; it is not known whether it exists. The proposed generators are all based on the assumed difficulty of some "famous" problem like, for instance, the discrete log [9], quadratic residuosity [10], and inverting the Rivest–Shamir–Adleman (RSA) encryption algorithm, [1,80]. Proving the perfectness of these generators would require superpolynomial lower bounds on the complexity of the associated problems. Since this has been impossible, researchers have resorted to heuristic arguments, called intractability hypotheses, to prove that a generator is "perfect." We will reserve the word *perfect* (without quotation marks) for the ideal device, which is not known to exist, and will use "perfect" (with quotation marks) whenever a generator is "perfect" modulo some intractability hypothesis. The broader question also remains of how much evidence is provided by an asymptotic analysis, inasmuch as all implementations of cryptosystems necessarily have a finite size.

4.1 Basic Notions and Concepts

A *bit generator* G is a sequence $\{G_n : n \geq 1\}$ of polynomial time algorithms G_n. Each $G_n : \{0, 1\}^n \to \{0, 1\}^l$ stretches a random key x^n of length n into a pseudorandom sequence z^l of length $l(n)$, where l is a polynomial function of n. Let $z^l = G_n(x^n)$ denote the keystream produced by G_n on input x^n.

Let $\mu_{R,l}$ denote the uniform probability distribution on the set of l bit sequences, that is, $\mu_{R,l}(s^l) = 2^{-l}$. Correspondingly, let $\mu_{G,l(n)}$ denote the probability distribution of keystream sequences z^l as generated by G_n for randomly selected keys (that is, for keys chosen according to $\mu_{R,n}$). The probability of a distinguished pseudorandom sequence z^l then is

$$\mu_{G,l}(z^l) = 2^{-n} \cdot \#\{x^n : G_n(x^n) = z^l\}$$

Let μ_G be the sequence of probability distributions $\{\mu_{G,l(n)} : n \geq 1\}$ as produced by G ; μ_G is said to be induced by G.

A practical security requirement for a keystream generator is its unpredictability. Given a segment z^i of i bits it must be unfeasible to extend the sequence beyond i. Otherwise, a captured segment of keystream would allow one to successfully decrypt a portion of the ciphertext without knowledge of the key. The concept of unpredictability can be formalized through the notion of a next-bit test (or predictor) [9,125].

A *predictor (next-bit test)* $C = \{C_n : n \geq 1\}$ is a sequence of probabilistic polynomial size circuits C_n with $i_n < l(n)$ input gates and one binary output gate. Loosely speaking, a predictor C is capable of predicting a pseudorandom generator G if the fraction of times C_n's output bit b agrees with z_{i+1} is "significantly" different from $\frac{1}{2}$. More precisely, a predictor C is said to predict a pseudorandom generator G (G is said to fail the next-bit test C) if there exists a polynomial $P(n)$ such that for infinitely many n

$$Pr(C_n(z^i) = z_{i+1}) \geq \frac{1}{2} + \frac{1}{P(n)}$$

where z^l is drawn according to $\mu_{G,l(n)}$.
Conversely, G is said to pass C if for all but a finite number of n and for all polynomials $P(n)$,

$$Pr(C_n(z^i) = z_{i+1}) < \frac{1}{2} + \frac{1}{P(n)}$$

G is said to be unpredictable if for all next-bit tests C, for all but a finite number of n, for all polynomials $P(n)$, and for all $i < l(n)$

$$Pr(C_n(z^i) = z_{i+1}) < \frac{1}{2} + \frac{1}{P(n)}$$

A keystream generator attempts to efficiently simulate randomness. If the pseudorandom sequences it generates were efficiently distinguishable from purely random sequences one could not claim that the generator simulates randomness. On the other hand, the generator's output can always be distinguished from purely random sequences by simply trying all the keys and comparing the resulting pseudorandom sequences with the sequence at hand. But this method takes exponential time and is considered unfeasible. The ability of distinguishing a pseudorandom sequence from a random sequence seems to be the most fundamental step preceding any other step in analyzing the pseudorandom sequence. The concept of indistinguishability can be formalized through the notion of a statistical test [125].

A *statistical test* $T = \{T_n : n \geq 1\}$ for the bit generator is a sequence of probabilistic polynomial-size circuits T_n with $l(n)$ input gates and one binary output gate.

Loosely speaking, a test T is able to distinguish a pseudorandom generator G if the fraction of times it puts out a 1 when it is confronted with sequences r^l drawn uniformly from $\mu_{R,l}$ is "significantly" different from the fraction of times it puts out a 1 when it is confronted with sequences z^l drawn according to the generator's output distribution $\mu_{G,l(n)}$. More formally, a test T is said to distinguish G if there exists a polynomial $P(n)$ such that for infinitely many n

$$\left| p_n^{T,G} - p_n^{T,R} \right| \geq \frac{1}{P(n)}$$

where $p_n^{T,G}$ and $p_n^{T,R}$ denote the probabilities that T emits a 1 on input of sequences drawn according to $\mu_{G,l(n)}$ and $\mu_{R,l}$.

Conversely, a generator G is said to pass the statistical test T if for all polynomials $P(n)$ and all but a finite number of n

$$\left| p_n^{T,G} - p_n^{T,R} \right| < \frac{1}{P(n)}$$

Yao succeeded in linking together the notions of predictor and statistical test, proving their equivalence [125].

Theorem: *A bit generator* G *passes all next-bit tests* C *if and only if it passes all statistical tests* T.

One then defines: A bit generator G is *perfect* if it passes all polynomial size statistical tests T.

Based on Yao's result, it is (at least in principle) possible to establish that a generator is perfect by proving that there is no polynomial-size predictor for the resulting pseudorandom sequences. But unfortunately, none of the proposed generators could be proved to be perfect. Indeed, it is not known if perfect generators exist.

Motivated by Yao's result, Blum and Micali [9] developed a general scheme to construct bit generators. Let $f = \{f_n : X_n \to X_n\}$ be a one-way permutation to be used as next-state function for the generator. Let $B = \{B_n : X_n \to \{0, 1\}\}$ be a binary predicate with domain X_n to be used as output function. Randomly select an element $x \in X_n$ as seed, iterate f_n on x, and output $z_i = B_n(f_n^i(x))$ for $1 \leq i \leq l(n)$. If B is an unpredictable predicate for f then the keystream z^l is unpredictable (by next-bit tests) to the left, and by Yao's result, is indistinguishable by any statistical test T (in particular, it is also unpredictable to the right). Concrete applications of this scheme are:

1. Blum-Micali generator [9]

$$f_n : Z_p^* \ni x \mapsto y = \alpha^x \bmod p \in Z_p^*$$

$$B_n : Z_p^* \ni y \mapsto \operatorname{half}_p(x) \in \{0, 1\}.$$

2. Quadratic residue generator [10]

$$f_n : QR_N \ni x \mapsto y = x^2 \bmod N \in QR_N$$

$$B_n : QR_N \ni y \mapsto \operatorname{lsb}(x) \in \{0, 1\}.$$

3. RSA generator [1]

$$f_n : Z_N^* \ni x \mapsto y = x^e \mod N \in Z_N^*$$

$$B_n : Z_N^* \ni y \mapsto \mathrm{lsb}(x) \in \{0, 1\}.$$

These proposals are all based on the assumed difficulty of some "famous" problem such as the discrete log, quadratic residuosity, and inverting RSA. To prove the "perfectness" of each generator then makes it necessary to introduce heuristic intractability hypotheses with respect to the difficulty of the underlying problem. In the sequel these generators will be discussed in detail.

4.2 Generators

4.2.1 Shamir's pseudorandom number generator.
The first proposal to use RSA in a pseudorandom number generator is due to Shamir [108]. His interest was in determining if for a given ciphertext Y the additional knowledge of an arbitrary number of decryptions mod N under various secret exponents $d_i = 1/e_i \mod \Phi(N)$ helps to find the decryption mod N under a specific secret exponent $d = 1/e \mod \Phi(N)$. Let e_1, $e_2, \ldots, e_l$ be a fixed sequence of public key exponents such that all the e_i are pairwise relatively prime. Choose an RSA-modulus $N = pq$ of size n such that all the e_i are relatively prime to $\Phi(N)$ and choose a public seed S from Z_N. The secret key of the generator consists only of the factorization $\{p, q\}$ of N.

Shamir's generator [108]:

Input: parameters: the modulus N,
  the (public) seed S randomly selected from Z_N
  the sequence $e_1, \ldots, e_l$,
  key: the factorization p, q.

1. For $i = 1, \ldots, l$ decrypt S under the secret exponents $d_i = (1/e_i) \mod \Phi(N)$

$$Z_i = S^{1/e_i} \mod N$$

Output: the sequence of Z_i, $i = 1, \ldots, l$

Note that this generator puts out whole n-bit numbers, not only bits. In [108] the following result is proved.

Theorem: *There exists a fixed polynomial* P(n) *such that any circuit* C_n *which, when given as input a modulus* N *of size* n *and a seed* S $\in$ Z$_N$, *is able to predict* Z_1 *from a number of known roots* $Z_2, \ldots, Z_l$ *for at least a fraction* δ(n) *of the instances, can be transformed into another circuit* A$_n$ *of size at most* $|C_n|$ + P(n) *that inverts RSA on input* N, S *for at least a fraction* δ(n) *of the instances.*

The proof exhibits the following circuit A_n and shows that it will recover $X = Y^{1/e_1} \mod N$ with the help of C_n.

Circuit A_n : simulates $D_N(Y) \mod N$

Input: Y, N

1. Define $S = Y^{e_2 \cdots e_l} \bmod N$ and $E = \prod e_i$; compute for $i = 2, \ldots, l$

$$Z_i = S^{1/e_i} = Y\left(\frac{E}{e_1 e_i}\right) \bmod N$$

by successive exponentiations.

2. Call C_n with input $N, S, Z_2, \ldots, Z_l$ and retrieve Z_1.

3. Use Euclid's extended gcd-algorithm to compute

$$\gcd\left(\frac{E}{e_1}, \ldots, \frac{E}{e_l}\right) = a_1 \frac{E}{e_1} + \ldots + a_l \frac{E}{e_l} = 1$$

4. Since $Z_i = X^{E/e_i} \bmod N$ multiply

$$Z_1^{a_1} Z_2^{a_2} \ldots Z_l^{a_l} = (X^{E/e_1})^{a_1} (X^{E/e_2})^{a_2} \ldots (X^{E/e_l})^{a_l} = X$$

Output: $X = Y^{1/e_1} \bmod N$

If C_n is successful for a fraction $\delta(n)$ of the instances of size n so must be A_n since the transformation $S = Y^{E/e_1} \bmod N$ is a permutation. Thus, predicting Shamir's pseudo-random number generator is provably equivalent to inverting RSA. If the RSA-function is almost everywhere secure, the random number generator must also be almost every-where secure. But in reality, when RSA is used for encrypting plaintext messages, the distribution of ciphertexts will not be uniform, and consequently, the success rates may be quite different. In [9] it was pointed out that the numbers of Shamir's generator could be unpredictable as a whole yet they could have a special form. For instance, individual bits could be heavily biased or predictable with high probability; as an illus-tration let N have the form $2^n + k$, then for uniform Z_i the most significant bit will be 1 with probability k/N and 0 with probability $2^n/N$.

4.2.2 Blum-Micali generator.

Let p be an odd prime and let α be a generator for Z_p^*, the ring of units modulo p. The discrete logarithm of an element $y \in Z_p^*$ with respect to α, denoted as $\text{index}_{p,\alpha}(y)$, is defined as the unique integer $0 \le x \le p - 2$ such that $y = \alpha^x \bmod p$. The discrete logarithm problem with input p, α, y consists of finding $\text{index}_{p,\alpha}(y)$. In general, no efficient algorithm is known that can solve the dis-crete logarithm problem. If $y \in QR_p$, that is, y is a quadratic residue modulo p, then $\text{index}_{p,\alpha}(y) = 2t$ for some integer $t < (p - 1)/2$. The two square roots of $y \in QR_p$ then are $\alpha^t \bmod p$, which is called the principal square root, and $\alpha^{t+(p-1)/2} \bmod p$, which is called the nonprincipal square root. For each p and α an element $x \in Z_p^*$ is a principal square root if and only if $\text{index}_{p,\alpha}(x) < (p - 1)/2$. Blum and Micali [9] showed that the existence of an efficient algorithm to decide if a given element is a principal square root with respect to p and α implies the existence of an efficient algo-rithm to compute the discrete logarithm. Define the binary predicate

$$\text{half}_p(x) = \begin{cases} 1 & \text{if } x < \dfrac{p - 1}{2} \\ 0 & \text{otherwise} \end{cases}$$

Blum-Micali generator [9]:

Input: parameters: (p, α)
       key: a randomly selected seed $x_1 \in Z_p^*$.

1. For $i = 1, \ldots, l$ do

   a. $z_i = \text{half}_p(x_i)$

   b. $x_{i+1} \leftarrow \alpha^{x_i} \bmod p$

Output: the sequence $z_1, z_2, \ldots, z_l$

The security of the index generator is based on the difficulty of computing discrete logarithms modulo p. In [9] it is shown:

Theorem: *If there is a probabilistic polynomial-size circuit that, when given as input* p, α, *and* y $= \alpha^x$ *mod* p, *guesses* half$_p$(x) *for at least a fraction* $1/\text{P}'(\text{n})$ *of the primes* p *of size* n *with advantage* $\geq 1/2 + 1/\text{P}(\text{n})$ *then, for any polynomial* Q(n), *there is another circuit that, when given as input* p, α, y *computes* index$_{\text{p},\alpha}$(y) *for at least a fraction* $1/\text{P}'(\text{n})$ *of the primes* p *of size* n *with advantage* $\geq 1 - 1/\text{Q}(\text{n})$.

If there is a predictor that, when given as input the keystream $z_1, \ldots, z_l$, is able to guess z_0 with probability $\geq 1/2 + \varepsilon(n)$, then it is possible to determine the predicate half$_p(x)$ from p, α, $y = \alpha^x \bmod p$ with the same probability of success. Use y as the seed for the index generator and produce the sequence $z_1, \ldots, z_l$; call the predictor to obtain z_0 which equals half$_p(x)$ with probability $\geq 1/2 + \varepsilon(n)$. By the above theorem, the existence of such a predictor with advantage $\geq 1/2 + 1/\text{P}(n)$ would then imply that the discrete logarithm problem could be solved in probabilistic polynomial time. To prove the "perfectness" of the index generator Blum and Micali introduced the following intractability hypothesis [9, 56].

Discrete logarithm assumption: For any polynomials P and P', any polynomial size circuit C, and all but a finite number of $n \geq 1$, the fraction of primes p of size n for which C is able to solve the discrete logarithm problem with probability $\geq 1 - 1/P(n)$ is upperbounded by $1/P'(n)$.

This hypothesis is true if and only if the index generator is unpredictable (to the left). As a last step, apply Yao's theorem [125] to conclude that the index generator is indistinguishable by probabilistic polynomial time statistical tests, and thus is "perfect."

The index generator is only one concrete implementation of a general scheme also proposed in [9]. Subsequently it was shown [53,62] that the discrete exponentiation function has $\log\log p$ simultaneous secure bits. This result implies a modified index generator with $\log\log p$ bits per each iteration of the discrete exponentiation.

4.2.3   RSA generators. The basic RSA generator [1] is another implementation of the general scheme proposed in [9], with $f(x) = x^e \bmod N$ as one-way permutation, and $B(x) = \text{lsb}(x)$ as unpredictable predicate for f. Let e be an integer ≤ 3. Let N be an integer of size $|n|$ which is the product of two primes p, q, such that $\gcd(e, \phi(N)) = 1$. The RSA generator accepts as input the triplet (N, e, x).

RSA generator:

Input: parameters: (N, e)

         key: a randomly selected seed $x \in Z_N^*$.

1. $x_0 \leftarrow x$
2. For $i = 1, \ldots, l$ do
   a. compute $x_i \leftarrow x_{i-1}^e \mod N$
   b. extract $z_i \leftarrow \text{lsb}(x_i)$

Output: the sequence $\{z_i\}_1^l$

The security of this generator is based on the difficulty of inverting RSA. The following theorem is proved in [1].

Theorem: *Every probabilistic algorithm which, when given as input* $y = x^e$ *mod N, e, and, N, is able to guess the least significant bit of* x *in expected time* $T_B(n)$ *with probability at least* $\frac{1}{2} + \varepsilon(n)$ *can be transformed into an algorithm that inverts RSA ciphertexts in expected time* $T(n)$ *in* $O(\varepsilon^{-8}(n)n^3 T_B(n))$.

If $T_B(n)$ is a polynomial function of n so is $T(n)$ and RSA ciphertexts can be inverted in polynomial time. The ability of predicting z_0 from the keystream sequence $z_1, \ldots, z_l$ implies the ability of deciding $\text{lsb}_N(x)$ from $y = x^e \mod N$. To see this, generate in a first step $z_1, \ldots, z_l$ using the RSA generator with y as x_1. In a second step call the predictor with $z_1, \ldots, z_l$ as input and obtain a bit b that equals z_0 with probability $1/2 + \varepsilon(n)$. Since z_0 equals $\text{lsb}_N(x_0)$ the probability that b equals $\text{lsb}_N(x)$ is also $1/2 + \varepsilon(n)$. The existence of a predictor with advantage $\varepsilon(n) = 1/P(n)$ would imply a polynomial-time algorithm for inverting RSA. Under the assumed intractability of inverting RSA there cannot be a probabilistic polynomial time algorithm that guesses the least significant bit of the plaintext x with advantage $1/P(n)$, which implies that there cannot be a polynomial time statistical test T that is able to distinguish the RSA generator. Thus, the RSA generator is "perfect."

Every probabilistic algorithm that, when given for input $x^e \mod N$, N, and e, can guess the least significant bit of x with probability at least $1/2 + \varepsilon(n)$ is equivalent to a statistical test T that is able to distinguish the following distributions

1. The uniform distribution on $[1, N]$
2. The distribution of $x^e \mod N$ for random, even $x \in [1, N]$

with advantage $\varepsilon(n)$. Instead of assuming that there is no polynomial-time algorithm that can invert a non-negligible fraction of RSA instances, one may directly start with the hypothesis that the above two distributions, for randomly selected RSA-moduli N of size n, are indistinguishable by polynomial-time statistical tests. A general result on the construction of pseudorandom functions presented in [36] then allows one to conclude that the iterated application of $x^e \mod N$ in the RSA generator yields a keystream z^l which is also indistinguishable by polynomial-time statistical tests and thus, that the RSA generator is "perfect." It was shown in [1] that the arguments extend to the log n least significant bits of x. These bits are indistinguishable if RSA is assumed to be secure. Hence, without losing "perfectness" log n bits could be extracted per modular exponentiation in the RSA generator.

Generating 1 bit (or log n bits) of keystream per modular exponentiation is too slow for most applications. To overcome this drawback Micali and Schnorr [80] introduced the following much stronger hypothesis:

RSA hypothesis [80]: Let $e \geq 3$ be an odd integer. For random moduli N of size n (which are the product of two primes each having size $n/2$) such that $\gcd(e, \phi(N)) = 1$ and all M proportional to $N^{2/e}$ the following distributions on $[1, N]$ are indistinguishable by polynomial-time statistical tests:

1. The uniform distribution on $[1, N]$
2. The distribution induced by $x^e \bmod N$ for random $x \in [1, M]$

As before the general result on the construction of pseudorandom functions presented in [36] now allows one to conclude that the iterated application of $x^e \bmod N$ in the RSA generator yields a keystream z^l which is also indistinguishable by polynomial-time statistical tests and thus, that the modified RSA generator is "perfect." But now a fraction $(e - 2)/e$ of the bits can be put out per exponentiation, and only a fraction $2/e$ is needed for the next iteration. This efficiency improvement is a simple consequence of the changed hypothesis.

4.2.4   Quadratic residue generator. Let N be the product of two distinct odd primes p and q. An element $y \in Z_N^*$ is called a quadratic residue modulo N if $y = x^2 \bmod N$ for some $x \in Z_N^*$. Denote the set of quadratic residues modulo N by QR_N. Every element $y \in QR_N$ has exactly four square roots. If $p = q = 3 \bmod N$ then exactly one of these four square roots belongs to QR_N. As a consequence, the mapping

$$QR_N \ni x \mapsto x^2 \bmod N \in QR_N$$

is one-to-one and onto, and has associated the inverse mapping

$$QR_N \ni y \mapsto \sqrt{y} \bmod N \in QR_N.$$

Rabin has shown [92] that factoring N and computing square roots are equivalent problems in the sense that an efficient algorithm for one of these problems implies an efficient algorithm for the other problem.

Quadratic residue generator [10]:

Input: parameters: modulus N of size n
        key: a randomly selected $x_1 \in QR_N$.

For $i = 1, 2, \ldots, l$ do

1. $z_i = \mathrm{lsb}(x_i)$
2. $x_{i+1} = x_i^2 \bmod N$

Output: the sequence $z_1, z_2, \ldots, z_l$.

If $p = q = 3 \bmod 4$ then exactly half the elements of Z_N^* have Jacobi symbol $+1$, and the other half have Jacobi symbol -1. Denote the corresponding sets by $Z_N^*(+1)$ and $Z_N^*(-1)$. None of the elements of $Z_N^*(-1)$ and exactly half the elements of $Z_N^*(+1)$ are quadratic residues modulo N. The quadratic residuosity problem with input N and $x \in Z_N^*(+1)$ consists of deciding if x is a quadratic residue modulo N. It is an open problem to find an efficient procedure that solves the quadratic residuosity problem. The security of the quadratic residue generator is based on the difficulty of deciding quadratic residuosity. In [10] it is shown that deciding quadratic residuosity can be efficiently reduced to determining the least significant bit of $x = \sqrt{y} \bmod N$ on input N and $y \in Z_N^*(+1)$. Combining this result with a result by Goldwasser and Micali [35] yields the following theorem [10]:

Theorem: *If there is a probabilistic polynomial-size circuit which, when given as input N and* $y \in QR_N$, *determines the least significant bit of* $x = \sqrt{y} \bmod N$ *for at least a fraction* $1/P'(n)$ *of the moduli N of size n with advantage* $\geq 1/2 + 1/P(n)$ *then, for any polynomial Q(n), there is another circuit that, when given as input N and* $x \in Z_N^*(+1)$, *determines quadratic residuosity for at least a fraction* $1/P'(n)$ *of the moduli N of size n with advantage* $\geq 1 - 1/Q(n)$.

If there is a predictor that, when given as input the keystream $z_1, \ldots, z_l$, is able to guess z_0 with probability $\geq \frac{1}{2} + \varepsilon(n)$, then it is possible to determine the lsb of $x = \sqrt{y} \bmod N$ with the same probability of success. Use y as the seed and generate the sequence $z_1, \ldots, z_l$, call the predictor to obtain z_0 which equals the lsb of x with probability $\geq \frac{1}{2} + \varepsilon(n)$. By the above theorem, the existence of such a predictor with advantage $\geq \frac{1}{2} + 1/P(n)$ would then imply that the quadratic residuosity problem could be solved in probabilistic polynomial time. To prove "perfectness" of the quadratic residue generator Blum, Blum, and Shub introduced the following intractability hypothesis [10,56].

Quadratic residuosity assumption: For any polynomials P and P', any circuit C, and all but a finite number of $n \geq 1$, the fraction of moduli N of size n for which C is able to determine quadratic residuosity with probability $\geq 1 - 1/P(n)$ is upper-bounded by $1/P'(n)$.

This hypothesis is true if and only if the quadratic residue generator is unpredictable (to the left). As a last step, apply Yao's theorem [125] to conclude that the quadratic residue generator is indistinguishable by probabilistic polynomial time statistical tests, and thus is "perfect."

5 RANDOMIZED STREAM CIPHERS

In general, it is difficult to prove lower bounds on the computational effort to solve all, or almost all, instances of a given problem. As illustrated in the complexity-theoretic approach one typically has to resort to heuristic arguments about the computational difficulty of a problem (called intractability hypotheses). A different approach is to focus on the size of the cryptanalytic problem, instead of the work effort. The objective is to increase the number of bits the cryptanalyst has to examine in the cryptanalytic process while keeping the secret key small. This can be done by making use of a large publicly accessible random string in the encryption and decryption process. The key then specifies which parts of the large randomizer are to be used, whereas the opponent, not knowing the secret key, is forced to search through all the random data.

The security level of randomized stream cipher systems may be expressed by the average number of bits the cryptanalyst has to examine before his chances of determining the key or the plaintext improve over pure guessing. The design objective is to establish a provable lower bound on the expected number of bittests the cryptanalyst must perform in order to have non-negligible probability of success. Different interpretations of such a result are possible. The expected number of bittests is a lower bound on the number of steps that any algorithm breaking the system must perform, and thus leads to a notion of computational security. But the expected number of bittests is also a lower bound on the number of bits the opponent has to observe before his a posteriori probabilities of the various keys or messages improve, and thus leads to a notion of information-theoretic security. These possible interpretations are the reason why we have treated randomized stream cipher systems in this section separately.

The following probabilistic cipher was proposed by Diffie [26] in unpublished work (see also [72]):

Diffie's randomized stream cipher:

Input: message $x = x_1, x_2, \ldots$
  key k: a random n-bit string.

1. Flip 2^n random sequences $r_1, r_2, \ldots, r_{2^n}$.
2. Use the kth random string r_k as the one-time pad to encrypt x.

Output: send $y = x \oplus r_k$ and $r_1, r_2, \ldots, r_{2^n}$ over $2^n + 1$ telephone lines.

In this randomized stream cipher the plaintext is encrypted with a one-time pad. The corresponding one-time pad is sent publicly over a telephone line, but disguised by a huge number of unrelated random strings also sent publicly. The cryptanalyst has no choice but to examine the random sequences one at a time until he finds the correct one-time pad. Thus, any attack must examine an expected number of bits which is in $O(2^n)$. It appears that a comparable security level is achievable if, instead of 2^n only n random strings are sent, and the key is used to specify a linear combination of those random strings.

Massey and Ingemarsson [68] proposed a different randomized stream cipher which, for reasons that soon will become clear, they called the Rip van Winkle cipher.

Rip van Winkle cipher:

Input: plaintext sequence $\bar{x} = x_1, x_2, \ldots$
key: a random n-bit number, $0 \le k < 2^n$.

1. Flip a random k-bit preamble $r_1, r_2, \ldots, r_k$
2. Flip a random keystream $\bar{z} = z_1, z_2, \ldots$
3. For $i = 1, 2, \ldots$ do
   a. Encrypt the plaintext with the random keystream into the ciphertext sequence $\bar{y}^{(1)}$:

$$y_i^{(1)} = x_i \oplus z_i$$

   b. Form a second sequence $\bar{y}^{(2)}$ by concatenating the random preamble with the keystream:

$$y_i^{(2)} = \begin{cases} r_i & 1 \le i \le k \\ z_{k-i} & k < i \end{cases}$$

4. Send alternatingly a bit of $\bar{y}^{(1)}$ and $\bar{y}^{(2)}$.

Output: the sequence of bitpairs $(y_i^{(1)}, y_i^{(2)})$, $i = 1, 2, \ldots$

$y_i^{(1)}$ contains the encrypted message and $y_i^{(2)}$ contains the keystream only disguised by a random preamble $r_1, \ldots, r_k$ of unknown length k. In order to decrypt, the receiver simply has to wait k time units until the random preamble has passed by. For the Rip van Winkle cipher a lower bound on the expected number of bittests of any attack can be proved [68]:

Theorem: *Any algorithm that wants to determine the key* k *in a known plaintext attack (with probability at least* $\delta \geq 2^{-n}$ *of being correct) satisfies*

$$E(B) \geq 2^{n/2}\sqrt{1 - 2^{-n}\lfloor \delta^{-1}\rfloor}\,(1 + 2^{-n}(\lfloor \delta^{-1}\rfloor - 1))$$

where B *denotes the number of bittests.*

Thus, by randomizing the ciphertext it is possible to guarantee that any opponent has to spend an exponential effort in the size of the key: about $2^{n/2}$ bittests are necessary before the opponent can sharpen the probability distribution of possible key values. Unfortunately, there is a trade-off since the receiver also has to wait until an exponential number of bits, on the average until 2^n bits of ciphertext have arrived before he can start decryption. In Massey's words, "One can easily guarantee that the enemy cryptanalyst will need thousands of years to break the cipher, if one is willing to wait millions of years to read the plaintext." The Rip van Winkle cipher is completely impractical.

Motivated by the ideas contained in the Rip van Winkle cipher, Maurer [75] developed a randomized stream cipher for which one can prove that the enemy obtains no information in Shannon's sense about the plaintext with probability close to 1 unless he accesses an infeasible number of bits (performs an infeasible computation). This approach provides an information-theoretic notion of security under a computational restriction of the opponent (whereas typically, information-theoretic security implies that the opponent has infinite computing power).

Maurer's randomized stream cipher [75]:

Input: plaintext $x^N = (x_1, \ldots, x_N) \in \mathbb{F}_2^N$
public randomizer $R[s, t]$, $1 \leq s \leq S$, $0 \leq t \leq T - 1$
key $k^S = (k_1, \ldots k_S) \in \mathbb{Z}_T^S$

1. Compute keystream

$$z_i = \sum_{s=1}^{S} R[s, (k_s + i - 1) \mod T] \mod 2 \quad 1 \leq i \leq N$$

2. Encrypt

$$y^N = x^N \oplus z^N$$

Output: ciphertext y^N.

The enemy attacking this system may have knowledge of the plaintext statistics and may also have some other a priori information about the plaintext. This situation is modeled by introducing an additional random variable V, jointly distributed with the plaintext according to $P(X^N, V)$, that collects all a priori information that the enemy may have about the plaintext. The enemy is allowed to use an arbitrary (possibly probabilistic) sequential strategy for selecting the addresses $A_i = (B_i, C_i)$ of the randomizer bits to be examined. Let $O_i = R[B_i, C_i]$ be the observed value of the randomizer bit at the address A_i in the ith step. The total information available to the enemy before execution of the ith step in his attack is the cryptogram Y^N, the values O^{i-1} of all previously examined randomizer bits together with the corresponding addresses A^{i-1}, and the a priori information V. Thus, the enemy's strategy is completely specified by the sequence of conditional probability distributions $P(A_i \mid Y^N V A^{i-1} O^{i-1})$, $i \geq 1$. For the described model Maurer was able to prove [75].

Theorem: *There exists an event $\mathscr{E}$ such that, for all joint probability distributions* $P(X^N, V)$ *and for all strategies of examining bits* O^M *at addresses* A^M *of the randomizer* R

$$I(X^N; Y^N A^M O^M \mid V, \mathscr{E}) = 0 \quad \text{and} \quad P(\mathscr{E}) \geq 1 - N\delta^S$$

where $\delta = M/ST$ is the fraction of randomizer bits examined.

The theorem states that if the event $\mathscr{E}$ occurs, then the enemy's total observation (Y^N, A^M, O^M) gives no additional information about the plaintext X^N beyond what he already knew before he started the attack. The chance that the enemy can learn something new about the plaintext is given by the probability that the event $\mathscr{E}$ does not occur. This probability is upperbounded by $P(\overline{\mathscr{E}}) < N\delta^S$. Hence, to have a substantial chance to obtain new information about the plaintext, the enemy is forced to examine a substantial fraction of the randomizer bits.

ACKNOWLEDGMENTS

First of all, I would like to thank Jim Massey for his continuous encouragement and help. His friendship deeply influenced my research, and my life.

Three recent workshops have greatly inspired me and have contributed to the present form of this manuscript. In October 1989, Jim Massey and Hansjürg Mey organized the Monte Verità Seminar on "Future Directions in Cryptography" in Ascona, Switzerland. In September 1989, Andy Odlyzko, Claus Schnorr, and Adi Shamir organized a Workshop on Cryptography in Oberwolfach, Germany. And early in 1989, Thomas Beth and Fred Piper organized a Workshop on Stream Ciphers in Karlsruhe, Germany. I would like to thank the organizers for their invitations, and the participants for the many stimulating discussions.

I would like to thank Lennart Brynielsson, Bill Chambers, Yvo Desmedt, Dieter Gollmann, Christoph Günther, Ueli Maurer, Harald Niederreiter, Kaisa Nyberg, Andy Odlyzko, Andrea Sgarro, Thomas Siegenthaler, Gus Simmons, and Othmar Staffelbach whose valuable comments and helpful discussions greatly improved the paper.

Finally, I would like to thank Crypto AG for their support in the initial phase of preparing this paper.

REFERENCES

[1] W. Alexi, B. Chor, O. Goldreich, and C. P. Schnorr, "RSA and Rabin functions: Certain parts are as hard as the whole," *SIAM J. Comput.*, vol. 17, pp. 194–209, April 1988.

[2] H. Beker and F. Piper, *Cipher Systems: the Protection of Communications*, London: Northwood Books, 1982.

[3] B. Benjauthrit and I. S. Reed, "Galois switching functions and their applications," *IEEE Trans. Comput.*, vol. C-25, pp. 78–86, Jan. 1976.

[4] E. R. Berlekamp, *Algebraic Coding Theory*, New York: McGraw-Hill, 1968.

[5] J. Bernasconi and C. G. Günther, "Analysis of a nonlinear feedforward logic for binary sequence generators," *BBC Tech. Rep.*, 1985.

[6] T. Beth and F. Piper, "The stop-and-go generator," in *Lecture Notes in Computer Science 209; Advances in Cryptology: Proc. Eurocrypt '84*, T. Beth, N. Cot, and I. Ingemarsson, Eds., Paris, France, April 9–11, 1984, pp. 88–92. Berlin: Springer-Verlag, 1985.

[7] R. E. Blahut, "Transform techniques for error-control codes," *IBM J. Res. Develop.* vol. 23, pp. 299–315, 1979.

[8] W. Blaser and P. Heinzmann, "New cryptographic device with high security using public key distribution," *Proc. IEEE Student Paper Contest 1979–80*, pp. 145–153, 1982.

[9] M. Blum and S. Micali, "How to generate cryptographically strong sequences of pseudo-random bits," *SIAM J. Comput.*, vol. 13, pp. 850–864, 1984.

[10] L. Blum, M. Blum, and M. Shub, "A simple unpredictable pseudo-random number generator," *SIAM J. Comput.*, vol. 15, pp. 364–383, 1986.

[11] J. O. Bruer, "On pseudo random sequences as crypto generators," in *Proc. Int. Zurich Seminar on Digital Communication*, Switzerland, 1984.

[12] L. Brynielsson, "On the linear complexity of combined shift register sequences," in *Lecture Notes in Computer Science 219; Advances in Cryptology: Proc. Eurocrypt '85*, F. Pichler, Ed., Linz, Austria, April 1985, pp. 156–166. Berlin: Springer-Verlag, 1986.

[13] L. Brynielsson, "Wie man den richtigen Schlüssel in einem Heuhaufen findet," *Kryptologie Aufbauseminar J.*, Kepler Universität, Linz, Austria, 1987.

[14] L. Brynielsson, "Below the unicity distance," Workshop on Stream Ciphers, Karlsruhe, Germany 1989.

[15] C. M. Campbell, "Design and specification of cryptographic capabilities," *IEEE Commun. Soc. Mag.*, vol. 16, pp. 15–19, 1978.

[16] W. G. Chambers and S. M. Jennings, "Linear equivalence of certain BRM shift-register sequences," *Electron. Lett.*, vol. 20, Nov. 1984.

[17] W. G. Chambers and D. Gollmann, "Generators for sequences with near-maximal linear equivalence," *IEE Proc. E.*, vol. 135, pp. 67–69, 1988.

[18] A. H. Chan and R. A. Games, "On the linear span of binary sequences obtained from finite geometries," in *Lecture Notes in Computer Science 263; Advances in Cryptology: Proc. Crypto '86*, A. M. Odlyzko, Ed., Santa Barbara, CA, Aug. 11–15, 1986, pp. 405–417. Berlin: Springer-Verlag, 1987.

[19] A. H. Chan, M. Goresky, and A. Klapper, "Correlation functions of geometric sequences," *Proc. Eurocrypt 90*, I. Damgård, Ed., Springer Verlag (in press).

[20] D. Chaum and J. H. Evertse, "Cryptanalysis of DES with a reduced number of rounds," in *Lecture Notes in Computer Science 218; Advances in Cryptology: Proc. Crypto '85*, H. C. Williams, Ed., Santa Barbara, CA, Aug. 18–22, 1985, pp. 192–211. Berlin: Springer-Verlag, 1986.

[21] Zong-duo Dai, "Proof of Rueppel's linear complexity conjecture," *IEEE Trans. Inform. Theory*, vol. 32, pp. 440–443, May 1986.

[22] D. E. Denning, *Cryptography and Data Security*, Reading, MA: Addison-Wesley, 1983.

[23] Y. Desmedt, J. J. Quisquater, and M. Davio, "Dependence of output on input of DES: Small avalanche characteristics," in *Lecture Notes in Computer Science 196; Advances in Cryptology: Proc. Crypto '84*, G. R. Blakley and D. Chaum, Eds., Santa Barbara, CA, Aug. 19–22, 1984, pp. 359–376. Berlin: Springer-Verlag, 1985.

[24] Y. G. Desmedt, "Cryptanalysis of conventional and public key cryptosystems," *Proc. SPRCI'89*, Rome, Nov. 23–24, 1989.

[25] W. Diffie and M. Hellman, "Privacy and authentication: An introduction to cryptography," *Proc. IEEE*, vol. 67, pp. 397–427, 1979.

[26] W. Diffie, Private communication, July 1984.

[27] J. F. Dillon, "Elementary Hadamard difference sets," *Proc. 6th Southeastern Conf. Combinatorics, Graph Theory, and Computing*, Boca Raton, FL, pp. 237–249, 1975; in Congressus Numerantium No. XIV, Utilitas Math., Winnipeg, Manitoba, 1975.

[28] J. H. Evertse, "Linear structures in block cyphers," in *Lecture Notes in Computer Science 304; Advances in Cryptology: Proc. Eurocrypt '87*, D. Chaum and W. L. Price, Eds., Amsterdam, The Netherlands, April 13–15, 1987, pp. 249–266. Berlin: Springer-Verlag, 1988.

[29] R. Forré, "The strict avalanche criterion: Spectral properties of boolean functions and an extended definition," in *Lecture Notes in Computer Science 403; Advances in Cryptology: Proc. Crypto '88*, S. Goldwasser, Ed., Santa Barbara, CA, Aug. 21–25, 1987, pp. 450–468. Berlin: Springer-Verlag, 1990.

[30] R. Forré, "A fast correlation attack on nonlinearly feedforward filtered shift-register sequences," in *Lecture Notes in Computer Science 434; Advances in Cryptology; Proc. Eurocrypt '89*, J.-J. Quisquater and J. Vandewalle, Eds., Houthalen, Belgium, April 10–23, 1989, pp. 586–595. Berlin: Springer-Verlag, 1990.

[31] J. Gait, "A new nonlinear pseudorandom number generator," *IEEE Trans. Software Eng.*, vols. S E3, no. 5, pp. 359–363, Sept. 1977.

[32] R. G. Gallager, "Low-density parity-check codes," Cambridge, MA: MIT Press 1963.

[33] M. R. Garey and D. S. Johnson, *Computers and Intractability*, New York: W. H. Freeman, 1979.

[34] P. R. Geffe, "How to protect data with ciphers that are really hard to break," *Electronics*, Jan. 4, 1973.

[35] S. Goldwasser and S. Micali, "Probabilistic encryption and how to play mental poker keeping secret all partial information," *J. Comput. Sys. Sci.*, vol. 28, no. 2, Apr. 1984.

[36] O. Goldreich, S. Goldwasser, and S. Micali, "How to construct random functions," *J. ACM*, vol. 33, no. 4, pp. 792–807, 1986.

[37] J. Golić and M. V. Zivković, "On the linear complexity of nonuniformly deci- mated pn-sequences," *IEEE Trans. Inform. Theory*, vol 34, pp. 1077–1079, Sept. 1988.

[38] J. D. Golić, "On the linear complexity of functions of periodic GF(q)- sequences," *IEEE Trans. Inform. Theory,* vol. IT-35, pp. 69–75, Jan. 1989.

[39] D. Gollman, "Pseudo random properties of cascade connections of clock con- trolled shift registers," in *Lecture Notes in Computer Science 209; Advances in Cryptology: Proc. Eurocrypt '84*, T. Beth, N. Cot, and I. Ingemarsson, Eds., Paris, France, April 9–11, 1984, pp. 93–98. Berlin: Springer-Verlag, 1985.

[40] D. Gollman and W. G. Chambers, "Lock-in effect in cascades of clock- controlled shift-registers," in *Lecture Notes in Computer Science 330; Advances in Cryptology: Proc. Eurocrypt '88*, C. G. Günther, Ed., Davos, Switzerland, May 25–27, 1988, pp. 331–343. Berlin: Springer-Verlag, 1988.

[41] D. Gollmann and W. G. Chambers, "Clock-controlled shift registers: A review," *IEEE J. Selected Areas Commun.*, vol. 7, pp. 525–533, May 1989.

[42] S. W. Golomb, "Deep space range measurements," Jet Propulsion Laboratory, Pasadena, CA Research Summary, No. 36-1, 1960.

[43] S. W. Golomb, *Shift Register Sequences*, San Francisco: Holden Day, 1967.

[44] E. J. Groth, "Generation of binary sequences with controllable complexity," *IEEE Trans. Inform. Theory*, vol. IT-17, no. 3, May 1971.

[45] C. G. Günther, "On some properties of the sum of two pseudorandom se- quences," paper presented at Eurocrypt'86, Linköping, Sweden, May 20–22, 1986.

[46] C. G. Günther, "Alternating step generators controlled by de Bruijn sequences," in *Lecture Notes in Computer Science 304; Advances in Cryptology: Proc. Eurocrypt'87*, D. Chaum and W. L. Price, Eds., Amsterdam, The Netherlands, April 13–15, 1987, pp. 5–14. Berlin: Springer-Verlag, 1988.

[47] C. G. Günther, "A universal algorithm for homophonic coding," in *Lecture Notes in Computer Science 330; Advances in Cryptology: Proc. Eurocrypt'88*, C. G. Günther, Ed., Davos, Switzerland, May 25–27, 1988, pp. 405–414. Ber- lin: Springer-Verlag, 1988.

[48] T. Herlestam, "On the complexity of functions of linear shift register se- quences," *Int. Symp. Inform. Theory*, Les Arc, France, 1982.

[49] T. Herlestam, "On functions of linear shift register sequences," in *Lecture Notes in Computer Science 219; Advances in Cryptology: Proc. Eurocrypt'85*, F. Pilcher, Ed., Linz, Austria, April 1985, pp. 119–129. Berlin: Springer-Verlag, 1986.

[50] C. J. Jansen, Investigations on nonlinear streamcipher systems: Construction and evaluation methods, Ph.D. thesis, Eindhoven University of Technology, The Netherlands, 1989.

[51] S. M. Jennings, "Multiplexed sequences: Some properties of the minimum poly- nomial," in *Lecture Notes in Computer Science 149; Cryptography: Proc. Work- shop Cryptography*, T. Beth, Ed., Burg Feuerstein, Germany, March 29–April 2, 1982, pp. 189–206. Berlin: Springer-Verlag, 1983.

[52] S. M. Jennings, "Autocorrelation function of the multiplexed sequence," *IEE Proc.*, vol. 131, no. 2, pp. 169–172, Apr. 1984.

[53] B. Kaliski, A pseudo random bit generator based on elliptic logarithms, M. Sc. thesis, Massachusetts Institute of Technology, 1987.

[54] E. L. Key, "An analysis of the structure and complexity of nonlinear binary sequence generators," *IEEE Trans. Inform. Theory*, vol. IT-22, no. 6, pp. 732–763, Nov. 1976.

[55] K. Kjeldsen and E. Andresen, "Some randomness properties of cascaded sequences," *IEEE Trans. Inform. Theory*, vol. IT-26, pp. 227–232, March 1980.

[56] E. Kranakis, *Primality and Cryptography*, Stuttgart: Teubner, Wiley, 1986.

[57] P. V. Kumar and R. A. Scholtz, "Bounds on the linear span of bent sequences," *IEEE Trans. Inform. Theory*, vol. IT-29, pp. 854–862, Nov. 1983.

[58] P. V. Kumar, R. A. Scholtz, and L. R. Welch, "Generalized bent functions and their properties," *J. Combinatorial Theory*, Ser. A 40, pp. 90–107, 1985.

[59] A. Lempel and M. Cohn, "Maximal families of bent sequences," *IEEE Trans. Inform. Theory*, vol. IT-28, pp. 865–868, Nov. 1982.

[60] R. Lidl and H. Niederreiter, "Finite Fields," in *Encyclopedia of Mathematics and Its Applications, Vol. 20*, Reading, MA: Addison-Wesley, 1983.

[61] S. Lloyd, "Counting functions satisfying a higher order strict avalanche criterion," in *Lecture Notes in Computer Science 434; Advances in Cryptology; Proc. Eurocrypt'89*, J.-J. Quisquater and J. Vandewalle, Eds., Houthalen, Belgium, April 10–23, 1989, pp. 63–74. Berlin: Springer-Verlag, 1990.

[62] D. L. Long and A. Wigderson, "How discrete is the discrete log?" in *Proc. 15th ACM Symposium on Theory of Computation*, Apr. 1983.

[63] M. Luby and C. Rackoff, "How to construct pseudorandom permutations from pseudorandom functions," *SIAM J. Comput.* vol. 17, pp. 373–386, 1988.

[64] F. J. MacWilliams and N. J. A. Sloane, "The theory of error correcting codes," Amsterdam: North-Holland, 1977.

[65] J. L. Massey, "Shift-register synthesis and BCH decoding," *IEEE Trans. Inform. Theory*, vol. IT-15, pp. 122–127, Jan. 1969.

[66] J. L. Massey, A. Gubser, A. Fischer, P. Hochstrasser, B. Huber, and R. Sutter, "A self-synchronizing digital scrambler for cryptographic protection of data," in *Proceedings of International Zurich Seminar*, March, 1984.

[67] J. L. Massey and R. A. Rueppel, "Linear ciphers and random sequence generators with multiple clocks," in *Lecture Notes in Computer Science 209; Advances in Cryptology: Proc. Eurocrypt'84*, T. Beth, N. Cot, and I. Ingemarsson, Eds., Paris, France, April 9–11, 1984, pp. 74–87. Berlin: Springer-Verlag, 1985.

[68] J. L. Massey and I. Ingemarsson, "The Rip van Winkle cipher—a simple and provably computationally secure cipher with a finite key," *IEEE Int. Symp. Inform. Theory*, Brighton, England, June 24–28, 1985.

[69] J. L. Massey, "Delayed-decimation/square sequences," *Proc. 2nd Joint Swedish-Soviet Workshop on Information Theory*, Gränna, Sweden, Apr. 14–19, 1985.

[70] J. L. Massey, "Cryptography and System Theory," *Proc. 24th Allerton Conf. Commun., Control, Comput.*, Oct. 1–3, 1986.

[71] J. L. Massey, "Probabilistic encipherment," *Elektrotechnik und Maschinenbau*, vol. 104, no. 12, Dec. 1986.

[72] U. Maurer and J. L. Massey, "Perfect local randomness in pseudo-random sequences," in *Lecture Notes in Computer Science 435; Advances in Cryptology: Proc. Crypto'89*, G. Brassard, Ed., Santa Barbara, CA, Aug. 20–24. 1989, pp. 110–112. Berlin: Springer-Verlag, 1990.

[73] J. L. Massey, "Applied digital information theory," Lecture Notes, Swiss Federal Institute of Technology, Zurich.

[74] U. Maurer, "A provable-secure strongly-randomized cipher," in *Lecture Notes in Computer Science 473; Advances in Cryptology: Proc. Eurocrypt'90*, I. Damgård, Ed., Aarhus, Denmark, May 21–24. 1990, pp. 361–373. Berlin: Springer-Verlag.

[75] R. L. McFarland, "A family difference sets in non-cyclic groups," *J. Combinatorial Theory*, Ser. A 15, pp. 1–10, 1973.

[76] W. Meier and O. Staffelbach, "Fast correlation attacks on stream ciphers," *J. Cryptol.*, vol. 1, no. 3, pp. 159–176, 1989.

[77] W. Meier and O. Staffelbach, "Nonlinearity criteria for cryptographic functions," in *Lecture Notes in Computer Science 434; Advances in Cryptology; Proc. Eurocrypt'89*, J.-J. Quisquater and J. Vandewalle, Eds., Houthalen, Belgium, April 10–23, 1989, pp. 549–562. Berlin: Springer-Verlag, 1990.

[78] W. Meier and O. Staffelbach, "Correlation properties of combiners with memory in stream ciphers," in *Lecture Notes in Computer Science 473; Advances in Cryptology: Proc. Eurocrypt'90*, I. Damgård, Ed., Aarhus, Denmark, May 21–24. 1990, pp. 204–213. Berlin: Springer-Verlag.

[79] S. Micali and C. P. Schnorr, "Efficient, perfect random number generators," preprint, Massachusetts Institute of Technology, University of Frankfurt, 1988

[80] L. M. Milne-Thomson, "The calculus of finite differences," London: Macmillan and Co., 1951.

[81] H. Niederreiter, "Continued fractions for formal power series, pseudorandom numbers, and linear complexity of sequences," contributions to General Algebra 5, Proc. Conf. Salzburg, Teubner, Stuttgart, 1986.

[82] H. Niederreiter, "Sequences with almost perfect linear complexity profile," in *Lecture Notes in Computer Science 304; Advances in Cryptology: Proc. Eurocrypt'87*, D. Chaum and W. L. Price, Eds., Amsterdam, The Netherlands, April 13–15, 1987, pp. 37–51. Berlin: Springer-Verlag, 1988.

[83] H. Niederreiter, "Probabilistic theory of linear complexity," in *Lecture Notes in Computer Science 330; Advances in Cryptology: Proc. Eurocrypt'88*, C. G. Günther, Ed., Davos, Switzerland, May 25–27, 1988, pp. 191–209. Berlin: Springer-Verlag, 1988.

[84] H. Niederreiter, "Keystream sequences with a good linear complexity profile for every starting point," in *Lecture Notes in Computer Science 434; Advances in Cryptology; Proc. Eurocrypt'89*, J.-J. Quisquater and J. Vandewalle, Eds., Houthalen, Belgium, April 10–23, 1989, pp. 523–532. Berlin: Springer-Verlag, 1990.

[85] K. Nyberg, "Construction of bent functions and difference sets," in *Lecture Notes in Computer Science 473;Advances in Cryptology:Proc. Eurocrypt'90*, I. Damgård, Ed., Aarhus, Denmark, May 21–24. 1990, pp. 151–160. Berlin: Springer-Verlag.

[86] P. Nyffeler, Binäre Automaten und ihre linearen Rekursionen, Ph.D. thesis, University of Berne, 1975.

[87] Y. Ohnishi, A study on data security, Master thesis (in Japanese), Tohuku University, Japan, 1988.

[88] F. Piper, "Stream ciphers," *Elektrotechnik und Maschinenbau*, vol. 104, no. 12, pp. 564–568, 1987.

[89] V. S. Pless, "Encryption schemes for computer confidentiality," *IEEE Trans. Comput.*, vol. C-26, pp. 1133–1136, Nov. 1977.

[90] B. Preneel, W. Van Leewijk, L. Van Linden, R. Govaerts, and J. Vandewalle, "Propagation characteristics of boolean functions," in *Lecture Notes in Com-*

puter Science 473; Advances in Cryptology: Proc. Eurocrypt'90, I. Damgård, Ed., Aarhus, Denmark, May 21–24. 1990, pp. 161–173. Berlin: Springer-Verlag.

[91] M. O. Rabin, "Digital signatures and public-key functions as intractable as factorization," Massachusetts Institute of Technology Laboratory for Computer Science, TR-212, 1979.

[92] O. S. Rothaus, "On bent functions," *J. Combinatorial Theory*, vol. 20, pp. 300–305, 1976.

[93] F. Rubin "Decrypting a stream cipher based on *J-K* flip-flops," *IEEE Trans Comput.*, vol. C-28, no. 7, pp. 483–487, July 1979.

[94] R. A. Rueppel, "Linear complexity and random sequences," in *Lecture Notes in Computer Science 219; Advances in Cryptology: Proc. Eurocrypt'85*, F. Pilcher, Ed., Linz, Austria, April 1985, pp. 167–188. Berlin: Springer-Verlag, 1986.

[95] R. A. Rueppel and J. L. Massey, "The knapsack as a nonlinear function," *IEEE Int. Symp. Inform. Theory*, Brighton, UK, May 1985.

[96] R. A. Rueppel, "Correlation immunity and the summation combiner," in *Lecture Notes in Computer Science 218; Advances in Cryptology: Proc. Crypto'85*, H. C. Williams Ed., Santa Barbara, CA, Aug. 18–22, 1985, pp. 260–272. Berlin: Springer-Verlag, 1986.

[97] R. A. Rueppel, *Analysis and Design of Stream Ciphers*, Berlin: Springer-Verlag, 1986.

[98] R. A. Rueppel and O. Staffelbach, "Products of sequences with maximum linear complexity," *IEEE Trans. Inform. Theory*, vol. IT-33, no. 1, pp. 124–131, Jan. 1987.

[99] R. A. Rueppel, "When shift registers clock themselves," in *Lecture Notes in Computer Science 304; Advances in Cryptology: Proc. Eurocrypt'87*, D. Chaum and W. L. Price, Eds., Amsterdam, The Netherlands, April 13–15, 1987, pp. 53–64. Berlin: Springer-Verlag, 1988.

[100] R. A. Rueppel, "On the security of Schnorr's pseudo random sequence generator," in *Lecture Notes in Computer Science 434; Advances in Cryptology; Proc. Eurocrypt'89*, J.-J. Quisquater and J. Vandewalle, Eds., Houthalen, Belgium, April 10–23, 1989, pp. 423–428. Berlin: Springer-Verlag, 1990.

[101] R. A. Rueppel, "Security models and notions for stream ciphers," *Proc. 2nd IMA Conf. Cryptography and Coding*, Cirencester, England, Dec. 1989.

[102] J. E. Savage, Some simple self-synchronizing digital data scramblers, *Bell Sys. Tech. J.*, vol. 46, no. 2, pp. 449–487, Feb. 1967.

[103] T. Schaub, A linear complexity approach to cyclic codes, Ph.D. thesis, Swiss Federal Institute of Technology, Zurich, 1988.

[104] C. P. Schnorr, "On the construction of random number generators and random function generators," in *Lecture Notes in Computer Science 330; Advances in Cryptology: Proc. Eurocrypt'88*, C. G. Gunther, Ed., Davos, Switzerland, May 25–27, 1988, pp. 225–232. Berlin: Springer-Verlag, 1988.

[105] E. S. Selmer, Linear recurrence relations over finite fields, Lecture Notes, University of Bergen, Bergen, Norway, 1966.

[106] J. A. Serret, "Cours d'algèbre supérieure," Tome II, p. 154, Gauthier-Villars, Paris, 1886.

[107] A. Shamir, "On the generation of cryptographically strong pseudo-random sequences," *8th Int. Colloquium on Automata, Languages, and Programming*, Lecture Notes in Computer Science 62, Springer Verlag, 1981.

[108] C. E. Shannon, "Communication theory of secrecy systems," *Bell Syst. Tech. J.*, vol. 28, pp. 656–715, Oct. 1949.

[109] T. Siegenthaler, "Correlation-immunity of nonlinear combining functions for cryptographic applications," *IEEE Trans. Inform. Theory*, vol. IT-30, pp. 776–780, Oct. 1984.

[110] T. Siegenthaler, "Cryptanalyst's representation of nonlinearity filtered ml-sequences," in *Lecture Notes in Computer Science 219; Advances in Cryptology: Proc. Eurocrypt'85*, F. Pilcher, Ed., Linz, Austria, April 1985, pp. 103–110. Berlin: Springer-Verlag, 1986.

[111] T. Siegenthaler, "Decrypting a class of stream ciphers using ciphertext only," *IEEE Trans. Comput.*, vol. C-34, pp. 81–85, Jan. 1985.

[112] B. Smeets, "A note on sequences generated by clock-controlled shift registers," in *Lecture Notes in Computer Science 219; Advances in Cryptology: Proc. Eurocrypt'85*, F. Pilcher, Ed., Linz, Austria, April 1985, pp. 40–42. Berlin: Springer-Verlag, 1986.

[113] B. Smeets, "The linear complexity profile and experimental results on a randomness test of sequences over the field F_q," *IEEE Int. Symp. Inform. Theory*, Kobe, Japan, June 19–24, 1988.

[114] O. Staffelbach and W. Meier, "Cryptographic significance of the carry for ciphers based on integer addition," in *Advances in Cryptology, Proc. Crypto 90*, S. Vanstone, Ed. (in press).

[115] R. C. Titsworth, "Optimal ranging codes," *IEEE Trans. Space Electron. Telemetry*, pp. 19–30, March 1964.

[116] S. A. Tretter, "Properties of PN^2 sequences," *IEEE Trans. Inform. Theory*, vol. IT-20, pp. 295–297, March 1974.

[117] G. S. Vernam, "Cipher printing telegraph systems for secret wire and radio telegraphic communications," *J. Amer. Inst. Elec. Eng.*, vol. 45, pp. 109–115, 1926.

[118] R. Vogel, "On the linear complexity of cascaded sequences," in *Lecture Notes in Computer Science 209; Advances in Cryptology: Proc. Eurocrypt'84*, T. Beth, N. Cot, and I. Ingemarsson, Eds., Paris, France, April 9–11, 1984, pp. 99–109. Berlin: Springer-Verlag, 1985.

[119] M. Z. Wang and J. L. Massey, "The characteristics of all binary sequences with perfect linear complexity profiles," paper presented at Eurocrypt'86, Linkoping, Sweden, May 20–22, 1986.

[120] M. Wang, "Linear complexity profiles and continued fractions," in *Lecture Notes in Computer Science 434; Advances in Cryptology; Proc. Eurocrypt'89*, J.-J. Quisquater and J. Vandewalle, Eds., Houthalen, Belgium, April 10–23, 1989, pp. 571–585. Berlin: Springer-Verlag, 1990.

[121] A. F. Webster and S. E. Tavares, "On the design of S-boxes," in *Lecture Notes in Computer Science 218; Advances in Cryptology: Proc. Crypto'85*, H. C. Williams, Ed., Santa Barbara, CA, Aug. 18–22, 1985, pp. 523–534. Berlin: Springer-Verlag, 1986.

[122] S. Wolfram, "Cryptography with cellular automata," in *Lecture Notes in Computer Science 218; Advances in Cryptology: Proc. Crypto'85*, H. C. Williams, Ed., Santa Barbara, CA, Aug. 18–22, 1985, pp. 429–432. Berlin: Springer-Verlag, 1986.

[123] G. Z. Xiao and J. L. Massey, "A spectral characterization of correlation-immune functions," *IEEE Trans. Inform. Theory*, vol. 34, no. 3, pp. 569–571, May 1988.

[124] A. C. Yao, "Theory and applications of trapdoor functions," *Proc. 25th IEEE Symp. Foundations Comput. Sci.*, New York, 1982.

[125] R. Yarlagadda and J. E. Hershey, "Analysis and synthesis of bent sequences," *Proc. IEE*, vol. 136, pt. E., pp. 112–123, March 1989.

[126] L. E. Zegers, "Common bandwidth transmission of data signals and wide-band pseudonoise synchronization waveforms," *Philips Res. Reports Suppl.*, no. 4, 1972.

[127] K. Zeng and M. Huang, "On the linear syndrome method in cryptanalysis," in *Lecture Notes in Computer Science 403; Advances in Cryptology: Proc. Crypto'88*, S. Goldwasser, Ed., Santa Barbara, CA, Aug. 21–25, 1987, pp. 469–478. Berlin: Springer-Verlag, 1990.

[128] Y. Zheng, T. Matsumoto, and H. Imai, "Impossibility and optimality results on constructing pseudorandom permutations," in *Lecture Notes in Computer Science 434; Advances in Cryptology; Proc. Eurocrypt'89*, J.-J. Quisquater and J. Vande-walle, Eds., Houthalen, Belgium, April 10–23, 1989, pp. 412–422. Berlin: Springer-Verlag, 1990.

[129] N. Zierler, "Linear recurring sequences," *J. Soc. Indust. Appl. Math.*, vol. 7, no. 1, pp. 31–48, March 1959.

[130] N. Zierler and W. H. Mills, "Products of linear recurring sequences," *J. Algebra*, vol. 27, no. 1, pp. 147–157, Oct. 1973.

CHAPTER 3

The First Ten Years of Public Key Cryptology

WHITFIELD DIFFIE
Bell-Northern Research

1. Initial Discoveries
2. Exponential Key Exchange
3. Trapdoor Knapsacks
4. The Rivest-Shamir-Adleman System
5. The McEliece Coding Scheme
6. The Fall of the Knapsacks
7. Early Responses to Public Key Cryptosystems
8. Application and Implementation
9. Multiplying, Factoring, and Finding Primes
10. Directions in Public Key Cryptography Research
11. Where Is Public Key Cryptography Going?

Abstract—Public key cryptosystems separate the capacities for encryption and decryption so that (1) many people can encrypt messages in such a way that only one person can read them or (2) one person can encrypt messages in such a way that many people can read them. This separation allows important improvements in the management of cryptographic keys and makes it possible to "sign" a purely digital message.

Public key cryptography was discovered in the spring of 1975 and has followed a surprising course. Although diverse systems were proposed early on, the ones that appear both practical and secure today are all very closely related and the search for new and different ones has met with little success. Despite this reliance on a limited mathematical foundation, public key cryptography is revolutionizing communication security by making possible secure communication networks with hundreds of thousands of subscribers.

Equally important is the impact of public key cryptography on the theoretical side of communication security. It has given cryptographers a systematic means of addressing a broad range of security objectives and pointed the way toward a more theoretical approach that allows the development of cryptographic protocols with proven security characteristics.

1 INITIAL DISCOVERIES

Public key cryptography was born in May 1975, the child of two problems and a misunderstanding.

First came the problem of key distribution. If two people who have never met before are to communicate privately using conventional cryptographic means, they must somehow agree in advance on a key that will be known to themselves and to no one else.

The second problem, apparently unrelated to the first, was the problem of signatures. Could a method be devised that would provide the recipient of a purely digital electronic message with a way of demonstrating to other people that it had come from a particular person, just as a written signature on a letter allows the recipient to hold the author to its contents?

On the face of it, both problems seem to demand the impossible. In the first case, if two people could somehow communicate a secret key from one to the other without ever having met, why could they not communicate their message in secret? The second is no better. To be effective, a signature must be hard to copy. How then can a digital message, which can be copied perfectly, bear a signature?

The misunderstanding was mine and prevented me from rediscovering the conventional key distribution center (KDC). The virtue of cryptography, I reasoned, was that, unlike any other known security technology, it did not require trust in any party not directly involved in the communication, only trust in the cryptographic systems. What good would it do after all to develop impenetrable cryptosystems; if their users were forced to share their keys with a KDC that could be compromised by either burglary or subpoena.

The discovery consisted not of a solution, but of the recognition that the two problems, each of which seemed unsolvable by definition, could be solved at all and that the solutions to both problems came in one package.

First to succumb was the signature problem. The conventional use of cryptography to authenticate messages had been joined in the 1950s by two new applications, whose functions when combined constitute a signature.

Beginning in 1952, a group under the direction of Horst Feistel at the Air Force Cambridge Research Center in Massachusetts began to apply cryptography to the military problem of distinguishing friendly from hostile aircraft. In traditional *Identification Friend or Foe* (IFF) systems, a fire control radar determines the identity of an aircraft by challenging it, much as a sentry challenges a soldier on foot. If the airplane returns the correct identifying information, it is judged to be friendly; otherwise it is thought to be hostile or at best neutral. To allow the correct response to remain constant for any significant period of time, however, is to invite opponents to record a legitimate friendly response and play it back whenever they themselves are challenged. The approach taken by Feistel's group, and now used in the MK XII IFF system, is to vary the exchange cryptographically from encounter to encounter. The radar sends a randomly selected challenge and judges the aircraft by whether it receives a correctly encrypted response. Because the challenges are never repeated, previously recorded responses will not be judged correct by a challenging radar.

Later in the decade, this novel authentication technique was joined by another, which seems first to have been applied by Roger Needham of Cambridge University [112]. This time the problem was protecting computer passwords. Access control systems often suffer from the extreme sensitivity of their password tables. The tables gather all of the passwords together in one place and anyone who gets access to this information can impersonate any of the system's users. To guard against this possibility, the password table is filled not with the passwords themselves, but with the images of the passwords under a *one-way function*. A one-way function is easy to compute, but difficult to invert. For any password, the correct table entry can be calculated easily. Given an output from the one-way function, however, it is exceedingly difficult to find any input that will produce it. This reduces the value of the password table to an intruder tremendously, since its entries are not passwords and are not acceptable to the password verification routine.

Challenge and response identification and one-way functions provide protection against two quite different sorts of threats. Challenge and response identification resists the efforts of an eavesdropper who can spy on the communication channel. Since the challenge varies randomly from event to event, the spy is unable to replay it and fool the challenging radar. There is, however, no protection against an opponent who captures the radar and learns its cryptographic keys. This opponent can use what he has learned to fool any other radar that is keyed the same. In contrast, the one-way function defeats the efforts of an intruder who captures the system password table (analogous to capturing the radar) but succumbs to anyone who intercepts the login message because the password does not change with time.

I realized that the two goals might be achieved simultaneously if the challenger could pose questions that it was unable to answer, but whose answers it could judge for correctness. I saw the solution as a generalization of the one-way function: *a trapdoor one-way function*, which allowed someone in possession of secret information to go backwards and compute the function's inverse. The challenger would issue a value in the range of the one-way function and demand to know its inverse. Only the person who knew the trapdoor would be able to find the corresponding element in the domain, but the challenger, in possession of an algorithm for computing the one-way function, could readily check the answer. In the applications that later came to seem most important, the role of the challenge was played by a message and the process took on the character of a signature, a *digital signature*.

It did not take long to realize that the trapdoor one-way function could also be applied to the baffling problem of key distribution. For someone in possession of the forward form of the one-way function to send a secret message to the person who knew the trapdoor, he had only to transform the message with the one-way function. Only the holder of the trapdoor information would be able to invert the operation and recover the message. Because knowing the forward form of the function did not make it possible to compute the inverse, the function could be made freely available. It is this possibility that gave the field its name: *public key cryptography*.

The concept that emerges is that of a *public key cryptosystem:* a cryptosystem in which keys come in inverse pairs [36] and each pair of keys has two properties:

- Anything encrypted with one key can be decrypted with the other.
- Given one member of the pair, the *public key,* it is infeasible to discover the other, the *secret key.*

This separation of encryption and decryption makes it possible for the subscribers to a communication system to list their public keys in a ''telephone directory'' along with their names and addresses. This done, the solutions to the original problems can be achieved by simple protocols:

- One subscriber can send a private message to another simply by looking up the addressee's public key and using it to encrypt the message. Only the holder of the corresponding secret key can read such a message; even the sender, should he lose the plaintext, is incapable of extracting it from the ciphertext.

- A subscriber can sign a message by encrypting it with his own secret key. Anyone with access to the public key can verify that it must have been encrypted with the corresponding secret key, but this is of no help to him in creating (forging) a message with this property.

The first aspect of public key cryptography greatly simplifies the management of keys, especially in large communication networks. In order for a pair of subscribers to communicate privately using conventional end-to-end cryptography, they must both have copies of the same cryptographic key and this key must be kept secret from anyone they do not wish to take into their confidence. If a network has only a few subscribers, each person simply stores one key for every other subscriber against the day he will need it, but for a large network, this is impractical.

In a network with n subscribers there are $[n(n-1)]/2$ pairs each of which may require a key. This amounts to five thousand keys in a network with only a hundred subscribers, half a million in a network with one thousand, and twenty million billion in a network the size of the North American telephone system. It is unthinkable to distribute this many keys in advance and undesirable to postpone secure communication while they are carried from one party to the other by courier.

The second aspect makes it possible to conduct a much broader range of normal business practices over a telecommunication network. The availability of a signature that the receiver of a message cannot forge and the sender cannot readily disavow makes it possible to trust the network with negotiations and transactions of much higher value than would otherwise be possible.

It must be noted that both problems can be solved without public key cryptography, but that conventional solutions come at a great price. Centralized key distribution centers can on request provide a subscriber with a key for communicating with any other subscriber and protocols for this purpose will be discussed later on. The function of the signature can also be approximated by a central registry that records all transactions and bears witness in cases of dispute. Both mechanisms, however, encumber the network with the intrusion of a third party into many conversations, diminishing security and degrading performance.

At the time public key cryptography was discovered, I was working with Martin E. Hellman in the Electrical Engineering Department at Stanford University. It was our immediate reaction, and by no means ours alone, that the problem of producing public key cryptosystems would be quite difficult. Instead of attacking this problem in earnest, Marty and I forged ahead in examining the consequences.

The first result of this examination to reach a broad audience was a paper entitled "Multiuser cryptographic techniques" [35], which we gave at the National Computer Conference (NCC) in 1976. We wrote the paper in December 1975 and sent preprints around immediately. One of the preprints went to Peter Blatman, a Berkeley graduate student and friend since childhood of cryptography's historian David Kahn. The result was to bring from the woodwork Ralph Merkle, possibly the single most inventive character in the public key cryptography saga.

1.1 Merkle's Puzzles

Ralph Merkle had registered in the fall of 1974 for Lance Hoffman's course in computer security at the University of California, Berkeley. Hoffman wanted term papers and required each student to submit a proposal early in the term. Merkle addressed the problem of public key distribution or as he called it "Secure Communication over Insecure Channels" [70]. Hoffman could not understand Merkle's proposal. He demanded that it be rewritten, but alas found the revised version no more comprehensible than the original. After one more iteration of this process, Merkle dropped the course, but he did not cease working on the problem despite continuing failure to make his results understood.

Although Merkle's original proposal may have been hard to follow, the idea is quite simple. Merkle's approach is to communicate a cryptographic key from one person to another by hiding it in a large collection of puzzles. Following the tradition in public key cryptography the parties to this communication will be called Alice and Bob rather than the faceless A and B, X and Y, or I and J common in technical literature.

Alice manufactures a million or more puzzles and sends them over the exposed communication channel to Bob. Each puzzle contains a cryptographic key in a recognizable standard format. The puzzle itself is a cryptogram produced by a block cipher with a fairly small key space. As with the number of puzzles, a million is a plausible number. When Bob receives the puzzles, he picks one and solves it, by the simple expedient of trying each of the block cipher's million keys in turn until he finds one that results in plaintext of the correct form. This requires a large but hardly impossible amount of work.

To inform Alice which puzzle he has solved, Bob uses the key it contains to encrypt a fixed text message, which he transmits to Alice. Alice now tries her million keys on the test message until she finds the one that works. This is the key from the puzzle Bob has chosen.

The task facing an intruder is more arduous. Rather than selecting one of the puzzles to solve, he must solve on average half of them. The amount of effort he must expend is therefore approximately the square of that expended by the legitimate communicators.

The n to n^2 advantage the legitimate communicators have over the intruder is small by cryptographic standards, but sufficient to make the system plausible in some circumstances. Suppose, for example, that the plaintext of each puzzle is 96 bits, consisting of 64 bits of key together with a 32-bit block of zeros that enables Bob to recognize the right solution. The puzzle is constructed by encrypting this plaintext using a block cipher with 20 bits of key. Alice produces a million of these puzzles and Bob requires about half a million tests to solve one. The bandwidth and computing power required to make this feasible are large but not inaccessible. On a DS1 (1.544 Mbit) channel it would require about a minute to communicate the puzzles. If keys can be tried on the selected puzzle at about ten thousand per second, it will take Bob another minute to solve it. Finally, it will take a similar amount of time for Alice to figure out, from the test message, which key has been chosen.

The intruder can expect to have to solve half a million puzzles at half a million tries apiece. With equivalent computational facilities, this requires 25 million seconds or about a year. For applications such as authentication, in which the keys are no longer of use after communication is complete, the security of this system might be sufficient.

When Merkle saw the preprint of "Multiuser cryptographic techniques" he immediately realized he had found people who would appreciate his work and sent us copies of the paper he had been endeavoring unsuccessfully to publish. We in turn realized that Merkle's formulation of the problem was quite different from mine and, because Merkle had isolated one of the two intertwined problems I had seen, potentially simpler.

Even before the notion of putting trapdoors into one-way functions had appeared, a central objective of my work with Marty had been to identify and study functions that were easy to compute in one direction, but difficult to invert. Three principal examples of this simplest and most basic of cryptographic phenomena occupied our thoughts.

- John Gill, a colleague in the Electrical Engineering Department at Stanford, had suggested discrete exponentiation because the inverse problem, discrete logarithm, was considered very difficult.
- I had sought suitable problems in the chapter on NP-complete functions in Aho, Hopcroft, and Ullman's book on computational complexity [3] and selected the knapsack problem as most appropriate.
- Donald Knuth of the Stanford Computer Science Department had suggested that multiplying a pair of primes was easy, but that factoring the result, even when it was known to have precisely two factors, was exceedingly hard. All three of these one-way functions were shortly to assume great importance.

2 EXPONENTIAL KEY EXCHANGE

The exponential example was tantalizing because of its combinatorial peculiarities. When I had first thought of digital signatures, I had attempted to achieve them with a scheme using tables of exponentials. This system failed, but Marty and I continued twisting exponentials around in our minds and discussions trying to make them fit. Marty eventually made the breakthrough early one morning in May 1976. I was working at the Stanford Artificial Intelligence Laboratory on the paper that we were shortly to publish under the title "New directions in cryptography" [36] when Marty called and explained exponential key exchange in its unnerving simplicity. Listening to him, I realized that the notion had been at the edge of my mind for some time, but had never really broken through.

Exponential key exchange takes advantage of the ease with which exponentials can be computed in a Galois (finite) field GF(q) with a prime number q of elements (the numbers $\{0, 1, \ldots, q - 1\}$ under arithmetic modulo q) as compared with the difficulty of computing logarithms in the same field. If

$$Y = \alpha^X \bmod q, \quad \text{for } 1 < X < q - 1$$

where α is a fixed primitive element of GF(q) (that is, the powers of α produce all the nonzero elements $1, 2, \ldots, q - 1$ of GF(q)), then X is referred to as the logarithm of Y to the base α, over GF(q):

$$X = \log_\alpha Y \quad \text{over GF}(q), \quad \text{for } 1 < Y < q - 1$$

Calculation of Y from X is easy: Using repeated squaring, it takes at most $2 \times \log_2 q$ multiplications. For example,

$$\alpha^{37} = \alpha^{32+4+1}$$

$$= \left(\left(\left((\alpha^2)^2 \right)^2 \right)^2 \right)^2 \times (\alpha^2)^2 \times \alpha$$

Computing X from Y, on the other hand is typically far more difficult [29,83,104]. If q has been chosen correctly, extracting logarithms modulo q requires a precomputation proportional to:

$$L(q) = e^{\sqrt{\ln q \times \ln \ln q}}$$

although after that individual logarithms can be calculated fairly quickly. The function $L(q)$ also estimates the time needed to factor a composite number of comparable size and will appear again in that context.

To initiate communication Alice chooses a random number X_A uniformly from the integers $1, 2, \ldots, q - 1$. She keeps X_A secret, but sends

$$Y_A = \alpha^{X_A} \mod q$$

to Bob. Similarly, Bob chooses a random number X_B and sends the corresponding Y_B to Alice. Both Alice and Bob can now compute:

$$K_{AB} = \alpha^{X_A X_B} \mod q$$

and use this as their key. Alice computes K_{AB} by raising the Y_B she obtained from Bob to the power X_A:

$$\begin{aligned} K_{AB} &= Y_B^{X_A} \mod q \\ &= (\alpha^{X_B})^{X_A} \mod q \\ &= \alpha^{X_B X_A} = \alpha^{X_A X_B} \mod q \end{aligned}$$

and Bob obtains K_{AB} in a similar fashion:

$$K_{AB} = Y_A^{X_B} \mod q$$

No one except Alice and Bob knows either X_A or X_B so anyone else must compute K_{AB} from Y_A and Y_B alone. The equivalence of this problem to the discrete logarithm problem is a major open question in public key cryptography. To date no easier solution than taking the logarithm of either Y_A or Y_B has been discovered.

If q is a prime about 1000 bits in length, only about 2000 multiplications of 1000-bit numbers are required to compute Y_A from X_A, or K_{AB} from Y_A and X_B. Taking logarithms over GF(q), on the other hand, currently demands more than 2^{100} (or approximately 10^{30}) operations.

The arithmetic of exponential key exchange is not restricted to prime fields; it can also be done in Galois fields with 2^n elements, or in prime product rings [68,103]. The "2^n" approach has been taken by several people [56,64,117] because arithmetic in these fields can be performed with linear shift registers and is much faster than arithmetic over large primes. It has turned out, however, that discrete logarithms can also be calculated much more quickly in "2^n" fields [10,28,29] and so the sizes of the registers must be about 50% greater.

Marty and I immediately recognized that we had a far more compact solution to the key distribution problem than Merkle's puzzles and hastened to add it to both the upcoming NCC presentation and to "New directions in cryptography" [36]. The latter now contained a solution to each aspect of the public key problem, though not the combined solution I had envisioned. It was sent off to the *IEEE Transactions on*

Information Theory prior to my departure for NCC and like all of our other papers was immediately circulated in preprint.

3 TRAPDOOR KNAPSACKS

Later in the same year, Ralph Merkle began work on his best-known contribution to public key cryptography: building trapdoors into the knapsack one-way function to produce the trapdoor knapsack public key cryptosystem.

The knapsack problem is fancifully derived from the notion of packing gear into a knapsack. A shipping clerk faced with an odd assortment of packages and a freight container will naturally try to find a subset of the packages that fills the container exactly with no wasted space. The simplest case of this problem, and the one that has found application in cryptography is the one-dimensional case: packing varying lengths of fishing rod into a tall thin tube.

Given a *cargo vector* of integers $\mathbf{a} = (a_1, a_2, \ldots, a_n)$ it is easy to add up the elements of any specified *subvector*. Presented with an integer S, however, it is not easy to find a subvector of $\mathbf{a}$ whose elements sum to S, even if such a subvector is known to exist. This *knapsack problem* is well known in combinatorics and is believed to be extremely difficult in general. It belongs to the class of NP-complete problems, problems thought not to be solvable in polynomial time on any deterministic computer.

I had previously identified the knapsack problem as a theoretically attractive basis for a one-way function. The cargo vector $\mathbf{a}$ can be used to encipher an n-bit message $\mathbf{x} = (x_1, x_2, \ldots, x_n)$ by taking the dot product $S = \mathbf{a} \cdot \mathbf{x}$ as the ciphertext. Because one element of the dot product is binary, this process is easy and simply requires n additions. Inverting the function by finding a binary vector $\mathbf{x}$ such that $\mathbf{a} \cdot \mathbf{x} = S$ solves the knapsack problem and is thus believed to be computationally infeasible if $\mathbf{a}$ is randomly chosen. Despite this difficulty in general, many cases of the knapsack problem are quite easy and Merkle contrived to build a trapdoor into the knapsack one-way function by starting with a simple cargo vector and converting it into a more complex form [71].

If the cargo vector $\mathbf{a}$ is chosen so that each element is larger than the sum of the preceding elements, it is called *superincreasing* and its knapsack problem is particularly simple. (In the special case where the components are 1, 2, 4, 8, etc., this is the elementary operation of binary decomposition.) For example, if $\mathbf{a}' = (171, 197, 459, 1191, 2410)$ and $S' = 3798$ then x_5 must equal 1. If it were 0 then even if $x_1, x_2, x_3,$ and x_4 were all equal to 1, the dot product $\mathbf{a} \cdot \mathbf{x}$ would be too small. Since $x_5 = 1$, $S' - a_5' = 3797 - 2410 = 1387$ must be a sum of a subset of the first four elements of a'. The fact that $1387 > a_4' = 1191$ means that x_4 too must equal 1. Finally $S' - a_5' - a_4' = 197 = a_2'$ so $x_3 = 0$, $x_2 = 1$, and $x_1 = 0$.

The simple cargo vector $\mathbf{a}'$ cannot be used as a public enciphering key because anyone can easily recover a vector $\mathbf{x}$ for which $\mathbf{x} \cdot \mathbf{a}' = S'$ from $\mathbf{a}'$ and S' by the process described above. The algorithm for generating keys therefore chooses a random superincreasing cargo vector $\mathbf{a}'$ (with a hundred or more components) and keeps this vector secret. It also generates a random integer m, larger than $\Sigma \, \mathbf{a}'$, and a random integer w, relatively prime to m, whose inverse w^{-1} mod m will be used in decryption. The public cargo vector or enciphering key $\mathbf{a}$ is produced by multiplying each component of $\mathbf{a}'$ by w mod m:

$$\mathbf{a} = w\mathbf{a}' \mod m$$

Alice publishes a permuted version of $\mathbf{a}$ as her public key, but keeps the permutation, the simple cargo vector $\mathbf{a}'$, the multiplier w and its inverse, and the modulus m secret as her private key.

When Bob wants to send the message $\mathbf{x}$ to Alice he computes and sends

$$S = \mathbf{a} \cdot \mathbf{x}$$

Because

$$
\begin{aligned}
S' &= w^{-1}S \mod m \\
&= w^{-1} \sum a_i x_i \mod m \\
&= w^{-1} \sum (wa_i' \mod m)x_i \mod m \\
&= \sum (w^{-1}wa_i' \mod m)x_i \mod m \\
&= \sum a_i' x_i \mod m \\
&= \mathbf{a}' \cdot \mathbf{x}
\end{aligned}
$$

when $m > \Sigma \, a_i'$, Alice can use her secret information, w^{-1} and m, to transform any message S that has been enciphered with her public key into $S' = w^{-1} \times S$ and solve the easy knapsack problem $S' = \mathbf{a}' \cdot \mathbf{x}$ to obtain $\mathbf{x}$.

For example, for the secret vector $\mathbf{a}'$, above, the values $w = 2550$ and $m = 8443$, result in the public vector $\mathbf{a} = (5457, 4213, 5316, 6013, 7439)$, which hides the structure present in $\mathbf{a}'$.

This process can be iterated to produce a sequence of cargo vectors with more and more difficult knapsack problems by using transformations (w_1, m_1), (w_2, m_2), etc. The overall transformation that results is not, in general, equivalent to any single (w, m) transformation.

The trapdoor knapsack system does not lend itself readily to the production of signatures because most elements S of the ciphertext space $\{0 \le S \le \Sigma \, a_i\}$, do not have inverse images. This does not interfere with the use of the system for sending private messages, but requires special adaptation for signature applications [71,98].

Merkle had great confidence in even the single-iteration knapsack system and posted a note on his office door offering a $100 reward to anyone who could break it.

4 THE RIVEST–SHAMIR–ADLEMAN SYSTEM

Unknown to us at the time we wrote "New directions in cryptography"[36] were the three people who were to make the single most spectacular contribution to public key cryptography: Ron Rivest, Adi Shamir, and Len Adleman. Ron Rivest had been a graduate student in computer science at Stanford while I was working on proving the correctness of programs at the Stanford Artificial Intelligence Laboratory. One of my colleagues in that work was Zohar Manna, who shortly returned to Israel and supervised the doctoral research of Adi Shamir, at the Weitzman Institute. Len Adleman was a native of San Franciscan with both undergraduate and graduate degrees from the University of California at Berkeley. Despite this web of near connections, not one of the three had previously crossed our paths and their names were unfamiliar.

When the "New directions in cryptography" paper [36] reached Massachusetts Institute of Technology (MIT) in the fall of 1976, the three took up the challenge of producing a full-fledged public key cryptosystem. The process lasted several months during which Ron Rivest proposed approaches, Len Adleman attacked them, and Adi Shamir recalls doing some of each.

In May 1977 they were rewarded with success. After investigating a number of possibilities, some of which were later put forward by other researchers [1,67], they had discovered how a simple piece of classic number theory could be made to solve the problem. The resulting paper [91] also introduced Alice and Bob, the first couple of cryptography [53].

The Rivest–Shamir–Adelman (RSA) cryptosystem is a block cipher in which the plaintexts and ciphertexts are integers between 0 and $N - 1$ for some N. It resembles the exponential key exchange system described above in using exponentiation in modular arithmetic for its enciphering and deciphering operations but, unlike that system, RSA must do its arithmetic not over prime numbers, but over composite ones.

Knowledge of a plaintext M, a modulus N, and an exponent e are sufficient to allow calculation of M^e mod N. Exponentiation, however, is a one-way function with respect to the extraction of roots as well as logarithms. Depending on the characteristics of N, M, and e, it may be very difficult to invert.

The RSA system makes use of the fact that finding large (e.g., 200-digit) prime numbers is computationally easy, but that factoring the product of two such numbers appears computationally infeasible. Alice creates her secret and public keys by selecting two very large prime numbers, P and Q, at random, and multiplying them together to obtain a *bicomposite* modulus N. She makes this product public together with a suitably chosen enciphering exponent e, but keeps the factors, P and Q, secret.

The enciphering process of exponentiation modulo N can be carried out by anyone who knows N, but only Alice, who knows the factors of N, can reverse the process and decipher.

Using P and Q, Alice can compute the Euler totient function $\phi(N)$, which counts the numbers of integers between 1 and N that are relatively prime to N and consequently invertible in arithmetic modulo N. For a bicomposite number this is:

$$\phi(N) = (P - 1)(Q - 1)$$

The quantity $\phi(N)$ plays a critical role in Euler's theorem, which says that for any number x that is invertible modulo N (and for large N that is almost all of them):

$$x^{\phi(N)} \equiv 1 \pmod{N}$$

or slightly more generally

$$x^{k\phi(N)+1} \equiv x \pmod{N}$$

Using $\phi(N)$ Alice can calculate [60] a number d such that

$$e \times d \equiv 1 \pmod{\phi(N)}$$

which is equivalent to saying that:

$$e \times d = k \times \phi(N) + 1$$

When the cryptogram M^e mod N is raised to the power d the result is:

$$(M^e)^d = M^{ed} = M^{k\phi(N)+1} \equiv M \quad (\text{mod } N)$$

the original plaintext M.

As a very small example, suppose $P = 17$ and $Q = 31$ are chosen so that $N = PQ = 527$ and $\phi(N) = (P - 1)(Q - 1) = 480$. If $e = 7$ is chosen then $d = 343$. ($7 \times 343 = 2401 = 5 \times 480 + 1$). And if $M = 2$ then

$$C = M^e \text{ mod } N = 2^7 \text{ mod } 527 = 128$$

Note again that only the public information (e, N) is required for enciphering M. To decipher, the private key d is needed to compute

$$
\begin{aligned}
M &= C^d \text{ mod } N \\
&= 128^{343} \text{ mod } 527 \\
&= 128^{256} \times 128^{64} \times 128^{16} \times 128^4 \times 128^2 \times 128^1 \text{ mod } 527 \\
&= 35 \times 256 \times 35 \times 101 \times 47 \times 128 \text{ mod } 527 \\
&= 2 \text{ mod } 527
\end{aligned}
$$

Just as the strength of the exponential key exchange system is not known to be equivalent to the difficulty of extracting discrete logarithms, the strength of RSA has not been proven equivalent to factoring. There might be some method of taking the eth root of M^e without calculating d and thus without providing information sufficient to factor. While at MIT in 1978, Michael O. Rabin [86] produced a variant of RSA, subsequently improved by Hugh Williams of the University of Manitoba [113], that is equivalent to factoring. Rivest and I have independently observed [38, 92], however, that the precise equivalence Rabin has shown is a two-edged sword.

5 THE McELIECE CODING SCHEME

Within a short time yet another public key system was to appear, this due to Robert J. McEliece of the Jet Propulsion Laboratory at Cal Tech [69]. McEliece's system makes use of the existence of a class of error-correcting codes, the Goppa codes, for which a fast decoding algorithm is known. His idea was to construct a Goppa code and disguise it as a general linear code, whose decoding problem is NP-complete. There is a strong parallel here with the trapdoor knapsack system in which a superincreasing cargo vector, whose knapsack problem is simple to solve, is disguised as a general cargo vector whose knapsack problem is NP-complete.

In a knapsack system, the secret key consists of a superincreasing cargo vector $\mathbf{v}$, together with the multiplier w and the modulus m that disguise it; in McEliece's system, the secret key consists of the generator matrix G for a Goppa code together with a nonsingular matrix S and a permutation matrix P that disguise it. The public key appears as the encoding matrix $G' = SGP$ of a general linear code.

- To encode a data block u into a message x, Alice multiplies it by Bob's public encoding matrix G' and adds a locally generated noise block z.
- To decode, Bob multiplies the received message x by P^{-1}, decodes xp^{-1} to get a word in the Goppa code and multiplies this by S^{-1} to recover Alice's data block.

McEliece's system has never achieved wide acceptance and has probably never even been considered for implementation in any real application. This may be because the public keys are quite large, requiring on the order of a million bits; it may be because the system entails substantial expansion of the data; or it may be because McEliece's system bears a frightening structural similarity to the knapsack systems whose fate we will discover shortly.

6 THE FALL OF THE KNAPSACKS

The year 1982 was the most exciting time for public key cryptography since its spectacular first 3 years. In March, Adi Shamir sent out a research announcement: He had broken the single-iteration Merkle-Hellman knapsack system [101,102]. By applying new results of Lenstra at the Mathematische Centrum in Amsterdam, Shamir had learned how to take a public cargo vector and discover a w' and m' that would convert it back into a superincreasing "secret" cargo vector—not necessarily the same one the originator had used, but one that would suffice for decrypting messages encrypted with the public cargo vector.

Shamir's original attack was narrow. It seemed that perhaps its only consequence would be to strengthen the knapsack system by adding conditions to the construction rules for avoiding the new attack. The first response of Gustavus J. Simmons, whose work will dominate a later chapter, was that he could avoid Shamir's attack without even changing the cargo vector merely by a more careful choice of w and m. He quickly learned, however, that Shamir's approach could be extended to break a far larger class of knapsack systems [16].

Crypto '82 revealed that several other people had continued down the trail Shamir had blazed. Shamir himself had reached the same conclusions. Andy Odlyzko and Jeff Lagarias at Bell Laboratories were on the same track and Len Adleman had not only devised an attack but had also programmed it on an Apple II. The substance of the attacks will not be treated here since it is central to another chapter in this volume (Ernest F. Brickell and Andrew M. Odlyzko, "Cryptanalysis: A Survey of Recent Results"). The events they engendered, however, will be discussed.

I had the pleasure of chairing the cryptanalysis session at Crypto '82 in which the various results were presented. Ironically, at the time I accepted the invitation to organize such a session, Shamir's announcement stood alone and knapsack systems were only one of the topics to be discussed. My original program ran into very bad luck, however. Of the papers initially scheduled only Donald Davies's talk on "The Bombe at Bletchley Park" was actually presented. Nonetheless, the lost papers were more than replaced by presentations on various approaches to the knapsack problem.

Last on the program were Len Adleman and his computer, which had accepted a challenge on the first night of the conference. The hour passed; various techniques for attacking knapsack systems with different characteristics were heard; and the Apple II sat on the table waiting to reveal the results of its labors. At last Adleman rose to speak, mumbling something self-deprecatingly about "the theory first, the public humiliation later" and beginning to explain his work. All the while the figure of Carl Nicolai was moving silently in the background setting up the computer and copying a sequence of numbers from its screen onto a transparency. At last another transparency was drawn from a sealed envelope and the results placed side by side on the projector. They were identical. The public humiliation was not Adleman's, it was the knapsack system's.

Ralph Merkle was not present, but Marty Hellman, who was, gamely rose to make a concession speech on their behalf. Merkle, always one to put his money where his mouth was, had long since paid Shamir the $100 in prize money that he had placed on the table nearly 6 years before.

The press wrote that knapsacks were dead. I was skeptical but ventured that the results were sufficiently threatening that I felt "nobody should entrust anything of great value to a knapsack system unless he had a much deeper theory of their functioning than was currently available." Nor was Merkel's enthusiasm dampened. He promptly raised his bet and offered $1000 to anyone who could break a multiple-iteration knapsack [72].

It took 2 years, but in the end, Merkle had to pay [42]. The money was finally claimed by Ernie Brickell in the summer of 1984 when he announced the destruction of a knapsack system of 40 iterations and a hundred weights in the cargo vector in about an hour of Cray-1 time [17]. That fall I was forced to admit that "knapsacks are flat on their back."

Closely related techniques have also been applied to make a dramatic reduction in the time needed to extract discrete logarithms in fields of type $GF(2^n)$. This approach was pioneered by Blake, Fuji-Hara, Vanstone, and Mullin in Canada [10] and refined by Coppersmith in the United States [28]. A comprehensive survey of this field was given by Andy Odlyzko at Eurocrypt '84 [79].

7 EARLY RESPONSES TO PUBLIC KEY CRYPTOSYSTEMS

A copy of the MIT report [90] on the RSA cryptosystem was sent to Martin Gardner, Mathematical Games editor of *Scientific American*, shortly after it was printed. Gardner promptly published a column [48] based on his reading of both the MIT report and "New directions in cryptography" [36]. Bearing the title "A new kind of cryptosystem that would take millions of years to break," it began a confusion that persists to this day between the two directions explored by the "New directions" paper: public key cryptography and the problem of proving the security of cryptographic systems. More significant, however, was the prestige that public key cryptography got from being announced in the scientific world's most prominent lay journal more than 6 months before its appearance in the *Communications of the ACM*.

The excitement public key cryptosystems provoked in the popular and scientific press was not matched by corresponding acceptance in the cryptographic establishment, however. In the same year that public key cryptography was discovered, the National Bureau of Standards, with the support of the National Security Agency (NSA), proposed a conventional cryptographic system, designed by International Business Machines (IBM), as a federal *Data Encryption Standard* (DES) [44]. Marty Hellman and I criticized the proposal on the grounds that its key was too small [37], but manufacturers were gearing up to support the proposed standard and our criticism was seen by many as an attempt to disrupt the standards-making process to the advantage of our own work. Public key cryptography in its turn was attacked, in sales literature [74] and technical papers [59,76] alike, more as though it were a competing product than a recent research discovery. This, however, did not deter NSA from claiming its share of the credit. Its director, in the words of the *Encyclopedia Britannica* [110], "pointed out that two-key cryptography had been discovered at the agency a decade earlier," although no evidence for this claim was ever offered publicly.

Far from hurting public key cryptography, the attacks and counterclaims added to a ground swell of publicity that spread its reputation far faster than publication in scientific journals alone ever could. The criticism nonetheless bears careful examination, because the field has been affected as much by discoveries about how public key cryptosystems should be used as by discoveries about how they can be built.

In viewing public key cryptography as a new form of cryptosystem rather than a new form of key management, I set the stage for criticism on grounds of both security and performance. Opponents were quick to point out that the RSA system ran about one-thousandth as fast as DES and required keys about ten times as large. Although it had been obvious from the beginning that the use of public key systems could be limited to exchanging keys for conventional cryptography, it was not immediately clear that this was necessary. In this context, the proposal to build *hybrid* systems [62] was hailed as a discovery in its own right.

At present, the convenient features of public key cryptosystems are bought at the expense of speed. The fastest RSA implementations run at only a few thousand bits per second, while the fastest DES implementations run at many million. It is generally desirable, therefore, to make use of a hybrid in which the public key systems are used only during key management processes to establish shared keys for employment with conventional systems.

No known theorem, however, says that a public key cryptosystem must be larger and slower than a conventional one. The demonstrable restrictions mandate a larger minimum block size (though perhaps no larger than that of DES) and preclude use in stream modes whose chunks are smaller than this minimum. For a long time I felt that "high-efficiency" public key systems would be discovered and would supplant both current public key and conventional systems in most applications. Using public key systems throughout, I argued, would yield a more uniform architecture with fewer components and would give the best possible damage limitation in the event of a key distribution center compromise [38]. Most important, I thought, if only one system were in use, only one certification study would be required. As certification is the most fundamental and most difficult problem in cryptography, this seemed to be where the real savings lay.

In time I saw the folly of this view. Theorems or not, it seemed silly to expect that adding a major new criterion to the requirements for a cryptographic system could fail to slow it down. The designer would always have more latitude with systems that did not have to satisfy the public key property and some of these would doubtless be faster. Even more compelling was the realization that modes of operation incompatible with the public key property are essential in many communication channels.

To date, the "high-efficiency public key systems" that I had hoped for have not appeared and the restriction of public key cryptography to key management and signature applications is almost universally accepted. More fundamental criticism focuses on whether the public key system actually makes any contribution to security, but before examining this criticism, we must undertake a more careful study of key distribution mechanisms.

7.1 Key Management

The solution to the problem of key management using conventional cryptography is for the network to provide a KDC: a trusted network resource that shares a key with each subscriber and uses these in a bootstrap process to provide additional keys to the subscribers as needed. When one subscriber wants to communicate securely with another, he first contacts the KDC to obtain a *session key* for use in that particular conversation.

Key distribution protocols vary widely depending on the cost of messages, the availability of multiple simultaneous connections, whether the subscribers have synchronized clocks, and whether the KDC has authority not only to facilitate, but also to allow or prohibit, communications. The following example is typical and makes use of an important property of cryptographic authentication. Because a message altered by anyone who does not have the correct key will fail when tested for authenticity, there is no loss of security in receiving a message from the hands of a potential opponent. In so doing, it introduces, in a conventional context, the concept of a *certificate*—a cryptographically authenticated message containing a cryptographic key—a concept that plays a vital role in modern key management.

1. When Alice wants to call Bob, she first calls the KDC and requests a key for communicating with Bob.
2. The KDC responds by sending Alice a pair of certificates. Each contains a copy of the required session key, one encrypted so that only Alice can read it and one so that only Bob can read it.
3. When Alice calls Bob, she presents the proper certificate as her introduction. Each of them decrypts the appropriate certificate under the key that he shares with the KDC and thereby gets access to the session key.
4. Alice and Bob can now communicate securely using the session key.

Alice and Bob need not go through all of this procedure on every call; they can instead save the certificates for later use. Such *cacheing* of keys allows subscribers to avoid calling the KDC every time they pick up the phone, but the number of KDC calls is still proportional to the number of distinct pairs of subscribers who want to communicate securely. A far more serious disadvantage of the arrangement described above is that the subscribers must share the secrecy of their keying information with the KDC and if it is penetrated they too will be compromised.

A big improvement in both economy and security can be made by the use of public key cryptography. A certificate functions as a letter of introduction. In the protocol above, Alice has obtained a letter that introduces her to Bob and Bob alone. In a network using public key encryption, she can instead obtain a single certificate that introduces her to any network subscriber [62].

What accounts for the difference? In a conventional network, every subscriber shares a secret key with the KDC and can only authenticate messages explicitly meant for him. If one subscriber has the key needed to authenticate a message meant for another subscriber, he will also be able to create such a message and authentication fails. In a public key network, each subscriber has the public key of the KDC and thus the capacity to authenticate any message from the KDC, but no power to forge one.

Alice and Bob, each having obtained a certificate from the KDC in advance of making any secure calls, communicate with each other as follows:

1. Alice sends her certificate to Bob.

2. Bob sent his certificate to Alice.

3. Alice and Bob each check the KDC's signature on the certificates they have received.

4. Alice and Bob can now communicate using the keys contained in the certificates. When making a call, there is no need to call the KDC and little to be gained by cacheing the certificates. The added security arises from the fact that the KDC is not privy to any information that would enable it to spy on the subscribers. The keys that the KDC dispenses are public keys and messages encrypted with these can only be decrypted by using the corresponding secret keys, to which the KDC has no access.

The most carefully articulated attack came from Roger Needham and Michael Schroeder [76], who compared conventional key distribution protocols with similar public key ones. They counted the numbers of messages required and concluded that conventional cryptography was more efficient than public key cryptography. Unfortunately, in this analysis, they had ignored the fact that security was better under the public key protocol they presented than the conventional one.

To compromise a network that employs conventional cryptography, it suffices to corrupt the KDC. This gives the intruders access to information sufficient for recovering the session keys used to encrypt past, present, and perhaps future messages. These keys, together with information obtained from passive wiretaps, allow the penetrators of the KDC access to the contents of any message sent on the system.

A public key network presents the intruder with a much more difficult problem. Even if the KDC has been corrupted and its secret key is known to opponents, this information is insufficient to read the traffic recorded by a passive wiretap. The KDC's secret key is useful only for signing certificates containing subscribers' public keys; it does not enable the intruders to decrypt any subscriber traffic. To be able to gain access to this traffic, the intruders must use their ability to forge certificates as a way of tricking subscribers into encrypting messages with phony public keys.

To spy on a call from Alice to Bob, opponents who have discovered the secret key of the KDC must intercept the message in which Bob sends Alice the certificate for his public key and substitute one for a public key they have manufactured themselves and whose corresponding secret key is therefore known to them. This will enable them to decrypt any message that Alice sends to Bob. If such a misencrypted message actually reaches Bob, however, he will be unable to decrypt it and may alert Alice to the error. The opponents must therefore intercept Alice's messages, decrypt them, and re-encrypt them in Bob's public key in order to maintain the deception. If the opponents want to understand Bob's replies to Alice, they must go through the same procedure with Bob, supplying him with a phony public key for Alice and translating all the messages he sends her.

The procedure above is cumbersome at best. Active wiretaps are in principle detectable and the number the intruders must place in the net to maintain their control grows rapidly with the number of subscribers being spied on. Over large portions of many networks—radio broadcast networks, for example—the message deletions essential to this scheme are extremely difficult. This forces the opponents to place their taps very close to the targets and recreates the circumstances of conventional wiretapping, thereby denying the opponents precisely those advantages of communications intelligence that make it so attractive.

It is worth observing that the use of a hybrid scheme diminishes the gain in security a little because the intruder does not need to control the channel after the session key has been selected. This threat, however, can be countered, without losing the advantages of a session key, by periodically (and unpredictably) using the public keys to exchange new session keys [40].

Public key techniques also make it possible to conquer another troubling problem of conventional cryptographic security, the fact that compromised keys can be used to read traffic taken at an earlier date [5]. At the trial of Jerry Whitworth, a spy who passed U. S. Navy keying information to the Russians, the judge asked the prosecution's expert witness [27]: "Why is it necessary to destroy yesterday's . . . [key] . . . list if it's never going to be used again?" The witness responded in shock: "A used key, Your Honor, is the most critical key there is. If anyone can gain access to that, they can read your communications."

The solution to this problem is to be found in a judicious combination of exponential key exchange and digital signatures, inherent in the operation of a secure telephone currently under development at Bell-Northern Research [41,81] and intended for use on the Integrated Services Digital Network (ISDN).

Each ISDN secure phone has an operating secret key/public key pair that has been negotiated with the network's key management facility. The public key portion is embodied in a certificate signed by the key management facility along with such identifying information as its phone number and location. In the call setup process that follows, the phone uses this certificate to convey its public key to other phones.

1. The telephones perform an exponential key exchange to generate session keys unique to the current phone call. These keys are then used to encrypt all subsequent transmissions in a conventional cryptosystem.

2. Having established an encrypted (though not yet authenticated) channel, the phones begin exchanging credentials. Each sends the other its public key certificate.

3. Each phone checks the signature on the certificate it has received and extracts from it the other phone's public key.

4. The phones now challenge each other to sign test messages and check the signatures on the responses using the public keys from the certificates. Once the call setup is complete, each phone displays for its user the identity of the phone with which it is in communication.

The use of the exponential key exchange creates unique session keys that exist only inside the phones and only for the duration of the call. This provides a security guarantee whose absence in conventional cryptography is at the heart of many spy cases: once a call between uncompromised ISDN secure phones is completed and the session keys are destroyed, no compromise of the long-term keys that still reside in the phones will enable anyone to decrypt the recording of the call. Using conventional key management techniques, session keys are always derivable from a combination of long-term keying material and intercepted traffic. If long-term conventional keys are ever compromised, all communications, even those of earlier date, encrypted in derived keys, are compromised as well.

In the late 1970s, a code clerk named Christopher Boyce, who worked for a Central Intelligence Agency (CIA)-sponsored division of TRW, copied keying material that was supposed to have been destroyed and sold it to the Russians [66]. More recently, Jerry Whitworth did much the same thing in the communication center of the Alameda Naval Air Station [8]. The use of exponential key exchange would have rendered such previously used keys virtually worthless.

Another valuable ingredient of modern public key technology is the *message digest*. Implementing a digital signature by encrypting the entire document to be signed with a secret key has two disadvantages. Because public key systems are slow, both the signature process—encrypting the message with a secret key—and the verification process—decrypting the message with a public key—are slow. There is also another difficulty. If the signature process encrypts the entire message, the recipient must retain the ciphertext for however long the signed message is needed. To make any use of it during this period, he must either save a plaintext copy as well or repeatedly decrypt the ciphertext.

The solution to this problem seems first to have been proposed by Donald Davies and Wyn Price of the National Physical Laboratory in Teddington, England. They proposed constructing a cryptographyically compressed form or digest of the message [33] and signing by encrypting this with the secret key. In addition to its economy, this method has the advantage of allowing the signature to be passed around independently of the message. This is often valuable in protocols in which a portion of the message that is required in the authentication process is not actually transmitted because it is already known to both parties.

Most criticism of public key cryptography came about because public key management has not always been seen from the clear, certificate-oriented, view described above. When we first wrote about public key cryptography, we spoke either of users looking in a public directory to find each other's keys or simply of exchanging them in the course of communication. The essential fact that each user had to authenticate any public key he received was glossed over. Those with an investment in traditional cryptography were not slow to point out this oversight. Public key cryptography was stigmatized as being weak on authentication and, although the problems the critics saw have long been solved, this criticism is heard to this day.

8 APPLICATION AND IMPLEMENTATION

While arguments about the true worth of public key cryptography raged in the late 1970s, it came to the attention of one person who had no doubt: Gustavus J. Simmons, head of the mathematics department at Sandia National Laboratories. Simmons was responsible for the mathematical aspects of nuclear command and control and digital signatures were just what he needed. The applications were limitless: A nuclear weapon could demand a digitally signed order before it would arm itself; a badge admitting someone to a sensitive area could bear a digitally signed description of the person; a sensor monitoring compliance with a nuclear test ban treaty could place a digital signature on the information it reported. Sandia began immediately both to develop the technology of public key devices [89,107,108,109] and to study the strength of the proposed systems [16,34,105]. To this end, they produced a 336-bit RSA encryption/decryption board using discrete components (Fig. 1), which was applied to personnel

Figure 1   Sandia 336-bit RSA board. (Courtesy of Sandia National Laboratories.)

identity verification at the Idaho Falls Zero Power Plutonium Reactor (ZPPR) facility. This was followed shortly thereafter by a VLSI implementation (Fig. 2) for the same size modulus.

The application about which Simmons spoke most frequently, test ban monitoring by remote seismic observatories [106] is the subject of another chapter in this volume; Gustavus J. Simmons, "How to Insure That Data Acquired to Verify Treaty Compliance Are Trustworthy". If the United States and the Soviet Union could put seismometers on each other's territories and use these seismometers to monitor each other's nuclear tests, the rather generous 150 kiloton upper limit imposed on underground nuclear testing by the Limited Nuclear Test Ban Treaty of 1963 could be tightened considerably—perhaps to 10 kilotons or even 1 kiloton. The problem is this: A *monitoring* nation must assure itself that the *host* nation is not concealing tests by tampering with the data from the monitor's observatories. Conventional cryptographic authentication techniques can solve this problem, but in the process create another. A host nation wants to assure itself that the monitoring nation can monitor only total yield and does not employ an instrument package capable of detecting staging or other aspects of the weapon not covered by the treaty. If the data from the remote seismic observatory are encrypted, the host country cannot tell what they contain.

Digital signatures provided a perfect solution. A digitally signed message from a remote seismic observatory cannot be altered by the host, but can be read. The host country can assure itself that the observatory is not exceeding its authority by comparing the data transmitted with data from a nearby observatory conforming to its own interpretation of the treaty language.

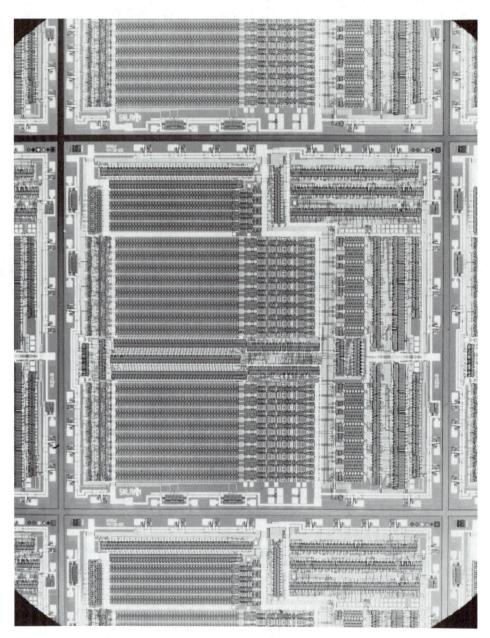

Figure 2   Wafer photo: Sandia low-speed chip. (Courtesy of Sandia National Laboratories.)

Sandia was not the only hardware builder. Ron Rivest and colleagues at MIT, ostensibly theoretical computer scientists, learned to design hardware and produced a board (Fig. 3) at approximately the same time as Sandia. The MIT board would carry out an RSA encryption with a 100 digit modulus in about a twentieth of a second. It

Figure 3   MIT RSA board. (Courtesy of Ron Rivest.)

was adequate ''proof of concept'' but too expensive for the commercial applications Rivest had in mind.

No sooner was the board done than Rivest started studying the recently popularized methods for designing large-scale integrated circuits. The result was an experimental nMOS chip that operated on approximately 500-bit numbers and should have been capable of about three encryptions per second [93]. This chip was originally intended as a prototype for commercial applications. As it happened the chip was never gotten to work correctly, and the appearance of a commercially available RSA chip was to await the brilliant work of Cylink Corporation in the mid-eighties [31].

As the 1990s dawned, public key technology began the transition from esoteric research to product development. Part of AT&T's response to a Carter Administration initiative to improve the overall security of American telecommunications, was to develop a specialized cryptographic device for protecting the Common Channel Interoffice Signaling (CCIS) on telephone trunks. The devices were link encryptors that used exponential key exchange to distribute DES keys [16,75].

Although AT&T's system was widely used within its own huge network, it was never made available as a commercial product. At about the same time, however, Racal-Milgo began producing the Datacryptor II (Fig. 4), a link encryption device that offered an RSA key exchange mode [87]. One device used exponential key exchange, the other RSA, but overall function was quite similar. When the public key option of the Datacryptor is initialized, it manufactures a new RSA key pair and communicates the public portion to the Datacryptor at the other end of the line. The device that receives this public key manufactures a DES key and sends it to the first Datacryptor encrypted with RSA. Unfortunately, the opportunity for sophisticated digital signature based authentication that RSA makes possible was missed.

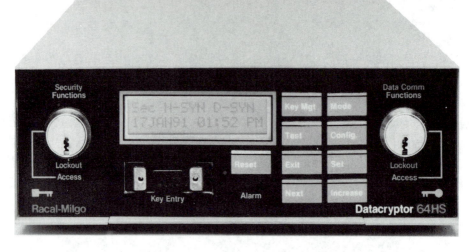

Figure 4   Racal–Milgo Datacryptor II. (Courtesy of Racal–Milgo.)

8.1 Future Secure Voice System

As the early eighties became the mid-eighties, public key cryptography finally achieved official, if nominally secret, acceptance. In 1983, NSA began feasibility studies for a new secure phone system. There were fewer than 10,000 of their then latest system the Secure Telephone Unit II or STU-II and already the key distribution center for the principal network was overloaded, with users often complaining of busy signals. At $12,000 or more apiece, 10,000 STU-IIs may have been all the government could afford, but it was hardly all the secure phones that were needed. In its desire to protect far more than just explicitly classified communications, NSA was dreaming of a million phones, each able to talk to any of the others. They couldn't have them all calling the key distribution center every day.

The system to be replaced employed electronic key distribution that allowed the STU-II to bootstrap itself into direct end-to-end encryption with a different key on every call. When a STU-II made a secure call to a terminal with which it did not share a key, it acquired one by calling a key distribution center using a protocol similar to one described earlier.

Although the STU-II seemed wonderful when first fielded in the late 1970s, it had some major shortcomings. Some cacheing of keys was permitted, but calls to the KDC entailed significant overhead. Worse, each network had to be at a single clearance level, because there was no way for a STU-II to inform the user of the clearance level of the phone with which it was talking. These factors, as much as the high price and large size, conspired against the feasibility of building a really large STU-II network.

The STU-III (Fig. 5) is the size of a large conventional telephone and, at about $3000 apiece, substantially cheaper than its predecessor. It is equipped with a two-line display that, like the display of the ISDN secure phone, provides information to each party about the location, affiliation, and clearance of the other. This allows one phone

Figure 5   Motorola STU-III secure telephone. (Courtesy of Motorola, Inc.)

to be used for the protection of information at various security levels. The phones are also sufficiently tamper resistant that, unlike earlier equipment, the unkeyed instrument is unclassified. These elements will permit the new system to be made much more widely available with projections of the number in use by the early nineties running from half a million to three million [18,43].

To make a secure call with a STU-III, the caller first places an ordinary call to another STU-III, then inserts a key-shaped device containing a cryptographic variable and pushes a "go secure" button. After an approximately 15-second wait for cryptographic setup, each phone shows information about the identity and clearance of the other party on its display and the call can proceed.

In an unprecedented move, Walter Deeley, NSA's deputy director for communications security, announced the STU-III or Future Secure Voice System in an exclusive interview given to the *New York Times* [18]. The objective of the new system was primarily to provide secure voice and low-speed data communications for the U.S. Defense Department and its contractors. The interview didn't say much about how it was going to work, but gradually the word began to leak out. The new system was using public key.

The new approach to key management was reported early on [88] and one article [6] spoke of phones being "reprogrammed once a year by secure telephone link," a turn of phrase strongly suggestive of a certificate passing protocol, similar to that described earlier, that minimizes the need for phones to talk to the key management center. Recent reports have been more forthcoming, speaking of a key management system called *FIREFLY* that [95] "evolved from public key technology and is used to establish pair-wise traffic encryption keys." Both this description and testimony submitted to the U.S. Congress by Lee Neuwirth of Cylink [78] suggest a combination of key exchange and certificates similar to that used in the ISDN secure phone and it is plausible that FIREFLY too is based on exponentiation.

Three companies, AT&T, Motorola, and RCA are manufacturing the instruments in interoperable versions and GTE is building the key management system. So far, contracts have been issued for an initial 75,000 phones and deliveries began in November 1987.

8.2 Current Commercial Products

Several companies dedicated to developing public key technology have been formed in the 1980s. All have been established by academic cryptographers endeavoring to exploit their discoveries commercially.

The first was RSA Data Security, founded by Rivest, Shamir, and Adleman, the inventors of the RSA cryptosystem, to exploit their patent on RSA and develop products based on the new technology. RSA produces a stand-alone software package called *Mailsafe* for encrypting and signing electronic mail. It also makes the primitives of this system available as a set of embeddable routines called *Bsafe* that has been licensed to major software manufacturers [9].

Cylink Corporation of Sunnyvale, California has chalked up the most impressive engineering record in the public key field. Its first product was the CIDEC HS [32,63] (Fig. 6), a high-speed (1.544 M byte) data encryptor for protecting DS1 telephone trunks. Like AT&T's CCIS encryptor, it uses exponential key exchange to establish DES session keys [77].

Cylink was also first to produce a commercially available RSA chip [7,31] (Figs. 7–8). The CY1024 is, despite its name, a 1028-bit exponential engine that can be cascaded to perform the calculations for RSA encryptions on moduli more than 16,000 bits long. A single CY1024 does a 1000-bit encryption in under half a second—both modulus size and speed currently being sufficient for most applications.

Figure 6   Cylink CIDEC-HS. (Courtesy of Cylink Corporation.)

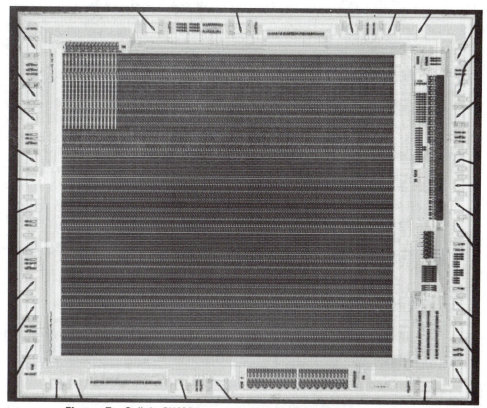

Figure 7   Cylink CY1024 exponentiator. (Courtesy of Cylink Corporation.)

The cryptography group at Waterloo University in Ontario have brought the fruits of their labors to market through a company called Cryptech. Their initial inroads into the problem of extracting logarithms over finite fields with 2^n elements [10] made it necessary to employ larger fields. Hewlett-Packard abandoned a VLSI chip designed to implement Diffie-Hellman by exchange in $GF(2^{127})$ as a direct consequence [117]. This

Figure 8   Chip photo: Cylink CY1024. (Courtesy of Cylink Corporation.)

in turn inspired them to develop high-speed exponentiation algorithms. The result is a system providing both exponential key exchange and half-megabit data encryption with the same system [56].

9 MULTIPLYING, FACTORING, AND FINDING PRIMES

The successes of the RSA system and of exponential key exchange over prime fields have led to significant developments in three areas: multiplying, factoring, and finding prime numbers.

Factoring the modulus has remained the front runner among attacks on the RSA system. As factoring has improved, the modulus size required for security has more than doubled, requiring the system's users to hunt for larger and larger prime numbers to operate the system securely. As the numbers grow larger, faster and faster methods for doing modular arithmetic are required. The result has been not only the development of a technical base for public key cryptography, but an inspiration and source of support for number theory [61,65].

9.1 Factoring

In addressing the question of how large the primes in the RSA system should be, Rivest, Shamir, and Adleman's original memo spoke of a number d such that

> determining the prime factorization of a number n which is the product of just two prime numbers of length d (in digits) is "computationally impossible." When MIT/LCS/TM-82 first appeared, it contained the statement "Choosing $d = 40$ seems to be satisfactory at present." [60] In a second printing the recommended value of d was changed to 50 and in a third took a sharp leap to 100. This escalation is symbolic of the direction of factoring in the late seventies and early eighties.

In 1970, the factoring of a 41-digit number [73] constituted a landmark. The advent of the RSA system, however, was to usher in a decade of rapid progress in this field. By the end of that decade, numbers twice as long could be factored, if not with ease, at least with hours of Cray-1 time [34]. These factorizations confirmed, by actual computer implementation, the number theorists' predictions about factoring speed.

Several factoring techniques of comparable performance have become available in recent years [85]. All factor in time proportional to:

$$L(n) = e^{\sqrt{\ln n \times \ln \ln n}} ,$$

a figure that has already been seen in connection with discrete logarithms. The one that has been most widely applied is called quadratic sieve factoring [34] and lends itself well to machine implementation. One of factoring's gurus, Marvin Wunderlich, wrote a paper in 1983 [116] that examined the way in which quadratic sieve factoring could exploit parallel processing to factor a 100-digit number in 2 months. In the same lecture, Wunderlich also explained the importance of uniformity in factoring methods applied in cryptanalysis. To be used in attacking RSA, a factoring method must be uniform, at least over the class of bicomposite numbers. If it is only applicable to numbers of some particular form, as many methods used by number theorists have been, the cryptographers will simply alter their key production to avoid numbers of that form.

More recently, Carl Pomerance [85] has undertaken the design of a modular machine employing custom chips and specialized to factoring. The size of the numbers you can factor is dependent on how much of such a machine you can afford. He has begun building a $25,000 implementation that he expects to factor 100-digit numbers in 2 weeks [96]. Ten million dollars worth of similar hardware would be able to factor 150-digit numbers in a year, but Pomerance's analysis does not stop there. Fixing 1 year as a nominal upper limit on our patience with factoring any one number, he is prepared to give a dollar estimate for factoring a number of any size. For a 200-digit number, often considered unapproachable and a benchmark in judging RSA systems, the figure is one hundred billion dollars. This is a high price to be sure, but not beyond human grasp.

9.2 Prime Finding

Prime finding has followed a somewhat different course from factoring. This is in part because there are probabilistic techniques that identify primes with sufficient certainty to satisfy all but perhaps the pickiest of RSA users and in part because primality is not in itself a sufficient condition for numbers to be acceptable as RSA factors.

Fermat's Little Theorem guarantees that if n is prime then for all $0 < b < n$

$$b^{n-1} \equiv 1 \pmod{n}$$

and any number that exhibits this property for some b is said to pass the pseudoprime test to base b. Composite numbers that pass pseudoprime tests to all bases exist, but they are rare and a number that passes several pseudoprime tests is probably a prime.

The test can be refined by making use of the fact that if n is an odd prime only the numbers 1 and -1 are square roots of 1, whereas if n is the product of distinct odd primes, the number of square roots of unity grows exponentially in the number of factors. If the number n passes the pseudoprime test to base b, it can be further examined to see if

$$b^{\frac{n-1}{2}} \equiv \pm 1 \pmod{n}$$

Tests of this kind are called strong pseudoprime tests to base b and very few composite numbers that pass strong pseudoprime tests to more than a few bases are known.

Although there has been extensive work in the past decade on giving genuine proofs of primality [2,51,84], the strong pseudoprime tests take care of the primality aspect of choosing the factors for RSA moduli. Another aspect arises from the fact that not all prime numbers are felt to be equally good. In many RSA implementations, the factors of the modulus are not random large primes p, but large primes chosen for particular properties of the factors of $p - 1$ [52,91].

9.3 High-Speed Arithmetic

Because of the progress in factoring during the decade of public key's existence, the size of the numbers used in RSA has grown steadily. In the early years, talk of 100-digit moduli was common. One hundred-digit numbers, 322 bits, did not seem likely to be factored in the immediate future and, with the available computing techniques, systems with bigger moduli ran very slowly. Today, 100-digit numbers seem only just out of reach and there is little discussion of moduli smaller than 512 bits. Two hundred digits, 664 bits, is frequently mentioned, and Cylink has not only chosen to make its chip a comfortable 1028 bits, but also to allow up to 16 chips to be used in cascade. If this expansion has been pushed by advances in factoring, it has been made possible by advances in arithmetic.

Most of the computation done both in encryption and decryption and in the ancillary activity of manufacturing keys is exponentiation and each exponentiation, in turn, is made up of multiplications. Because, as discussed in Section 2 on exponential key exchange, numbers can be raised to powers in a small number of operations by repeated squaring, it is the speed of the underlying multiplication operation that is crucial.

According to Rivest [94] multiplication on a fixed-word-length processor takes time to proportional to the square of the length of the operands or $O(k^2)$. If dedicated serial/parallel hardware is constructed for the purpose, this time can be reduced to $O(k)$. In this case, the number of gates required is also proportional to the lengths of the operands, $O(k)$. The fastest implementations [15] run in time $O(\log k)$, but here the hardware requirements grow sharply to $O(k^2)$ gates.

10 DIRECTIONS IN PUBLIC KEY CRYPTOGRAPHY RESEARCH

Public key cryptography has followed a curious course. In its first 3 years, three systems were invented. One was broken; one has generally been considered impractical; and the third reigns alone as the irreplaceable basis for a new technology. Progress in producing new public key cryptosystems is stymied as is the complementary problem of proving the one system we have secure, or even of proving it equivalent to factoring in a useful way.

Stymied though it may be in its central problems, however, the theoretical side of public key cryptography is flourishing. This is perhaps because the public key problem changed the flavor of cryptography. It may be difficult to produce good conventional cryptosystems, but the difficulty is all below the surface. It is typically easier to construct a transformation that appears to satisfy the requirements of security than it is to show that a proposed system is no good. The result is a long development cycle ill suited to the give and take of academic research. Systems that even appear to exhibit the public key property, however, are difficult to find and this sort of difficulty is something the theoretical computer scientists can get their teeth into. The early taste of success that came with the development of RSA has inspired the search for solutions to other seemingly paradoxical problems and led to active exploration of a variety of new cryptographic disciplines.

This is not to say that contemporary cryptographic research is not motivated by application. A constant caution in conventional cryptography, for example, is that the strength of a cryptosystem in one mode of operation does not guarantee its strength in another. It is widely felt, for example, that a conventional block cryptosystem such as DES is a suitable component with which to implement other modes of operation, but no proofs have been offered. This burdens anyone who chooses the system as a building block with a separate certificational examination of every configuration in which it is to be used. One objective of research in public key cryptography has been to demonstrate the equivalence of many secondary cryptographic problems to those that define the strength of the system. Substantial progress has been made in proving that the strength of cryptographic protocols is equivalent to the strength of the RSA system and that the protection provided by RSA is uniform [4].

There is another sort of applied flavor to even the purest of cryptographic research—a search for ways of transplanting our current social and business mechanisms to a world in which communication is primarily telecommunication. The digital signature was the first great success in this direction, which can be characterized as asking: What can we do with paper, pencil, coins, and handshakes that would be hard to do without them? And, how can we do it without them?

In 1977, I gave a talk on the problem of developing a purely electronic analog of the registered mail receipt, in the current topics session of the International Symposium on Information Theory at Cornell. My message was pessimistic, arguing for both the importance and the intractability of the problem, but fortunately my pessimism was premature. A year and a half later, the MIT group published a paper entitled "Mental poker" [99]. It did not solve the problem of receipts for registered mail, but did show how to do something just as surprising: gamble over the telephone in a way that prevented either party from cheating without being discovered. This, as it turned out, was just the beginning.

To my delight, the problem of registered mail was rediscovered in Berkeley in 1982 as part of a larger category of problems that could be solved by *ping-pong protocols* and the emergence of this subject was one of the highlights of Crypto '82 [20]. Despite problems with protocols that were either broken or impossibly expensive [55], progress has been sufficient to provide hope that registered mail, contract signing, and related problems will one day have practical solutions.

In separate 1979 papers, G. R. Blakley at the University of Texas and Adi Shamir at MIT [11,100] opened yet another direction of investigation: how secret information can be divided among several people in such a way that any *k* of them, but no fewer, can recover it. Although this field of *secret sharing*, unlike that of ping-pong protocols, emerged full grown with provably correct and easily implementable protocols, it has been the subject of continuing examination [5,26,45,58] and is the subject of a chapter in this volume, "An Introduction to Shared Secret and/or Shared Control Schemes and Their Application," by G. J. Simmons.

David Chaum, currently at the Center for Mathematics and Computer Science in Amsterdam, has applied public key technology to a particularly challenging set of problems [21,22]. In a society dominated by telecommunication and computers, organizations ranging from credit bureaus to government agencies can build up dossiers on private citizens by comparing notes on the credentials issued to the citizens. This dossier building occurs without the citizens' knowledge or consent and, at present, the only protection against abuses of this power lies in legal regulation. Chaum has developed technical ways of permitting an individual to control the transfer of information about him from one organization to another. Without action on the part of an individual to whom credentials have been issued, no organization is able to link the information it holds about the individual with information in the databanks of any other organization. Nonetheless, the systems guarantee that no individual can forge organizational credentials. Chaum's techniques address problems as diverse as preventing spies from tracing messages through electronic mail networks [19,24] and protecting the privacy of participants in transactions with systems that recapture in electronic media both the assurance and the anonymity of cash [21].

The work drawing most attention at present is probably the field best known under the name of *zero-knowledge proofs* [49,50], though similar theories, based on different assumptions about the capabilities of the participants, have been developed independently [13,14,23]. One of the idea's originators, Silvio Micali at MIT, described it as "the inverse of a digital signature." A zero-knowledge proof permits Alice to demonstrate to Bob that she knows something, but gives him no way of conveying this assurance to anybody else. In the original example, Alice convinced Bob that she knew how to color a map with three colors, but gave him no information whatever about what the coloring was.

The view that a zero-knowledge proof is the inverse of a digital signature now seems ironic, because a form of challenge and response authentication, applicable to the signature problem, has become the best-known outgrowth of the field. In this system, the responder demonstrates to the challenger his knowledge of a secret number, without revealing any information about what the number is. Amos Fiat and Adi Shamir have recently brought forth an identification system of this sort, and announced a proof that breaking it is equivalent to factoring [47].

A purist might respond to all this by saying that having failed to solve the real problems in public key cryptography, cryptographers have turned aside to find other things about which to write papers. It is a situation that has been seen before in mathematics. At the end of the last century, mathematical analysis ground to a halt against intractable problems in Fourier theory, differential equations, and complex analysis. What many mathematicians did with their time while not solving the great problems was viewed with scorn by critics who spoke of the development of point set topology and abstract algebra as "soft mathematics." Only at mid-century did it become clear what had happened. In the abstractions a great hammer had been forged and through the fifties and sixties the classic problems began to fall under its blows. Perhaps cryptography will be equally lucky.

11 WHERE IS PUBLIC KEY CRYPTOGRAPHY GOING?

In just over 10 years public key cryptography has gone from a novel concept to a mainstay of cryptographic technology. It is soon to be implemented in hundreds of thousands of secure telephones and efforts are under way to apply the same mechanisms to data communications on a similar scale [97]. The outlook in the commercial world is equally bright. Digital signatures have been applied in retail electronic funds transfer technology in a British experiment with point of sale terminals [57]. The demand for public key is exemplified by a recent conference on smart cards in Vienna [111], where one question was heard over and over again: When will we have an RSA card?

Now that it has achieved acceptance, public key cryptography seems indispensable. In some ways, however, its technological base is disturbingly narrow. With the exception of the McEliece scheme and a cumbersome knapsack system devised explicitly to resist the known attacks [25], virtually all surviving public key cryptosystems and most of the more numerous signature systems employ exponentiation over products of primes. They are thus vulnerable to breakthroughs in factoring or discrete logarithms. Key exchange systems are slightly better off since they can use the arithmetic of primes, prime products, or Galois fields with 2^n elements and are thus sensitive to progress on the discrete logarithm problem only.

From the standpoint of conventional cryptography, with its diversity of systems, the narrowness bespeaks a worrisome fragility. This worry, however, is mitigated by two factors:

- The operations on which public key cryptography currently depends—multiplying, exponentiating, and factoring—are all fundamental arithmetic phenomena. They have been the subject of intense mathematical scrutiny for centuries and the increased attention that has resulted from their use in public key cryptosystems has on balance enhanced rather than diminished our confidence.
- Our ability to carry out large arithmetic computations has grown steadily and now permits us to implement our systems with numbers sufficient in size to be vulnerable only to a dramatic breakthrough in factoring, logarithms, or root extraction.

It is even possible that RSA and exponential key exchange will be with us indefinitely. The fundamental nature of exponentiation makes both good candidates for eventual proof of security and if complexity theory evolves to provide convincing evidence of the strength of either, it will establish a new paradigm for judging cryptographic mechanisms. Even if new systems were faster and had smaller keys, the current systems might never be superceded altogether.

Such proofs have yet to be found, however, and proposed schemes are continually presented at the cryptographic conferences [12,30,80,82,114]. Approaches include generalizing RSA to other rings and various attempts to replace exponentials with polynomials, but in general they have not fared well and some of their fates are discussed elsewhere in this book (see Ernest F. Brickell and Andrew M. Odlyzko, ''Cryptanalysis: A Survey of Recent Results''). So far, the goal of improving on the performance of RSA without decreasing its security has yet to be achieved.

An appealing idea that has been put forward by Stephen Wolfram and studied by Papua Guam [54] is the use of cellular automata. Guam's system is too new to have received careful scrutiny and superficial examination suggests that it may suffer a weakness similar to one seen in other cases [46]. Even should this effort fail, however, the cellular automaton approach is attractive. Cellular automata differ from such widely accepted cryptographic mechanisms as shift registers in that, even if they are invertible, it is not possible to calculate the predecessor of an arbitrary state by simply reversing the rule for finding the successor. This makes them a viable vehicle for trapdoors. Cellular automata also lend themselves to study of the randomness properties required of strong cryptographic systems [115].

What will be the outcome of such research? In an attempt to foresee the future of cryptography in 1979, I wrote [39]:

> Prospects for development of new and more efficient public key cryptographic systems by the latter part of the eighties are quite good. Public key cryptography is more successful today than algebraic coding theory was at the age of four. The major breakthroughs in that field did not begin till the latter part of its first decade, but then progressed rapidly. The similarity of the two fields is reason for optimism that . . . public key cryptography will follow a similar course.
>
> Increasing use of the available public key systems in the 1980s will spread awareness of both their advantages and the performance shortcomings of the early examples. The research response to this awareness will probably produce better public key systems in time for use during the first half of the nineties.

My schedule was clearly too optimistic. If there are public key cryptosystems with better performance or greater security waiting in the wings, they are proprietary systems that have yet to make even their existence known. Other aspects of the argument are closer to the mark, however. The use of public key cryptosystems has increased dramatically and with it awareness of their advantages. Judicious use of hybrid systems and improved arithmetic algorithms have reduced the "performance shortcomings" to the status of a nuisance in most applications and the biggest motivation for seeking new systems today is probably the desire not to have all our eggs in one basket. Unless the available systems suffer a cryptanalytic disaster, moreover, the very success of public key cryptography will delay the introduction of new ones until the equipment now going into the field becomes outmoded for other reasons.

For a discipline just entering its teens, the position of public key cryptography should be seen not as fragile, but as a strong one.

DEDICATION

To my wife Mary Lynn Fischer, my first discovery on the 2-year odyssey that led me to public key cryptography and without whom the others would probably never have been made.

REFERENCES

[1] L. M. Adleman and R. L. Rivest, "How to break the Lu-Lee (COMSTAT) public key cryptosystem," MIT Laboratory for Computer Science, July 24, 1979.

[2] L. M. Adleman, C. Pomerance, and R. S. Rumley, "On distinguishing prime numbers from composite numbers." *Ann. Math.*, vol. 117, no. 2, pp. 173–206, 1983.

[3] Alfred V. Aho, John E. Hopcroft, and Jeffrey D. Ullman, *The Design and Analysis of Computer Algorithms*, Reading, MA: Addison-Wesley, 1974.

[4] W. Alexi, B. Chor, O. Goldreich, and C. P. Schnor, "RSA/Rabin bits are $1/2 + 1/\text{poly}(\log N)$ secure," in *Proceedings of 25th Ann. IEEE Symp. Foundations Computer Sci.*, Singer Island, FL, Oct. 24–26, 1984, pp. 449–457. Los Angeles: IEEE Computer Society Press, 1984.

[5] C. Asmuth and J. Blum, "A modular approach to key safeguarding" *IEEE Trans. Inform. Theory*, vol. IT-29, pp. 208–210, March 1983.

[6] "Contractors ready low-cost, secure telephone for 1987 service start," *Aviat. Week Space Technol.* pp. 114–115, Jan. 20, 1986.

[7] C. Barney, "Cypher chip makes key distribution a snap," *Electronics,* vol. 59, no. 28, pp. 31–32, Aug. 7, 1986.

[8] J. Barron, *Breaking the Ring*, Boston: Houghton Mifflin, 1987.

[9] D. ben-Aaron, "Mailsafe signs, seals, and delivers files," *Information Week*, pp. 19, 22, Sept. 15, 1986.

[10] I. F. Blake, R. Fuji-Hara, R. C. Mullin, and S. A. Vanstone, "Computing logarithms in finite fields of characteristic two" *SIAM J. Alg. Disc. Methods,* vol. 5, no. 2, pp. 276–285, June 1984.

[11] G. R. Blakley, "Safeguarding cryptographic keys," in *Proceedings of AFIPS National Computer Conference,* New York, June 4–7, 1979, vol. 48, pp. 313–317.

[12] G. R. Blakley and D. Chaum (eds.), *Lecture Notes in Computer Science 196; Advances in Cryptology: Proc. Crypto'84,* Santa Barbara, CA, Aug. 19–22, 1984. Berlin: Springer-Verlag, 1985.

[13] G. Brassard and C. Crépeau, "Non-transitive transfer of confidence: A perfect zero-knowledge interactive protocol for SAT and beyond," in *Proceedings of 27th Ann. IEEE Symp. Foundations Computer Sci.,* Toronto, Canada, Oct. 27–29, 1986, pp. 188–195. Los Angeles: IEEE Computer Society Press, 1986.

[14] G. Brassard, C. Crépeau, and D. Chaum, "Minimum disclosure proofs of knowledge," *J. Computer and Sys. Sci.,* vol. 37, no. 2, pp. 156–189, Oct. 1988.

[15] E. F. Brickell, "A fast modular multiplication algorithm with application to two key cryptography," in *Advances in Cryptology: Proc. Crypto'82,* D. Chaum, R. L. Rivest, and A. T. Sherman, Eds., Santa Barbara, CA, Aug. 23–25, 1982, vol. 20, pp. 51–60. New York: Plenum Press, 1983.

[16] E. F. Brickell and G. J. Simmons, "A status report on knapsack based public key cryptosystems," *Congressus Numerantium,* vol. 7, pp. 3–72, 1983. The CCIS encryptor is mentioned on pp. 4–5.

[17] E. F. Brickell, "Breaking iterated knapsacks," in *Lecture Notes in Computer Science 196; Advances in Cryptology: Proc. Crypto'84,* G. R. Blakley and D. Chaum, Eds., Santa Barbara, CA, Aug. 19–22, 1984, vol. 12, pp. 342–358. Berlin: Springer-Verlag, 1985.

[18] D. Burnham, "NSA seeking 500,000 'secure' telephones," New York Times News Service, news release pp. 1–6, Oct. 6, 1984.

[19] D. L. Chaum, "Untraceable electronic mail, return addresses, and digital pseudonyms," *CACM,* vol. 24, no. 2, pp. 84–88, Feb. 1981.

[20] D. Chaum, R. L. Rivest, and A. T. Sherman, (eds.), *Advances in Cryptology: Proc. Crypto'82,* Santa Barbara, CA, Aug. 23–25, 1982. New York: Plenum Press, 1983.

[21] D. Chaum, "Security without identification: Transaction systems to make Big Brother obsolete," *CACM,* vol. 28, no. 10, pp. 1030–1044, Oct. 1985.

[22] D. Chaum and J.-H. Evertse, "A secure and privacy-protecting protocol for transmitting personal information between organizations," in *Lecture Notes in Computer Science 263; Advances in Cryptology: Proc. Crypto'86,* A. M. Odlyzko, Ed., Santa Barbara, CA, Aug. 11–15, 1986, vol. 80, pp. 118–167. Berlin: Springer-Verlag, 1987.

[23] D. Chaum, "Demonstrating that a public predicate can be satisfied without revealing any information about how," in *Lecture Notes in Computer Science 263; Advances in Cryptology; Proc. Crypto'86,* A. M. Odlyzko, Ed., Santa Barbara, CA, Aug. 11–15, 1986, pp. 195–199. Berlin: Springer-Verlag, 1987.

[24] D. Chaum, "The dining cryptographer's problem: Unconditional sender untraceability," *J. Cryptology,* vol. 1, no. 1, pp. 65–75, 1988.

[25] B. Chor and R. L. Rivest, "A knapsack type public-key cryptosystem based on arithmetic in finite fields," in *Lecture Notes in Computer Science 196; Advances in Cryptology; Proc. Crypto'84,* G. R. Blakley and D. Chaum, Eds., Santa Barbara, CA, Aug. 19–22, 1984, vol. 12, pp. 54–65. Berlin: Springer-Verlag, 1985.

[26] B. Chor, S. Goldwasser, S. Micali, and B. Awerbuch, "Verifiable secret sharing and achieving simultaneity in the presence of faults," in *Proceedings of 26th*

Ann. IEEE Symp. Foundations Computer Sci., Portland, OR, Oct. 21–23, 1985, pp. 383–395. Los Angeles; IEEE Computer Society Press, 1985.

[27] Testimony of David Earl Clark at the trial of Jerry Alfred Whitworth before Judge J. P. Vukasin, Jr., in the United States District Court, Northern District of California, Mar. 25, 1986. Reported by Vivian Pella Balboni, pp. 11–1345.

[28] D. Coppersmith, "Fast evaluation of logarithms in fields of characteristic two," *IEEE Trans. Inform. Theory,* vol. IT-30, pp. 587–594, 1984.

[29] D. Coppersmith, A. M. Odlyzko, and R. Schroeppel, "Discrete logarithms in GF(p)," *Algorithmica,* vol. 1, no. 1, pp. 1–16, 1986.

[30] T. Beth, N. Cot, and I. Ingemarsson (eds.), *Lecture Notes in Computer Science 209; Advances in Cryptology: Proc. Eurocrypt'84,* Paris, France, April 9–11, 1984. Berlin: Springer-Verlag, 1985.

[31] "Cidec-HS high speed DES encryption for digital networks," product description, Cylink Corporation, Sunnyvale, CA.

[32] "Key management development package," product description, Cylink Corporation, Sunnyvale, CA.

[33] D. W. Davies and W. L. Price, "The applications of digital signatures based on public key cryptosystems," National Physical Laboratory Report DNACS 39/80, Dec. 1980.

[34] J. A. Davis, D. B. Holdridge, and G. J. Simmons, "Status report on factoring (at the Sandia National Laboratories)," in *Lecture Notes in Computer Science 209; Advances in Cryptology: Proc. Eurocrypt'84,* T. Beth, N. Cot, and I. Ingemarsson, Eds., Paris, France, April 9–11, 1984, vol. 30, pp. 183–215. Berlin: Springer-Verlag, 1985.

[35] W. Diffie and M. E. Hellman, "Multiuser cryptographic techniques," in *Proceedings of AFIPS National Computer Conference,* New York, June 7–10, 1976, pp. 109–112.

[36] W. Diffie and M. E. Hellman, "New directions in cryptography," *IEEE Trans. Inform. Theory,* vol. IT-22, pp. 644–654, Nov. 1976.

[37] W. Diffie and M. E. Hellman, "Exhaustive cryptanalysis of the NBS data encryption standard," *Computer,* vol. 10, no. 6, pp. 74–84, June 1977.

[38] W. Diffie, "Conventional versus public key cryptosystems," in *Secure Communications and Asymmetric Cryptosystems,* AAAS Selected Symposium 69, G. J. Simmons, Ed., pp. 41–72. Boulder, CO: Westview Press, 1982. Rabin's system is discussed on p. 70; the relative strength of conventional and public key distribution, on pp. 64–66.

[39] W. Diffie, "Cryptographic technology: Fifteen-year forecast," in *Secure Communications and Asymmetric Cryptosystems,* AAAS Selected Symposium 69, G. J. Simmons, Ed., pp. 301–327. Boulder, CO: Westview Press, 1982.

[40] W. Diffie, "Securing the DoD transmission control protocol," in *Lecture Notes in Computer Science 218; Advances in Cryptology: Proc. Crypto'85,* H. C. Williams, Ed., Santa Barbara, CA, Aug. 18–22, 1985, pp. 108–127. Berlin: Springer-Verlag, 1986.

[41] W. Diffie, L. Strawczynski, B. O'Higgins, and D. Steer, "An ISDN secure telephone unit," in *Proceedings, National Communications Forum 1987,* Rosemont, IL, Sept. 28–30, vol. 41, book 1, pp. 473–477. Chicago: National Engineering Consortium, 1987.

[42] E. Dolnick, "N.M. scientist cracks code, wins $1000," *The Boston Globe*, Nov. 6, 1984.

[43] Electronic Industries Association, "Comsec and Compusec market study," Jan. 14, 1987.

[44] "Encryption algorithm for computer data protection" *Fed. Register,* vol. 40, no. 52, pp. 12134–12139, March 17, 1975.

[45] P. Feldman, "A practical scheme for non-interactive verifiable secret sharing," in *Proceedings of 28th Ann. IEEE Symp. Foundations Computer Sci.,* Los Angeles, CA, Oct. 12–14, 1987, pp. 427–437. Los Angeles: IEEE Computer Society Press, 1987.

[46] H. Fell and W. Diffie, "Analysis of a public key approach based on polynomial substitution," in *Lecture Notes in Computer Science 218; Advances in Cryptology: Proc. Crypto'85,* H. C. Williams, Ed., Santa Barbara, CA, Aug. 18–22, 1985, pp. 108–127. Berlin: Springer-Verlag, 1986.

[47] A. Fiat and A. Shamir, "How to prove yourself: Practical solutions to identification and signature problems," in *Lecture Notes in Computer Science 263; Advances in Cryptology: Proc. Crypto'86,* A. M. Odlyzko, Ed., Santa Barbara, CA, Aug. 11–15, 1986, pp. 186–212. Berlin: Springer-Verlag, 1987.

[48] M. Gardner, "A new kind of cipher that would take millions of years to break," *Scientific American,* pp. 120–124, Aug. 1977.

[49] O. Goldreich, S. Micali, and A. Wigderson, "Proofs that yield nothing but their validity and a methodology of cryptographic protocol design," in *Proceedings of 27th Ann. IEEE Symp. Foundations Computer Sci.,* Toronto, Canada, Oct. 27–29, 1986, pp. 174–187. Los Angeles: IEEE Computer Society Press, 1986.

[50] S. Goldwasser, S. Micali, and C. Rackoff, "Knowledge complexity of interactive proofs," in *Proceedings of 17th Symp. Theory of Computing,* Providence, RI, May 6–8, 1985, pp. 291–304. New York: Association for Computing Machinery, 1985.

[51] S. Goldwasser and J. Killian, "All primes can be quickly certified," in *Proceedings of 18th Symp. Theory of Computing,* Berkeley, CA, May 28–30, 1986, pp. 316–329. New York: Association for Computing Machinery, 1985.

[52] J. Gordon, "Strong primes are easy to find," in *Lecture Notes in Computer Science 209; Advances in Cryptology: Proc. Eurocrypt'84,* T. Beth, N. Cot, and I. Ingemarsson, Eds., Paris, France, April 9–11, 1984, pp. 215–223. Berlin: Springer-Verlag, 1985.

[53] J. Gordon, afterdinner speech at the Zurich Seminar, 1984. In this lecture, which has unfortunately never been published, Gordon assembled the facts of Alice and Bob's precarious lives, which had previously been available only as scattered references in the literature.

[54] P. Guam, "Cellular automaton public key cryptosystem," *Complex Systems,* vol. 1, pp. 51–56, 1987.

[55] J. Hastad and A. Shamir, "The cryptographic security of truncated linearly related variables," in *Proceedings of 17th Symp. Theory of Computing,* Providence, RI, May 6–8, 1985, pp. 356–362. New York: Association for Computing Machinery, 1985.

[56] D. Helwig, "Coding chip devised in Waterloo," *The Globe and Mail,* Toronto, Canada, Jan. 1, 1987.

[57] "National EFT POS to use public key cryptography," *Information Security Monitor,* vol. 2, no. 12, p. 1, Nov. 1987.

[58] M. Ito, A. Saito, and T. Nishizeki, "Secret sharing scheme realizing general access structure," in *Proceedings, Globecom '87,* Tokyo, Nov. 15–18, 1987, pp. 361–364. New York: IEEE.

[59] C. S. Kline and G. J. Popek, "Public key vs. conventional key encryption," in *Proceedings of AFIPS National Computer Conference,* New York, June 4–7, 1979, pp. 831–837.

[60] D. Knuth, "Semi-numerical algorithms," in *The Art of Computer Programming,* 2d ed., vol. 2, pp. 316–336. Reading, MA: Addison-Wesley, 1981.

[61] N. Koblitz, *A Course in Number Theory and Cryptography,* New York: Springer-Verlag, 1987.

[62] L. M. Kohnfelder, "Toward a practical public key cryptosystem," bachelor's thesis, MIT Department of Electrical Engineering, May 1978.

[63] R. Kopeck, "T1 encryption plan protects data," *PC Week,* vol. 4, no. 9, pp. C9–C10, March 3, 1987.

[64] J. Kowalchuk, B. P. Schanning, and S. Powers, "Communication privacy: Integration of public and secret key cryptography," in *Proceedings of National Telecommunications Conf.,* Houston, TX, Nov. 30–Dec. 4, 1980, pp. 49.1.1–5. New York: IEEE Press.

[65] E. Kranakis, *Primality and Cryptography.* Stuttgart: B. G. Teubner; New York: Wiley, 1986.

[66] R. Lindsey, *The Falcon and the Snowman,* New York: Simon and Schuster, 1979.

[67] S. Lu and L. Lee, "A simple and effective public key cryptosystem," *Comsat Tech. Rev.,* vol. 9, no. 1, Spring 1979.

[68] K. S. McCurley, "A key distribution system equivalent to factoring," Department of Mathematics, University of Southern California, June 3, 1987.

[69] R. J. McEliece, "A public key cryptosystem based on algebraic coding theory" JPL DSN Progress Report 42–44, pp. 114–116, Jan.–Feb. 1978.

[70] R. Merkle, "Secure communication over insecure channels" *CACM,* pp. 294–299, April 1978.

[71] R. C. Merkle and M. E. Hellman, "Hiding information and signatures in trap door knapsacks" *IEEE Trans. Inform. Theory,* vol. IT-24, pp. 525–530, Sept. 1978.

[72] R. Merkle, Letters to the editor, *Time Magazine,* vol. 120, no 20, p. 8, Nov. 15, 1982.

[73] M. A. Morrison and J. Brillhart, "A method of factoring and the factorization of F_7" *Math. Comp.,* vol. 29, pp. 18–205, 1975.

[74] "Advanced techniques in network security," Motorola Government Electronics Division, Scottsdale, AZ, 1977.

[75] F. H. Myers, "A data link encryption system" in *Proceedings of National Telecommunications Conf.,* Washington, D.C., Nov. 27–29, 1979, pp. 4.5.1–4.5.8.

[76] R. M. Needham and M. D. Schroeder, "Using encryption for authentication in large networks of computers," *CACM, vol. 21, pp. 993–999, Dec. 1978.*

[77] L. Neuwirth, "A comparison of four key distribution methods," *Telecommunications,* vol. 20, no. 7, pp. 110–111, 114–115, 135, July 1986.

[78] "Statement of Lee Neuwirth of Cylink on HR145," submitted to Congressional committees considering HR145, Feb. 1987.

[79] A. M. Odlyzko, "Discrete logarithms in finite fields and their cryptographic significance," in *Lecture Notes in Computer Science 209; Advances in Cryptology: Proc. Eurocrypt'84*, T. Beth, N. Cot, and I. Ingemarsson, Eds., Paris, France, April 9–11, 1984, pp. 225–314. Berlin: Springer-Verlag, 1985.

[80] A. M. Odlyzko (ed.), *Lecture Notes in Computer Science 263; Advances in Cryptology: Proc. Crypto'86*, Santa Barbara, CA, Aug. 11–15, 1986. Berlin: Springer-Verlag, 1987.

[81] B. O'Higgins, W. Diffie, L. Strawczynski, and R. de Hoog, "Encryption and ISDN—a natural fit," in *Proceedings International Switching Symp.*, Phoenix, AZ, March 16–20, 1987, pp. A11.4.1–7.

[82] F. Pichler (ed.), *Lecture Notes in Computer Science 219; Advances in Cryptology: Proc. Eurocrypt'85*, Linz, Austria, April 1985. Berlin: Springer-Verlag, 1986.

[83] S. C. Pohlig and M. E. Hellman, "An improved algorithm for computing logarithms in GF(p) and its cryptographic significance," *IEEE Trans. Inform. Theory*, vol. IT-24, pp. 106–110, Jan. 1978.

[84] C. Pomerance, "Recent developments in primality testing," in *Mathematical Intelligencer*, vol. 3, no. 3, pp. 97–105, 1981.

[85] C. Pomerance, J. W. Smith, and R. Tuler, "A pipe-line architecture for factoring large integers with the quadratic sieve algorithm," *SIAM J. Computing*, vol. 17, no. 2 (special issue on cryptography), pp. 387–403, April 1988.

[86] M. O. Rabin, "Digitalized signatures and public-key functions as intractable as factorization," MIT Laboratory for Computer Science, MIT/LCS/TR-212, Jan. 1979.

[87] "Datacryptor II, public key management option," Racal-Milgo, Sunrise, FL. 1981.

[88] "AT&T readying new spy-proof phone for big military and civilian markets," *The Report on AT&T*, pp. 6–7, June 2, 1986.

[89] R. F. Rieden, J. B. Snyder, R. J. Widman, and W. J. Barnard, "A two-chip implementation of the RSA public encryption algorithm," GOMAC (Government Microcircuit Applications Conference), Orlando, FL, pp. 24–27, Nov. 1982.

[90] R. L. Rivest, A. Shamir, and L. Adleman, "On digital signatures and public key cryptosystems," MIT Laboratory for Computer Science, MIT/LCS/TR-212, Jan. 1979.

[91] R. L. Rivest, A. Shamir, and L. Adleman, "A method for obtaining digital signatures and public key cryptosystems," *CACM*, vol. 21 no. 2, pp. 120–126, Feb. 1978.

[92] R. Rivest, personal communication with H. C. Williams, in H. C. Williams, "A modification of the RSA public-key cryptosystem," *IEEE Trans. Inform. Theory*, Vol. IT-26, no. 6, p. 729, Nov. 1980.

[93] R. L. Rivest, "A description of a single-chip implementation of the RSA cipher," *Lambda*, vol. 1, no. 3, pp. 14–18, Fall 1980.

[94] R. Rivest, "RSA chips (past/present/future)," in *Lecture Notes in Computer Science 209; Advances in Cryptology: Proc. Eurocrypt'84*, T. Beth, N. Cot, and I. Ingemarsson, Eds., Paris, France, April 9–11, 1984, pp. 159–165. Berlin: Springer-Verlag, 1985.

[95] H. L. Rogers, "An overview of the caneware program," paper 31 in *Proceedings of 3rd Ann. Symp. Physical/Electronic Security*, August 1987. Philadelphia: Armed Forces Communications and Electronics Association, 1987.

[96] "Toward a new factoring record," *Science News,* vol. 133, no. 4, p. 62, Jan. 23, 1988.

[97] "SDNS: A network on implementation," in *Proceedings 10th National Computer Security Conf.,* Baltimore, MD, Sept. 21–24, 1987, pp. 150–174. Session containing six papers on the Secure Data Network System.

[98] A. Shamir, "A fast signature scheme," MIT Laboratory for Computer Science, Technical Memorandum, MIT/LCS/TM-107, July 1978.

[99] A. Shamir, R. L. Rivest, and L. M. Adleman, "Mental poker," MIT Laboratory for Computer Science, Technical Memorandum, MIT/LCS/TM-125, Jan. 29, 1979.

[100] A. Shamir, "How to share a secret," *CACM,* vol. 22, no. 11, pp. 612–613, Nov. 1979.

[101] A. Shamir, "A polynomial time algorithm for breaking Merkle–Hellman cryptosystems (extended abstract)," research announcement, preliminary draft, Applied Mathematics, Weizmann Institute, Rehovot, Israel, April 20, 1982. This paper appears with a slightly different title: "A polynomial time algorithm for breaking the basic Merkle–Hellman cryptosystem (extended abstract)," in *Advances in Cryptology: Proc. Crypto'82,* D. Chaum, R. L. Rivest, and A. T. Sherman, Eds., Santa Barbara, CA, Aug. 23–25, 1982, pp. 279–288. New York: Plenum Press, 1983.

[102] A. Shamir, "A polynomial time algorithm for breaking the basic Merkle-Hellman cryptosystem," *IEEE Trans. Inform. Theory,* vol. IT-30, no. 5, pp. 699–704, Sept. 1984.

[103] Z. Shmuely, "Composite Diffie-Hellman public-key generating systems are hard to break," Computer Science Department, Technion, Haifa, Israel, Technical Report 356, Feb. 1985.

[104] R. Silver, "The computation of indices modulo P" Mitre Corporation, Working Paper WP-07062, p. 3, May 7, 1964.

[105] G. J. Simmons and M. J. Norris, "Preliminary comments on the M.I.T. public key cryptosystem," *Cryptologia,* vol. 1, 1977, pp. 406–414.

[106] G. J. Simmons, "Authentication without secrecy: A secure communications problem uniquely solvable by asymmetric encryption techniques," IEEE EASCON '79, Washington, DC, Oct. 9–11, 1979, pp. 661–662.

[107] G. J. Simmons and M. J. Norris, "How to cipher faster using redundant number systems" Sandia National Laboratories, SAND-80-1886, Aug. 1980.

[108] G. J. Simmons, "High speed arithmetic utilizing redundant number systems," in *Proceedings of National Telecommunications Conf.,* Houston, TX, Nov. 30–Dec. 4, 1980, pp. 49.3.1–2.

[109] G. J. Simmons (ed.), *Secure Communications and Asymmetric Cryptosystems,* AAAS Selected Symposium 69, Boulder, CO: Westview Press, 1982.

[110] G. J. Simmons, "Cryptology," *Encyclopedia Britannica,* 16th ed., 1986, pp. 913–924B.

[111] *Proceedings of Smart Card 2000,* Vienna, Oct. 19–20, 1988.

[112] M. V. Wilkes, *Time-Sharing Computer Systems,* New York: American Elsevier, 1972.

[113] H. C. Williams, "A modification of the RSA public-key cryptosystem," *IEEE Trans. Inform. Theory,* vol. IT-26 no. 6, pp. 726–729, Nov. 1980.

[114] H. C. Williams (ed.), *Lecture Notes in Computer Science 218; Advances in Cryptology: Proc. Crypto'85,* Santa Barbara, CA, Aug. 18–22, 1985. Berlin: Springer-Verlag, 1986.

[115] S. Wolfram, "Cryptography with cellular automata," in *Lecture Notes in Computer Science 218; Advances in Cryptology: Proc. Crypto'85,* H. C. Williams, Ed., Santa Barbara, CA, Aug. 18–22, 1985, vol. 114, pp. 429–432. Berlin: Springer-Verlag, 1986.

[116] M. C. Wunderlich, "Recent advances in the design and implementation of large integer factorization algorithms," in *Proceedings of 1983 Symp. Security and Privacy,* R. Blakley and D. Denning, Eds., Oakland, CA, April 25–27, 1983, pp. 67–71. Los Angeles: IEEE Computer Society Press, 1983.

[117] K. Yiu and K. Peterson, "A single-chip VLSI implementation of the discrete exponential public key distribution system," in *Proceedings of GOMAC (Government Microcircuit Applications Conference),* Orlando, FL, Nov. 1982, pp. 18–23. Washington, D.C.: Department of Defense.

CHAPTER 4

Public Key Cryptography

JAMES NECHVATAL
National Institute of Standards and Technology
Gaithersburg, Maryland 20899

INTRODUCTION

This chapter presents a state-of-the-art survey of public key cryptography circa 1988–1990. In doing so, it covers a number of different topics including:

- The theory of public key cryptography
- Comparisons to conventional (secret key) cryptography
- A largely self-contained summary of relevant mathematics
- A survey of major existing public key systems
- An exploration of hash functions
- A survey of public key implementations in networks
- An introduction to zero-knowledge computing and probabilistic encryption
- An exploration of security issues and key sizes

In some of these areas there is significant overlap with material covered by other contributors to this volume, notably the coverage of the basic concepts of public key cryptography appearing in the chapters by J. L. Massey and by W. Diffie and of hashing functions in the chapter on digital signatures by C. Mitchell, F. Piper and P. Wild. To make each author's presentation be as complete and cohesive as possible, however, a deliberate decision was made by the editor to tolerate a reasonable degree of repetition of essential notions between the various chapters to make it possible for a reader to refer to any chapter as an essentially self-contained treatise on its subject.

It should be pointed out that the treatment of public key cryptography given in this chapter is more comprehensive than any of the expositions listed in the references. Most of the latter treat either the theory or specific systems and implementations, but not both. The viewpoint here is that the theory and practice are inseparable.

The selection of cryptosystems and hash functions mentioned in this chapter serve only to provide examples. The focus is on issues such as criteria for systems and protocols for use. These are presumably long-term, in contrast, for example, to the set of existing public key systems which is more volatile. Thus we provide information that will hopefully be of use to implementors of systems, but the frameworks we develop are versatile enough to be relevant in a variety of settings. The latter may include, for example, both electronic mail systems and electronic funds transfer systems.

The core of this exposition is Sections 1–5. Sections 1–3 cover the fundamentals of public key cryptography and the related topics of hash functions and digital signatures. The reader may also wish to refer to the chapter on these same topics by C. Mitchell, F. Piper and P. Wild for additional information about these very important (to applications) topics. Section 4 gives some examples of public key systems and hash functions. Section 5 gives some examples of actual or proposed implementations of public key cryptography.

Section 6 gives a sample proposal for a local area network (LAN) implementation of public key cryptography; it draws heavily on the work of the International Standards Organization (ISO) and the Internet Activities Board (IAB) Privacy Task Force which is summarized in Section 5. Section 7 discusses some theoretical issues related to security and choice of key size. Section 8 gives a very brief introduction to zero-knowledge protocols. Section 9 treats probabilistic encryption and identity-based public key systems.

A variety of topics is covered in the appendices, including a summary of relevant mathematics and algorithms. In the following sections, letters refer to appendices; for example, Lemma G.2.1 refers to a lemma appearing in Section 2 of Appendix G.

The author wishes to thank Dr. Ronald L. Rivest for providing many comments and suggestions, and Dr. Burton S. Kaliski, Jr., for providing information on implementations of the Rivest–Shamir-Adleman (RSA) public key system.

1 CRYPTOSYSTEMS AND CRYPTANALYSIS

Cryptography deals with the transformation of ordinary text (plaintext) into coded form (ciphertext) by encryption, and transformation of ciphertext into plaintext by decryption. Normally these transformations are parameterized by one or more keys. The motive for encrypting text is security for transmissions over insecure channels.

Three of the most important services provided by cryptosystems are secrecy, authenticity, and integrity. *Secrecy* refers to denial of access to information by unauthorized individuals. *Authenticity* refers to validating the source of a message; that is, that it was transmitted by a properly identified sender and is not a replay of a previously transmitted message. *Integrity* refers to assurance that a message was not modified accidentally or deliberately in transit, by replacement, insertion, or deletion. A fourth service that may be provided is nonrepudiation of origin, that is, protection against a sender of a message later denying transmission. Variants of these services, and other services, are discussed in [ISO-87].

Classic cryptography deals mainly with the secrecy aspect. It also treats keys as secret. In the past 15 years two new trends have emerged:

1. Authenticity as a consideration that rivals and sometimes exceeds secrecy in importance.
2. The notion that some key material need not be secret.

The first trend has arisen in connection with applications such as electronic mail systems and electronic funds transfers. In such settings an electronic equivalent of the handwritten signature may be desirable. Also, intruders into a system often gain entry by masquerading as legitimate users; cryptography presents an alternative to password systems for access control.

The second trend addresses the difficulties that have traditionally accompanied the management of secret keys. This may entail the use of couriers or other costly, inefficient, and not really secure methods. In contrast, if keys are public the task of key management may be substantially simplified.

An ideal system might solve all of these problems concurrently, that is, using public keys, providing secrecy, and providing authenticity. Unfortunately no single technique proposed to date has met all three criteria. Conventional systems such as the Data Encryption Standard (DES) require management of secret keys; systems using public key components may provide authenticity but are inefficient for bulk encryption of data due to low bandwidths.

Fortunately, conventional and public key systems are not mutually exclusive; in fact they can complement each other. Public key systems can be used for signatures and also for the distribution of keys used in systems such as DES. Thus it is possible to construct hybrids of conventional and public key systems that can meet all of the above goals: secrecy, authenticity, and ease of key management.

For surveys of the preceding and related topics see, for example, ([BRAS88], [COPP87], [DENN83], [DIFF82], [DIFF79], [KLIN79], [KONH81], [LEMP79], [MASS88], [MERK82], [POPE78], [POPE79], [SALO85], [SIMM79]). More specialized discussions of public key cryptography are given, for example, in ([DIFF88], [LAKS83], [MERK82b]). Mathematical aspects are covered, for example, in ([KRAN86], [PATT87]).

In the following, E and D represent encryption and decryption transformations, respectively. It is always required that $D(E(M)) = M$. It may also be the case that $E(D(M)) = M$; in this event E or D can be employed for encryption. Normally D is assumed to be secret, but E may be public. In addition it may be assumed that E and D are relatively easy to compute when they are known.

1.1 Requirements for Secrecy

Secrecy requires that a cryptanalyst (i.e., a would-be intruder into a cryptosystem) should not be able to determine the plaintext corresponding to given ciphertext, and should not be able to reconstruct D by examining ciphertext for known plaintext. This translates into two requirements for a cryptosystem to provide secrecy:

1. A cryptanalyst should not be able to determine M from $E(M)$; that is, the cryptosystem should be immune to ciphertext-only attacks.

2. A cryptanalyst should not be able to determine D given $\{E(M_i)\}$ for any sequence of plaintexts $\{M_1, M_2, \ldots\}$; that is, the cryptosystem should be immune to known-plaintext attacks. This should remain true even when the cryptanalyst can choose $\{M_i\}$ (chosen-plaintext attack), including the case in which the cryptanalyst can inspect $\{E(M_1), \ldots, E(M_j)\}$ before specifying M_{j+1} (adaptive chosen-plaintext attack).

To illustrate the difference between these two categories, we use two examples. First, suppose $E(M) = M^3 \mod N$, $N = p * q$, where p and q are large secret primes. Then (see Section 4) it is infeasible for a cryptanalyst to determine D, even after inspecting numerous pairs of the form $\{M, E(M)\}$. However, an eavesdropper who intercepts $E(M) = 8$ can conclude $M = 2$. Thus a ciphertext-only attack may be feasible in an instance where known- or chosen-plaintext attack is not useful.

On the other hand, suppose $E(M) = 5M \mod N$ where N is secret. Then interception of $E(M)$ would not reveal M or N; this would remain true even if several ciphertexts were intercepted. However, an intruder who learns that $E(12) = 3$ and $E(16) = 4$ could conclude $N = 19$. Thus a known- or chosen-plaintext attack may succeed where a ciphertext-only attack fails.

Deficiencies in (1), that is, vulnerability to ciphertext-only attack, can frequently be corrected by slight modifications of the encoding scheme, as in the $M^3 \mod N$ encoding above. Adaptive chosen-plaintext is often regarded as the strongest attack.

Secrecy ensures that decryption of messages is infeasible. However, the enciphering transformation E is not covered by the above requirements; it could even be public. Thus secrecy, per se, leaves open the possibility that an intruder could masquerade as a legitimate user, or could compromise the integrity of a message by altering it. That is, secrecy does not imply authenticity/integrity.

1.2 Requirements for Authenticity and Integrity

Authenticity requires that an intruder should not be able to masquerade as a legitimate user of a system. Integrity requires that an intruder should not be able to substitute false ciphertext for legitimate ciphertext. Two minimal requirements should be met for a cryptosystem to provide these services:

1. It should be possible for the recipient of a message to ascertain its origin.
2. It should be possible for the recipient of a message to verify that it has not been modified in transit.

These requirements are independent of secrecy. For example, a message M could be encoded by using D instead of E. Then assuming D is secret, the recipient of $C = D(M)$ is assured that this message was not generated by an intruder. However, E might be public; C could then be decoded by anyone intercepting it.

A related service which may be provided is nonrepudiation; that is, we may add a third requirement if desired:

3. A sender should not be able to deny later that he sent a message.

We might also wish to add:

4. It should be possible for the recipient of a message to detect whether it is a replay of a previous transmission.

1.3 Conventional Systems

In a conventional cryptosystem E and D are parameterized by a single key K, so that we have $D_K(E_K(M)) = M$. It is often the case that the algorithms for obtaining D_K and E_K from K are public, although both E_K and D_K are secret. In this event the security of a conventional system depends entirely on keeping K a secret. Then secrecy and authenticity are both provided: If two parties share a secret K, they can send messages to one another that are both private (since an eavesdropper cannot compute $D_K(C)$) and authenticated (since a would-be masquerader cannot compute $E_K(M)$). In some cases (e.g., transmission of a random bit string), this does not assure integrity; that is, modification of a message en route may be undetected. Typically integrity is provided by sending a compressed form of the message (a message digest) along with the full message as a check.

Conventional systems are also known as one-key or symmetric systems [SIMM79].

1.4 Example of a Conventional Cipher: DES

The most notable example of a conventional cryptosystem is DES. It is well-documented (e.g., [DENN83], [EHRS78], [NATI77], [NATI80], [NATI81], [SMID81], [SMID88b]) and will not be discussed in detail here, except to contrast it with other ciphers. It is a block cipher, operating on 64-bit blocks using a 56-bit key. Essentially the same algorithm is used to encipher or decipher. The transformation employed can be written $P^{-1}(F(P(M)))$, where P is a certain permutation and F is a certain function that combines permutation and substitution. Substitution is accomplished via table lookups in so-called S-boxes.

The important characteristics of DES from the standpoint of this exposition are its one-key feature and the nature of the operations performed during encryption and decryption. Both permutations and table lookups are easily implemented, especially in hardware. Thus encryption rates exceeding 40M bits/sec have been obtained (e.g., see [BANE82], [MACM81]). This makes DES an efficient bulk encryptor, especially when implemented in hardware.

The security of DES is produced in classic fashion: alternation of substitutions and permutations. The function F is obtained by cascading a certain function $f(x, y)$, where x is 32 bits and y is 48 bits. Each stage of the cascade is called a round. A sequence of 48-bit strings $\{K_i\}$ is generated from the key. Let $L(x)$ and $R(x)$ denote the left and right halves of x, and let XOR denote exclusive-or. Then if M_i denotes the output of stage i, we have

$$L(M_i) = R(M_{i-1})$$
$$R(M_i) = L(M_{i-1}) \text{ XOR } f(L(M_i), K_i)$$

The hope is that after 16 rounds, the output will be statistically flat; that is, all patterns in the initial data will be undetectable.

1.5 Another Conventional Cipher: Exponentiation

Pohlig and Hellman [POHL78] noted a type of cipher that deviated from the classic methods such as transposition and substitution. Their technique was conceptually much simpler. Let GCD denote greatest common divisor (Appendix G). Suppose $p > 2$ is a prime and suppose K is a key in $[1, p - 1)$ with $GCD(K, p - 1) = 1$ (i.e., K is relatively prime to $p - 1$). If M is plaintext in $[1, p - 1]$, an encryption function E may be defined by

$$E(M) = M^K \bmod p$$

Now the condition $GCD(K, p - 1) = 1$ implies (Lemma G.1) that we can find I with $I * K \equiv 1 \pmod{p - 1}$. Note that I is not a separate key; I is easily derived from K or vice versa (Appendix H). We may set

$$D(C) = C^I \bmod p$$

It may then be shown (Corollary G.2.3) that $D(E(M)) = M$, as required. Furthermore, $E(D(C)) = C$ as well; that is, E and D are inverse functions. This makes exponentiation a versatile cipher; in particular, we can encipher with D as well as E. Later we will note that this can be useful in authentication. However, Pohlig and Hellman used the equation above only as an example of a conventional cryptosystem. That is, since I is easily derived from K, both E and D are generated by the single, secret, shared key K. In Section 4.1 we will see that if p were replaced by a product of two primes, derivation of I from K would be nontrivial. This would cause the key material to be split into two portions.

Despite the relative simplicity of the definitions of E and D, they are not as easy to compute as their counterparts in DES. This is because exponentiation mod p is a more time-consuming operation than permutations or table lookups if p is large.

Security of the system in fact requires that p be large. This is because K should be nondeterminable even in the event of a known-plaintext attack. Suppose a cryptanalyst knows p, M, C, and furthermore knows that $C = M^K \bmod p$ for some K. He should still be unable to find K; that is, he should not be able to compute discrete logarithms modulo p. At present there are no efficient algorithms for the latter operation: the best techniques now available take time (e.g., [ADLE79], [COPP86])

$$\exp(c((\log p)(\log \log p))^{1/2})$$

Proceeding mod p is equivalent to using the Galois field $GF(p)$; it is possible to use $GF(p^n)$ instead (Appendix G). There are both advantages and disadvantages to the extension. Arithmetic in $GF(p^n)$ is generally easier if $n > 1$, and especially so in $GF(2^n)$. On the other hand, taking discrete logarithms in $GF(p^n)$ is also easier. The case of small n is treated in [ELGA85b]. The greatest progress has been made for the case $p = 2$ (e.g., [BLAK84], [COPP84]). In [COPP84] it is shown that discrete logarithms in $GF(2^n)$ can be computed in time

$$\exp(c * n^{1/3} * (\log n)^{2/3})$$

For a survey of the discrete logarithm problem see, for example, [ADLE86], [COPP87], [MCCU89], [ODLY84b].

1.6 Public Key Cryptosystems

The notion of public key cryptography was introduced by Diffie and Hellman [DIFF76b]; for a history see [DIFF88]. Public key systems, also called two-key or asymmetric systems [SIMM79], differ from conventional systems in that there is no longer a single secret key shared by a pair of users. Rather, each user has his own key material. Furthermore, the key material of each user is divided into two portions, a private component and a public component. The public component generates a public transformation E, and the private component generates a private transformation D. In analogy to the conventional case, E and D might be termed encryption and decryption functions, respectively. However, this is imprecise: In a given system we may have $D(E(M)) = M$, $E(D(M)) = M$, or both.

A requirement is that E must be a trapdoor one-way function. *One-way* refers to the fact that E should be easy to compute from the public key material but hard to invert unless one possesses the corresponding D, or equivalently, the private key material generating D. The private component thus yields a "trapdoor" which makes the problem of inverting E seem difficult from the point of view of the cryptanalyst, but easy for the (sole legitimate) possessor of D. For example, a trapdoor may be the knowledge of the factorization of an integer (see Section 4.1).

We remark that the trapdoor functions employed as public transformations in public key systems are only a subclass of the class of one-way functions. The more general case will be discussed in Section 3.2.2.

We note also that public/private dichotomy of E and D in public key systems has no analogue in a conventional cryptosystem: In the latter, both E_K and D_K are parameterized by a single key K. Hence if E_K is known then it may be assumed that K has been compromised, whence it may also be assumed that D_K is also known, or vice versa. For example, in DES, both E and D are computed essentially by the same public algorithm from a common key; so E and D are both known or both unknown, depending on whether the key has been compromised.

1.6.1 Secrecy and authenticity.
To support secrecy, the transformations of a public key system must satisfy $D(E(M)) = M$. Suppose A wishes to send a secure message M to B. Then A must have access to E_B, the public transformation of B (note that subscripts refer to users rather than keys in this context). Now A encrypts M via $C = E_B(M)$ and sends C to B. On receipt, B employs his private transformation D_B for decryption; that is, B computes $D_B(C) = D_B(E_B(M)) = M$. If A's transmission is overheard, the intruder cannot decrypt C since D_B is private. Thus secrecy is ensured. However, presumably anyone can access E_B; B has no way of knowing the identity of the sender per se. Also, A's transmission could have been altered. Thus authenticity and integrity are not assured.

To support authentication and integrity, the transformations in a public key system must satisfy $E(D(M)) = M$. Suppose A wishes to send an authenticated message M to B. That is, B is to be able to verify that the message was sent by A and was not altered. In this case A could use his private transformation D_A to compute $C = D_A(M)$ and send C to B. That is, A employs D_A as a de facto encryption function. Now B can use A's public transformation E_A to find $E_A(C) = E_A(D_A(M)) = M$; that is, E_A acts as a de facto decryption function. Assuming M is valid plaintext, B knows that C was in fact sent by A, and was not altered in transit. This follows from the one-way nature of E_A: if a cryptanalyst, starting with a message M, could find C' such that $E_A(C') = M$, this would imply that he can invert E_A, a contradiction.

If M, or any portion of M, is a random string, then it may be difficult for B to ascertain that C is authentic and unaltered merely by examining $E_A(C)$. Actually, however, a slightly more complex procedure is generally employed: an auxiliary public function H is used to produce a much smaller message $S = D_A(H(M))$ that A sends to B along with M. On receipt B can compute $H(M)$ directly. The latter may be checked against $E_A(S)$ to ensure authenticity and integrity, since once again the ability of a cryptanalyst to find a valid S' for a given M would violate the one-way nature of E_A. Actually H must also be one-way; we return to this subject in Section 3.2.

Sending C or S above ensures authenticity, but secrecy is nonexistent. In the second scheme M was sent in the clear along with S; in the first scheme, an intruder who intercepts $C = D_A(M)$ presumably has access to E_A and hence can compute $M = E_A(C)$. Thus in either case M is accessible to an eavesdropper.

It may be necessary to use a combination of systems to provide secrecy, authenticity, and integrity. However, in some cases it is possible to employ the same public key system for these services simultaneously. We note that for authenticity and integrity purposes, D is regarded as an encryptor; for secrecy, E is the encryptor. If the same public key system is to be used in both cases, then $D(E(M)) = M$ and $E(D(M)) = M$ must both hold; that is, D and E are inverse functions. A requirement is that the plaintext space (i.e., the domain of E) must be the same as the ciphertext space (i.e., the domain of D).

Suppose that in addition to E and D being inverses for each user, for each pair of users A and B the functions E_A, D_A, E_B, and D_B all have a common domain. Then both secrecy and authenticity can be accomplished with a single transmission: A sends $C = E_B(D_A(M))$ to B; then B computes $E_A(D_B(C)) = E_A(D_A(M)) = M$. An intruder cannot decrypt C since he lacks D_B; hence secrecy is assured. If the intruder sends C' instead of C, C' cannot produce a valid M since D_A is needed to produce a valid C. This assures authenticity.

In actuality there are no common systems versatile enough for the last usage. In fact there is only one major system (Rivest–Shamir–Adelman[RSA]) that satisfies $E(D(M)) = D(E(M)) = M$. The lack of a common domain between two users creates a technical problem in using such a system for secrecy and authenticity. We discuss some approaches to this problem in Section 4.1.

1.6.2 Applicability and limitations. The range of applicability of public key systems is limited in practice by the relatively low bandwidths associated with public key ciphers, compared to their conventional counterparts. It has not been proven that time or space complexity must necessarily be greater for public key systems than for conventional systems. However, the public key systems that have withstood cryptanalytic attacks are all characterized by relatively low efficiency. For example, some are based on modular exponentiation, a relatively slow operation. Others are characterized by high data expansion (ciphertext much larger than plaintext). This inefficiency, under the conservative assumption that it is in fact inherent, seems to preclude the use of public key systems as replacements for conventional systems utilizing fast encryption techniques such as permutations and substitutions. That is, using public key systems for bulk data encryption is not feasible, at least for the present.

On the other hand, there are two major application areas for public key crypto-systems:

1. Distribution of secret keys
2. Digital signatures

The first involves using public key systems for secure and authenticated exchange of data-encrypting keys (DEKs) between two parties (Section 2). DEKs are secret shared keys connected with a conventional system used for bulk data encryption. This permits users to establish common keys for use with a system such as DES. Classically, users have had to rely on a mechanism such as a courier service or a central authority for assistance in the key exchange process. Use of a public key system permits users to establish a common key that does not need to be generated by or revealed to any third party, providing both enhanced security and greater convenience and robustness.

Digital signatures are a second major application (Section 3). They provide authentication, nonrepudiation, and integrity checks. As noted above, in some settings authentication is a major consideration; in some cases it is desirable even when secrecy is not a consideration (e.g., [SIMM88]). We have already seen an example of a digital signature, that is, when A sent $D_A(M)$ to B. This permitted B to conclude that A did indeed send the message. As we will note in Section 3, nonrepudiation is another property desirable for digital signatures. Public key cryptosystems provide this property as well.

No bulk encryption is needed when public key cryptography is used to distribute keys, since the latter are generally short. Also, digital signatures are generally applied only to outputs of hash functions (Section 3). In both cases the data to be encrypted or decrypted are restricted in size. Thus the bandwidth limitation of public key is not a major restriction for either application.

2 KEY MANAGEMENT

Regardless of whether a conventional or public key cryptosystem is used, it is necessary for users to obtain other users' keys. In a sense this creates a circular problem: to communicate securely over insecure channels, users must first exchange key information. If no alternative to the insecure channel exists, then secure exchange of key information presents essentially the same security problem as subsequent secure communication.

In conventional cryptosystems this circle can be broken in several ways. For example, it might be assumed that two users can communicate over a supplementary secure channel, such as a courier service. In this case it is often the case that the secure channel is costly, inconvenient, low-bandwidth, and slow; furthermore, use of a courier cannot be considered truly secure. An alternative is for the two users to exchange key information via a central authority. This presumes that each user individually has established a means of communicating securely with the central authority. Use of a central authority has several disadvantages as noted below.

In public key systems the key management problem is simpler because of the public nature of the key material exchanged between users, or between a user and a central authority. Also, alternatives to the insecure channel may be simpler; for example, a physical mail system might suffice, particularly if redundant information is sent via the insecure (electronic) channel.

2.1 Secret Key Management

In a conventional (one-key) system, security is dependent on the secrecy of the key that is shared by two users. Thus two users who wish to communicate securely must first securely establish a common key. As noted above, one possibility is to employ a third party such as a courier; but as remarked there are various disadvantages to this implementation. The latter is the case even for a single key exchange. In practice, it may be necessary to establish a new key from time to time for security reasons. This may make use of a courier or similar scheme costly and inefficient.

An alternative is for the two users to obtain a common key from a central issuing authority (e.g., [POPE79]). Security is then a major consideration: a central authority having access to keys is vulnerable to penetration. Due to the concentration of trust, a single security breach would compromise the entire system. In particular, a central authority could engage in passive eavesdropping for a long period of time before the practice was discovered; even then it might be difficult to prove.

A second disadvantage of a central issuing authority is that it would probably need to be online. In large networks it might also become a bottleneck, since each pair of users needing a key must access a central node. If the number of users is n then the number of pairs of users wishing to communicate could theoretically be as high as $n(n - 1)/2$, although this would be unlikely. Also, each time a new key is needed, at least two communications involving the central authority are needed for each pair of users. Furthermore, such a system is not robust: failure of the central authority disrupts the key distribution system.

Some of the disadvantages of a central authority can be mitigated by distributing key distribution via a network of issuing authorities. One possibility is a hierarchical (tree-structured) system, with users at the leaves and key distribution centers at intermediate nodes. However, distributing key management creates a new security problem, since a multiplicity of entry points for intruders is created. Furthermore, such a modification might be inefficient unless pairs of users communicating frequently were associated to a common subtree; otherwise the root of the tree would be a bottleneck as before.

2.2 Public Distribution of Secret Keys

It was noted by Diffie [DIFF76], Diffie and Hellman [DIFF76b], and Merkle [MERK78] that users can publicly exchange secret keys. This is not related a priori to public key cryptography; however, a public key system can be used to implement public distribution of secret keys (often referred to as public key distribution). The underlying public key system must support secrecy, for use in key encryption.

The notion of public distribution of secret keys was originated in 1974 by Merkle, who proposed a "puzzle" system to implement it [MERK78]. However, even before this work was published it was superseded by a public key scheme supporting public key distribution [DIFF76b]. This has come to be known as the Diffie–Hellman exponential key exchange. To begin with, users A and B are assumed to have agreed on a common prime p and a common primitive root $g \bmod p$ (Appendix K). Then (Lemma

K.2) each number in $[1, p)$ can be expressed as $g^x \bmod p$ for some x. Both p and g may be public. Now A chooses a random $x(A)$ in $[0, p - 1]$ and B chooses a random $x(B)$ in $[0, p - 1]$; these are the private components of A and B, respectively. Then A computes

$$y(A) = g^{x(A)} \bmod p$$

while B computes

$$y(B) = g^{x(B)} \bmod p$$

These are the public components (along with the common p and g) of A and B, respectively. Finally A sends $y(A)$ to B and B sends $y(B)$ to A. Since A knows $x(A)$ and $y(B)$ he can compute

$$K = y(B)^{x(A)} \bmod p$$

We note

$$K = g^{x(A)x(B)} \bmod p$$

Now B knows $x(B)$ and $y(A)$ and can compute

$$y(A)^{x(B)} \bmod p = K$$

The secrecy of this scheme depends, as in Section 1.5, on the difficulty of computing discrete logarithms. That is, knowing p, g, y, and $y = g^x \bmod p$ for some x does not yield x. Thus, if an intruder intercepts $y(A)$ or $y(B)$ or both he cannot find $x(A)$ or $x(B)$ and hence cannot compute K.

A disadvantage of this scheme is lack of support for authenticity. If an intruder C intercepts $y(B)$, sends $y(C) = g^{x(C)} \bmod p$ to B and B thinks he is receiving $y(A)$, B will inadvertently establish a secret key with C. That is, the underlying public key cryptosystem supports secrecy but not authentication. Thus it may be desirable to augment a secrecy-providing system with one that provides authentication. Examples of the latter will be given later.

2.3 Management of Public Components in a Public Key System

Prior to using a public key cryptosystem for establishing secret keys, signing messages, etc. users A and B must exchange their public transformations (E_A and E_B in Section 1.6), or equivalently their public components. The latter may be regarded as a key management problem. It is a simpler problem than establishment of secret keys, since public components do not require secrecy in storage or transit. Public components can, for example, be managed by an online or offline directory service; they can also be exchanged directly by users. However, authenticity is an issue as in the previous section. If A thinks that E_C is really E_B then A might encrypt using E_C and inadvertently allow C to decrypt using D_C. A second problem is integrity: any error in transmission of a public component will render it useless. Hence some form of error detection is desirable.

Regardless of the scheme chosen for public component distribution, at some point a central authority is likely to be involved, as in conventional systems. However, it may not be necessary for the central authority to be online; exchange of public components between users need not involve the central authority. An example of how this can be implemented is given in Section 2.5. We also note there that the implications of compromise of the central authority are not as severe as in the conventional case.

Validity is an additional consideration: A user's public component may be invalidated because of compromise of the corresponding private component, or for some other reason such as expiration. This creates a stale data problem in the event that public components are cached or accessed through a directory. Similar problems have been encountered in management of multi-cache systems and distributed databases, and various solutions have been suggested. For example, users could be notified immediately of key compromises or invalidations. This may be impractical, in which case either users should be informed within a given time period of changes needed for cached information on public components, or users should periodically check with a central authority to update validity information.

2.3.1   Use of certificates.   A technique to gain a partial solution to both authenticity and integrity in distribution of public components is use of certificates [KOHN78]. A certificate-based system assumes a central issuing authority S as in the secret key case. Again it must be assumed that each user can communicate securely with S. This is relatively simple in the present instance: it merely requires that each user possess E_S, the public transformation of S. Then each user A may register E_A with S. Since E_A is public, this might be done via the postal service, an insecure electronic channel, a combination of these, etc.

Normally A will follow some form of authentication procedure in registering with S. Alternatively, registration can be handled by a tree-structured system: S issues certificates to local representatives (e.g., of employing organizations), who then act as intermediaries in the process of registering users at lower levels of the hierarchy. An example of such a system is given in Section 5.4.

In any case, in return A receives a certificate signed by S (see Section 3 for a more thorough discussion of signatures) and containing E_A. That is, S constructs a message M containing E_A, identification information for A, a validity period, and so on. Then S computes $C_A = D_S(M)$ which becomes A's certificate. The latter is then a public document that both contains E_A and authenticates it, since the certificate is signed by S. Certificates can be distributed by S or by users, or used in a hierarchical system that we return to later. The inclusion of the validity period is a generalization of timestamping. The latter was not treated in [KOHN78], but Denning [DENN81] notes the importance of timestamping in guarding against the use of compromised keys.

In general, the problem of stale data is not wholly solved by timestamping: a certificate may be invalidated before its expiration date, because of compromise or administrative reasons. Hence if certificates are cached by users (as opposed to being redistributed by S each time they are used), S must periodically issue lists of invalidated certificates. Popek and Kline [POPE79] have noted that certificates are analogous to capabilities in operating systems. Management of certificates creates analogous problems, such as selective revocation of capabilities. The public nature of certificates, however, a priori precludes an analogue of a useful feature of a capability system: Users cannot be restricted to communicating with a subset of other users. If a selective capability feature is desired it must be implemented through a controlled distribution of certificates, which would be difficult and contrary to the spirit of public key.

2.3.2 Generation and storage of component pairs. Generation of public/private component pairs is an important consideration. If this service is offered by a central facility, it may result in certain advantages; for example, keys might be longer or chosen more carefully. However, this introduces the same problem noted in Section 2.1: A central authority holding users' private components would be vulnerable to penetration. Also, the central facility would have to be trusted not to abuse its holdings by monitoring communications between users. Both problems are circumvented if users generate and store their own private/public components.

2.3.3 Hardware support for key management. If users store their own components, a management problem is created. One possibility is software storage, for example, in a file with read protection. If greater security is desired, a second option is hardware keys ([DENN79], [FLYN78], [RIVE78]).

In the classic version of this scheme, each public/private key pair is stored on a pair of read-only memory (ROM) chips. The public key is revealed to the owner of the keys. However, the private key need not be known to any user, including the owner.

There are two advantages to this mode of storage, namely, the fact that neither the user nor a central authority need know the private component. We have previously noted the advantage of not having users reveal private keys to a central authority. Also, as noted in [FLYN78], it may also be desirable to protect keys from disloyal employees. Entering keys from a keyboard rather than from ROM is insecure. However, there is also an obvious disadvantage to the classic mode of hardware key storage, namely, the fact that the key manufacturer has access to the user's keys, and may retain copies. An alternative, but still classic, scheme is for the user to be provided with a means of writing to a memory chip and then sealing it from further writing.

A more modern and more secure hardware adjunct is a token, smart card, or other device that contains both memory and a microprocessor. Such a device can both generate and store user keys. Ideally, it should be implemented via a single chip that stores both cryptographic algorithms and data.

In the classic case, encryption and decryption are handled by separate units in a hardware device acting as an interface between a user's computer and the network. As noted in [DENN79], this creates a security risk: An intruder may rig the device. Also, all data entering or exiting a user's computer are automatically decrypted or encrypted. Hence all other users must know a user's public key to communicate with him. On the other hand, a user can easily store encrypted data in a central facility: By setting a toggle switch, a user enciphers data exiting his computer using his own private component. Furthermore, a form of Trojan horse attack is thwarted: If a user runs a program from an external source that attempts to have data illicitly sent back to it, the data will be encrypted and hence worthless.

If a token or smart card is used, it may be possible in the near future to generate and store keys on the same device that executes encryption and decryption algorithms. Again, security is optimal if all of these functions are combined onto one or a minimal number of chips. However, use of such auxiliary devices entails some of the same problems as in the classic case. For example, a token or card reader is needed; this device must be secure. Also, a token or smart card may be stolen, permitting the thief to masquerade as the user. Passwords (personal identification numbers [PINs]) used to access devices can substantially lessen this threat.

A classic hardware system can support signatures (see Section 3), but this requires an additional data path from a user's computer through the deciphering unit and back to the computer. This creates an additional security risk. A second path is needed through the enciphering unit, but this is more secure.

Advanced capabilities such as signature generation should become feasible on devices such as smart cards in the next few years. This would provide an excellent solution to the problem of using private components without exposing them.

2.4 Using Public Key Systems for Secret Key Distribution

As noted above, one of the major applications of public key cryptosystems is in regard to public distribution of secret keys. Thus, a public key system can be used in conjunction with a conventional system such as DES. We have already seen an example of a public key cryptosystem implementing public key distribution in Section 2.2. Both secrecy and authentication can be provided simultaneously in the distribution process via public key systems. This goal may be achieved using a single public key system if its encryption and decryption functions are inverses, or via a hybrid of two public key systems providing secrecy and authentication separately.

The essence of this usage is very simple: A secret key is merely a particular form of message. Assuming that a system has been established for distribution of public components of users as discussed in the previous section, users can then establish secret keys at will by employing the public system for encryption and signing of secret keys. Thus the latter can be exchanged or changed without difficulty. This is in contradistinction to a system in which secret keys must be established via couriers or a central authority.

If users in a public key system generate their own private/public component pairs, it is in fact never necessary for them to use a "secure" auxiliary mechanism such as a courier service, or to share their private components with a central authority. Users may normally register their public components without use of a secure channel; and two users may use each other's public components to establish common keys for use in message encryption, again without resort to a secure channel. In contrast, in a secret key system a secure communication may be required each time a user initiates communication with a new user or a central authority, or when two users change a master key.

2.4.1 A protocol for key exchange. Suppose A and B wish to establish a shared secret key. Suppose further that they have obtained each other's public components; some ways by which this may be effected are discussed below. In any case, A may now generate a secret key K. For example, this may be a key-encrypting key to be used to exchange session keys and possibly initialization vectors if, for example, DES is used in cipher feedback mode or cipher block-chaining mode [NATI80]. Then A might form an expanded message that contains K and further data, which could include an identification for A, a timestamp, a sequence number, a random number, and so on. This added material is needed for A to authenticate himself to B in real time, and thus prevent replays. For example, Needham and Schroeder [NEED78] suggest a three-way handshake protocol:

First, A may send to B the message $M = E_B(R_A, ID_A)$, where E_B is B's public transformation, ID_A is the identification of A, and R_A is a random number. Now B can decrypt M and obtain ID_A. As noted above, M could also contain other information such as a timestamp or sequence number. Now B generates a random number R_B and sends $M' = E_A(R_A, R_B)$ to A. On decryption of M', A can verify in real time that B has received R_A, since only B could decrypt M. Finally A sends $M'' = E_B(R_B)$ to B; when B decrypts M'' he can verify in real time that A has received R_B, since only A could decrypt M'. Thus A and B are authenticated to each other in real time; that is, each knows that the other is really there.

Now A sends $E_B(D_A(K))$ to B, who can decrypt to obtain K. This ensures both secrecy and authenticity in exchange of K.

2.5 Protocols for Certificate-Based Key Management

There are a number of different approaches to public key component management and the application to secret key distribution (e.g., [KLIN79], [NEED78]). For simplicity we assume a certificate-based system. We also assume that a system has been established for a central authority to create certificates.

2.5.1 Certificate management by a central authority. In a certificate-based public key system, one possibility is to have the central issuing authority manage certificate distribution. Suppose the authority S has created certificates C_A and C_B for A and B, respectively. These contain E_A and E_B (or equivalently, the public components of A and B) as noted in Section 2.3.1.

We may assume that A and B have cached C_A and C_B, respectively. If A has exchanged certificates with B previously, A and B may have each other's certificates cached. If not, then there are two ways in which A could obtain B's certificate. The simplest is for A to request B's certificate directly from B; this is explored in the next section. The advantage of this simple approach is that online access to a central authority is avoided.

On the other hand, if A requests B's certificate directly from B, or obtains it from a directory listing, security and integrity considerations arise. For example, the certificate A obtains may have been recently invalidated.

Thus A may wish to access B's certificate through S, assuming that the latter is online. In this event A requests C_A and C_B from S. We recall that each C has the form $D_S(M)$ where M contains an E. Thus when A receives the requested certificates, he can validate C_A by computing $E_S(C)$ and recovering E_A from the decrypted form. This validates C_B as well. Hence A may retrieve a duly authenticated E_B from C_B. The disadvantage is the requirement of an online central node; the advantage is that A is assured of the instantaneous validity of the certificate. Later we will note that this has implications for nonrepudiation.

Caching of certificates implies that authentication is not done regularly. Thus the procedure above is only necessary when two users first communicate, or when components are compromised or certificates invalidated. As long as the public components involved are valid, secret keys can be exchanged at will, although a handshake protocol may still be used to ensure real-time authenticity and integrity.

An advantage of this scheme is that the entire process is conducted over insecure channels with excellent security. A second advantage is that distribution of certificates by a central authority assures that certificates are valid at time of receipt.

A disadvantage is that, as in Section 2.1, the central authority may be a bottleneck. The severity is mitigated by the fact that a secure channel is not needed, but the essential problem is similar.

A more significant disadvantage arises from the concentration of trust in one entity [KLIN79]. The security of the central authority might be compromised, for example, by penetration. The fact that the central authority does not generally access users' private components means that existing messages are not compromised, as might be the case in a secret key system [DIFF82]. If an intruder accesses the central authority's private component, however, the intruder could now forge certificates.

Diffie [DIFF88] notes that the latter situation is not catastrophic. Suppose A requests B's certificate from the penetrated central authority. The intruder might substitute his own public transformation for E_B, then forge the central authority's signature. The intruder could then decrypt messages sent to B by A. However, if the intruder's eavesdropping is passive, B will be unable to decrypt, and the masquerade will be short-lived. To continue the deception, the intruder must intercept, decrypt, and re-encrypt the message using E_B and then pass it along to B. When B replies to A the procedure must be repeated. This involves wiretaps and message deletions that, according to Diffie, will most likely expose the intruder.

A more blatant approach would be for the intruder merely to intercept messages intended for B. In either event, the issuance of false certificates is a practice that presumably would be exposed in a short period of time, exposing the intruder and limiting the damage. Also, forged certificates would provide proof of criminal activity. In contrast, we noted above that in a conventional system, compromise of the central authority would permit passive eavesdropping which could proceed undetected for a long period and might be difficult to prove.

A more detailed discussion of recovery from compromised keys may be found in [DENN83b]. There Denning also suggests that the central authority keep a log of all transactions to aid in recovery. In passing she also notes that a potential weak link in the certificate system is that for the central authority to properly identify users, it must assume that users' hosts are secure. Denning also observes that hosts may be vulnerable to Trojan horse attacks in the form of subversion of their operating systems or network interfaces.

Merkle [MERK82b] has suggested a tree-structured system as a means of coping with the compromise of a central authority. However, this scheme is not practical for large networks since the tree must be restructured each time the user population changes.

2.5.2 Decentralized management.
Users may be responsible for managing their own certificates. In this event the protocol is much simpler. When A wishes to initiate communication (e.g., exchange of a secret key) with B, he sends a message to B containing A's certificate, A's identification, and other information such as a date, random number, and so on, as described in the protocol in the previous section. This message also requests B's certificate. On completion of the certificate exchange, employing some protocol such as the handshake above, A and B will possess each other's authenticated certificates. A can validate the certificate C_B by computing $E_S(C_B)$ as usual. Then E_B may be retrieved. The certificate must contain information properly identifying B to A, so that an intruder cannot masquerade as B. The certificate must also have a validity period. In turn B may proceed similarly.

The central authority must periodically issue lists of certificates that have become invalid before their expiration dates due to key compromise or administrative reasons. It is likely that in a large system this would be done, for example, on a monthly basis. Hence a user receiving a certificate from another user would not have complete assurance of its validity, even though it is signed by S. Thus this system trades higher efficiency for some security loss, compared to the previous scheme.

For greater assurance of validity, a user could access a centrally managed list of invalid certificates; presumably these would be very current. This would require online access to a central facility, which would create the same type of bottleneck we have noted previously. However, the accesses would be very quick, since presumably only a certificate sequence number would be transmitted.

Yet another online possibility would be for the central authority to enforce coherence by a global search of cached certificates each time a certificate is invalidated. Again there is a trade-off of validity assurance and efficiency.

2.5.3 A phone book approach to certificates.   Some of the features of the previous schemes could be combined in a phone-book approach, using an electronic equivalent such as a floppy disk containing certificates. This would optimize ease of use since a user could communicate securely with another by accessing the latter's certificate very rapidly. However, again the central authority would have to issue "hot lists." Periodic distribution of the compilations of certificates would be a separate management process.

Security of such a scheme is clearly in question [KLIN79]. Phone book entries might contain errors, and if phone books are public entries these errors could be altered.

3 DIGITAL SIGNATURES AND HASH FUNCTIONS

Digital signatures were introduced by Diffie and Hellman [DIFF76b]; for surveys see, for example, [AKL-83], [POPE79], [RABI78]. A digital signature is the electronic analogue of a handwritten signature. A common feature is that they must provide the following properties:

1. A receiver must be able to validate the sender's signature.
2. A signature must not be forgeable.
3. The sender of a signed message must not be able to repudiate it later.

We have already seen an example of the usage of digital signatures, namely, when a central authority signs certificates. In this section we are concerned with the signing of arbitrary messages.

A major difference between handwritten and digital signatures is that a digital signature cannot be a constant; it must be a function of the document that it signs. If this were not the case then a signature, due to its electronic nature, could be attached to any document. Furthermore, a signature must be a function of the entire document; changing even a single bit should produce a different signature. Thus a signed message cannot be altered.

There are two major variants of implementation:

1. True signatures
2. Arbitrated signatures

In a true signature system, signed messages are forwarded directly from signer to recipient. In an arbitrated system, a witness (human or automated) validates a signature and transmits the message on behalf of the sender. The use of an arbitrator may be helpful in event of key compromise as noted below.

The notion of authentication by some form of signature can be traced back as far as 1952, as noted by Diffie [DIFF88]. Cryptography was used by the Air Force to identify friendly aircraft through a challenge/response system. The protocols utilized influenced the development of public key cryptography by Diffie and Hellman [DIFF76b].

As we note below, hash functions are useful auxiliaries in this context, that is, in validating the identity of a sender. They can also serve as cryptographic checksums, thereby validating the contents of a message. Use of signatures and hash functions can thus provide authentication and verification of message integrity at the same time.

Numerous digital signature schemes have been proposed. As noted in [DAVI83], a major disadvantage of signature schemes in conventional systems is that they are generally one-time schemes. A signature is generated randomly for a specific message, typically using a large amount of key material, and is not reusable. Furthermore, later resolution of disputes over signed documents requires written agreements and substantial bookkeeping on the part of the sender and receiver, making it more difficult for a third party to adjudicate. A conventional scheme was noted in [DIFF76b] (the Lamport–Diffie one-time signature) and is refined in [MERK82]. Some other conventional schemes are DES-based.

Public key schemes supporting authentication permit generation of signatures algorithmically from the same key repeatedly, although the actual signatures are of course different. That is, in a public key scheme, signatures are a function of the message and a long-term key. Hence key material can be reused many times before replacement. This obviates the necessity of special written agreements between individual senders and receivers. It also makes it easy for a third party to resolve disputes (e.g., involving repudiation) later, and simplifies storage and bookkeeping. As we note below, the bandwidth limitation of public key systems is unimportant, due to the use of hash functions as auxiliaries.

In passing we remark that Goldwasser, Micali, and Rivest [GOLD88] have proposed a nondeterministic public key signature scheme that incorporates an aspect of conventional schemes: A message does not have a unique signature. This scheme is intended to deflect adaptive chosen-text attacks, in which an attacker is permitted to use a signer as an oracle. In such an attack, the attacker specifies a sequence of messages to be signed, and is able to study all previous signatures before requesting the next. In the Goldwasser–Micali–Rivest (GMR) scheme, it can be proved that an attacker can gain no information by studying any number of signatures no matter how they are generated.

This security gain is offset, however, to some extent by data expansion (high ratio of signature to message size) and a negative effect on nonrepudiation. As in the case of the one-time schemes discussed above, the signatures generated in nondeterministic public key schemes tend to be large in comparison to signatures produced by deterministic public key schemes (i.e., those that produce a unique signature for a given message). Adjudication of disputes is made difficult by increased bookkeeping and the necessity of storing large databases of messages and their signatures.

In the following sections, we discuss only deterministic public key signature schemes, in which a message has a unique signature.

3.1 Public Key Implementation of Signatures

We have noted above that a public key system can be used for authentication. There are several distinctive features of a public key implementation of signatures, including:

1. The possibility of incorporating both secrecy and authenticity simultaneously, assuming that the system supports both services.
2. The reuse of key material in generating signatures algorithmically.
3. Nonrepudiation of sending is essentially a built-in service.

These features make public key implementation of signatures very efficient and versatile.

3.1.1 Signing messages. We recall that A signs a document M by sending $S = D_A(M)$ to B; B validates A's signature by computing $M = E_A(S)$. That is, A's signature can be validated. We also noted above that because E_A is a trapdoor one-way function, it should not be possible for an intruder to find S' such that $M = E_A(S')$ for a given message M. Thus A's signature cannot be forged. Also, if A attempts to repudiate the M sent to B above, B may present M and S to a judge. The judge can access E_A and hence can verify $M = E_A(S)$; assuming the trapdoor D_A is indeed secret, only A could have sent S. Thus a priori we have nonrepudiation (but see Section 3.1.2, below).

For both authenticity and secrecy, as before A may send

$$S = E_B(D_A(M))$$

to B; B computes

$$M = E_A(D_B(S))$$

For nonrepudiation, B retains $D_B(S) = D_A(M)$. Again a judge can use E_A to find M. This again assumes common domains for the E's and D's; in Section 4.1 we will note an example of how this point may be treated in practice. In passing we remark that the preceding schemes satisfy another desirable property: In the adjudication process, they do not compromise security by exposing private components to a judge.

3.1.2 The issue of nonrepudiation. Nonrepudiation, that is, preventing senders from denying they have sent messages, is contingent on users keeping their private components secret (e.g., [NEED78], [POPE79]). In the discussion above, if D_A should be compromised, then A might be able to repudiate messages sent even before the compromise. This is an administrative issue, and ultimately a matter for litigation; hence it is beyond the scope of cryptography per se. However, some partial solutions have been proposed that can be incorporated into protocols. Most of these involve use of some form of arbitrator, as noted in [DEMI83].

For example, in [POPE79], notary public machines are employed as arbitrators. These sign messages on top of the senders' signatures. A sender then cannot claim that a message sent via the arbitrator was forged. However, a sender can still claim that his private component was compromised in the past without his knowledge, and a forged message was notarized. A partial solution is for the notary to send a copy of a signed message back to the sender; in this case the notary must know the current address of the sender.

A general protocol for usage of an arbitrator A when U is sending a message M to R is given in [AKL-83]:

1. U computes $S_1 = E_R(D_U(M))$.
2. U computes a header m = identifying information.
3. U computes $S_2 = D_U(m, S_1)$.
4. U sends m and S_2 to A.
5. A decrypts S_2 and validates the identifying information in m, thereby verifying the origin of M.
6. A computes $M' = (m, S_1, \text{timestamp})$.
7. A computes $S' = D_A(M')$.
8. A sends S' to R.

As noted above, a copy of S' may also be sent to U.

The obvious disadvantages of an arbitrator system are inconvenience and loss of security. In particular, as noted in [POPE78], this type of scheme violates a desirable criterion for network protocols, namely, point-to-point communication. It also requires trusted software, which is undesirable.

In [BOOT81] the use of a central authority is suggested for this purpose. In this scheme, the receiver of a message sends a copy to the central authority. The latter can attest to the instantaneous validity of the sender's signature; that is, it has not been reported that the sender's private component has been compromised at the time of sending. The value of such testimony is mitigated by the necessity of users rapidly reporting key compromise to the central authority. Also, the central authority becomes a bottleneck.

Another partial solution involves timestamps ([DENN81], [MERK82b]). This again may involve a network of automated arbitrators, but very simple in nature, since they need merely timestamp messages. In contrast to the notary publics above, however, in [MERK82b] it is suggested that receivers obtain timestamps. If a receiver needs to be sure of the validity of a signature, he may check the validity of the sender's private component by checking with a central authority. As long as the received message is timestamped before the validity check, the receiver is assured of nonrepudiation. To be legally cohesive, however, this system requires the sender to be responsible for signing until a compromise of his private component is reported to the central authority. In analogy to lost credit cards, this may not be a desirable system. Also, it requires an online central authority and real-time validity checks and timestamps, which may again create bottlenecks. Furthermore, such schemes require a networkwide clock and are vulnerable to forgery of timestamps, as noted in [BOOT81].

If users are permitted to change their components, a central authority should retain past components for use in disputes that may arise later. Neither compromise nor change of components of a user should cause catastrophic problems in practice, since credit card systems have established precedents for protocols, both administrative and legal. For example, as noted above, conventions have been established for treatment of cases of lost or stolen credit cards; presumably analogous procedures could be established for component compromise.

3.1.3 The issue of proof of delivery. The literature on public key protocols concentrates on the issues of validity of signatures and the effects of key compromise on nonrepudiation. However, DeMillo and Merritt [DEMI83] note that the inverse of a signature, namely, protection against the recipient of a message denying receipt, may also be a desired feature of a system. It is mentioned as an optional security service in [ISO-87].

Both nonrepudiation and proof of delivery must be implemented via adjudicable protocols, that is, protocols that can be verified later by a third party. In [DEMI83] it is noted that nonarbitrated adjudicable reception protocols are difficult to design. A simple handshaking protocol might proceed as follows:

1. A computes $X = E_B(D_A(M))$.
2. A sends X to B.
3. B computes $M = E_A(D_B(X))$.
4. B computes $Y = E_A(D_B(M))$.
5. B acknowledges receipt of M by sending Y to A.

If this protocol is standard procedure then A can assume nonreception unless he receives acknowledgment from B. However, to serve as an adjudicable proof-of-reception protocol, B is required to acknowledge anything A sends. Then an intruder C could proceed as follows:

1. C intercepts $Y = E_A(D_B(M))$.
2. C sends Y to A.
3. A thinks this is a legitimate message from C, unrelated to his communication with B. Following protocol:
   a. A computes $M' = E_C(D_A(Y))$.
   b. A computes $Z = E_C(D_A(M'))$.
   c. A acknowledges receipt of M' by sending Z to C.
4. C computes $E_B(D_C(E_A(D_C(Z)))) = E_B(D_C(M')) = E_B(D_A(Y)) = M$.

This of course occurs because of step 3, in which A is required by protocol to acknowledge $M' =$ gibberish. This shows that such an automatic protocol may be undesirable. On the other hand, selective acknowledgment might have an adverse effect on adjudicability.

3.2 Hash Functions and Message Digests

We noted that public key systems generally encrypt more slowly than conventional ciphers such as DES. Other digital signature schemes are also relatively slow in general. Furthermore, some schemes produce signatures comparable in size to, and in some cases larger than, the messages they sign. This results in data expansion and effectively lower bandwidth of transmission. Thus it is usually not desirable to apply a digital signature directly to a long message. On the other hand, we remarked that the entire message must be signed. This is seemingly contradictory, but a heuristic solution can be obtained by using a hash function as an intermediary.

A hash function H accepts a variable-size message M as input and outputs a fixed-size representation $H(M)$ of M, sometimes called a message digest [DAVI80]. In general, $H(M)$ is much smaller than M; for example, $H(M)$ might be 64 or 128 bits, whereas M might be a megabyte or more. A digital signature may be applied to $H(M)$ in relatively quick fashion. That is, $H(M)$ is signed rather than M. Both M and the signed $H(M)$ may be encapsulated in another message which may then be encrypted for secrecy. The receiver may validate the signature on $H(M)$ and then apply the public function H directly to M and check to see that it coincides with the forwarded signed version of $H(M)$. This validates both the authenticity and integrity of M simultaneously. If $H(M)$ were unsigned only integrity would be assured.

A hash function can also serve to detect modification of a message, independent of any connection with signatures. That is, it can serve as a cryptographic checksum (also known as a manipulation detection code [MDC] or message authentication code [MAC]). This may be useful in a case where secrecy and authentication are unimportant but accuracy is paramount. For example, if a key is sent in unencrypted form, even if the key is only a few hundred bits it might be useful to transmit a checksum along with it. Another instance where this case arises is when secrecy is provided by a system such as DES that does not provide a signature capability. An important distinction is that a hash function such as a MAC used in connection with a conventional system is typically parameterized by a secret shared key, although the latter may be distinct from the session key used in transmitting the message and its MAC. In contrast, hash functions used in connection with public key systems should be public and hence keyless.

We referred above to the use of hash functions as a "heuristic" solution to the problem of signing long messages. This refers to the fact that the signature for a message should be unique, but it is theoretically possible that two distinct messages could be compressed into the same message digest (a collision). The security of hash functions thus requires collision avoidance. Collisions cannot be avoided entirely, since in general the number of possible messages exceeds the number of possible outputs of the hash function. However, the probability of collisions must be low. If a function has a reasonably random output, the probability of collisions is determined by the output size.

A hash function must meet at least the following minimal requirement to serve the authentication process properly: It must not be computationally feasible to find a message that hashes to the same digest as a given message. Thus, altering a message will change the message digest. This is important to avoid forgery.

In many settings this minimal requirement is regarded as too weak. Instead, the requirement is added that it should not be possible to find any two strings that hash to the same value. We return to the distinction between these two criteria in Section 3.2.3.

3.2.1 Use of hash functions. In a public key system augmented by a hash function H, A might send a message M to B as follows: For simplicity ignore secrecy considerations. Then A sends M and $X = D_A(H(M))$ to B. The latter uses E_A to retrieve $H(M)$ from X, then computes $H(M)$ directly and compares the two values for authentication. For nonrepudiation, B retains M, $H(M)$, and X. If A attempts to repudiate M, a judge can use the three items to resolve the dispute as before: He computes $H(M) = E_A(X)$ and recomputes $H(M)$ from M. If B could find M' with $H(M') = H(M)$, B could claim A sent M'. A judge receiving M', $H(M)$ and X would reach a false conclusion.

3.2.2 Relation to one-way functions. Merkle [MERK82] defines a hash function F to be a transformation with the following properties:

1. F can be applied to an argument of any size.

2. F produces a fixed-size output.

3. $F(x)$ is relatively easy to compute for any given x.

4. For any given y it is computationally infeasible to find x with $F(x) = y$.

5. For any fixed x it is computationally infeasible to find $x' \neq x$ with $F(x') = F(x)$.

The most important properties are (4) and (5). In particular, (5) guarantees that an alternative message hashing to the same value as a given message cannot be found. This prevents forgery and also permits F to function as a cryptographic checksum for integrity. Actually (5) can be strengthened, as noted below.

Property (4) states that F is a one-way function (e.g., [DIFF76b]). One-way functions are used in various other contexts as well. For example, we noted above that the security of public key systems depends on the fact that the public transformations are trapdoor one-way functions. Trapdoors permit decoding by recipients. In contrast, hash functions are one-way functions that do not have trapdoors.

The concept of (nontrapdoor) one-way functions originated in connection with log-in procedures ([WILK68], p. 91): Needham [NEED78] noted that if passwords were stored encrypted under a one-way function, a list of encrypted passwords would be of no use to an intruder since the original passwords could not be recovered. When a user logged in, the string entered would be encrypted and compared to the stored version for authenticity. Essentially the same system was rediscovered later in [EVAN74].

Use of one-way functions to produce message digests or encrypt passwords is very different from use of trapdoor one-way functions to generate encryption functions $\{E_A\}$ in a public key cryptosystem. In the latter, (4) above becomes a dichotomy: it is computationally infeasible for anyone except A to find M from $C = E_A(M)$, but it is easy for A, the unique holder of the trapdoor D_A, to compute $M = D_A(C)$.

An example of functions that are not one-way are the private transformations in public key systems: anyone can solve $S = D(M)$ for M; that is, $M = E(S)$. This shows that signature schemes are not one-way functions; that is, they do not satisfy (4). On the other hand, a deterministic signature function S satisfies (5); that is, messages have unique signatures. For example, if a signature is generated via $S = D(M)$ where D is a decryption function of a public key system, then $D(M) = D(M')$ implies $E(D(M)) = M = E(D(M')) = M'$, so (5) is trivially satisfied.

Merkle's requirements are that a hash function must be both one-way (4) and effectively collisionless (5). To satisfy (1) and (2) concurrently, collisions must exist; (5) requires that a cryptanalyst should not be able to find them. This notion is amenable to further refinement.

3.2.3 Weak and strong hash functions.

The security of hash functions can be refined further: We may distinguish between weak and strong hash functions. A function satisfying items (1)–(5) in Section 3.2.2, may be termed a *weak hash function*. Property (5) characterizes weak hash functions; it states that a cryptanalyst cannot find a second message producing the same message digest as a fixed message. On the other hand, H may be termed a *strong hash function* if items (1)–(4) still hold, but (5) is modified as follows: It is computationally infeasible to find any $\{x_1, x_2\}$ with $H(x_1) = H(x_2)$.

Strong and weak functions may often be obtained from the same class of functions. Strong functions are then characterized by larger message digests, which reduce the probability of collisions. Strong functions are thus more secure, but the longer message digests are likely to increase time needed to hash.

Although the two definitions above are superficially similar, computationally they are quite different. For example, suppose H is a hash function with an 80-bit output. Suppose a cryptanalyst starts with a fixed message M and wishes to find a second message M' with $H(M) = H(M')$. Assuming that the 2^{80} outputs of H are totally random, any candidate M' has a probability of only 2^{-80} of meeting the required condition. More generally, if the cryptanalyst tries k candidates, the probability that at least one candidate satisfies $H(M) = H(M')$ is $1 - (1 - 2^{-80})^k$ which is about $2^{-80}k$ by the binomial theorem if the latter is small. For example, the cryptanalyst will probably have to compute H for about $k = 2^{74} = 10^{22}$ values of M' to have even one chance in a million of finding one M' that collides with M. Thus H is secure in the weak sense.

Suppose for the same H the cryptanalyst merely seeks any values $\{M_1, M_2\}$ with $H(M_1) = H(M_2)$. By Example F.1 (see Appendix F), if he examines $H(M)$ for at least $1.17 * 2^{40} < 2 * 10^{12}$ random values of M, the probability of at least one collision exceeds $\frac{1}{2}$. A supercomputer could probably find M_1 and M_2 with $H(M_1) = H(M_2)$ in at most a few days. Thus H is not secure in the strong sense.

The latter attack is called a birthday attack (Appendix F). It should be noted that finding a collision $H(x) = H(y)$ gives the cryptanalyst no valuable information if x and y are random bit strings. For a purpose such as forgery an adversary may need to generate a large number (e.g., 10^{12} above) of variations of a message to find two that collide. Inclusion of timestamps or sequence numbers in messages, according to a fixed format, may make it computationally infeasible to find a collision of use to an adversary.

3.3 Digital Signatures and Certificate-Based Systems

We noted above that digital signatures can be employed for authentication in the process of distributing public components in public key systems. In particular, if the system is certificate-based, the central issuing authority can sign certificates containing public components.

The notion of a central authority can be extended to a hierarchical (tree) structure. The central authority serves as the root of the tree; leaves represent users. Intermediate nodes may also be users. Typically the intermediate nodes will be arranged in several levels representing, for example, organizations and organizational units. Each node of the tree is responsible for authenticating its children. Thus an authentication path is created for each node, ending at the root. For example, the central authority may certify an organization; the organization may certify a unit; the unit may certify an individual user. Certification may be accomplished by having a parent node sign a certificate for the child node. To validate another user's certificate, a user may request the entire authentication path.

It is also possible for the tree to be replaced by a forest. For example, in a multinational system there may be a different tree for each country. In this event the roots must certify each other. An authentication path may then pass through two or more roots.

More generally, an authentication structure can be an arbitrary directed graph, with a directed edge from A to B if A certifies B. Then authentication paths may be constructed by conjoining directed paths from two users to a common trusted node; an example of this more complex structure is given in Section 5.3.

4 EXAMPLES OF PUBLIC KEY SYSTEMS AND HASH FUNCTIONS

Numerous public key systems have been proposed. These may be divided into several categories:

1. Systems that have been broken
2. Systems that are considered secure
   a. systems that are of questionable practicality
   b. systems that are practical
      (1) systems suitable for key distribution only
      (2) systems suitable for digital signatures only
      (3) systems suitable for key distribution and digital signatures

From the standpoint of sheer numbers, most systems have proven to be insecure. This is unfortunate, because some of the techniques producing insecure systems have relatively high bandwidths, where bandwidth refers to rates of encryption and decryption. Knapsack ciphers are an example (Section 4.2.1; Appendix E) of high-bandwidth but generally insecure systems. Of the systems that have not been broken, many are regarded as impractical for reasons such as large key sizes or large data expansion (ciphertext much larger than plaintext).

Only a relative handful of systems are widely regarded as secure and practical. In particular, a cryptosystem is usually regarded as secure if breaking it is essentially equivalent to solving a long-standing mathematical problem that has defied all attempts at solution.

Of the well-known systems that are generally regarded as secure and practical, some are limited to digital signatures and hence unsuitable for public key distribution (e.g., Section 4.2.2). On the other hand, in instances such as the Diffie–Hellman exponential exchange scheme (Section 2.2), public key distribution is supported but authentication is not. Such systems may need augmentation by a system that supports signatures. The only well-known system discovered to date that is secure, practical, and suitable for both secrecy and authentication is described in Section 4.1.

Category (2.b.3) in the list above could in theory be further subdivided into systems with relatively low and high bandwidths, but there are no well-known examples with high bandwidths. There does not exist a secure, practical system suitable for key distribution and digital signatures, and with bandwidth high enough to support bulk data encryption. Prospects for creating radically new systems seem dim; in fact, as noted in [DIFF88], most of the extant systems are based on a small number of hard mathematical problems:

- Discrete exponentiation
- Knapsack problems
- Factoring

According to Diffie, discrete exponentiation was suggested by J. Gill, and the use of the knapsack problem by Diffie. A suggestion to use factoring was apparently made by Knuth to Diffie during the early years of public key cryptography, but factoring was not incorporated into a cryptosystem until several years later (Section 4.1).

The knapsack problem is noted briefly in Section 4.2.1. The other two mathematical problems above are easy to state (although also hard to solve):

1. If p is a given prime and g and M are given integers, find x such that $g^x \equiv M$ (mod p).
2. If N is a product of two secret primes:
   a. Factor N.
   b. Given integers M and C, find d such that $M^d \equiv C$ (mod N).
   c. Given integers e and C, find M such that $M^e \equiv C$ (mod N).
   d. Given an integer x, decide whether there exists an integer y such that $x \equiv y^2$ (mod N).

Gill's suggestion was based on (1), that is, on the difficulty of computing discrete logarithms modulo a prime. This has generated systems (e.g., Section 2.2) suitable for key distribution, and others suitable for signatures (e.g., Section 4.2.2).

Exploitation of the difficulty of factoring has proven to be the most versatile approach to design of public key systems. Factoring is widely regarded as very difficult. If a modulus has unknown factorization, various problems in modular arithmetic become difficult. For example, discrete logarithm (2b) is difficult, as is taking roots (2c) and deciding quadratic residuosity (2d). These problems have generated the most widely known public key system (Section 4.1) and various others. We remark in passing that the probabilistic scheme mentioned in Sections 3 and 9.1 are based on (2d); more generally, quadratic residuosity forms the basis of many zero-knowledge schemes.

Diffie's knapsack suggestion is one of many attempts to utilize nondeterministic polynomial (NP)-complete problems for cryptosystems. Such attempts have met with very limited success. We return to this topic and further discussion of the above problems later. In the meantime we discuss a few of the most well-known public key systems along with some hash functions.

4.1 The RSA Public Key Scheme

Rivest, Shamir, and Adleman [RIVE78] obtained the best-known and most versatile public key cryptosystem. It supports both secrecy and authentication, and hence can provide complete and self-contained support for public key distribution and signatures.

A user chooses primes p and q and computes $n = p * q$ and $m = (p - 1)$ $(q - 1)$. He then chooses e to be an integer in $[1, m - 1]$ with GCD(e, m) = 1. He finds the multiplicative inverse of e, modulo m; that is, he finds (see Appendix H) d in $[1, m - 1]$ with $e * d \equiv 1$ (mod m). Now n and e are public; d, p, q, and m are secret. That is, for a user A, n_A and e_A constitute the public component, and d_A, p_A, and q_A the private component.

After a user has computed p, q, e, and d the private transformation D and public transformation E are defined by (see Appendix M)

$$E(M) = M^e \bmod n$$
$$D(C) = C^d \bmod n$$

In the equation above, M and C are in $[0, n - 1]$. As in Section 1.5, we have $D(E(M)) = M$ and $E(D(C)) = C$; that is, D and E are inverses. Since d is private, so is D; and since e and n are public, so is E. This constitutes a cryptosystem that can be used for both secrecy and authentication. That is, for secrecy, A sends $E_B(M)$ to B as usual; for authentication, A sends $D_A(M)$ as usual. For both secrecy and authentication, suppose first that message digests are not employed. Assuming $n_A \leq n_B$, A computes

$$C = E_B(D_A(M))$$

and sends C to B. Then B recovers M as usual by

$$M = E_A(D_B(E_B(D_A(M))))$$

As noted in Section 3.1.1, for nonrepudiation B may retain $D_A(M)$.

This implementation only works because the range of D_A is a subset of the domain of E_B; that is, $[0, n_A - 1]$ is a subset of $[0, n_B - 1]$. In the event that $n_A \geq n_B$, Kohnfelder [KOHN78b] notes that A can instead transmit

$$C' = D_A(E_B(M))$$

Then B can recover M as

$$M = D_B(E_A(D_A(E_B(M))))$$

This works since the range of E_B is a subset of the domain of D_A. For adjudication of possible disputes, B must retain C' and M. Then to prove that A sent M, the judge can compute $E_A(C')$ and $E_B(M)$ and test for equality.

However, in [DAVI83] and [DENN83b] a disadvantage to this solution is noted: A signs $E_B(M)$ rather than M. In Section 3.1.1 the judge was able to apply E_A to the stored quantity $D_A(M)$ and M is obtained. In Kohnfelder's protocol both C' and M must be stored, doubling the storage requirement.

The use of message digests eliminates the preceding problem. Suppose H is a hash function. Then to send M to B, A can create a new message M' containing M and $D_A(H(M))$ and send $C = E_B(M')$ to B. This gives both secrecy and authentication; also, C may be retained for nonrepudiation. Moreover, M and $D_A(H(M))$ are reblocked in forming M', so that no problem arises from the sizes of n_A and n_B.

4.1.1   Choice of p and q.   To use the RSA system, a user must first choose primes p and q. As noted in Section 4.1.3, p and q should be at least 75–80 decimal digits; for the sake of discussion we assume p and q to be on the order of about 100 digits. The user begins by choosing a random odd b of around 100 digits; b is a candidate for p. Now b is acceptable only if b is prime. There are two approaches to this problem. The deterministic approach decides with certainty whether b is prime (see below). An alternative is to use the probabilistic approach as implemented, for example, by Solovay and Strassen [SOLO77]: Randomly choose a set of about 100 numbers in $[1, b - 1]$. For each number a in the set, test to see if GCD$(a, b) = 1$, where GCD = greatest common divisor. If this condition holds, compute the Jacobi symbol (a/b) (Appendix J). Both GCDs and Jacobi symbols are easy to compute via well-known algorithms (Appendices H and J). Now check to see if

$$(a/b) \equiv a^{(b-1)/2} \pmod{b}$$

If b is not prime, this check will fail with probability at least $\frac{1}{2}$ (Appendix L). Thus, if 100 random a's are tested and all pass the test above, the probability of b not being prime is about 1 in 2^{100}. Hence it is possible to test in polynomial time whether b is probably a prime. The probability of b being prime is about 1 in 115; hence repeated applications of the preceding algorithm will probably quickly yield b which is probably a prime. This can be taken to be p; a second application of the algorithm will yield q.

The Solovay–Strassen algorithm is discussed further in Appendix L, which also mentions alternative algorithms of Lehman [LEHM82] and Miller and Rabin ([MILL76], [RABI80]).

The advantage of such probabilistic algorithms is that the algorithms run in polynomial time. They do not guarantee a correct answer, but the possibility of error is negligible as noted above. If absolute certainty were required, deterministic algorithms for primality testing could be used ([ADLE83], [COHE87], [COHE84]). These guarantee primality, but have runtimes of the form

$$\exp(c(\log \log n)(\log \log \log n))$$

If a sufficient amount of computing power is available, primality of 100-digit numbers can usually be established deterministically in minutes. However, the code needed is complicated; moreover, most users do not have access to high-performance computers. The probabilistic algorithms have acceptable runtimes even on small computers, particularly if hardware support is available.

4.1.2 Further notes on implementation. Once p and q have been chosen, e is chosen to be relatively prime to $m = (p - 1)(q - 1)$; that is, GCD$(e, m) = 1$. For example, e can be chosen to be a random small prime > 2. If e is relatively small, $E(M)$ can be computed relatively rapidly; that is, only a small number of modular multiplications are needed. The condition GCD$(e, m) = 1$ presents no problem; in fact the probability that two random numbers are relatively prime is about $\frac{3}{5}$ ([KNUT81], p. 324). Now d can be found in polynomial time; however, d may be large. Hence a version of the Chinese remainder theorem is employed for decryption (Appendix M).

There are some restrictions on p and q in addition to the size specification mentioned above. Some of these are noted in Section 4.1.3.1. Also, the time and space requirements of RSA are considerably larger than, for example, DES. Storage for each of p, q, and d requires at least 300 bits; n is about 600 bits. Only e can be chosen to be small.

Exponentiation mod n is a relatively slow operation for large n, especially in software. Efficient algorithms for computing products mod n have been given, for example, in [BLAK83], [BRIC82], [MONT85], and [QUIS82]. In particular, [BRIC82] shows how multiplication mod n can be done in m + 7 clock pulses if n is m bits. In [MONT85] it is shown how modular multiplication can avoid division.

At the present time RSA decryption (encryption is slower) with a 508-bit modulus has been performed at about 225K bits/sec in hardware [SHAN90]. Decryption with a 512-bit modulus has been effected at about 11K bits/sec in software [DUSS90].

4.1.3 Security of RSA. The security of the private components in the RSA cryptosystem depends on the difficulty of factoring large integers. No efficient algorithms are known for this problem. Consequently, knowing $n = p * q$ does not yield p and q. Thus it is computationally infeasible to find $(p - 1)(q - 1) = m$, and hence the relation $e * d \equiv 1 \pmod{m}$ is useless for determining d from e.

The difficulty of computing discrete logarithms (see Section 1.5) deflects known-plaintext attacks. Suppose an intruder knows M, C and $M = D(C)$ for some plaintext/

ciphertext pair $\{M, C\}$. To find d he must solve $M = C^d$ mod n; that is, he must find a discrete logarithm.

Ciphertext-only attacks are equivalent to taking roots modulo a composite number with unknown factorization, that is, solving $C = M^e$ mod n for M. This is probably about as difficult as discrete logarithms, even in the case of square roots (e.g., [ADLE86]). In fact (Lemma N.3.2; see [KNUT81], p. 389; [RABI79]), taking square roots modulo such n is, loosely speaking, as hard as factoring n.

It should be noted that the security of RSA is not provably equivalent to the difficulty of factoring or the other problems mentioned here. Also, security of RSA or other public key systems does not preclude breaking specific protocols for their use; see, for example, [MOOR88] or [DOLE81]. An example of a protocol failure is given in [MOOR88]: Suppose two users use a common modulus n and encryption exponents e and e' with $GCD(e, e') = 1$. If the same message M is sent to both, then let $C = M^e$ mod n and $C' = M^{e'}$ mod n. If both C and C' are intercepted, an intruder can find (Appendix H) r and s with $r * e + s * e' = 1$; now $M = C^r * C'^s$ mod n.

4.1.3.1 Restrictions on p and q.

There are two major restrictions on p and q in order to foil attempts at factoring n = $p * q$:

1. p and q should be about the same length.
2. p and q should be at least 75 digits in length.

Present factoring methods will be foiled by such a choice. For longer-term security, p and q should be, for example, around 100 digits. Condition (1) is needed against the elliptic curve method (see below).

An attempt to predict the period of time for which a given modulus length will remain secure is necessarily guesswork. Intelligent guesses are the best that can be hoped for by implementors; these are connected with the status of progress in factoring, which is highly dynamic.

4.1.3.2 Notes on factoring.

Choosing n to be about 200 digits will foil any present attempt at factoring n. The most powerful general-purpose factoring algorithm is Pomerance's quadratic sieve [POME84] and its variations. Using a parallel version of the quadratic sieve, Caron and Silverman [CARO87] factored 90-digit integers in several weeks on a network of several dozen Sun-3's. Recently, 106-digit integers have been factored on a larger network [LENS89], using the multiple polynomial variant of the quadratic sieve, and 111- and 116-digit composite integers have been factored by Lenstra and Manasse [BRILL88, update 91] using the quadratic sieve and the two-large-prime method.

Pomerance, Smith, and Tuler [POME88] discuss a prototype of a special pipelined architecture, optimized for the quadratic sieve, which they claim is extensible to a machine that might factor 125-digit integers in a year. However, this presents no threat to 200-digit moduli, since all of the best general-purpose factoring algorithms known, including the quadratic sieve, run in time roughly

$$\exp(((\log n)(\log \log n))^{1/2})$$

The fastest present computers only execute on the order of 10^{10} operations per second, at a cost of $10 million or more. Even if a future computer reached 10^{12} operations per second (and such a machine would presumably have a price tag of $1 billion or more) it would take on the order of 1000 years to factor a 200-digit number using existing algorithms.

The strongest results on factoring obtained to date have utilized networks of computers. The power of such networks is more difficult to extrapolate than for supercomputers, since they are expandable and the power of personal computers and workstations has been rising at a higher rate than at the supercomputer level. In theory it would be preferable to know that a cryptosystem would be immune to an attack by an assemblage of every computer in the universe. It seems difficult to estimate the total amount of computing power this would bring to bear; furthermore, the question in practice is what fraction of this total amount can be realistically allotted to a given factoring problem (Appendix B).

It follows that implementors may have a high level of confidence in 200-digit moduli at the present. An interesting issue at the time of this writing is whether a modulus with a length between 150 and 200 digits will provide security for a given length of time. It seems certain that 154-digit (i.e., 512-bit) integers will be factored eventually, but the question is when. If the preceding runtime estimate is used, the conclusion is that 154 digits are likely to be secure until the late 1990s, barring major algorithmic advances (see Appendix B). RSA moduli of around 512 bits are generally preferable to 200 digits (around 660 bits) from a hardware-implementation standpoint.

A potential qualifier to the preceding analysis is that the runtime estimate above is generic; in particular it assumes that runtime is essentially independent of the integer being factored. There are, however, algorithms that have the runtime estimate above as their worst-case time [DIXO81]. For example, the Schnorr–Lenstra algorithm [SCHN84] factors n by using quadratic forms with discriminant $-n$. The equivalence classes of such forms form a group (the class group) whose order is denoted by $h(-n)$. The algorithm factors n quickly if $h(-n)$ is free of large prime divisors. The algorithm is Monte Carlo in the sense that the n's that will be factored quickly are evidently fairly random. No algorithms are known to generate class numbers with large prime divisors, although interestingly RSA comes close: Class numbers $h(-n)$ of discriminants n with one or two prime factors have a higher probability of having large prime factors than the class numbers of discriminants with only small prime factors.

The net result is that, for example, perhaps 1 in 1000 n's will factor 1000 times more quickly than average. If this method were practicable it would argue for the 200-digit modulus in RSA. However, the Monte Carlo phenomenon has not produced noticeably lower factoring times in practice as yet.

Some other factoring methods are applicable to numbers that do not have the form required for an RSA modulus. For example, the elliptic curve method [LENS87] is oriented to integers whose second-largest prime factor is considerably smaller than the largest prime factor. This motivates the condition that an RSA modulus should have factors roughly equal in size. A recent algorithm, the number field sieve [LENS90], applies to integers of the form $r^d + s$ or $r^d - s$ where r and s are small. Again this will not affect properly chosen RSA moduli.

4.1.4   Low-exponent versions of RSA.

A priori the encryption exponent e in the RSA scheme is arbitrary. However, it need not be random. It has been suggested that e be chosen to have a common value by all users of an RSA-based system. Using $e = 2$ is not feasible per se, since 2 is not relatively prime to $p - 1$ or $q - 1$. However, extensions of this type have been made. These are of some theoretical interest, since Rabin [RABI79] and Williams [WILL80] have noted modifications of RSA with exponent 2 for which successful ciphertext-only attacks are essentially as difficult as factoring the modulus (Appendix N).

The exponent 3 can be used directly with RSA. It has the advantage of greatly simplifying the encryption (but not the decryption) process. In fact Knuth [KNUT81], Rivest [RIVE84], and others recommend its use. However, 3 and other very low exponents suffer from some serious flaws. The case $e = 3$ is an illustration. For one thing, as Knuth notes, users must make certain that each block M to be encrypted satisfies $M^3 \gg n$. This is because $C = M^3 \bmod n$ becomes simply $C = M^3$ if $M^3 < n$; in this event finding M reduces to the trivial problem of taking ordinary cube roots of integers. Thus the use of $e = 3$ causes the domain of E to be a subset of $[0, n)$ rather than the whole interval as would be preferable.

A related but more subtle flaw has also been noted if, for example, $e = 3$ [HAST88]. Suppose A sends the same message M to each of $\{B_i\}$, $i = 1, 2, 3$. Suppose B_i uses modulus n_i, and $M < n_i$ for each i (this will be true, for example, if the sender uses a modulus smaller than each of the $\{n_i\}$). Assuming that each of $\{n_1, n_2, n_3\}$ is generated as a product of two random primes, the probability of duplicates among the six primes used is near zero. Hence it may be safely assumed that the $\{n_i\}$ are pairwise relatively prime. Let $C_i = M^3 \bmod n_i$. Suppose all three $\{C_i\}$ are intercepted. Using the Chinese remainder theorem (Appendix I), the intruder can find x in $[0, n')$, $n' = n_1 *$ $n_2 * n_3$, with $x \equiv C_i \pmod{n_i}$, $i = 1, 2, 3$. Thus $x \equiv M^3 \pmod{n'}$. But $M^3 < n'$, so $M^3 = x$, so $M = x^{1/3}$ (ordinary cube root). Hence the plaintext M can be easily recovered from the three ciphertexts. Thus the use of $e = 3$ or other small exponents makes RSA vulnerable to ciphertext-only attacks. The sender may attempt to modify M for each recipient to deflect this attack, but Hastad shows that more generally, a low exponent is vulnerable to linear dependencies in portions of messages.

4.2 Other Public Key Systems

Several public key systems other than RSA have been proposed [MERK78b, MCEL78, ELGA85]. These are briefly surveyed here. As noted earlier, most of these are either insecure, of questionable practicality, or limited in scope. In some cases their security and/or practicality has not been established. None of these systems rivals RSA if a combination of versatility, security, and practicality is the criterion. However, this does not preclude their use for specific applications such as digital signatures.

4.2.1 Knapsack systems.   These systems were proposed by Merkle and Hellman [MERK78b]. The essence, however, is the use of NP-complete problems (e.g., [GARE79], [HORO78]) that had been suggested for use in designing public key systems in [DIFF76b]. The Merkle–Hellman approach was to employ a superincreasing sequence $\{a_i\}$, that is, a sequence of positive integers for which

$$a_i > a_{i-1} + \ldots + a_1$$

The associated knapsack problem [HORO78] is to find nonnegative integers $\{M_i\}$ such that for a given Y,

$$Y = a_1 * M_1 + \ldots + a_n * M_n$$

The above is an instance of integer programming. In the present context the $\{M_i\}$ are binary; thus the above reduces to sum-of-subsets. Solving the above is easy for superincreasing $\{a_i\}$; in fact the solution is unique. However, even deciding existence of a solution of sum-of-subsets, or more generally integer programming, is NP-complete ([GARE79], pp. 223, 245). Now for any w and u satisfying

$$u > a_1 + \ldots + a_n$$
$$\text{GCD}(u, w) = 1$$

a disguised knapsack problem may be produced from $\{a_i\}$ by defining

$$b_i = w * a_i \mod u$$

Then the associated knapsack problem

$$C = b_1 * M_1 + \ldots + b_n * M_n$$

appears to a cryptanalyst to be hard since the $\{b_i\}$ appear random. In actuality, it is easily solved using the connection with the $\{a_i\}$: since $\text{GCD}(w, u) = 1$, there exists W (Appendix H) such that

$$w * W \equiv 1 \pmod{u}$$

Then

$$W * C \equiv a_1 * M_1 + \ldots + a_n * M_n \pmod{u}$$

Since $0 \leq M_i \leq 1$,

$$0 \leq a_1 * M_1 + \ldots + a_n * M_n < u$$

from which

$$W * C = a_1 * M_1 + \ldots + a_n * M_n$$

As noted above, the latter knapsack problem has an easily found unique solution, from which C may be retrieved. This is called a trapdoor; it permits a legitimate user to easily solve what appears to be a hard problem. The modular disguise above can be iterated; that is, new w' and u' chosen and the $\{b_i\}$ disguised, etc.

The trapdoor knapsack yields a public key system: if the binary representation of plaintext M is $M_n \ldots M_1$, let

$$E(M) = b_1 * M_1 + \ldots + b_n * M_n$$

If $C = E(M)$ then decrypting C is equivalent to solving what appears to be a general knapsack problem, although decryption is easy using the trapdoor. The public component is $\{b_i\}$ and the private component is u, w, and $\{a_i\}$ (see [MERK78b] for details).

The advantage is that the arithmetic is much quicker than in RSA: about 200 additions are needed, as opposed to about 500 squarings and 150 multiplications in RSA. In fact knapsacks may rival DES in speed [HENR81]. Unfortunately the trapdoor knapsack above and most variations on the above have been broken (Appendix E).

It should also be noted that even if a knapsack approach proves to be secure and practical, such schemes are generally limited to support for either secrecy or authentication but not both. In contrast to RSA, the encryption and decryption functions are not inverses, although $D(E(M)) = M$. A trivial example suffices to show this: Suppose

$$E(M) = 2M_1 + 3M_2$$

To be invertible a function must be both injective and surjective. However, E is not surjective (onto). For example, there is no M such that $E(M) = 1$; hence $D(1)$ is undefined. Thus we could not employ D for signatures as in RSA, since, for example, 1 could not be signed by computing $D(1)$.

4.2.2 The El Gamal signature scheme.   A signature scheme derived from a modification of exponentiation ciphers was proposed by El Gamal [ELGA85].

First, a prime p and a base a which is a primitive root modulo p (Appendix K) are public. Now suppose A wishes to send a signed message to B. The private component of A consists of two portions, one fixed and the other message-dependent. The fixed portion is a random $x(A)$ from $[1, p - 2]$. The public component of A also has fixed and message-dependent components. The fixed portion is

$$y(A) = a^{x(A)} \bmod p$$

Now for a given message M in $[0, p - 1]$, A chooses a secret k in $[0, p - 1]$ with $GCD(k, p - 1) = 1$. Thus k is the message-dependent portion of A's private component. Also, A finds (Appendix H) I with

$$k * I \equiv 1 \pmod{(p - 1)}$$

The message-dependent portion of A's public component consists of r and s, where

$$r = a^k \bmod p$$
$$s \equiv I * (M - r * x(A)) \qquad (\bmod(p - 1))$$

Now A sends M, r, and s to B. For authentication B computes

$$C = a^M \bmod p$$
$$C' = y(A)^r * r^s \bmod p$$

We note that

$$M \equiv r * x(A) + k * s \pmod{(p - 1)}$$

Hence if M is authentic and valid,

$$C = (a^{x(a)})^r * (a^k)^s \bmod p = C'$$

A different k must be chosen for each M; using any k twice determines $x(A)$ uniquely. The security of the method then depends mainly on the difficulty of computing discrete logarithms in $GF(p)$: suppose an intruder intercepts M, r, and s; a and p are public. Let $d = a^r \bmod p$ and $u * r^s \equiv 1 \pmod p$. Then the intruder knows

$$a^M \equiv y(A)^r * r^s \pmod p$$

Hence

$$u * a^M \equiv y(A)^r \quad (\text{mod } p)$$

Thus

$$d^{x(A)} \equiv (a^{x(A)})^r \equiv y(A)^r \quad (\text{mod } p)$$

Finally

$$d^{x(A)} \equiv u * a^M \quad (\text{mod } p)$$

Finding $x(A)$, which permits the intruder to masquerade as A, is thus equivalent to the above discrete logarithm problem.

It is easy to solve this problem if $p - 1$ has only small factors [POHL78], hence $p - 1$ should have a large prime factor as in RSA. An alternative approach is to seek k and m satisfying

$$M = r * x(A) + s * k + (p - 1) * m$$

but this is underdetermined, so that there are an exponential number of possible solutions to check. A third approach at cryptanalysis seeks r and s satisfying

$$a^M \equiv y(A)^r * r^s \quad (\text{mod } p)$$

This is easier than discrete logarithm since r and s can both be varied, but it is not clear how much easier.

It is also not clear whether this scheme has a substantial advantage over RSA. Both employ exponentiation, so their speeds of encryption and decryption should be comparable. Generating keys in the two methods is similar; finding appropriate primes is the main step in either.

Cryptanalytically, the last attack (finding r and s simultaneously) may have a complexity substantially lower than factoring or discrete logarithm; hence security of El Gamal is at best no better than RSA, and possibly much inferior, with respect to secrecy of components. The use of message-dependent portions of components is a plus and minus: It increases bookkeeping, and makes it more difficult for a third party to adjudicate a dispute. On the other hand, it may produce greater security against certain types of attacks. In fact the k above could be chosen differently for different blocks of one message.

In passing we note that, like knapsacks, this scheme goes in one direction only: We have in effect

$$M = E(r, s) = (x(A) * r + k * s) \text{ mod } (p - 1)$$

This maps the ordered pair (r, s) to a unique M. The condition $GCD(k, p - 1) = 1$ guarantees that an (r, s) can always be found; that is, E is surjective. But it is not injective (one-to-one): for example, if $p = 7$, $k = 5$, $x(A) = 5$, then $E(1, 2) = E(2, 1) = 3$. Thus text M cannot be represented uniquely by a pair (r, s), which would lead to ambiguity in an attempt to use this scheme in two directions.

Also, El Gamal notes that extension to $GF(p^n)$ (Appendix G) is feasible. However, it is not clear that extensions to finite fields are desirable (see Section 1.5); gains in efficiency may be offset by lower security for comparable key sizes.

4.3 Examples of Hash Functions

To be useful in conjunction with public key systems, a hash function should ideally be keyless. This is in contradistinction to message authentication codes used in connection with secret key systems. Several hash functions have been proposed for use in public key settings. A few are mentioned here, but it seems to be difficult to produce secure, keyless, efficient hash functions. Thus we include some examples of functions that have keys, are insecure, or both.

4.3.1 Merkle's metamethod.

Merkle ([MERK82], [MERK89]) has proposed a general technique for obtaining hash functions and digital signatures from existing conventional cryptosystems such as DES. Although secret key systems are used as adjuncts, the resulting hash functions are keyless; keys needed for DES are generated from the message to be hashed. The hash functions are precertified in the sense that their security can often be proven to be the same as that of the underlying conventional function.

Merkle assumes that encryption in a system such as DES defines a function E_K, where K is a key, that produces a random output. This is technically impossible, and in fact DES is not a random function since it is a permutation. Nonetheless, small deviations from randomness may be tolerable. Another requirement is that for a given key/plaintext (or ciphertext) pair, the ciphertext (or plaintext) returned will not vary with time. Normally a cryptosystem would meet this criterion.

Assume that a function $E_K(M)$ is available that satisfies these requirements. In deference to the potential use of DES to define E, assume K is 56 bits, M is 64 bits, and $E_K(M)$ is 64 bits. Since E may be assumed to be an encryption function, it may be assumed that there exists a decryption function with $E_K(D_K(C)) = C$. This is undesirable in producing one-way hash functions. This problem may be mitigated as follows: given a 120-bit input x, write $x = \text{CAT}(K, M)$ where CAT denotes concatenation, K is the first 56 bits of x, and M is the last 64 bits. Let

$$f(x) = E_K(M) \ \text{XOR} \ M$$

where XOR is exclusive-or. Then f hashes 120 bits to 64 bits. Furthermore, Merkle claims that f produces a random output even if x is chosen nonrandomly. A strong one-way hash function, as defined in Section 3.2.3, meets this criterion. Thus f might be termed a fixed-size strong one-way hash function, with the term "fixed-size" referring to the fact that f does not accept arbitrary inputs.

There are various ways in which f can be extended to a one-way hash function accepting arbitrary inputs (see [MERK89]). Furthermore, it is desirable to have a function that outputs more than 64 bits, to deflect birthday or square root attacks (Appendix F), which are based on the statistical phenomenon that the probability of collisions produced when hashing becomes significant when the number of items hashed is around the square root of the total number of hash function values.

In Section 3.2.3 we noted how such attacks affect the security of hash functions. For example, a 64-bit output (i.e., 2^{64} hash values) would be vulnerable to an attack using only $2^{32} = 4$ billion messages. On the other hand, a 112-bit output would require exhaustive search of on the order of 2^{56} values, producing a security level comparable to DES. Merkle discusses several constructions for producing an output exceeding 64 bits from f. Only the simplest construction will be discussed here. Given an input x of

119 bits, let

$$\text{CAT}(f(\text{CAT}(``0", x)), f(\text{CAT}(``1", x))) = \text{CAT}(y, z)$$

where y is 112 bits. Then define $F_0(x) = y$. In the equation above, " " denotes a constant bit string. We note that F_0 hashes 119 bits to 112 bits. This is not very efficient; however, the 112-bit output will deter a birthday attack. The latter would require $2^{56} = 7 * 10^{16}$ messages, which is computationally infeasible; more precisely, adequate computing power to effect a birthday attack would also break DES, thereby obliterating DES-based hash functions.

This produces F_0 which is also a fixed-size strong one-way hash function. Finally F_0 is extended to a one-way hash function F accepting inputs of arbitrary size by cascading copies of F_0. Given an input x, suppose

$$x = \text{CAT}(x_1, \ldots, x_n)$$

where n is arbitrary and each x_i is 7 bits (pad x with a few zeros if need be). Let

$$R_0 = 0$$
$$R_i = F_0(\text{CAT}(R_{i-1}, x_i)) \qquad (1 \le i \le n)$$

where in the first equation the right side is a string of 112 zeros. Let $F(x) = R_n$.

Now Merkle claims that F is as secure as F_0. That is, if $x \ne x'$ can be found with $F(x') = F(x)$, then F_0 can also be broken. The proof is inductive: if $n = 1$ we have $F(x) = F_0(\text{CAT}(0, x))$. If $F(x) = F(x')$ and $x \ne x'$ let $z = \text{CAT}(0, x)$ and $z' = \text{CAT}(0, x')$; then $F_0(z) = F_0(z')$ and $z \ne z'$, so that F_0 is broken. Now suppose the claim is true for n. Suppose

$$z_i = \text{CAT}(x_1, \ldots, x_i) \qquad (1 \le i \le n + 1)$$
$$x = z_{n+1}$$
$$z_i' = \text{CAT}(x_1', \ldots, x_i') \qquad (1 \le i \le n + 1)$$
$$x' = z_{n+1}'$$
$$R_i = F_0(\text{CAT}(R_{i-1}, x_i)) \qquad (1 \le i \le n + 1)$$
$$R_i' = F_0(\text{CAT}(R_{i-1}', x_i')) \qquad (1 \le i \le n + 1)$$

where each x_i and x_i' is 7 bits. Suppose $x \ne x'$ but $F(x) = F(x')$, that is, $R_{n+1} = R_{n+1}'$. Let $y = \text{CAT}(R_n, x_{n+1})$ and $y' = \text{CAT}(R_n', x_{n+1}')$. Then $F_0(y) = F_0(y')$. If $x_{n+1} \ne x_{n+1}'$ then $y \ne y'$ and F_0 is broken. Suppose $x_{n+1} = x_{n+1}'$. Then $R_n = R_n'$; but $z_n \ne z_n'$ since $x \ne x'$. Now we observe that by definition $R_i = F(z_i)$ and $R_i' = F(z_i')$. Hence $F(z_n) = F(z_n')$. By the induction hypothesis, F_0 can be broken, completing the proof.

A disadvantage of this F is that it hashes 7 bits per stage of the cascade, and each stage involves two DES calculations. Thus 3.5 bits are hashed per DES calculation. Merkle gives other constructions that are more complicated but hash up to 17 bits per application of DES.

4.3.2 Coppersmith's attack on Rabin-type functions.

The Merkle meta-method is an attempt to use a conventional system as a generator for hash functions. This type of approach has its origin in some predecessors that have been broken by combinations of birthday attacks and meet-in-the-middle attacks as formulated by Coppersmith [COPP85].

An early attempt to construct a hash function in support of signature schemes was given by Rabin [RABI78]. Let H_0 be random (Rabin uses 0) and suppose a message M is divided into fixed-size blocks $M_1, \ldots, M_n$. Suppose $E(K,N)$ represents encryption of block N with key K using a conventional system such as DES. For $i = 1, \ldots, n$ let

$$H_i = E(M_i, H_{i-1})$$

Then a hash value is given by (H_0, H_n). Coppersmith's attack assumes that an opponent knows H_0 and H_n. He then constructs a bogus message whose hash value is also (H_0, H_n). Assume blocksize is 64 bits. The opponent begins by specifying $M_1, \ldots, M_{n-2}$ using H_0. He then generates 2^{32} trial values X. For each X he computes a trial H_{n-1} of the form

$$H(n - 1, X) = E(X, H_{n-2})$$

He sorts these 2^{32} values and stores them. Now he generates 2^{32} trial values Y. For each Y he computes a trial H_{n-1} of the form

$$H'(n - 1, Y) = D(Y, H_n)$$

where D is the decryption function corresponding to E. We note that $H(n - 1, X)$ is the value that H_{n-1} would have if $M_{n-1} = X$, and $H'(n - 1, Y)$ is the value H_{n-1} would have if $M_n = Y$. Each time the opponent tries a Y he searches through the sorted H-list (about 32 operations per search) and tries to find an X and Y with $H(n - 1, X) = H'(n - 1, Y)$. By the birthday phenomenon (Appendix F), the probability of finding X and Y is at least $\frac{1}{2}$. If X and Y are found, the opponent has obtained a bogus message with the proscribed hash value; furthermore, this takes at most about 2^{33} encryptions, 2^{32} storage, and $64 * 2^{32}$ comparison operations. This effectively breaks the scheme. Actually Coppersmith's attack was directed against a refinement by Davies and Price [DAVI80]; Rabin's original scheme had been broken earlier.

4.3.3   Quadratic congruential hash functions.

Another attempt to create hash functions, differing from both Merkle- and Rabin-type constructions, was made by Jueneman [JUEN82]. Unlike the Merkle metamethod, this method uses no external system as an adjunct. It is not keyless; an initialization vector is used. This limits the scope of use; in particular, such a function is useless for signature-only schemes. Nonetheless, quadratic congruential constructions are simple and efficient, and hence would be useful in some contexts if secure. Unfortunately it seems as though Coppersmith's attack applies to these as well.

Jueneman uses the same partition into fixed-size blocks as above, and again begins by choosing $H_0 = 0$. He also chooses a secret seed M_0, however, which is changed every message and transmitted as a prefix to the message. Thus assuming 32-bit blocks (to use the Mersenne prime $2^{31} - 1$), for $i = 0, \ldots, n$ he computes

$$H_{i+1} = (H_i + M_i)^2 \ \mod(2^{31} - 1)$$

Then the hash value is H_n. Coppersmith broke this first scheme. A revised scheme was proposed in [JUEN86]. The text is split into 128-bit blocks, which are further divided into four words. Recent results, as described in detail in the chapter on Digital Signatures by Mitchell, Piper, and Wild in this volume, suggest that this scheme may be insecure also. This is especially unfortunate in light of the fact that a hash function given in Annex D of X.509 [CCIT87] is defined similarly, via

$$H_{i+1} = M_i + H_i^2 \quad (\text{mod } n)$$

4.4 Hardware and Software Support

As noted at the beginning of Section 4, RSA and similar exponentiation-based algorithms suffer from relatively low bandwidths. At a minimum this implies that supporting software needs to be carefully designed; hardware support is probably a necessity in many settings. As noted by Sedlak [SEDL87], bandwidths of less than 64K bits/sec will degrade performance in many networks. Furthermore, arithmetic modulo numbers of an adequate number of bits (at least 512) should be supported. Essentially the requirement is a chip (or chip set) that can quickly compute quantities such as $a \bmod b$, $a^i \bmod b$, and perhaps multiplicative inverses or greatest common divisors. As Sedlak further notes, a single chip is preferable for security reasons. Since off-chip communication is slower than on-chip, single-chip implementations should also yield higher bandwidths.

4.4.1 Design considerations for RSA chips. Some general trade-offs in such design schemes are discussed by Rivest [RIVE84]. He classifies architectures as sequential (S), serial/parallel (S/P), or parallel/parallel (P/P). He then notes the following time and hardware costs:

1. Multiplying two k-bit numbers modulo a k-bit number:
   a. $O(k^2)$ time and $O(1)$ gates on an S
   b. $O(k)$ time and $O(k)$ gates on an S/P
   c. $O(\log k)$ time and $O(k^2)$ gates on a P/P
2. Exponentiating a k-bit number modulo a k-bit number:
   a. $O(k^3)$ time and $O(1)$ gates on an S
   b. $O(k^2)$ time and $O(k)$ gates on an S/P
   c. $O(k \log k)$ time and $O(k^2)$ gates on a P/P
3. Key generation, k-bit primes:
   a. $O(k^4)$ time and $O(1)$ gates on an S
   b. $O(k^3)$ time and $O(k)$ gates on an S/P
   c. $O(k^2 \log k)$ time and $O(k^2)$ gates on a P/P

This is a somewhat oversimplified version presented only for comparison purposes. Also, there are two basic assumptions used above: Exponentiation is an inherently sequential process; and the 100 or so primality tests used in key generation are done sequentially. The first assumption seems intrinsic. However, the second is pragmatic: The tests all use the same hardware, and replicating the latter 100-fold for parallel execution would be highly cost-ineffective. Rivest uses the time and cost estimates above to give some sample timings:

1. Decryption rate, 200-digit modulus:
   a. 0.005K bits/sec on an S
   b. 1.3K bits/sec on an S/P
   c. 95K bits/sec on a P/P

2. Key generation, 100-digit primes:
   a. 1,200,000 msec = 20 minutes on an S
   b. 5000 msec = 5 seconds on an S/P
   c. 70 msec on a P/P

It may be seen that reasonably high bandwidths can be achieved, but the time/cost trade-off is significant.

4.4.2 Proposed designs for RSA chips.   Various authors have proposed designs for chips supporting RSA. Of course, such chips could also support other exponential-based algorithms.

Orton et al. [ORTO86] discuss an implementation of RSA in 2-μm CMOS that should encrypt at 40K bits/sec for 512-bit moduli.

Sedlak [SEDL87] discusses a highly optimized chip which makes substantial use of lookahead. Thus the number of cycles required for exponentiation is not fixed; analysis of expected time is performed using a probabilistic finite-state machine (Appendix D). Support is provided not only for RSA but also for the ISO hash function (see above). Sedlak claims a deciphering rate of nearly 200K bits/sec is achievable for 780-bit moduli using a single 160,000-transistor chip with dual ciphering units, in 1.5 μm* CMOS. He also claims a key generation time of 2 seconds. A 5000-transistor, 5-μm CMOS prototype has been realized.

In [BRIC89] it is noted that chips capable of up to 450K bits/sec are being designed.

5 IMPLEMENTATIONS OF PUBLIC KEY CRYPTOGRAPHY

We examine here some existing implementations of public key cryptography, some implementations that are in progress, and some that have been proposed.

5.1 MITRENET

One of the earliest implementations of public key cryptography was in the MITRE encrypted mail office (MEMO) system, a secure electronic mail system for MITRENET ([SCHA82], [SCHA80]). MITRENET is a broadband cable system with a bandwidth of 1M bits/sec. It uses a carrier-sense protocol (e.g., [TANE81]); that is, each station can sense transmissions of all other stations. In fact the protocol requires all parties to monitor all communication, in effect requiring passive eavesdropping. A priori this provides no secrecy, authentication, or integrity services. Furthermore, it employs distributed switching, creating a potential for active intrusion. This is a good setting for a privacy enhancement testbed.

The MEMO system is a hybrid public/private cryptosystem. DES is used for data encryption. The Diffie–Hellman exponential key exchange of Section 2.2 is used for

---

* μm = micro meter, much more commonly called a micron. VLSI design rules are always stated in microns; 1 micron or 1 μm etc.

establishment of secret keys. To implement the Diffie–Hellman system, use of GF(2^n), with $2^n - 1$ a prime (called a Mersenne prime), was chosen for efficiency of implementation via linear feedback shift registers. Unfortunately the MEMO implementors used $n = 127$; the work of Adleman [ADLE79] rendered this choice insecure even before the system was implemented. This is noted by Schanning; in fact the use of $n = 521$ is recommended in [SCHA82], suggesting that the MEMO system was intended mainly for experimental purposes.

In the MEMO system, a public key distribution center (PKDC) is a separate network node containing public components in erasable programmable read-only memory (EPROM). Private components can be generated by users or by the system.

Each user workstation establishes secure communication with the PKDC. A session begins with the user requesting the file of public keys from the PKDC. The request is honored if it passes an identification test involving the user's private component. The file of public keys is then downloaded to the user's workstation, encrypted with DES to ensure integrity.

When the user sends a message to another user, the workstation generates a random document key. The latter is used to DES-encrypt the message. The public key of the recipient is used to generate a key-encrypting key which is used to DES-encrypt the document key.

There is no provision for lost keys. Some provision is made for detecting modifications to messages, using checksums. The use of Diffie–Hellman alone, however, does not permit authentication to be introduced.

5.2 Integrated Services Digital Network

In [DIFF87] a testbed secure Integrated Services Digital Network (ISDN) terminal developed at Bell-Northern Research is described. It can carry voice or data at 64K bits/sec. Public key cryptography is used for both key exchange and authentication. Reference to the Diffie–Hellman exponential key exchange is made. Evidently it is used in conjunction with DES. The article also alludes to signatures, but implementation is unclear.

As noted in [DIFF88], the use of public key cryptography in conjunction with DES provides good security. In particular, the exponential key exchange permits a session key unique to each call. Thus if long-term keys are compromised, recordings of calls made prior to compromise cannot be decoded. In conventional systems the compromise of a long-term key may compromise communications made previously with derived keys.

Public key computations in the terminal are implemented via a digital signal processing (DSP) chip, while an off-the-shelf integrated circuit implements DES, which is used for data encryption. Key management involves keys installed in phones, carried by users, and exchanged electronically.

Each Bell-Northern Research, Inc. (BNR) terminal incorporates a Northern Telecom M3000 Touchphone.

5.2.1 Keys. A public/private component pair is embedded in the phone; this pair is called the *intrinsic key*. The private component is contained in a tamper-resistant compartment of the phone. The public component serves as the name of the phone. These cannot be altered.

A second long-term public key stored in the phone is the *owner key*. This key is used to authenticate commands from the owner. It can be changed by a command signed with the current owner key, thus permitting transfer to a new owner.

A third long-term public key in the phone is the *network key*. This identifies the network with which the phone is associated. It validates commands signed with the private component of the network's key management facility (KMF). It can authenticate calls from network users, and it can be changed by a command signed by the owner key.

A short-term component pair stored in the phone is the working pair. This component is embodied in a certificate signed by the key management facility. During the setup process for a call, phones exchange certificates. The network key is used to authenticate certificates. This permits station-to-station calls.

Further information is needed for authenticated person-to-person calls. This information is contained on a hardware "ignition key" which must be inserted into the phone. This key contains the user's private component encrypted under a secret password known only to the user. It also contains a certificate signed by the KMF which contains the user's public component and identifying information. The latter is encrypted as well. Decryption of the information on the ignition key is effected via a password typed on the telephone touchpad.

Further certificates authorizing users to use particular phones are acquired by the phone from the key management facility; these are cached and replaced in first in, first out (FIFO) fashion.

5.2.2 Calling.   A person-to-person call begins as follows: The caller inserts his ignition key and dials the number. The phone interrogates the ignition key to check the caller's identity. The phone then checks its cache for an authorization certificate for the caller. If not found it is acquired by the phone from the KMF. The phone then dials the number of the other phone.

The two phones then set up. This begins with an exponential key exchange. The calling phone then transmits its certificate and user authorization. The receiving phone authenticates the signatures on both of the latter using the network key. The receiving phone then employs challenges. It demands real-time responses to time-dependent messages; the responses must be signed. This ensures that certificates are not played back from a previous conversation. One response is signed by the calling phone; another with the user's ignition key. A further level of security may be obtained via the ISDN D channel, which makes calling party identification available from the network without interrupting the primary call.

Now the called phone sends its certificates and responds to challenges as above. If the called party is home he inserts his ignition key. If the called phone does not have authorization for the called party to use the phone, it must obtain a certificate. Again this can be accomplished via the D channel without interrupting the call. The called party then undergoes challenge and response. Finally the conversation ensues.

5.3 ISO Authentication Framework

Public key cryptography has been recommended for use in connection with the ISO authentication framework, X.509 [CCIT87]. This is based on the directory, a collection of services and databases that provide an authentication service. Authentication refers to verification of the identity of a communicating party. Strong authentication uses cryptography. The credentials of a communicating party may be obtained from the directory.

No specific cryptosystem or hash function is endorsed in support of strong authentication; however, it is specified in [CCIT87] that a cryptosystem should be usable for both secrecy and authenticity. That is, the encryption and decryption transformations should be inverses, as is the case with RSA. Multiple cryptosystems and hash functions may be used in the system.

A user must possess a distinguished name. Naming authorities are responsible for assigning unique names. The crux of the authentication framework is the binding of user names and public components. Assuming for the moment that such binding has occurred, subsequent authentication in the ISO framework consists of locating a chain of trusted points within the directory. Such a chain exists if there is a common point of trust between two authenticating parties.

5.3.1   Use of certificates.   The X.509 public component management system is certificate-based. Binding of a user's name and public component occurs when a certification authority (CA) issues a certificate to a user. The certificate contains the user's public component and distinguished name, and the certificate's validity period. It is generated offline and signed by the CA using the CA's private component. Normally a user would generate his own public/private component pair and transmit the public component to the CA for inclusion in the certificate. At the user's request the CA may also generate the user's public/private component pair.

The CA vouches for the binding of the user's name and public component. The CA must not issue multiple certificates with one name. Also, a CA must keep his private component secret; compromise would affect not only the CA but also the integrity of communications involving users certified by the CA.

Obtaining a certificate from a CA requires a secure communication between the user and CA. In particular, the integrity of the user's public component must be assured. This communication can be online, offline, or both for redundancy.

The user receives a copy of the certificate obtained from the CA. This can be cached; for example, a component pair could be stored on a smart card along with the user's certificate and the CA's certificate. Additionally, certificates are entered in the directory. The user may place a copy of the certificate in the directory, or the CA may be authorized to do this. A user's directory entry may contain multiple certificates. A CA's entry in the directory information tree (DIT) contains the certificates issued for it by other CAs, as well as certificates it issues for other nodes.

Semiformally a certificate may be defined as follows:

> certificate : : =
> {
> signature algorithm identifier;
> issuer name;
> validity period;
> subject name;
> subject information
> }
> validity period : : =
> {

```
        start date;
        finish date
        }
subject information :: =
        {
        subject public key;
        public key algorithm identifier
        }
```

This format permits use of different algorithms. For a more formal description of relevant formats using ASN.1 see Annexes G and H of [CCIT87].

5.3.2 Certification paths.   An associated data structure is the directory information tree (DIT). Certification authorities are otherwise undistinguished users who are nodes in the DIT. A user may have certificates issued by several CAs. Thus the authentication structure, despite the use of the term DIT, is not tree-structured. Instead it may be modeled as a directed graph.

A certification path consists of certificates of nodes in the DIT. The public component of the first node is used to initiate a domino-type process that ultimately unlocks the whole path, leading to recovery of the public component of the final node. The simplest path is of the form (A, B), where A is a CA for B. Then a knowledge of the public component of A permits recovery of the public component of B from the certificate issued by A for B, since the latter is signed using the private transformation of A. This is readily extended by recursion: In a certification path $A_1, \ldots, A_n$, knowing the public component of A_i permits recovery of the public component of A_{i+1} from the certificate for A_{i+1} issued by A_i.

For a user A to obtain B's certificate involves finding a common trusted point, that is, a node that has certification paths to the two users individually. Joining these two paths produces a certification path between A and B. The paths, if they exist, may be obtained from the directory.

Although there is no restriction placed on the structure of the authentication graph, an important special case is when it is tree-structured. In this event the CAs are arranged hierarchically; hence each node (except the root) has a unique CA (its parent in the tree). Then each CA stores the certificate obtained from its superior CA, as well as various certificates issued by it. The common trusted point for a pair of users is the root of the DIT. User A may cache the certificates of nodes along the path from A to the root. The other half of the path to another user B is obtained by conjoining the path from the root to B.

More generally, a user A who communicates frequently with users certified by a particular CA could store paths to and from that CA (these may be different in the nonhierarchical case). Then to communicate with a given user B certified by that CA, A need only consult the directory entry for the CA and obtain the certificate of B.

A related concept is cross-certification: If two CAs C_1 and C_2 have certified users who communicate frequently, C_1 and C_2 may certify each other. Then the certificate issued by C_1 for C_2 can be stored in the directory entry of C_1 and vice versa. We note that in a hierarchical system, a priori a directory entry for a node contains only the certificate of a node's superior and the reverse certificate; if cross-certification is permitted then entries contain an indefinite number of certificates.

A user may cache certificates of other users. A CA may add another's certificate to its directory entry.

5.3.3 Expiration and revocation of certificates.
When a certificate expires it should be removed from the directory. Expired certificates should be retained by the issuing CAs for a period of time in support of the nonrepudiation service.

Revocation of certificates may occur because of component compromise involving either the user or the issuing CA. Revocation may also be necessary for administrative reasons, for example, when a CA is no longer authorized to certify a user. A CA keeps a time-stamped list of revoked certificates which the CA had issued, and a list of revoked certificates issued by other CAs certified by the first CA. Entries for CAs in the directory should contain revocation lists for users and CAs.

5.3.4 Authentication protocols.
Suppose A wishes to engage in communication with an authenticated B. Suppose further that A has obtained a certification path from A to B, for example, by accessing the directory, and has utilized the path to obtain B's public component. Then A may initiate one-, two-, or three-way authentication protocols.

One-way authentication involves a single communication from A to B. It establishes the identities of A and B and the integrity of any communicated information. It also deflects replay in communication of authentication information.

Two-way authentication adds a reply from B. It establishes that the reply was sent by B and that information in the reply had integrity and was not replayed. It also establishes secrecy of the two communications. Both one-way and two-way authentication use timestamps. Three-way authentication adds a further message from A to B, and obviates the necessity of timestamps.

Let

I_A = identity of A

I_B = identity of B

D_A = private transformation of A

E_A = public transformation of A

D_B = private transformation of B

E_B = public transformation of B

T_A = timestamp by A

T_B = timestamp by B

R_A = random number generated by A

R_B = random number generated by B

C_A = certification path from A to B

Identities refer to the distinguished names of A and B. A timestamp included in a message M includes an expiration date for M. Optionally, it also may include the time of generation of M. Random numbers may be supplemented with sequence numbers; they should not be repeated within the expiration period indicated by a timestamp in the same communication.

The one-way protocol is as follows:

1. A:
   a. generates an R_A.
   b. constructs message $M = (T_A, R_A, I_B, <data>)$ where $<data>$ is arbitrary. The latter may include data encrypted under E_B for secrecy, for example, when A is sending a DEK to B.
   c. sends $(C_A, D_A(M))$ to B.

2. B:
   a. decrypts C_A and obtains E_A. Also checks certificate expiration dates.
   b. uses E_A to decrypt $D_A(M)$, verifying both A's signature and the integrity of the signed information.
   c. checks the I_B contained in M for accuracy.
   d. checks the T_A in M for currency.
   e. optionally, checks the R_A in M for replay.

The two-way protocol is:

1. as above.

2. as above.

3. B:
   a. generates an R_B.
   b. constructs $M' = (T_B, R_B, I_A, R_A, <data>)$ where R_A was obtained previously and $<data>$ may include data encrypted using E_A.
   c. sends $D_B(M')$ to A.

4. A:
   a. decrypts $D_B(M')$, verifying B's signature and the integrity of the enclosed information.
   b. checks the I_A contained in M' for accuracy.
   c. checks the T_B in M' for currency.
   d. optionally checks R_B for replay.

The three-way protocol is:

1. A:
   a. generates an R_A.
   b. constructs $M = (T_A, R_A, I_B, <data>)$. Unlike the previous cases, T_A may be zero.
   c. sends $(C_A, D_A(M))$ to B.

2. B:
   a. decrypts C_A and obtains E_A. Also checks certificate expiration dates.
   b. uses E_A to decrypt $D_A(M)$, verifying both A's signature and the integrity of the signed information.
   c. checks the I_B contained in M for accuracy.
   d. optionally, checks the R_A in M for replay.
   e. generates an R_B.
   f. constructs $M' = (T_B, R_B, I_A, R_A, <data>)$ where R_A was obtained previously; T_B may be zero.
   g. sends $D_B(M')$ to A.

3. A:

   a. decrypts $D_B(M')$, verifying B's signature and the integrity of the enclosed information.

   b. checks the I_A contained in M' for accuracy.

   c. optionally checks R_B for replay.

   d. checks the received version of R_A against the version sent to B.

   e. sends $D_A(R_B)$ to B.

4. B:

   a. decrypts $D_A(R_B)$, verifying the signature and data integrity.

   b. checks the R_B received against the value sent to A.

Remarks: It has been noted that there are some potential problems with these protocols. For example, in the three-way version, it would be better to have A send $D_A(R_B, I_B)$ to B rather than just $D_A(R_B)$. This would prevent a third party from intercepting random numbers and using them to subvert the protocol.

Also, since in both of the multipass protocols authentication tokens may be of the form $M = (T_A, R_A, I_B, R_B, E_B(\text{encdata}))$. Then encrypted data are signed, which may cause problems. For example, suppose C intercepts $D_A(M)$ on its way to B. Then C may decrypt M and construct $(T_C, R_C, I_B, R_B, E_B(\text{encdata}))$. In a communication between B and C, B may think that $E_B(\text{encdata})$ was generated by C and inadvertently reveal encdata to C. In particular, encdata may have included a data-encrypting key to be used by A and B. A solution to this problem is to let $M = (T_A, R_A, I_B, R_B, \text{encdata})$, $M' = (T_A, R_A, I_B, R_B, E_B(\text{encdata}))$, and $M'' = (M', D_A(M))$, then send M''.

5.3.5 Further notes.   Authentication uses a hash function. Signed information includes identifications of the hash function used to produce the message digest and the decryption function used to generate the digital signature. Timestamps and random numbers are included in messages to prevent replays and forgery.

Annex C of [CCIT87] mentions RSA as an example of a public key system. A recommendation of a 512-bit modulus is made. It is also recommended that a common public exponent of $e = 2^{16} + 1$ be used (the fourth Fermat number). In particular it is noted that a smaller e would be vulnerable to ciphertext-only attacks.

Other annexes specify a strong hash function as defined in Section 4, and the Abstract Syntax Notation 1 (ASN.1) syntax for the authentication framework. Algorithm identifiers are included in ASN.

5.4 Defense Advanced Research Projects Agency (DARPA)-Internet

The IAB Privacy Task Force is in the process of developing a proposal for privacy-enhanced electronic mail for the DARPA-Internet community ([IAB-88], [IAB-88b], [LINN89]). The summary presented here is of work in progress.

Public key cryptography has been recommended for distribution of secret keys and in support of authentication of communicating parties (i.e., for signature generation). It is anticipated that a conventional secret key cryptosystem will be used for encryption of messages and for message integrity check (MIC) computation. RSA has been recommended for key distribution; it may also be used for signatures, although other possibilities have been suggested. A universal hash function is not currently supported. Fields are included in messages to identify cryptographic algorithms and hash functions. It is specified that all implementations should support RSA and DES; the role of other systems has not been finalized at the time of this writing.

Much of the authentication framework is borrowed from a subset of X.509. In particular, the recommended public component management system is certificate-based.

5.4.1   Services.   Services to be provided in the DARPA-Internet system include:

- Data confidentiality
- Data origin authentication
- Message integrity assurance
- Nonrepudiation of origin

The framework for providing these services is X.400. End-to-end service is supported between user agent (UA) processes. To send mail, a user calls on a UA, which is an application program that interacts with the message transfer system (MTS). The MTS delivers the message to one or more UAs. No special requirements are made on the MTS.

Privacy enhancement occurs at the application layer only, with integration at the UA level or above. End-to-end encryption is employed. Encapsulation of privacy-enhanced messages within an enclosing layer of headers interpreted by MTS is supported. Lack of involvement of lower layers implies a corresponding limitation of services. For example, access control and routing control are not addressed; nor are security services aimed at attacks such as traffic analysis. Also, no provision is made for nonrepudiation of receipt.

Message text may or may not be encrypted, at the sender's option; encryption of only subsets of message text is also supported. Authentication always uses the whole message.

5.4.2   Key management.   DEKs are shared secret keys that are used for text encryption. They are used in connection with secret key data encryption systems such as DES. They are also employed when secret key cryptography is used to compute MICs. New DEKs are generated for each message. They can be generated by a KDC or at endpoints. No form of predistribution is required for these keys.

Interchange keys (IKs) are used to encrypt DEKs and MICs. The same IKs can be used by two users for a period of time. Assuming that public key cryptography is used for key distribution, the recipient's public component is used as an IK to encrypt DEKs; the sender's private component is used as an IK to sign MICs. If secret key cryptography is employed for key distribution, an IK functions as a master key.

A message contains X-Sender-ID and X-Recipient-ID control fields which are used to identify the IK used to encrypt the DEKs or MICs for a given recipient. If public key cryptography is employed, and a message is intended for several recipients, multiple X-Recipient-ID fields are required to hold the IKs (recipient public keys) used for encrypting DEKs. Again in the case of public key cryptography, a single IK (the originator's private component) may be employed directly in MICs.

5.4.3   Encapsulation and encoding.   A semiformal description of message format is as follows:

```
message :: =
    {
    enclosing header;
    encapsulated message
    }
encapsulated message :: =
    {
    encapsulation boundary;
    encapsulated header;
    blank line;
    encapsulated text;
    encapsulation boundary
    }
```

The enclosing header contains RFC-822 header fields. It also contains user-supplied data, for example, "subject." These fields pertain only to message transport, and not privacy or authentication.

The encapsulated header contains encryption control fields such as DEKs and IKs. It also contains user address information needed for encryption. These fields, however, pertain only to privacy and/or authentication, and not to message transport. The division of fields between the enclosing and encapsulated headers is related to the end-to-end encryption employed; intermediate nodes and gateways need to access unencrypted routing information.

Encapsulated text contains encrypted user message text. Optionally, it may contain encrypted copies of protected fields of the enclosing header, for purposes of redundancy in case the enclosing header is accidentally or deliberately modified. Separation of the two headers ensures that modification of header fields used for transport will not affect privacy.

A message is accepted in local form, using the host's character set and line representation. The local form is then converted to canonical form. The latter is padded per the encryption mode, and a MIC is computed. Assuming that data confidentiality is desired, the padded canonical representation is encrypted. The encrypted form is encoded into printable form, universal across all sites. The printable form is combined with header fields. Finally the result is passed to the mail system for encapsulation as the text portion.

5.4.4 Use of certificates. Certificate-based key management is supported. This is compatible with a subset of X.509 [CCIT87]. Certificates can be distributed through insecure channels. A distributed directory service has been suggested, but implementation prospects are unclear at the moment. A sender must obtain a certificate for each recipient of a privacy-enhanced message.

Semiformally, a certificate has the following format:

```
certificate :: =
    {
    serial number;
    issuer name;
    user name;
```

```
validity period;
user public component / algorithm identifier;
issuer signature / algorithm identifier
}
```

The issuer's signature is computed on a hash value based on certificate contents; the certificate also contains an identifier for the hash algorithm used. Components of the subject (user) name include country, administration domain (DARPA-Internet), personal name, and a domain-defined attribute to identify the subject's mailbox address. Organization and organizational unit affiliations are optional. Issuer names follow a similar format. Validity period is at most 2 years.

5.4.5 Authentication framework.

The authentication structure here deviates somewhat from X.509 in that minimal assumptions are made regarding a directory system infrastructure. Thus, for example, it could not be assumed that certification paths through an arbitrary directed graph could be located by users. This dictates the use of a hierarchical (tree-structured) certification framework, with users at the leaves. Each node (except the root) then possesses a unique parent node and a unique path leading to the root. The root thus functions as a common point of trust for all users.

Each user has exactly one certificate. The certificate for node A is issued by the parent node of A, or other nodes lying between A and the root of such a hierarchical system. The latter typically has three or four levels, so that certification paths are correspondingly short. An issuing node is called a certification authority (CA). The latter generates a certificate from a subject's public component and identification and signs a message digest computed on the certificate. Thus a CA vouches for the binding between user ID information and public component. The CA for a user may be the root itself.

A simple hierarchy has users at the leaves, with only one intermediate level between leaves and root, consisting of organizations. In this case a CA may represent an organization; then users affiliated with that organization can obtain certificates through it. The binding between the user's ID and public component implicitly affirms the user's affiliation with the organization.

It is also possible to add a layer to the hierarchy by having organizational units or organizational notaries (ONs) within organizations. Further refinement is discouraged to shorten certification paths. Whether or not or organizational units are present, the root certifies organizations. Organizations may certify users, or they may certify organizational divisions or notaries which in turn certify users. Organizational units and notaries may possess a second certificate issued by the root.

This tree-structured framework may be extended to a forest by permitting multiple roots. For example, the U.S. government might have its own certifying facility. Other roots may exist in other countries as well. If multiple roots are present they should cross-certify.

Public components of CAs may be obtained recursively from certification paths as usual; that is, each certificate in the path is used to decrypt the next, until finally the user's public component is recovered.

5.4.6 Obtaining certificates.

The process by which a user obtains a certificate is independent of whether the certificate is acquired through affiliation with an organization, or directly from the root. In any case the root does the actual certificate generation. Intermediate nodes merely vouch for the binding between IDs and public components.

When a user obtains a certificate through an organization, it is assumed that the user properly identifies himself to the organizational notary for the organization. The user may generate his own public/private pair.

To obtain a certificate, the following information pertaining to the user is sent to the root:

- Distinguished name
- Postal address
- Electronic mail address
- Message digest

Distinguished names are formed from legal names, organizations, and so on; they should be unique. The message digest is computed on the four items above concatenated with the user's public key. Canonical forms are needed for these items. The form containing these items is taken to an organizational notary if the user is affiliated with an organization possessing such a notary. Otherwise the user must take the information to a notary public.

In either case the user information is then sent to the root via the postal service. The file containing the user's public component and the information above, including the message digest, is sent concurrently to the root via electronic mail. A receipt and message digest computed on the user's public component and identifying information is returned to the user by the root via the postal service; the certificate is sent via electronic mail. The public component of the root is also sent to the user.

For an organization to issue certificates it must register with the root, obtaining a certificate as above. One organizational official represents the organization; this is merely a distinguished user. The organizational official becomes the first organizational notary (ON) for the organization; he may authorize others. Then there is one ON per organizational unit. Any ON may obtain certificates for users. The ON may vouch for only a portion of the user's information, for example, name.

Private components of organizations, rather than those of ONs, are used to sign certificates. That is, there is a distinction between ONs and CAs: Although ONs are certified to perform gathering and validating of data, they are otherwise undistinguished users. Only the root and organizations are certified to act as CAs. The root acts as CA for nonaffiliated users and users affiliated with organizations lacking an ON.

5.4.7 Revocation. Revocation of certificates may be necessitated by compromise of a user's private component or a change of distinguished name caused by a change of the user's affiliation with an organization or organizational unit.

The root should maintain a hot list of invalidated certificates or compromised keys. This may be available online. Hot lists might be issued about once per month.

In addition, each CA should maintain a time-stamped list of revoked certificates for users and other CAs. It is assumed that certificates contain serial numbers, unique for a given CA. Lists of revoked certificates can contain only serial numbers. It is recommended that each CA publish hot lists periodically, with each hot list containing the date of release of the next one.

5.4.8 Authentication and key exchange. Validation of certificates involves verification that the affixed signature is valid; that is, the hash value computed on certificate contents matches the value obtained by decrypting the signature field using the issuer's public component. This assumes that the latter has already been validated, that is, by possession of a validated certificate issued to the issuer in the event that the issuer is not the root. More generally, this involves recursive decryption of the certificates in a certification path; however, the authentication framework is assumed to have at most four levels, meaning that at most four certificates are involved.

To communicate with another user with whom the sender has not corresponded previously, the sender forwards a message to the recipient, with the X-certificate field of the message header containing the sender's certificate C_1. The message header may also contain the certificate C_2 of the issuer I of C_1 (see below). Assuming for simplicity that C_2 is present and was issued by the root, the recipient uses the root's public component to validate C_2 as above. The recipient then extracts the public component of I from C_2 and uses it to validate C_1 as above. The recipient in turn extracts the public component of the sender which is now validated.

The recipient now uses the sender's public component to decrypt the MIC, contained in the X-MIC-Info field of the message header. This permits verification of the authenticity and integrity of the message, by comparing the recovered MIC to the locally calculated MIC. We recall that the sender signed the MIC with his private component.

Once a user has validated the public component of another user, this may be used to encrypt a DEK which can in turn be used to encrypt messages. An encrypted DEK is placed in the X-Key-Info field of the message header. It is decrypted with the private component of the recipient.

The preceding process can be speeded up by caching of validated public components. Also, it has been suggested that multiple issuer fields could be contained in certificate headers; then entire certification paths could be represented, as assumed above in the case where the root is the issuer of the sender's certificate. This could serve as a substitute for a directory service (possibly on an interim basis).

5.4.9 Further notes.   The X.509 Annex C recommendation of using the Fermat number $F_4 = 65,537$ as the common public exponent of RSA has received partial support. However, the use of 3 has also been suggested as a user option. Modulus size of 512–1200 bits has been suggested. Suggested hash functions include MD4 [RIVE90]. It must be emphasized that this is work in progress, however.

Uniqueness of the root has not been decided as yet. It is possible that several roots may be needed; these could cross-certify each other.

6 A SAMPLE PROPOSAL FOR A LAN IMPLEMENTATION

We present here a sample proposal for the implementation of privacy enhancement in a packet-switched LAN, using public key cryptography for key management and authentication. The main purpose is to explore the relevant decision-making process. In particular, some services will be needed in some settings but not others, and hence a monolithic structure incorporating all conceivable services would be inappropriate. One approach to the design of a generic structure is layered, with a kernel consisting of the most basic services and outer layers added as desired. This is the paradigm we adopt here.

A hybrid of public key cryptography and conventional cryptography is recommended. The former is used for signatures and for distribution of secret keys used in the latter; the latter is used for bulk data encryption. In addition, a hash function is needed so that only a compressed form of long text need be signed. We do not endorse specific public key systems or hash functions. The conventional system may be taken to be DES.

6.1 Integration into a Network

There are basically two modes of implementation of encryption in a network (e.g., [DIFF84], [DIFF86], [TANE81]), namely, link and end-to-end.

Link encryption provides good protection against external threats such as traffic analysis because all data flowing on links can be encrypted, including addresses. Entire packets, including addresses, are encrypted on exit from, and decrypted on entry to, a node. Link encryption is easy to incorporate into network protocols.

On the other hand, link encryption has a major disadvantage: A message is encrypted and decrypted several times. If a node is compromised, all traffic flowing through that node is also compromised. A secondary disadvantage is that the individual user loses control over algorithms used.

In end-to-end encryption, a message is encrypted and decrypted only at endpoints, thereby largely circumventing problems with compromise of intermediate nodes. However, some address information (data link headers) must be left unencrypted to allow nodes to route packets. Also, high-level network protocols must be augmented with a separate set of cryptographic protocols.

Here we assume end-to-end encryption. In terms of the OSI model, encryption can occur at various levels, including application, presentation, network, or transport. As noted in [ISO-87], the appropriate level depends on desired granularity of protection. In particular, high granularity refers to separate keys for each application or user and assurance of integrity. For this granularity the presentation or application layers are most appropriate. These two layers will be assumed here. In particular, integration at the application layer gives the individual user complete control over the algorithms used.

6.2 Security Threats

Some basic security threats (see Annex A of [CCIT87]) include:

- Masquerade
- Replay
- Interception of data
- Manipulation of messages
- Repudiation

Masquerade refers to users representing themselves as other users. *Replay* refers to recording and resending a message at a later time. *Interception of data* refers to passive eavesdropping on communications. *Manipulation* refers to unauthorized insertions, deletions, or other changes to messages. *Repudiation* refers to denial of sending (or possibly receipt) of a message.

There are other threats that are outside the present scope per se. For example, misrouting of packets naturally occurs in OSI layers 1–3, and we are restricting ourselves to higher layers.

6.3 Security Services

Again from Annex A of [CCIT87], some of the most basic security services that can be provided include:

- Authentication
- Secrecy
- Integrity
- Nonrepudiation

Authentication refers to verification of the identity of the sender or receiver of a communication. It can protect against masquerade. It may also provide protection against replay, depending on implementation. *Secrecy* refers to protection against interception of data. *Integrity* refers to protection against manipulation (including accidental) of data. *Nonrepudiation* refers to protection against denial of sending (or possibly receipt) of a message.

The kernel of a proposed system would normally support at least authentication, secrecy, and integrity. Nonrepudiation is somewhat more complex as we have noted. A public key signature system implements basic nonrepudiation of sending automatically, but does not a priori protect against repudiation of sending due to compromised private components, nor does it provide proof of receipt. Thus nonrepudiation is an example of a service (beyond the basic) which is most appropriately treated as an outer layer of an implementation, whose existence and structure depends on considerations such as contractual agreements that cannot be incorporated into a generic structure.

There are other services that could also be provided. One example is access control with differing capabilities for different classes of users. A public key system, however, does not provide this type of service per se, since it would require restricted access to public components. This type of service is thus another example of a layer which could be added in a given implementation, for example, by centralized key distribution which might be considered undesirable in many settings.

6.4 Security Mechanisms

Once more we follow Annex A of [CCIT87], which identifies some of the basic mechanisms that can implement security services. These include:

- Encryption
- Manipulation detection codes
- Signatures
- Authentication framework

Encryption refers to application of cryptographic transformations to messages or keys; it implements secrecy. *Manipulation detection* can be effected via hash functions producing a compressed version of the text; this implements integrity. *Signature* refers to application of private components to message digests (output of hash functions); it implements authentication and basic nonrepudiation, in concert with an authentication framework. The authentication framework is the system protocol and procedures within which authentication is achieved. Its form is implementation-dependent. For example, a directory service might be provided, along with hot lists of compromised keys.

The four mechanisms above may be regarded as the kernel of an implementation. Various other mechanisms that could be provided are noted in [ISO-87]. For example, to guard against replay, timestamps using synchronized clocks might be provided. Another possibility is the addition of a handshaking protocol. For enhancement of the basic nonrepudiation mechanism (the signature), a notary system could be used. Again these auxiliary services are best added at the discretion of implementors. For example, synchronized clocks may not be present, and notaries violate the desirable feature of point-to-point communication, and hence should not be included in standard specifications.

6.5 Criteria for Cryptosystems

There are various criteria that could be employed in selection of a cryptosystem (or systems) to implement key distribution and signatures. Logically these are separate functions and a hybrid of two separate systems could be used. Some relevant criteria (including suggestions from Dr. Dennis Branstad of the National Institute of Standards and Tests (NIST) and Dr. Ronald Rivest) are:

- Security
- Versatility
- Bandwidth
- Data expansion
- Key size
- Key generation time
- Patent restrictions/license fees
- Software support
- Hardware support
- Number of pseudorandom bits needed
- Interoperability

Most of these are self-explanatory. The reference to pseudorandom bits, above, is related to key generation, which requires a stream of pseudorandom bits generated from a random seed. *Interoperability* refers to current use and endorsement within and outside of the United States.

6.5.1   Security.   The first and foremost criterion is of course security. It may take 5 years or more for a given method to be thoroughly cryptanalyzed, starting from the time it receives widespread public exposure.

One subcriterion in this regard is that a system should be published in an outlet such as a refereed journal with a reasonably large circulation, or a book or conference proceeding that is present in libraries at larger academic institutions and research laboratories. This subcriterion is somewhat vague and not intrinsic, but may serve to avoid future embarrassment on the part of implementors.

A second subcriterion connected with security deals with mathematical and computational infrastructure. For example, the security of systems such as RSA is connected with the problem of factoring. It is safe to assume that thousands (or perhaps millions) of hours have been spent on this problem. This does not preclude major advances over the next decade or so, but at least guarantees that such advances will not occur simply because experts have suddenly become aware of a problem. In particular, systems based on long-standing problems such as factoring, discrete logarithm, and discrete root extraction have a certain degree of security in this regard. In contrast, for example, the El Gamal signature scheme can be broken if the "transcendental" equation $c = b^r r^s$ (mod n) can be solved for some r and s. This is easier than solving the logarithm or root problem $y = x^a$ (mod n), and furthermore in all likelihood the El Gamal equation has not been studied as extensively.

6.5.2 Numeric criteria. Quantitative criteria for cryptosystems include bandwidth (encryption and decryption rates), data expansion (relative sizes of plaintext and ciphertext), key size, and key generation time. We have already noted the phenomenon that systems that appear to be secure are also characterized by low bandwidths. Exponentiation-based methods such as RSA and the Diffie–Hellman exponential key exchange are examples. Other systems suffer from large data expansion, large key size, or long key generation time.

There seem to be some inherent trade-offs in this regard. That is, it does not seem possible to construct a system that is secure and also scores well on all numeric counts. The classic example is key size; for example, small key size in RSA would produce a high bandwidth, but would also produce insecurity. That is, in this case high bandwidth would produce low cryptanalysis time. Data expansion seems to have a similar impact. For example, in Appendix E it is noted that Shamir's knapsack attack runs in time which rises as n^d where d = expansion factor. Again high bandwidth produced insecurity. It would be interesting to know whether this trade-off notion could be formalized.

6.5.3 Other criteria. Versatility is an important criterion. The RSA system is distinguished in this regard since it supports both key distribution and signatures. All other major systems are limited to one or the other.

Patent restrictions and license fees may be a major factor in practice. Software and hardware support is another important practical matter.

6.6 Criteria for Hash Functions

In Section 3.2 we discussed some of the characterizations that have been proposed for hash functions, including measures of security. Some of the discussion in Section 6.5 applies here as well. For example, a hash function should be widely publicized for a period of time before it is trusted. Bandwidth is also an important criterion. Software and hardware support is relevant as well.

6.7 Some Recommendations

Finally we give a brief outline of a framework for incorporating public key cryptography into a LAN.

6.7.1 Key management.   The recommended framework for key management is compatible with a subset of [CCIT87]. Public components of receivers are used as key-encrypting keys; private components of senders are used to encrypt message digests. DEKs are generated for individual sessions. These keys may be associated to an arbitrary conventional cryptosystem, although DES is recommended. The public and private components may be associated to different public key systems if different algorithms are used for key encryption and signatures.

Certificate-based key management is recommended. Since we are focusing on LANs, a simple tree structure is proposed, consisting of a root and one additional level containing all users. In particular, the root issues all certificates. Certification paths are thus trivial.

6.7.2 Component generation and storage.   It is recommended that users generate their own public/private component pairs, using trusted software or hardware supplied by the system. Key pairs could be stored on smart cards [HAYK88] or tokens, along with the user's certificate. Such kernel mechanisms could be augmented by involvement of a central key distribution facility. However, we have noted this would negate a major advantage of public key cryptography, since compromise of the central facility would compromise all keys of users who obtained their keys from it.

6.7.3 Secret key generation.   Numerous schemes exist for generation of data-encrypting keys. For example, in [MATY78] it is noted that keys can be generated by a conventional system such as DES. A master key might be used to encrypt DEKs or other key-encrypting keys. The master key, presumably long-term, is generated randomly. Other key-encrypting keys can be generated using DES as a pseudorandom generator. These are then encrypted under the master, which can also be involved in generation of other key-encrypting keys. A whole set of key-encrypting keys can be generated from a random 64-bit seed, with every eighth bit adjusted for parity. DEKs can be produced dynamically by pseudorandom number generators that are time-variant.

An example of a generator for DEKs, as well as initializing vectors (IVs), is given in Appendix C of [ANSI85]. Let $E(K, Y)$ be encryption by data encryption algorithm (DEA) (essentially equivalent to DES) with key K. Let K be a DEA key reserved for use in key generation and let V_0 be a 64-bit secret seed. Let T be a date/time vector, updated on each key or IV generation. Let

$$R_i = E(K, E(K, T_i) \text{ XOR } V_i)$$
$$V_{i+1} = E(K, R_i \text{ XOR } E(K, T_i))$$

Then R_i may be employed as an IV. To obtain a DEA key from R, reset every eighth bit to odd parity.

Routines for generating DES keys are supplied with various products. Schemes for pseudorandom number generation include [AKL-84], [BLUM84], [BRIG76]. The last reference gives a pseudorandom bit generator whose security is equivalent to discrete logarithm.

6.7.4 Issuance and distribution of certificates.   Public components are registered with the root. The root generates a certificate containing the user's public component and identifying information, and a validity period. Distribution of certificates by users is recommended; certificates may be cached. This eliminates the necessity of having the root be online.

However, a user may wish to send a privacy-enhanced message to a user with whom he has not previously communicated, and who is currently unavailable. Thus it may be desirable to augment this kernel mechanism with a supplementary source of certificates. There are disadvantages to any augmentation. For example, if a phone book approach to certificates is used, entries may be inaccurate or altered. If a central directory mechanism is involved in key distribution it must be online. On the other hand, a central mechanism can provide instantaneous validity checks of public components and certificates.

The root should maintain a list of old public components for a period of time in event of disputes, for example, over-attempted repudiation.

6.7.5 Compromised or invalidated certificates.

Assuming that certificates are cached, the root must periodically issue hot lists of invalidated certificates. This kernel service may be augmented in various ways to provide more current validity information. Again, however, additional mechanisms have drawbacks. As noted above, a central directory service could provide real-time validity checks, but it must be online. Furthermore, such checks a priori do not account for compromised private components, during the period following compromise but before a report is filed with the root. Even if users are required to report known compromises within a specified time period, a compromise may not become known to the user until later. As we noted, this creates an administrative and legal problem, since a user could disavow a signature on the basis of the latter type of compromise. A notary system can be used to add a layer of validity by having secondary signatures attest to the validity of senders' signatures, but this violates the desired criterion of point-to-point communication.

This is clearly an area in which solutions must be customized to some degree. For example, authentication in a system used for financial transactions may require more stringent controls than a kernel provides.

The root should maintain a time-stamped list of revoked certificates.

6.7.6 Authentication.

A hash function is used to produce a message digest. This digest is signed with the private component of the sender. Timestamps and random numbers may also be included to prevent replay. If more than one hash function is permitted, identification of the function used should be included.

Privacy-enhanced communication between two users begins when A requests the certificate of B. This initial message contains A's certificate. As noted above, it is desirable if this request can be met directly by B. In this event, B first validates A's certificate. This uses the public component of the root, which is assumed to be known to all users. Then B forwards his certificate to A, who validates it. Each may cache the certificate of the other.

Now A and B may communicate securely. If A sends a message to B, B uses A's validated public component, extracted from A's certificate, to decipher the message digest. Then B may decrypt the key used for data encryption, using B's private component. Now B may decrypt the message text using this session key, then recompute the message digest and compare to the transmitted form.

7 MATHEMATICAL AND COMPUTATIONAL ASPECTS

We discuss here some issues related to the computational complexity of public key cryptography. The foundation of the security of such systems is the computational infeasibility of performing cryptanalysis, in contradistinction to the relative ease of encryption/decryption. We analyze some of the issues that arise in this context. Also included is some background theoretical computer science needed to discuss the issue of computational complexity of cryptography and cryptanalysis, as well as zero-knowledge proofs and related schemes.

This discussion may aid in understanding the security basis of public key cryptography. It may also shed some light on the more practical matter of choosing key sizes, that is, the number of bits in private and public components. This section may safely be skipped by readers who do not wish to gain an in-depth understanding of such issues.

7.1 Computational Complexity and Cryptocomplexity

An ideal cryptosystem would have the property that encryption and decryption are easy, but cryptanalysis is computationally infeasible. In practice, this is commonly interpreted (e.g., [DIFF76b]) to mean that encryption/decryption should be executable in polynomial time, while ideally cryptanalysis should take exponential time. More generally, cryptanalytic time should be an exponential function of encryption/decryption time.

Unfortunately, it is difficult to determine when this criterion holds in practice. This is because it is usually very difficult to determine a nonpolynomial lower bound for the time required for cryptanalysis (or computations in general), even in instances when this process reduces to simple and well-known problems. In some cases the latter problems have been studied for centuries, but their computational complexity is still unknown.

In fact, whenever we try to determine the relative complexity of cryptanalysis and encryption, we encounter the problem of determining accurate lower bounds on computations.

Also, an analysis of security and efficiency of public key systems should take into account not only encryption/decryption time versus anticipated time for cryptanalysis, but also other parameters such as key size, key generation time, and data expansion. Developing a theory to characterize both security and practicality of cryptosystems seems very challenging.

7.2 Classic Complexity Theory

Here we briefly introduce some notions from the theory of computation. Some of these notions are given a more formal treatment in Appendix C.

One attempt to formalize the study of hard problems is the theory of NP-completeness (e.g., [GARE79], [HORO78]). The class P is defined (loosely) to be the class of decision problems solvable in polynomial time via a deterministic algorithm. The latter is essentially an algorithm executable on an ordinary sequential computer. A nondeterministic algorithm is a more ephemeral concept: It is essentially executable on a machine with unbounded parallelism. This means that an unlimited number of possibilities can be explored in parallel. For example, suppose a set of n items is to be checked for the presence or absence of a given item. The worst-case deterministic complexity is n, the number of operations needed by a sequential machine to check all n items. In contrast, the nondeterministic complexity is 1 since a machine with unbounded parallelism could check all n items in parallel regardless of n.

The class NP is (loosely) the class of decision problems solvable in polynomial time via a nondeterministic algorithm. Perhaps the single most important problem in computer science is to decide whether P = NP. The NP-complete problems are the hardest subclass of NP, having the property that if one instance of the subclass is in P then P = NP. Many classic combinatorial search problems can be given NP-complete formulations. The traveling salesman and knapsack problems are examples (see [GARE79] for a long list).

The class of NP-hard problems are those problems, decision or otherwise, that are at least as hard as NP-complete problems. Some NP-hard problems are so difficult that no algorithm will solve them; they are undecidable. Such problems cannot be in NP. An example is the halting problem (e.g., [HORO78]).

A more formal treatment of the classes P and NP requires a framework such as Turing machines (Appendix C).

7.3 Public Key Systems and Cryptocomplexity

The use of NP-complete problems to generate public key cryptosystems was suggested in [DIFF76b]. Later this approach materialized in the form of trapdoor knapsacks [MERK78b]. As we have noted, however, most knapsacks have been broken, despite the continuing intractability of the underlying NP-complete problem (integer programming). There are two separate explanations for this phenomenon. First of all, in most cases it has not been proven that the security of the public key system is equivalent to the intractability of the underlying problem.

Second, it is important to note that the classic theory of computational complexity has been founded around the cornerstone of worst-case complexity, with average-case complexity as a secondary criterion. As Rabin [RABI76] notes, these measures are of limited relevance in analyzing the complexity of cryptanalysis, since a cryptosystem must be immune to attack in almost all instances.

Attempts to formalize the notion of "almost-everywhere hardness" have been made (e.g., [GROL88]). An early characterization was suggested by Shamir [SHAM79], who quantifies this notion by defining a complexity measure $C(n, a)$ for algorithms as follows: $C(n, a)$ is a measure of an algorithm if at least a fraction a of problem instances of size n can be solved by the algorithm in time $C(n, a)$. For example, an algorithm for finding one factor of n could have complexity $C\left(n, \frac{1}{2}\right) = 0(1)$, since n has a 50% chance of being even. However, such investigations have thus far failed to produce a comprehensive framework, although they may be useful in a few special instances.

Another simple example illustrates the difference between worst-case and average-case complexity: As noted in [RABI76], algorithms to sort n items are often predicated on the assumption that the items are arranged in random order; that is, all $n!$ arrangements are equally likely. If the file has a real-world origin it is more likely that the file is already partially sorted. An optimized algorithm might anticipate this and utilize an adaptive strategy.

In practice, a cryptosystem for which cryptanalysis has an average-case polynomial complexity is generally worthless; in fact this remains true if any measurable fraction of all instances permits polynomial-time cryptanalysis (this is difficult to make precise, since the fraction may vary with key size). Thus there is no a priori connection between the breaking of a system and the difficulty of the underlying problem, since the latter is characterized in terms of worst-case complexity. For example, there often exists a heuristic technique that yields an approximate solution to an NP-complete problem. Such techniques may not converge in polynomial time to an exact solution, but may break the corresponding cryptosystem.

In passing we remark that some authors (e.g. [WAGN84]) have attempted to go beyond the Diffie–Hellman notion of utilizing NP-complete problems to construct systems by using the even more intractable class of undecidable problems instead. It would be interesting to know if such systems can be made practical.

7.4 Probabilistic Algorithms

Another important distinction between the classic theory of complexity and cryptocomplexity is that the classic theory of algorithms does not encompass probabilistic algorithms (e.g., [RABI76]), which again may produce rapid solutions to problems in many instances but not even terminate in a few instances. They may also produce answers that are probably but not necessarily correct. Such algorithms are inappropriate in many contexts, for example, real-time control settings in which a response must be guaranteed in a fixed period of time, or where provable solutions are required. However, they are powerful tools in both cryptography and cryptanalysis.

As noted in [RABI76], probabilistic algorithms cannot be measured by classic criteria, which focus on worst-case runtime. Instead, a probabilistic algorithm may involve a trade-off between execution time and confidence. For example, most numbers have all small factors. If an algorithm exploits this fact it may terminate quickly for most inputs but take exponential time when large primes appear as factors.

Gill [GILL77] models probabilistic computation by extending the classic Turing machine model (e.g., [HORO78]) to the probabilistic Turing machine (Appendix D). This has proven valuable in the analysis of cryptographic systems, and probabilistic encryption schemes in particular (see Section 9.1).

7.5 Status of Some Relevant Problems

The security of several major cryptosystems mentioned here depends on the hardness of problems whose complexity status is unresolved. This includes factoring and discrete logarithm. The importance of the subclass of NP-complete problems emerges in this regard: If a problem is known to be in NP but is not known to be NP-complete, a solution to the problem in polynomial time (placing the problem in P) would not imply that problems such as traveling salesman are in P. The latter proposition is widely disbelieved.

A second relevant class is co-NP, the complements of problems in NP. For example, the complement of deciding whether n is prime is deciding if n is composite. In fact, primality and compositeness are in both NP and co-NP. Some experts have speculated (e.g., [GARE79]) that P is the intersection of NP and co-NP. For example, linear programming was known to be in both of the latter long before its status was resolved; eventually it was shown to be in P, via the ellipsoid method [KHAC79]. This illustrates that it would indeed be desirable to have available cryptosystems based on NP-complete problems (but see below).

Good surveys of the status of many problems important in security of public key systems are given in [ADLE86] and [POME86]. Let

$$L(n) = \exp((1 + o(1))(\ln n * \ln \ln n)^{1/2})$$

Then Adleman notes that many algorithms for factoring n have probabilistic execution times believed to be of the form $L(n)^c$ (e.g., [SCHN84]). However, only the algorithm of Dixon [DIXO81] has been rigorously analyzed. Similarly, various algorithms for discrete logarithms mod p are believed to take time $L(p)$. These results would be viewed with mistrust in the classic theory due to their probabilistic nature and lack of rigorous upper bounds for worst-case or average-case times. In contrast, Adleman [ADLE83] gives a deterministic algorithm for primality testing which executes in time

$$O((\ln n)^{c \ln \ln \ln n})$$

In the classic theory this result would be valuable, but we have noted that probabilistic algorithms are more efficient and far less complex, hence much more relevant in the present setting. Again this is indicative of the divergence between classic complexity and cryptocomplexity.

The status of the problem of taking roots mod n, that is, solving $x^e \equiv c \pmod{n}$, depends on the parameters (e.g., [SALO85]). A polynomial-time probabilistic algorithm to find x exists if e, c, and n are fixed and n is prime. If n is composite with unknown factors, no polynomial-time algorithm exists, even probabilistic. If $e = 2$ the complexity is essentially equivalent to factoring n (Lemma N.3.2). If $e > 2$ the status is less clear; as a consequence it is not clear that the security of RSA is equivalent to factoring.

Brassard [BRAS79] has shown that the discrete logarithm has an associated decision problem which lies in the intersection of NP and co-NP. Thus, if either factoring or taking discrete logarithms is NP-hard, it follows from the definition of this class that the associated decision problem is NP-complete and in co-NP. In turn this would imply NP = co-NP, which is widely disbelieved.

More generally, in [BRAS83] a function f is called restricted one-way if f is easily computable, f^{-1} is not, and f is injective and polynomially bounded. It is shown that if restricted one-way functions exist then P is not the intersection of NP and co-NP as many believe; and if f is restricted one-way and f^{-1} is NP-hard then NP = co-NP. Discrete logarithm is thought to be restricted one-way.

We conclude that there is little hope for fulfilling the Diffie–Hellman quest for public key systems based on NP-complete problems. The same may be true of the search for one-way functions to employ as hash functions. Prospects for use of NP-hard problems seem difficult to assess at this time.

8 AN INTRODUCTION TO ZERO-KNOWLEDGE

The notion of zero-knowledge proofs was introduced in [GOLD89]. The essence of zero-knowledge is that one party can prove something to another without revealing any additional information. In deference to the depth and scope of this area, we give only an illustrative example in this section. However, there is a close connection between zero-knowledge schemes and probabilistic public key encryption. Furthermore, some zero-knowledge schemes that have drawn attention because of applicability to smart card implementations, such as the one used as the basis of the example in this section, use Shamir's notion of identity-based public key systems. Thus the reader desiring a more formal introduction to these topics may wish to skip to Section 9.

Suppose that Alice knows a fact P. She wants to convince Bob that she knows P, but she does not trust Bob. Thus, Alice does not want to reveal any more knowledge to Bob than is necessary. What Alice needs is a zero-knowledge proof of P.

For example, suppose that Alice wants to prove to Bob that she really is Alice. Suppose for convenience that there is some authority that verifies identities. One possibility is that the authority could issue Alice an identification. If this were contained on a device such as a smart card, Alice could simply show it to Bob. However, if Alice and Bob are communicating over a network, then Alice's identifying information would have to be transmitted to Bob over the network. On receiving it, Bob could use it to impersonate Alice. Even if Bob were trusted, an eavesdropper such as Alice's adversary Carol could do the same.

This situation also arises commonly in computer access control: Bob might then be a host computer or network server, and Alice's identification might be a password. If Alice uses her password to identify herself, her password is exposed to the host software as well as eavesdroppers; anyone who knows this password can impersonate Alice.

It is thus desirable for Alice to be able to prove her identity without revealing any private information. More generally, we need a scheme through which Alice can prove to Bob that she possesses something (e.g., a password) without having to reveal it. Such a scheme is an example of a zero-knowledge proof. In fact, this example is the major practical use of zero-knowledge that has been suggested to date.

Here is one way that such a system could be organized: The authority decides on a number N used for everyone; for example, take $N = 77$. Everyone knows this number. The authority may then choose, for example, two numbers that form an ID for Alice. Suppose these are $\{58, 67\}$. Everyone knows Alice's ID. The authority then computes two other numbers $\{9, 10\}$ that are given to Alice alone; she keeps these private. The latter numbers were chosen because $9^2 * 58 \equiv 1 \pmod{77}$ and $10^2 * 67 \equiv 1 \pmod{77}$.

Now Alice can identify herself to Bob by proving that she possesses the secret numbers $\{9, 10\}$ without revealing them. Each time she wishes to do this she can proceed as follows: She can choose some random numbers such as $\{19, 24, 51\}$ and compute

$$19^2 \equiv 53 \pmod{77}$$
$$24^2 \equiv 37 \pmod{77}$$
$$51^2 \equiv 60 \pmod{77}$$

Alice then sends $\{53, 37, 60\}$ to Bob. Bob chooses a random 3 by 2 matrix of 0's and 1's, for example,

$$E = \begin{matrix} 0 & 1 \\ 1 & 0 \\ 1 & 1 \end{matrix}$$

Bob sends E to Alice. On receipt, Alice computes

$$19 * 9^0 * 10^1 \equiv 36 \pmod{77}$$
$$24 * 9^1 * 10^0 \equiv 62 \pmod{77}$$
$$51 * 9^1 * 10^1 \equiv 47 \pmod{77}$$

Alice sends {36, 62, 47} to Bob. Finally, Bob can check to see that Alice is who she says she is. He does this by checking that

$$36^2 * 58^0 * 67^1 \equiv 53 \pmod{77}$$
$$62^2 * 58^1 * 67^0 \equiv 37 \pmod{77}$$
$$47^2 * 58^1 * 67^1 \equiv 60 \pmod{77}$$

The original numbers {53, 37, 60} that Alice sent reappear. Actually, this doesn't really prove Alice's identity; she could have been an impersonator. But the chances of an impersonator succeeding would have been only 1 in 64.

In an actual system, the number N would have been much larger (e.g., 160 digits). Also, Alice would have been assigned an ID consisting of more numbers, for example, 4, by the authority, with a secret also consisting of four numbers. Furthermore, Alice would have generated more random numbers, for example, 5, to send to Bob. The ID numbers, secret numbers, and random numbers would have been about as large as N. This would have reduced an impersonator's chances of cheating successfully to about 1 in a million (more precisely 2^{-20}) if 4 and 5 are the parameters, which certainly would have convinced Bob of Alice's identity.

Why does this work? Because the authority chose {58, 67} and {9, 10} so that

$$9^2 * 58 \equiv 1 \pmod{77}$$
$$10^2 * 67 \equiv 1 \pmod{77}$$

This says that 9^2 and 58 are multiplicative inverses modulo 77, as are 10^2 and 67.

Thus

$$
\begin{aligned}
36^2 * 58^0 * 67^1 &\equiv 19^2 * 9^{2*0} * 58^0 * 10^{2*1} * 67^1 \\
&\equiv 19^2 * (9^2 * 58)^0 * (10^2 * 67)^1 \\
&\equiv 19^2 \equiv 36 \pmod{77} \\
62^2 * 58^1 * 67^0 &\equiv 24^2 * 9^{2*1} * 58^1 * 10^{2*0} * 67^0 \\
&\equiv 24^2 * (9^2 * 58)^1 * (10^2 * 67)^0 \\
&\equiv 24^2 \equiv 37 \pmod{77} \\
47^2 * 58^1 * 67^1 &\equiv 51^2 * 9^{2*1} * 58^1 * 10^{2*1} * 67^1 \\
&\equiv 51^2 * (9^2 * 58)^1 * (10^2 * 67)^1 \\
&\equiv 47^2 \equiv 53 \pmod{77}
\end{aligned}
$$

Thus the checks that Bob uses serve their purpose properly; that is, Alice is identified. Also, Bob has learned nothing that would permit him to masquerade as Alice; nor has Carol, who may have been eavesdropping. Either would need to know that 9^2 and 10^2 are the multiplicative inverses of 58 and 67, respectively, modulo 77. To see this, suppose Carol tries to convince Bob that she is really Alice; that is, Carol pretends that {58, 67} is her own ID. For simplicity, suppose Carol tries to identify herself as Alice by generating the same random {19, 24, 51}. Then she sends {53, 37, 60} to Bob. Again for simplicity suppose Bob generates the same E and sends it to Carol. Now Carol is in trouble; she doesn't know the numbers 9 and 10, and can only guess them. The protocol requires Carol to send three numbers, say {x, y, z}, to Bob. Then Bob will check:

$$x^2 * 58^0 * 67^1 \equiv 53 \pmod{77}$$
$$y^2 * 58^1 * 67^0 \equiv 37 \pmod{77}$$
$$z^2 * 58^1 * 67^1 \equiv 60 \pmod{77}$$

Carol will have succeeded in her masquerade if she chose $\{x, y, z\}$ to make these checks come out right. But $67^{-1} \equiv 10^2 \equiv 23 \pmod{77}$ and $58^{-1} \equiv 9^2 \equiv 4 \pmod{77}$, so $x^2 \equiv 53 * 23 \equiv 64 \pmod{77}$, $y^2 \equiv 37 * 4 \equiv 71 \pmod{77}$, and $z^2 \equiv 60 * 23 * 4 \equiv 53 \pmod{77}$. Could Carol solve these quadratic equations to find, for example, $x = 36$, $y = 62$, $z = 47$? For a small value of N such as $N = 77$, she could indeed. However, if N is a product of two appropriately chosen large primes (e.g., each 80 digits or more), and if these primes are kept secret by the authority, then the answer is no. That is, computing square roots modulo a composite is computationally infeasible (Appendix N). Thus, anyone who does not know Alice's secret ($\{9, 10\}$ in the example above) cannot impersonate her when N is large.

Another possibility is that Alice's interaction with Bob might give Bob information that could allow impersonation of Alice, at least on one occasion, by replay. Suppose Bob tries to convince Carol he is really Alice. Bob might try to imitate Alice by sending $\{53, 37, 60\}$ to Carol to start the protocol. Now Carol doesn't necessarily select the E above. Suppose she selects

$$F = \begin{matrix} 1 & 1 \\ 0 & 0 \\ 0 & 1 \end{matrix}$$

and sends F to Bob. Now Bob is in trouble; he might try to imitate Alice again by sending $\{36, 62, 47\}$ to Carol. Then Carol will check

$$36^2 * 58^1 * 67^1 \equiv 71 \pmod{77}$$

Since $71 \neq 53 \pmod{77}$, Carol knows Bob is a fraud even without the other two checks. Can Bob send some $\{r, s, t\}$ instead of $\{36, 62, 47\}$? This will only work if

$$r^2 * 58^1 * 67^1 \equiv 53 \pmod{77}$$
$$s^2 * 58^0 * 67^0 \equiv 37 \pmod{77}$$
$$t^2 * 58^0 * 67^1 \equiv 60 \pmod{77}$$

As before, this gives $r^2 \equiv 58^{-1} * 67^{-1} * 53 \equiv 25 \pmod{77}$, and similarly $s^2 \equiv 37 \pmod{77}$, $t^2 \equiv 71 \pmod{77}$. As above, Bob can solve these quadratic equations to find r, s, t because $N = 77$ is small, but could not do so if N were large. Also, in the example above there is one chance in 64 that Carol would choose $F = E$, in which case Bob's deception would go undetected; but for larger parameters such as 4 and 5 instead of 2 and 3, this would again be improbable (one chance in a million (2^{-20}) instead of 64).

Another possibility is that when Alice identified herself to Bob, she may have inadvertently revealed some information that could enable Bob to learn her secret, that is, $\{9, 10\}$, that would permit impersonation. For this protocol to be zero-knowledge this should not be possible. Can Bob deduce the numbers $\{9, 10\}$ from the two sets of numbers $\{53, 37, 60\}$ and $\{36, 62, 47\}$, and E? He knows Alice started with three numbers $\{a, b, c\}$, but he doesn't know these were $\{19, 24, 51\}$. He knows Alice's secret is

$\{u, v\}$ but doesn't know these are $\{9, 10\}$. He knows the authority computed u and v from

$$u^2 \equiv 58^{-1} \equiv 4 \pmod{77}$$
$$v^2 \equiv 67^{-1} \equiv 23 \pmod{77}$$

He also knows Alice computed

$$a^2 \equiv 53 \pmod{77}$$
$$b^2 \equiv 37 \pmod{77}$$
$$c^2 \equiv 60 \pmod{77}$$
$$a * u^0 * v^1 \equiv 36 \pmod{77}$$
$$b * u^1 * v^0 \equiv 62 \pmod{77}$$
$$c * u^1 * v^1 \equiv 47 \pmod{77}$$

This gives Bob eight equations from which to deduce $\{u, v\}$. However, the last three are redundant, and as we have noted, the first five cannot be solved when N is large.

The following list shows informally that the protocol above works:

1. It identifies Alice by proving she possesses the secret $\{9, 10\}$.
2. It identifies Alice uniquely: Anyone who doesn't know this secret cannot impersonate Alice.
3. Alice's secret is not revealed in the process of proving she possesses it.

Actually, (1) means the possessor of $\{9, 10\}$ is identified as the system user who has ID $= \{58, 67\}$ assigned by the authority. Also, no matter how many times Alice proves her identity, her secret will not be revealed (if N is sufficiently large); that is, (3) states loosely that the protocol is zero-knowledge. All this depends on the assumption that equations of the form $x^2 \equiv y \pmod{N}$ cannot be solved if N is the product of two large primes that are not known, but can be solved easily if the primes are known (Appendix N). Such equations lie at the heart of most concrete zero-knowledge schemes.

The formal definition of zero-knowledge is more complicated (see [GOLD89]), but the example above demonstrates its essence in a context which has been proposed as a candidate for actual implementation in smart card–based identification schemes.

9 ALTERNATIVES TO THE DIFFIE–HELLMAN MODEL

Thus far this chapter has been based on the work of Diffie and Hellman [DIFF76b]. This model of public key cryptography has received some criticism on two grounds:

- Security of most Diffie–Hellman-type systems is difficult to characterize formally.
- Security of Diffie–Hellman-type systems is dependent on a superstructure that binds user IDs and public components.

In Section 7 we noted that it has proven difficult to develop a comprehensive axiomatic framework in which to establish the security of Diffie–Hellman-type public key systems. For example, it is difficult to guarantee that partial information about plaintext (e.g., its least significant bit) cannot be recovered from the corresponding ciphertext even though the entire plaintext cannot be found.

In Section 5 we noted some examples of authentication frameworks (e.g., use of certificates) that bind user IDs and public keys. Without such a superstructure a public key system is useless, since a user would not be certain that he was employing the correct public key for encryption or decryption. The security of the public key system thus depends on proper functioning of the authentication framework, which is a priori unrelated to the underlying cryptosystem.

In this section we briefly examine two modifications of the basic Diffie–Hellman model. In one case the goal is to incorporate the binding between a user ID and public key directly into the cryptosystem, thereby eliminating the separation between the cryptosystem and the authentication framework. The other scheme addresses the subject of knowledge concealment; it is closely related to zero-knowledge proofs. Both schemes have received considerable attention not only because of possibly enhanced security, but also because of their potential relevance to smart card implementations of public key cryptography.

9.1 Probabilistic Encryption

Goldwasser and Micali [GOLD84] note that the public key systems of Diffie and Hellman are not provably secure. They observe that use of a trapdoor function f to generate such systems does not exclude the possibility that $f(x) = y$ may be solvable for x without the (original) trapdoor under certain conditions. Also, even if x cannot be found from $f(x)$, it may be possible to extract partial information about x. One problem is that such trapdoor systems proceed block by block. This makes it difficult to prove security with respect to concealment of partial information such as least significant bit.

In [GOLD84] Goldwasser and Micali suggest an alternative to trapdoor-based public key systems. Their procedure is to encrypt bit by bit. They call this probabilistic encryption. It introduces several advantages, namely, uniformly difficult decoding and hiding of partial information. Their scheme has the following properties:

1. Decoding is equivalent to deciding quadratic residuosity modulo a composite N, whose factorization is not known to an adversary.

2. Suppose a predicate P has probability p of being true in message space M. For $c > 0$, assuming intractability of quadratic residuosity, an adversary given ciphertext cannot decide with probability $> p + c$ whether the corresponding plaintext satisfies P; that is, the adversary does not have a c-advantage in guessing P.

In item (1) the reference is to the problem of deciding for a given x whether there is a y such that $y^2 \equiv x \pmod{N}$. As in the case of computing discrete logarithms or extracting roots, this is computationally infeasible if the factorization of N is unknown (Appendix N).

An example of item (2): If the messages consist of uniformly distributed bit strings and P is "least significant bit $= 0$" then $p = \frac{1}{2}$. If the Goldwasser–Micali scheme is employed, an adversary cannot guess the least significant bit of plaintext with probability greater than $\frac{1}{2} + c$. It may be very difficult to verify that traditional public key systems conceal such partial information.

The public key system proposed by Goldwasser and Micali is as follows: A user A chooses primes p and q and makes public $N = p * q$; p and q are private. Also, A selects a random y with $(y/N) = 1$ (Appendix N.2) with y a quadratic nonresidue modulo N; y is public. By Lemma N.4.6, y can be found in probabilistic polynomial time.

To send the binary string $m = (m_1, \ldots, m_k)$ to A, B randomly selects $\{x_1, \ldots, x_k\}$ in Z_N^* and computes

$$z_i = x_i^2 \mod N \qquad (i = 1, \ldots, k)$$

Then B sets $e_i = z_i$ if $m_i = 0$, $e_i = y * z_i$ otherwise. The encoding of m is $e = (e_1, \ldots, e_k)$, which B sends to A. To decode e, A sets $m_i = 1$ if e_i is a quadratic residue modulo N, and $m_i = 0$ otherwise. This can be effected by A in polynomial time since A knows p and q (Lemmas J.4.1, N.1.1). It is correct because $y * z_i$, the product of a quadratic nonresidue and residue, is a quadratic nonresidue modulo N (Lemma N.2.2).

The security of partial information follows from the fact that each bit of the message m is encoded independently. A disadvantage of the technique is data expansion: if $|N| = n$, then n bits are transmitted for each bit of a message.

One application is coin flipping by telephone: A randomly chooses r in Z_N^* with $(r/N) = 1$, where (r/N) is the Jacobi symbol (Appendix N.2). Then B guesses whether or not r is a quadratic residue modulo N; B wins if and only if he guesses correctly. The probability of the latter is $\frac{1}{2}$; that is, exactly half of the elements of Z_N^* satisfying $(r/N) = 1$ are quadratic residues modulo N (Lemma N.2.1). The correctness of B's guess can be checked by A; the result can be verified by B if A releases the factorization of N to B.

A second application is to mental poker (e.g., [DENN83b] pp. 110–117). Goldwasser and Micali [GOLD82] show that their implementation corrects some deficiencies in hiding partial information in a scheme of Shamir, Rivest, and Adleman.

Quadratic residuosity modulo a composite is an example of a trapdoor function. Yao [YAO-82] showed that under certain conditions, trapdoor functions in general can be used to construct provably secure probabilistic public key cryptosystems. An important role is also played by probabilistic encryption in zero-knowledge proofs, especially for problems in NP.

We remark briefly that the scheme above is only one of a class of encryption schemes that Rivest and Sherman [RIVE82] term randomized encryption techniques; various other schemes are summarized there. We have characterized schemes such as the one above as bit-by-bit; more generally, a *randomized scheme* is any scheme in which a ciphertext for a given plaintext is randomly chosen from a set of possible ciphertexts. Rivest and Sherman also develop a classification for such schemes. A major aspect of randomized schemes is often significant data expansion. For example, the Goldwasser–Micali scheme would probably have ciphertext around 512 times as large as the plaintext. On the other hand, probabilistic schemes may be much faster than their deterministic counterparts, an attractive feature for smart card implementations.

9.2 Identity-Based Schemes

In [SHAM84], Shamir suggests yet another modification of traditional public key systems. In Shamir's framework a user's public key coincides with his system ID. This eliminates the need for any superstructure for distribution of public keys. It also trivializes the authentication of public keys. However, unlike a traditional public key scheme this modification requires a trusted key generation center.

The center issues a smart card to each user, containing the user's private key. Thus the "private" key is really a shared secret key. A card can generate the user's digital signature. The center does not maintain a database of keys; in fact it need not even keep information beyond a list of user IDs (needed to ensure that there are no duplicate IDs).

Security for such a system differs from the usual requirement that a user keep his private key a secret. The latter condition is retained, in that a user must guard his card. As in a certificate-based traditional system, the center must ascertain a user's identity before issuing a card. In addition, however, the cryptographic function used by the center to generate private keys from IDs must be kept secret. Typically this function would employ trapdoor information such as the factorization of a modulus. The requirement is that computing the private key from an ID is easy if the trapdoor information is known, but knowing any polynomial number of public/secret pairs should not reveal the trapdoor.

A weakness in such a scheme is that anyone (intruder or insider) possessing the trapdoor information can forge the signature of any user. This is in contradistinction, for example, to a credit card scheme in which a transaction is generally valid only if accompanied by a user's physical signature. Also, the center is not required to maintain any information on lost or stolen cards; thus the loss of a card is disastrous since the possessor can forge the legitimate user's signature indefinitely. Furthermore, the center may find itself involved in litigation produced by any such security problems, since it is providing the means by which users generate signatures. Again this is in contrast to the credit card situation: If a credit card is stolen, the thief cannot forge the legitimate holder's physical signature, providing a means of distinguishing between legal and illegal use of the card. Use of passwords (PINs) with cards could largely eliminate the problems accruing from lost or stolen cards, but compromise of the trapdoor would still be disastrous.

Shamir's notion was later ([FEIG87], [FIAT86]) incorporated into zero-knowledge schemes. One of these was used as the basis for the example in Section 8.

APPENDIX A
TIMINGS

We briefly summarize some timings related to RSA obtained by Keller at NIST [KELL89]. These were obtained by using either software or a combination of software and hardware. Tests were run on an International Business Machines (IBM) PC AT.

Time to generate RSA public/private key pairs was as follows:

- Software
  - Keysize = 32 bytes: 18 sec
  - Keysize = 64 bytes: 122 sec
- Software/hardware:
  - Keysize = 32 bytes: 4 sec
  - Keysize = 64 bytes: 12 sec

Rates of encryption and decryption can be found in [KELL89].

APPENDIX B
ALGORITHMS AND ARCHITECTURES

We briefly survey some of the considerations relevant to determining what type of hardware and software support for cryptanalysis, or to a lesser extent encryption, may be forthcoming from the creators of algorithms and architectures of the future. We also mention some examples already in existence. This represents an excursion into high-performance computing, only a small fraction of which is applicable to cryptography. Nonetheless, to evaluate security of methods or ascertain key sizes, it is necessary to make some educated guesses as to how this field will progress over the next few decades.

B.1 TECHNOLOGY

Computing at present is silicon-based. Thus all estimates of achievable computer performance are geared to this technology. The question arises as to whether a radically different technology such as superconductivity or optical computing will make silicon-based performance standards obsolete, and if so, when this might occur. We have no idea, and hence we ignore these questions.

Gallium arsenide (GaAs) technology is another matter. It has already been integrated into some existing supercomputers. Some of the differences between GaAs and silicon very large-scale integration (VLSI) are (e.g., [MILU88]):

- GaAs gates have a higher switching speed.
- Off-chip communication in GaAs pays a relatively higher penalty.
- GaAs chips have lower density.

Gate delays in GaAs (DCFL E/D-MESFET*) may be as low as 50 psec, as opposed to at least 1 nsec in silicon (NMOS). Similarly, on-chip memory access in GaAs may take as little as 500 psec, as opposed to at least 10 nsec in silicon. This indicates that performance of GaAs-based computers could theoretically be as much as 20 times greater than even the fastest silicon-based supercomputers. However, GaAs levels of integration are currently much lower: at most about 50,000 transistors per chip, as opposed to 1,000,000 in silicon. This is due to problems in GaAs of power dissipation, yield, and area limitations. Thus the number of chips necessary to build systems using GaAs is higher; minimizing chip count is important for high performance.

Off-chip communication in GaAs is another factor at the system level. The peak performance of a computer system is limited by the bandwidth of the slowest subsystem [HWAN84]. Interchip signal propagation is not significantly different for silicon or GaAs, but the relative effect is different: off-chip communication is more of a bottleneck in GaAs because of its ratio to on-chip speed. Silicon solutions to the problem of central processing unit (CPU) speed versus other subsystem speeds include cache and multilevel memory hierarchies; these may not carry over mutatis mutandis to GaAs.

---

*DCFL E/D-MESFET is the acronymn for direct coupled FET logic enhancement/depletion metal-semiconductor field effect transistor.

We conclude that at the moment, GaAs technology does not present a real threat to silicon performance standards. Furthermore, it does not appear likely that problems such as yield and integration will be solved in the near future. Thus, for the remainder of Appendix B we assume no radical change in technology from the present.

B.2 COMPUTING MODES

Classically, computing was dominated by the Von Neumann model, with a single processor executing a single instruction scheme on a single data stream. This paradigm reached its peak with computers such as the Cray-1 [CRAY80] and CYBER 205 [CONT80]. However, the performance of a single processing unit is limited by its clock speed. It does not seem feasible to reduce major cycles much below 1 nsec with silicon technology. Furthermore, at these speeds memory access becomes a bottleneck, not to mention input/output (I/O). Thus it is not likely that uniprocessors will exceed 10^9 operations per second, barring radical new technologies.

Two alternatives to the Von Neumann model are parallel and distributed computing (e.g., [HWAN84]). Parallel computing generally refers to processors in one box; distributed means each processor is in its own box. Parallel computers may (loosely) be subdivided into shared-memory, in which all processors share a common address space, and distributed memory, in which each processor has its own address space. These modes remove the restriction on the number of instruction and/or data streams that may be processed concurrently. In theory these modes of computing can produce unlimited computing power, by splicing together single processing nodes and memory units (or processor/memory pairs) into a unified system. However, there are three major limitations in this regard:

- Cost-effectiveness
- The interconnection network
- Parallel algorithms.

Cost-effectiveness alone has been a deterrent to the development of parallel systems. The Denelcor HEP (e.g., [KOWA85]), Floating Point T-Series [HAWK87], and ETA-10 [STEI86] are examples of parallel systems that have not proven to be commercially successful. Also, Cray and other supercomputer manufacturers have been reluctant to expand into large-scale parallelism, partially because of cost-related concerns. This creates an interesting situation from a cryptographic point of view: There may be cases where a computer powerful enough to break a given system could be built in theory. Security of the system might then rest on the assumption that no one will spend the money to build it, or that the only units built will belong to wealthy government agencies and will be subject to tight controls.

Even if computers with a thousand or more powerful processors are constructed, the question arises as to whether such a configuration can achieve computing power in proportion to the number of processors, that is, linear speedup. The mode of interconnecting the processors and memories is critical. If the shared-memory configuration is used, with a collection of processors accessing common memory partitioned into modules, interference in paths through the network or in simultaneous attempts to access a module will cause a bottleneck if a low-cost, low-bandwidth network such as a bus is used. If a high-bandwidth network such as a crossbar is used (i.e., with concurrent data transfers possible between many processors and memory modules), the number of units interconnected is limited by cost which rises as the square of the number of processors. Thus such systems seem to be inherently limited in the extent to which they can improve on uniprocessor performance.

An alternative is to connect processor/memory pairs using a network such as a hypercube [SEIT85]. Machines of this type with up to 65,536 weak processing elements (e.g., the Connection Machine [HILL85]) or up to 1024 powerful processors (e.g., the NCUBE/10 [HAYE86]) have been constructed, and systems with up to 32,000 Cray-1 level processors have been proposed. There is debate (and some controversy) over the speedups that such distributed-memory machines can achieve. The latter consideration is related to cost-effectiveness, which requires nearly linear speedup. Also, the cost of interconnection in a hypercube-based system rises as $O(n \log n)$, where n is the number of processor/memory pairs.

Another concern is algorithms [QUIN87]. A problem must be highly decomposable to be amenable to parallel or distributed computation. Furthermore, algorithms for non–Von Neumann machines must be tailored to individual architectures to a much greater extent than their Von Neumann counterparts, adding to the cost of software development.

Because of inherent trade-offs between performance and cost, there may be a divergence between the computing power attainable in theory and that which is practical (and in the event of commercial machines, marketable). Within 10 years it is conceivable that machines executing 10^{12} operations per second could be built with refinements of existing technology; the real question seems to be financing.

An alternative to the single-machine approach is the use of networks for completely distributed computing (e.g., [LENS89]). The class of problems amenable to solution via networks (and wide area networks in particular) is restricted, since the nodes must communicate infrequently (typically only at the beginning and end of a computation). Nonetheless, in Section 4 it was noted that some of the strongest cryptanalytically related results have been obtained by networks of computers. This is possible because many of the relevant cryptanalytic algorithms are fully decomposable into independent portions that do not need to interact. An example is the quadratic sieve.

It may be possible to assemble more computing power in such a network than is present in any single computer. Once again, the constraints are largely pragmatic. Designers of cryptosystems must therefore attempt to anticipate not only advances in algorithms and architectures, but also the greatest amount of computing power that might realistically be brought to bear against a given task.

B.3 SOME RELEVANT ALGORITHMS AND IMPLEMENTATION

Most of the powerful algorithms for factoring and discrete logarithm are fairly new. As we have noted, most of them have not been fully analyzed for runtimes. Nonetheless, some standards have emerged in this regard. The best guess for the future can only amount to anticipation of improvements in the present algorithms.

B.3.1 Quadratic Sieve Factoring Algorithm

The quadratic sieve [POME84] provides an alternative to the earlier continued fraction factorization algorithm [MORR75].

As noted in [DAVI83b], the continued fraction approach uses considerable multiple precision division. The quadratic sieve works with larger residues but involves mainly single-precision subtraction. Both have runtimes of exp(sqrt($c \ln n \ln \ln n$)) but $c = 2$ for continued fraction, $c = \frac{9}{8}$ for the quadratic sieve.

In [DAVI84] the results of implementing the quadratic sieve on a Cray X-MP [CRAY85] are reported. Factorization of 70-digit numbers takes about an hour; factorization of 100-digit numbers should take about a year. The key to this match of algorithm and architecture is the use of the vector capability of the Cray architecture. In particular, a steady stream of operands is necessary to take advantage of a pipelined, register-to-register machine. The loops in the sieve are amenable to streaming of operands, and about 75% efficiency was obtained. However, the multitasking capability of the X-MP was not utilized. Since Crays with up to 16 processors are expected in the near future, with even greater parallelism to be anticipated, it follows that the results of such experiments are probably too conservative. On the other hand, using both vector and parallel capabilities of a system is nontrivial, partially as a result of memory bank contention. Furthermore, adaption of the algorithm to exploit both capabilities may be nontrivial. Hence it is not clear to what extent the preceding efficiency can be maintained as the degree of parallelism grows.

An alternative is distributed computing: Silverman [SILV87] has used the quadratic sieve implemented via a network of nine SUN 3 workstations to factor numbers up to 80 digits in 8 weeks. Recently another network was used to factor 106-digit numbers [LENS89].

It should be noted that the results obtained via the quadratic sieve are directly relevant cryptanalytically, since the integers factored are general. Integers of up to 155 digits, but with special forms, have been factored, also by use of networks [SIAM90]. However, these results are only indirectly relevant to systems such as RSA, assuming moduli are properly chosen.

B.3.2 Computations in Finite Fields

Computations in finite fields is a subject that has been explored fairly thoroughly (e.g., [BART63]) and we will not review it here.

Multiplication of elements in GF(m) has classically been implemented at the circuit level using linear feedback shift registers. However, Laws [LAWS71] has noted the potential for using cellular arrays. These are highly amenable to VLSI implementation. A pipeline architecture for multiplication and inverses is proposed in [WANG85].

One class of algorithms of particular interest is for discrete logarithms in GF(p^n). Still more particularly, the case $n = 2$ has been explored considerably ([BLAK84], [BLAK84b], [COPP84]). Since finite logarithms are easy to compute in GF(m) if m has only small factors [POHL78], n should be chosen so that $2^n - 1$ is (a Mersenne) prime, for example, $n = 521$. However, Odlyzko [ODLY84b] notes that a next-generation supercomputer might break $n = 521$ in a year. A special-purpose computer might attack n on the order of 700 in a year.

Odlyzko also notes that with similar bounds on key size, $GF(p)$ may be preferable to $GF(2^n)$; for example, $n = 2000$ is roughly equivalent to p of about 750 bits. As noted above, this advantage may be counterbalanced by implementation efficiency.

B.3.3 Other Algorithms

Many of the most fundamental operations in this setting seem to be characterized by inherent sequentiality. An example is exponentiation; computing the nth power seems to take log n steps regardless of the number of processors available.

Another classic example is GCD. Although some partial results are noted in [ADLE86], major improvement over Euclid does not seem forthcoming regardless of advances in architecture.

Many of the powerful factoring algorithms are highly parallelizable (e.g., [SCHN84]). The main requirement is existence of appropriate architectures.

B.4 APPLICATION-SPECIFIC ARCHITECTURES

General-purpose architectures such as that of the Cray supercomputer are of interest in this setting because of their immediate availability. At the other extreme, architectures closely matching the algorithms of interest could be constructed, but their over-specialized nature would virtually preclude their commercial distribution. In between are classes of architectures that are more versatile but not truly general-purpose. We note several examples of partly specific architectures and a proposed highly-specific machine.

B.4.1 Systolic and Wavefront Arrays

The notion of systolic arrays ([KUNG82], [KUNG78]) is an extension of pipelining. The idea is to have a collection of processors operate on data that are pulsed rhythmically through the array. If the original requirement of a synchronous mode of operation is relaxed, the result is a wavefront array [KUNG82b]. Both types were originally targeted at applications such as signal processing that are compute-intensive and involve repetitions of operations on many operands.

A systolic array for the computation of GCDs is proposed in [BREN83]. It is very amenable to VLSI implementation because of its regular topology. Furthermore it is linear, limiting I/O requirements which can constitute a bottleneck in VLSI implementations. Supporting algorithms are noted in [BREN83b].

It would be of interest to know what a more general-purpose linear systolic array such as the Warp [ANNA87] (which curiously more closely resembles a wavefront array) could accomplish on some of the problems discussed in this chapter, and factoring in particular. The Warp is already in production.

B.4.2 Proposal for a Quadratic Sieve Machine

Pomerance, Smith, and Tuler [POME88] describe a pipeline architecture that would efficiently execute the quadratic sieve. This notion is of considerable interest to anyone implementing an algorithm such as RSA, since such a machine could presumably factor numbers beyond the reach of present-day supercomputers.

Cost-effectiveness was carefully analyzed by these authors, as is critical for an application-specific architecture. Pomerance et al. speculate that a $50,000 version of the architecture should factor 100-digit numbers in about 2 weeks, or 140-digit numbers in a year on a $10 million version, or 200-digit numbers in 1 year on a $100 billion version. It should be noted that the last example is clearly predicated on the notion that the architecture and algorithm will scale linearly with the number of processing units; however as we noted in Section B.2 various bottlenecks such as interprocessor communication and memory access make this assumption suspect. Nonetheless the possibility arises that moduli approaching 200 digits may be necessary in RSA because of the potential existence of machines such as these.

The crux of the architecture are the pipe and pipe I/O units. These should be custom-made and should be able to handle variable strides without the considerable loss of efficiency that usually accompanies nonconstant stride. The two units should be chained together. The pipe should consist of stages, all of which are bus-connected to the pipe I/O unit. The cycle time of memory should be the same as that of the processing elements.

It is interesting to note that this architecture bears a close resemblance to a linear systolic array.

B.4.3 Massively Parallel Machines

An alternative to pipelined machines is to configure a large number of primitive processing elements in an array and have them execute the same instruction stream synchronously, processing a large quantity of data in parallel [single instruction multiple data (SIMD) mode]. Such machines are generally applied to real-time image or signal processing, or to large-scale matrix operations.

The massively parallel processor (MPP) [BATC80] is an example. It consists of 16,384 processors in a square array. It was intended mainly for image processing of data from satellites; but in [WUND83] and [WILL87] it is used for an implementation of the continued-fraction factoring algorithm. Wunderlich [WUND85] also implements this algorithm on the distributed array processor (DAP) (e.g., [HOCK81]), another massively parallel machine with 4096 processors. Unlike the MPP, the DAP has been marketed.

APPENDIX C
THE CLASSIC THEORY OF COMPUTATION

In Section 7 a number of topics from the classic theory of computation were mentioned. Here we give a more precise treatment of some notions from the classic theory (e.g., [LEWI81]).

An *alphabet* is a finite set of symbols. A *string* is a sequence of symbols from an alphabet. A *language* on an alphabet is a set of strings from the alphabet. If S is an alphabet, S^* denotes the set of all strings from S, including the empty string. Concatenation of strings a and b is written ab. If A and B are subsets of S^*, AB is the set of strings formed by concatenating elements from A and B.

C.1 TURING MACHINES

A (one-tape, deterministic) Turing machine is a quadruple (K, S, D, s) where:

S is an alphabet that contains a blank $= \#$, but not the symbols L or R which are reserved for tape movement to the left or right

K is a set of states that does not include the halt state $= h$, which signals an end to computation

$s =$ initial state; it is in K

$D : K \times S \rightarrow (K + \{h\}) \times (S + \{L, R\})$ where $+$ denotes union

A Turing machine $= M$ may be interpreted semiphysically as consisting of a control unit and a tape. At any time M is in some state, and the read/write head of the tape is over some symbol on the tape. If $D(q, a) = (p, b)$ then initially the head is scanning a and M is in state q; its next move will be to enter state p. If b is in S the head will set $a := b$ without moving; otherwise $b = L$ or R in which case the head will move to the left or right. M halts when state h is entered, or M hangs (the left end of the tape is surpassed). The tape has no right end.

Input to M consists of a string. The string is padded by a $\#$ on each side and placed on the leftmost squares of the tape. The rest of the tape consists of $\#$'s. Initially the head is positioned over the $\#$ that marks the right end of the input. Initially M is in state s; its first move is thus $D(s, \#)$.

At a given time M is in configuration (q, w, a, u),

where:

$q =$ state (element of $K + \{h\}$)

$w =$ portion of the tape to the left of the head (element of S^*)

$a =$ symbol under the head (element of S)

$u =$ portion of the tape to the right of the head

If e denotes the empty string, u in the list above is required to be an element of $(S^*)(S - \{\#\}) + \{e\}$; that is, either $u = e$, meaning the tape has all $\#$'s to the right of the head, or the last symbol of u is the last nonblank symbol to the right of the head position. This gives the configuration a unique description. In particular, if the input to M is w then the initial configuration of M is $(s, \#w, \#, e)$.

M is said to halt on input w if some halted configuration (state $= h$) is reached from $(s, \#w, \#, e)$ in a finite number of steps; then it is said that $(s, \#w, \#, e)$ yields a halted configuration. If M halts on input w, M is said to accept w. The language accepted by M is the set of strings w accepted by M. Conversely, a language is said to be Turing-acceptable if it is accepted by some Turing machine.

Suppose S is an alphabet, T and W are subsets of S, and $f : W \rightarrow T$. Suppose there exists a Turing machine $M = (K, S, D, s)$ such that for any w in W, the configuration $(s, \#w, \#, e)$ yields $(h, \#u, \#, e)$, that is, u is the output of M on input w, and furthermore $u = f(w)$. Then f is said to be computed by M.

Suppose alphabet A does not contain $\#$, Y, or N. Suppose L is a language in A^* and X is its characteristic function, that is, for w in A^*, $X(w) = Y$ if w is in L, $X(w) = N$ otherwise. If X is computed by Turing machine M then M is said to decide (or recognize) L. Conversely, if X is the characteristic function of a language L and X is Turing-computable, that is, there exists a Turing machine that computes X, then L is said to be Turing-decidable.

An extension of the basic model is to permit the control unit to control (a finite number of) multiple tapes. However, a language accepted by a multitape Turing machine is also accepted by a one-tape Turing machine; that is, additional tapes do not increase the computing power of Turing machines in terms of expanding the class of languages accepted.

C.2 NONDETERMINISTIC TURING MACHINES

The preceding Turing machines were deterministic; that is, D was a function: if $D(q, a) = (p, b)$ then the next state p and scanned symbol b were uniquely determined by the present state q and symbol a. A nondeterministic Turing machine is defined similarly except that D is now a relation on

$$(K \times S) \times ((K + \{h\}) \times (S + \{L, R\}))$$

That is, D is now multiple-valued, so that the next configuration is no longer uniquely determined by the present. Instead, in a given number of steps a configuration may yield a number of configurations. Consequently an input may yield many outputs. A sequence of steps starting from a given input defines a computation; in general, many computations are possible on one input. A nondeterministic machine is said to accept an input if there exists a halting computation on it. Again the language accepted by a machine is the set of strings accepted by it. Trivially any language accepted by nondeterministic machines is also accepted by deterministic machines, which are merely special cases of the more general nondeterministic case.

It is also true (but not as trivial) that any language accepted by a nondeterministic Turing machine is also accepted by a deterministic Turing machine. That is, nondeterminism does not increase the power of Turing machines insofar as the class of languages is concerned.

Similar extensions hold for decidable languages.

C.3 COMPUTATIONAL COMPLEXITY

The time complexity of Turing machines is measured by the number of steps taken by a computation. If T is a function on the nonnegative integers, a deterministic Turing machine M is said to decide language L in time T if it decides in time $T(n)$ or less whether w is or is not in L, where w has length n. If T is a polynomial then L is said to be decided in polynomial time. If a language is decidable in polynomial time on a multitape deterministic Turing machine then it is decidable in polynomial time on a one-tape Turing machine.

A nondeterministic Turing machine M is said to accept w in time T if there is halting computation on w of $T(n)$ or fewer steps, where w has length n. M is said to accept language L in time T if it accepts each string in L in time T.

The class of languages decidable in polynomial time on some deterministic Turing machine is denoted by P. The class of languages acceptable in polynomial time on some nondeterministic Turing machine is denoted by NP.

It is not known whether P = NP.

APPENDIX D
THE THEORY OF PROBABILISTIC COMPUTING

Probabilistic algorithms are employed as adjuncts in cryptosystems for purposes such as finding primes. They have also produced virtually all major practical cryptanalytic algorithms for factoring, discrete logarithms, etc. Here we review an extension of the classic theory of computation which incorporates probabilistic computing. This extension has proven particularly valuable in the study of probabilistic cryptosystems.

An ordinary deterministic multitape Turing machine may be considered to have an input tape, an output tape, and read-write worktapes. A modification of the ordinary model (e.g., [GILL77]) is the probabilistic Turing machine. It has a distinguished state called the coin-tossing state, which permits the machine to make random decisions. In terms of languages recognized, these have the same power as deterministic machines. However, time considerations are more subtle. In particular, a notion of probabilistic runtime is needed, rather than measures such as maximum runtime used for ordinary machines.

A probabilistic Turing machine operates deterministically, except when it is in a special coin-tossing state (or states). In such a state the machine may enter either of two possible next states. The choice between these is made via the toss of an unbiased coin. The sequence of coin tosses may be considered to constitute the contents of an auxiliary read-only input tape, the random tape, which contains a binary string. Thus a computation by a probabilistic machine is a function of two variables, the ordinary input tape and the random tape.

If the random tape is unspecified, the output of the computation of probabilistic machine M, $M(x)$, is a random variable (e.g., [MCEL78]): M produces output y with probability $Pr\{M(x) = y\}$. For a given input x, there may exist a y such that $Pr\{M(x) = y\} > \frac{1}{2}$. Such a y is clearly unique if it exists, in which case we can write $q(x) = y$. This defines a partial function: q is undefined if no such y exists. The partial function q is said to be computed by M. The set accepted by M is the domain of q.

If x is accepted by M let $e(x) = Pr\{M(x) \, ! = y\}$. It is direct from definition that $e(x) < \frac{1}{2}$. Suppose there exists a constant $c < \frac{1}{2}$ such that $e(x) \leq c$ for all x in the domain of e. Then it may be said that M has bounded error probability (the error probability e is bounded away from $\frac{1}{2}$), another concept important in zero-knowledge frameworks.

Again leaving the random tape unspecified, $Pr\{M(x) = y$ in time $n\}$ is the probability that probabilistic Turing machine M with input x gives output y in some computation of at most n steps. The probabilistic runtime $T(x)$ of M is defined to be infinity if x is not accepted by M, that is, if x is not in the domain of the partial function q computed by M. If x is in the domain of q, $T(x)$ is the smallest n such that $Pr\{M(x) = q(x)$ in time $n\} > \frac{1}{2}$. This is somewhat analogous to defining the runtime of a nondeterministic Turing machine to be the length of the shortest accepting computation.

A function is probabilistically computable if it is computed by some probabilistic Turing machine. There are many important examples of functions that are probabilistically computable in polynomial probabilistic runtime and bounded error probability, but are not known to be computable in polynomial deterministic time, that is, in P. An example is the characteristic function of the set of primes, which is probabilistically computable in polynomial time (e.g., [SOLO77]), but for which no deterministic algorithm is known.

Let BPP be the class of languages recognized by polynomial-bounded probabilistic Turing machines with bounded error probability. Letting $<$ denote inclusion, it follows easily that $P < BPP < NP$. An important question, with implications for schemes such as RSA as well as zero-knowledge schemes, probabilistic encryption, and so on is whether either of these inclusions is proper.

APPENDIX E
BREAKING KNAPSACKS

We give a brief account of the demise of the Merkle–Hellman trapdoor knapsack public key system and some of its variants. A much more complete discussion is given in [BRIC88].

We recall from Section 4.2.1 that the security of this approach rests on the difficulty of solving knapsack problems of the form

$$C = b_1 * M_1 + \ . \ . \ . \ + b_n * M_n$$

where the $\{b_i\}$ are obtained from superincreasing $\{a_i\}$ by modular "disguising":

$$b_i = w * a_i \ \mathrm{mod} \ u$$

Around 1982, several authors (e.g., [DESM83]) made the observation that if $W * w \equiv 1 \pmod{u}$, where W may be found as in Appendix H, then for some $\{k_i\}$,

$$a_i = W * b_i - u * k_i$$

In particular

$$a_1 = W * b_1 - u * k_1$$

and hence

$$b_i * k_1 - b_1 * k_i = (b_1 * a_i - b_i * a_1)/u$$

Since $u > a_1 + \ . \ . \ . \ + a_n$, $b_i < u$, and $a_1 < a_i$,

$$\left| b_i * k_1 - b_1 * k_i \right| < ua_i/(a_1 + \ . \ . \ . \ + a_n)$$

Now it is easily shown that

$$a_{i+j} \geq 2^{j-1} * (a_i + 1)$$

and hence

$$\left| b_i * k_1 - b_1 * k_i \right| < 2^{i+1-n} * u$$

Thus

$$\left| k_1/k_i - b_1/b_i \right| < 2^{i+1-n} * u/k_i b_i < 2^{i+1-n} * u/b_i$$

This shows that the $\{k_i\}$ are in fact not random; they are determinable via the inequalities above if u is known. Shamir [SHAM84b] observed that an intruder merely seeks any trapdoor, as represented by any u, w, and $\{a_i\}$ that produce an easy knapsack. Thus the last inequality may be regarded as an integer programming problem. In this particular instance Shamir noted that the algorithm of Lenstra [LENS83] is applicable, together with classic methods of Diophantine approximation. This yields the $\{k_i\}$, and the system is then broken easily.

Lenstra's algorithm made use of the Lovasz lattice basis reduction algorithm [LENS82], one of the basic tools in "unusually good" simultaneous Diophantine approximation [LAGA84]. This approach was utilized by Adleman [ADLE82] to break the Shamir–Graham knapsack, and by Brickell [BRIC84] to break the iterated Merkle–Hellman knapsack. The multiplicative version was broken by Odlyzko [ODLY84]. In fact, all major proposed knapsacks based on modular disguises have been broken using this approach.

It should be noted that the low-density attacks ([BRIC83], [LAGA83]) are successful where finding trapdoors fails. These use a measure of density for a knapsack with coefficients $b_1, \ldots, b_n$ defined by density $= n/(\log \max \{b_i\})$. This type of attack is independent of whether the knapsack is a disguised version of another. Trapdoors, in contrast, are easiest to find for high-density knapsacks.

The concept of density is related to two important parameters of cryptosystems, namely, information rate and expansion factor d. In [LAGA84] the information rate is defined to be the ratio of the size of plaintext to the maximum size of the ciphertext. This is the reciprocal of d, that is, ciphertext/plaintext. Information rate is essentially the same as density, although for the above knapsack and modulus u it is defined slightly differently, namely as $n/(\log n * v)$. Both definitions are derived from approximations to the actual ciphertext size, which is $\log(b_1 * M_1 + \ldots + b_n * M_n)$. Lagarias [LAGA84] notes that the attack in [SHAM84b] runs in time $O(P(n) * n^d)$ for a polynomial P. Hence the attack is feasible if d is fixed but not if the expansion factor d is large. This illustrates the interrelation between security and practicality.

APPENDIX F
BIRTHDAY ATTACKS

In Section 4 we noted several uses of birthday attacks against hash functions. Here we give a brief summary of the relevant mathematics.

Suppose H is a function that has m possible outputs. Whenever H outputs a value it is totally random and independent of any previous values which have been output.

If H is evaluated k times, each output is akin to an object placed in one of m cells corresponding to the range of H. Since the k values of H are independent, any object could be placed in any of m cells; the total number of ways of distributing the k objects is m^k. If no two objects are to be placed in any cell (i.e., if there are to be no collisions in applying H k times), the first object can be placed anywhere; the second can go in any of the $m - 1$ remaining cells; the third in any of $m - 2$ cells, etc. The total number of ways of distributing the objects is $m(m - 1) \ldots (m - k + 1)$ (sometimes called a falling factorial). The probability of no collisions is $m(m - 1) \ldots (m - k + 1)/m^k$. Hence the probability of at least one collision is

$$P(m, k) = 1 - (m - 1) \ldots (m - k + 1)/m^{k-1}$$
$$= 1 - (1 - 1/m) \ldots (1 - (k - 1)/m)$$

This yields:

Lemma F.1:   Suppose the function H, with m possible outputs, is evaluated k times, where $m > k > (2cm)^{1/2}$ for some constant c. Then the probability of at least one collision (i.e., x and y with $H(x) = H(y)$ for some x, y) is at least $1 - e^{-c}$, $e = 2.718 \ldots$

Proof: for $0 < x < 1$,

$$1 - x < 1 - x + x^2(1 - x/3)/2 + x^4(1 - x/5)/24 + \ldots$$
$$= e^{-x}$$

For $k < m$ this gives

$$(1 - 1/m) \ldots (1 - (k - 1)/m) < e^{-1/m \ldots -(k-1)/m}$$
$$= e^{-k(k-1)/2m}$$

Thus

$$P(m, k) > 1 - e^{-k(k-1)/2m}$$

The lemma follows.   ∎

Example F.1:   Suppose the H of Lemma F.1 is evaluated k times where $k > (2(\ln 2)m)^{1/2} = 1.17 * m^{1/2}$. Then the probability of at least one collision is $> \frac{1}{2}$.

This suggests an attack on hash functions. Its name derives from the classic problem of computing the probability that two members of a group of people have the same birthday.

APPENDIX G
MODULAR ARITHMETIC AND GALOIS FIELDS

We give a brief introduction to arithmetic modulo n where n is a positive integer. A ring structure may be imposed on $Z_n = \{0, 1, \ldots, n - 1\}$ by doing addition, subtraction, and multiplication mod n (e.g., [DENN83], [HERS64] or any book on elementary number theory). We use GCD for greatest common divisor; for example, $GCD(x, y) = n$ if n is the largest positive integer dividing x and y. For $n > 1$ let

$$Z_n^* = \{x \in Z_n : x > 0 \quad \text{and} \quad GCD(x, n) = 1\}$$

For example, $Z_{10}^* = \{1, 3, 7, 9\}$. That is, Z_n^* consists of the nonzero elements of Z_n that are relatively prime to n.

If $a \in Z_n^*$ let

$$a * Z_n^* = \{a * x \mod n : x \in Z_n^*\}$$

For example, $2 * Z_{10}^* = \{2, 6, 14, 18\} \mod 10 = \{2, 6, 4, 8\}$, $3 * Z_{10}^* = \{3, 9, 21, 27\} \mod 10 = Z_{10}^*$. We have:

Lemma G.1:   Suppose $n > 1$ and $a \in Z_n^*$. Then

1. For any x and y: $a * x \equiv a * y \pmod{n}$ iff $x \equiv y \pmod{n}$.

2. $a * Z_n^* = Z_n^*$.
3. a has a multiplicative inverse modulo n, that is, there exists $b \in Z_n^*$ such that $b * a \equiv 1 \pmod n$.

Proof: If $x \equiv y \pmod n$ then trivially $a * x \equiv a * y \pmod n$. Conversely, suppose $a * x \equiv a * y \pmod n$. Then $n \mid (a * (x - y))$. But $\mathrm{GCD}(a, n) = 1$, so $n \mid (x - y)$, that is, $x \equiv y \pmod n$. Thus (1) holds.

For (2): if $x \in Z_n^*$ then $\mathrm{GCD}(x, n) = 1$, so $\mathrm{GCD}(a * x, n) = 1$. Thus $a * x$ mod $n \in Z_n^*$. Hence $a * Z_n^*$ is contained in Z_n^*. By (1), if $a * x \equiv a * y \pmod n$, then $x \equiv y \pmod n$. Thus $a * Z_n^*$ consists of n distinct elements, and cannot be a proper subset of Z_n^*; (2) follows. In particular, since $1 \in Z_n^*$, there exists some $b \in Z_n^*$ such that $a * b$ mod $n = 1$; (3) follows. ∎

G.1 THE EULER PHI FUNCTION

The cardinality of a set S is denoted by $|S|$. In particular, $|Z_n^*|$ is denoted by $\phi(n)$, the Euler totient function.

Lemma G.1.1:

1. If p is prime then $\phi(p) = p - 1$.
2. If $n = p * q$, p and q prime, then $\phi(n) = (p - 1)(q - 1)$.

Proof: (1) is trivial, since $Z_p^* = \{1, \ldots, p - 1\}$.
For (2): let $Y_p = \{p, \ldots, (q - 1)p\}$, that is, the nonzero elements of Z_n divisible by p. Let $Y_q = \{q, \ldots, (p - 1)q\}$, that is, the nonzero elements of Z_n divisible by q. If $a * p = b * q$, then $p \mid b$ and $q \mid a$; hence $a * p$ and $b * q$ cannot be in Y_p and Y_q, respectively. Thus Y_p and Y_q are disjoint. Letting $+$ denote disjoint union,

$$Z_n = \{0\} + Y_p + Y_q + Z_n^*$$

Taking cardinalities,

$$p * q = 1 + q - 1 + p - 1 + |Z_n^*|$$

Then (2) follows. ∎

G.2 THE EULER–FERMAT THEOREM

Lemma G.2.1 (Euler's theorem): Suppose $n > 0$ and $a \in Z_n^*$. Then

$$a^{\phi(n)} \equiv 1 \pmod n$$

Proof: If $Z_n^* = \{r_1, \ldots, r_m\}$ then a restatement of (2) of Lemma G.1 is

$$\{a * r_1 \bmod n, \ldots, a * r_m \bmod n\} = Z_n^*$$

Hence

$$a * r_1 * \ldots * a * r_m \equiv r_1 * \ldots * r_m \pmod{n}$$

Thus

$$a^m * r_1 * \ldots * r_m \equiv r_1 * \ldots * r_m \pmod{n}$$

By (1) of Lemma G.1, each r_i above may be canceled, leaving

$$a^m \equiv 1 \pmod{n}$$

Noting that by definition $m = \phi(n)$ gives Euler's theorem. ∎

Corollary G.2.1 (Fermat's theorem):   Suppose p is prime and $a \in Z_p^*$. Then

$$a^{p-1} \equiv 1 \pmod{p}$$

Proof: Use (1) of Lemma G.1.1 in Lemma G.2.1. ∎

Corollary G.2.2: Suppose $n > 0$, $a \in Z_n^*$, and $x \equiv 1 \pmod{m}$ where $m = \phi(n)$. Then $a^x \equiv a \pmod{n}$.

Proof: We have $x = m * y + 1$ for some y. Now by Lemma G.2.1,

$$a^x \equiv (a^m)^y * a \equiv 1^y * a \equiv a \pmod{n} \quad ∎$$

Corollary G.2.3: Suppose $n > 0$. Let $m = \phi(n)$. Suppose e and d are in Z_m^* and $e * d \equiv 1 \pmod{m}$. Let $E(M) = M^e \bmod n$ and $D(C) = C^d \bmod n$. Then $D(E(M)) = E(D(M)) = M$ for any M in $[0, n)$.

Proof:

$$D(E(M)) \equiv (M^e \bmod n)^d \bmod n \equiv M^{e*d} \bmod n \equiv M \pmod{n}$$

The last step uses Corollary G.2.2. Also $0 \le D(E(M)) < n$ and $0 \le M < n$, so $D(E(M)) = M$. Similarly $E(D(M)) = M$. ∎

G.3. GALOIS FIELDS

Lemma G.1 shows that Z_n^* is an Abelian group (e.g., [HERS64], p. 27) under the operation of multiplication modulo n; Z_n^* is called the multiplicative group of units of Z_n.

In particular, if p is prime then $Z_p^* = \{1, \ldots, p - 1\}$ is a group; that is, each nonzero element of Z_p has a multiplicative inverse modulo p. Thus the ring $Z_p = \{0, 1, \ldots, p - 1\}$ is in fact a finite field (e.g., [HERS64], p. 84) with p elements.

It can be shown (e.g., [HERS64], p. 314) that every finite field has p^n elements for some prime p. These are called the Galois fields, and denoted $GF(p^n)$. We have already noted that $GF(p) = Z_p$ is defined by doing arithmetic modulo p. The elements of Z_p are called the residues modulo p; that is, each integer x has a unique representation $x = q * p + r$ where $r \in Z_p$. To define $GF(p^n)$ we choose an irreducible polynomial $f(x)$ of degree n in the ring of polynomials modulo p (e.g., [DENN83], p. 49).

Now arithmetic may be defined on this ring of polynomials, modulo $f(x)$; that is, write $g(x) \equiv h(x)$ iff $f(x)$ is a divisor of $g(x) - h(x)$. Each polynomial $g(x)$ has a unique representation $g(x) = q(x)f(x) + r(x)$ for some polynomial $r(x)$ of degree at most $n - 1$. The residues modulo $f(x)$ are the elements of $GF(p^n)$. These consist of polynomials of degree $\le n - 1$ with coefficients from Z_p. Each of the n coefficients of a residue has p possible values, accounting for the p^n element count for $GF(p^n)$.

APPENDIX H
EUCLID'S ALGORITHM

On a number of occasions we referred to Euclid's algorithm, which can be used to find GCDs and multiplicative inverses. We give some versions of it here. These versions are recursive, and minimize storage requirements.

Suppose x and y are arbitrary positive integers. Their greatest common divisor $GCD(x, y)$ can be computed in $O(\log \max\{x, y\})$ steps by recursively employing $GCD(s, t) = GCD(s, t \bmod s)$. This is Euclid's algorithm:

```
function GCD(x, y) returns integer;
  /* return GCD of x > 0 and y > 0 */
  s := x; t := y;
  while (s > 0) do
      div := s; s := t mod s; t := div;
  end while;
  return(div);
end GCD;
```

This is readily generalized to

```
function Multi_GCD(m; x : array[0..m]);
  /* for m > 0 return GCD of
      x₁,...,xₘ > 0 */
  if (m = 1) return(x₁)
  else return(GCD(xₘ, Multi_GCD(m - 1, x)));
end Multi_GCD;
```

The above runs in $O(m * \log \max\{x_i\})$ time.

For $x \in Z_n^*$, with the latter as in Appendix G, a simple extension of GCD yields the multiplicative inverse of x modulo n, that is, u with $x * u \equiv 1 \pmod{n}$. With $[y]$ denoting the largest integer $\le y$, the extended Euclid algorithm to find u is:

```
function INVERSE(n, x) returns integer;
  /* for n > 0 and x ∈ Zₙ* return u ∈ Zₙ*
    with u * x ≡ 1 (mod n) */
  procedure Update(a, b);
    temp := b; b := a - y * temp; a := temp;
  end Update;
  g := n; h := x; w := 1; z := 0; v := 0; r := 1;
  while (h > 0) do
```

$y := [g/h]$; Update(g, h); Update(w, z); Update(v, r);
    end while;
    return$(v \bmod n)$;
  end INVERSE;

For example, for $n = 18$, $x = 7$, this algorithm gives the values

g:	18	7	4	3	1
h:	7	4	3	1	0
w:	1	0	1	-1	2
z:	0	1	-1	2	-7
v:	0	1	-2	3	-5
r:	1	-2	3	-5	18
y:		2	1	1	3

Finally $u = -5 \bmod 18 = 13$ is returned. This is equivalent to finding the sequence $\frac{7}{18}, \frac{2}{5}, \frac{1}{3}, \frac{1}{2}, \frac{0}{1}$ in which each neighboring pair is a pair of Farey fractions; $\frac{a}{b}$ and $\frac{c}{d}$ are Farey fractions (e.g., [RADE64]) if $a*d - b*c = \pm 1$. We have found $2*18 - 5*7 = 1$, so that $-5*7 \equiv 1 \pmod{18}$, or equivalently $13*7 \equiv 1 \pmod{18}$. Thus finding multiplicative inverses modulo n can also be done in $O(\log n)$ steps.

This will also find s with $u*x + s*n = 1$: take $s = (1 - u*x)/n$. More generally we have (with $GCD(x) = x$):

```
procedure Coeffs(m; x : in array[1..m]; u : out array[1..m]);
    /* given m > 0 and x₁,...,xₘ > 0 with GCD(x₁,...,xₘ) = 1,
       find u₁,...,uₘ with u₁x₁ + ... + uₘxₘ = 1 */
    if (m = 1) return(x₁)
    else
        g = Multi_GCD(m - 1, x);
        for i := 1 to m - 1 do yᵢ := xᵢ/g;
        coeffs(m - 1, y, u');
        uₘ := INVERSE(g, xₘ);
        b := (1 - uₘ * xₘ)/g;
        for i := 1 to m - 1 do uᵢ := b * uᵢ';
    end else;
end Coeffs;
```

This runs in $O(m * \log \max\{x_i\})$ time.

APPENDIX I
THE CHINESE REMAINDER THEOREM

The Chinese remainder theorem is useful for purposes such as simplifying modular arithmetic. In particular, we note in Appendix N that its use increases the efficiency of decryption in RSA.

We begin with a special case: with Z_n as in Appendix G, we have

Lemma I.1:   Suppose p and q are primes and $p < q$. Then:

1. For arbitrary $a \in Z_p$ and $b \in Z_q$, there exists a unique $x \in Z_{pq}$ with

$$x \equiv a \pmod{p}$$
$$x \equiv b \pmod{q}$$

2. If $u * q \equiv 1 \pmod{p}$, the x in (1) is given by

$$x = (((a - b) * u) \bmod p) * q + b$$

Proof: We note that for any y and s, $0 \leq y \bmod s \leq s - 1$. Thus if x and u are as in (2), $0 \leq x \leq (p - 1) * q + q - 1 = p * q - 1$. Hence $x \in Z_{pq}$. Trivially $x \equiv b \pmod{q}$. Also

$$x \equiv (((a - b) * u) \bmod p) * q + b$$
$$\equiv (a - b) * u * q + b$$
$$\equiv (a - b) * 1 + b$$
$$\equiv a \pmod{p}$$

Thus x is a solution to the simultaneous linear system in (1). To show that it is unique, suppose $x' \equiv a \pmod{p}$ and $x' \equiv b \pmod{q}$. Then for some k and m, $x - x' = k * p = m * q$. Thus $k = q * k'$ for some k'; hence $x - x' = p * q * k'$, that is, $x' = x - p * q * k'$. Since $0 \leq x \leq p * q - 1$, if $k' > 0$ then $x' < 0$; if $k' < 0$ then $x' \geq p * q$. Hence $x' \in Z_{pq}$ iff $k' = 0$, that is, $x' = x$. ∎

We note that in (2) above, u can be found via the INVERSE function of Appendix H.

The condition $p < q$ is arbitrary, but useful in noting

Corollary I.1: Suppose p and q are primes, $p < q$, $0 \leq x < p * q$, and $a = x \bmod p$, $b = x \bmod q$. Then

1. $x = (((a - (b \bmod p)) * u) \bmod p) * q + b$

$$(a \geq b \bmod p)$$

2. $x = (((a + p - (b \bmod p)) * u) \bmod p) * q + b$

$$(a < b \bmod p)$$

Proof: Immediate from Lemma I.1. ∎

Corollary I.1 provides an optimal framework for representing x via a and b, that is, the residues of x modulo p and q, respectively.

Although the most general case of the Chinese remainder theorem is not used in this exposition, we remark that the above can be extended:

Theorem I.1: Given pairwise relatively prime moduli $\{p_1, \ldots, p_n\}$ and arbitrary $\{a_1, \ldots, a_n\}$, there exists a unique $x \in [0, p_1 * \ldots * p_n)$ satisfying $x \equiv a_i \pmod{p_i}$ for each i.

Proof: A straightforward generalization of the above (e.g., [DENN83], p. 47).

APPENDIX J
QUADRATIC RESIDUES AND THE JACOBI SYMBOL

Quadratic residues play a role in primality testing as we note in Appendix N. In Sections 8 and 9 we also noted their use in encryption.

Let Z_n^* be as in Appendix G. For a positive integer n and $a \in Z_n^*$, a is called a quadratic residue modulo n if $x^2 \equiv a \pmod{n}$ for some x; otherwise a is a quadratic nonresidue.

Lemma J.1: Suppose $n > 0$ and a is a quadratic residue modulo n. Then every y with $y^2 \equiv a \pmod{n}$ has the form $y = x + k * n$ where k is an integer and $x \in Z_n^*$.

Proof: Suppose $y^2 \equiv a \pmod{n}$. Let $x = y \bmod n$. Then $x \in [0, n)$. Also, $x^2 \equiv a \pmod{n}$. Now $x^2 = a + j * n$ for some j. Suppose for some m we have $m > 0$, $m \mid x$ and $m \mid n$. Then $m \mid a$, so $m = 1$ since $GCD(a, n) = 1$. Hence $x \in Z_n^*$; also $y = x + k * n$ for some k. ∎

Thus the square roots of a quadratic residue modulo n can be taken to be elements of Z_n^*; all other square roots are obtained by adding multiples of n to these.

J.1 QUADRATIC RESIDUES MODULO A PRIME

If p is prime and $0 < a < p$, a is a quadratic residue modulo p iff $x^2 \equiv a \pmod{p}$ for some x with $0 < x < p$. For example, the quadratic residues modulo 7 are

$$\{1^2, 2^2, 3^2, 4^2, 5^2, 6^2\} \bmod 7 = \{1, 2, 4\}$$

The quadratic nonresidues modulo 7 are $\{3, 5, 6\}$.
Given a prime p, let $s = (p - 1)/2$ and

$$S_p = \{x^2 \bmod p : 0 < x \leq s\}$$

Then S_p is a set of quadratic residues modulo p. In fact we have

Lemma J.1.1: Suppose $p > 2$ is prime. Let $s = (p - 1)/2$. Then

1. The elements of S_p are precisely the quadratic residues modulo p.
2. There are s quadratic residues modulo p.
3. There are s quadratic nonresidues modulo p.

Proof: As noted above, S_p is a subset of the set of quadratic residues modulo p. Furthermore, if $x^2 \equiv y^2 \pmod{p}$ then $p \mid x^2 - y^2$; hence $p \mid x - y$ or $p \mid x + y$. If x and y are in $(0, s]$ then $1 < x + y < p$; thus $p \mid x - y$; hence $x = y$. It follows that S_p contains distinct elements.

Now suppose a is a quadratic residue, $x^2 \equiv a \pmod{p}$, and $x \in Z_p^*$. We note

$$(p - x)^2 \equiv p^2 - 2 * p * x + x^2 \equiv a \pmod{p}$$

Since $0 < x < p$, either x or $p - x$ is in $(0, s]$. It follows that $a \in S_p$. Thus the set of quadratic residues modulo p is contained in S. Hence the two sets are identical, establishing (1). Since $|S_p| = s$, (2) follows. Also, the complement of S_p in Z_p^* is the set of quadratic nonresidues modulo p. Since $|Z_p^*| = 2s$, the complement of S_p also has cardinality s; (3) follows. ∎

J.2 THE JACOBI SYMBOL

If p is a prime > 2 and $0 < a < p$, the Legendre symbol (a/p) is a characteristic function of the set of quadratic residues modulo p (e.g., [RADE64]):

1. $(a/p) = 1$ if a is a quadratic residue mod p.
2. $(a/p) = -1$ if a is a quadratic nonresidue mod p.

More generally, if $k > 1$ is odd and h is in Z_k^*, the Jacobi symbol (h/k) may be defined as follows:

1. The Jacobi symbol (h/p) coincides with the Legendre symbol if p is prime.
2. If $k = p_1 * \ldots * p_m$ with p_i prime, $(h/k) = (h/p_1) * \ldots * (h/p_m)$.

An efficient mode for computing the Jacobi symbol is via the recursion:

1. $(1/k) = 1$
2. $(a * b/k) = (a/k)(b/k)$
3. $(2/k) = 1$ if $(k^2 - 1)/8$ is even, -1 otherwise
4. $(b/a) = ((b \mod a)/a)$
5. If $GCD(a, b) = 1$:
   a. $(a/b)(b/a) = 1$ if $(a - 1)(b - 1)/4$ is even
   b. $(a/b)(b/a) = -1$ if $(a - 1)(b - 1)/4$ is odd

The key step in the recursion is (5), which is Gauss's law of quadratic reciprocity (e.g., [RADE64]). The above list shows that the Jacobi symbol (a/n) can be computed in $O(\log n)$ steps.

J.3 SQUARE ROOTS MODULO A PRIME

Regarding solutions of $x^2 \equiv a \pmod p$, that is, the square roots of a modulo p, we have:

Lemma J.3.1: Suppose $p > 2$ is prime. Let $s = (p - 1)/2$. Suppose a is a quadratic residue modulo p. Then

1. a has exactly two square roots modulo p in Z_p^*. One of these lies in $(0, s]$ and the other in $(s, p - 1]$.

2. the square roots of a modulo p can be found in probabilistic polynomial time.

Proof: By Lemma J.1.1 we know that $a \in S_p$ as defined in Section J.1; hence there exists a unique x in $(0, s]$ with $x^2 \equiv a \pmod{p}$. Then $y = p - x$ satisfies $y^2 \equiv a \pmod{p}$, and y is in $(s, p - 1]$. Conversely, suppose y is in $(s, p - 1]$ and $y^2 \equiv a \pmod{p}$. Then $x = p - y$ is in $(0, s]$, and $x^2 \equiv a \pmod{p}$. Hence y is unique. Thus we have (1).

For (2) we invoke a probabilistic polynomial-time algorithm for finding square roots modulo a prime (e.g., [PERA86] or p. 22 of [KRAN86]). ∎

J.4 QUADRATIC RESIDUOSITY MODULO A PRIME

Deciding quadratic residuosity plays a role in both primality testing and probabilistic encryption. For a prime $p > 2$, to decide whether a given element $a \in Z_p^*$ is a quadratic residue modulo p, define

$$e_p(a) = a^s \bmod p \qquad (s = (p - 1)/2)$$

Then we have

Lemma J.4.1: If $p > 2$ is a prime and $a \in Z_p^*$, $e_p(a) = 1$ iff a is a quadratic residue modulo p; $e_p(a) = p - 1$ iff a is a quadratic nonresidue modulo p.

Proof: By the Euler–Fermat theorem (Corollary G.2.1), $e_p(a)^2 \equiv 1 \pmod{p}$. By Lemma J.3.1, 1 has exactly two square roots modulo p in Z_p^*, namely, 1 and $p - 1$. Hence $e_p(a) = 1$ or $p - 1$. If a is a quadratic residue modulo p, then $a = x^2 \bmod p$ for some x with $0 < x < p$. Then $e_p(a) = x^{p-1} \bmod p$. Again by the Euler–Fermat theorem, $e_p(a) = 1$. Thus all s quadratic residues a satisfy $a^s - 1 = 0 \pmod{p}$. An sth degree congruence modulo p has at most s solutions (e.g., [RADE64], p. 21). Thus all quadratic nonresidues b must have $e_p(b) = p - 1$. ∎

Also, e_p can be evaluated in O(log p) time. Thus quadratic residuosity modulo p can be decided in O(log p) time.

APPENDIX K
PRIMITIVE ROOTS AND DISCRETE LOGARITHMS

The use of primitive roots was noted in Sections 2.2 and 4.2.2. We also observed that the security of exponentiation-based cryptosystems depends in part on the difficulty of computing discrete logarithms. Here we briefly explore these topics.

Let Z_p^* be as in Appendix G. Suppose p is a prime and $a \in Z_p^*$. Suppose $m > 0$, $a^m \equiv 1 \pmod{p}$, but $a^r \not\equiv 1 \pmod{p}$ for any r with $0 < r < m$. Then we say that a belongs to the exponent m modulo p. The existence of m is guaranteed by the Euler–Fermat theorem (Corollary G.2.1), which shows $m < p$. Let

$$C_p(a) = \{a^x \bmod p : 0 \le x < m\}$$

Then we have

Lemma K.1: Suppose p is prime, $a \in Z_p^*$, and a belongs to the exponent m modulo p. Then:

1. $C_p(a)$ contains distinct elements, that is, $|C_p(a)| = m$.
2. $C_p(a)$ is a cyclic subgroup of Z_p^*.
3. $m \mid p - 1$.

Proof: Suppose $C_p(a)$ does not contain distinct elements. Then for some $\{k, j\}$ in $[0, m)$ with $k < j$ we have $a^k \equiv a^j \pmod{p}$. Thus $a^{j-k} \equiv 1 \pmod{p}$, which is a contradiction. This gives (1). Now suppose x and y are in $[0, m)$ and $x + y = k * m + r$ where r is in $[0, m)$. Since $a^m \equiv 1 \pmod{p}$, $a^x * a^y \bmod p = a^r \bmod p$. Thus $C_p(a)$ is closed under multiplication, and forms a subgroup of Z_p^*; this is called the cyclic subgroup generated by a (e.g., [HERS64]). This gives (2). The order of a subgroup divides the order ($= p - 1$) of the whole group Z_p^* (ibid.). This gives (3). ∎

If g belongs to the exponent $p - 1$ modulo prime p, g is called a primitive root modulo p.

Lemma K.2: Suppose p is a prime and g is a primitive root modulo p. Then $C_p(g) = Z_p^*$; that is, each $y \in [1, p)$ has a unique representation $y = g^x \bmod p$ for some $x \in [0, p - 1)$.

Proof: $C_p(g)$ is a subset of Z_p^*. The result follows from Lemma K.1, which shows $|C_p(g)| = p - 1 = |Z_p^*|$. ∎

The x in Lemma K.2 is the discrete logarithm, or index, of y modulo p with base $= g$. However, the range of the logarithm can be any interval of length $p - 1$. We have used $[0, p - 2]$, but, for example, we could also take x to lie in $[1, p - 1]$.

A restatement of Lemma K.2 is that the cyclic subgroup generated by the primitive root g modulo p is the whole group Z_p^*. Thus g is a generator of Z_p^*.

Lemma K.3: Suppose $p > 2$ is prime and $p - 1$ has k distinct prime factors which are known. Then:

1. The number of primitive roots modulo p is $\phi(p - 1)$, where ϕ is the phi function (Appendix G.1).
2. Testing a given a to determine if it is a primitive root modulo p can be done in time $O(k * \log p) = O((\log p)^2)$.

Proof: For (1) see, for example, [RADE64], p. 49. For (2), if the exponent of a modulo p is m, then $m \mid p - 1$ by Lemma K.1. Now $m < p - 1$ iff $m \mid (p - 1)/q$ for some prime factor q of p, whence $a^{(p-1)/q} \equiv 1 \pmod{p}$. Thus a is a primitive root modulo p iff $a^{(p-1)/q} \not\equiv 1 \pmod{p}$ for each prime factor q of p. This condition can be tested for each of the k factors in $O(\log p)$ time. Clearly $k = O(\log p)$. ∎

In particular, if p has the form $2q + 1$ for prime q, it is easy to find the exponent m modulo p for a given a, since $m = 2$, q, or $2q$. It is also easy to find primitive roots modulo p : $\phi(p - 1) = q - 1 = (p - 3)/2$; thus for large p, a random element of Z_p^* has about a $\frac{1}{2}$ probability of being a primitive root.

Lemma K.4:   Suppose $p > 2$ is prime and g is a primitive root modulo p. Let $s = (p - 1)/2$. Then:

1. $g^s \equiv -1 \pmod{p}$.
2. g is a quadratic nonresidue modulo p.

Proof: For (1): We note $g^0 \equiv 1 \pmod{p}$. Since g is a generator of $\{1, g, \ldots, g^{p-2}\}$ and $0 < s < p - 1$ we thus know $g^s \:!\equiv 1 \pmod{p}$. Also, $g^{2s} \equiv 1 \pmod{p}$, so that g^s is a square root of 1 modulo p. By Lemma J.3.1, 1 has exactly two square roots modulo p, namely, 1 and -1. Thus $g^s \equiv -1 \pmod{p}$.

For (2): Suppose $x^2 \equiv g \pmod{p}$. Now $x \equiv g^r \pmod{p}$ for some r; hence $g^{2r} \equiv g$ $\pmod{p}$, and $g^{2r-1} \equiv 1 \pmod{p}$. Thus $2r - 1 \equiv 0 \pmod{p - 1}$, which is impossible since $p - 1$ is even. ∎

APPENDIX L
PRIMALITY TESTING

Testing large integers to determine whether they are prime plays a major role in key generation in RSA and other public key systems. We give a few examples of algorithms to effect this.

It is well known (e.g., [KNUT81], p. 366) that if the number of primes between 1 and x is $f(x)$, then with ln denoting natural logarithm (to base $e = 2.718 \ldots$),

$$f(x) = x/(\ln x) + O(x/(\ln x)^2)$$

The number of primes between x and $x + h$ is $f(x + h) - f(x)$, and hence the probability that a number between x and $x + h$ is prime is roughly

$$(f(x + h) - f(x))/h = f'(x) = 1/(\ln x) - 1/(\ln x)^2$$

That is, roughly ln x numbers must be tested before a prime is found near x; but the even numbers may be ignored, so that roughly $(\ln x)/2$ odd numbers need be tested. For example, about $(\ln 10^{100})/2 = 115$ odd numbers must be tested to find a prime of about 100 digits. If multiples of small primes > 2 are eliminated, even fewer numbers need be tested before a prime is found (e.g., [GORD84, GORD84b]).

A number of tests have been given to establish the primality of a candidate. Proving primality deterministically (i.e., with certainty) is less difficult than factoring or computing discrete logarithms, but is nonetheless nontrivial. For example, the algorithms of [ADLE83] or [COHE84] could be used; but they have a runtime on the order of

$$(\log n)^{c \: \log \log \log n}$$

Such deterministic algorithms are computationally infeasible or inadvisable unless high-performance computers are used for key generation. This may be undesirable even if a high-performance computer is available, since it may not constitute a cryptographically secure environment. For key generation on, for example, personal computers or smart cards, which is preferable from a security standpoint, more efficient algorithms are desirable. Thus, probabilistic tests may be used in practice to make an "educated guess" as to whether a candidate is prime.

Often a Monte Carlo approach is employed. In this event there is a slight possibility of an erroneous conclusion; however, the error probability can be made arbitrarily small. Typically, a sequence of "witnesses" attest to the primality or compositeness of a number. Agreement among a group of about 100 witnesses is generally sufficient to reach a conclusion beyond any reasonable doubt, although evidently the legality of this procedure has not been tested in the courts.

L.1 THE SOLOVAY–STRASSEN TEST

One example of a probabilistic polynomial-time primality test is the Solovay–Strassen test [SOLO77] mentioned in Section 4.1.1 (although this approach is commonly attributed to Solovay and Strassen, it was noted implicitly by Lehmer [LEHM76]; see [LENS86]). If n is an odd positive integer and $a \in Z_n^*$, with the latter as in Appendix G, let

$$e(a, n) \equiv a^{(n-1)/2} \pmod{n}, \quad -1 \le e(a, n) \le n - 2$$

For convenience we have $e(a, n)$ take on the value -1 instead of the usual $n - 1$.

If p is prime we have $e(a, p) \equiv e_p(a) \pmod{p}$ with e_p as in Appendix J.4. Lemma J.4.1 shows that $e(a, p) = (a/p)$, where the latter is the Jacobi symbol (Appendix J.3). The latter equality thus provides evidence of primality. That is, if p is prime and $a \in Z_n^*$ then $e(a, p) = (a/p)$. Per se this is useless, but the contrapositive provides a proof of compositeness: if $(a/n) \ne e(a, n)$ then n is composite. Also, both (a/n) and $e(a, n)$ can be computed in $O(\log n)$ steps, so this proof of compositeness runs in deterministic polynomial time. The converse is more subtle. Suppose n is an odd positive integer and $0 < a < n$. If $GCD(a, n) > 1$ then n is certainly composite. Let

$$G = \{a \in Z_n^* : (a/n) = e(a, n)\}$$

Now the Jacobi symbol is multiplicative; that is,

$$(a * b/n) = (a/n) * (b/n)$$

Also

$$e(a * b, n) \equiv e(a, n) * e(b, n) \pmod{n}$$

That is, e is also multiplicative. It follows that G is a subgroup of Z_n^*. The order of G must divide the order of Z_n^* (e.g., [HERS64], p. 35). Thus if $G \ne Z_n^*$, G has cardinality at most $(n - 1)/2$. Solovay and Strassen showed that indeed $G \ne Z_n^*$ if n is composite [SOLO77]. Hence if n is composite and a is chosen at random, the probability that a is in G is at most $\frac{1}{2}$. We thus test as follows:

```
function Solovay_Strassen (a, n) returns charstring;
/* for n > 2, n odd, a ∈ Zn, decides probabilistically
```

```
            whether n is prime */
        if (GCD(a, n) > 1) return("composite")
        else
            if ((a/n) = e(a, n)) return("prime")
            else return("composite");
    end Solovay_Strassen;
```

We will certainly reach a correct conclusion in any of the following cases:

1. n is prime.

2. n is composite and GCD$(a, n) > 1$.

3. n is composite, GCD$(a, n) = 1$, and $(a/n) \neq e(a, n)$.

In the remaining case, that is, n is composite, GCD$(a, n) = 1$, and $(a/n) = e(a, n)$, n "masquerades" as a prime due to the perjury of a. But such an a is in G above; hence the probability of false testimony from an a is at most $\frac{1}{2}$ in this case. If one hundred random a's are used as witnesses, we conclude n is prime or composite with probability of error zero if n is prime, and at most 2^{-100} otherwise. At most $6 * \log n$ operations are needed ([SOLO77]).

L.2 LEHMAN'S TEST

Lehman [LEHM82] noted that the Jacobi function is not needed in Monte Carlo testing for primality. He defines

$$e'(a, n) = a^{(n-1)/2} \mod n$$
$$G = \{e'(a, n) : a \in Z_n^*\}$$

We note that e' differs only slightly from e, taking on the value $p - 1$ instead of -1.

Lehman shows that if n is odd, $G = \{1, p - 1\}$ iff n is prime. Again one hundred a's are tested, but only $a^{(n-1)/2} \mod n$ is computed. If anything except 1 or -1 is found, n is composite. If only 1's and -1's are found, we conclude n is prime if any -1's are found; otherwise we conclude n is composite. Again the probability of error is at most 2^{-100}.

L.3 THE MILLER–RABIN TEST

Another Monte Carlo test was noted by Miller [MILL76] and Rabin [RABI80]. If $n - 1 = u * 2^k$ where u is odd and $k > 0$, let exp$(i) = 2^i$, $0 \leq i \leq k - 1$. If $a \in Z_n^*$ then a is a witness to the compositeness of n if $a^u \not\equiv 1 \pmod{n}$ and $a^{u*\exp(i)} \not\equiv -1 \pmod{n}$, $0 \leq i \leq k - 1$. Again, for example, a hundred witnesses may be tested; if none attest to compositeness of n then n may be assumed prime.

When $n \equiv 3 \pmod 4$ this test is similar to Lehman's.

APPENDIX M
MATHEMATICS OF RSA AND OTHER
EXPONENTIAL SYSTEMS

In Section 1.5 the basic exponential cipher was introduced; that is, $E(M) = M^k \bmod p$, $D(C) = C^I \bmod p$, where $K * I \equiv 1$ (modulo $p - 1$). We recall (Appendix H) that I can be computed from K using Euclid's algorithm in $O(\log p)$ time. Also, by Lemma G.1.1 we have $\phi(p) = p - 1$. Thus, by Corollary G.2.3 we have $D(E(M)) = M$ and $E(D(C)) = C$.

The RSA system is analogous: we have $E(M) = M^e \bmod n$, $D(C) = C^d \bmod n$, $n = p * q$, $e * d \equiv 1 \pmod{m}$, $m = (p - 1)(q - 1)$. By Lemma G.1.1 we have $\phi(n) = m$, so once again Corollary G.2.3 gives $D(E(M)) = M$ and $E(D(C)) = C$.

We have noted in Appendix L how the Solovay–Strassen or Lehman tests can be used to choose p and q efficiently, that is, in $O(\log n)$ steps. However, these involve multiple-precision arithmetic [KNUT81]. Hence the total time can be significant, especially in software.

Once p and q have been chosen, e is chosen easily; then d is computed via $e * d \equiv 1 \pmod{m}$. This can be done in $O(\log n)$ steps using Euclid's algorithm. Thus key material can be generated in linear time.

Encryption is easy since e can be chosen small. However, efficient decryption requires the Chinese remainder theorem. In general we have

$$(y \bmod n) \bmod p = y \bmod p$$
$$(y \bmod n) \bmod q = y \bmod q$$

Suppose we are given ciphertext C; we wish to efficiently compute

$$M = C^d \bmod n$$

To use the machinery of Appendix I, let $y = C^d$ and $x = M = y \bmod n$. Then

$$a = C^d \bmod p$$
$$b = C^d \bmod q$$

Also, suppose $d = k * (p - 1) + r$. Then by the Euler–Fermat theorem,

$$a = (C^{p-1})^k C^r \bmod p = 1^k C^r \bmod p = (C \bmod p)^r \bmod p$$

Similarly if $d = j * (q - 1) + s$,

$$b = (C \bmod q)^s \bmod q$$

Also $r = d \bmod p - 1$ and $s = d \bmod q - 1$. Thus by Corollary I.1, an algorithm for decryption [QUIS82] is as follows:

1. Compute
   a. $a = (C \bmod p)^{d \bmod (p-1)} \bmod p$
   b. $b = (C \bmod q)^{d \bmod (q-1)} \bmod q$
2. Find u with $0 < u < p$ and

$$u * q \equiv 1 \pmod{p}$$

3. Use one of

    a. $M = (((a - (b \mod p)) * u) \mod p) * q + b$

                                                   $(a \geq b \mod p)$

    b. $M = (((a + p - (b \mod p)) * u) \mod p) * q + b$

                                                   $(a < b \mod p)$

These represent roughly optimal formulae for deciphering. Again u is found easily using Euclid's algorithm, and M is easy to compute once a and b have been found. Finding a and b requires at most 2 log p and 2 log q multiplications, respectively ([KNUT81], p. 442). Thus both encryption and decryption can be done in O(log n) operations, that is, linear time. Nonetheless, these operations are relatively time-consuming (though elementary), since they involve multiplications and divisions of numbers of about 100 digits. For efficiency this requires hardware support.

APPENDIX N
QUADRATIC RESIDUOSITY MODULO A COMPOSITE

Deciding quadratic residuosity modulo a composite played a major role in Sections 8 and 9. In particular, suppose $N = p * q$ where p and q are large primes. Deciding quadratic residuosity modulo such N when p and q are unknown is regarded as computationally intractable. This forms the basis of many probabilistic encryption and zero-knowledge schemes.

Also, it was noted by Rabin [RABI79] that finding square roots modulo N in this event has essentially the same complexity as factoring N, the intractability of which forms the basis for RSA. This was exploited by Rabin (see Section 4.1.4) to obtain a modification of RSA which is provably equivalent to factoring. Essentially his scheme was to encrypt via

$$E_N(x) = x^2 \mod N.$$

Let Z_N^* be as in Appendix G.

N.1 CHARACTERIZING QUADRATIC RESIDUES

The basic extension of quadratic residuosity modulo a prime to the composite modulus case is:

Lemma N.1.1: If $N = p * q$, p and q primes > 2, then:

1. Suppose $z \in Z_N^*$. Then z is a quadratic residue modulo N iff $z \mod p$ is a quadratic residue modulo p and $z \mod q$ is a quadratic residue modulo q.
2. A quadratic residue z modulo N has exactly four square roots of the form $\{x, N - x, y, N - y\}$ in Z_N^*.
3. If z is a quadratic residue modulo N, and p and q are known, then the square roots of z modulo N can be found in probabilistic polynomial time.

Proof: Suppose $z \in Z_N^*$, $z \mod p$ is a quadratic residue modulo p, and $z \mod q$ is a quadratic residue modulo q. Then for some r and t in Z_p^* and Z_q^*, respectively, we have $r^2 \equiv z \pmod{p}$ and $t^2 \equiv z \pmod{q}$. By the Chinese remainder theorem (Appendix

I) there exists w in Z_N with $w \equiv r \pmod{p}$ and $w \equiv t \pmod{q}$. Then $w^2 \equiv z \pmod{p \text{ or } q}$, so $w^2 \equiv z \pmod{N}$. Thus z is a quadratic residue modulo N. This proves one direction of (1).

Now suppose z is a quadratic residue modulo N. Let $z_p = z \bmod p$ and $z_q = z \bmod q$. Then $z \in Z_N^*$, and hence $z_p \in Z_p^*$ and $z_q \in Z_q^*$. Also, $w^2 \equiv z \pmod{N}$ for some w in Z_N^*. Thus $w^2 \equiv z \pmod{p \text{ or } q}$ and hence $w^2 \equiv z_p \pmod{p}$ and $w^2 \equiv z_q \pmod{q}$. Thus z_p and z_q are quadratic residues modulo p and q, respectively, proving (1) in the other direction.

Furthermore, by Lemma J.3.1, z_p has exactly two square roots $\{x_1, x_2\}$ in Z_p^*, and z_q has exactly two square roots $\{y_1, y_2\}$ in Z_q^*. Hence $w \equiv x_i \pmod{p}$ and $w \equiv y_j \pmod{q}$ for some i and j. There are four possible pairs ($i = 1, 2$ and $j = 1, 2$). The Chinese remainder theorem shows that w is uniquely determined in Z_N by a given i and j; hence z has at most four square roots modulo N in Z_N^*. Conversely, by the Chinese remainder theorem once again, w can be found for each (i, j) pair. Thus z has exactly four square roots modulo N in Z_N^*. Let x denote one root. Then $N - x$ is another. If y is a third, then $N - y$ is the fourth. This proves (2).

By Lemma J.3.1, $\{x_i\}$ and $\{y_j\}$ can be found in probabilistic polynomial time; the corresponding w's can then be found in polynomial time. This proves (3). ∎

Corollary N.1.1: Suppose $N = p * q$, p and q primes > 2. Then:

1. E_N is a four-to-one function.
2. If p and q are known and z is a quadratic residue modulo N, the four values of x for which $E_N(x) = z$ can be found in probabilistic polynomial time.

Proof: Immediate from the previous lemma. ∎

N.2 THE JACOBI SYMBOL ONCE MORE

Suppose $N = p * q$ where p and q are primes > 2. Then for x in Z_N^*, the Jacobi symbol (x/N) is given by

$$(x/N) = (x/p)(x/q)$$

If the Jacobi symbol $(x/N) = -1$, then (x/p) or $(x/q) = -1$. Hence $x \bmod p$ and $x \bmod q$ are quadratic nonresidues modulo p and q, respectively. By Lemma N.1.1, x is a quadratic nonresidue modulo N. Since (x/N) can be evaluated in $O(\log N)$ time, $(x/N) = -1$ is a deterministic polynomial-time test for quadratic nonresiduosity of x modulo N, $N = p * q$. The interesting case is $(x/N) = 1$, whence no conclusion can be drawn regarding quadratic residuosity of x modulo N. To study this further, Z_N^* may be partitioned into four subclasses, according to the Jacobi symbols (x/p) and (x/q):

(x/p)	(x/q)	Class	(x/N)
1	1	$Q_{00}(N)$	1
-1	-1	$Q_{11}(N)$	1
1	-1	$Q_{01}(N)$	-1
-1	1	$Q_{10}(N)$	-1

Lemma N.2.1: Suppose $N = p * q$, p and q primes > 2. Then:

1. The $\{Q_{ij}(N)\}$ partition Z_N^* into disjoint classes.
2. $Q_{00}(N)$ is the set of quadratic residues modulo N; the other Q's contain nonresidues.
3. For $i = 0$ or 1 and $j = 0$ or 1, $|Q_{ij}(N)| = (p - 1)(q - 1)/4$.

Proof: (1) Is trivial, and (2) is immediate from Lemma N.1.1. For (3), let g and h be generators for Z_p^* and Z_q^*, respectively. For $w \in Z_N^*$, suppose $w \equiv g^r \pmod{p}$ and $w \equiv h^s \pmod{q}$, where $0 \leq r < p - 1$ and $0 \leq s < q - 1$. Then r and s are unique. Conversely, any such pair (r, s) uniquely determines w, by the Chinese remainder theorem. Let $f(w) = (r, s)$. Then f is a bijection from Z_N^* to $Z_p \times Z_q$. Also, if $f(w) = (r, s)$ then

$$(w/p) = (g^r/p) = (g/p)^r$$
$$(w/q) = (h^s/q) = (h/q)^s$$

By Lemma K.4, g and h are quadratic nonresidues modulo p and q, respectively. Thus $(w/p) = (-1)^r$ and $(w/q) = (-1)^s$. Let $i = r \bmod 2$ and $j = s \bmod 2$. Then $(w/p) = (-1)^i$ and $(w/q) = (-1)^j$. Hence w is in $Q_{ij}(N)$. For $k = 0$, 1 let Z_p^k and Z_q^k be the subsets of Z_p^* and Z_q^*, respectively, which contain elements $= k \pmod 2$. Then for $i = 0$, 1 and $j = 0$, 1, f is a bijection from $Q_{ij}(N)$ to $Z_p^i \times Z_q^j$ The latter has cardinality $(p - 1)(q - 1)/2$; this gives (3). ∎

Thus exactly half of the elements of Z_N^* have $(x/N) = -1$; these are nonresidues modulo N. The other half have $(x/N) = 1$. Of the latter, half (i.e., $Q_{00}(N)$) are quadratic residues and half (i.e., $Q_{11}(N)$) are nonresidues. The quadratic residuosity conjecture is that determining which of the elements of Z_N^* with $(x/N) = 1$ are quadratic residues modulo N is not solvable in probabilistic polynomial time, for N a product of two primes. This is in contradistinction to the case of one prime p, for which we have noted that quadratic residuosity is decidable in deterministic polynomial time.

Lemma N.2.2: Suppose $N = p * q$, p and q primes > 2. Then:

1. For $x, y \in Z_N^*$, $x * y$ is a quadratic residue modulo N iff x and y are in the same $Q_{ij}(N)$.
2. The product of quadratic residues is a quadratic residue.
3. The product of a quadratic residue and a nonresidue is a nonresidue.

Proof: Suppose $x \in Q_{ij}(N)$ and $y \in Q_{rt}(N)$. Then $(x/p) = (-1)^i$, $(x/q) = (-1)^j$, $(y/p) = (-1)^r$, $(y/q) = (-1)^t$. Hence $(x * y/p) = (-1)^{i+r}$ and $(x * y/q) = (-1)^{j+t}$. Now $x * y \bmod N$ will be a residue iff $i = r$ and $j = t$. This gives (1). In particular, $x * y \bmod N$ is a residue if both are in $Q_{00}(N)$, yielding (2). If x is a residue and y a nonresidue, x is in $Q_{00}(N)$ but y is not; (3) follows. ∎

Thus the quadratic residues modulo N form a subgroup of Z_N^*.

N.3 QUADRATIC RESIDUOSITY AND FACTORING

We note the connection between quadratic residuosity and factoring observed by Rabin [RABI79].

Lemma N.3.1:   Suppose $N = p * q$, p and q prime, x and y are in Z_N^*, $x^2 \equiv y^2$ (mod N), and $y \,! \equiv x$ or $-x$ (mod N). Then possession of x and y permits factoring N in deterministic polynomial time.

Proof: We have $x - y \,! \equiv 0$ (mod N) and $x + y \,! \equiv 0$ (mod N), so $GCD(N, y + x) < N$ and $GCD(N, y - x) < N$. Now $x^2 - y^2 \equiv (x - y)(x + y) \equiv 0$ (mod N). Hence $(x - y)(x + y) \equiv 0$ (mod p). Thus $p \mid y - x$ or $p \mid y + x$; also $p \mid N$. If $p \mid y - x$, $p \mid GCD(N, y - x) < N$, so $p = GCD(N, y - x)$. The latter can be found via Euclid's algorithm (Appendix H). If $p \mid y + x$, $p \mid GCD(N, y + x) < N$, so $p = GCD(N, y + x)$.   ∎

Lemma N.3.2:   Suppose $N = p * q$, p and q prime. Suppose there exists an algorithm to find in probabilistic polynomial time a square root of a quadratic residue modulo N, which works for at least a fraction $1/\log^c N$ of quadratic residues, c a constant positive integer. Then N can be factored in probabilistic polynomial time.

Proof: Let $m = \log^c N$. Choose $x_1, \ldots, x_m$ randomly in Z_N^* and compute $z_i = x_i^2 \bmod N$, $i = 1, \ldots, m$. If any z_i has a square root w computable in probabilistic polynomial time via the hypothetical algorithm, find w. The probability that $w \,! \equiv x_i$ or $-x_i$ (mod N) is $\frac{1}{2}$. In this event possession of x_i and w factors N by Lemma N.3.1. The probability that some z_i has a computable square root is at least $\frac{1}{2}$. Thus the probability of factoring N is at least $\frac{1}{4}$. If this procedure is repeated m times, the probability of success is at least $1 - \left(\frac{3}{4}\right)^m$. Also, the problem size is $\log N$, so m is a polynomial in the problem size.   ∎

Corollary N.3.2:   If $N = p * q$, p and q prime, and the factorization of N is computationally intractable, then the Blum encryption function E_N above is a trapdoor one-way function, with (p, q) forming the trapdoor.

Proof: By the previous theorem, an algorithm to invert E_N would yield an algorithm to factor N.   ∎

N.4 QUADRATIC RESIDUOSITY AND BLUM INTEGERS

Suppose $N = p * q$ where $p \equiv 3$ (mod 4) and $q \equiv 3$ (mod 4). Then N is called a Blum integer (normally p and q are specified to have roughly the same size). For example, $N = 77$ was used in Section 8.

Lemma N.4.1:   If $p, q \equiv 3$ (mod 4) then $(-1/p) = (-1/q) = -1$.

Proof: In Lemma J.4.1 take $a = -1$. We recall that $(-1/p) = 1$ iff -1 is a quadratic residue modulo p. Thus $(-1/p) = (-1)^{(p-1)/2}$ (mod p) and similarly $(-1/q) = (-1)^{(q-1)/2}$ (mod q). The result follows.   ∎

Lemma N.4.2: If $N = p * q$ is a Blum integer then $(-x/p) = -(x/p)$, $(-x/q) = -(x/q)$, $(-1/N) = 1$, and $(-x/N) = (x/N)$.

Proof: The first two are immediate from the previous lemma; also $(-1/N) = (-1/p)(-1/q)$. ∎

Lemma N.4.3: Suppose $N = p * q$ is a Blum integer, x and y are in Z_N^*, $x^2 \equiv y^2 \pmod{N}$, and $x \: !\equiv y$ or $-y \pmod{N}$. Then $(x/N) = -(y/N)$.

Proof: $x^2 \equiv y^2 \pmod{p}$, so $(y - x)(y + x) \equiv 0 \pmod{p}$, so $y \equiv d * x \pmod{p}$ where $d = 1$ or -1. Similarly $y \equiv e * x \pmod{q}$ where $e = 1$ or -1. Now $(x/N) = (x/p)(x/q)$ and $(y/N) = (y/p)(y/q) = (d * x/p)(e * x/q) = (d/p)(e/q)(x/N)$. Since $(1/p) = (1/q) = 1$, by Lemma N.4.1, $(y/N) = e * d * (x/N)$. If $e = d$ then $y \equiv d * x \pmod{q \text{ or } p}$, so $y \equiv d * x \pmod{N}$. But $y \: !\equiv x$ or $-x \pmod{N}$, which is a contradiction. Thus $e \neq d$ and $e * d = -1$. ∎

Lemma N.4.4: If $N = p * q$ is a Blum integer and z is a quadratic residue modulo N, then z has a square root in each $Q_{ij}(N)$, $i = 0, 1, j = 0, 1$.

Proof: Let $\{x, N - x, y, N - y\}$ be the square roots of z in Z_N^*. Since $N - x \equiv -x \pmod{p}$, by Lemma N.4.2, $(N - x/p) = -(x/p)$. Similarly $(N - x/q) = -(x/q)$, so if $x \in Q_{ij}(N)$, $N - x \in Q_{1-i, 1-j}(N)$; similarly for y and $N - y$. Now $x^2 \equiv y^2 \pmod{N}$ and $y \: !\equiv x$ or $-x \pmod{N}$. Since by Lemma N.4.3, $(y/N) = -(x/N)$, y cannot be in the same Q as x. By Lemma N.4.2, $((N - y)/N) = (y/N)$, so $N - y$ cannot be in the same Q as x. Thus y and $N - y$ must be in $Q_{i, 1-j}(N)$ and $Q_{1-i, j}(N)$ in some order. ∎

For Blum integers we can restrict the function E_N to $Q_{00}(N)$; that is, let $B_N : Q_{00}(N) \to Q_{00}(N)$ by $B_N(x) = x^2 \bmod N$. Then we have

Lemma N.4.5: If $N = p * q$ is a Blum integer then:

1. B_N is a permutation on $Q_{00}(N)$, that is, on the set of quadratic residues modulo N.
2. If the factorization of N is computationally intractable then B_N is a trapdoor one-way permutation, with (p, q) constituting the trapdoor.

Proof: By Corollary N.1.1 we know that if p and q are known and y is in $Q_{00}(N)$, the equation $E_N(x) = y$ has exactly four solutions which can be found in probabilistic polynomial time. By Lemma N.4.4, exactly one of these, x_{00}, lies in $Q_{00}(N)$; x_{00} is easily extracted from the set of four since only it satisfies $(x/p) = (x/q) = 1$. Hence x_{00} is the unique solution of $B_N(x) = y$. Thus B_N is a permutation on $Q_{00}(N)$ whose inverse may be computed in probabilistic polynomial time with knowledge of the trapdoor (p, q). On the other hand, Lemma N.3.2 shows that an algorithm to invert B_N can be converted to an algorithm to factor N. ∎

Lemma N.4.6: If $N = p * q$, p and q prime, then it is possible to find x in $Q_{11}(N)$, that is, $x \in Z_N^*$ such that x is a quadratic nonresidue modulo N and $(x/N) = 1$, in probabilistic polynomial time.

Proof: Choose a in Z_p^* and evaluate $a^{(p-1)/2}$; if the latter is not 1 then choose a new a, etc. The probability of failure each time is $\frac{1}{2}$, so that the probability of not finding an a with $a^{(p-1)/2} = 1 \pmod{p}$ in n tries is 2^{-n}. Hence a can be found in probabilistic polynomial time (in $n = $ length of p). Now $(a/p) = 1$. Similarly, find b with $(b/q) = 1$, also in probabilistic polynomial time. Use the Chinese remainder theorem to find y in Z_N^* with $y \equiv a \pmod{p}$ and $y \equiv b \pmod{q}$, in polynomial time (in length of N). ∎

REFERENCES

[ADLE79] L. Adleman, "A subexponential algorithm for the discrete logarithm problem with applications to cryptography," in *20th Annu. Symp. Found. Comput. Sci.,* San Juan, Puerto Rico, October 29–31, 1979, pp. 55–60. Silver Spring, MD: IEEE Computer Society Press, 1979.

[ADLE82] L. M. Adleman, "On breaking the iterated Merkle–Hellman public-key cryptosystems" in *Advances in Cryptology: Proc. Crypto'82,* D. Chaum, R. L. Rivest, and A. T. Sherman, Eds., Santa Barbara, CA, Aug. 23–25, 1982, pp. 303–308. New York: Plenum Press, 1983.

[ADLE86] L. M. Adleman and K. S. McCurley, "Open problems in number theoretic complexity," in *Discrete Algorithms and Complexity, Proc. Japan–U.S. Joint Seminar,* D. S. Johnson, T. Nishizeki, A. Nozaki, and H. S. Wilf, Eds., June 4–6, 1986, pp. 237–262. Orlando, FL: Academic Press, 1987.

[ADLE83] L. M. Adleman, C. Pomerance, and R. S. Rumely, "On distinguishing prime numbers from composite numbers," *Ann. Math.,* vol. 117, pp. 173–206, 1983.

[AKL-83] S. G. Akl, "Digital signatures: A tutorial survey," *Computer,* vol. 16, no. 2, pp. 15–24, Feb. 1983.

[AKL-84] S. G. Akl and H. Meijer, "A fast pseudo random permutation generator with applications to cryptology," in *Lecture Notes in Computer Science 196; Advances in Cryptology: Proc. Crypto'84,* G. R. Blakley and D. Chaum, Eds., Santa Barbara, CA, Aug. 19–22, 1984, pp. 269–275. Berlin: Springer-Verlag, 1985.

[ANSI85] American National Standard X9.17-1985, Financial Institution Key Management (Wholesale), American Bankers Association, Washington, DC, 1985.

[ANNA87] M. Annaratone, E. Arnould, T. Gross, H. T. Kung, M. Lam, O. Menzilcioglu, and J. A. Webb, "The Warp computer: Architecture, implementation and performance," *IEEE Trans. Comput.,* vol. C-36, no. 12, pp. 1523–1538, Dec. 1987.

[BANE82] S. K. Banerjee, "High speed implementation of DES," *Computers and Security,* vol. 1, no. 3, pp. 261–267, Nov. 1982.

[BART63] T. C. Bartee and D. I. Schneider, "Computation with finite fields, *Inform. Contr.,* vol. 6, no. 2, pp. 79–98, June 1963.

[BATC80] K. E. Batcher, "Design of a massively parallel processor," *IEEE Trans. Comput.,* vol. C-29, no. 9, pp. 836–840, Sept. 1980.

[BLAK84] I. F. Blake, R. Fuji-Hara, R. C. Mullin, and S. A. Vanstone, "Computing logarithms in finite fields of characteristic two," *SIAM J. Algebraic Discrete Methods,* vol. 5, no. 2, pp. 276–285, June 1984.

[BLAK84b] I. F. Blake, R. C. Mullin, and S. A. Vanstone, "Computing logarithms in GF(2^n)," in *Lecture Notes in Computer Science 196; Advances in Cryptology: Proc. Crypto'84,* G. R. Blakley and D. Chaum, Eds., Santa Barbara, CA, Aug. 19–22, 1984, pp. 73–82. Berlin: Springer-Verlag, 1985.

[BLAK83] G. R. Blakley, "A computer algorithm for calculating the product AB modulo M," *IEEE Trans. Comput.,* vol. C-32, no. 5, pp. 497–500, May 1983.

[BLUM84] M. Blum and S. Micali, "How to generate cryptographically strong sequences of pseudo-random bits," *SIAM J. Comput.,* vol. 13, no. 4, pp. 850–864, Nov. 1984.

[BOOT81] K. S. Booth, "Authentication of signatures using public key encryption," *Commun. ACM,* vol. 24, no. 11, pp. 772–774, Nov. 1981.

[BRAS79] G. Brassard, "A note on the complexity of cryptography," *IEEE Trans. Inform. Theory,* vol. IT-25, no. 2, pp. 232–233, March 1979.

[BRAS83] G. Brassard, "Relativized cryptography," *IEEE Trans. Inform. Theory,* vol. IT-29, no. 6, pp. 877–894, Nov. 1983.

[BRAS88] G. Brassard, *Modern Cryptology: a Tutorial: Lecture Notes in Computer Science Vol. 325.* Berlin/New York: Springer-Verlag, 1988.

[BREN83] R. P. Brent and H. T. Kung, "Systolic VLSI arrays for linear-time GCD computation," in *VLSI 83, Proceedings of the IFIP TC 10/WG 10.5 International Conference on VLSI,* F. Anceau and E. J. Aas, Eds., Trondheim, Norway, August 16–19, 1983, pp. 145–154. Amsterdam/New York: North-Holland, 1983.

[BREN83b] R. P. Brent, H. T. Kung, and F. T. Luk, "Some linear-time algorithms for systolic arrays," in *IFIP Congress Series Vol. 9: Information Processing 83, Proceedings of the IFIP 9th World Congress,* R. E. A. Mason, Ed., Paris, Sept. 19–23, 1983, pp. 865–876. Amsterdam/New York: North-Holland, 1983.

[BRIC82] E. F. Brickell, "A fast modular multiplication algorithm with application to two key cryptography," in *Advances in Cryptology: Proc. Crypto'82,* D. Chaum, R. L. Rivest, and A. T. Sherman, Eds., Santa Barbara, CA, Aug. 23–25, 1982, pp. 51–60. New York: Plenum Press, 1983.

[BRIC83] E. F. Brickell, "Solving low density knapsacks," in *Advances in Cryptology: Proc. Crypto'83,* D. Chaum, Ed., Santa Barbara, CA, Aug. 22–24, 1983, pp. 25–37. New York: Plenum Press, 1984.

[BRIC84] E. F. Brickell, "Breaking iterated knapsacks," in *Lecture Notes in Computer Science 196; Advances in Cryptology: Proc. Crypto'84,* G. R. Blakley and D. Chaum, Eds., Santa Barbara, CA, Aug. 19–22, 1984, pp. 342–358. Berlin: Springer-Verlag, 1985.

[BRIC89] E. F. Brickell, "A survey of hardware implementations of RSA," in *Lecture Notes in Computer Science 435; Advances in Cryptology: Proc. Crypto'89,* G. Brassard, Ed., Santa Barbara, CA, Aug. 20–24, 1989, pp. 368–370. Berlin: Springer-Verlag, 1990.

[BRIC88] E. F. Brickell and A. M. Odlyzko, "Cryptanalysis: A survey of recent results," *Proc. IEEE,* vol. 76, no. 5, pp. 578–593, May 1988.

[BRIG76] H. S. Bright and R. L. Enison, "Cryptography using modular software elements," in *AFIPS Conference Proceedings Vol. 45: National Computer Conference,* S. Winkler, Ed., New York, June 7–10, 1976, pp. 113–123. Montvale, NJ: AFIPS Press, 1976.

[CCIT87] CCIT, Draft Recommendation X.509: *The Directory—Authentication Framework.* Geneva: Consultation Committee, International Telephone and Telegraph, Nov. 1987.

[CARO87] T. R. Caron and R. D. Silverman, "Parallel implementation of the quadratic scheme," *J. Supercomput.,* vol. 1, no. 3, 1987.

[COHE87] H. Cohen and A. K. Lenstra, "Implementation of a new primality test," *Math. Comput.,* vol. 48, no. 177, pp. 103–121, Jan. 1987.

[COHE84] H. Cohen and H. W. Lenstra, Jr., "Primality testing and Jacobi sums," *Math. Comput.,* vol. 42, no. 165, pp. 297–330, Jan. 1984.

[CONT80] Control Data Corporation, *CDC CYBER 200 Model 205 Computer System.* Minneapolis, MN, 1980.

[COPP84] D. Coppersmith, "Fast evaluation of logarithms in fields of characteristic two," *IEEE Trans. Inform. Theory,* vol. IT-30, no. 4, pp. 587–594, July 1984.

[COPP85] D. Coppersmith, "Another birthday attack," in *Lecture Notes in Computer Science 218; Advances in Cryptology: Proc. Crypto'85,* H. C. Williams, Ed., Santa Barbara, CA, Aug. 18–22, 1985, pp. 14–17. Berlin: Springer-Verlag, 1986.

[COPP87] D. Coppersmith, "Cryptography," *IBM J. Res. and Develop.,* vol. 31, no. 2, pp. 244–248, March 1987.

[COPP86] D. Coppersmith, A. M. Odlyzko, and R. Schroeppel, "Discrete logarithms in GF(p)," *Algorithmica,* vol. 1, no. 1, pp. 1–15, 1986.

[CRAY80] Cray Research, Inc., *Cray 1-S Series Hardware Reference Manual.* Mendota Heights, MN: Cray Research, Inc., June 1980.

[CRAY85] Cray Research, Inc., *Cray X-MP Series of Computer Systems.* Mendota Heights, MN: Cray Research, Inc., 1985.

[DAVI83] D. W. Davies, "Applying the RSA digital signature to electronic mail," *Computer,* vol. 16, no. 2, pp. 55–62, Feb. 1983.

[DAVI80] J. A. Davies and W. L. Price, "The application of digital signatures based on public key cryptosystems," *NPL Report DNACS 39/80.* Teddington, Middlesex, England: National Physics Laboratory, Dec. 1980.

[DAVI83b] J. A. Davis and D. B. Holdridge, "Factorization using the quadratic sieve algorithm," in *Advances in Cryptology: Proc. Crypto'83,* D. Chaum, Ed., Santa Barbara, CA, Aug. 22–24, 1983, pp. 103–113. New York: Plenum Press, 1984.

[DAVI84] J. A. Davis, D. B. Holdridge, and G. J. Simmons, "Status report on factoring," in *Lecture Notes in Computer Science 209; Advances in Cryptology: Proc. Eurocrypt'84,* T. Beth, N. Cot, and I. Ingemarsson, Eds., Paris, France, April 9–11, 1984, pp. 183–215. Berlin: Springer-Verlag, 1985.

[DEMI83] R. DeMillo and M. Merritt, "Protocols for data security," *Computer,* vol. 16, no. 2, pp. 39–51, Feb. 1983.

[DENN79] D. E. Denning, "Secure personal computing in an insecure network," *Commun. ACM,* vol. 22, no. 8, pp. 476–482, Aug. 1979.

[DENN83b] D. E. Denning, "Protecting public keys and signature keys," *Computer,* vol. 16, no. 2, pp. 27–35, Feb. 1983.

[DENN81] D. E. Denning and G. M. Sacco, "Timestamps in key distribution protocols," *Commun. ACM,* vol. 24, no. 8, pp. 533–536, Aug. 1981.

[DENN83] D. E. R. Denning, *Cryptography and Data Security.* Reading, MA: Addison-Wesley, 1983.

[DESM83] Y. Desmedt, J. Vandewalle, and R. J. M. Govaerts, "A critical analysis of the security of knapsack public key algorithms," *IEEE Trans. Inform. Theory,* vol. IT-30, no. 4, pp. 601–611, July 1984.

[DIFF82] W. Diffie, "Conventional versus public key cryptosystems," in *Secure Communications and Asymmetric Cryptosystems,* G. J. Simmons, Ed., pp. 41–72. Boulder, CO: Westview Press, 1982.

[DIFF84] W. Diffie, "Network security problems and approaches," *Proceedings of the National Electronics Conference,* vol. 38, pp. 292–314, 1984.

[DIFF86] W. Diffie, "Communication security and national security business, technology, and politics," *Proc. National Communications Forum,* vol. 40, pp. 734–751, 1986.

[DIFF88] W. Diffie, "The first ten years of public-key cryptography," *Proc. IEEE,* vol. 76, no. 5, pp. 560–577, May 1988.

[DIFF76] W. Diffie and M. E. Hellman, "Multiuser cryptographic techniques," in *AFIPS Conference Proceedings, vol. 45: National Computer Conference,* S. Winkler, Ed., New York, June 7–10, 1976, pp. 109–112. Montvale, NJ: AFIPS Press, 1976.

[DIFF76b] W. Diffie and M. E. Hellman, "New directions in cryptography," *IEEE Trans. Inform. Theory,* vol. IT-22, no. 6, pp. 644–654, Nov. 1976.

[DIFF79] W. Diffie and M. E. Hellman, "Privacy and authentication: An introduction to cryptography," *Proc. IEEE, vol. 67, no. 3, pp. 397–427, March 1979.*

[DIFF87] W. Diffie, B. O'Higgins, L. Strawczynski, and D. Steer, "An ISDN secure telephone unit," *Proc. National Communication Forum,* vol. 41, no. 1, pp. 473–477, 1987.

[DIXO81] J. D. Dixon, "Asymptotically fast factorization of integers," *Math. Comput.,* vol. 36, no. 153, pp. 255–260, Jan. 1981.

[DOLE81] D. Dolev and A. C. Yao, "On the security of public key protocols," in *22nd Annual Symposium on Foundations of Computer Science,* Nashville, TN, Oct. 28–30, 1981, pp. 350–357. Silver Spring, MD: IEEE Computer Society Press, 1981.

[DUSS90] S. R. Dusse and B. S. Kaliski, Jr., "A cryptographic library for the Motorola DSP 56000," in *Lecture Notes in Computer Science, 473; Advances in Cryptology: Proc. Eurocrypt'90,* I. Damgård, Ed., May 21–24, pp. 230–244. Berlin: Springer-Verlag 1990.

[EHRS78] W. F. Ehrsam, S. M. Matyas, C. H. Meyer, and W. L. Tuchman, "A cryptographic key management scheme for implementing the Data Encryption Standard," *IBM Syst. J.,* vol. 17, no. 2, pp. 106–125, 1978.

[ELGA85] T. El Gamal, "A public key cryptosystem and a signature scheme based on discrete logarithms," *IEEE Trans. Inform. Theory,* vol. IT-31, no. 4, pp. 469–472, July 1985.

[ELGA85b] T. El Gamal, "On computing logarithms over finite fields," in *Lecture Notes in Computer Science 218; Advances in Cryptology: Proc. Crypto'85*, H. C. Williams, Ed., Santa Barbara, CA, Aug. 18–22, 1985, pp. 396–402. Berlin: Springer-Verlag, 1986.

[EVAN74] A. Evans, Jr., and W. Kantrowitz, "A user authentication scheme not requiring secrecy in the computer," *Commun. ACM*, vol. 17, no. 8, pp. 437–442, Aug. 1974.

[FEIG87] U. Feige, A. Fiat, and A. Shamir, "Zero knowledge proofs of identity," in *Proceedings of the Nineteenth Annual ACM Symposium on Theory of Computing*, New York, May 25–27, 1987, pp. 210–217. New York: ACM, 1987.

[FIAT86] A. Fiat and A. Shamir, "How to prove yourself: Practical solutions to identification and signature problems," in *Lecture Notes in Computer Science 263; Advances in Cryptology: Proc. Crypto'86*, A. M. Odlyzko, Ed., Santa Barbara, CA, Aug. 11–15, 1986, pp. 186–194. Berlin: Springer-Verlag, 1987.

[FLYN78] R. Flynn and A. S. Campasano, "Data dependent keys for selective encryption terminal," in *Am. Fed. Inform. Process. Socs. Conference Proc., Vol 47: National Computer Conf.*, S. P. Ghosh and L. Y. Liu, Eds., Anaheim, CA, June 5–8, 1978, pp. 1127–1129. Montvale, NJ: AFIPS Press, 1978.

[GARE79] M. R. Garey and D. S. Johnson, *Computers and Intractability*. New York: W. H. Freeman, 1979.

[GILL77] J. Gill, "Computational complexity of probabilistic Turing machines," *SIAM J. Comput.*, vol. 6, no. 4, pp. 675–695, Dec. 1977.

[GOLD82] S. Goldwasser and S. Micali, "Probabilistic encryption and how to play mental poker keeping secret all partial information," in *Proc. 14th ACM Symp. Theory of Computing*, San Francisco, May 5–7, 1982, pp. 365–377. New York: Association for Computing Machinery, 1982.

[GOLD84] S. Goldwasser and S. Micali, "Probabilistic encryption," *J. Comput. Syst. Sci.*, vol. 28, no. 2, pp. 270–299, April 1984.

[GOLD89] S. Goldwasser, S. Micali, and C. Rackoff, "The knowledge complexity of interactive proof systems," *SIAM J. Comput.*, vol. 18, no. 1, pp. 186–208, Feb. 1989.

[GOLD88] S. Goldwasser, S. Micali, and R. L. Rivest, "A digital signature scheme secure against adaptive chosen-message attacks," *SIAM J. Comput.*, vol. 17, no. 2, pp. 281–308, Apr. 1988.

[GORD84b] J. Gordon, "Strong RSA keys," *Electron. Lett.*, vol. 20, no. 12, pp. 514–516, June 7, 1984.

[GORD84] J. A. Gordon, "Strong primes are easy to find," in *Lecture Notes in Computer Science 209; Advances in Cryptology: Proc. Eurocrypt'84*, T. Beth, N. Cot, and I. Ingemarsson, Eds., Paris, France, April 9–11, 1984, pp. 216–223. Berlin: Springer-Verlag, 1985.

[GROL88] J. Grollman and A. L. Selman, "Complexity measures for public-key cryptosystems," *SIAM J. Comput.*, vol. 17, no. 2, pp. 309–335, Apr. 1988.

[HAST88] J. Hastad, "Solving simultaneous modular equations of low degree," *SIAM J. Comput.*, vol. 17, no. 2, pp. 336–341, Apr. 1988.

[HAWK87] S. Hawkinson, "The FPS T Series: A parallel vector supercomputer," in *Multiprocessors and Array Processors*, W. J. Karplus, Ed., pp. 147–155. San Diego, CA: Simulation Councils, Inc., 1987.

[HAYE86] J. P. Hayes, T. Mudge, Q. F. Stout, S. Colley, and J. Palmer, "A microprocessor-based hypercube supercomputer," *IEEE Micro*, vol. 6, no. 5, pp. 6–17, Oct. 1986.

[HAYK88] M. E. Haykin and R. B. J. Warnar, *Smart Card Technology: New Methods for Computer Access Control*, NIST Special Publication 500-157. Washington, D.C.: National Institute of Standards and Technology, Sept. 1988.

[HENR81] P. S. Henry, "Fast decryption algorithm for the knapsack cryptographic system," *Bell Syst. Tech. J.*, vol. 60, no. 5, pp. 767–773, May–June 1981.

[HERS64] I. N. Herstein, *Topics in Algebra*. Waltham, MA: Blaisdell, 1964.

[HILL85] D. Hillis, *The Connection Machine*. Cambridge, MA: MIT Press, 1985.

[HOCK81] R. W. Hockney and C. R. Jesshope, *Parallel Computers: Architecture, Programming and Algorithms*. Bristol, England: Adam Hilger, 1981.

[HORO78] E. Horowitz and S. Sahni, *Fundamentals of Computer Algorithms*. Rockville, MD: Computer Science Press, 1978.

[HWAN84] K. Hwang and F. A. Briggs, *Computer Architecture and Parallel Processing*. New York: McGraw-Hill, 1984.

[IAB-88] IAB (Internet Activities Board) Privacy Task Force, "Privacy enhancement for Internet electronic mail, Part I: Message encipherment and authentication procedures," *Request for Comments (RFC) 1040B*, Dec. 22, 1988.

[IAB88b] IAB (Internet Activities Board) Privacy Task Force, "Privacy enhancement for Internet electronic mail, Part II: Certificate-based key management," *Request for Comments (RFC) KM-RFC*, Dec. 23, 1988.

[ISO-87] International Organization for Standardization, *Draft International Standard ISO/DIS 7498-2, Information Processing Systems—Open Systems Interconnection Model, Part 2: Security Architecture*. Geneva: ISO, 1987.

[JUEN82] R. R. Jueneman, "Analysis of certain aspects of output feedback mode," in *Advances in Cryptology: Proc. Crypto'82*, D. Chaum, R. L. Rivest, and A. T. Sherman, Eds., Santa Barbara, CA, Aug. 23–25, 1982, pp. 99–127. New York: Plenum Press, 1983.

[JUEN86] R. R. Jueneman, "A high speed manipulation detection code," in *Lecture Notes in Computer Science 263; Advances in Cryptology: Proc. Crypto'86*, A. M. Odlyzko, Ed., Santa Barbara, CA, Aug. 11–15, 1986, pp. 327–346. Berlin: Springer-Verlag, 1987.

[KELL89] S. S. Keller, "Comparison of data rate timings using BSAFE versus BSAFE with CYLINK," July 1989.

[KHAC79] L. G. Khachian, "A polynomial algorithm in linear programming," *Dokl. Akad. Nauk. SSSR*, vol. 244, pp. 1093–1096. English translation in *Sov. Math. Dokl.*, vol. 20, pp. 191–194.

[KLIN79] C. S. Kline and G. J. Popek, "Public key vs. conventional key encryption," in *Am. Fed. Inform. Process. Socs. Conference Proc., Vol 48: National Computer Conf.*, R. E. Merwin, Ed., New York, June 4–7, 1979, pp. 831–837. Montvale, NJ: AFIPS Press, 1979.

[KNUT81] D. E. Knuth, *The Art of Computer Programming, Vol. 2: Seminumerical Algorithms.* Reading, MA: Addison-Wesley, 1981.

[KOHN78b] L. M. Kohnfelder, "On the signature reblocking problem in public-key cryptosystems," *Commun. ACM*, vol. 21, no. 2, p. 179, Feb. 1978.

[KOHN78] L. M. Kohnfelder, "A method for certification," *MIT Laboratory for Computer Science*, Cambridge, MA: MIT Press, May 1978.

[KONH81] A. G. Konheim, *Cryptography: A Primer.* New York: John Wiley & Sons, 1981.

[KOWA85] J. Kowalik, Ed., *Parallel MIMD Computation: The HEP supercomputer and Its Applications.* Cambridge, MA: MIT Press, 1985.

[KRAN86] E. Kranakis, *Primality and Cryptography.* Chichester/New York: John Wiley & Sons, 1986.

[KUNG82] H. T. Kung, "Why systolic architectures," *Computer,* vol. 15, no. 1, pp. 37–46, Jan. 1982.

[KUNG78] H. T. Kung and C. Leiserson, "Systolic arrays (for VLSI)," in *Sparse Matrix Proceedings,* I. S. Duff and G. W. Stewart, Eds., pp. 245–282. Philadelphia: SIAM, 1978.

[KUNG82b] S. Y. Kung, K. S. Arun, R. J. Gal-Ezer, and D. V. B. Rao, "Wavefront array processor: Language, architecture, and applications," *IEEE Trans. Comput.,* vol. C-31, no. 11, pp. 1054–1066, Nov. 1982.

[LAGA84] J. C. Lagarias, "Performance analysis of Shamir's attack on the basic Merkle-Hellman knapsack system," in *Lecture Notes in Computer Science, Vol. 172: Automata, Languages and Programming: 11th Colloquium,* J. Paredaens, Ed., Antwerp, Belgium, July 16–20, 1984, pp. 312–323. Berlin/New York: Springer-Verlag, 1984.

[LAGA83] J. C. Lagarias and A. M. Odlyzko, "Solving low-density subset sum problems," in *24th Annual Symposium on Foundations of Computer Science,* Tucson, AZ, November 7–9, 1983, pp. 1–10. Silver Spring, MD: IEEE Computer Society Press, 1983. Revised version in *J. ACM,* vol. 32, no. 1, pp. 229–246, Jan. 1985.

[LAKS83] S. Lakshmivarahan, "Algorithms for public key cryptosystems: Theory and application," *Advances in Computers,* vol. 22, pp. 45–108, 1983.

[LAWS71] B. A. Laws, Jr., and C. K. Rushforth, "A cellular-array multiplier for $GF(2^m)$," *IEEE Trans. Comput.,* vol. 20, no. 12, pp. 1573–1578, Dec. 1971.

[LEHM82] D. J. Lehmann, "On primality tests," *SIAM J. Comput.,* vol. 11, no. 2, pp. 374–375, May 1982.

[LEHM76] D. H. Lehmer, "Strong Carmichael numbers," *J. Austr. Math. Soc.,* vol. 21 (Ser. A), pp. 508–510, 1976.

[LEMP79] A. Lempel, "Cryptology in transition," *ACM Comput. Surveys,* vol. 11, no. 4, pp. 285–303, Dec. 1979.

[LENS82] A. K. Lenstra, H. W. Lenstra, Jr., and L. Lovasz, "Factoring polynomials with rational coefficients," *Mathematische Annalen,* vol. 261, pp. 515–534, 1982.

[LENS82] A. K. Lenstra, H. W. Lenstra, Jr., M. S. Manasse, and J. M. Pollard, "The number field sieve," *Proc. 22nd Annual ACM Symp. Theory of Computing,* Baltimore, May 14–16, 1990, pp. 564–572. New York: Association for Computing Machinery, 1990.

[LENS89] A. K. Lenstra and M. S. Manasse, "Factoring by electronic mail," in *Lecture Notes in Computer Science 434; Advances in Cryptology; Proc. Eurocrypt'89*, J.-J. Quisquater and J. Vandewalle, Eds., Houthalen, Belgium, April 10–23, 1989, pp. 355–371. Berlin: Springer-Verlag, 1990.

[LENS83] H. W. Lenstra, Jr., "Interger programming with a fixed number of variables," *Math. Oper. Res.*, vol. 8, no. 4, pp. 538–548, Nov. 1983

[LENS86] H. W. Lenstra, Jr., "Primality testing," in *Mathematics and Computer Science, CWI Monographs* Vol. I (CWI Symp., Nov. 19, 1983), J. W. de Bakker et al., Eds., pp. 269–287. Amsterdam/New York: North Holland, 1986.

[LENS87] H. W. Lenstra, Jr., "Factoring integers with elliptic curves," *Ann. Math.*, vol. 126, pp. 649–673, 1987.

[LEWI81] H. R. Lewis and C. H. Papadimitriou, *Elements of the Theory of Computation*. Englewood Cliffs, NJ: Prentice Hall, 1981.

[LINN89] J. Linn and S. T. Kent, "Privacy for DARPA-Internet mail," in *Proc. 12th Nat. Computer Security Conf.*, Baltimore, MD, pp. 215–229, October 1989.

[MACM81] D. MacMillan, "Single chip encrypts data at 14 Mb/s," *Electronics*, vol. 54, no. 12, pp. 161–165, June 16, 1981.

[MASS88] J. L. Massey, "An introduction to contemporary cryptology," *Proc. IEEE*, vol. 76, no. 5, pp. 533–549, May 1988.

[MATY78] S. M. Matyas and C. H. Meyer, "Generation, distribution, and installation of cryptographic keys," *IBM Syst. J.*, vol. 17, no. 2, pp. 126–137, 1978.

[MCCU89] K. S. McCurley, "The discrete logarithm problem," in *AMS Proc. Symp. Appl. Math., Vol. 42: Cryptology and Computational Number Theory*, C. Pomerance, Ed., pp. 49–74. Providence, RI: American Mathematical Society, 1991.

[MCEL78] R. J. McEliece, "A public-key cryptosystem based on algebraic coding theory," *DSN Progress Report 42–44*, Jet Propulsion Laboratory, Pasadena, CA, pp. 114–116, 1978.

[MERK78] R. C. Merkle, "Secure communications over insecure channels," *Commun. ACM*, vol. 21, no. 4, pp. 294–299, Apr. 1978.

[MERK82] R. C. Merkle, *Secrecy, Authentication, and Public Key Systems*. Ann Arbor: UMI Research Press, 1982.

[MERK82b] R. C. Merkle, "Protocols for public key cryptosystems," in *Secure Communications and Asymmetric Cryptosystems*, G. J. Simmons, Ed., pp. 73–104. Boulder, CO: Westview Press, 1982.

[MERK89] R. C. Merkle, "One way hash functions and DES," in *Lecture Notes in Computer Science 435; Advances in Cryptology: Proc. Crypto'89*, G. Brassard, Ed., Santa Barbara, CA, Aug. 20–24, 1989, pp. 428–446. Berlin: Springer-Verlag, 1990.

[MERK78b] R. C. Merkle and M. E. Hellman, "Hiding information and signatures in trapdoor knapsacks," *IEEE Trans. Inform. Theory*, vol. 24, no. 5, pp. 525–530, Sept. 1978.

[MILL76] G. L. Miller, "Riemann's hypothesis and tests for primality," *J. Comput. Syst. Sci.*, vol. 13, no. 3, pp. 300–317, Dec. 1976.

[MILU88] V. M. Milutinovic, *Computer Architecture: Concepts and Systems.* New York: North-Holland, 1988.

[MONT85] P. L. Montgomery, "Modular multiplication without trial division," *Math. Comput.,* vol. 44, no. 170, pp. 519–521, Apr. 1985.

[MOOR88] J. H. Moore, "Protocol failures in cryptosystems," *Proc. IEEE,* vol. 76, no. 5, pp. 594–602, May 1988.

[MORR75] M. A. Morrison and J. Brillhart, "A method of factoring and the factorization of F_7," *Math. Comput.,* vol. 29, no. 129, pp. 183–205, Jan. 1975.

[NATI77] National Bureau of Standards, *Federal Information Processing Standards Publication 46: Data Encryption Standard,* Washington, D.C.: U.S. Dept. Commerce, Jan. 15, 1977.

[NATI80] National Bureau of Standards, *Federal Information Processing Standards Publication 81: DES Modes of Operation,* Washington, D.C.: U.S. Dept. Commerce, Dec. 2, 1980.

[NATI81] National Bureau of Standards, *Federal Information Processing Standards Publication 74: Guidelines for Implementing and Using the NBS Data Encryption Standard,* Washington, D.C.: U.S. Dept. Commerce, Apr. 1, 1981.

[NEED78] R. M. Needham and M. D. Schroeder, "Using encryption for authentication in large networks of computers," *Commun. ACM,* vol. 21, no. 12, pp. 993–999, Dec. 1978.

[ODLY84] A. M. Odlyzko, "Cryptanalytic attacks on the multiplicative knapsack cryptosystem and on Shamir's fast signature scheme," *IEEE Trans. Inform. Theory,* vol. IT-30, no. 4, pp. 594–601, July 1984.

[ODLY84b] A. M. Odlyzko, "Discrete logarithms in finite fields and their cryptographic significance," in *Lecture Notes in Computer Science 209; Advances in Cryptology: Proc. Eurocrypt'84,* T. Beth, N. Cot, and I. Ingemarsson, Eds., Paris, France, April 9–11, 1984, pp. 224–314. Berlin: Springer-Verlag, 1985.

[ORTO86] G. A. Orton, M. P. Roy, P. A. Scott, L. E. Peppard, and S. E. Tavares, "VLSI implementation of public-key encryption algorithms," in *Lecture Notes in Computer Science 263; Advances in Cryptology: Proc. Crypto'86,* A. M. Odlyzko, Ed., Santa Barbara, CA, Aug. 11–15, 1986, pp. 277–301. Berlin: Springer-Verlag, 1987.

[PATT87] W. Patterson, *Mathematical Cryptology for Computer Scientists and Mathematicians.* Totowa, NJ: Rowman & Littlefield, 1987.

[PERA86] R. C. Peralta, "A simple and fast probabilistic algorithm for computing square roots modulo a prime number," *IEEE Trans. Inform. Theory,* vol. 32, no. 6, pp. 846–847, Nov. 1986.

[POHL78] S. C. Pohlig and M. E. Hellman, "An improved algorithm for computing logarithms over GF(p) and its cryptographic significance," *IEEE Trans. Inform. Theory,* vol. IT-24, no. 1, pp. 106–110, Jan. 1978.

[POME84] C. Pomerance, "The quadratic sieve factoring algorithm," in *Lecture Notes in Computer Science 209; Advances in Cryptology: Proc. Eurocrypt'84,* T. Beth, N. Cot, and I. Ingemarsson, Eds., Paris, France, April 9–11, 1984, pp. 169–182. Berlin: Springer-Verlag, 1985.

[POME86] C. Pomerance, "Fast, rigorous factorization and discrete logarithm algorithms," in *Discrete Algorithms and Complexity, Proceedings of the Japan-US Joint Seminar,* D. S. Johnson, T. Nishizeki, A. Nozaki, and H. S. Wilf, Eds., Kyoto, June 4–6, 1986, pp. 119–143. Orlando, FL: Academic Press, 1987.

[POME88] C. Pomerance, J. W. Smith, and R. Tuler, "A pipeline architecture for factoring large integers with the quadratic sieve algorithm," *SIAM J. Comput.,* vol. 17, no. 2, pp. 387–403, Apr. 1988.

[POPE78] G. J. Popek and C. S. Kline, "Encryption protocols, public key algorithms and digital signatures in computer networks," in *Foundations of Secure Computation,* R. A. DeMillo, D. P. Dobkin, A. L. Jones, and R. J. Lipton, Eds., pp. 133–153. New York: Academic Press, 1978.

[POPE79] G. L. Popek and C. S. Kline, "Encryption and secure computer networks," *ACM Comput. Surveys,* vol. 11, no. 4, pp. 331–356, Dec. 1979.

[QUIN87] M. J. Quinn, *Designing Efficient Algorithms for Parallel Computers.* New York: McGraw-Hill, 1987.

[QUIS82] J.-J. Quisquater and C. Couvreur, "Fast decipherment algorithm for RSA public-key cryptosystem," *Electron. Lett.,* vol. 18, no. 21, pp. 905–907, Oct. 14, 1982.

[RABI76] M. O. Rabin, "Probabilistic algorithms," in *Algorithms and Complexity: New Directions and Recent Results, Proceedings of a Symposium,* J. F. Traub, Ed., Pittsburgh, PA, April 7–9, 1976, pp. 21–39. New York: Academic Press, 1976.

[RABI78] M. O. Rabin, "Digitalized signatures," in *Foundations of Secure Computation,* R. A. DeMillo, D. P. Dobkin, A. K. Jones, and R. J. Lipton, Eds., pp. 155–168. New York: Academic Press, 1978.

[RABI79] M. O. Rabin, *Digitalized Signatures and Public-Key Functions as intractable as factorization,* Cambridge, MA: MIT Laboratory for Computer Science, Tech. Rept. LCS/TR-212, Jan. 1979.

[RABI80] M. O. Rabin, "Probabilistic algorithms for testing primality," *J. Number Theory,* vol. 12, pp. 128–138, 1980.

[RADE64] H. Rademacher, *Lectures on Elementary Number Theory.* New York: Blaisdell, 1964.

[RIVE84] R. L. Rivest, "RSA chips (past/present/future)," in *Lecture Notes in Computer Science 209; Advances in Cryptology: Proc. Eurocrypt'84,* T. Beth, N. Cot, and I. Ingemarsson, Eds., Paris, France, April 9–11, 1984, pp. 159–165. Berlin: Springer-Verlag, 1985.

[RIVE90] R. L. Rivest, "The MD4 message digest algorithm," paper presented at Crypto'90, Santa Barbara, CA, Aug. 11–15, 1990; to appear in *Advances in Cryptology,* S. Vanstone, Ed. Berlin: Springer-Verlag (in press).

[RIVE78] R. L. Rivest, A. Shamir, and L. Adleman, "A method for obtaining digital structures and public-key cryptosystems," *Commun. ACM,* vol. 21, no. 2, pp. 120–127, Feb. 1978.

[RIVE82] R. L. Rivest and A. T. Sherman, "Randomized encryption techniques," in *Advances in Cryptology: Proc. Crypto'82,* D. Chaum, R. L. Rivest, and A. T. Sherman, Eds., Santa Barbara, CA, Aug. 23–25, 1982, pp. 23–25. New York: Plenum Press, 1983.

[SIAM90] "Number field sieve produces factoring breakthrough," *SIAM News*, vol. 23, no. 4, July 1990.

[SALO85] A. Salomaa, "Cryptography," in *Encyclopedia of Mathematics and its Applications, Vol. 25: Computation and Automata*, A. Salomaa, pp. 186–230. Cambridge, United Kingdom: Cambridge University Press, 1985.

[SCHA82] B. P. Schanning, "Applying public key distribution to local area networks," *Computers and Security*, vol. 1, no. 3, pp. 268–274, Nov. 1982.

[SCHA80] B. P. Schanning, S. A. Powers, and J. Kowalchuk, "MEMO: Privacy and authentication for the automated office," in *Proceedings of the 5th Conference on Local Computer Networks, Minneapolis, MN, Oct. 6–7, 1980*, pp. 21–30. Silver Spring, MD: IEEE Computer Society Press, 1980.

[SCHN84] C. P. Schnorr and H. W. Lenstra, Jr., "A Monte Carlo factoring algorithm with linear storage," *Math. Comput.*, vol. 43, no. 167, pp. 289–311, July 1984.

[SEDL87] H. Sedlak, "The RSA cryptography processor," in *Lecture Notes in Computer Science 304; Advances in Cryptology: Proc. Eurocrypt'87*, D. Chaum and W. L. Price, Eds., Amsterdam, The Netherlands, April 13–15, 1987, pp. 95–105. Berlin: Springer-Verlag, 1988.

[SEIT85] C. L. Seitz, "The Cosmic Cube," *Commun. ACM*, vol. 28, no. 1, pp. 22–33, Jan. 1985.

[SHAM79] A. Shamir, "On the cryptocomplexity of knapsack systems," in *Proceedings of the Eleventh Annual ACM Symposium on Theory of Computing, Atlanta, GA, April 30–May 2, 1979*, pp. 118–129. New York: ACM, 1979.

[SHAM84b] A. Shamir, "Identity-based cryptosystems and signature schemes," in *Lecture Notes in Computer Science 196; Advances in Cryptology: Proc. Crypto'84*, G. R. Blakley and D. Chaum, Eds., Santa Barbara, CA, Aug. 19–22, 1984, pp. 47–53. Berlin: Springer-Verlag, 1985.

[SHAM84] A. Shamir, "A polynomial-time algorithm for breaking the basic Merkle-Hellman cryptosystems," *IEEE Trans. Inform. Theory*, vol. IT-30, no. 5, pp. 699–704, Sept. 1984.

[SHAN90] M. Shand, P. Bertin, and J. Vuillemin, "Hardware speedups in long integer multiplication," in *Proc. 2nd ACM Symposium on Parallel Algorithms and Architectures*, Island of Crete, Greece, July 2–6, 1990, pp. 138–145. New York: Association for Computing Machinery, 1991.

[SILV87] R. D. Silverman, "The multiple polynomial quadratic sieve," *Math. Comput.*, vol. 48, no. 177, pp. 329–339, Jan. 1987.

[SIMM79] G. J. Simmons, "Symmetric and asymmetric encryption," *ACM Comput. Surveys*, vol. 11, no. 4, pp. 305–330, Dec. 1979.

[SIMM88] G. J. Simmons, "How to ensure that data acquired to verify treaty compliance are trustworthy," *Proc. IEEE*, vol. 76, no. 5, pp. 621–627, May 1988.

[SMID81] M. E. Smid, "Integrating the data encryption standard into computer networks," *IEEE Trans. Comput.*, vol. COM-29, no. 6, pp. 762–772, June 1981.

[SIMM88b] M. E. Smid and D. K. Branstad, "The Data Encryption Standard: Past and future," *Proc. IEEE*, vol. 76, no. 5, pp. 550–559, May 1988.

[SOLO77] R. Solovay and V. Strassen, "A fast Monte-Carlo test for primality," *SIAM J. Comput.*, vol. 6, no. 1, pp. 84–85, March 1977. Erratum: *Ibid.*, vol. 7, no. 1, p. 118, Feb. 1978.

[STEI86] L. K. Steiner, "The ETA-10 supercomputer system," in *Proceedings of the 1986 IEEE International Conference on Computer Design, Port Chester, NY, Oct. 6–9,* p. 42. Washington, DC: IEEE Computer Society Press, 1986.

[TANE81] A. S. Tanenbaum, *Computer Networks.* Englewood Cliffs, NJ: Prentice Hall, 1981.

[WAGN84] N. R. Wagner and M. R. Magyarik, "A public-key cryptosystem based on the word problem," in *Lecture Notes in Computer Science 196; Advances in Cryptology: Proc. Crypto'84,* G. R. Blakley and D. Chaum, Eds., Santa Barbara, CA, Aug. 19–22, 1984, pp. 19–36. Berlin: Springer-Verlag, 1985.

[WANG85] C. C. Wang, T. K. Truong, H. M. Shao, L. J. Deutsch, J. K. Omura, and I. S. Reed, "VLSI architectures for computing multiplications and inverses in $GF(2^m)$," *IEEE Trans. Comput.*, vol. C-34, no. 8, pp. 709–717, Aug. 1985.

[WILK68] M. V. Wilkes, *Time-Sharing Computer Systems.* New York: Elsevier, 1968.

[WILL80] H. C. Williams, "A modification of the RSA public-key encryption procedure," *IEEE Trans. Inform. Theory,* vol. IT-26, no. 6, pp. 726–729, Nov. 1980.

[WILL87] H. C. Williams and M. C. Wunderlich, "On the parallel generation of the residues for the continued fraction factoring algorithm," *Math. Comput.*, vol. 48, no. 177, pp. 405–423, Jan. 1987.

[WUND83] M. C. Wunderlich, "Factoring numbers on the Massively Parallel computer," in *Advances in Cryptology: Proc. Crypto'83,* D. Chaum, Ed., Santa Barbara, CA, Aug. 22–24, 1983, pp. 87–102. New York: Plenum Press, 1984.

[WUND85] M. C. Wunderlich, "Implementing the continued fraction factoring algorithm on parallel machines," in *Math. Comput.*, vol. 44, no. 169, pp. 251–260, Jan. 1985.

[YAO-82] A. C. Yao, "Theory and applications of trapdoor functions," in *Proceedings of 23rd Ann. IEEE Symposium Foundations Computer Science,* Chicago, Nov. 3–5, 1982, pp. 80–91. Los Angeles: IEEE Computer Society Press, 1982.

CHAPTER 5

A Comparison of Practical Public Key Cryptosystems Based on Integer Factorization and Discrete Logarithms*

PAUL C. VAN OORSCHOT
Bell-Northern Research

---

*Partial support for this work was provided by the University of Waterloo, Waterloo, Ontario, and by Newbridge Microsystems (a division of Newbridge Networks Corporation), Kanata, Ontario.

Abstract—based on the current literature, this survey carries out a detailed analysis of a version of the multiple polynomial quadratic sieve integer factorization algorithm, and of the Coppersmith algorithm for computing discrete logarithms in $GF(2^n)$. This is used for a practical security comparison between the Rivest–Shamir–Adleman (RSA) cryptosystem and the El Gamal cryptosystem in fields of characteristic 2. Other aspects of the cryptosystems are also compared. In addition, the security of elliptic curve cryptosystems over $GF(2^n)$ is discussed, and related to that of the previously mentioned cryptosystems.

1 INTRODUCTION

Since its inception in the mid 1970s, public key cryptography has flourished as a research activity, and significant theoretical advances have been made. In more recent years, many public key concepts have gained acceptance in the commercial world. Without question, the best-known public key cryptosystem is the RSA cryptosystem of Rivest, Shamir, and Adleman [46]. Although not as well known, another public key cryptosystem of practical interest is that of El Gamal [17]. The latter system and its variations use a basic extension of Diffie-Hellman key exchange [15] for encryption, together with an accompanying signature scheme.

The security of the RSA and El Gamal cryptosystems is generally equated to the difficulty of integer factorization and that of the computation of discrete logarithms in finite fields, respectively. Each of these problems has been the subject of extensive research in recent years, and significant progress has been made. An early survey of progress in integer factorization is given by Davis, Holdridge, and Simmons [14]. Since then, the quadratic sieve algorithm has been carefully examined with an eye to specialized hardware [45] and networks of distributed processors [11,28]. More recently, dramatic theoretical progress has been made using elliptic curves [30] and algebraic number fields [27]. Each of the factoring algorithms resulting from these latter two advances is most efficient when applied to specific classes of composite integers as will be discussed below—although the number field sieve may eventually prove to be a competitive factoring technique for general integers of a size to be of cryptographic interest. Regarding discrete logarithms, for reasons discussed below we restrict our attention primarily to fields of characteristic 2. Early work on the problem of computing logarithms in $GF(2^n)$ resulted in the acknowledgment that the field $GF(2^{127})$ is totally inadequate for cryptographic security [5]; 127 bits, which corresponds to 38 digits, is simply insufficient. Subsequent work [12,41] has led to further improvements in the subexponential-time index-calculus techniques for computing discrete logarithms in $GF(2^n)$. Progress in solving large sparse linear systems over finite fields applies to both problems [24,52].

Significant advances have also been made, in theory and in practice, on techniques for efficient implementation of these cryptosystems, including custom very large-scale integration (VLSI) chips and very efficient digital signal processor software implementations for modular exponentiation for RSA [7], and custom VLSI chips for arithmetic operations in the El Gamal cryptosystem in $GF(2^n)$ [47].

Asymptotic running times for many integer factorization algorithms are often given in the form $L(N)^c$, where $L(N)$, $= \exp(\sqrt{\log N \log\log N})$, with analysis carried out with sufficient accuracy to derive the correct value of c (see [43]). Such analysis is very important, and allows one to gauge the relative difficulty of problems of different sizes using a fixed algorithm; furthermore, given an actual running time obtained by applying an implementation to one problem instance, this permits reasonable extrapolations yielding running time estimates for similar implementations on larger problem instances. However, while asymptotic formulas such as $L(N)^c$ by themselves suffice for theoreticians, they leave much to be desired by those interested in estimating the best achievable running time of an algorithm in practice; and without further investigation, they are of limited use in comparing the difficulty of a particular instance of one problem (e.g., integer factorization) with a particular instance of another problem (e.g., the discrete logarithm problem).

Earlier surveys discussing both integer factorization and the discrete logarithm problem include work by Lenstra and Lenstra [26], Blake, Van Oorschot, and Vanstone [6], and Bach [1]. As noted by Bach, little work has been done on attaining more precise running time formulas for algorithms for these problems. Two notable exceptions are the discrete analysis by Odlyzko [41] of the Coppersmith algorithm for computing discrete logarithms in $GF(2^n)$, and a detailed analysis of a version of the quadratic sieve integer factorization algorithm by Pomerance, Smith, and Tuler [45] in a paper that proposes a hardware design for an integer factorization machine. Regarding the former, fields of characteristic 2 have traditionally been of much interest in practice, as arithmetic in such fields is particularly amenable to efficient hardware implementation.

The computation of discrete logarithms in odd prime fields $GF(p)$ is discussed in [13] and [25]. The latter paper concludes, based on empirical results—the computation of discrete logarithms in a prime field $GF(p)$ of size 192 bits—that the computation of discrete logarithms in $GF(p)$, using the best currently known techniques, is slightly harder than factorization of integers N (where $N \approx p$) via the multiple polynomial quadratic sieve. This conclusion is in line with earlier asymptotic analysis of these algorithms, and gives an indication of the relative security of RSA versus that of cryptosystems whose security relies on the difficulty of computing discrete logarithms in prime fields $GF(p)$.

This raises the question of the relative difficulty of integer factorization and the computation of discrete logarithms in fields of characteristic 2. This issue has apparently not been addressed in the literature. Asymptotic running time formulas for the best currently known algorithms for each of these problems are well known, but as noted above are of limited use in "head-to-head" comparisons of particular instances of different problems. For example, a cryptographer might ask, "How large a field $GF(2^n)$ must be chosen to make the computation of discrete logarithms in that field as difficult as the factorization of an m-bit integer?" While large-scale implementations provide valuable information from which to base extrapolations, answers to such questions cannot in general be obtained empirically, since problem instances of cryptographic interest should by definition be well beyond computational capabilities; furthermore, distinct implementations may (due to quality, software, and architecture details) vary dramatically with respect to running times, and thus provide only upper bounds on the difficulty of a problem instance even for a given algorithm. A worthwhile endeavor would be the refinement of existing asymptotic formulas, incorporating running time constants and currently missing factors to bring these formulas into agreement with actual (absolute) running times observed for the largest problem instances solved to date; however, this has not been done to date (nor is it done herein).

In this chapter, based on the current literature we carry out a practical comparison of the relative difficulty of integer factorization and the computation of discrete logarithms in $GF(2^n)$. By "practical" here we mean a comparison suitable for dealing with particular problem instances of practical interest, rather than dwelling exclusively on asymptotic complexities. This facilitates a security comparison between the El Gamal cryptosystem in $GF(2^n)$ and RSA. In addition to the relative security, we consider the practical efficiency of the systems, discuss further aspects, and report on recent advances. We also briefly consider the related elliptic curve cryptosystems, including the cryptographic significance of recent progress on the problem upon which their security rests—the computation of elliptic curve logarithms. Elliptic curve cryptosystems over $GF(2^n)$ have recently received considerable attention as cryptographic alternatives to other candidate cryptosystems, offering greater security at shorter key lengths [22,34]. The recent reduction [35] of the elliptic curve logarithm problem, for certain elliptic curves, to the discrete logarithm problem in extension fields of the underlying field, has made an understanding of the relative difficulty of the discrete logarithm problem in $GF(2^n)$ and that of integer factorization of even greater importance.

The analysis of the Coppersmith algorithm for computing discrete logarithms here is largely based on that of Odlyzko [41], with suitable modifications made to facilitate comparison. Similarly, our practical analysis of the quadratic sieve factorization algorithm is based on the analysis in [45], again with suitable modifications and subject to several assumptions. We emphasize that for our purposes, we are interested only in the relative difficulty of these different problems, and make no estimates as to absolute running times (e.g., number of days or weeks on a particular machine). As our model, we count the number of primitive machine instructions (such as addition and exclusive-or) on a single-processor machine such as a modern workstation; operation counts for instances of the different problems can then be compared to determine relative running times within this model. We base the security comparison on the best currently known algorithms for factorization and logarithms; future algorithmic advances cannot be predicted. Advances in technology (e.g., processor speeds) are not taken into account; these should affect the running times of the underlying problems similarly. To simplify the task, advanced architectures are not taken into account either, although these may or may not affect both algorithms similarly—for example, both can take advantage of large networks of loosely coupled workstations during the relation collection phase, but the quadratic sieve algorithm appears better suited to take full advantage of vectorized supercomputers [14]. The analysis we carry out involves several assumptions and estimates (see Sections 2 and 3). Regarding absolute running times, which are often strongly influenced by implementation, processor and architecture details, we refer the reader to [11], [27–29], and [51] for factorization, and to [12] for logarithms in $GF(2^n)$; however, meaningful extrapolations of absolute running times from the latter may be difficult, since the largest implementation carried out to date is $n = 127$.

The remaining sections of this chapter are organized as follows. In Section 2 we review the index-calculus techniques currently available for computing discrete logarithms in $GF(2^n)$. As is well known, these techniques consist of two stages, the first of which dominates the running time and memory requirements, involving construction of a large database through the generation and solution of a large sparse linear system. The costly operations in this first stage are the testing of a large number of pairs of polynomials for smoothness with respect to a particular degree bound, and the solution of the linear system over $\mathbb{Z}_M$, $M = 2^n - 1$. We review a discrete estimate of the running time (in number of shifts and adds) of the algorithm, and also consider the size of the linear system and the resulting database. Estimates of algorithmic parameters, the size of the resulting system, and running times are tabulated for a small number of fields of sizes representative of systems of practical interest.

In Section 3 we consider the fastest practical general purpose method for factoring integers N currently available—the multiple polynomial quadratic sieve algorithm. Analogous to the smoothness testing required in computing discrete logarithms, the quadratic sieve algorithm requires that a large number of residues (mod N) be generated and then tested for smoothness with respect to an appropriate integer bound. In this case, the smoothness testing is done by "sieving." A sufficient number of smooth residues provide a sufficient number of relations, from which a linear dependency in a large binary sparse linear system is sought. We estimate the running time (in number of single precision adds) of the algorithm, and also consider the size of the linear system. Tabulations are given estimating algorithmic parameters, system size, and running time for integers N of 100 and 155 digits. (The latter corresponds to 512 bits, a number arising as often as any other in discussions of RSA modulus size.) The currently best implementations and systems of quadratic sieve factoring can handle numbers at the low end of this range, and we restrict our examination to the specified upper limit as we feel that extrapolations beyond this point are not possible (for us) to make with any confidence, and would have little practical meaning. Furthermore, it is not clear what the range of practical utility of the quadratic sieve algorithm is.

In Section 4, we first briefly review the RSA and El Gamal cryptosystems. The estimates of Sections 2 and 3 are used to compare the running times for integer factorization and the computation of discrete logarithms in $GF(2^n)$, for a limited set of bitlengths. This facilitates a first attempt at estimation of the relative security of the El

Gamal cryptosystem in $GF(2^n)$, and RSA, at particular bitlengths of practical interest. We examine many further aspects of these respective cryptosystems and the underlying problems, including currently available throughput for implementations of exponentiation in $\mathbb{Z}_N$ and exponentiation in $GF(2^n)$.

In Section 5 we survey recent work in the area of elliptic curve cryptosystems as mentioned above, and consider the cryptographic impact of recent progress in computing elliptic curve logarithms.

2 DISCRETE LOGARITHMS IN FIELDS OF CHARACTERISTIC 2

As a measure of the security of the El Gamal cryptosystem in $GF(2^n)$, we consider the difficulty of computing discrete logarithms in fields of characteristic 2. In this section, we review the current techniques available for this problem [12,24,41,52]. We rely heavily on the authoritative analysis of Odlyzko [41]. Other surveys discussing the computation of discrete logarithms include [1], [26], and [32].

Elements of $GF(2^n)$ are represented as polynomials over $GF(2)$ of degree at most $n - 1$, with polynomial arithmetic done modulo a fixed binary irreducible polynomial $f(x) = x^n + f_1(x)$, $\deg f_1(x) \leq n - 1$. A simple heuristic argument indicates that we may expect to find such $f(x)$ with $\deg f_1(x) \approx \log(n)$, and henceforth we assume the degree of $f_1(x)$ to be about $\log(n)$.* Let $g(x)$ be a primitive element of $GF(2^n)$. Then given any polynomial $b(x)$ in $GF(2^n)$, the *discrete logarithm problem* in this setting is to find the unique integer y, $0 \leq y \leq 2^n - 2$, such that $b(x) \equiv g(x)^y \pmod{f(x)}$. We begin by reviewing the basic algorithm, outline the improvements that have been made, and review the running time and size of the resulting linear system.

2.1 Basic Algorithm

The standard index-calculus approach consists of two stages. The first is the precomputation (carried out only once for a given field) of a large database containing the logarithms of all irreducible polynomials of degree at most m, for appropriate m. The second stage then reuses this database each time the logarithm of a particular field element $b(x)$ is required.

Stage 1. In the first stage of the basic algorithm, an appropriate bound m ($m \approx n^{1/2} (\ln n)^{1/2}$) is selected, and then S is defined to be the set of all irreducible polynomials over $GF(2)$ of degree at most m. A polynomial all of whose irreducible factors have degree at most m is said to be *smooth with respect to m* (or simply *smooth,* when m is understood). The objective of stage 1 is to compute the logarithms (with respect to $g(x)$) of the elements in S. This is done by obtaining approximately $|S|$ linear equations relating $|S|$ unknowns, the unknowns being the required logarithms, and then solving this system. To obtain the equations, select a random integer c, $1 \leq c \leq 2^n - 1$, and

---

*We use $\ln x$ to denote the natural logarithm of the number x, and $\log(x)$ to denote the logarithm of x to the base 2. For discrete logarithms of finite field elements, we also use $\log$ to denote the logarithm with respect to a primitive element $g(x)$, where $g(x)$ is implicitly assumed to be known and fixed.

compute $h(x)$, where deg $h(x) < n$ and

$$h(x) \equiv g(x)^c \pmod{f(x)} \tag{2.1}$$

Check if $h(x)$ is smooth. If so, factor $h(x)$ to obtain the irreducible polynomial decomposition

$$h(x) = \prod_{s \in S} s^{i_s}, \qquad i_s \geq 0$$

(s is a binary polynomial here.) Taking logs with respect to $g(x)$ then yields

$$c \equiv \sum_{s \in S} i_s \cdot \log(s) \pmod{2^n - 1} \tag{2.2}$$

This equation relates the (unknown) logarithms of elements in S. $|S|$ such equations are obtained (or slightly more, in the case there are linear dependencies—but the resulting equations are largely expected to be independent here, and this expectation is confirmed in practice), and the linear system over the integers modulo $2^n - 1$ is solved to obtain the unknown logs.

  Stage 2. In the second stage of the basic algorithm, we start with the knowledge of the logarithms of (almost) all elements in S. Given $b(x)$, we seek y, $0 \leq y \leq 2^n - 2$, such that $b(x) \equiv g(x)^y \pmod{f(x)}$. To find y, select random c, $1 \leq c \leq 2^n - 1$, and compute

$$h(x) \equiv b(x)g(x)^c \pmod{f(x)} \tag{2.3}$$

Check if the reduced polynomial $h(x)$ factors over S, and if so, factor $h(x)$ to obtain

$$h(x) = \prod_{s \in S} s^{i_s}, \qquad i_s \geq 0$$

Then

$$\log b(x) + c \equiv \sum_{s \in S} i_s \cdot \log(s) \pmod{2^n - 1}$$

and one can determine $y = \log b(x)$ from the known quantities $\log(s)$.

2.2 Coppersmith Algorithm

The probability that a polynomial $h(x)$ is smooth (e.g., in Eqs. (2.1) and (2.3)) increases as the degree of $h(x)$ decreases. Blake, Fuji-Hara, Mullin, and Vanstone [5] noted that one way to take advantage of this is to make use of the extended Euclidean algorithm. This, together with another observation of Blake et al., concerning "systematic equations," motivated Coppersmith to conceive a far more powerful method, resulting in a dramatic decrease in the expected running time [12]. This latter variation is now described.

  Stage 1. Coppersmith's algorithm obtains stage 1 equations more rapidly via the following technique. Positive integers k and B are carefully selected, and k is used to define h, such that

$$2^k \approx n^{1/3}(\ln n)^{-1/3}, \qquad B \approx n^{1/3}(\ln n)^{2/3}, \qquad h = \lceil n/2^k \rceil$$

Here h is the least integer greater than or equal to $n/2^k$ ($h \approx n^{2/3}(\ln n)^{1/3}$). Next select a pair of coprime polynomials $u_1(x)$, $u_2(x)$ of degrees at most B. Define

$$w_1(x) = x^h \cdot u_1(x) + u_2(x)$$
$$w_2(x) \equiv w_1(x)^{2^k} \pmod{f(x)} \tag{2.4}$$

Note deg $w_1(x) \approx n^{2/3}$, and

$$w_2(x) \equiv x^{h2^k} \cdot u_1(x)^{2^k} + u_2(x)^{2^k} \pmod{f(x)}$$
$$= x^{h2^k - n} \cdot f_1(x) \, u_1(x^{2^k}) + u_2(x^{2^k}) \tag{2.5}$$

and hence deg $w_2(x) \approx n^{2/3}$ also (since $h2^k - n = O(2^k) \approx O(n^{1/3})$, and $B2^k \approx n^{2/3}$). Now check if both $w_1(x)$ and $w_2(x)$ are smooth. (This is far more likely than $h(x)$ being smooth in Eq. (2.1), since $h(x)$ has degree $\approx n$ there.) If so, factor $w_1(x)$ and $w_2(x)$, and note

$$\log (w_2(x)) \equiv 2^k \cdot \log (w_1(x)) \pmod{2^n - 1} \tag{2.6}$$

is an equation that can be used in place of Eq. (2.2).

Whereas $m \approx n^{1/2}(\ln n)^{1/2}$ is optimal for the basic algorithm, $m \approx n^{1/3}(\ln n)^{2/3}$ is optimal for the Coppersmith algorithm, and the database is significantly smaller.

Stage 2. Stage 2 of the Coppersmith algorithm falls into two steps. Given a polynomial $b(x)$ whose logarithm is to be obtained, the first step reduces the problem to that of finding several logarithms of smaller-degree polynomials. The second step recursively decomposes each of these subproblems into that of finding the logarithms of a number of still smaller degree polynomials, until eventually the logs of these can be looked up from the stage 1 database. The time required to determine the log of a given polynomial $b(x)$ via this second stage is far less than that required for the first stage; we do not consider the details further here.

2.3 Faster Generation of Coppersmith Equations (Due to Odlyzko)

Odlyzko noted that the equation generation in the first stage of Coppersmith's variation can be further expedited (significantly in practice, but not asymptotically), as follows. Instead of selecting $u_1(x)$, $u_2(x)$, for use in Eq. (2.4), to be any coprime pair of polynomials as above, construct candidate pairs $u_1(x)$, $u_2(x)$ more carefully so as to increase the probability that $w_1(x)$ and $w_2(x)$ are smooth. This is done as follows. Select (possibly from a previously prepared table) a pair of polynomials $v_1(x)$, $v_2(x)$ each of degree at most (but near) $B - 1$, each known to be smooth with respect to m. The pair $v_1(x)$, $v_2(x)$ is used to determine (again for use in Eq. (2.4)) a pair of polynomials $u_1(x)$, $u_2(x)$, each of degree at most B, such that

$$v_1(x) \mid w_1(x) \quad \text{and} \quad v_2(x) \mid w_2(x) \tag{2.7}$$

These two conditions define at most $2B - 2$ ($= \deg v_1(x) + \deg v_2(x)$) linear equations in $2B$ unknowns (the latter being the defining coefficients of $u_1(x)$ and $u_2(x)$). Solving this homogeneous binary system of order $2B - 2 \times 2B$ (e.g., by standard Gaussian elimination) yields three or more nontrivial solution pairs $u_1(x)$, $u_2(x)$. (Note it is essential here that the solution of this system does not take more time than that necessary to test the resulting $w_1(x)$, $w_2(x)$ pairs for smoothness, otherwise the purpose of the exercise will have been defeated.)

Using this technique, however, several pairs $v_1(x)$, $v_2(x)$ will result in the same pair of smooth polynomials $w_1(x)$, $w_2(x)$, and hence yield the same equation. To compensate for this, we increase the expected requirement from the previous value of $|S|$ equations to $\alpha|S|$ equations (with $\alpha > 1$), that $|S|$ of these be linearly independent. It can be shown that $|S|$ independent equations are expected if the number of smooth $w_1(x)$, $w_2(x)$ pairs obtained by this method is $\alpha|S|$, where $\alpha \approx 1.6$. To accommodate this, it may be necessary to increase the degree bound B by 1.

2.4 Smoothness Testing of Polynomials

As discussed above, it is necessary to determine whether certain polynomials (e.g., $w_i(x)$) are smooth with respect to some degree bound m. Once such smoothness is determined, the factorization of a polynomial can be achieved by standard techniques (e.g., via a square-free decomposition followed by a distinct-degree factorization [21] and then Berlekamp's classic Q-matrix method, or via the probabilistic method of Cantor and Zassenhaus [10]). Since far more polynomials fail to be smooth than require factorization, the cost of smoothness testing dominates that of factorization.

Regarding testing the smoothness of a polynomial $w(x) = \prod y_i(x)^{e_i}$ with distinct irreducible factors $y_i(x)$, Odlyzko notes the following technique. (For an alternative method of smoothness testing, see [12].) Use standard techniques (e.g., see [31]) to compute the square-free component

$$w^{(0)}(x) = \prod_i y_i(x)$$

of $w(x)$. (This operation itself reveals partial information about the factorization of $w(x)$, including possibly individual factors $y_i(x)$ of degree m or less, which may be excluded from $w^{(0)}(x)$ right away.) Given $w^{(0)}(x)$, consider calculating

$$w^{(i)}(x) = \frac{w^{(i-1)}(x)}{(w^{(i-1)}(x), x^{2^i} - x)}, \quad 1 \le i \le m \tag{2.8}$$

It is easily seen that $w^m(x) = 1$ if and only if $w^{(0)}(x)$ is smooth with respect to m, since the ith iteration removes from consideration all irreducible factors whose degrees divide i. Furthermore, the effect is more efficiently achieved by replacing Eq. (2.8) by the sequence

$$w^{(M)}(x) = \frac{w^{(0)}(x)}{(w^{(0)}(x), x^{2^M} - x)}$$

and for $M + 1 \le M + i \le m$, where $M = \lfloor (m - 1)/2 \rfloor + 1$

$$w^{(M+i)}(x) = \frac{w^{(M+i-1)}(x)}{(w^{(M+i-1)}(x), x^{2^{M+i}} - x)} \tag{2.9}$$

Within Eq. (2.9), the polynomials $x^{2^{M+i}} - x$ can be computed efficiently noting that if

$$r_i(x) \equiv x^{2^i} - x \pmod{w^{(i-1)}(x)},$$

then

$$x^{2^{i+1}} \equiv (r_i(x) + x)^2, \text{ so } x^{2^{i+1}} - x \equiv r_i(x^2) + x^2 - x \pmod{w^{(i-1)}(x)},$$

and since $w^{(i)}(x) \mid w^{(i-1)}(x)$, this last congruence is true modulo $w^{(i)}(x)$ also.

2.5 Solution of Linear System

In stage 1, once a sufficient number of linear equations have been determined via Eq. (2.6), the linear system over $\mathbb{Z}_{2^n-1}$ must be solved. (We overlook issues that arise from the fact that $2^n - 1$ may be composite.) The stage 1 database, containing the logarithms of irreducible polynomials of degree at most m, has $|S| \approx 2^{m+1}/m$ elements. To solve for these logarithms, $|S|$ independent equations are required. Using Odlyzko's method for generating stage 1 equations as noted above, about $N = 1.6|S|$ equations (in $|S|$ unknowns) must be generated in order to obtain a system of full rank.

Standard Gaussian elimination, at $O(N^3)$ operations, would dominate all other costs in the algorithm; for this reason, it is bypassed in favor of faster solution techniques which take advantage of the fact that the resulting linear system is sparse. These techniques include the conjugate gradient and Lanczos algorithms [13,24,41], and Wiedemann's technique [52], which have expected running times not much greater than $O(N^2)$. Structured or so-called *intelligent* Gaussian elimination can be used to reduce an original sparse system to a smaller system, before applying one of these techniques. Unfortunately, it is not known how to efficiently distribute the linear algebra among a large number of processors, as can be done with the equation collection phase; a single machine appears necessary. Due to memory constraints, access to a supercomputer appears desirable for carrying out this linear algebra for large problem instances. The recent work of LaMacchia and Odlyzko [24] shows that the linear algebra stages arising in both integer factorization and the discrete logarithm problem no longer appear to be a running-time bottleneck in practice.

2.6 Practical Analysis of Coppersmith Algorithm

In our analysis, we estimate the number of operations in terms of the number of shifts and adds required on a 32-bit single-processor machine.

Stage 1 of the Coppersmith variation, with generation of equations using Odlyzko's technique, is summarized below.

1. Carefully choose $m \approx n^{1/3}(\ln n)^{2/3}$, implying $|S| \approx 2^{m+1}/m$.
2. While less than $\approx 1.6|S|$ equations have been generated,
   a. Generate pairs $w_1(x)$, $w_2(x)$ of the form Eq. (2.4), via Eq. (2.7);
   b. Test each pair for smoothness; if smooth, obtain an equation of form Eq. (2.6).
3. Solve the large sparse linear system to obtain the logarithms of the elements in S.

Running time. Since stage 1 takes far more time than stage 2, the running time for computing discrete logs in $GF(2^n)$ is that of stage 1. It is assumed that the time required for the solution of the linear system in stage 1 is less than that required for the equation generation phase of stage 1. This assumption seems reasonable, in light of progress made in solving similar large sparse systems [24]. However, there is limited practical experience with large systems of cryptographic interest ($n \approx 600$ and greater). The largest sparse system solved to date is that by LaMacchia and Odlyzko [25] as previously mentioned, resulting from a 192-bit prime field.

Although intelligent Gaussian elimination works better when there are many more equations than unknowns [24], we consider only the time required to generate enough equations to allow for solution of the system. Generating additional equations would apparently make the linear algebra stage run faster, at the expense of the equation generation phase.

Regarding the time required for the generation of equations, following the analysis in [41] we neglect the time required to generate the $w_1(x)$, and equate the running time of the equation generation part of stage 1 to that of testing smoothness for the resulting pairs $w_1(x)$, $w_2(x)$. To determine this cost, let deg $u_i(x) \approx d_i$ and deg $v_i(x) \approx d_i - 1$ (where $d_i \approx B$; minor variations between d_1 and d_2 can result in a small savings). From Eqs. (2.4) and (2.5) note that deg $w_1(x)/v_1(x) \approx h + 1$, and deg $w_2(x)/v_2(x) \approx M$ where

$$M = \max(h2^k - n + \deg f_1(x) + d_1 2^k - d_2 + 1, d_2(2^k - 1) + 1)$$

To satisfy the requirement of a sufficient number of independent equations, d_1 and d_2 must be sufficiently large such that

$$3 \cdot 2^{d_1 + d_2} (1 - \varepsilon)^2 \cdot p(h + 1, m) \cdot p(M, m) \geq (1.6) \cdot 2^{m+1}/m$$

where ε depends on how far we allow the degrees of $v_i(x)$ to fall below $B - 1$; we simplify this to

$$p(h + 1, m) \cdot p(M, m) \cdot 2^{d_1 + d_2 + 1} \geq 2^{m+2}/m \qquad (2.10)$$

Here $p(r, m)$ is the probability that all irreducible factors of a binary polynomial of degree exactly r have degree at most m, that is, the probability of smoothness with respect to m. Formulas for $p(r, m)$ are readily available (e.g., see [5], [41, Appendix A]), and these exact probabilities are easily computed.

Now testing $w_i(x)$ for smoothness requires about $m/2$ operations of the form Eq. (2.9), involving the squaring of one polynomial modulo another of about the same degree (say t), a greatest common divisor (GCD) of two polynomials of this degree, and a division (the cost of the division being negligible since the denominator is typically unity). The squaring and GCD can be performed in about $6t$ t-bit vector shifts and adds (exclusive-ors), or $6t \cdot (t/32)$ simple machine operations assuming the work is done on a machine with 32-bit shifts and adds. The use of 32-bit machines, as opposed to larger-wordsize supercomputers, is a reasonable assumption for our purposes, in light of recent projects involving the networking of hundreds of workstations to carry out various pieces of the computations in quadratic sieve, elliptic curve, and number field sieve factorizations [27,28]. Since in the vast majority of cases $w_1(x)$ will fail to be smooth and $w_2(x)$ then need not even be tested, only the cost of testing one polynomial from each pair is charged. This leads to an operation estimate for smoothness testing of

$$m/2 \cdot 6t(t/32) \cdot (1.6) \cdot 2^{m+1}/m \cdot p(h + 1, m)^{-1} \cdot p(M, m)^{-1}$$

We have deg $w_i(x) \approx h$, and use $t = h$ here, although the degrees of the polynomials decrease during a smoothness test, and on average are thus somewhat less than h. Our estimate becomes about

$$h^2 \cdot 2^{m-2} \cdot p(h + 1, m)^{-1} \cdot p(M, m)^{-1} \qquad (2.11)$$

32-bit shifts/adds. Tables 1, 2, and 3 illustrate sample parameters and estimated operation counts (Eq. (2.11)) for the smoothness testing in stage 1 of the algorithm, for $n = 400$, 700, and 850, respectively. (The reason for choosing to focus on these particular bitlengths will become clear in Section 4. If one is willing to work with a million or more equations, then $2^k = 8$ is worth considering.)

Memory requirements. The tables below include figures indicating the size of the linear systems arising from various parameter choices. The resulting stage 1 database itself contains about $b = 2^{m+1}/m$ entries, each being the n-bit logarithm of an element in $GF(2^n)$ (e.g., for $n = 1000$, each entry is 125 bytes). Hence roughly bn bits are required to store the completed stage 1 database. During the stage 1 computation the linear system itself requires storage on the order of the number of nonzero entries in the system (assuming a sparse representation is used), each nonzero entry being a coefficient in a linear relation (Eq. (2.6)); storage is required for the coefficient itself plus overhead associated with the sparse representation. Working memory required by the sparse solution techniques is of about the same order as that required for the representation of the system.

The sparseness of the relations depends on the number of distinct irreducible factors of $w_1(x)$ and $w_2(x)$ in Eq. (2.6). These are polynomials of degree about $n^{2/3}$ which are smooth with respect to m. The number of irreducible factors in each polynomial is thus $O(n^{2/3})$, and in some cases as small as $O(n^{2/3}/m)$.

Further speedups in the computation of discrete logarithms. We note below two further methods by which the Coppersmith variation, as modified by Odlyzko, can be improved. These improvements have no asymptotic impact, but can speed things up in practice. Details are given in [41].

TABLE 1.   ESTIMATES FOR SMOOTHNESS TESTING ($n = 400$)

			deg $f_1(x) = 10$,	$2^k = 4$,	$h = 100$
m	d_1	d_2	$b = 2^{(m+1)}/m$	bn	Operations
14	20	23	2.3406e + 03	9.3623e + 05	8.8509e + 16
15	18	21	4.3691e + 03	1.7476e + 06	6.7538e + 15
16	17	19	8.1920e + 03	3.2768e + 06	1.1959e + 15
17	16	18	1.5420e + 04	6.1681e + 06	3.4882e + 14
18	17	17	2.9127e + 04	1.1651e + 07	3.4324e + 14
19	15	17	5.5188e + 04	2.2075e + 07	8.8542e + 13
20	15	17	1.0486e + 05	4.1943e + 07	6.6796e + 13
21	16	16	1.9973e + 05	7.9892e + 07	8.8178e + 13
22	15	16	3.8130e + 05	1.5252e + 08	5.7013e + 13
23	15	16	7.2944e + 05	2.9178e + 08	5.7346e + 13
24	15	16	1.3981e + 06	5.5924e + 08	6.1942e + 13
25	16	16	2.6844e + 06	1.0737e + 09	9.2897e + 13
26	16	16	5.1622e + 06	2.0649e + 09	1.1056e + 14

TABLE 2.   ESTIMATES FOR SMOOTHNESS TESTING ($n = 700$)

m	d_1	d_2	$b = 2^{(m+1)}/m$	bn	Operations
			$\deg f_1(x) = 10,$	$2^k = 4,$	$h = 175$
20	25	28	$1.0486e + 05$	$7.3400e + 07$	$6.3195e + 20$
21	26	26	$1.9973e + 05$	$1.3981e + 08$	$3.1137e + 20$
22	25	25	$3.8130e + 05$	$2.6691e + 08$	$8.2279e + 19$
23	23	25	$7.2944e + 05$	$5.1061e + 08$	$1.6711e + 19$
24	23	24	$1.3981e + 06$	$9.7867e + 08$	$9.9167e + 18$
25	23	23	$2.6844e + 06$	$1.8790e + 09$	$6.5894e + 18$
26	22	23	$5.1622e + 06$	$3.6136e + 09$	$3.2556e + 18$
27	22	23	$9.9421e + 06$	$6.9594e + 09$	$2.4706e + 18$
28	21	23	$1.9174e + 07$	$1.3422e + 10$	$1.5647e + 18$
29	22	22	$3.7026e + 07$	$2.5918e + 10$	$1.9366e + 18$
30	22	22	$7.1583e + 07$	$5.0108e + 10$	$1.8354e + 18$
31	22	22	$1.3855e + 08$	$9.6983e + 10$	$1.8454e + 18$
32	22	22	$2.6844e + 08$	$1.8790e + 11$	$1.9555e + 18$
33	22	22	$5.2060e + 08$	$3.6442e + 11$	$2.1720e + 18$
34	21	23	$1.0106e + 09$	$7.0741e + 11$	$1.9733e + 18$
35	22	23	$1.9634e + 09$	$1.3744e + 12$	$2.8893e + 18$
36	22	23	$3.8177e + 09$	$2.6724e + 12$	$3.6044e + 18$

1. *Large irreducible factor method.* Motivated by the "large prime" variations of the quadratic sieve and earlier integer factorization algorithms, a "large irreducible factor" variation of the index-calculus algorithms for computing discrete logarithms follows. The basic idea is that when testing a polynomial $w_i(x)$ for smoothness with respect to a degree-bound m, if it fails smoothness due to a single irreducible factor

TABLE 3.   ESTIMATES FOR SMOOTHNESS TESTING ($n = 850$)

m	d_1	d_2	$b = 2^{(m+1)}/m$	bn	Operations
			$\deg f_1(x) = 10,$	$2^k = 4,$	$h = 213$
21	30	33	$1.9973e + 05$	$1.6977e + 08$	$9.2274e + 23$
22	30	31	$3.8130e + 05$	$3.2411e + 08$	$2.6795e + 23$
23	28	30	$7.2944e + 05$	$6.2003e + 08$	$3.5360e + 22$
24	27	29	$1.3981e + 06$	$1.1884e + 09$	$9.3776e + 21$
25	27	28	$2.6844e + 06$	$2.2817e + 09$	$4.3878e + 21$
26	27	27	$5.1622e + 06$	$4.3879e + 09$	$2.3381e + 21$
27	25	27	$9.9421e + 06$	$8.4507e + 09$	$6.7622e + 20$
28	26	26	$1.9174e + 07$	$1.6298e + 10$	$6.2617e + 20$
29	25	26	$3.7026e + 07$	$3.1472e + 10$	$3.2172e + 20$
30	24	26	$7.1583e + 07$	$6.0845e + 10$	$1.8677e + 20$
31	24	26	$1.3855e + 08$	$1.1777e + 11$	$1.5459e + 20$
32	25	25	$2.6844e + 08$	$2.2817e + 11$	$1.8362e + 20$
33	25	25	$5.2060e + 08$	$4.4251e + 11$	$1.6989e + 20$
34	25	25	$1.0106e + 09$	$8.5899e + 11$	$1.6618e + 20$
35	25	25	$1.9634e + 09$	$1.6689e + 12$	$1.7090e + 20$
36	25	25	$3.8177e + 09$	$3.2451e + 12$	$1.8390e + 20$
37	25	25	$7.4291e + 09$	$6.3148e + 12$	$2.0613e + 20$
38	25	25	$1.4467e + 10$	$1.2297e + 13$	$2.3977e + 20$

whose degree exceeds m but is at most $m + d$ (for a suitably small integer d), then retain the equation resulting from this polynomial where it otherwise would have been discarded. If, over the entire equation collection phase, $t \geq 2$ such equations result from the same large irreducible factor, then these can be combined to yield $t - 1$ "ordinary" equations (involving only factors whose degrees are less than the original bound m). This technique may lead to improvements in running time for equation collection by a factor of at most 3, for fields of cryptographic interest.

2. *Early abort strategy.* This technique is similarly motivated by an analogous technique used in integer factorization—the continued fractions algorithm. The reasoning here is that most of the polynomial pairs tested for smoothness will fail; if failure can be detected at an earlier stage, then running time will be saved. The method relies on the fact that a polynomial is unlikely to be smooth unless it has some smaller-degree factors. The idea is to abort the smoothness test (guessing it is destined to fail) once it is found that the sum of the degrees of irreducible factors of degree at most k, in the polynomial $w_i(x)$ being tested, is less than R, for appropriately chosen bounds k and R. The abort decision process can be efficiently combined with the smoothness testing technique of Section 2.4. This technique may lead to improvements in running time of the equation collection by a factor of about 2, for fields of cryptographic interest.

3 INTEGER FACTORIZATION

For the purpose of comparative analysis, the method we focus on for integer factorization is the quadratic sieve [43] and its variations. This is a logical choice since the quadratic sieve is the most efficient general purpose (i.e., applicable to composite integers of no special form) factoring algorithm in 1990. Other algorithms, such as variations of Lenstra's elliptic curve technique [30] and the number field sieve [27], have factored numbers of special form much larger than can be factored at present using the quadratic sieve, but such numbers can be easily avoided in cryptographic applications. Hence for our purpose in comparing the relative cryptographic security of schemes based on the difficulty of factoring to schemes based on the difficulty of extracting discrete logarithms, the comparison should be based on the quadratic sieve. General descriptions of the algorithm are readily available in the literature, and much practical experience has been documented (e.g., [11,14,28,29,44,51]). We include here an outline of the basic ideas for convenience, followed by a practical analysis of running time.

3.1 Outline of the Quadratic Sieve Algorithm

Let N be the composite integer to be factored. Define $FB = \{p_1, p_2, \ldots, p_b\}$, the *factor base*, to consist of $p_1 = -1$, $p_2 = 2$, and all odd primes $p_i < B$ (where B is a suitably chosen bound) for which N is a quadratic residue modulo p_i. Consider pairs of integers a_i, r_i such that $a_i^2 \equiv r_i \pmod{N}$, with $-N/2 \leq r_i \leq N/2$, where r_i factors over the factor base, that is,

$$r_i = \prod_{j=1}^{b} p_j^{\alpha_{ij}}, \qquad \alpha_{ij} \geq 0$$

We say that such integers are *smooth* (with respect to p_b). A sufficient number of such pairs (a_i, r_i) are found, such that there exists a product of a subset of the r_i's that is a perfect square. To find such a product, since only the parity of the exponents α_{ij} is of import, associate with each r_i a binary vector

$$v_i = (v_{i1}, v_{i2}, \ldots, v_{ib}) \quad \text{where} \quad v_{ij} \equiv \alpha_{ij} \pmod{2}$$

After at most $b + 1$ (a_i, r_i) pairs are found, a linear dependency will exist among the corresponding b-tuples v_i, in which case there exists a subset S of indexes such that $\sum_{s \in S} v_i \equiv (0, 0, \ldots, 0)$, implying $\prod_{i \in S} r_i$ is a perfect square (call it x^2). Then letting $y^2 = \prod_{i \in S} a_i^2$ yields the relation $x^2 \equiv y^2 \pmod{N}$. Now if $x \not\equiv \pm y \pmod{N}$, which would be expected at least half the time for composite N and randomly chosen such x, y, then $GCD(x + y, N)$ is a proper divisor of N.

To obtain such (a_i, r_i) pairs with r_i smooth, the quadratic sieve algorithm generates r_i candidates as follows. Carefully choose a quadratic polynomial $r(x)$, such that for $x = i$ in a suitable range, say $-M \leq i \leq M$, $r(i)$ is small in absolute value. For example, for the (ordinary) quadratic sieve as described in [43], Pomerance originally suggested $r(x) = ([\sqrt{N}] + x)^2 - N$. (Note that the only odd primes p that the factor base need contain are those for which N is a quadratic residue mod p, for if p divides $r(i)$, then $([\sqrt{N}] + x)^2 \equiv N \pmod{p}$.) Consider an array $R[]$ of $2M + 1$ cells, with the cell indexed by i corresponding to the value $r_i = r(i)$. To check whether r_i is smooth, proceed as follows. Solve the quadratic congruence

$$r(x) \equiv 0 \pmod{q}, \quad \text{for each prime power} \quad q = p^\alpha < N, p \in FB \quad (3.1)$$

There are two solutions x_j for each odd p in the factor base; the case $p = 2$ requires special consideration (e.g., see [14]). Since $r(x + kq) \equiv r(x) \pmod{q}$, given one solution x_j to Eq. (3.1), $(x_j + kq)$ is also a solution for all k, implying that every qth cell is divisible by q. To efficiently record this, initialize all cells in $R[]$ to 0, and add $\log(p)$ to cell $R[i]$ whenever it is found that $q = p^\alpha$ divides $r(i)$. For each solution x_j to Eq. (3.1), this is done by "sieving" the array $R[]$—modifying every qth cell. Once this process (via Eq. (3.1)) has been followed for all primes and prime powers, those cells $R[i]$ with value near $\log|r_i|$ are likely smooth. This is verified by factoring the corresponding r_i explicitly.

3.2 Practical Analysis of the Quadratic Sieve

In our analysis, we count what we refer to as *simple operations*—a wordsize (16- or 32-bit) arithmetic operation, equivalent to an add, shift, or move, on a single-processor machine. (In actual implementations, sieving operations may in fact be byte operations; this saves memory, but not time.) We emphasize that the analysis below does not attempt to establish the actual real time required to carry out the algorithm, but rather serves to derive an estimated operation count to which a similar analysis of discrete logarithm algorithms can be compared.

In practice, the condition $q = p^\alpha < N$ in Eq. (3.1) is relaxed, for example, to $q = p^\alpha < B$. It is not expected that this results in the loss of many smooth residues [45, Section 3]. Since for each solution $x_j, j = 1, 2$ to Eq. (3.1) the sieve processes the interval using stride q, we are interested in the sum

$$S = \sum_{q=p^a<B;\ p\in FB} \frac{1}{q} \tag{3.2}$$

(We shall use S in the analysis which follows.) S can be approximated for our purposes by

$$S_1 = \sum_{p<B;\ p\in FB} \frac{1}{p} \tag{3.3}$$

It is well known [19, Section 22.7] that $\sum_{p\le B}p^{-1} = \ln\ln(B) + S_2 + o(1)$, where $S_2 = \gamma + \sum_p\{\ln(1 - p^{-1}) + p^{-1}\}$, $\gamma = 0.57721566 \ldots$; for our purposes $S_2 \approx 0.26$ more than suffices for this convergent sum. Since about half the primes less than B are in the factor base, we use $S_1 \approx (1/2)\ln\ln(B)$.

A disadvantage of using a single polynomial $r(x)$ is that the magnitude of $r_i = r(i)$ increases linearly with i, and this is undesirable since the probability that a number is smooth decreases as its size increases. For the *multiple polynomial* version of the quadratic sieve, many distinct polynomials $r(x)$ are constructed and employed, allowing a smaller interval size M to be used on each of these. The penalty paid is the cost of determining the new polynomial, the solution of the congruences (Eq. (3.1)) for each new polynomial, and the reinitialization of the sieve interval. In practice the payoff more than compensates for these costs; furthermore, multiple polynomials facilitate parallelization.

While other factors bear on the asymptotic running time of the algorithm (notably finding the linear dependency), it appears in practice that the sieving step dominates. Caron and Silverman [11, Section 3] report that sieving takes 80–85% of the running time on a machine with 32 × 32-bit arithmetic, but the time required for changing polynomials dominates on machines with only 16 × 16-bit arithmetic. Pomerance, Smith, and Tuler [45, Section 5] suggest the algorithm parameters be adjusted so that the time for sieving matches the time for preparing a polynomial and sieve initialization data. Once in possession of a sufficient number of smooth residues, finding a linear dependency/postprocessing does not appear to be a time bottleneck in practice ([11, Section 1], [45, Section 5], and most recently [24]—see Section 2.5). Again, while equation collection can be done using a loosely coupled network of workstations, use of an appropriately large machine is necessary to effectively handle the resulting large sparse linear system. In light of these observations, we shall take as a measure of the running time of the quadratic sieve algorithm the time required for the sieving process. This appears reasonable provided an appropriately large interval size M is selected (since the cost of changing polynomials is reduced by staying with each polynomial for a longer period of time), however, this assumption requires further examination, which we will not pursue here. If the cost of changing polynomials is as great as sieving itself, the operation count for the algorithm will be two or three times that of the sieving stage, depending on the effort required to deal with the linear system. Regarding algorithmic parameters, asymptotic estimates for the optimal size of the factor base, as a function of N, can be established; in practice, B and M are best determined and refined by experimentation, and are strongly influenced by available memory resources (e.g., see [28,51]).

In our analysis here, we consider the version of the quadratic sieve described in [45, Section 8], using multiple polynomials as suggested by Peter Montgomery [51], and multiplier 1. (The use of an appropriate "multiplier" (e.g., see [45, Section 4], [38, Sections 4.5, and 5.3]) can be used to skew the factor base in favor of more small primes, at the cost of residues of somewhat larger magnitude.) The residues then have maximum size $M\sqrt{N/2}$, and for the purpose of estimating the probability of smoothness, it appears reasonable to assume "typical" residues obtained have magnitude $Z = (1/3)M\sqrt{N/2}$. Let $r(u)$ be the probability that a residue is smooth with respect to B (notation to be justified shortly). We require an estimate for $r(u)$. Let $\Psi(x, y)$ denote the number of positive integers $\leq x$ that are free of prime divisors $> y$. $\Psi(Z, Z^{1/u})/Z$ is essentially the probability we require, for $B = Z^{1/u}$. This function has received much attention in the literature. It is well known that for fixed $u \geq 1$, $\lim_{x\to\infty}\Psi(x, x^{1/u})/x = \rho(u)$, where $\rho(u)$, is *Dickman's function* (see [40, Section 3] for a comprehensive survey on $\Psi(x, y)$). Knuth and Trabb Pardo [20] provide an error bound on the estimation of $\Psi(x, x^{1/u})/x$ by $\rho(u)$, and tabulate Dickman's function for discrete values in the range $1 \leq u \leq 10$. The bound on this error term is somewhat larger than desirable for larger values of u, but is apparently difficult to tighten. Canfield, Erdös, and Pomerance [9, Corollary to Theorem 3.1] have established that if $\varepsilon > 0$ is arbitrary and $3 \leq u \leq (1 - \varepsilon)\ln x/\ln\ln(x)$, then

$$\Psi(x, x^{1/u})/x = e^{-u(\ln u + (\ln\ln u - 1)(1 + 1/\ln u) + E(x,\, u))}, \text{ where } |E(x, u)| \leq c_\varepsilon \left(\frac{\ln\ln u}{\ln u}\right)^2$$

and the constant c_ε depends only on ε. The tabulations of Dickman's function have been used by others to estimate smoothness probability (e.g., [45, 54]). We use as our approximation simply $r(u) \approx e^{-u\ln u}$, which appears to be as good an estimate as any currently available, agrees reasonably well with Dickman's function in the range $3 \leq u \leq 10$, and is a crude approximation to the result of Canfield et al. for $u > 10$. To increase the accuracy of our estimates, we would like to obtain better estimates to $r(u)$ for discrete values of x and u, but these are unavailable at present. Bach [1, Section 4] also comments on this situation, and offers suggestions.

For fixed N and a chosen value M, selecting $u > 1$ determines $B = Z^{1/u}$, and then we have $b \approx B/2\ln B$ from the prime number theorem and the fact that about half the primes less than B are in the factor base. (Better approximations for $\pi(x)$ are available, for example the Chebychev formula $\pi(x) \approx \int_2^x dx/\ln x$, but are unnecessary for our purposes.) Note that Z, and thus B and b, depend on M here.

We now estimate the number of simple operations required for the sieving phase. We do not factor in the time required to access data structures, etc. For each solution x_j, $j = 1, 2$ to Eq. (3.1) the sieve processes the interval using stride q. We charge three operations for the process of updating a cell each time it is "hit" during the sieving (as this typically will involve a read-add-write). Fix M and let K be the number of intervals of size $2M$ (= the number of polynomials) expected to be required to acquire b smooth residues. The number of operations is then roughly $12KMS$, where S is given by Eq. (3.2). Now to obtain b smooth residues, approximately $b/r(u)$ cells need be processed (meaning $K \approx r(u)^{-1} b/2M$). From this and Eqs. (3.2)–(3.3), the number of simple machine operations for sieving is estimated to be

$$3\ln\ln B \cdot b \cdot r(u)^{-1} \tag{3.4}$$

Recall that b here is the number of primes in the factor base, and hence the approximate number of relations required in the linear system; from this, an estimate of the memory requirements can be derived. Lenstra and Manasse [29] have used a factor base of 50,000 primes for typical 100-digit numbers; they have also employed a factor base of 65,500 elements to factor a 107-digit integer, using the "two-large-primes" variation.

Numeric determinations of optimal parameters for a fixed-size problem have been published by Wunderlich [54] for the continued fractions algorithm, but as noted by Morrison and Brillhart already in [38], optimizing parameters seems to be mainly a matter of experience. The situation has not changed (Silverman makes a similar comment [51]) and in fact appears more difficult for the multiple polynomial quadratic sieve, as the latter involves a greater number of parameters. Tables 4 and 5 illustrate sample parameters and estimated sieving operation counts (Eq. (3.4)) for the quadratic sieve algorithm as outlined above, for implementations for 100- and 155-digit composite integers N. Although crude, and based on a hypothetical model, they give a rough indication as to what parameter choices are possible, and their effects. From the tables, the sizes of factor bases that minimize the running time of the sieving stage, and/or are practical, can be estimated.

We note that the figures under the heading "Operations" in Tables 4 and 5 are estimates for the sieving phase only. For the larger values of b in these tables, the running time of the sieving phase will be dominated by the time taken (which is $O(b^2)$) to deal with the larger sparse linear systems; in these cases the figures under the "operations" heading are not attainable for the algorithm as a whole, and hence would not be used. Practical considerations (e.g., memory constraints, machine addressing limitations, etc.) also rule out the use of unreasonably large factor bases.

Speedups to quadratic sieve factorization. Other techniques are known for speeding up the basic version of the multiple polynomial quadratic sieve. As in the case for other similar integer factorization algorithms, and in the computation of discrete

TABLE 4.   ESTIMATES FOR SIEVING PHASE $N \approx 10^{100}$ ($M = 10^7$)

u	b	B	Estimated $r(u)$	K	Operations
10.00	1.6708e + 04	4.3375e + 05	1.00000e − 10	8.3540e + 06	1.2849e + 15
9.75	2.2723e + 04	6.0504e + 05	2.27617e − 10	4.9916e + 06	7.7532e + 14
9.50	3.1430e + 04	8.5888e + 05	5.14785e − 10	3.0527e + 06	4.7892e + 14
9.25	4.4273e + 04	1.2425e + 06	1.15662e − 09	1.9139e + 06	3.0332e + 14
9.00	6.3610e + 04	1.8348e + 06	2.58117e − 09	1.2322e + 06	1.9731e + 14
8.75	9.3380e + 04	2.7705e + 06	5.72044e − 09	8.1619e + 05	1.3207e + 14
8.50	1.4033e + 05	4.2859e + 06	1.25875e − 08	5.5742e + 05	9.1169e + 13
8.25	2.1635e + 05	6.8079e + 06	2.74951e − 08	3.9343e + 05	6.5053e + 13
8.00	3.4302e + 05	1.1131e + 07	5.96046e − 08	2.8775e + 05	4.8110e + 13
7.75	5.6085e + 05	1.8787e + 07	1.28207e − 07	2.1873e + 05	3.6986e + 13
7.50	9.4856e + 05	3.2833e + 07	2.73552e − 07	1.7338e + 05	2.9659e + 13
7.25	1.6654e + 06	5.9635e + 07	5.78828e − 07	1.4386e + 05	2.4903e + 13
7.00	3.0478e + 06	1.1303e + 08	1.21427e − 06	1.2550e + 05	2.1988e + 13
6.75	5.8405e + 06	2.2463e + 08	2.52464e − 06	1.1567e + 05	2.0519e + 13
6.50	1.1783e + 07	4.7062e + 08	5.20072e − 06	1.1329e + 05	2.0352e + 13
6.25	2.5185e + 07	1.0461e + 09	1.06108e − 05	1.1868e + 05	2.1600e + 13
6.00	5.7442e + 07	2.4854e + 09	2.14335e − 05	1.3400e + 05	2.4717e + 13
5.75	1.4101e + 08	6.3663e + 09	4.28460e − 05	1.6455e + 05	3.0773e + 13
5.50	3.7633e + 08	1.7763e + 10	8.47238e − 05	2.2209e + 05	4.2126e + 13

TABLE 5.   ESTIMATES FOR SIEVING PHASE $N \approx 10^{155}$ ($M = 10^9$)

u	b	B	Estimated $r(u)$	K	Operations
12.25	$3.1698e + 05$	$1.0233e + 07$	$4.68094e - 14$	$3.3858e + 09$	$5.6503e + 19$
12.00	$4.3462e + 05$	$1.4323e + 07$	$1.12157e - 13$	$1.9376e + 09$	$3.2574e + 19$
11.75	$6.0426e + 05$	$2.0337e + 07$	$2.67335e - 13$	$1.1302e + 09$	$1.9143e + 19$
11.50	$8.5263e + 05$	$2.9320e + 07$	$6.33833e - 13$	$6.7260e + 08$	$1.1479e + 19$
11.25	$1.2222e + 06$	$4.2963e + 07$	$1.49463e - 12$	$4.0887e + 08$	$7.0323e + 18$
11.00	$1.7819e + 06$	$6.4059e + 07$	$3.50494e - 12$	$2.5419e + 08$	$4.4062e + 18$
10.75	$2.6451e + 06$	$9.7303e + 07$	$8.17258e - 12$	$1.6183e + 08$	$2.8274e + 18$
10.50	$4.0032e + 06$	$1.5077e + 08$	$1.89458e - 11$	$1.0565e + 08$	$1.8608e + 18$
10.25	$6.1861e + 06$	$2.3867e + 08$	$4.36597e - 11$	$7.0844e + 07$	$1.2580e + 18$
10.00	$9.7755e + 06$	$3.8658e + 08$	$1.00000e - 10$	$4.8877e + 07$	$8.7519e + 17$
9.50	$2.6292e + 07$	$1.0945e + 09$	$5.14785e - 10$	$2.5537e + 07$	$4.6512e + 17$
9.00	$7.9162e + 07$	$3.4784e + 09$	$2.58117e - 09$	$1.5335e + 07$	$2.8427e + 17$
8.50	$2.7224e + 08$	$1.2666e + 10$	$1.25875e - 08$	$1.0814e + 07$	$2.0418e + 17$
8.00	$1.0966e + 09$	$5.4206e + 10$	$5.96046e - 08$	$9.1987e + 06$	$1.7703e + 17$
7.50	$5.3409e + 09$	$2.8161e + 11$	$2.73552e - 07$	$9.7620e + 06$	$1.9165e + 17$
7.00	$3.2770e + 10$	$1.8513e + 12$	$1.21427e - 06$	$1.3494e + 07$	$2.7050e + 17$

logarithms, savings in running time (at the expense of memory) are possible due to a "large prime variation." A speedup by a factor of about 2.5 appears possible (see [28], Fig. 1). A "two-large-prime" variation appears to give a further speedup by about the same factor, at the expense of a denser linear system—making its solution more time-consuming, and significantly more memory for the storage of relations [29]. A "small prime variation" may slightly reduce the time spent sieving, perhaps by 20% [45]. Other variations, including use of a small multiplier (as briefly mentioned earlier), and use of different classes of polynomials, may offer further advantages.

3.3 Recent Advances

Very recently, dramatic theoretical progress has been made on the problem of integer factorization; for integers of special form, this has resulted in a very efficient factorization algorithm. The algorithm, originated by Pollard and refined by H. W. Lenstra, has been implemented by A. K. Lenstra and Manasse [27]. It makes use of algebraic number fields, and is referred to as the (special) *number field sieve*. It applies to numbers of the form $N = r^e \pm s$ for small integers r and s (and in fact, to a somewhat larger class of "special" numbers), and runs in heuristic expected time $\exp((c + o(1))(\ln N)^{1/3}(\ln\ln N)^{2/3})$, where $c \approx 1.526$. In mid-June 1990, the factorization of a special-form 155-digit number (the ninth Fermat number, $2^{512} + 1$) was completed using this algorithm. As in previous factorizations (see [28]), this factorization employed hundreds of workstations communicating by electronic mail, and involved a sparse linear system of over 200,000 equations. The theory has been extended to the factorization of general integers, with a larger running time constant of $c = 3^{2/3} \approx 2.08$ (see further comments below). While asymptotically significantly faster than all previous general factoring algorithms, it does not currently appear that this (general) number field sieve is practical for integers of cryptographic interest; for numbers within the reach of present computational power, the multiple polynomial quadratic sieve reigns. In summary, this work in its present state does not impact the cryptographic security of RSA, although it now appears that integer factorization is asymptotically an easier problem than previously believed.

Of related theoretical interest is the work of Gordon [18], which argues that the ideas of number field sieve factorization can be applied to the computation of discrete logarithms in odd prime fields GF(p), resulting in a heuristic expected asymptotic running time for the latter problem which is of the same form as that of the former. To record asymptotic running times succinctly, we define

$$L_x[v, c] = e^{(c+o(1))x^v(\ln x)^{1-v}}$$

The heuristic expected running times for the best algorithms for the problems of interest can then be summarized as follows:

Factoring an interger N—(general) number field sieve: $L_{\ln N}[1/3, 2.08]$

Discrete logarithms in GF(p)—via number field sieve: $L_{\ln p}[1/3, 2.08]$

Discrete logarithms in GF(2^n)—Coppersmith algorithm:

$$L_n[1/3, c], \qquad 1.3507 \le c \le 1.4047$$

Note that the critical parameter v is $1/3$ in all cases, although only the latter algorithm—for extracting discrete logarithms in GF(2^n)—is of practical use at the current time.

The constant $c = 2.08$ for the (general) number field sieve noted above has recently been improved to $c = 1.923$, by work due to J. Buhler, H. W. Lenstra, and C. Pomerance, with contributions from Len Adleman, and further reduced to $c = 1.902$ by Coppersmith [12a]. The theoretical crossover for the quadratic sieve and the (general) number field sieve is then in the general area of the often-suggested 512-bit moduli for RSA implementations. This is, however, presently only of theoretical interest since neither algorithm is currently capable of factoring integers this large. Also of theoretical interest, following a massive precomputation on the order of $L_{\ln N}[1/3, 2.01]$, Coppersmith's work allows number field sieve factorization with a per-factorization running time of $L_{\ln N}[1/3, 1.639]$.

Of great interest is the degree to which theoretical factorization advances have actually been of use in practice. M. Manasse and A. Lenstra have carried out extensive implementations over the past few years. As of mid-summer 1990, the state of the art in practice is as follows [31a]. Using only the idle time on a network of 200 loosely coupled engineering workstations over one month, any of the following tasks is currently feasible:

1. (Special) number field sieve factorization of 155-digit Cunningham numbers (of form $r^e \pm s$; r, s small)
2. (Two-large-prime) multiple polynomial quadratic sieve factorization of 116-digit general integers
3. Elliptic curve factorization to extract up to 40-digit factors from up to 200-digit general integers

We emphasize that, similar to the (special) number field sieve discussed above, the remarkable power of elliptic curve factorization is limited to a special class of integers—in this case, integers containing moderately large prime factors. Neither of these algorithms apply directly to RSA moduli, which should be carefully chosen in light of such known techniques.

4 COMPARING EL GAMAL IN GF(2^n) VERSUS RSA

In this section, we first briefly review the RSA and El Gamal cryptosystems, and then consider their relative security based on the discussions in Sections 2 and 3, above. We then compare other aspects of the cryptosystems.

4.1 Review of RSA and El Gamal Cryptosystems

In the RSA cryptosystem [46], each user has three integer parameters e, d, and n, where $n = pq$ with p and q large, suitably chosen primes, and the pair e, d satisfying the relation $ed \equiv 1 \pmod{\phi(n)}$, ϕ being the Euler totient function. Suppose we have two users, Alice and Bob. Alice has a public key consisting of the pair (e_A, n_A), and private key d_A; likewise, Bob has (e_B, n_B) and d_B. If Alice wishes to send the encrypted message M to Bob, she encrypts it using Bob's public key, computing $C = M^{e_B} \pmod{n_B}$. If Alice wishes to send the signed message M to Bob, she signs it using her own private key, computing $C = M^{d_A} \pmod{n_A}$. Encryption and signing of messages can be composed. By considering M to be a key K, it is clear that the system can be used for key exchange.

To describe the El Gamal system [17] in the field GF(q), where q is a prime or prime power (with $q = 2^n$ of interest here), let α be a primitive element in the field. Alice now has a private key a and a public key α^{-a}; likewise Bob has b and α^{-b}. To send the message M to Bob encrypted, Alice selects a random integer r, and sends the pair $(\alpha^r, \alpha^{-br}m)$. Here m is the field element with integer representation M. Bob computes $(\alpha^r)^b \cdot \alpha^{-br}m$ to recover m. If M is a key K chosen by Alice, then this serves as a key exchange mechanism. (This is not an *authenticated* key exchange, since an imposter Charlie could masquerade as Alice to Bob, but the protocol can be modified to prevent this.) To send a signed message M to Bob, Alice again selects a random integer r and computes α^r. Let R be an integer representation of α^r, with $0 \leq R \leq q - 1$. Alice also computes $s = (M + aR)r^{-1} \pmod{q - 1}$, and sends the message M along with the signature (α^r, s). Bob verifies the signature by computing $(\alpha^{-a})^R(\alpha^r)^s$, and checking this is equal to α^M. For security reasons (see [17]), a new value r must be chosen for each signature. To avoid the cost of computing r^{-1} each time, one might alter the protocol slightly as follows: Rather than as above, have Alice compute $s = -(M - rR)a^{-1} \pmod{q - 1}$, and as verification have Bob check that $(\alpha^r)^R(\alpha^{-a})^s$ is equal to α^M. Recently, a new signature protocol has been proposed by Schnorr [48], resulting in a shorter signature which is more efficient to construct and verify.

Since their inception, the difficulty with using most public key systems for encryption has been their throughput—until now, available implementations have operated at bit rates too low to accommodate even standard 64K bits/sec speech, let alone megabit and higher data rates. While throughputs are increasing (see Section 4.3), and elliptic curve systems offer hope (see Section 5), symmetric systems still apparently dominate for the encryption function, with public key systems of primary use for key exchange and signatures and authentication.

4.2 Security Comparison: Discrete Logs in GF(2^n) Versus Integer Factorization

In this section, we compare the running times for fixed sizes of integer factorization and the computation of discrete logarithms in GF(2^n). For our model of comparison, we refer back to Section 1. We focus on a relative comparison based on the present level of security offered by 512-bit integer factorization.

Asymptotic analysis for the quadratic sieve factorization of an integer N [43] and the Coppersmith discrete logarithm algorithm in GF(2^n) [41] gives respective (heuristic expected) asymptotic running times proportional to

$$e^{\sqrt{\ln N \ln \ln N}} \quad \text{and} \quad e^{c \cdot n^{1/3}(\ln n)^{2/3}}, \quad c \approx 1.35 \tag{4.1}$$

(For the Coppersmith algorithm, the asymptotic running time actually exhibits periodic oscillations with $1.3507 \leq c \leq 1.4047$ [41]; being cryptographically conservative we use 1.35.) Without additional knowledge, simply plugging numbers into guideline functions such as these and expecting to obtain accurate absolute running times is unreasonable; trying to compare algorithms for different problems based on the use of two such unrelated expressions is most unreasonable. Such formulas do not take into account running time constants, which might vary wildly for the different algorithms, nor do they reflect the difficulty of one "operation." However, before consulting the more careful analysis of previous sections, out of interest we pause to check what these guideline functions would suggest if interpreted literally. For 512-bit numbers N (Eq. (4.1)) gives a guideline for the difficulty of factoring as 6.69×10^{19} "operations" (where a factoring operation here is not clearly defined). To match this using the expression in Eq. (4.1) for discrete logarithms literally, requires $n \approx 850$. Similarly, 332-bit integer factorization in Eq. (4.1) yields 2.34×10^{15} "operations," which is matched by $n \approx 475$ for discrete logarithms in Eq. (4.1).

We now resort to the discrete analysis carried out in Sections 2 and 3, from which a more reliable comparison is expected since there the algorithms are analyzed on a more equal footing. Recall that the analysis in those sections estimated operation counts for the smoothness testing and sieving stages only, respectively; to facilitate comparison, we presume that these are the overall running times of the algorithms. Table 5 is not far off the value of Eq. (4.1) for the difficulty of factoring, with the analysis indicating that the sieving process as analyzed in our model could be carried out perhaps an order of magnitude faster than by Eq. (4.1) taken literally. However, this depends on the choice of database size used; note that dealing with a linear system involving over a million unknowns would appear to be formidable. On the other hand, Tables 2 and 3 suggest that Eq. (4.1) taken literally underestimates the difficulty of taking logarithms within the 700- to 850-bit range, as compared to our model. Tables 3 and 5 together suggest that n in the range of 700+ bits would make the smoothness testing in the computation of discrete logs about as difficult as the sieving phase in quadratic sieve factoring of 512 bits. This analysis suggests that it is reasonable to estimate that the GF(2^n) discrete logarithm problem of equivalent difficulty to 512-bit integer factorization (with respect to running time) is somewhere in the range of 700–800 bits. Similarly, Tables 1 and 4 indicate that 400-bit discrete logarithms appear to be of approximately the same difficulty as 100-digit factorization. Thus the analysis in Sections 2 and 3 suggests that for equivalent levels of security to integer factorization at

100 and 155 digits, the discrete logarithm problem in GF(2^n) requires somewhat smaller bitlengths n than those that result from aligning the expressions of Eq. (4.1) literally. Table 6 summarizes this discussion.

TABLE 6.   ESTIMATED BITLENGTHS FOR EQUIVALENT SECURITY

Factoring	Discrete Logs (Characteristic 2)
332	400 +
512	700 +

The similarities between integer factorization and discrete logarithm algorithms have been discussed by many. For example, the sieving of residues is analogous to the smoothness testing in Coppersmith's algorithm; in both cases, the objective is to find a sufficient number of linear relations, and in both cases, large sparse linear systems must be dealt with. We now consider and contrast a number of items regarding these algorithms.

Largest problem instances to date. The current champion among general integers factored by the quadratic sieve algorithm is 116 digits ($\approx$ 386 bits), using the two-large-prime method [8a, 29]. The largest field for which the Coppersmith algorithm has been applied for extracting discrete logarithms is GF(2^{127})—that is, 127 bits. While discrete logarithm problems in far larger fields GF(2^n) are tractable with this algorithm and current technology, computational number theorists have apparently not found it as fashionable a problem to pursue, one possible reason being the absence of a motivational analogue of the Cunningham tables [8], which have stimulated much of the work in integer factorization. However, with renewed interest in fields of characteristic 2, it is anticipated that researchers will begin to explore larger-scale implementations of the Coppersmith algorithm. For the harder problem of discrete logarithms in prime fields, a larger problem instance has been solved—a 192-bit field by LaMacchia and Odlyzko, as noted earlier [25]. This problem, which was motivated by a cryptographic scheme in actual use by a major computer corporation, provided among other things, experience solving a system of 300,000 equations in 100,000 unknowns.

Modifications to "basic" algorithms. The running time of the quadratic sieve algorithm can be improved by a factor of about 2.5 by the large prime variation, and another factor of about 2.5 by the two-large-primes variation, as noted in Section 3.2. The running time of the Coppersmith algorithm (modified as described in Section 2.3) for computing logs in GF(2^n) can be improved by a factor of up to 3 by the large irreducible factor method, and further by a factor of up to 2 by early abort strategies, as noted in Section 2.6. As these factors would appear to balance out with respect to running time, they are not factored into our analysis, as our objective is a relative comparison. In addition, for very large problem instances, it is unclear which, if any, of the large prime variations would be used, as they result in somewhat denser linear systems, and substantially increased memory requirements.

Attention given to problems. Both the problem of integer factorization and that of extracting discrete logarithms have been well-studied, the former having received more attention in both theory and in practice. In light of this, it might be suggested that current algorithms for computing discrete logarithms—particularly in GF(2^n)—may afford more room for improvement than those for integer factorization. While this might be true with respect to implementation (as opposed to asymptotic) speedups, it is interesting to note that the number field sieve for factoring [27] evolved from the Gaussian integer method proposed for computing discrete logarithms in GF(p) [13], and in turn led to an algorithm for computing discrete logarithms in GF(p) by Gordon [18]. Unfortunately, for neither the factorization nor the logarithm problem is there definitive evidence at hand that any of the currently known algorithms are even close to optimal. The inherent complexity of these problems appears most difficult to establish, and accordingly the optimality of the best current algorithms is most difficult to judge.

Difficulty of solving additional problem instances. In integer factorization and RSA, once the factors of the modulus N are determined, the secret RSA key corresponding to the public key can be found directly, and messages signed with this RSA key pair are then vulnerable with essentially no additional work. For discrete logarithm-based cryptosystems, once the stage 1 database is built, individual logarithms can be computed in a relatively short time (compared to the time required to compute the database), but the database itself, which is nontrivial in size for larger problem instances, must be retained and accessed during the computation of each individual logarithm.

Solution of the linear system. In the quadratic sieve algorithm, a linear dependency must be found in a binary linear system. This can be efficiently implemented by packing, for example, 32 coefficients in one word on a 32-bit machine, and making use of exclusive-or machine instructions. For the Coppersmith algorithm in GF(2^n), the linear system must actually be solved, modulo a prime (or several primes) on the order of n bits. The latter requires more time and space for systems with the same number of unknowns.

Large memory requirements arise in both algorithms in relation to the large sparse linear systems. Caron and Silverman [11] point out that in practice, most of the memory requirement arises from storing the large-prime factorizations in the large-prime variation of the quadratic sieve. The situation is worse in the two-large-prime variation [29]. Similar results would be expected for similar variations of the Coppersmith algorithm. In [41], Odlyzko notes that in the Coppersmith algorithm, for large problems the equation solution phase is limited more by memory than by running time. Thus for very large problems, memory becomes a real barrier affecting both the variation of an algorithm that is used, and the parameters chosen.

For the algorithms in question, note that running time and memory differ in one important respect. The running time is dominated by the sieving and smoothness testing stages in these algorithms, which can be performed using very large numbers of loosely coupled processors, as noted earlier; using more machines of the same size (perhaps through the help of additional anonymous friends with idle workstations) can compensate for increased running time demands of larger problems. However, for the linear algebra stage, one cannot similarly pool memory resources, since a loosely coupled system does not work for that stage—larger problems apparently require machines with larger memory capabilities.

4.3 El Gamal in GF(2^n) Versus RSA

Having examined the difficulty of the underlying number-theoretic problems on which the security of the cryptosystems in question is based, we now compare other aspects of the cryptosystems.

Exponentiation throughput. As modular exponentiation and discrete exponentiation in the field are required to implement the RSA and El Gamal cryptosystems, the throughput that is presently attainable for these operations is of interest. We consider this below.

Regarding modular exponentiation as used in RSA, a 1989 survey of hardware implementations is given by Brickell [7]. The RSA chips cited range up to 17K bits/sec for 512-bit exponentiation at 14 MHz. Recently, Shand et al., [50] have achieved a 226K bits/sec implementation of 508-bit RSA modular exponentiation. This timing is for RSA exponentiation using the Chinese remainder theorem (CRT). The technology used is *programmable active memory* (PAM) [4], and their system consists of a host processor driving three distinctly configured PAM boards, each PAM being on a 25×25 cm^2 printed circuit board; two of these are noted as 38-ns units. To achieve the cited throughput, two of the PAM boards are "programmed"—via an automated process taking about 30 min—with data customized for the particular RSA modulus being employed. This implementation gives a dramatic order of magnitude speedup over previous RSA implementations, although its nature makes comparison somewhat difficult, particularly in real-life cryptosystems where RSA moduli may change periodically. Other recent work includes [37].

Software implementations of modular exponentiation on digital signal processors rival custom-hardware speeds. An implementation on the Motorola DSP56000 by Michael Wiener of BNR achieves 13.4K bits/sec for 512-bit exponentiation using CRT, at a clock rate of 20.48 MHz; without the CRT, estimated throughput is 5.4K bits/sec. On the same processor, Dussé and Kaliski report software achieving 11.6K bits/sec for 512-bit exponentiation using CRT, and 4.6K bits/sec without CRT [16]. Shand, Bertin, and Vuillemin [50] report a 10.3K bits/sec software implementation of 512-bit RSA on a 25 MHz RISC-based workstation.

For exponentiation in GF(2^n) with $n = 593$, a single-chip 2-micron complementary metal-oxide-semiconductor (CMOS) implementation is available with average throughput of 150K bits/sec at 10 MHz; the chip has also been run at 20 MHz yielding 300K bits/sec [47]. These timings are for limited Hamming weight exponentiation, using exponents of maximum weight 30 to decrease the number of multiplications required in an exponentiation. The hardware implementation makes use of a so-called *optimal normal basis* [39] to reduce the complexity of multiplication in the field GF(2^{593}). It is projected that using 1-micron technology, a clock rate of 33 MHz, and a one-pass multiply (instead of a two-pass multiply), an average throughput of over 1M bits/sec, and perhaps 1.5M bits/sec, is achievable for such limited Hamming weight exponentiation.

Both modular exponentiation and discrete exponentiation in GF(2^n) can be implemented using hardware which performs n-bit multiplication (respectively, modular and in the field) in $O(n)$ clock cycles—in fact, in $n + c$ clocks for small constants c. Exponentiation in GF(2^n) can take advantage of the fact that, using a normal basis representation of fields elements, squaring an element is simply a circular shift of that element's coefficient vector, which cuts down the average number of multiplications in the standard "square-and-multiply" exponentiation technique from $3n/2$ to $n/2$; there is no analogy to this in modular exponentiation. To be fair, it should be noted that to facilitate the hardware implementation of normal basis multiplication, fields GF(2^n) should be sought which exhibit optimal normal bases (see [39] for further details). The more dramatic reason that throughputs mentioned for exponentiation in $GF(2^n)$ are high is due to the use of limited Hamming weight exponents. The use of this technique for encryption in El Gamal–like cryptosystems appears safe—provided the Hamming weight is sufficiently large to rule out exhaustive-search attacks—and significantly cuts down on the number of multiplications required in an exponentiation.

Of course, in evaluating the speed of cryptosystems, one must consider the "types" of exponentiation that are required for both encryption and the application of signatures, and correspondingly, for decryption and the checking of signatures. For example in RSA, encryption is typically carried out with a small public exponent—$2^{16} + 1$ has been suggested by many—but this weight-2 exponent is not used to determine the throughput of RSA, since decryption is more costly. The corresponding secret exponent is in general full-length, and of expected weight $n/2$ for an n-bit modulus. Fortunately, decryption time can be reduced here through use of the Chinese remainder theorem, as the holder of the secret key can exploit knowledge of the factors p and q of his modulus N. (We note the work of Wiener [53] warning against the use of short decrypting exponents.) Thus in evaluating throughput, one should take into account the types and number of exponentiations required, and the weights of the exponents involved, as dictated by each stage of the particular encryption and signature protocols used.

Message expansion. For RSA encryption, there is no message expansion; the encrypted message is the same size as the original. Similarly, signed RSA messages have no message expansion. For the El Gamal system as described in [17] and above, encryption results in message expansion by a factor of 2, and signing a message results in message expansion by a factor of 3. As noted earlier, more recently proposed signature schemes for El Gamal–like systems have smaller message expansion [48].

Choice of arithmetic systems. It has been suggested as an option by some that the duration of use of each RSA key pair be restricted by a fixed "cryptoperiod," after which time the key pair be changed. This results in a new RSA modulus N, moving future computations into a new arithmetic system, giving an opponent less incentive to devote considerable resources to attacking a particular modulus N and key pair. Of course, the system is either impervious to a factoring attack or it is not; if a particular modulus size is insecure, moving to a new modulus does not prevent an attacker from continuing to factor the previous modulus, but does limit the damage to the lifetime of that modulus. Altering the length of the cryptoperiod may also provide system administrators with some degree of flexibility for adjusting the security of the system. A change of RSA key pair and modulus is easily accommodated by most modular exponentiation implementations (i.e., by those that handle general n-bit moduli), and there are many n-bit RSA moduli N offering the same degree of security.

The situation differs for the El Gamal cryptosystem in $GF(2^n)$. Custom hardware is typically built to handle a particular representation of a field of a fixed dimension n. But more importantly, from the attacker's point of view, there is effectively only a single n-bit arithmetic system to attack for each value n, since all fields with 2^n elements are isomorphic and mappings between these fields are easily established. The lack of choice of arithmetic systems here might be viewed as a weakness relative to RSA, or El Gamal in $GF(q)$—for a fixed value n, a single database suffices for computing logarithms in $GF(2^n)$. On the other hand, if n is chosen sufficiently large to resist the most concerted discrete logarithm attack, then there is no need to move to a new system. If, for reasons of implementation efficiency, it is desired to use fields $GF(2^n)$ that admit optimal normal basis representations, then the choice of values n is also restricted (see [39]).

Use of extension fields of $GF(2^n)$. To increase the security of the El Gamal cryptosystem in $GF(2^n)$, larger fields are used. Given hardware for multiplication (exponentiation) in $GF(2^n)$, the quadratic extension $GF(2^{2n})$ is appealing because the hardware for the smaller field can be reused. The multiplicative group here has $2^{2n} - 1 = (2^n - 1)(2^n + 1)$ elements. To guard against a combined discrete logarithm attack using a generalized Pohlig–Hellman algorithm [42] on the cyclic group of $2^n + 1$ elements, and index-calculus on the subgroup of $2^n - 1$ elements, it should be ensured that $2^n + 1$ has a large prime factor. Although it has been suggested that the presence of the subfield $GF(2^n)$ in $GF(2^{2n})$ may make computation of discrete logarithms in $GF(2^{2n})$ easier than in a similar-sized field that does not contain a subfield, aside from the above-mentioned idea no algorithm has been reported that gives weight to this suggestion. Presuming there is no weakness introduced, other extension fields of $GF(2^n)$ are similarly available. Analogously, the security of RSA can be increased by resorting to moduli N of greater bitlength, through use of appropriate modular multiplication (exponentiation) hardware or software for the larger bitlength.

5 RECENT WORK REGARDING ELLIPTIC CURVE CRYPTOSYSTEMS

In this section we survey recent work related to elliptic curve cryptosystems, and consider the cryptographic impact of recent progress in the computation of discrete logarithms over elliptic curves.

We first briefly review some background. Given a finite field F, the points of an elliptic curve E over F form an abelian group. A well-known theorem of Hasse states that the number of points k on an elliptic curve E over the field $F = GF(q)$ is of the form $k = q + 1 - t$, where $|t| \leq 2\sqrt{q}$, that is, the order of the group is $k \approx q$. The Diffie–Hellman key exchange technique, and the related El Gamal cryptosystem, are based on exponentiation, which is simply the repeated application of a group operation (multiplication in the original case). Therefore the elliptic curve group may be used to construct cryptosystems, as outlined by Miller [36] and Koblitz [22]. In the case of the elliptic curve group, the group is additive, with the addition operation requiring a small number of operations in underlying field F; the additive analogue of exponentiation is multiplication, that is, repeated addition in the group.

In an arbitrary group G, the *general discrete logarithm problem* is as follows: Given a specified element $g \in G$ of (typically maximum) order r, and any other group element $b \in G$, determine the unique y, $0 \leq y \leq r - 1$, such that $b = g^y$, if such an integer exists. The *elliptic curve logarithm problem* then follows: Given a point P of (typically maximum) order r in the elliptic curve group, and another point R in the group, determine unique y, $0 \leq y \leq r - 1$, such that $R = yP$, if such an integer exists. The elliptic curve analogues of the Diffie–Hellman key exchange and the El Gamal cryptosystem are insecure if the elliptic curve logarithm problem is easy.

Until recently, the best discrete logarithm attacks known which applied to elliptic curves were general methods applicable in any group, having running time $O(\sqrt{k})$ in a group of order k (e.g., see [1,26,32]). To avoid vulnerability to a Pohlig–Hellman attack [42], elliptic curves should be chosen such that the order of the relevant group has a large prime divisor. The order of the group in question can be determined via a polynomial time, but not very practical, algorithm due to Schoof [49] (see [23] for a further discussion); alternatively, subclasses of elliptic curves are known whose group orders can be easily determined, thus avoiding the need to use Schoof's algorithm ([3, 34]). The above-mentioned "square root" attacks are very weak, and the apparent absence of stronger general attacks has resulted in the belief that elliptic curve cryptosystems with relatively short keylengths may afford greater security than alternative cryptosystems with larger keylengths. To reiterate the point with an example, suppose it *were* true that the best elliptic curve logarithm attack on a particular curve having an elliptic curve group of order about 2^{132} (i.e., the curve is over a field GF(2^n) with $n \approx 132$) *was* one of running time precisely $\sqrt{k}$ "operations"; then extracting elliptic logarithms would take about 7.4×10^{19} such operations. Shorter keylengths translate into simpler implementations of arithmetic, and smaller bandwidth and memory requirements. Among other applications, this significantly impacts the design and feasibility of smart card systems.

Practical implementation of elliptic curve cryptosystems, in particular the elliptic curve analogue of the El Gamal cryptosystem, has recently been studied by Menezes and Vanstone [34]. They have explored the feasibility of hardware implementation of an arithmetic processor for carrying out elliptic curve computations over fields of characteristic 2. The class of curves of so-called *zero* j-*invariant* have received special attention as they offer significant additional computational advantages in implementing the group operation. Furthermore, these are convenient in that the order of the elliptic curve groups of curves over GF(2^n) of zero j-invariant are known [33]: for n odd, the order is one of $q + 1$, $q + 1 \pm \sqrt{2q}$; for n even, the order is one of $q + 1$, $q + 1 \pm \sqrt{q}$, $q + 1 \pm 2\sqrt{q}$. Of interest in what follows, it is also known that there are precisely three isomorphism classes of elliptic curves with j-invariant 0 when n is odd, and precisely seven such classes for n even (see [33] for a representative from each class). Finally, the case when n is odd is preferred because a point on the curve, which is in general an ordered pair (x, y) of elements of GF(2^n), can in that case be represented by the x-component and a single bit of the y-component. This can be used to reduce message expansion for encryption in the elliptic El Gamal system to a factor of $3/2$ [34, Section 5].

More recently, Menezes, Okamoto, and Vanstone [35] have made the first significant progress on the elliptic curve logarithm problem. For certain classes of curves over fields GF(q), they show how to reduce the elliptic curve logarithm problem in a curve E over GF(q) to the discrete logarithm problem in an extension field GF(q^k) of GF(q). In general, k is exponentially large and the reduction takes exponential time. However, for *supersingular* elliptic curves, k is small ($k = 1, 2, 3, 4,$ or 6) and the reduction is probabilistic polynomial time, yielding a (probabilistic) subexponential-time elliptic curve logarithm algorithm. (If E is an elliptic curve over a field GF(q) where $q = p^m$, and the elliptic curve group has order $q + 1 - t$, then E is supersingular if p divides t. Hence for $p = 2$, the supersingular curves are precisely those curves with j-invariant equal to 0.) For curves over GF(2^n) of zero j-invariant, when n is odd the technique applies with $k = 2$ for one of the isomorphism classes of curves, and for $k = 4$ for the other two classes; when n is even the technique also applies to each of the seven classes of curves with $k = 3, 2,$ or 1. Hence among curves over GF(2^n) of zero j-invariant, the classes of curves recommended for cryptographic use are those for which k is exactly 4, and in this case computing elliptic curve logarithms is essentially no more difficult than computing discrete logarithms in the quartic extension $GF(2^{4n})$—which would be done via the Coppersmith algorithm. As a very rough guideline, to match the level of security offered by 512-bit RSA, if computing discrete logarithms in $GF(2^n)$ for $n = 700$ or 800 is as difficult as factoring 512-bit numbers, then for an elliptic curve over $GF(2^m)$ which has $k = 4$ as discussed above, one would require $m \approx 175$ or 200.

    To summarize, in light of recent results, in using elliptic curve cryptosystems over $GF(2^n)$, one must now (i) take special care in the particular choice of elliptic curve, and (ii) compensate for the attack by using appropriately larger fields to preserve the security. These larger fields may still be smaller than those required for equivalent security in other types of cryptosystems, so that elliptic curve cryptosystems remain attractive in practice, albeit to a somewhat lesser degree than prior to discovery of this new technique. Arithmetic operations in elliptic curve systems are easier due to the fact that smaller fields can be used, but this is partially offset by the fact that several arithmetic operations in the underlying field are required for one "elliptic curve operation." Ironically, the classes of curves susceptible to the new attack include many of those that have been recommended for use, including curves suggested in [3], [22], [34], and [36].

6 CONCLUDING REMARKS

It is well known that the computation of discrete logarithms in $GF(2^n)$ is easier than the factorization of n-bit integers N, using the best currently known algorithms for each problem. Hence, for example, the El Gamal cryptosystem in $GF(2^n)$ is less secure than RSA using an n-bit modulus N. This can be compensated for by using larger bitlengths n for the $GF(2^n)$ system. Indeed, fields $GF(2^n)$ can be carefully chosen that are larger and yet admit efficient arithmetic resulting in high throughput; the price is somewhat larger key sizes, and greater bandwidth and storage requirements. Asymptotic running times for the best practical algorithms for integer factorization and discrete logarithms in $GF(2^n)$ reveal that for similar levels of security, the relative bitlength by which the $GF(2^n)$ system must exceed the bitlength in RSA increases with the bitlength of the RSA system. For extremely secure (by present measures) RSA systems, such as 1024 bits, the corresponding bitlength for El Gamal in $GF(2^n)$ becomes problematic. To overcome problems due to larger key sizes, serious consideration should be given to elliptic curve systems over $GF(2^n)$. However, in light of recent advances, one must now exercise caution in the choice of elliptic curves, and in the size of the underlying fields. Further experience and large-scale implementations for computing discrete logarithms in larger fields $GF(2^n)$ will provide further empirical knowledge leading to a better understanding and more accurate practical estimates of the relative difficulty of integer factorization and the computation of discrete logarithms in $GF(2^n)$.

ACKNOWLEDGMENTS

The author would like to thank Arjen Lenstra, Alfred Menezes, Gus Simmons, Scott Vanstone, and Michael Wiener for discussions and comments, and for bringing various research to his attention.

REFERENCES

[1] E. Bach, "Intractable problems in number theory," in *Lecture Notes in Computer Science 403; Advances in Cryptology: Proc. Crypto'88*, S. Goldwasser, Ed., Santa Barbara, CA, Aug. 21–25, 1987, pp. 77–93. Berlin: Springer-Verlag, 1990.

[2] E. Bach, "Number-theoretic algorithms," *Ann. Rev. Comput. Sci.*, vol. 4, pp. 119–172, 1990.

[3] A. Bender and G. Castagnoli, "On the implementation of elliptic curve cryptosystems," in *Lecture Notes in Computer Science 435; Advances in Cryptology: Proc. Crypto'89*, G. Brassard, Ed., Santa Barbara, CA, Aug. 20–24, 1989, pp. 186–192. Berlin: Springer-Verlag, 1990.

[4] P. Bertin, D. Roncin, and J. Vuillemin, "Introduction to programmable active memories," in *Systolic Array Processors*, J. McCanny, J. McWhirter, and E. Swartzlander, Eds., Englewood Cliffs, NJ: Prentice Hall, pp. 301–309, 1989.

[5] I. F. Blake, R. Fuji-Hara, R. C. Mullin, and S. A. Vanstone, "Computing logarithms in finite fields of characteristic two," *SIAM J. Alg. Disc. Meth.* vol. 5, no. 2, pp. 276–285, June 1984.

[6] I. F. Blake, P. C. van Oorschot, and S. A. Vanstone, "Complexity issues for public key cryptography," in *Performance Limits in Communication Theory and Practice*, J. K. Skwirzynski, Ed., Amsterdam: Kluwer Academic Publishers, pp. 75–97, 1988.

[7] E. F. Brickell, "A survey of hardware implementations of RSA (abstract)," in *Lecture Notes in Computer Science 435; Advances in Cryptology: Proc. Crypto'89*, G. Brassard, Ed., Santa Barbara, CA, Aug. 20–24, 1989, pp. 368–370. Berlin: Springer-Verlag, 1990.

[8] J. Brillhart, D. H. Lehmer, J. L. Selfridge, B. Tuckerman, and S. S. Wagstaff, Jr., *Factorizations of $b^n \pm 1$, b = 2,3,5,6,7,10,11,12 Up to High Powers*, 2nd ed., *Contemporary Mathematics*, vol. 22, Amer. Math. Soc., 1988.

[8a] Update, dated March 12, 1991, by S. Wagstaff.

[9] E. R. Canfield, P. Erdös, and C. Pomerance, "On a problem of Oppenheim concerning 'Factorisatio Numerorum'," *J. Number Theory*, vol. 17, no. 1, pp. 1–28, Aug. 1983.

[10] D. G. Cantor and H. Zassenhaus, "A new algorithm for factoring polynomials over finite fields," *Math. Comp.*, vol. 36, pp. 587–592, 1981.

[11] T. T. Caron and R. D. Silverman, "Parallel implementation of the quadratic sieve," *J. Supercomput.*, vol. 1, pp. 273–290, 1988.

[12] D. Coppersmith, "Fast evaluation of logarithms in fields of characteristic two," *IEEE Trans. Inform. Theory*, vol. IT-30, no. 4, pp. 587–594, July 1984.

[12a] D. Coppersmith, "Modifications to the number field sieve," IBM Research Report #RC 16264 (Nov. 1990; updated Mar. 1991).

[13] D. Coppersmith, A. M. Odlyzko, and R. Schroeppel, "Discrete logarithms in GF(p)," *Algorithmica*, vol. 1, no. 1, pp. 1–15, 1986.

[14] J. A. Davis, D. B. Holdridge, and G. J. Simmons, "Status report on factoring," in *Lecture Notes in Computer Science 209; Advances in Cryptology: Proc. Eurocrypt'84*, T. Beth, N. Cot, and I. Ingermarsson, Eds., Paris, France, April 9–11, 1984, pp. 183–215. Berlin: Springer-Verlag, 1985.

[15] W. Diffie and M. Hellman, "New directions in cryptography," *IEEE Trans. Inform. Theory*, vol. IT-22, no. 6, pp. 644–654, Nov. 1976.

[16] S. R. Dusse and B. S. Kaliski, Jr., "A cryptographic library for the Motorola DSP56000," in *Lecture Notes in Computer Science 493; Advances in Cryptology, Proc. Eurocrypt '90*, Aarhus, Denmark, May 21–24, 1990; I. Damgård, Ed., pp. 230–244. Berlin: Springer-Verlag, 1991.

[17] T. El Gamal, "A public key cryptosystem and a signature scheme based on discrete logarithms," *IEEE Trans. Inform. Theory*, vol. IT-31, no. 4, pp. 469–472, July 1985.

[18] D. M. Gordon, "Discrete logarithms in GF(p) using the number field sieve," presented at the *Workshop on Number Theory and Algorithms*, Mathematical Sciences Research Institute (Berkeley, CA), March 26–29, 1990.

[19] G. H. Hardy and E. M. Wright, *An Introduction to the Theory of Numbers*, Oxford: Oxford University Press, 5th ed. (reprinted with corrections), 1983.

[20] D. E. Knuth and L. Trabb Pardo, "Analysis of a simple factorization algorithm," *Theoretical Computer Science*, vol. 3, pp. 321–348, 1976.

[21] D. E. Knuth, *The Art of Computer Programming*, Vol. 2: *Semi-numerical Algorithms*, 2nd ed., Reading, MA: Addison-Wesley, 1981.

[22] N. Koblitz, "Elliptic curve cryptosystems," *Math. Comp.*, vol. 48, pp. 203–209, 1987.

[23] N. Koblitz, "Constructing elliptic curve cryptosystems in characteristic 2," paper presented at Crypto'90, Santa Barbara, CA, Aug. 11–15, 1990; to appear in *Advances in Cryptology*, S. Vanstone, Ed. Berlin: Springer-Verlag (in press).

[24] B. A. LaMacchia and A. M. Odlyzko, "Solving large sparse linear systems over finite fields," paper presented at Crypto'90, Santa Barbara, CA, Aug. 11–15, 1990; to appear in *Advances in Cryptology*, S. Vanstone, Ed. Berlin: Springer-Verlag (in press).

[25] B.Ä. LaMacchia and A. M. Odlyzko, "Computation of discrete logarithms in prime fields," in Designs, Codes, and Cryptography, vol. 1, pp. 46–62, 1991.

[26] A. K. Lenstra and H. W. Lenstra, Jr., "Algorithms in number theory," in *Handbook of Theoretical Computer Science*, A. Meyer, M. Nivat, M. Paterson, and D. Perrin, eds., North Holland, Amsterdam (in press).

[27] A. K. Lenstra, H. W. Lenstra, Jr., M. S. Manasse, and J. M. Pollard, "The number field sieve," *Proc. 22nd ACM Symp. Theory of Computing*, pp. 464–572, 1990.

[28] A. K. Lenstra and M. S. Manasse, "Factoring by electronic mail," in *Lecture Notes in Computer Science 434; Advances in Cryptology; Proc. Eurocrypt'89*, J.-J. Quisquater and J. Vandewalle, Eds., Houthalen, Belgium, April 10–23, 1989, pp. 355–371. Berlin: Springer-Verlag, 1990.

[29] A. K. Lenstra and M. S. Manasse, "Factoring with two large primes," in *Lecture Notes in Computer Science 473; Advances in Cryptology, Proc. Eurocrypt'90*, Aarhus, Denmark, May 21–24, 1990; I. Damgård, Ed., pp. 72–82. Berlin: Springer-Verlag, 1991.

[30] H. W. Lenstra, Jr., "Factoring with elliptic curves," *Ann. Math.*, vol. 126, pp. 649–673, 1987.

[31] R. Lidl and H. Niederreiter, *Finite Fields, Encyclopedia of Mathematics and Its Applications*, Reading MA: Addison-Wesley, vol. 20, 1983.

[31a] M. Manasse, seminar given at Sandia National Laboratories, Albuquerque, NM, June 27, 1990, communicated by Gustavus J. Simmons.

[32] K. S. McCurley, "The discrete logarithm problem," in *Cryptology and Computational Number Theory; Proc. Symp Appl. Math.*, vol. 42, C. Pomerance, Ed.,

Boulder, CO, Aug. 6–7, 1989, pp. 145–166. Providence, RI: American Mathematical Society (1990).

[33] A. Menezes and S. Vanstone, "Isomorphism classes of elliptic curves over finite fields," Research Report 90-91 (Jan. 1990), Faculty of Mathematics, University of Waterloo, revised manuscript (Aug 1990).

[34] A. Menezes and S. Vanstone, "The implementation of elliptic curve cryptosystems," *Proceedings of Auscrypt'90,* (in press).

[35] A. Menezes, T. Okamoto, and S. Vanstone, "Reducing elliptic curve logarithms to logarithms in a finite field," (Unpublished manuscript, Sept. 1990).

[36] V. Miller, "Uses of elliptic curves in cryptography," in *Lecture Notes in Computer Science 218; Advances in Cryptology: Proc. Crypto'85,* H. C. Williams, Ed., Santa Barbara, CA, Aug. 18–22, 1985, pp. 417–426. Berlin: Springer-Verlag, 1986.

[37] H. Morita, "A fast modular-multiplication algorithm based on a radix 4 and its application," *Trans. IEICE,* vol. E 73, no. 7, special issue on Cryptography and Information Security, pp. 1081–1086, July 1990.

[38] M. A. Morrison and J. Brillhart, "A method of factoring and the factorization of F_7," *Math. Comp.,* vol. 29, no. 129, pp. 183–205, Jan. 1975.

[39] R. Mullin, I. Onyszchuk, S. Vanstone, and R. Wilson, "Optimal normal bases in $GF(p^n)$," *Discrete Applied Mathematics,* vol. 22, pp. 149–161, 1988/1989.

[40] K. K. Norton, "Numbers with small prime factors, and the least kth power non-residue," *Memoirs of the AMS,* no. 106, 1971.

[41] A. M. Odlyzko, "Discrete logarithms in finite fields and their cryptographic significance," in *Lecture Notes in Computer Science 209; Advances in Cryptology: Proc. Eurocrypt'84,* T. Beth, N. Cot, and I. Ingemarsson, Eds., Paris, France, April 9–11, 1984, pp. 224–314. Berlin: Springer-Verlag, 1985.

[42] S. C. Pohlig and M. E. Hellman, "An improved algorithm for computing logarithms over *GF(p)* and its cryptographic significance," *IEEE Trans. Inform. Theory,* vol. IT-24, no. 1, pp. 106–110, Jan. 1978.

[43] C. Pomerance, "Analysis and comparison of some integer factoring algorithms," in *Computational Methods in Number Theory,* H. W. Lenstra, Jr., and R. Tijdeman, Eds., *Math. Centrum Tract,* vol. 154, pp. 89–139, 1982.

[44] C. Pomerance, "The quadratic sieve factoring algorithm," in *Lecture Notes in Computer Science 209; Advances in Cryptology: Proc. Eurocrypt'84,* T. Beth, N. Cot, and I. Ingemarsson, Eds., Paris, France, April 9–11, 1984, pp. 169–182. Berlin: Springer-Verlag, 1985.

[45] C. Pomerance, J. W. Smith, and R. Tuler, "A pipeline architecture for factoring large integers with the quadratic sieve algorithm," *SIAM J. Computing,* vol. 17, no. 2, pp. 387–403, April 1988.

[46] R. Rivest, A. Shamir, and L. Adleman, "A method for obtaining digital signatures and public-key cryptosystems," *Commun. ACM,* vol. 21, pp. 120–126, 1978.

[47] T. Rosati, "A high speed data encryption processor for public key cryptography," in *Proceedings of the IEEE Custome Integrated Circuits Conference,* San Diego, May 15–18, 1989, pp. 12.3.1–12.3.5.

[48] C. P. Schnorr, "Efficient identification and signatures for smart cards," in *Lecture Notes in Computer Science 435; Advances in Cryptology: Proc. Crypto'89,* G. Brassard, Ed., Santa Barbara, CA, Aug. 20–24, 1989, pp. 239–351. Berlin: Springer-Verlag, 1990.

[49] R. Schoof, "Elliptic curves over finite fields and the computation of square roots mod p," *Math. Comp.*, vol. 44, pp. 483–494, 1985.

[50] M. Shand, P. Bertin, and J. Vuillemin, "Hardware speedups in long integer multiplication," *Proceedings of the 2nd ACM Symposium on Parallel Algorithms and Architectures*, Crete, July 2–6, 1990, (in press).

[51] R. D. Silverman, "The multiple polynomial quadratic sieve," *Math. Comp.*, vol. 48, pp. 329–339, 1987.

[52] D. H. Wiedemann, "Solving sparse linear equations over finite fields," *IEEE Trans. Inform. Theory*, vol. IT-32, no. 1, pp. 54–62, Jan. 1986.

[53] M. J. Wiener, "Cryptanalysis of short RSA secret exponents," *IEEE Trans. Inform. Theory*, vol. IT-36, no. 3, pp. 553–558, May 1990.

[54] M. C. Wunderlich, "Implementing the continued fraction factoring algorithm on parallel machines," *Math. Comp.*, vol. 44, no. 169, pp. 251–260, Jan. 1985.

SECTION 2

Authentication

CHAPTER 6

Digital Signatures

C. J. MITCHELL
Royal Holloway and Bedford New College
Department of Computer Science
Egham, TW20 OEX, England

F. PIPER AND P. WILD
Royal Holloway and Bedford New College
Department of Mathematics
Egham, TW20 OEX, England

1 INTRODUCTION

Since the original description of the concept of a digital signature in Diffie and Hellman's 1976 paper [Diffie 76], there has been a great deal of public interest shown in applying this idea to practical security problems. The publication of the Rivest–Shamir–Adleman (RSA) algorithm in 1978 [Rivest 78] showed that practical schemes for realizing the concept do exist. As a result many experts have been expecting digital signatures, and public key cryptography in general, to have a major impact on the design of secure systems.

However, it is only now, some 15 years later, that this revolution can truly be said to be taking place; even as recently as 5 years ago, digital signature techniques still had not had much impact on the marketplace for security products. Why should this be happening now, and why did it not take place earlier? One possible reason is that only in the last 5 years or so has general commercial awareness of the need for security grown to the point at which serious money is being spent. The traditional marketplaces for security products, namely the military and national governments, are not the places in which digital signature services are likely to be very important. One the other hand, resolution of disputes and protection against repudiation of messages are of fundamental importance in the commercial arena, and are often much more important than issues such as confidentiality of data.

The purpose of this chapter is to provide a user guide for digital signatures. In doing so it covers four major topics: (i) what a digital signature is (and how one is used in practice), (ii) techniques for generating digital signatures, (iii) techniques for cryptographic hash functions, and (iv) applications for digital signatures.

The second section, Fundamental Concepts, discusses the nature of digital signature techniques, and what distinguishes them from other methods for providing data authentication and integrity protection services. General approaches for constructing digital signature algorithms are described, prefatory to the material covered in Section 3. A discussion of the legal implications of digital signatures is included, together with a general discussion of commitments and of disputes and their resolution. Finally, the use of hash functions is motivated by considering practical aspects of the use of digital signature algorithms.

The third and fourth sections, Techniques for Digital Signatures and Techniques for Hashing, respectively, concentrate on specific examples of digital signature and hashing algorithms. The aim is to cover some of the most important algorithms, while not attempting to be encyclopedic. Full descriptions of each algorithm are given, with some discussion of possible attacks against these types of technique.

The chapter concludes in the fifth section, Applications for Digital Signatures, with a description of some key areas for the application of digital signatures, including a discussion of key certification.

2 FUNDAMENTAL CONCEPTS

2.1 Digital Signatures

There are many situations where the only security requirement is that the receiver should be confident that the contents of a message have not been altered since it left the sender and that the identity of the sender is not misrepresented. In other words, the receiver needs assurances concerning the authenticity of the message and of its origin.

For the moment we will ignore the problem of identifying the sender and concentrate on the problem of ensuring the authenticity of the message, that is, the message integrity. If data are transmitted over a noisy channel then, provided that the noise level is not too great, there are a number of well-known techniques for ensuring that the receiver can, at worst, detect that the message has been distorted. In fact, in many situations it is even possible to ensure that the distortion is "removed" so that the correct message is received. These techniques involve the use of a suitable error detecting (or correcting) code. Unfortunately, as we will illustrate with a trivial example, they do not offer any protection against deliberate alterations.

The simplest type of error detecting code is probably the parity check code. If the message is a string of bits then, if a parity check code is used, the transmitted signal contains an additional bit which is chosen to ensure that the total number of 1's in the transmitted signal is even. If a single bit, or any odd number of bits, is changed during the transmission then the received message will contain an odd number of 1's and the receiver will know that the received message is different to the message that was sent. Thus this code is guaranteed to detect a single error.

The reason why such a code offers no protection against deliberate alteration should be clear; anyone wishing to change some bits would merely adjust the parity bit (if necessary) so that the number of 1's in the altered message was still even. The alteration would then be undetectable and the receiver would accept the message as authentic. Although this particular example is ridiculously simple, it illustrates the fact that if the technique that is used to provide the authenticity is publicly known and implementable by everyone then anyone who knows the technique can alter the content of a message and then use the technique to ''adjust'' the altered message so that it appears authentic. Thus, if an authentication technique is to protect against deliberate alteration it must be such that the receiver is confident that only the sender could have applied it to the message. This suggests the use of some form of (secret) identifying information by the sender.

One possible solution involves the use of a conventional symmetric key cryptographic algorithm. In this scenario the sender uses the algorithm with a secret key, known only to the sender and receiver, to produce a ''cryptographic checksum'' from the message. This checksum, whose value should depend on each bit of the message, is then appended to the message prior to transmission. Thus the process is somewhat analogous to the parity check code described above. There is one crucial difference, however. A cryptographic checksum for a given message can only be computed by someone who knows the secret key and thus, since anyone who alters the message will be unable to make the appropriate change to the cryptographic checksum, the participants can be confident that, so long as they manage to keep the key value secret, no third party can alter their messages without detection. (This technique is frequently used by financial institutions and one very common application is the protection of transmissions between an automatic teller machine (ATM) and the institution host computer. A particular technique for computing the cryptographic check, which is often called a message authentication code (MAC), is described in American National Standards Institute (ANSI) X9.9; see sections 2.8 and 4, below.)

This solution to the message authentication problem is perfectly acceptable provided that the participants have faith in their ability to keep the key value secret and provided there is never a dispute between the sender and receiver. Furthermore, it may also be sufficient to convince the receiver of the sender's identity. There is nothing, however, to prevent the receiver from changing the received message and then altering the cryptographic checksum so that it is authentic.

Similarly, there is no way to prevent the sender from ''proving'' that he sent a message that was different to the one received. Thus there is nothing in this solution that prevents either party from trying to cheat the other and, in the case of a dispute, no evidence to enable a third party to settle the disagreement. The reason for this is that the cryptographic checksum depends on secret information that is shared by both parties and, therefore, either party can ''imitate'' the other.

We must stress here that if the two parties trust each other then the above solution is probably sufficient. If they do not, however, as will certainly be the case if they do not even know each other, then the ability of the receiver to impersonate the sender must be removed. Instead we need the sender to be able to ''sign'' the message in such a way that if anyone changes the message during transmission then the ''signature'' will reveal the fact that an alteration has occurred. Furthermore, if the signature is such that it cannot be ''forged'' then this will also authenticate the sender. We have deliberately used the term ''signature'' here because of the obvious analogy with written signatures.

In paper-based transactions the validity of a document is authenticated by a written signature. This signature then serves as evidence of the signer's agreement to the authenticity of the information on the document and, furthermore, can be presented in court if the signer ever tries to deny agreeing to the statements on that document. The emergence of computer-based message systems has led to the need to find a digital equivalent of the written signature.

The concept of a "digital signature" was first discussed by Diffie and Hellman in their classic paper "New Directions in Cryptography" [Diffie 76] and has since been the subject of numerous research papers, conference lectures, etc. [Akl 83]. In general terms, a digital signature should comprise some data that the receiver can keep as evidence that a particular message was sent and that the signer was the originator. It should thus prevent two types of fraud; the forging of a signature by the receiver (or any third party) and the repudiation of the transmission of a message by the sender. Moreover, it must be such that the receiver can save it as evidence which can be presented to a referee who can then check the validity of the signature and settle any dispute.

2.2 Properties of Signatures

If digital signatures are to be as widely used as written signatures and to be accepted for such important tasks as signing legal contracts, then they must have those properties of the written signature that make it such a workable and reliable form of authentication. But what are those basic properties? The crucial properties of a written signature are that it is easy to produce, easy to recognize but difficult to forge. It is this latter property that means that it cannot be repudiated and, as a consequence, written signatures are accepted as providing lasting evidence of precisely what has been authenticated and who authenticated it.

Thus, in an electronic message system we require that each user be able to produce with ease a digital signature whose authenticity is easily checked by a recipient. Moreover, such a digital signature must have the property that it could not have been produced by anyone else and can be used by a referee to resolve disputes.

A digital signature is a digital signal, usually represented as a string of bits. It may be appended to a message or the message may form an integral part of it. Thus, by its very nature, it differs from a written signature. Producing a written signature is a physical process, considered to be intrinsic to the person who is signing, and it is this uniqueness that identifies the signer and means that a signer cannot later repudiate his signature. In contrast, a digital signature is produced by machine. Instead of the signer performing a physical process, the signer merely provides some input to the process and it is this input that determines what particular pattern of bits makes up the digital signature. The aim is that no other person should be able to produce the same pattern of bits which, of course, means that they must not know the signer's input. Thus it is the use of the signer's input that identifies him. The signer's input may be some (secret) information that only he knows or a physical characteristic such as a fingerprint. Since it is the use of the signer's input that identifies him, if the input is secret information then the signature only identifies the signer as someone who knows the secret information. It is clearly possible that someone else may be able to obtain that information and if they do then they will be able to impersonate the signer. If, on the other hand, the signature is dependent on a physical characteristic then a forger must successfully copy that physical characteristic. However, it must not be forgotten that if, during the signing process, that physical characteristic is converted into a sequence of bits then the signer may be able to obtain that bit string without actually needing to copy the physical characteristic.

Another fundamental property of a person's written signature is that it is the same on all documents. Although it may be a difficult task, a forger may, therefore, be able to learn from studying examples of the signature and so duplicate it (without detection).

The security of a written signature lies in the difficulty of producing undetectable forgeries. In the case of a written signature, which is physically attached to a paper document, it is the ability to detect whether or not the document has been doctored that guarantees that the document is one that was signed. On the other hand, since there is no physical way of determining how a digital signal was produced or what input was given and since a digital signal is easily replicated, a digital signature must be different for each message. It is the particular pattern of bits of each individual digital signature that guarantees what message was signed and therefore, to prevent the substitution of an (altered) message to correspond to a signature, the signature must be a function which is dependent on all of the message. A forger, having seen many examples of a person's digital signature, should be no better informed as to how to produce a new (forged) digital signature of another message.

2.3 Signing and Verifying Transformations

Since a digital signature is a message-dependent bit-pattern, the signing process must transform the message into the signed message (or signature). Furthermore, since the signature can only be computed by the sender, this transformation must use some information that is unique to the sender. As we saw in the last section, it is then the use of this secret information that is accepted as proof of the identity of the signer, that is, as establishing the origin of the message. We must now express those concepts more formally.

We will let $\mathcal{M}$ denote the set of all messages which the sender may sign and $\mathcal{S}$ be the set of all possible signatures. The signing process is now nothing more than a transformation from $\mathcal{M}$ to $\mathcal{S}$. If we let S_X be the (secret) signing transformation used by signer X then, for any $M \in \mathcal{M}$, $S_X(M)$ is the signature on message M by user X. This signature $S = S_X(M)$ can then be concatenated to the message M to give (M, S), which we call the signed version of M. If, as is frequently the case, M can be recovered from S then there may be no need to transmit M. In this situation we abuse our notation slightly and regard S as both the signature and the signed version of the message.

Before accepting a message as authentic the recipient must verify the signature. However, since he does not know S_X, the receiver cannot compute $S_X(M)$ from M. Instead, during the verification process he must use public information to check that the user's secret transformation was used. Thus, since the verification process must also authenticate the message, we may view this process as the application of a (public) transformation V_X from $\mathcal{M} \times \mathcal{S}$ to the set {True, False}. (Thus V_X is the transformation that verifies signatures by X.)

Clearly the first requirement of V_X is that it should identify those pairs for which the signature does not "belong" to the message. More formally we require:

1. $V_X(M, S) = $ True if, and only if, $S = S_X(M)$.
However, it is also necessary for the sender to have confidence that, whenever $V_X(M, S) = $ True, the message M originated from X and that the signature is not a forgery. Thus we also require:

2. It is not computationally feasible for anyone, except user X using S_X, to construct $M \in \mathcal{M}$ and $S \in \mathcal{S}$ such that $V_X(M, S) = $ True.

Concern is often expressed about the special case of (2) when someone might be able to change the message without invalidating the signature. It is clearly desirable to require that it is impossible to find two distinct messages M, M' and a single signature S such that $V_X(M, S) = V_X(M', S) =$ True. This is equivalent to requiring that $S_X(M) \neq S_X(M')$ for all messages $M \neq M'$ and is automatically satisfied if a message is recoverable from the signature. Unfortunately, as we shall see when we discuss specific examples, in many practical situations $|\mathcal{M}| \gg |\mathcal{S}|$ so that this stronger requirement is unachievable. However, condition (2) implies the weaker property that, given signature S on message M, it is computationally infeasible to find another message M' such that $V_X(M', S) =$ True.

Another clearly desirable property is that the probability of guessing (or randomly generating) $S_X(M)$ for a given M should be minimal. This imposes size restrictions on $|\mathcal{S}|$.

Yet another important observation about V_X is that, although V_X can only use public information, it is crucial that knowledge of this public verifying information should not enable anyone to deduce the secret signing transformation S_X or, in any other way, allow false signatures to be generated.

We end this section by noting that if the message were recoverable from the signature then we could have given an alternative definition of the verifying process as one determined by a (public) transformation V_X from $\mathcal{S}$ to $\mathcal{M}$ such that $V_X(S) = M$ belongs to $\mathcal{M}$ if and only if $S = S_X(M)$.

2.4 Initial Agreement and Legality of Signatures

It is important to realize that conditions (1) and (2) are both necessary. The need for condition (1) is obvious. It is condition (2), however, that guarantees that forgery is difficult. Since condition (2) is so crucial each user must agree to the verifying transformation that is to be used to validate his signatures. Furthermore, the receiver requires a guarantee that the verifying transformation he uses to validate a signature is one to which the signer has agreed. Thus, before any signed messages are transmitted, there must be an agreement about what verifying transformation is valid. This agreement may necessarily need to be the subject of a paper document with handwritten signatures.

The sender and recipient must also agree on how disputes are to be resolved. Usually a dispute will be resolved by a third party, the referee. When a dispute arises it usually means that a receiver, user B, has a signature S that he claims to be the message M signed by the sender, user A, while A claims he did not sign M. The essence of a digital signature is that a referee can determine whether B is putting forward an S that he knows A did not generate or whether A is repudiating a signature he has generated. We have already noted that if messages are sent authenticated by a conventional (shared key) cryptosystem, a third party is not able to discover these ruses. (This is, basically, because the verifying transformation V_X involves the use of S_X. Indeed, to determine whether $V_X(M, S)$ is True or False the receiver computes $S_X(M)$ and compares its value with S.)

When B claims that S is the signature of M by A, his claim is presumably based on the fact that $V_A(M, S) =$ True where V_A is the verifying transformation for signatures by user A. Therefore B presents to the referee the evidence S, M, and V_A. The referee verifies for himself whether or not $V_A(M, S) =$ True and thereby determines whether B is making a false claim or A is attempting to repudiate his signature. In such a dispute the referee must also rule on whether V_A is the verifying transformation for

A's signature, agreed to by A and B. This process is also likely to include verifying that M is a valid message, that is, it belongs to the agreed message set $\mathcal{M}$.

Although in some circumstances the referee may be trusted by A and B to resolve disputes fairly in his own way, there are many situations where A and B may have a signed written agreement with the referee on how disputes between them will be resolved. In any case the protocol for resolving disputes must be defined and agreed to by A and B in advance of any messages being signed. Different protocols determine different signature schemes even though the signing and verifying transformations may be the same. This problem is discussed further in Section 2.7.

Whereas both the laws and legal precedent regarding signed paper documents are well established, the situation regarding digital signatures is not so clear. It seems that, at present, if digital signatures are to have any standing in law, then an initial written agreement between the parties concerned, that is, the sender, receiver, and referee, is required. This agreement should define the procedures for obtaining a digital signature, how it would be recorded, what each party's commitment would be and how disputes would be resolved. If one party to this agreement felt that another party had not honored his commitment then he could test this in court.

This may seem unsatisfactory and, as digital signatures become more widely accepted, it is possible that legislation dealing with digital signatures may be enacted. Experience with the use of digital signatures will undoubtedly help determine what form these laws may take. There will, of course, be many problems that need addressing. These include: the means of authenticating the verifying information, the obligation of a user to ensure the security of his signing transformation, the means of resolving disputes, penalties for the misuse of digital signatures, the use of stolen keys, signing false statements or misdating signed messages, and the action required in case of compromised keys. It is likely to be a few years before adequate legislation is introduced.

2.5 Methods for Digital Signatures

Whether or not users can reach an agreement to accept digital signatures will depend largely on their belief that transformations S_X and V_X satisfying conditions (1) and (2) above (see Section 2.3) can be constructed. When the concept of digital signature was first discussed by Diffie and Hellman, [Diffie 76], they suggested a solution to the problem by use of public key cryptosystems. Since then there have been many proposed implementations of digital signatures. Some suggested schemes make use of conventional (or symmetric) key cryptosystems, some of public (or asymmetric) key cryptosystems, and others rely on transformations that are independent of cryptosystems.

Two categories of digital signatures may be identified: true signatures and arbitrated signatures. In a true signature scheme, signed messages are sent directly by the signer to the receiver who verifies their validity. Unless there is a dispute there is no need for the signature to be referred to a third party. The situation for an arbitrated signature scheme is totally different and signed messages can be sent only via a trusted third party called the arbitrator. The recipient is unable to verify the sender's signature directly, but is assured of its validity through the mediation of the arbitrator. The arbitrator is trusted by both parties to play his role correctly and this gives the sender assurance that his signature will not be forged and the receiver assurance that the signatures he receives are valid. The arbitrator also plays the role of referee. He is trusted to resolve disputes (fairly) should they arise.

2.5.1   Arbitrated signatures.   Two users, X and Y, who trust each other can use a conventional cryptosystem to exchange authenticated messages. By some secure means they share a secret key k, known only to them, which determines both an enciphering transformation E_k and a deciphering transformation D_k. They trust each other not to reveal this key to any other user. For X to send an authenticated message M to Y, the cryptogram $C = E_k(M)$ is formed and transmitted. User Y decrypts C to obtain $M = D_k(C)$ and checks that M is valid, that is, that M belongs to the agreed message set. If M belongs to the agreed set then Y can be confident that C came from X provided that the probability that someone, who doesn't know k, can find a cryptogram that decrypts to a valid message is small. If this probability is too high then the problem can be overcome by transmitting M concatenated with C, that is, (M, C).

We have already discussed this method of authentication in detail and seen that it does not provide a signature as a third party cannot determine whether X or Y produced $C = E_k(M)$ and cannot resolve a dispute between X and Y about whether X sent message M to Y or not. However, if X and Y both trust a third party A, then this form of authentication may be used to allow X, with the cooperation of A, to send a signed message to Y. In this scheme each user X shares with A, called the arbitrator, a secret key k_X which determines enciphering and deciphering transformations E_{k_x} and D_{k_x}. (This key k_X is known only to X and A.)

One such scheme involves user X sending messages to user Y via A. User X signs message M by encrypting it with key k_X. Thus $S_X(M) = E_{k_x}(M)$. User X sends $E_{k_x}(M)$ to A. The arbitrator A has key k_X and so can decrypt this to obtain M. Provided M is a valid message for user X, A now appends X and M to the signature $E_{k_x}(M)$, encrypts the result with key k_y, and sends $E_{k_y}(X, M, E_{k_x}(M))$ to user Y. User Y decrypts this to obtain X, M, and $E_{k_x}(M)$. Provided the identity X and message M are acceptable to him, Y accepts $E_{k_x}(M)$ as the signature of M by X. Thus $V_X(M, S) = $ True if and only if Y receives $E_{k_y}(X, M, S)$ from A (which occurs if and only if $S = E_{k_x}(M)$). In case of dispute, Y, who claims that he has the signature $S = S_X(M)$, sends $E_{k_y}(X, M, S)$ to A who uses k_Y to obtain X, M and S and then uses k_X to check that $E_{k_x}(M) = S$. Thus in this scheme Y cannot actually check X's signature directly. It is there solely to settle disputes. Thus in this arbitrated digital signature scheme, both users X and Y must trust the arbitrator A. Signatures can be sent only via A and only A can verify signatures and resolve disputes. User X must trust A not to reveal key k_X and not to generate false signatures $E_{k_x}(M)$. Similarly, user Y must trust A to send $E_{k_y}(X, M, S)$ only if $E_{k_x}(M) = S$. Furthermore, both users must trust A to resolve disputes fairly. If the arbitrator acts as expected then X has an assurance that no one can forge his signature and Y has the assurance that X cannot disavow his signature.

In a slight variation of this method, user X sends $E_{k_x}(M)$ directly to Y who then sends $E_{k_y}(X, M, E_{k_x}(M))$ to A to have the signature validated.

There is another arbitrated signature method in which user X and the arbitrator A produce the signature between them. In this scheme the arbitrator A also has another key k_A known only to him. To sign message M, user X sends $E_{k_x}(M)$ to A. Now A uses k_X to obtain M, adjoins the identity X, and enciphers the result with key k_A. This he sends back to X and it is this that forms the signature of M by X, that is, $S_X(M) = E_{k_A}(X, M)$. Now X can send the message M and signature $S = E_{k_A}(X, M)$ to Y. To validate S, user Y sends M and S to A who uses k_A to obtain $D_{k_A}(S)$ which he enciphers with key k_y. Thus A returns $E_{k_y}(D_{k_A}(S))$ to Y. Now Y uses k_y to obtain $D_{k_A}(S)$ and accepts S as a valid signature only if $D_{k_A}(S) = (X, M)$. Thus $V_X(M, S) = $ True if and only if $D_{k_y}(E_{k_y}(D_{k_A}(S))) = D_{k_A}(S) = (X, M)$, that is, if and only if $E_{k_A}(X, M) = S$. In case of dispute user Y sends $E_{k_y}(X, M, S)$ to A, who rules the signature valid if and only if $S = E_{k_A}(X, M)$.

In this scheme the assistance of A is required both to generate and to validate signatures and, as before, only A can resolve disputes. Thus both users X and Y must trust A, and if A is trustworthy they have an assurance that signatures cannot be forged nor disavowed.

2.5.2 True signatures.

A true signature scheme is one in which the signer sends the signature directly to the receiver who is able to verify the signature without recourse to a third party. In this case the transformation S_X acts as a broadcast cipher and all users can use V_X to verify the origin of a message. However, only one user, namely X, can generate S from M. Diffie and Hellman [Diffie 76] describe this property as one-way authentication.

The verifying transformation V_X allows the receiver to recognize an S as the signature $S = S_X(M)$ on message M by user X, but knowledge of V_X must not allow him to form the signature of a message. In arbitrated signature schemes the receiver needs the cooperation of the arbitrator to perform V_X and it is the trustworthiness of the arbitrator that ensures that user X, and only X, is able to generate signatures. True signature schemes rely on properties of the signing and verifying transformations to provide this assurance.

Clearly this reliance necessitates the use of a special type of function for the signing and verifying transformations. These functions, whose importance was recognized by Diffie and Hellman, are called one-way functions. Their fundamental property is that they are "easy" to implement but "hard" to invert. More precisely, a *one-way function* is a function f such that, given any x in its domain, $y = f(x)$ is easy to evaluate, but, for almost all y (in its range), it is difficult to find an x such that $y = f(x)$. One particularly important family of such functions are the *trapdoor one-way functions*, that is, one-way functions that allow someone in possession of specific secret information (the trapdoor) to compute an inverse.

In fact, in any true signature scheme the verifying transformation V_X determines a function with the properties of a trapdoor one-way function. To see this we define a function $f : \mathcal{M} \times \mathcal{S} \to \mathcal{M} \cup \{\infty\}$ by $f(M, S) = M$ if $V_X(M, S) = $ True and $f(M, S) = \infty$ otherwise. Then, clearly, f is easy to evaluate since, otherwise, checking signatures would be difficult. However, for any given $M \in \mathcal{M}$ it is difficult to find $(M, S) \in \mathcal{M} \times \mathcal{S}$ such that $f(M, S) = M$ without knowledge of S_X. (This is condition (2) in our requirements for the signing and verifying transformations.) This knowledge of S_X is the trapdoor that enables the signer to compute the inverse.

Several methods for obtaining a digital signature scheme based on one-way functions and their trapdoors have been published. When Diffie and Hellman introduced the concept of a public key cryptosystem based on a trapdoor one-way function, they showed how it may be used to obtain a digital signature scheme. Other asymmetric transformations (with trapdoors), however, and conventional (symmetric key) cryptosystems have also been used to provide digital signature schemes.

2.6 Some General Settings for Signature Schemes

In this section we list a number of general settings for true signature schemes.

According to ISO the term "digital signature" is used "to indicate a particular authentication technique used to establish the origin of a message in order to settle disputes of what message (if any) was sent." Unfortunately there is not always universal agreement on precisely what is acceptable as indisputable evidence for settling such disputes and thus there is often debate as to whether or not a specific proposal constitutes a true digital signature scheme, according to the ISO definition. We do not attempt to act as judge as to whether or not a given scheme warrants the term "digital signature" but merely list all those proposals that we know are being seriously considered for implementation.

2.6.1   Public key cryptography.   A public key cryptosystem is a multiaccess cipher with the fundamental property that the encryption and decryption transformations are separated. Thus whereas the encryption transformation is public and anyone can encipher a message, decryption is only possible to the holder of the (secret) decryption key.

In a public key cryptosystem each user X has two keys: a public key p_X which determines an enciphering transformation and a private (secret) key S_X which determines a deciphering transformation. A message M may be sent to X in secret by enciphering it as the cryptogram $C = E_{p_x}(M)$. User X can decipher the cryptogram using his deciphering transformation D_{s_x} which must satisfy $D_{s_x}(E_{p_x}(M)) = M$ for all messages M. Both transformations E_{p_x} and D_{s_x} must be easy to perform and it must be computationally infeasible for anyone, from the knowledge of E_{p_x}, to derive D_{s_x}. Thus, with p_X made public, anyone can send to X a secret message that only X can decipher (using D_{s_x}).

The public key cryptosystems considered by Diffie and Hellman that can be used to provide digital signatures are those for which the transformations E_{p_x} and D_{s_x} both act on the same set. Then E_{p_x} and D_{s_x} are inverses of one another and $E_{p_x}(D_{s_x}(M)) = M$ for all messages M. When this is the case, user X may sign a message by forming the signature $S = D_{s_x}(M)$. The sender and receiver agree on a subset $\mathcal{M}$ of all possible messages to be the set of valid messages. The signing transformation S_X is then the secret deciphering transformation D_{s_x} restricted to this set. The verifying transformation V_X is defined by $V_X(M, S) =$ True if and only if $E_{p_x}(S) = M$ and $M \in \mathcal{M}$. Once again we stress that the set $\mathcal{M}$ of valid messages must be chosen so that, unless D_{s_x} is known, it is computationally infeasible to construct M and S such that $V_X(M, S) =$ True. So, for example, for randomly chosen S, the probability that $E_{p_x}(S) \in \mathcal{M}$ must be negligible.

Public key cryptosystems for which the range of the enciphering transformation E_{p_x} is not the set of messages have also been proposed. These systems are not immediately suitable for signature schemes as in such a system $D_{s_x}(M)$ may not be defined for all messages M or $E_{p_x}(D_{s_x}(M))$ may not necessarily be equal to M. However, Merkle and Hellman [Merkle 78] and Schöbi and Massey [Schöbi 83] have shown that, by suitable modifications, a signature scheme may be obtained from certain of these public key cryptosystems provided $E_{p_x}(D_{s_x}(M)) = M$ for a suitable proportion of the set of messages M. In these signature schemes the signature of a message M is obtained by applying D_{s_x} to a message M' obtained from M (in a defined way) such that $E_{p_x}(D_{s_x}(M')) = M'$. Thus $S_X(M) = D_{s_x}(M')$ and $V_X(M, S) =$ True if and only if $E_{p_x}(S) = M'$.

2.6.2   Other trapdoor signature schemes.   A public key cryptosystem has two transformations E_{p_x} and D_{s_x} satisfying $D_{s_x}(E_{p_x}(M)) = M$ for all messages M. Furthermore it is not possible to determine the secret deciphering transformation D_{s_x} from knowledge of the public enciphering transformation E_{p_x}. Suppose instead that we have two transformations E_{p_x}, D_{s_x} satisfying $D_{s_x}(E_{p_x}(M)) = M$ for all messages M with the extra property that it is not possible to determine E_{p_x} from knowledge of D_{s_x}. Then a signature scheme may be obtained by making the key S_X public while keeping the key p_X secret. Thus the signature by X on message M is $S_X(M) = E_{p_x}(M)$ and the signature S is verified by checking that $D_{s_x}(S) = M$ is a valid message.

Thus, given transformations f, g satisfying $f(g(M)) = M$ for all messages $M \in \mathcal{M}$, we obtain a public key cryptosystem if g is a one-way function for which f provides the trapdoor information (and then $E_{p_x} = f$ and $D_{s_x} = g$) and we obtain a signature scheme if f is a one-way function for which g provides the trapdoor information (and then $S_X = f$ and $V_X(M, S) = $ True if and only if $g(S) = M$ is a valid message). A public key cryptosystem may or may not be suitable for a signature scheme and similarly a signature scheme may exist independently of any public key cryptosystem.

2.6.3 Symmetric key cryptography.

In a conventional (symmetric key) cryptosystem both the encryption transformation E_k and the decryption transformation D_k are determined by the same key k. Then $D_k(E_k(M)) = M$ for all messages. However, the transformation E_k is not a one-way function and if made public would make known the deciphering transformation D_k. For such a cryptosystem to be secure against a known plaintext attack it must be difficult to determine the key k given pairs of corresponding plaintext M and ciphertext $C = E_k(M)$. If this is so, then we may use the enciphering transformation to define a one-way function acting on the set of keys k. Fix a message X and define f_X by $f_X(k) = E_k(X)$ for all keys k. Provided the cryptosystem is secure against a known plaintext attack, f_X is a one-way function. This one-way function can be used to obtain a signature scheme provided it is possible to find a suitable trapdoor as the signing transformation S_X.

For most conventional cryptosystems the one-way function f_X may not have a trapdoor. However when a one-way function f is used to obtain a signature scheme, the trapdoor information is (merely) the means by which the signer finds a solution for S to an equation $f(S) = h(M)$ for some M. If two tables of pairs are constructed, one containing the pairs (M, S) and the other $(M, f(S))$, this solution can be found by comparing entries in the two tables with the same first entry M. Thus the table of pairs $(M, f(S))$ can be made public and used to evaluate $h(M)(= f(S))$ and the table of pairs (M, S) kept secret and used to produce the signature $S = S_X(M)$ by X on message M.

This is the approach taken in the Diffie–Lamport signature scheme [Diffie 76]. In its simplest form this scheme has a set of two messages $\{M_1, M_2\}$, a secret table containing pairs $(M_1, k_1),(M_2, k_2)$, and a public table containing pairs $(M_1, K_1),(M_2, K_2)$ and public one-way functions f_1, f_2 such that $f_1(k_1) = K_1$ and $f_2(k_2) = K_2$ (e.g., $f_i(k) = E_k(M_i)$).

This use of tables shows one way in which a conventional cryptosystem, secure against a known plaintext attack, may be used to provide a signature scheme. The method has the drawback that the number of possible messages must be small enough for the tables to be manageable. If the number of messages that can be signed is too small then this restriction can be overcome by representing long messages as a sequence of messages from the small set and signing each member of the sequence. To prevent deletion or replay of signatures of messages that have appeared in a sequence before, however, the tables of pairs (i.e., the one-way functions f_1, f_2) must be continually changing. In any scheme where the number of messages is small it is always feasible to keep a log of previous signatures and then replay them (in a different order) as a forged message at a later time. Signatures obtained using this type of tables are therefore called one-time signatures, since the tables must be continually being used and discarded. They cannot be used again.

To sign a message of n bits these one-time signatures require of the order of $2n$ bits of verification data to be made available in the tables and so, if a great many messages are to be signed, there may be difficulties in storing, handling, and authenticating this verification data. Merkle [Merkle90] has suggested ways in which the verification data may be generated (and authenticated) in a tree structure as the messages are signed. The storage requirement is reduced to the storage of the root of the tree. The essential idea is to use a succession of one-way functions (and corresponding tables) to successively authenticate the (next) tables and finally sign the message. The length of the resulting signature increases proportionally to the log of the number of messages signed (see Section 3.4).

2.6.4 Signatures by tamper-resistant modules. It is the fact that there is a (natural) separation between encryption and decryption in a public key cryptosystem that makes such systems candidates for digital signature schemes. Since each of the two functions has its own key it is possible to be able to verify a signature (using encryption) without being able to generate a signature (which requires decryption). For conventional systems there is not the same (natural) separation of keys. However, if the encryption and decryption transformations were successfully separated then it would be possible to provide signatures using such a system. One way of achieving this separation is by physical means using tamper-resistant modules (TRMs).

Suppose we have a system where each user has a TRM that contains a key k_X for a conventional cryptosystem and that k_X determines an encryption transformation E_{k_x} and decryption transformation D_{k_x}. Suppose the key k_X could be communicated securely to all other TRMs and the modules were constructed so that, while only X's TRM could perform E_{k_x}, all other TRMs could only perform D_{k_x}. Then, provided that the tamper resistance of each module could be trusted, only X's TRM could generate a signature $S_X(M) = E_{k_x}(M)$ but any other TRM could verify a signature by computing $D_{k_x}(S)$ and checking whether or not $D_{k_x}(S) = M$.

It should be noted that, although this is the first example where it is explicitly stated that the security relies on tamper resistance, most digital signature schemes rely on some form of physical security. If, for example, we consider a scheme based on public key cryptography then the security relies on the sender X's ability to keep his secret key S_X secret. Great care is taken to ensure that S_X cannot be computed from knowledge of the algorithm and X's public key p_X. However, this is not sufficient and X must also take steps to ensure that S_X is not exposed by physical means. Typical ways of achieving this are to use tamper-resistant devices and/or to store all devices that hold s_X in physically secure locations.

2.6.5 Composition of trapdoor permutations. In this case we assume that we have two trapdoor one-way permutations, f_0 and f_1, on the set of all users, and that messages are represented by binary strings. If $M = M_1 M_2 \ldots M_n$ is a sequence of 0's and 1's then the signature of X on M is

$$S_X(M) = f_{M_n}^{-1}(f_{M_{(n-1)}}^{-1}(\ldots (f_{M_1}^{-1}(X)) \ldots))$$

where f_0^{-1} and f_1^{-1} are the (trapdoor) inverses of f_0 and f_1, respectively, and are only known by user X. The rule for checking the signature should be clear, that is, $V_X(M, S) =$ True if and only if $X = f_{M_1}(f_{M_2}(\ldots (f_{M_n}(S)\ldots)))$.

Note that while computing $V_X(M, S)$ the receiver produces signatures for the messages $Mi = M_1, M_2 \ldots M_i$ for $i = 1, \ldots, n - 1$. Thus, this scheme can only be used if, for any $i < n$, the sequence Mi is not a valid message.

2.6.6 Schemes with implicit verification functions. So far we have considered signature schemes in which $V_X(M, S) =$ True if and only if $f(S) = g(M)$ where f is a one-way function and g is easily computed. More generally it may be preferable to consider verifying transformations defined by $V_X(M, S) =$ True if and only if $h(M, S) = 0$ for some implicit function h of M and S. Gibson [Gibson 88b] discusses an example modeled on the Fiat–Shamir scheme [Fiat 87] where an implicit function arises as a result of the introduction of a random element and the use of a family of trapdoor one-way functions.

For this signature scheme we use a publicly known one-way function h and a family of trapdoor one-way functions indexed by elements $e \in E$. User X has private functions g_e which provide the trapdoor inverses to the functions f_e. He signs message M in the following way.

X chooses a random element r and calculates $e = h(M, r)$ and $y = g_e(r)$. His signature $S_X(M)$ is then given by $S_X(M) = (y, e)$ which, clearly, depends on r. The receiver may verify the signature $S = (y, e)$ by checking that $h(M, f_e(y)) = e$.

2.6.7 Scheme with probabilistic verification. In arbitrated signatures the signer shares his secret signing information with the arbitrator. This allows the arbitrator to verify a signature and pass on this assurance to the recipient. Rabin [Rabin 78] has described a signature scheme in which, after signing a message, the signer reveals only part of his secret signing information to the recipient. This allows him to have confidence in the signature. However, only in the case of dispute does the signer reveal all his secret signing information to an arbitrator who can then verify the signature. The receiver's verification is a probabilistic verification in the sense that there is a (small) probability that a signature verified by the receiver will not be validated by the arbitrator.

Any signature scheme of this type must be a one-time signature scheme in the sense that the signer's secret information can be used to sign only one message. Thus the scheme must have a register of validating information used in the verifying process to validate the signer's revealed secret signing information.

In Rabin's scheme a user wishing to sign a message M uses an even number, say $2t$, of one-way functions $f_1, \ldots, f_{2t}$. (The f_i could, for example, arise from a conventional cipher system as enciphering transformations E_{k_i} for some choice of keys $k_1, \ldots, k_{2t}$.) The functions $f_1, \ldots, f_{2t}$ are the signer's secret signing information and, in order to enable verification of their use in the signing process, pairs (x_i, y_i) where $y_i = f_i(x_i)$ are published in a register whose authenticity is accepted by all parties. The signature on message M is then the tuple $(f_1(M), \ldots, f_{2t}(M))$.

To validate this signature the recipient chooses a subset I of t of the subscripts $1, \ldots, 2t$ and asks the signer to reveal f_i for $i \in I$. The recipient verifies that the revealed f_i are part of the signer's secret signing information by consulting the register of pairs (x_i, y_i) and then partially checks the signature $(S_1, \ldots, S_{2t})$ on message M by checking that $S_i = f_i(M)$ for $i \in I$. The recipient accepts the signature $(S_1, \ldots, S_{2t})$ if and only if all these checks are satisfied.

In case of dispute the recipient presents the message M and signature $(S_1, \ldots, S_{2t})$ and the signer reveals all of $f_1, \ldots, f_{2t}$ to an arbitrator. The arbitrator verifies that $f_1, \ldots, f_{2t}$ are the secret signing functions by checking that $f_i(x_i) = y_i$ for $i = 1, \ldots, 2t$ and validates $(S_1, \ldots, S_{2t})$ as the signature on M provided $S_i = f_i(M)$ for at least $t + 1$ functions f_i.

The problem of determining the precise number of pairs (f_i, S_i) to be checked by the arbitrator is interesting. If, for instance, we insisted that he checked all $2t$ of them then the sender might provide a "signature" with $2t - 1$ correct S_i and one incorrect one. This signature would be guaranteed to be declared invalid by the arbitrator but the chances of the receiver choosing t correct values, and thereby accepting the signature, would be high. If, on the other hand, the arbitrator checks only $t + 1$ then the signer might be able to create a "signature" with exactly t correct values which would be accepted by the receiver but, of course, rejected by the judge. However, the chance that the receiver would request these t functions, and so accept this signature (which an arbitrator would not validate) is only one in $\binom{2t}{t}$. For example, when $t = 18$, this results in only one chance in $\binom{36}{18} \sim 10^{10}$.

It is important to reemphasize that this signature scheme is a one-time signature scheme. The pairs (x_i, y_i) can only be used to sign one message because any repeated pair for which f_i was revealed could be used to forge a signature.

Note also that, since in case of dispute the signer shares all of his secret signing information with the arbitrator, it is possible for the arbitrator, after resolving a dispute, to forge a signature. The signer must therefore trust the arbitrator not to use this information or give it to anyone else.

2.7 Commitments and Disputes

We have already stressed the care needed to set up a signature scheme in which all the participants can have confidence. The basis of this confidence lies in the two transformations S_X and V_X. If they satisfy properties (1) and (2), above (see Section 2.3) then user X may feel happy about being committed to any message M for which a receiver holds the signature S such that $V_X(M, S) = $ True, knowing that the only (feasible) way for the receiver to obtain S is from X himself. User X, of course, must also be satisfied that no one else can discover the transformation S_X. It follows that S_X must be held securely. Provided user X is assured that no other person has access to the transformation S_X he can feel sure that no one can produce (forge) his signature. He may then be willing to commit himself to a signature S on a message such that $V_X(M, S) = $ True.

If a receiver has a (signed, written) agreement with X or some other trusted assurance that X is committed to signatures S verified by V_X then he may be happy to accept these signatures. Of course, the receiver may also require an assurance that the signer will be held to that commitment.

Unfortunately, no matter how much care is taken in setting up the scheme, it is still possible that a sender and receiver may end up in dispute. A receiver may (falsely) claim that S is the signature on message M by user X when no such message was signed by X. Similarly, a sender may (falsely) deny signing message M when in fact he did. In such cases the sender and/or receiver might refer their dispute to a referee for resolution. As we stressed in Section 2.4, it is crucial that both sender and receiver know [and agree (in writing)] the means which the referee will use to resolve a dispute. Furthermore, they must both trust the referee or have some commitment from the referee to abide by the agreed method of resolution.

2.7.1 Resolution of disputes. When the sender and receiver seek resolution of a dispute, the receiver sends to the referee X, M, and S, claiming that S is the signature on message M by user X. The referee requires some way of determining whether S is a valid signature on M by user X. One method of resolution might be for the arbiter to use the verifying transformation V_X agreed to by user X. Thus the referee rules in favor of the receiver if $V_X(M, S) = $ True.

The receiver will be happy with this method of resolution provided he is assured that he has access to the same verifying transformation V_X as the referee uses. The sender will be happy with this method of resolution provided that he is confident that the referee uses the agreed verifying transformation V_X, that V_X is a (trapdoor) one-way function, and that S_X has not been compromised (i.e., that he (user X) is the only person, except possibly the referee, with access to S_X). Both sender and receiver must then rely on the referee not to make a false ruling but to rule according to the outcome of $V_X(M, S)$.

Thus we have identified four necessary properties of a signature scheme for the resolution of disputes:

1. V_X and S_X have properties (1) and (2) (see Section 2.3).
2. V_X is authentic.
3. The referee is trustworthy.
4. S_X is secure.

Where a proof that a signature scheme satisfies (1) is not available, the validity of (1) is usually claimed by the scheme's successful resistance to cryptanalysis. Of course, while there is an incentive to do so, new cryptanalytic attacks will be sought which may provide the means of breaking a scheme. So this may be a volatile property and a scheme's security status may change with time.

Establishing (2) may be difficult and, indeed, the means of authenticating V_X may depend on the nature of V_X.

In an arbitrated signature scheme in which the receiver (and sender) must trust the arbitrator to verify a signature, it seems reasonable to assume that property (2) holds if property (3) holds. Of course, it would be the duty of a responsible trusted arbitrator to ensure that the verifying transformation V_X he uses is the one agreed to by user X (and has not been tampered with by anyone else).

In a true signature scheme, where more than one person requires access to (the same) V_X, there must be some formally agreed-on method of authenticating verification functions. Prior to setting up the scheme, user X must have agreed to the verifying transformation V_X and, in the final analysis, this may require some signed written document that details how to derive V_X. Furthermore, to make it widely available, it may be necessary to hold details of V_X in some trusted register. It may be possible for the referee to get a copy of V_X signed by the registry. This signed copy of V_X is often called a certificate and the registry is trusted to issue only valid certificates, that is, certificates containing only true copies of V_X. Of course, this creates a further requirement that the registry uses a secure signature scheme and keeps its signing transformation secret. However, the verifying transformation of the registry may be authenticated by wide publication (e.g., in the daily press) so that everyone can be reasonably sure of its validity.

Property (3) is essential—without a referee trusted by both sender and receiver there seems to be no possibility that a dispute can be resolved by a third party.

The method of resolution of a dispute described above makes it essential (for the sender) that S_X is kept secure. If the referee upholds any signature S that is verified by V_X then, in addition to being sure that it is infeasible for anyone to find S such that $V_X(M, S) =$ True for some message M without using S_X, the user must be confident that no one but himself has access to S_X.

We have already discussed the use of one-way functions to prevent anyone from calculating S_X. However, this is not sufficient and property (4), the security of S_X, will almost certainly also require physical security. One option is for user X to be the only person with physical access to the machine that performs S_X. Another is that some physical characteristic of user X that cannot be duplicated (e.g., a fingerprint) may be a necessary input before S_X can be performed. Alternatively, the security of S_X may be obtained by making something (a secret) that user X knows, but no one else does, a necessary input before S_X can be performed.

In the case of physical security user X would need to be confident that it was physically impossible for anyone else to perform S_X either due to lack of access to the machine or due to lack of the appropriate physical characteristic that was needed to activate the machine to perform S_X. In this latter case the machine would have to be tamper-resistant so that it is impossible to discover how to alter the machine or build another machine to perform S_X without the input of user X.

Similarly, when the input is some secret information, user X would want an assurance that no one else has access to the secret input that he (alone) knows. Secret information, however, can be lost or stolen and so, as the holder of this information, it may be the duty of user X to ensure its secrecy. As an "insurance policy" against compromise of this secret information, X might insist on being able to ensure that no more signatures be verified with V_X, that is, he might insist that V_X be withdrawn from the authentication registry. Unfortunately this type of requirement causes new problems. One obvious problem arises from the fact that a signature S may need to be verified long after the message was signed. Thus signatures signed before S_X was compromised may still need to be verified long after that compromise. This would not be possible if V_X were withdrawn.

Another problem, perhaps of even greater importance, arises from the fact that if V_X is withdrawn then no earlier signature of X can be verified. This is a fact that X may exploit for fraudulent purposes. Indeed, by falsely declaring the loss of his secret information or by revealing this information, user X could revoke messages he has signed. Clearly, signatures that could be repudiated in this way by the sender would not be acceptable to a recipient.

It seems reasonable to try to insist that if a signature could be identified as having been signed before S_X was compromised then V_X could still be used to verify it. However, there are practical difficulties with this notion. For instance, it is usually difficult to determine exactly when S_X may have been compromised. Even if we ignore this problem, V_X must now be authenticated as valid at a certain date and time. Furthermore a means, independent of S_X, must be used to authenticate the date and time of each signature to verify whether or not it was signed before or after the compromise of S_X.

The "obvious solution" of merely adding the date and time to a message and signing the resulting message using S_X is not sufficient since anyone who obtains S_X can predate a message and then sign it. Thus the receiver must obtain a time-stamp on the signature $S = S_X(M)$ from a third party. (This third party may be a registered timekeeper who adds the date and time to a signature S presented to him and signs the result to produce a signature certificate.) In addition, the receiver must also obtain from the registry of verifying transformations an authenticated copy of V_X which has been time-stamped by the timekeeper. The receiver would then check each message and would not accept a signature time-stamped with a date and time that did not precede that stamped on V_X. Of course, in this scenario the certificates issued by the registry and timekeeper need to be verified, but this can be achieved by using their respective verifying transformations which, in turn, can be authenticated by being published widely (e.g., in the daily press).

As soon as the "machinery" discussed above is established then we can determine our procedures for reporting compromises of secret information. If user X discovers (or suspects) that his signing transformation has been compromised, he notifies the registry. If this notification is signed using S_X, the registry knows that either the notification has come from user X or that, in any case, S_X has been compromised. Thus it can safely accept it as genuine. The registry then issues X with a certificate, time-stamped by the timekeeper, containing the statement that the compromise of S_X had been reported. User X could then repudiate any signatures that were verified by a verifying transformation which was time-stamped after that time.

In the case of dispute a referee would hold user X responsible for a signature only if (i) the receiver had certificates that verified that the signature was time-stamped at a date and time that preceded that at which the verifying transformation was time-stamped and validated; and (ii) the signer could not produce a certificate that verified that his signing transformation had been reported compromised at a date and time prior to the date and time stamped on the signature.

If (i) does not hold the referee dismisses the receiver's claim. If (i) holds but (ii) does not then the referee rules that the registry has issued a certificate recording the compromise of S_X with a date and time that precedes the date and time of a certificate that validates the verifying transformation V_X and holds the registry responsible.

Thus this protocol allows the signer to repudiate signatures when his secret information has been compromised by shifting the responsibility to the registry which issues the certificates. Now, however, the signer is not able to use this for fraudulent means, as the receiver is protected by having certificates with the signature and verifying transformation time-stamped and presumably the registry will protect itself by not issuing certificates for verifying transformations that are time-stamped with a date and time that does not precede the date and time stamped on a certificate that records the compromise of S_X.

The registry must also protect itself by safeguarding its signing transformation and keeping it secret so that no one can issue a forged certificate. The responsibility of user X to keep S_X secret has been shifted to the responsibility of the registry to keep its secret signing information secret. But, in view of their relative roles, this is probably more realistic. Note, however, that, to allow user X to repudiate signatures when S_X has been compromised, it has now become imperative to the registry that its signing transformation is kept secret. The reason is, quite simply, that the registry is likely to be held responsible for any signatures verified using forged certificates.

A disadvantage of this form of signature scheme is the fact that the recipient must communicate with the registry and timekeeper at the time of receiving the signature. However, unlike an arbitrated signature scheme, the recipient does not need to trust the intermediaries (the registry and timekeeper) as he can verify for himself whether a signature he obtains via the certificates will be upheld by the referee (assuming that he can verify the signatures of the registry and timekeeper). On the other hand, the signer does need to trust the registry and timekeeper not to issue certificates that are time-stamped with a date and time that precede the date and time at which the signer reported the compromise of his signing transformation. The signer must also trust the registry and timekeeper to keep their respective signing transformation secure since if they are also compromised then forged signatures and certificates may be generated. It is sufficient that the signer trust the timekeeper, as the registry's signing transformation alone is not enough to generate the required certificates. Conversely, it is necessary that the signer trust the timekeeper, since, with the aid of a copy of V_X time-stamped and authenticated before S_X has been reported compromised, the timekeeper could predate forged signatures to produce certificates which would be upheld by the referee.

Denning [Denning 83] describes a method of protection against the predating of forged signatures. This method involves the timekeeper keeping a log of all time-stamped certificates in order of issue on a write-once device. Only certificates appearing in the log would be valid. When user X reports the compromise of his signing transformation this is also recorded in the log. Everyone would have open read access to the log to ensure this. As the device cannot be overwritten or erased, the predating of signature certificates before the date of notification of the compromise would be detected. Of course, some means of maintaining the security and authenticity of the log is required. It seems that this would be best achieved by some suitable physical means.

2.7.2 Witnessed digital signatures.

The use of witnesses can also be used to relieve the sender of some of the burden of keeping his signing transformation secure. When user X signs message M, a witness, user W, physically confirms that user X agrees to message M and signs the message $M' =$ "User X agrees to M." The pair $S = S_X(M)$, $S' = S_W(M')$ are sent to the receiver and form the signature. The signature is verified using V_X and V_W (which, of course, must be authenticated). Thus the recipient (and the referee in case of dispute) accepts the pair S, S' as the signature on M by user X if and only if $V_X(M, S) =$ True and $V_W(M', S') =$ True.

The sender and receiver must agree in advance on which user may be an acceptable witness, and the referee must be able to verify this agreement. Now, if the signing transformation of user X is compromised, forged signatures still cannot be generated unless the signing transformation of a witness is also compromised. Of course the signer must trust the witness not to witness any signature that the signer has not produced.

2.8 Practical Use of Signature Schemes

In the definition of signature given above, it is assumed that the message to be signed is an element of a message space that is acted on by a signature function S_X. In practice, signature functions will take as input strings of bits of fixed length, and messages of any greater length will need to be processed in some way prior to signature. One possibility would be to divide the message into a sequence of "blocks" of appropriate size for the signature function, and then sign each piece individually, that is, if the message M is made up of the sequence

$$M_1, M_2, \ldots, M_t$$

where each M_i has the required length, then user X would sign the message by computing:

$$S_X(M_1), S_X(M_2), \ldots, S_X(M_t).$$

Unfortunately, this method has a number of disadvantages. The first, and probably most important, is that there is no linkage between different parts of the message, and so the recipient of a signed message will not know if the message components (M_i) have been reordered, replicated, or partially deleted during transmission. This can be remedied by the inclusion of redundancy in the message blocks, but this has the disadvantageous effect of increasing the number of signatures to be computed.

Second, even if the message only consists of one block, there is a problem caused by the public nature of the signature checking process. Suppose Y, a would-be forger of X's signature, knows X's public verification function V_X. Then Y chooses a value S at random and computes

$$M = V_X(S)$$

Then S will be a valid signature on M, although Y has no control on the contents of M. This problem, which we first encountered in Section 2.6.1, can be removed by the addition of redundancy to the message to be signed. However, this would reduce the efficiency of the system still further.

A third important defect is that the above process requires the signature algorithm to be applied t times. Many proposed signature functions are relatively difficult to compute (i.e., they run slowly) and therefore this could be very inefficient.

These deficiencies are sufficient to deter the use of this type of signature and, as a result, Rabin [Rabin 78], [Rabin 79], Merkle [Merkle 79], and Davies and Price [Davies 80] introduced the concept of a "one-way hash function." A one-way hash function is a public function h (which should be simple and fast to compute) that satisfies three main properties:

- H1. It must be capable of converting a message M of arbitrary length into a fixed-length "digest," $h(M)$, which has the length required to be input to the secret signature transformation.
- H2. It must be *one-way*, that is, given an arbitrary value y in the domain of h, it must be computationally infeasible to find a message M such that $h(M) = y$.
- H3. It must be *collision-free*, that is, it must be computationally infeasible to construct two messages M, M' with the property that $h(M) = h(M')$. (As we discuss below, $H3$ implies that h should normally be chosen to have a domain of size at least 2^{100}.)

Given an appropriate hash function h then, in order to compute a signature on a message M, user X first computes the digest (or "hashed version") $h(M)$ and then signs this digest using his secret transformation S_X. Thus the signed digest for the message M is:

$$S_X(h(M))$$

X then transmits M concatenated with the signed digest $S_X(h(M))$. Note that a further motive for using one-way hash functions is provided by Denning [Denning 84]. Briefly, if a signature scheme is used without first applying a hash function, and if a crypt-analyst can persuade a user to sign certain selected messages, then in some cases sig-natures can be forged on other messages.

The verification of this signature requires both prior knowledge of the public hash function h and X's public transformation V_X. Using this knowledge $h(M)$ and $V_X(h(M), S)$ can be computed. If $V_X(h(M), S) =$ True then the message is accepted as valid.

In the discussion above we introduce three properties ($H1$, $H2$, $H3$) that a hash function should have; we now show why these are necessary. The need for $H1$ should be clear. $H2$ is present to prevent the interceptor of a message and its signature (M and $S_X(h(M))$ say) replacing the valid message M with a fraudulent one M' with the property that $h(M) = h(M')$. The need for $H3$ is less obvious, as is the fact that $H3$ requires the domain of h to be so large. Before discussing this point further we observe that, in practice, $H3$ implies $H2$.

To demonstrate the need for $H3$ we will assume that we have a hash function h that does not satisfy $H3$. If this is so then it may be possible for a malicious party, say Z, to construct two messages M, M' with the following properties:

1. $h(M) = h(M')$.
2. User X would be happy to sign M.
3. User X would not be happy to sign M' but Z would like to obtain X's signature on this message.

Z could then persuade X to sign M and send this signed message to some third party, that is, X sends M accompanied by $S_X(h(M))$. It is then only necessary for Z to attach this signature to M' to achieve his fraudulent aim.

Having seen why collision freedom is so important, we now indicate why a hash function with a domain of size significantly less than 2^{100} cannot be collision-free. In general, if the hash function has a domain of size 2^n, then it is only necessary to generate approximately $2^{n/2}$ variants of each message to have a reasonable chance of finding two variants with the same digest. (This is the so-called "birthday attack" and is discussed in detail in Section 4.1.2.) Thus, to make such an approach completely impractical requires n to be large and $n = 100$ has been suggested as a practical mini-mum, giving a suitable margin for error.

An alternative approach, allowing the use of 64-bit hash functions, has been proposed by Davies and Price [Davies 84]. They suggest that the originator of a mes-sage should always modify it in some way prior to signature, typically by appending a random value to the message. Unfortunately, as pointed out by Akl [Akl 84] and Mitchell et al. [Mitchell 89], this procedure does not prevent potential frauds by the originator, and it would seem prudent to avoid use of any hash functions producing digests of less than 100 bits when producing a digital signature. However, we must point out that hash functions producing 64-bit digests are still of value for other types of authentication function.

Having given three properties that a hash function must satisfy, (i.e., $H1$–$H3$, above), we observe that satisfying all three properties does not mean that a hash function can be used in a secure way with all signature algorithms. In certain circumstances apparently strong signature and hashing algorithms, when used together, reveal weaknesses. This is precisely what occurs when the standard RSA signature technique is combined with the X.509 hash function, as has recently been shown by Coppersmith [Coppersmith 1989]. (This topic is discussed further in Section 4.2.2.)

In all our discussions so far we have assumed that the message M is sent in "clear" form along with the signature $S_X(h(M))$. However, there may be situations where secrecy is also required. Of course, if M is confidential then it can be encrypted before transmission. But it is important to note that this encryption should always take place after the signature process, that is, signatures should always be computed on "clear" data. This is not only established good practice (see, e.g., Davies and Price [Davies 84],) but also of great significance in protecting against various subtle forms of attack. This is exemplified by weaknesses recently revealed in certain of the security facilities in the 1988 CCITT X.400 and X.500 recommendations, in which construction of "tokens" can involve signing encrypted data; these standards and the attacks on them are discussed further in Section 5.

A second question that naturally arises, when considering encryption of signed messages, is whether or not the signature itself needs to be encrypted. In general this depends on the entropy of the message space. If there are a very large number of possible messages, all with small associated probabilities of occurrence, then the signature does not need to be encrypted. However, if the number of possible messages is small, or a certain small number of messages are significantly more probable than the others, and the signature is unencrypted, then the value of an intercepted signed message could be revealed by an interceptor hashing each possible (or each likely) message in turn until one is found which matches the signature. The only alternative to encrypting the signature in this case is for all messages to be expanded by the addition of random padding prior to signature and encryption. This latter process may be mandated where standards dictate unencrypted signatures (e.g., the International Telegraph and Telephone Consultative Committee (CCITT) X.500 series recommendations—see Section 5).

Apart from their use with digital signatures, cryptographic hash functions can be used to provide authentication and integrity protection for sets of data in other ways. For example, data files on a computer can be protected against change by computing a digest and then storing this digest in a secure place. In this case, property $H3$ would not normally be necessary, and so a 64-bit digest would suffice.

Alternatively, a message digest could be appended to a message prior to encryption to provide origin verification and integrity protection for the message; again, in this case a 64-bit digest would normally suffice. Generalizing this application to the case where a message is to be sent to more than one recipient, much cryptoprocessing can be saved by computing a single message digest and then encrypting it differently for each intended recipient (see, e.g., the Internet electronic mail scheme [RFC1113], [RFC1114]). In this latter application $H3$ is necessary to prevent certain kinds of attack (see, e.g., [Mitchell 90]).

We conclude this general discussion of hash functions by considering *keyed* hash functions, also called *parameterized* hash functions by Merkle [Merkle 89a], [Merkle 89b]. In all the above discussion we have considered hash functions that are completely public, that is, no secret knowledge is required to compute a message digest. However, in some applications it may be desirable to use a hash function that requires use of a secret key to compute the digest for a message. Of course, this means that the digital signature can only be checked by someone knowing the appropriate secret key, but it will also help prevent attacks by malicious third parties without knowledge of the secret key.

For example, we noted above that if the number of possible messages is small and message confidentiality is required, then an unencrypted signature could reveal the message; use of a keyed hashing function would prevent this. In addition, where straightforward origin authentication and integrity protection is required, a keyed message digest can often be used on its own without signature to provide the desired service. This is the basis of the standardized message authentication codes (MACs) based on the Data Encryption Standard (DES) block cipher [ANSI X9.9], [ANSI 9.19]; these functions were mentioned in Section 2.1 and are discussed further in Section 4.

3 TECHNIQUES FOR DIGITAL SIGNATURES

We have already seen the importance of one-way functions for digital signature schemes and how trapdoors are exploited to achieve the objective that, although it is computationally infeasible (without knowledge of S_X) to find a pair M and S with $V_X(M, S) =$ True, it must be easy to check whether $V_X(M, S)$ is True or False.

In Section 3.1 we discuss some of the one-way functions that have assumed central roles in modern cryptography. Then in Section 3.2 we discuss how they may be used for digital signature schemes.

3.1 One-Way Functions

3.1.1 Multiplication of two large primes. If p and q are large primes then computing their product $n = pq$ is easy. However, unless at least one of the primes has special properties that facilitate factorization, it is extremely difficult to determine p and q from n. Within the last decade considerable resources have been directed toward the factorization problem and considerable progress has been made. A number of general-purpose factoring algorithms have been discovered. The most practical one seems to be the multiple polynomial quadratic sieve (MPQS) [Silverman 87] which has an expected running time: $L(n) = \exp((1 + 0(1))(\ln(n)\ln(\ln(n)))^{1/2})$, independent of the size of the factors of n. This algorithm is well suited to parallel implementation and has been used to factor 102- and 106-decimal digit numbers in about 1 and 4 months, respectively.

In addition to the general-purpose factoring algorithms, there are a number of other algorithms that are much faster but which only "work" if either n itself or (at least one of) the factors of n have special properties. The most recent of these algorithms, the number field sieve [Lenstra 89], is particularly interesting and warrants discussion.

The number field sieve (NFS) algorithm factors Cunningham numbers, that is, numbers of the form $r^e \pm s$ with r and s small, in heuristic expected running time:

$$T(n, c) = \exp(c + 0(1))\ln(n)^{1/3}\ln(\ln(n))^{2/3}) \text{ with } c \cong 1.526.$$

This is asymptotically faster than any other known algorithm applicable to numbers of this form and it has been used to factor several numbers with more than 100 digits in a matter of weeks*. Work is in progress on generalizations of the number field sieve algorithm, applicable to general integers. It is suspected that these algorithms will have expected running time $T(n, c)$ with a larger value of c. It remains to be seen whether c will be small enough to give a general-purpose factoring algorithm that is significantly faster than MPQS for numbers of the size likely to be used in digital signature schemes.

3.1.2 Exponentiation modulo $n = pq$.   If $n = pq$ is the product of two large primes and e is chosen so that $(e, (p - 1)(q - 1)) = 1$, then the modular exponentiation function E, defined by $E(m) = m^e(\text{mod } n)$, is a trapdoor one-way function. If we are given a value c such that $c = m^e(\text{mod } n)$ for some unknown m then the only known practical method of determining m is to use the exponent d with $ed = 1(\text{mod}(p - 1)(q - 1))$ and calculate $c^d(\text{mod } n)$. (The fact that $m = c^d(\text{mod } n)$ is, of course, the basis of the RSA public key system, see [Beker 82]). However, the determination of d from n and e requires knowledge of the factors p and q and it is, therefore, this knowledge that provides the trapdoor. It is also true that knowledge of one pair of values for e and d is sufficient to determine p and q, that is, to factor n. Thus inverting E without the trapdoor appears to be as difficult as factoring n.

3.1.3 Exponentiation in a finite field.   For any prime power q the multiplicative group of the finite field $GF(q)$ of order q is cyclic. If q is a large prime power and a is a primitive element of $GF(q)$ (i.e., a generator of the cyclic multiplicative group) then for any nonzero element y in $GF(q)$, $y = a^x$ with $0 \le x \le q - 1$.

The integer x is called the discrete logarithm of y to the base a in $GF(q)$. When q is a prime $GF(q)$ is isomorphic to the integers modulo q and thus we may write $y = a^x(\text{mod } q)$. In practice it appears that the only fields that are used for implementations are $GF(p)$, where p is a large prime, or $GF(2^n)$ for large n. For a large prime p the function defined on the integers modulo p by $x \rightarrow a^x(\text{mod } p)$ is a one-way function. The computation of $a^x(\text{mod } p)$ requires at most $2 \log_2 x$ multiplications while the best general algorithms known for extracting logarithms modulo p require a precomputation of the order of $\exp((\ln p)^{1/2}(\ln \ln p)^{1/2})$ operations.

This expression is similar to the time complexity function $L(n)$ for the MPQS factoring algorithm. However, just as with factorization, we must point out that, although it represents the best results known for arbitrary primes p, it is possible to do considerably better if the primes have special properties. For instance, Pohlig and Hellman [Pohlig 78] have described an efficient algorithm for computing discrete logarithms mod p when $p - 1$ has only small prime factors. Thus if we are to regard $x \rightarrow a^x(\text{mod } p)$ as a one-way function then the prime must be chosen very carefully.

Arithmetic in $GF(2^n)$ for large n can be performed faster than arithmetic over large primes, so there is some reason to prefer the one-way function $x \rightarrow a^x$ in $GF(2^n)$. However, fast algorithms for evaluating discrete logarithms in $GF(2^n)$ have been developed [Coppersmith 84] so care must be taken to ensure that the exponent n is large

---

*In mid-June 1990, the ninth Fermat number ($F_9 = 2^{2^9} + 1$)—having 155 digits—was factored by using the NFS [Lenstra90,90a,91]. The calculations were done over a four-month period on approximately 700 workstations scattered worldwide with the final computation being done on a supercomputer.

enough. (Coppersmith's algorithm has an expected asymptotic complexity function of $\exp(n^{1/3}\ln(n)^{2/3})$).

3.1.4   Squaring modulo n.   If $n = pq$ where p and q are two large primes then the function $x \rightarrow x^2(\bmod\ n)$ is a trapdoor one-way function, where the trapdoor information is knowledge of the prime factors.

For any modulus m, an element y is called a quadratic residue modulo m if there exists an x such that $y = x^2(\bmod\ m)$. Any value x such that $y = x^2(\bmod\ m)$ is then called a square root of y modulo m.

If the modulus m is a prime then Adleman, Manders, and Miller [Adleman 77] have described a probabilistic polynomial time algorithm for finding square roots modulo m. Furthermore, if m is composite with known factors then this algorithm can be "combined" with the Chinese remainder theorem to extract square roots modulo m. Thus, extracting square roots modulo m is easy provided the factors of m are known. However, Rabin has shown that any algorithm that extracts square roots modulo m can be used to factor m [Kranakis 86]. Thus if $n = pq$, where p and q are unknown, it is only as feasible to invert the one-way function $x \rightarrow x^2(\bmod\ n)$ as it is to factor n.

3.1.5   The knapsack function.   Let $A = \{a_1, \ldots, a_n\}$ be a set of n distinct integers. If $\mathbf{a} = (a_1, \ldots, a_n)$ and $\mathbf{x} = (x_1, \ldots, x_n)$ is a binary n-tuple we write

$$\mathbf{a} \cdot \mathbf{x} = \sum_{i=1}^{n} a_i x_i$$

Then we may define the function $\mathbf{x} \rightarrow \mathbf{a} \cdot \mathbf{x}$ from the set of binary n-tuples to the set of integers. Clearly this function is easy to compute. However, for general $\mathbf{a}$, it is difficult to determine $\mathbf{x}$ from $\mathbf{a}$ and $\mathbf{a} \cdot \mathbf{x}$. (In fact, this is an NP-complete problem.) Thus, for many $\mathbf{a}$, the function $\mathbf{x} \rightarrow \mathbf{a} \cdot \mathbf{x}$ is one-way. Of course if $\mathbf{a}$ is chosen "badly" then this function is easy to invert and it was this observation that formed the basis of the Merkle–Hellman public key system [Merkle 78].

3.1.6   DES.   Let $E_k(m)$ denote DES encipherment of message m under key k. Apart from a few weak keys, the only known way to determine the key k from a pair c, m with $c = E_k(m)$ is by an exhaustive key search. Hence, for any given plaintext block x, the mapping $k \rightarrow E_k(x)$ is a one-way function from the set of binary 56-tuples to the set of binary 64-tuples.

3.2 Implementations of Digital Signature Schemes

There have been many proposals for implementing digital signature schemes. In this section we look at some of them and discuss the various ways in which the appropriate one-way function is used in the design.

3.2.1   Digital signatures using RSA.   The RSA public key scheme uses the one-way function $E(m)$ defined by $E(m) = m^e(\bmod\ n)$, where n is the product of two (suitably chosen) large primes p and q and $(e, (p - 1)(q - 1)) = 1$. Its security relies on the infeasibility of determining the deciphering exponent d from the public information e and n and so, since it is only knowledge of d that distinguishes the genuine receiver from everyone else, it seems reasonable to regard d as identifying the genuine receiver of the public key (e, n).

For RSA, the enciphering and deciphering transformations act on the same domain (integers modulo n) and are inverses of each other. Thus if $s = m^d(\text{mod } n)$ then $s^e = m^{de} = m(\text{mod } n)$. Furthermore, given $m = s^e(\text{mod } n)$, n and e then it is difficult to compute s (unless d is known). Thus as user X holds the secret key d, he may sign a given message block m by forming the signature $s = m^d(\text{mod } n)$. Anyone who knows the public key e can compute $s^e(\text{mod } n)$ and thus decide whether $V_X(m, s) = $ True or False.

By using this technique user X may sign any message from the set of messages $\mathcal{M}$ of integers between 0 and $n - 1$. However, if we are to have a secure signature scheme then we must restrict our message set. In particular, we need $\mathcal{M}$ to have the following two properties:

1. There must be only a small probability that an integer s chosen at random from 0 to $n - 1$ satisfies $s^e(\text{mod } n) \in \mathcal{M}$.

2. If $m_1, m_2 \in \mathcal{M}$ then $m_1 m_2 \notin \mathcal{M}$.

The reason for property (1) has already been discussed in Section 2. The reason for property (2) is that if X signs two messages m_1 and m_2, with signatures s_1 and s_2, respectively, then anyone can forge his signature on the message $m_1 m_2$ by computing $s_1 s_2$. (*Note:* $(s_1 s_2)^e = s_1{}^e s_2{}^e = m_1 m_2(\text{mod } n)$.)

The usual solution to properties (1) and (2) is to require that messages be hashed before they are signed. In this case the hash function must satisfy the following property:

3. It is infeasible to find $m_1, m_2 \in \mathcal{M}$ such that $h(m_1 m_2) = h(m_1)h(m_2)$.

If property (3) is not satisfied then $(m_1 m_2, s_1 s_2)$ can be constructed from (m_1, s_1) and (m_2, s_2).

3.2.2 El Gamal's signature scheme.

El Gamal [El Gamal 85] has proposed a public key cryptosystem and a digital signature scheme based on discrete logarithms. These schemes have the advantage that the public enciphering key and the public information used in the verifying transformation are the same. However, the two transformations are distinct. Furthermore, it appears that the public key cryptosystem cannot be adapted (as RSA was) to provide digital signatures and the digital signature scheme cannot be adapted to provide encryption. El Gamal's digital signature scheme makes use of the difficulty of computing discrete logarithms over $GF(p)$ where p is a large prime.

In this scheme, a large prime number p is chosen such that $p - 1$ has a large prime factor. The secret information that user X has as input to the signing transformation is an integer x and the public information is a primitive element a of $GF(p)$ together with the integer y such that $y = a^x \pmod p$. Thus, recovering the secret information from the public information requires the computation of a discrete logarithm modulo p. A message block m is now an integer between 0 and $p - 1$ and to sign m user X does the following:

1. Randomly chooses an integer k, $1 < k < p - 1$, such that $(k, p - 1) = 1$ and k has not been used to sign a previous message

2. Computes $r = a^k \pmod p$

3. Finds s such that $m = xr + ks \pmod{(p - 1)}$.

The calculation in (3) is straightforward because, since $(k, p - 1) = 1$, Euclid's algorithm can be used to find k^{-1} such that $kk^{-1} = 1 (\mathrm{mod}\ (p - 1))$ and then $s = k^{-1}(m - xr)(\mathrm{mod}\ (p - 1))$.

Once these calculations have been performed, X's signature on m is the pair (r, s).

Since $m = xr + ks\ (\mathrm{mod}\ (p - 1))$, $a^m = a^{xr+ks} = a^{xr}a^{ks} = y^r r^s\ (\mathrm{mod}\ p)$. Thus the recipient can verify that $(m, r, s) = S_X(m)$ by checking that $a^m = y^r r^s\ (\mathrm{mod}\ p)$.

Clearly, for any fixed $r = a^k\ (\mathrm{mod}\ p)$, finding s such that $a^m = y^r r^s\ (\mathrm{mod}\ p)$ is equivalent to computing a discrete logarithm modulo p. However, it is not clear that the more general problem of simultaneously finding a pair r, s with $a^m = y^r r^s\ (\mathrm{mod}\ p)$ is also equivalent to finding discrete logarithms modulo p. Nevertheless, it is believed to be difficult and, as yet, no one has found an efficient algorithm to solve this more general problem.

In this scheme user X should only be prepared to sign all messages in the message set $\mathcal{M}$ provided that it is computationally infeasible for anyone else to find $m \in \mathcal{M}$ and integers r, s satisfying $a^m = y^r r^s\ (\mathrm{mod}\ p)$. As for the systems based on RSA, this places restrictions on the choice of $\mathcal{M}$. For instance, if b and c are arbitrary integers with $(c, p-1) = 1$ then we can construct a triple (m, r, s) satisfying $a^m = y^r r^s\ (\mathrm{mod}\ p)$ as follows: $r = a^b y^c\ (\mathrm{mod}\ p)$, $s = -r/c\ (\mathrm{mod}\ (p - 1))$ and $m = rb/c\ (\mathrm{mod}\ (p - 1))$. Thus a necessary condition is that

1. If b, c are integers with $(c, p - 1) = 1$ and if $r = a^b y^c\ (\mathrm{mod}\ p)$ then it should be unlikely that $-rb/c\ (\mathrm{mod}\ (p - 1)$ is an element in $\mathcal{M}$.

Along similar lines we have that if (r, s) is a valid signature for $m \in \mathcal{M}$ and if z, b, c are arbitrary integers with $(zr - cs, p - 1) = 1$ then the following triple (m', r', s') satisfies $a^{m'} = y^{r'} r'^{s'}(\mathrm{mod}\ (p - 1))$:

$$r' = r^z a^b y^c\ (\mathrm{mod}\ p), s' = (sr^1/(zr - cs))(\mathrm{mod}\ (p - 1)), m'$$

$$= \frac{r'(zm - bs)}{zr - cs}\ (\mathrm{mod}\ (p - 1))$$

Thus we also require:

2. The probability that $m' \in \mathcal{M}$ for an arbitrarily chosen r, z, b, c should be negligible.

Again, the usual solution to (1) and (2) is to require that the messages be hashed before they are signed.

The El Gamal scheme can be generalized to use arithmetic in the field $GF(q)$. For practical implementations q is usually either a prime or a power of 2.

3.2.3 The Fiat–Shamir signature scheme.

The Fiat–Shamir scheme [Fiat 87] exploits the difficulty of extracting square roots modulo n where n is the product of two large primes p and q. User X holds k integers $s_1, \ldots, s_k$ in the range $0-n$ such that, for each i, $(s_i, n) = 1$ and they are the secret information that he uses in the signing transformation. The values $v_1 = s_1^{-2}\ (\mathrm{mod}\ n), \ldots, v_k = s_k^{-2}(\mathrm{mod}\ n)$ are made public and are used in the verifying transformation. For this scheme to be secure, it is crucial that $s_1, \ldots, s_k$ cannot be computed from $v_1, \ldots, v_k$. Thus, since being able to find square roots implies being able to factor n, this means that n must be large enough that it is difficult to factor.

To sign a message M, user X

1. chooses random $r_1, \ldots, r_t$ between 0 and $n - 1$ and computes $x_i = r_i^2 (\mathrm{mod}\ n)$, $i = 1, \ldots t$.

2. computes $f(M, x_1, \ldots, x_t)$ where f is a public hashing function and uses the first kt bits as entries e_{ij} of a t by k binary matrix E.

3. computes $y_i = r_i \prod_{e_{ij}=1} s_j$ (modulo n) for $i = 1, \ldots, t$.

The signature on M by user X is then (y, E) where y is the vector $(y_1, \ldots, y_t)$ and E is the matrix (e_{ij}).

To verify this signature, the recipient

1. computes $z_i = y_i^2 \prod_{e_{ij}=1} v_j\ (\mathrm{mod}\ n)$ for $i = 1, \ldots, t$.

2. checks that the first kt bits of $f(M, z_1, \ldots, z_t)$ are the entries e_{ij} of E.

A signature generated by user X will always be verified since

$$z_i = y_i^2 \prod_{e_{ij}=1} v_j = r_i^2 \prod_{e_{ij}=1} s_j^2 \prod_{e_{ij}=1} v_j = r_i^2 \prod_{e_{ij}=1} s_j^2 v_j = r_i^2 = x_i\ (\mathrm{mod}\ n)$$

A forger would need to be able to find a message M, a vector y, and a matrix E such that the first kt bits of $f(M, z_1, \ldots, z_t)$ are the entries of E where

$$z_i = y_i^2 \prod_{e_{ij}=1} v_j\ (\mathrm{mod}\ n)\ \text{for}\ i = 1, \ldots, t.$$

Fiat and Shamir [Fiat 87] suggest that if a forger were able to do this then he could, with non-negligible probability, find a pair of signatures (y, E) and (y', E') such that

$$y_i^2 \prod_{e_{ij}=1} v_j = y'^2_i \prod_{e'_{ij}=1} v_j.$$

But this would then enable him to determine a square root modulo n of some product $\prod_{j=1}^k v_j^{c_j}\ (\mathrm{mod}\ n)$ where for $j = 1, \ldots, k$, $c_j = -1, 0,$ or 1, not all zero. This would then allow the forger to factor n. Thus, provided n is difficult to factor and f is a suitable hashing function, it is computationally infeasible to forge signatures in this scheme.

3.2.4 The Goldwasser–Micali–Rivest signature scheme.

In [Goldwasser 84] Goldwasser, Micali, and Rivest define claw-free permutations and a signature scheme that relies on the fact that squaring modulo n (where $n = pq$ with p and q secret large primes) is a trapdoor one-way function.

Before we can describe their scheme we must introduce some notation.

For any integer n we let Z_n^* denote the integers between 1 and n which are coprime to n. If a is an integer that is coprime to the prime p, the Legendre symbol $(a/p) = 1$ if a is a quadratic residue modulo p, and $(a/p) = -1$ if a is a quadratic nonresidue modulo p. If $n = pq$ then the Jacobi symbol $(a/n) = (a/p)(a/q)$ for $a \in Z_n^*$. Thus a is a quadratic residue modulo n if and only if $(a/p) = (a/q) = 1$. We note that $(-1/n) = 1$ but -1 is not a quadratic residue and $(2/n) = -1$.

If we put $D_n = \{x \in Z_n^* \mid (x/n) = 1 \text{ and } 0 < x < n/2\}$ then we can define two functions $g_0 : D_n \to D_n$ and $g_1 : D_n \to D_n$:

$$g_0(x) \left\}\begin{array}{l} = x^2 \ (\text{mod } n) \text{ if } x^2 \ (\text{mod } n) < n/2 \\ = -x^2 \ (\text{mod } n) \text{ if } x^2 \ (\text{mod } n) > n/2 \end{array}\right.$$

$$g_1(x) \left\}\begin{array}{l} = 4x^2 \ (\text{mod } n) \text{ if } 4x^2 \ (\text{mod } n) < n/2 \\ = -4x^2 \ (\text{mod } n) \text{ if } 4x^2 \ (\text{mod } n) > n/2 \end{array}\right.$$

Both are trapdoor claw-free permutations with knowledge of the primes p and q as the trapdoor. If the factorization of n is known then it is easy to determine whether or not an element of D_n is a quadratic residue or a nonresidue. Furthermore, given this knowledge, the computation of the square root (in D_n) of a quadratic residue is also straightforward. Finally, before describing their scheme we note that if x and y in D_n are such that $g_0(x) = g_1(y)$, that is, $x^2 = 4y^2 (\text{mod } n)$ then $(x - 2y)(x + 2y) = 0(\text{mod } n)$ but $x - 2y \neq 0(\text{mod } n)$. Thus the greatest common divisor of $x - 2y$ and n is a nontrivial factor of n. This, of course, means that finding a pair with $g_0(x) = g_1(y)$ would result in being able to factor n.

The signature scheme of Goldwasser, Micali, and Rivest has a set of messages $\mathcal{M}$ consisting of (finite) sequences with terms from the set $\{0, 1\}$. The secret information used in signing a message is the factorization of n which is such that $p = 3(\text{mod } 4)$ and $q = 7(\text{mod } 8)$. It is this information that allows g_0 and g_1 to be inverted. The public information used in verifying a signature consists of the functions g_0 and g_1 and a verification parameter r from D_n.

To sign a message $m = m_1 m_2, \ldots, m_k$ where $m_i \in \{0, 1\}$ for $i = 1, \ldots, k$ the signature $S = S_X(m) = g_{<m>}^{-1}(r) = g_1^{-1}(g_0^{-1} (g_{m_1}^{-2}(\ldots (g_{m_k}^{-2}(r)) \ldots))) $ is computed. To verify a signature S on message m it is checked that $g_{<m>}(S) = g_{m_1}^2(\ldots (g_{m_k}^2(g_0(g_1 (S)))) \ldots) = r$. Thus $V_X(m, S) = $ True if and only if $g_{<m>}(S) = r$.

Provided that some means of authenticating them can be established, different values for the verification parameter r may be used for different messages. This is certainly feasible and, as an illustration, we now show how a tree structure is used to generate authenticated verification parameters.

Suppose that f_0 and f_1 are a pair of trapdoor claw-free permutations and let $r_0 \in D_n$ be a public, authenticated verification parameter. Then we can use r_0 to authenticate two (randomly chosen) values r_{00}, r_{01}. Let $<r_{00}>$, $<r_{01}>$ be encodings of r_{00}, r_{01} as sequences of terms from $\{0, 1\}$. Put $T_0 = f^{-1}_{<r00><r01>}(r_0)$. Each of r_{00}, r_{01} can now be used to authenticate an authentication parameter r (chosen at random from D_n) which itself can be used to sign a message m. Put $T_r = f^{-1}_{<r>}(r_{00})$ and $S = g^{-1}_{<m>}(r)$. Then the signature on m becomes $(S, T_r, r, T_0, r_{00}, r_{01})$ which is verified by checking that $f_{<r00><r01>}(T_0) = r_0$ (the published verification parameter), $f_{<r>}(T_r) = r_{00}$ and $g_{<m>}(S) = r$.

Clearly r_{00} and r_{01} could then be used to authenticate four values $r_{000}, r_{001}, r_{010},$ and r_{011} in a similar way. By generating a tree of values "r" in this way any number of verification parameters r may be authenticated and used to sign messages.

3.2.5 Digital signatures using Rabin's public key cryptosystem. In [Rabin 79] Rabin proposed a public key cryptosystem based on the one-way function $x \to x^2$ (mod n) where $n = pq$ as the product of two large primes. This cryptosystem is provably equivalent to factoring n.

Each user X has a public key p_X consisting of $n(= pq)$ and an integer $B < n$ chosen by user X. The secret key s_X is knowledge of p and q.

The message set $\mathcal{M}$ is a subset of the integers between 0 and $n - 1$. A message $m \in \mathcal{M}$ is encrypted as the cryptogram $C = E_{p_X}(M)) = M(M + B) \pmod{n}$. (Note that if w is any square root of 1 modulo n then $E_p(M') = C$ where $M' = -\frac{1}{2}B + w(M + \frac{1}{2}B)$. Thus, for C to be uniquely decipherable it is necessary that whenever $M \in \mathcal{M}$, $M' - \frac{1}{2}B + w(M + \frac{1}{2}B) \notin \mathcal{M}$ for w satisfying $w^2 = 1\pmod{n}$, $w \neq 1$.

Given a cryptogram C, it may be deciphered by finding a solution $x = M$ to the equation $x^2 + Bx = C\pmod{n}$ which is equivalent to finding a solution $y = M + \frac{1}{2}B$ to the equation $y^2 = C + \frac{1}{4}B^2\pmod{n}$. This latter equation has a solution only if $C + \frac{1}{4}B^2$ is a quadratic residue modulo n. Furthermore, when this condition is satisfied, there are four solutions for y. However only one satisfies $M = y - \frac{1}{2}B \in \mathcal{M}$. With the knowledge of the factors p and q of n (i.e., the secret key s_X) it is possible to extract the square-root of $C + \frac{1}{4}B^2\pmod{n}$ and thus decipher C, that is, to compute $M = D_{s_X}(C)$. Moreover, since being able to extract square root modulo n implies being able to factor n, this public key cryptosystem is secure provided n is difficult to factor.

Rabin's public key cryptosystem is not immediately applicable as a digital signature scheme since it is only possible to decipher a value C if $C + \frac{1}{4}B^2$ is a quadratic residue modulo n. So not all messages could be signed using D_{s_X}. However, it is possible to combine this cryptosystem with a suitably chosen hashing function h to provide digital signatures. The hashing function is used to map a subset of the set of messages $\mathcal{M}$ onto the set of quadratic residues modulo n. The guarantee that the hashed version is a quadratic residue is achieved in the following way.

A random string U (of fixed size) is concatenated with the message $M \in \mathcal{M}$ to be signed and $h(M, U)$ is computed. User X, in possession of his secret key s_X (i.e., the factors p and q of n) can determine whether or not $h(M, U)$ is a quadratic residue modulo n. If it is not, then another random string is chosen. The process is repeated until the value $h(M, U)$ is a quadratic residue modulo n. The signature $S_X(M)$ on message M is then the pair (S, U) where $S = D_{s_X}(h(M, U)$. A signature (S, U) is verified by checking that $h(M, U) = E_{p_X}(S) = S(S + B)\pmod{n}$.

3.4 Merkle's Tree Signature Scheme

In [Merkle90] Merkle proposed a digital signature scheme, based on a symmetric encryption function that provides an infinite tree of one-time signatures.

The basic building block of Merkle's scheme is a one-way function f based on a symmetric encryption function such as the DES. Thus, if the encryption transformation under key k is denoted E_k and X is a fixed (public) plaintext value then $f(k) = E_k(X)$. Merkle proposed the use of f to provide verification parameters in an adaptation of the Diffie–Lamport one-time signature scheme. Let h be a (public) hashing function which maps a message M to a string $h(M)$ of n bits. The hashing function h may be constructed from the one-way function f. If $c(M)$ is a count of the number of zero bits in $h(M)$, then $(h(M), c(M))$ is a string $(a_1, \ldots, a_t)$ of $t = n + \log n$ bits. The signer generates $x_1, \ldots, x_t$ at random and puts $y_i = f(x_i)$ for $i = 1, \ldots, t$. Suppose that $y_1, \ldots, y_t$ can be authenticated. Then the message M can be signed by revealing $y_1, \ldots, y_t$ and those x_i for which $a_i = 1$. A signature $S = (y_1, \ldots, y_t, x_{j_1}, \ldots, x_{j_l})$ is verified by checking that $y_1, \ldots, y_t$ are authentic and that $a_i = 1$ for $i = j_1, \ldots, j_l$ and $a_i = 0$ for $i \neq j_1, \ldots, j_l$. Only the signer, who knows $x_1, \ldots, x_t$

can generate this signature. Moreover, no other message M' can be signed using the knowledge of the revealed $x_{j_1}, \ldots, x_{j_i}$ since either $h(M')$ will have a "1" where $h(M)$ has a "0" (requiring a "new x_j" to be revealed) or $c(M')$ will be greater than $c(M)$ and so $c(M')$ will have a "1" where $c(M)$ has a "0".

The verification parameters $y_1, \ldots, y_t$ could be authenticated by publishing them in a registry. However, since the Diffie-Lamport signature is a one-time signature, to produce a large number of signatures would then require the publication of a large amount of data.

To avoid this complication, verification parameters are generated in a tree structure. Each node of this tree consists of the verification parameters that are used to sign a message and authenticate those verification parameters of subsequent nodes.

The signer generates $u_1, \ldots, u_t, x_1, \ldots, x_t$ and $w_1, \ldots, w_t$ at node 1 and puts $v_i = f(u_i)$, $y_i = f(x_i)$, and $z_i = f(w_i)$ for $i = 1, \ldots, t$. The value $R = h(h(v_1, \ldots, v_t), h(y_1, \ldots, y_t), h(z_1, \ldots, z_t))$ is published in a registry and is used to authenticate the verification parameters $v_1, \ldots, v_t, y_1, \ldots, y_t$, and $z_1, \ldots, z_t$.

The values $x_1, \ldots, x_t$ and $y_1, \ldots, y_t$ are used to sign the first message as above. (Note that the values $V = h(v_1, \ldots, v_t)$ and $Z = h(z_1, \ldots, z_t)$ must be included in the signature so that the values $y_1, \ldots, y_t$ may be authenticated by checking that $h(V, h(y_1, \ldots, y_t), W) = R$.) Values $u'_1, \ldots, u'_t, x'_1, \ldots, x'_t$, and $w'_1, \ldots, w'_t$ and values $u''_1, \ldots, u''_t, x''_1, \ldots, x''_t$, and $w''_1, \ldots, w''_t$ are generated for the next two nodes, nodes 2 and 3, and the corresponding images under f computed. The value $R' = h(h(v'_1, \ldots, v'_t), h(y'_1, \ldots, y'_t), h(z'_1, \ldots, z'_t))$ is signed using $u_1, \ldots, u_t$ and $v_1, \ldots, v_t$. Again the values $Y = h(y_1, \ldots, y_t)$ and $Z = h(z_1, \ldots, z_t)$ are included in this signature so that $v_1, \ldots, v_t$ can be authenticated by checking that $h(h(v_1, \ldots, v_t), Y, Z) = R$.

The signature on the second message M' consists of this signature on R' and a signature on M' computed as above using $x'_1, \ldots, x'_t$, $y'_1, \ldots y'_t$, $V' = h(v'_1, \ldots, v'_t)$, $Z' = h(z'_1, \ldots, z'_t)$ and R'.

Similarly the value $R'' = h(h(v''_1, \ldots, v''_t), h(y''_1, \ldots, y''_t), h(z''_1, \ldots, z''_t))$ is signed using $w_1, \ldots, w_t$ and $z_1, \ldots, z_t$ and a third message M'' may be signed.

Now the nodes 2 and 3 can be used to generate authenticated verification parameters at two more nodes each. In this way any number of messages may be signed. The signature on the message at node 2^i has length $i + 1$ times that of the signature on the first message. Only the value R is stored in the registry, all other verification parameters are authenticated by including signatures on them in the signature on the message. Thus the problem of one-time signatures requiring large amounts of verification data is overcome by allowing the size of the signature to grow as the number of messages signed increases.

4 TECHNIQUES FOR HASHING

Many proposals have been made for one-way hash functions, and we do not attempt to list all such proposals here; instead we describe some of the more significant. Unfortunately, a high proportion of the existing proposals have turned out to be flawed in some way, perhaps indicating that designing easily implemented and secure hash functions is not as easy as it appears.

It is interesting to note that most of the published proposals for hash functions fall into one of two categories: those based on use of a block cipher, and those based on modular arithmetic. We use this classification to survey some of the more important proposed schemes and, where relevant, their weaknesses.

4.1 Block Cipher–Based Hash Functions

In this section we consider hash functions built using a block cipher. By *block cipher* we mean here a symmetric key cryptosystem that transforms message (plaintext) blocks of fixed length (say, n bits) into ciphertext blocks of the same length, under the control of a key k (of, say, m bits). Thus, if M is an n-bit message block then

$$C = E_k(M)$$

is the corresponding n-bit ciphertext block when key k is used. In addition, we have

$$M = D_k(C) = D_k(E_k(M))$$

A widely used example is provided by the U.S. Standard DES algorithm, [ANSI3.92], which has $n = 64$ and $m = 56$.

To be secure against known-plaintext attacks, block ciphers require the following property. Given a pair (M, C) with the property that $C = E_k(M)$, or a set of such pairs, no means of discovering k exists that is significantly more efficient than searching through all 2^m possibilities for k. It is also normally assumed that the block cipher will resist chosen ciphertext/plaintext attacks, that is, where the value(s) of M or C can be chosen by the cryptanalyst. We assume here that the block cipher used to construct a hash function has all such desirable properties.

4.1.1 Cipher block chaining-message authentication code. Probably the most obvious way of using a block cipher to construct a message digest is based on the standard mode of use for a block cipher called cipher block chaining (CBC) [ANSI3.106], [ISO8372]. Note that this gives an example of a keyed hash function. In this mode of use the message is divided into a sequence of n-bit blocks,

$$M_1, M_2, \ldots, M_t$$

say. The sequence (C_i) is then derived, where

$$C_i = E_k(C_{i-1} \oplus M_i)$$

where $\oplus$ denotes bit-wise ex-or and C_0 is the all-zero block. The derived message digest is then simply C_t (or part of it), and we call C_t the CBC-MAC of the message. The CBC-MAC method of computing message digests is standardized for banking authentication applications, [ANSI9.9], [ANSI9.19], [ISO8731–1], and is also a draft standard for general authentication purposes, [DIS9797]. Note that, in these simple authentication and integrity applications, the CBC-MAC provides the desired security without subsequent signature; the key used to produce the digest is sufficient to guarantee the integrity and authenticity of the message.

Although satisfactory for simple authentication applications, the CBC-MAC is not, however, a secure means of producing a digital signature, as we now show. First observe that the recipient of a signed message must be equipped with the key k used to produce the digest. Given knowledge of k, the above hashing function does not satisfy $H2$, that is, it is not one-way. In fact, given any n-bit block, C say, and any sequence of n-bit blocks

$$N_1, N_2, \ldots, N_w$$

it is possible to choose a further n-block N_{w+1} such that the sequence

$$N_1, N_2, \ldots, N_{w+1}$$

has CBC-MAC C. This can be done by setting N_{w+1} to $D_k(C) \oplus C_w$ where C_w is the CBC-MAC derived from $N_1, N_2, \ldots, N_w$.

Another problem with the CBC-MAC is that, when used with an n-bit block cipher it gives a digest of at most n bits. The most well-known block cipher is DES [ANSI3.92] which has $n = 64$, which therefore gives digests of 64 bits—too small for use for computing digital signatures.

4.1.2 Bidirectional message authentication code.

Many attempts have been made to overcome the above problems by using a block cipher in different ways, and we consider some of them below. We start by considering a closely related scheme, called the bidirectional message authentication code (BMAC), originally described in Internet RFC 1040 [RFC1040]. The BMAC is again based on use of CBC; this has significant practical advantages since existing hardware implementations of DES can perform CBC very fast. As we shall see, the BMAC is again unsuitable for use in producing digital signatures—the attack that can be made against it is of significance since very similar attacks can be made against many other proposed hash functions.

The BMAC produces a message digest of $2n$ bits (given that the block cipher used to produce it produces ciphertext blocks of length n bits). Hence, when based on DES, the BMAC produces 128-bit digests, thereby avoiding one of the problems of the CBC-MAC. Suppose, as before, that the BMAC is to be computed for the sequence of n-bit blocks

$$M_1, M_2, \ldots, M_t$$

Let C_t be the CBC-MAC derived from this sequence and let B_1 be the CBC-MAC derived from the sequence

$$M_t, M_{t-1}, \ldots, M_1$$

The BMAC is then simply the concatenation of C_t and B_1.

Unfortunately, given knowledge of the key k used to compute it, and given n is of the order of 64, the BMAC is not one-way, that is, given any pair of n-bit blocks a message can be found having BMAC equal to this pair of blocks. Indeed, the method we now describe (previously given in [Mitchell90]) allows the message to be chosen arbitrarily, except that it must contain two 64-bit blocks which are essentially random in nature.

Suppose that the supplied pair of blocks is (C, B) and the given key is k. Suppose also that the message to be "matched" to the BMAC (C, B) is M, where M can be divided into two parts M_1 and M_2. In addition, choose at random an n-bit block X.

First prepare $2^{n/2}$ variants of M_1, each variant consisting of a whole number of n-bit blocks (although the number of blocks in each variant may vary). Davies and

Price ([Davies84], p. 278) illustrate a simple technique by which this may be done so that each variant is valid English and each variant is semantically the same. Basically, if q positions are identified within the message at which two possible wordings have the same meaning, then 2^q different variants of the message may be derived.

For each variant do the following. Suppose that

$$N_2, N_3, \ldots, N_s$$

is the decomposition of the variant into n-bit blocks (note that the first block of the variant is not defined yet). We now choose N_1 so that the "reverse" CBC-MAC of

$$N_1, N_2, \ldots, N_s, X$$

is B. We also let Y be the CBC-MAC of $N_1, N_2, \ldots, N_s$. More formally, for every i $(i = s, s - 1, \ldots, 2)$ let

$$F_i = E_k(N_i \oplus F_{i+1})$$

where $F_{s+1} = X$. Then let

$$N_1 = D_k(B) \oplus F_2$$

N_1 then constitutes the first block of this variant of M_1. For every i $(i = 1, 2, \ldots, s)$ let

$$G_i = E_k(N_i \oplus G_{i+1})$$

where G_0 is the all-zero block. Finally let $Y = G_s$. Each variant $(N_1, N_2, \ldots, N_s)$ is then stored along with its corresponding value of Y.

Second, prepare $2^{n/2}$ variants of M_2, each variant consisting of a whole number of n-bit blocks (again, the number of blocks in each variant may vary). For each variant,

$$P_{s+1}, P_{s+2}, \ldots, P_{r-1}$$

say, we now choose P_r so that the "reverse" CBC-MAC of

$$P_{s+1}, P_{s+2}, \ldots, P_r$$

is X. We also let Z be the value required to ensure that the sequence of blocks Z, $P_{s+1}, P_{s+2}, \ldots, P_r$ has CBC-MAC equal to C. More formally, for every i $(i = s + 2, s + 3, \ldots, r)$ let

$$H_i = D_k(H_{i-1}) \oplus P_{i-1}$$

where $H_{s+1} = X$. Then let

$$P_r = D_k(H_r)$$

P_r then constitutes the last block of this variant of M_2. For every i $(i = r - 1, r - 2, \ldots, s)$ let

$$L_i = D_k(L_{i+1}) \oplus P_{i+1}$$

where $L_r = C$. Finally let $Z = L_s$. This value of Z is then compared with all the stored values of Y resulting from the $2^{n/2}$ variants of M_1. There is a good chance that, before all $2^{n/2}$ variants of M_2 have been processed, a match, that is, a pair of values Y, Z with $Y = Z$, will be found (we justify this claim below).

Now suppose that the sequences (N_i) $(1 \le i \le s)$ and (P_j) $(s + 1 \le j \le r)$ give such a match. Then the sequence of n-bit blocks

$$N_1, N_2, \ldots, N_s, P_{s+1}, P_{s+2}, \ldots, P_r$$

will have BMAC (C, B), as desired.

The above attack does require a nontrivial amount of processing time and storage, although neither of these two requirements make the attack infeasible unless n is chosen to be substantially larger than 64. The attack requires the processing of some $2^{n/2+1}$ part-messages, and a number of block cipher encryptions/decryptions are required for each such part-message. However, most block ciphers are designed to run fast and if DES is used as an example (which has $n = 64$), then the amount of computation required to perform the above attack is not excessive, even without massive computing resources.

Finally, note that we asserted that the probability of a match being found was good, without giving any justification for such a claim. In general, if samples of size u and v are drawn independently, at random, and with replacement from a population of size N, then the probability that there will be a match between the two samples is approximately

$$1 - e^{uv/N}$$

given that u and v are small compared with N. Thus given n is reasonably large, the probability of a match in the above argument is approximately $1 - 1/e$ which is around 0.63.

In conclusion, it should be clear that the BMAC is not one-way when used with a 64-bit block cipher such as DES. Moreover, given a 128-bit block cipher (which would defeat the above attack) then other options appear more attractive, such as the Davies–Meyer (DM) scheme that we now describe.

Before proceeding note that the above attack is just one example of a well-known class of attacks against hash functions, namely the so-called ''birthday attacks.'' All these attacks rely on arguments relating to the probability of a match being found within a set of samples from a large set.

4.1.3 The DM scheme.

We now describe a different and apparently more secure way of using a block cipher to construct a hash function. This scheme apparently dates back to the early 1980s and is variously attributed to Davies [Winternitz84a] and Meyer [Davies85]; we therefore follow Quisquater and Girault [Quisquater89] and refer to it as the DM scheme. This scheme is also the subject of a draft proposal for an international standard [DP10118].

As before, the message to be signed is first divided into a series of fixed length blocks; this time, however, the block length is m (the key length for the block cipher) rather than n. Suppose the message to be signed is

$$M_1, M_2, \ldots, M_t$$

where each M_i contains m bits. Then let H_i $(1 \le i \le t)$ be defined by

$$H_i = E_{M_i}(H_{i-1}) \oplus H_{i-1}$$

where $H_0 = I$, an "initializing value" which may be prearranged (e.g., to 0, the m-bit block of all zeros) or randomly chosen and therefore forms part of the digest, and, as before, $\oplus$ represents bitwise exclusive-or. The message digest is then simply H_t, an n-bit block. At this point note that, if a random initializing value is used then, to prevent certain types of attack, it must be signed along with the digest; this warning applies to all the schemes described below.

This hash function is of particular interest since, if it is assumed that no special properties of the underlying block cipher are used, then it can be proved to be one-way [Winternitz84b]. However, it is still not collision-free unless n (the plaintext/ciphertext block size for the cipher) is sufficiently large, say 128 bits. Thus, for digital signature applications, the DM scheme is not appropriate for use with 64-bit block ciphers such as DES.

Because of the widespread use of 64-bit block ciphers (in particular, DES), efforts have recently been made to modify the DM scheme so as to produce $2n$-bit message digests when using an n-bit block cipher. The following scheme appears in the latest draft standard DP10118 [DP10118].

The message to be signed is again divided into a series of m-bit blocks; this time, however, the number of blocks must be even (as with previous schemes, appropriate padding rules need to be devised for messages that do not divide conveniently). Note also that m and n must satisfy $m \leq n$. Label the sequence of blocks

$$M_1, M_2, \ldots, M_{2t}$$

Then let H_i and G_i $(1 \leq i \leq 2t)$ be defined by

$$G_{2j-1} = E_{M_{2j-1}}(H_{2j-3} \oplus e(M_{2j})) \oplus e(M_{2j}) \oplus H_{2j-2} \ (1 \leq j \leq t)$$
$$G_{2j} = E_{M_{2j}}(G_{2j-1} \oplus e(M_{2j-1})) \oplus e(M_{2j-1}) \oplus H_{2j-3} \ (1 \leq j \leq t)$$
$$H_{2j-1} = G_{2j} \oplus H_{2j-2} \ (1 \leq j \leq t)$$
$$H_{2j} = G_{2j-1} \oplus H_{2j-3} \ (1 \leq j \leq t)$$

where H_{-1} and H_0 are initializing values (possibly m-bit blocks of all zeros), $e(\ )$ is an "expanding function" mapping m-bit values onto n-bit values (e.g., by padding with zeros), and, as before, $\oplus$ represents bitwise exclusive-or. The message digest is then simply H_{2t-1} concatenated with H_{2t}, a $2n$-bit value.

A similar proposal has been published by Quisquater and Girault [Quisquater89]. Both of these schemes should be treated with great caution; the logic behind their construction is by no means obvious, and they need thorough review before practical use.

4.1.4 Some insecure schemes. We now consider a variety of other schemes based on block ciphers which have been first proposed and then found wanting. Their existence attests to the difficulty of finding secure hash functions that are usable in practice.

The DM scheme described above is a modification of a scheme described by Rabin [Rabin78], Matyas [Matyas79], and Davies and Price [Davies80]; we refer to this earlier scheme as the Rabin–Matyas–Davies–Price (RMDP) scheme. In the RMDP system the message to be signed is first divided into a series of m-bit blocks:

$$M_1, M_2, \ldots, M_t$$

Then let H_i $(1 \leq i \leq t)$ be defined by

$$H_i = E_{M_i}(H_{i-1})$$

where H_0 is an initializing value. The message digest is then simply H_t, an n-bit block. Merkle showed the RMDP function is not one-way when n is of the order of 64 (see, e.g., [Davies80] or Winternitz84a]) using a "birthday attack" argument very similar to that described in Section 4.1.2, above. We now briefly outline Merkle's attack. Before proceeding note that the basic idea behind this attack was given by Yuval in 1979 [Yuval79]; interestingly, Yuval also pointed out the need for a hash function to satisfy the collision-freedom property.

Consider any digest H with initializing value I. Choose any message and divide it into two parts. Devise $2^{n/2}$ variations of the first part and $2^{n/2}$ variations of the second part. Starting with I compute the digest for the $2^{n/2}$ variations of the first part of the message. Using the invertibility of E, start from H and work back to $2^{n/2}$ values using the variants of the second part of the message. As in Section 4.1.2 there will be a better than 50% probability of finding a first and second part that "match" and a message will have been found to fit the given digest. Thus the RMDP scheme is not one-way unless n is chosen to be substantially larger than 64.

It is straightforward to see that the DM scheme is a simple derivative of the RMDP system which avoids the obvious "birthday attack." Other methods of modifying the RMDP function have been proposed, and we now briefly consider two of them.

Davies [Davies83] describes a scheme attributed to Bitzer, where the message digest

$$M_1, M_2, \ldots, M_t$$

(M_i contains m bits) is derived by computing

$$H_i = E_{M_i'}(H_{i-1})$$

where $M_i' = M_i + s(H_{i-1})$, $s(\ )$ is a "selection function" reducing an n-bit block to an m-bit block, and H_0 is an initializing value. The message digest is then H_t, an n-bit block. As Winternitz has pointed out [Winternitz84a], this technique is vulnerable to a very similar attack to that described above for the RMDP scheme.

Davies and Price [Davies80] (quoted by Denning [Denning83]) suggest using the RMDP scheme twice. In 1983, Winternitz ([Winternitz84a], [Akl84]) showed how this method could be attacked if the DES block cipher is used, based on special properties of the DES algorithm. Subsequently, in 1985 Coppersmith [Coppersmith86] gave a general birthday attack applicable regardless of the block cipher employed, this time a little more complex but perfectly feasible for 64-bit block ciphers. Coppersmith also considered the case where the RMDP scheme is iterated three times and again exhibits a feasible birthday attack. Coppersmith's method was then generalized by Girault, Cohen, and Campana [Girault89], who described how to attack a p-times-iterated version of RMDP, which when $p = 4$ and $n = 64$ is just about feasible.

A number of other possibilities have been described; we consider one of them. This is a variant of CBC described by Meyer and Matyas [Meyer82]. Let the message be

$$M_1, M_2, \ldots, M_t$$

(M_i contains n bits). The digest, H, is derived by first computing

$$W = M_1 \oplus M_2 \oplus \ldots \oplus M_t$$

and then computing H as the CBC-MAC of the sequence

$$M_1, M_2, \ldots, M_t, W$$

Unfortunately, as described by Akl [Akl84] this method is yet again susceptible to a birthday attack when n is of the order of 64. Moreover this scheme has a number of other serious weaknesses (see, e.g., [Jueneman83b], [Jueneman83], [Meijer83], [Akl84]). Although Jueneman et al. [Jueneman83] and Akl [Akl84] have described a number of improvements to this scheme, including using addition modulo 2^n instead of bit-wise exclusive-or and the possibility of multiple iterations, it still does not appear an attractive option, since undesirable properties remain.

4.1.5 Merkle's methods.   Merkle has described a number of methods for deriving one-way hash functions from block ciphers [Merkle89a]. He starts by describing a "metamethod" (also described in his Ph.D. thesis [Merkle79], [Merkle82]) which builds a hash function h producing k-bit digests using a function h_0 mapping from s-bit values to k-bit values for some fixed $s > k$. It operates as follows. Divide the message to be hashed into a sequence of blocks, each of $s - k$ bits, say,

$$M_1, M_2, \ldots, M_t$$

Then let

$$H_i = h_0(H_{i-1}; M_t)$$

where the semicolon denotes concatenation of data and where H_0 is an initializing value. The message digest is then H_t. This metamethod forms a convenient general description for most of the methods described above; for example, in the DM scheme of Section 4.1.3,

$$h_0(H_{i-1}; M_i) = E_{M_i}(H_{i-1}) \oplus H_{i-1}$$

where $s = m + n$. Note also that this metamethod is the basis of recent theoretical work on hash functions by Damgard [Damgard89].

In [Merkle89a], a modified form of the DM scheme is described which, in the above notation and for some chosen d, uses an h_0 having $s = m + n - 1$ and $k = m + n - d$, that is, it processes a message $d - 1$ bits at a time. It operates as follows.

Suppose, as before, that

$$M_1, M_2, \ldots, M_t$$

is a decomposition of a message into $(d - 1)$-bit blocks. Let h_0^* be the DM metafunction, that is,

$$h_0^*(A; B) = E_B(A) \oplus A$$

where A is an n-bit block, B is an m-bit block, and h_0^* maps $(m + n)$-bit blocks into n-bit blocks. Then define h_0 by

$$h_0(H_{i-1}; M_i) = Tr_d(h_0^*(H_{i-1}; 0; M_i); h_0^*(H_{i-1}; 1; M_i))$$

where Tr_d denotes truncation by omitting d bits of the block and 0 and 1 denote single bits fixed to 0 and 1. The hash function obtained from this metamethod will clearly produce an $(m + n - d)$-bit result.

As Merkle points out, d needs to be chosen so that the value $m + n - d$ is still sufficiently large to rule out collision attacks. For DES (where $m = 56$ and $n = 64$), Merkle suggests using $d = 8$, that is, the message digests will contain 112 bits, while the message is processed in 7-bit blocks.

Unfortunately this method is rather inefficient (two DES encryptions are required to process 7 bits of message), and Merkle [Merkle89a] goes on to describe two other, rather more sophisticated, means for deriving hash functions from block ciphers which are more efficient. These other methods also have the virtue of retaining the ability to transform block ciphers with relatively small m and n (e.g., DES) into apparently collision-free hash functions.

Before proceeding we briefly describe another rather interesting hashing scheme based on block ciphers (see Merkle [Merkle79], [Merkle82] and also Akl [Akl84]). In this scheme the metamethod h_0 is given by

$$h_0(H_{i-1}; M_i) = E_{H_{i-1}; M_i}(I)$$

where I is initializing value. For this method to work, m (the key size) must be larger than n (the plaintext/ciphertext block size), and the message is then divided into $(m - n)$-bit blocks M_i. For this method to be collision-free it is also necessary that n be sufficiently large, that is, at least 100 bits. In the absence of block ciphers satisfying these properties the method is of rather academic interest, but it is nevertheless of importance in suggesting other ways in which practical and secure hash functions may be constructed.

4.1.6   Conclusions.

The use of existing block ciphers to construct one-way hash functions is often appealing. However, as should be clear from the above discussion, such an approach requires great care. However, the following recommendations can be made.

If a 128-bit block cipher (i.e., one with $n = 128$) is available then the DM scheme appears secure given that the block cipher has no regularities that a cryptanalyst could exploit (the proof of the one-way property for the DM scheme assumes a "perfect" block cipher). Unfortunately, although designing a practically secure 128-bit block cipher is probably not difficult (e.g., by modifying DES appropriately), no such algorithm exists either as a standard or even a de facto standard. The only widely used and trusted block cipher algorithm is the DES, which has $n = 64$. Therefore, in practice the real problem is to derive a secure hash function from a 64-bit cipher.

If a 64-bit block cipher is to be used, then, as can be seen from the many failures described above, the problem is much more difficult. The only candidate systems currently available are the scheme given in ISO DP10118 [DP10118] (see Section 4.1.3) and the schemes of Merkle (see Section 4.1.5). None of these schemes have been subjected to prolonged public scrutiny, and use of them at this stage would be rather risky. Other possibilities include multiple iterations of existing hash functions, although this approach can also be flawed.

In the context of existing knowledge, Winternitz's comments from 1984 [Winternitz84a] remain extremely appropriate:

System implementors would be well-advised to use as much overkill as they can afford. They should go through the message several times and use a hash value as long as possible. Given our ignorance, safety requires a system several times as complicated as the simplest system yet unbroken.

4.2 Hash Functions Based on Modular Arithmetic

Another widely discussed method for constructing hash functions is the use of modular exponentiation. This is especially attractive if the signature function itself is based on modular exponentiation (e.g., the RSA algorithm), since software or hardware must in any case be present for performing the desired operations.

4.2.1 Jueneman's methods. We start by considering various methods proposed by Jueneman. In 1982, Jueneman proposed the following hash function, which he called the quadratic congruential manipulation detection code (QCMDC) [Jueneman83b], [Jueneman83]. The message is first divided into m-bit blocks

$$M_1, M_2, \ldots, M_t$$

where M_i is regarded as a number between 0 and $2^m - 1$. Then

$$H_i = (H_{i-1} + M_i)^2 \ (\text{modulo } C)$$

where $+$ denotes integer addition, H_0 is an initializing value (which Jueneman suggests should be randomly chosen for each message and kept secret), and C is a prime satisfying $C \geq 2^m - 1$. Jueneman goes on to suggest using $m = 16$ and $C = 2^{31} - 1$ (a Mersenne prime).

The fact that H_0 needs to be secret implies that QCMDC is a keyed hash function. Note also that Jueneman, Matyas, and Meyer, [Jueneman83] suggest that an additional secret key could be added on to the front of the message as M_1 (the actual message starting at M_2). In fact, the original proposed use of QCMDC was not for digital signature but for simple message authentication; however, the requirements for the function remain very similar.

Because of a variety of birthday attacks that they had discovered in conjunction with Coppersmith, by 1985 Jueneman, Matyas, and Meyer [Jueneman85] were suggesting computing the QCMDC four times for a message (using the digest obtained from the ith iteration as the initializing value for the $(i + 1)$th iteration) and then concatenating the results to obtain a 128-bit digest. However, as discussed in [Jueneman87b], the four-times-iterated version of the QCMDC is still subject to the type of attack devised by Coppersmith in 1985 [Coppersmith86] to attack the doubly iterated RMDP hash function (see Section 4.1.4, above).

As a result of Coppersmith's work, Jueneman, Matyas, and Meyer proposed a revised function called QCMDC Version 4 (QCMDCV4) ([Jueneman83], [Jueneman87b], [Jueneman87]). This is essentially a more complex version of the four-times-iterated QCMDC, although this time the four iterations are done in parallel and are cross-linked. Although designed to resist attack, Merkle [Merkle89a] reports that this function has also been broken by Coppersmith; details of Coppersmith's latest attack do not appear to have been published as yet.

4.2.2 The TeleTrust/Open Shop Information System method. Proposals similar to those of Jueneman's QCMDC were published as early as 1980 by Davies and

Price [Davies80]; an elaborated version is described in their 1985 paper [Davies85].
They suggest dividing the messages into blocks of $(m - d)$ bits, say

$$M_1, M_2, \ldots, M_t,$$

and then computing

$$H_i = (H_{i-1} \oplus M_i)^2 \; (\text{modulo } C)$$

where $h_0 = 0$ and C is of the order of 2^m. The message digest is then H_t. Note that this
differs from Jueneman's QCMDC in two significant ways. First, exclusive-or is used
instead of integer addition, and second, Davies and Price suggest the use of a much
larger modulus. They suggest choosing m to be 512 and d to be 64.

The reason for choosing message blocks significantly smaller than the modulus is
to "add redundancy" and thereby prevent the sort of attack that works on the CBC-
MAC (see Section 4.1.1). Thought of in these terms, the effect of requiring M_i to be at
most $2^{m-d} - 1$ ensures that the most significant d bits of M_i are set to zero. More
specifically, if this redundancy was not present, then, given a prespecified digest, all
but one block of the message could be chosen arbitrarily and the remaining block chosen to make the digest "come out right."

Unfortunately, choosing the redundancy in this way does not make the hash function collision-free, as shown by an attack due to Jung and given in Girault [Girault88].
This attack uses ideas previously published by De Jonge and Chaum [DeJonge86]. Alternative simple methods for adding redundancy such as requiring a fixed number of
zeros to be at the least significant end of M_i or requiring fixed numbers of ones at either
end of M_i also fail to achieve the desired objective ([DeJonge86] and [Girault88]).

As a result a more complex method of adding redundancy was devised as part of
the European Open Shop Information System (OSIS)-TeleTrust project, [OSIS85]. It
was also subsequently quoted in a non-normative annex of CCITT Recommendation
X.509 [X.509] and was proposed for adoption as an international standard [DP10118].
The hash function operates as follows.

First suppose that C (the modulus) satisfies $2^{8m} \leq C$, that is, C contains at least
$8m + 1$ bits. The message is divided into blocks of $4m$ bits

$$M_1, M_2, \ldots, M_t$$

and the message digest is then computed by

$$H_i = (H_{i-1} \oplus R(M_i))^2 \; (\text{modulo } C)$$

where $R(M_i)$ denotes the $8m$-bit block obtained from M_i by adding blocks of four 1's
between every set of 4 bits of M_i, and $H_0 = 0$. The message digest is then H_t. Note that
the draft International Standard DP10118 [DP10118] allows the exclusive-or operation to
be replaced with integer addition, as in the Jueneman QCMDC hash function.

Note also that the version specified in draft standard DP10118 allows H_0 to be set
to an initializing value I, which may be randomly chosen (instead of being fixed at 0).
As pointed out by Jefferies and Walker [Jefferies88], this allows the construction in a
very simple way of two different messages having the same hash (given that the factorization of the modulus is known), albeit that the two initializing values are different.
This means that if a random initializing value is used then it must be regarded as part of
the message digest, and signed with the digest.

However, all these efforts to standardize the OSIS hash function have been nullified by recent work of Coppersmith [Coppersmith89]. Coppersmith has been able to show that this hash function is not secure when used in conjunction with RSA signatures (and probably any other signature technique based on modular exponentiation). His cryptanalytic methods are based on the observation that the redundancy method used in the hash function is "almost invariant" under integer multiplication by 256, that is, left-shifting by 8-bit positions. This essentially discredits the hash function since it was always intended for use with RSA signatures. Indeed, the advantage of using modular arithmetic for computing the digest disappears when other signature techniques are in use.

Before proceeding note that Girault [Girault88] gives a brief review of other hash functions based on modular exponentiation, including the use of exponents other than 2. However, none of the possibilities given by Girault appears any more promising than the schemes discussed above.

4.2.3 Conclusions. As can be seen from the above discussion, hash functions based on modular squaring have a fairly awful history; indeed, there appear to be no public proposals for hash functions of this type that have not been discredited. One reason for wishing to construct hash functions using modular exponentiation, namely, the fact that the signature algorithm may also use modular exponentiation, may in fact be a good argument *against* their adoption. Coppersmith's attack on the OSIS hash function makes use of the fact that both the hash function and signature technique (namely, the RSA algorithm) are based on modular arithmetic.

It would therefore seem prudent to use hash functions whose method of computation has very little in common with the method of computation used in the signature function. This would mitigate against the use of hash functions based on modular squaring when used with either the RSA or Fiat–Shamir algorithms. Moreover, in other situations the relatively high complexity of computing modular exponents would appear to make the use of such hash functions an unlikely choice. Overall, such hash functions appear to have a very limited future.

4.3 Other Hash Functions

We conclude our discussion of practical examples of hash functions by considering some examples that do not fit so easily into the above two categories. However, on close examination, two of the schemes discussed, namely, the schemes of Rivest and Merkle, can be regarded as being based on the use of novel block ciphers.

We start by mentioning two schemes proposed by Rivest and RSA Data Security, Inc. Both these schemes produce a 128-bit digest from a message of arbitrary length. However, the two methods are very different in their mode of operation, although they are both designed to be implemented very fast in software. The first, called MD2, appears in Internet RFC 1115 [RFC1115]. MD2 has the property that it processes a message 1 byte at a time. The second, called MD4, was posted in February 1990 on the "usenet" electronic bulletin board; this function processes a message 512 bits at a time. Both algorithms are based on complex nonlinear functions which practical experience indicates are difficult to invert, similar perhaps to the philosophy used to design DES-like ciphers. Hence, while they appear secure, formal proofs of the security of these hash functions are unlikely to be forthcoming.

Merkle's "Snefru" algorithms share some of the same "intuitive" design characteristics. In 1989, Merkle published the first version of his Snefru algorithm [Merkle89b] which is a keyed hash function; however, it may also be used as an unkeyed hash function. Recently (November 1989) Merkle released Snefru Version 2.0 [Merkle89b], a modified version, and software implementations of this algorithm are available from Merkle at Xerox PARC. Snefru Version 2.0 has a number of versions, including the option of producing either 128- or 256-bit digests. In April 1990, Merkle announced on Internet that Eli Biham has found two different messages hashing to the same 128-bit value using one version of Snefru. Although other versions of Snefru remain unbroken, this new attack must cast some doubt on the security of all versions of Snefru (particularly while Biham's attack remains unpublished).

Given the demise of Snefru, MD4 appears the most promising candidate for future use; of course, before it is widely adopted it needs to withstand the test of time. It is interesting to note that another hash function of the Snefru/MD2/MD4 type already exists as an international standard! This is ISO 8731-1 [ISO8731-1], adopted as long ago as 1987. Unfortunately, this algorithm was not designed with digital signatures in mind and therefore only produces a 32-bit digest using a 64-bit key.

To complete this discussion we mention certain other work which is of theoretical interest. Damgard [Damgard88] and Gibson [Gibson88a] have discussed examples of hash functions that are in some sense provably collision-free, although at the expense of being rather impractical. Godlewski and Camion [Godlewski89] consider hash functions based on error-correcting codes and "random knapsacks," although their work appears of little immediate practical significance. As in other areas of cryptography, it remains to be seen whether the "provably secure" techniques will become the main practical option.

5 APPLICATIONS FOR DIGITAL SIGNATURES

We conclude this chapter by considering some of the more significant applications of digital signatures.

5.1 Public Key Certification

It has been recognized since their invention that public key cryptosystems offer considerable advantages over conventional cryptosystems when used for key management. Their use for general data encryption is less attractive, particularly as the most widely accepted public key cryptosystem, that is, the RSA algorithm, is nontrivial to implement, and on conventional personal computers (PCs) offers relatively low throughput speeds even when implemented in optimized software.

There is an obvious advantage to the use of public key cryptography for key management in a network of communicating entities that do not have prearranged pairwise keying relationships. If conventional cryptography is used, then one or more online key distribution centers (KDCs) are needed. On the other hand, if public key cryptography is used then every user's public key can be stored in one or more publicly available lists, and no active intervention by third parties is required.

This latter scenario fails to mention one very important requirement, namely, that the user of a public key must have some means of verifying its authenticity. One solution to this problem is the use of *key certification*, an idea that, according to Denning

[Denning83], was originally proposed by Kohnfelder [Kohnfelder78] in 1978. A discussion of key certification may also be found in Davies [Davies83].

In this system, every user's public key is signed by a certification authority (CA); this signed key is then stored in the public list of keys. This CA must be trusted to only sign valid keys, and its public verification transformation must be known (in a trusted sense) to all users of the system. The key together with the signature is then usually referred to as a certificate for that user.

In practice it is necessary for the certificate to contain the name of the key owner as well as the public key itself. In addition, the certificate may also contain an expiration date and/or an identifier for the algorithm with which the public key is to be used ([Denning83], [X.509]). A typical certificate might then have the form

$$A, p_A, T, I, S_C(A, p_A, T, I)$$

where A is the name of the user, p_A is A's public key, T is the expiration date for p_A, I is the identifier of the algorithm with which p_A is to be used, and S_C is the secret signature transformation of the CA C. Note that the importance of the use of one-way hash functions to prevent manipulation of signed data is pointed out in a paper of Gordon [Gordon85], in which an attack on certificates created without the use of a hash function is described.

Certification of public keys appears to be one of the most promising application areas for digital signatures. The 1988 version of the CCITT X.500 Directory Recommendations specifies how public key certificates may be stored in user directory entries; see, in particular, CCITT Recommendation X.509 [X.509]. The idea of digital signature-based public key certificates has also been adopted to provide key management for Internet electronic mail security. Internet RFC 1114 [RFC1114] specifies the use of RSA signatures for certifying RSA public keys.

CCITT Recommendation X.509 also makes provision for the case where more than one CA is used. This is done by allowing CAs to produce certificates for each other's public verification transformations. Sequences of such cross-certificates can be used to enable a user to obtain an authenticated copy of a CA's public verification transformation, and hence check a key certificate produced by that CA. For further discussion see, for example, [Mitchell89b].

It is intriguing to note that, unlike Internet RFC 1114, the method to be used for digital signatures is not completely specified in CCITT X.509 [X.509], although it is mandated that the secret signature process should be identical to the decryption process for a public key cryptosystem (see Section 2.6.1). As noted in [I'Anson90], this unnecessarily restricts the choice of signature algorithm.

5.2 Authentication Using Digital Signature

In computer networks it is often necessary for communicating parties to verify one another's identity. Traditionally this is done by the use of passwords; however the security offered by passwords used in the standard way is very limited.

One alternative is the use of cryptographic authentication protocols, standards for which are now emerging. Some of the most important of these protocols are based on the use of digital signatures. CCITT Recommendation X.509 [X.509] specifies three different protocols for authentication, all based on the use of digital signatures. All these protocols are based on the use of a cryptographic data structure called a *token*.

Like a certificate, a token is merely a series of data items with a signature appended. However, tokens are always specific to a single communication between two parties, that is, when required a token is generated by an originator for transmission to a single recipient. The general form of the token specified in X.509 for transmission from user A to user B is:

$$B, D, S_A(B, D)$$

where B designates the name of the recipient, D designates any data that are to be sent as part of the authentication protocol, and S_A designates A's secret signature transformation.

One of the protocols specified in X.509 (*three-way authentication*) has the following general form:

1. A sends to B: $B, R_A, S_A(B, R_A)$ where R_A is a random number.
2. B verifies the signature and checks that B's name is in the token.
3. B sends to A: $A, R_A, R_B, S_B(A, R_A, R_B)$ where R_B is another random number.
4. A verifies the signature, checks that A's name is in the token, and checks for the presence of R_A (protecting against replays).
5. A sends to B: $B, R_B, S_A(B, R_B)$.
6. B verifies the signature, checks that B's name is in the token, and checks for the presence of R_B (protecting against replays).

At the end of this process A and B are convinced of each other's identity. With small modifications the above protocol can also be used to exchange secret keys; these keys can then be used to protect any subsequent exchange of data. Note that the above protocol is a corrected form of the one given in X.509 [X.509]. Most significantly, in step (5) the form given in the standard does not include the name of B; apart from contravening the definition of token this seriously weakens the protocol, as has been pointed out by Burrows, Abadi, and Needham [Burrows89].

Modified forms of two of the three authentication protocols in X.509 are also under consideration for adoption as ISO standards [DP9798–3] (including the one listed above). Unfortunately, although the error in step (5) has been corrected, further shortcomings have been inherited from X.509 in the versions allowing secret key transfer ([Burrows89], [I'Anson90]).

5.3 Electronic Mail Security Based on Digital Signatures

The 1988 versions of the X.400 series of recommendations support a variety of security services, based to a considerable extent on the use of digital signatures and public key cryptosystems; see, for example, [Mitchell89b]. Key management is based on the use of public key certificates, as specified in X.509 (see Section 5.1). Provision of end-to-end security services are almost all based on the use of structures called message-tokens, whose general form is similar to that of the token described in Section 5.2. Tokens are also present in the authentication protocols used by pairs of communicating X.400 entities.

The X.400 approach to key management, namely, the use of public key certificates, has also been followed in the latest triplet of Internet request for comments (RFCs) providing a security option for Internet electronic mail ([RFC1113], [RFC1114], [RFC1115]).

5.4 Resolution of Disputes

A digital signature on a message provides lasting evidence of the content and origin of that message. It therefore finds application where a dispute may arise between sender and receiver over what message (if any) was sent. The digital signature may be presented as evidence to a referee, who can then use it to settle the dispute. Signatures also find application where the sender and receiver are not in dispute but the receiver may wish to save the message and use it at a later date with the assurance that it has not been modified in the interim.

Perhaps the most obvious case where a dispute might arise is when the message constitutes a transaction between sender and receiver, such as a bank and one of its customers, which might profit either side by falsification. Typically a bank might issue its customers with a card that can then be presented to the provider of a service. Information on the card is then used to formulate a request to the bank for a transfer of funds to pay for the service; the service is only provided if the bank responds with a signed message that authorizes the transfer. The signature on the message protects the provider of the service against later repudiation of the transaction.

Since the bank transfers funds out of the customer's account, it is also reasonable for the customer to expect some protection against falsification of transactions. This can be achieved if the customer has a smart card which can provide digital signatures. In this case the request to the bank must be signed by the card, and only when this is verified will the bank send a signed authorization of funds transfer to the service provider. It now would appear that all three parties to the transaction are protected. The customer is protected since only requests signed by his card will result in transfers from his account; the bank is protected since it has the customer's signature authorizing transfer of funds, and the service provider is protected since it has the bank's signed authorization.

There is, however, a loophole in the above analysis if the electronic device that reads the customer's card and sends the request to the bank is fraudulent. There is no way of guaranteeing that the value of the transaction appearing on the device's display is the same as the value of the transaction the smart card is requested to sign! A customer could then unwittingly transfer much larger sums to the service provider than has apparently been agreed. The only solution to this problem (apart from trusting all transaction devices) is to provide a very much more sophisticated smart card that can interface directly with its owner.

A different example of digital signatures being used to settle disputes is provided by the monitoring of remote seismic observatories for nuclear weapon test ban treaties (see Simmons [Simmons88] and the chapter by Simmons "How to Insure That Data Acquired to Verify Treaty Compliance Are Trustworthy," in this volume). In this scenario, two countries have agreed to limit (or stop) underground testing of nuclear weapons. Each country has seismic observatories in the other country which can detect any noncompliance with the agreement. These observatories send messages that either confirm or deny compliance. The messages are signed so that the receiver can be sure both of their origin and that they have not been tampered with, as well as providing evidence that can be presented to a neutral body in the event of disputes between the two countries.

5.5 A Secure Telephone System

Diffie [Diffie89] describes a secure telephone system for use with Integrated Services Digital Network (ISDN) which relies on digital signatures. To make a secure telephone call a user places a smart card in the telephone and dials the required number. When the receiving telephone answers, the two telephones perform a Diffie–Hellman exponential key exchange [Diffie76]. This exchange provides the telephones with a shared key which is a combination of secret pieces of information chosen at random by the two telephones. No eavesdropper can obtain this key from the information exchanged. This key is now used to encrypt all data subsequently exchanged between the two telephones. However, at this stage the two telephones have not verified each other's identity; to do this they use digital signatures.

The telephones now exchange public key certificates (see Section 5.1) and both parties check the signatures on the certificates, thereby obtaining verified copies of each other's public keys. To check that the other telephone and its user are the legitimate holders of these certificates, each telephone issues a challenge message. The challenge should be a piece of information that has previously been sent, such as the public information from the exponential key exchange, to ensure that the entities owning the secret keys are the same as those performing the encryption. The response to the challenge is a signed version of the challenge, which can then be checked using the public key in the certificate.

Each secure telephone must therefore contain its own secret signature key, together with a copy of the CA's public key and a certificate for its own public key (this certificate will also contain identification information for the telephone). The tamper-resistance of the telephone must therefore be sufficient to prevent extraction of the secret key, and ensure that none of the telephone secret key, the CA's public key, or the certificate can be changed. The telephone must also have a public key belonging to its owner which is used to verify commands from the owner (signed using the owner's smart card). The owner's public key may be changed by giving the phone two messages; one contains the new owner's public key signed by the old owner and the other signed by the new owner.

Before a telephone becomes operational it must be accepted by the network. The owner gives the telephone a signed command that identifies its characteristics and the identity of the network control center with which it will communicate. The telephone sends a request to become affiliated to the network along with a newly generated public key, all signed with its secret key. The network center verifies the signatures of the telephone and its owner, and if all is in order issues a certificate containing the telephone's newly generated public key. With the issue of this certificate the telephone is able to engage in authenticated key exchange with other telephones in the network. The network center saves the request and the corresponding signatures to give an audit trail of network activity.

5.6 Other Applications

Digital signatures can also be used for user authentication and identification. For this application anyone seeking access to a secure site or use of an information system could be issued with a challenge message. The required response is the user's signature on the challenge message, and only if the signature is verified is the user granted physical access or use of the system. All such responses could then be stored to provide an audit trail of authentication attempts.

Another example is in the distribution and validation of software [Merkle80]. Software updates produced at a central source can be signed and then distributed to individual sites. Each site can then verify the signature before using the new software, thereby protecting against use of erroneous or fraudulent software. Indeed, it would be possible for a signature on each program to be verified every time it is executed, thereby protecting against malicious or accidental changes.

A further application relates again to computer networks. Within a single multi-user computer system, control of access to data can normally be provided through careful design of the operating system and application software. However, when similar controls are to be applied to networks of computers, problems arise that require cryptographic solutions [Mitchell88, Mitchell89c]. In particular, service requests and/or objects passed from one machine to another may have associated access control information. In the case of service requests this information might indicate the privileges of the requestor, or in the case of a transferred object the information might list what types of user are allowed access to the object. Protection and validation of this information is therefore of great importance. Work is currently underway within ISO and other bodies to standardize access control procedures, and a proposal from the European Computer Manufacturers Association (ECMA) [ECMA89] calls for the use of access control certificates; such certificates would be lists of access control information with a digital signature appended to enable the authenticity of the information to be checked.

Note added in proof: On August 30, 1991, the U.S. National Institute of Standards and Technology (NIST) published "A Proposed Federal Information Processing Standard for Digital Signature Standard (DSS)" soliciting public comment prior to the adoption of the standard. The DSS is a variant of the El Gamal public key algorithm (Chapter 4, "Public Key Cryptography" of this volume) in a field GF(p), p a prime modulus 512 bits in size, but carrying out the signature and verification computations in a subfield GF(q), q a prime divisor of $p - 1$; $2^{159} < q < 2^{160}$. The complete specifications for the proposed DSS can be obtained from NIST at Gaithersburg, MD 20899.

REFERENCES

[Adleman79] L. M. Adleman, K. Manders, and G. Miller, "On taking roots in finite fields," in *Proc. 20th Ann. IEEE Symp. Foundations Computer Sci.,* pp. 175–178. Los Angeles, CA: IEEE Computer Society Press, 1979.

[Akl83] S. G. Akl, "Digital signatures: A tutorial survey," *IEEE Computer Magazine,* pp. 15–24, Feb. 1983.

[Akl84] S. G. Akl, "On the security of compressed encodings," in *Advances in Cryptology: Proc. Crypto'83,* D. Chaum, Ed., Santa Barbara, CA, Aug. 22–24, 1983, pp. 209–230. New York: Plenum Press, 1984.

[ANSI3.92] ANSI X3.92-1981, *Data encryption algorithm,* American National Standards Institute, New York, 1981.

[ANSI3.106] ANSI X3.106-1983, *American National Standard for Information Systems—Data Encryption Algorithm—Modes of Operation,* American National Standards Institute, New York, 1983.

[ANSI9.9] ANSI X9.9-1986, *Financial Institution Message Authentication (Wholesale),* American Bankers Association, Washington, D.C., 1986.

[ANSI9.19] ANSI X9.19-1985, *Financial Institution Retail Message Authentication*, American Bankers Association, Washington, D.C.

[Beker82] H. J. Beker and F. C. Piper, *Cipher Systems*, London: Van Nostrand Reinhold, 1982.

[Burrows89] M. Burrows, M. Abadi, and R. M. Needham, "A logic of authentication," *ACM Operating Systems Review*, vol. 23, no. 5, pp. 1–13, 1989.

[X509] CCITT X.509-1988, *The Directory—Authentication Framework*, Geneva: Consultation Committee, International Telephone and Telegraph, Dec. 1988.

[Coppersmith84] D. Coppersmith, "Fast evaluation of logarithms in fields of characteristic two," *IEEE Trans. Inform. Theory*, vol. IT-30, pp. 587–594, 1984.

[Coppersmith86] D. Coppersmith, "Another birthday attack," in *Lecture Notes in Computer Science 218; Advances in Cryptology: Proc. Crypto'85*, H. C. Williams, Ed., Santa Barbara, CA, Aug. 18–22, 1985, pp. 14–17. Berlin: Springer-Verlag, 1986.

[Coppersmith89] D. Coppersmith, "Analysis of ISO/CCITT Document X.509 Annex D," preprint, June 1989.

[Damgård88] I. B. Damgård, "Collision free hash functions and public key signature schemes," in *Lecture Notes in Computer Science 304; Advances in Cryptology: Proc. Eurocrypt'87*, D. Chaum and W. L. Price, Eds., Amsterdam, The Netherlands, April 13–15, 1987, pp. 203–216. Berlin: Springer-Verlag, 1988.

[Damgård89] I. B. Damgård, "Design principles for hash functions," in *Lecture Notes in Computer Science 435; Advances in Cryptology: Proc. Crypto'89*, G. Brassard, Ed., Santa Barbara, CA, Aug. 20–24, 1989, pp. 416–427. Berlin: Springer-Verlag, 1990.

[Davies80] D. W. Davies and W. L. Price, "The application of digital signatures based on public key cryptosystems," in *Proc. 5th Internat. Conf. Computer Commun.*, J. Salz, Ed., Atlanta, GA, Oct. 27–30, 1980, pp. 525–530. International Council of Computer Communications.

[Davies83] D. W. Davies, "Applying the RSA digital signature to electronic mail," *IEEE Computer Magazine*, vol. 16, no. 2, pp. 55–62, Feb. 1983.

[Davies84] D. W. Davies and W. L. Price, *Security for Computer Networks*, Chichester, England: Wiley, 1984.

[Davies85] D. W. Davies and W. L. Price, "Digital signatures—an update," in *Proc. 7th Internat. Conf. Computer Commun.*, J. M. Bennett and T. Pearcy, Eds., Sydney, Australia, Oct. 30–Nov. 2, 1984, pp. 843–847. Amsterdam: Elsevier/North Holland, 1985.

[DeJonge86] W. De Jonge and D. Chaum, "Attacks on some RSA signatures," in *Lecture Notes in Computer Science 218; Advances in Cryptology: Proc. Crypto'85*, H. C. Williams, Ed., Santa Barbara, CA, Aug. 18–22, 1985, pp. 18–27. Berlin: Springer-Verlag, 1986.

[Denning83] D. E. Denning, "Protecting public keys and signature keys," *IEEE Computer Magazine*, vol. 16, no. 2, pp. 27–35, 1983.

[Denning84] D. E. Denning, "Digital signatures with RSA and other public-key cryptosystems," *Commun. ACM,* vol. 27, pp. 388–392, 1984.

[Diffie76] W. Diffie and M. Hellman, "New directions in cryptography," *IEEE Trans. Inform. Theory,* vol. IT-22, pp. 644–654, 1976.

[Diffie89] W. Diffie, "Digital signatures, electronic negotiations and tamper resistant audit trails," paper given at *COMPSEC 89,* (Computer Security) London, Oct. 1989.

[ECMA89] ECMA, *Security in Open Systems—Data Elements and Service Definitions,* ECMA/TC32/TG9, final draft of July 1989 (output of 12th (Oslo) meeting). Geneva: European Computer Manufacturers Association.

[El Gamal85] T. El Gamal, "A public key cryptosystem and a signature scheme based on discrete logarithms," *IEEE Trans. Inform. Theory,* vol. IT-31, pp. 469–472, 1985.

[Fiat87] A. Fiat and A. Shamir, "How to prove yourself: Practical solutions to identification and signature problems," in *Lecture Notes in Computer Science 263; Advances in Cryptology: Proc. Crypto'86,* A. M. Odlyzko, Ed., Santa Barbara, CA, Aug. 11–15, 1986, pp. 186–194. Berlin: Springer-Verlag, 1987.

[Gibson88a] J. K. Gibson, "A collision free hash function and the discrete logarithm problem for a composite modulus," preprint, 1988.

[Gibson88b] J. K. Gibson, "Intertwining a hash function with a public key cryptosystem, Fiat-Shamir style," preprint, 1988.

[Girault88] M. Girault, "Hash-functions using modulo-N operations," in *Lecture Notes in Computer Science 304; Advances in Cryptology: Proc. Eurocrypt'87,* D. Chaum and W. L. Price, Eds., Amsterdam, The Netherlands, April 13–15, 1987, pp. 217–226. Berlin: Springer-Verlag, 1988.

[Girault89] M. Girault, R. Cohen, and M. Campana, "A generalised birthday attack," in *Lecture Notes in Computer Science 330; Advances in Cryptology: Proc. Eurocrypt'88,* C. G. Günther, Ed., Davos, Switzerland, May 25–27, 1988, pp. 129–156. Berlin: Springer-Verlag, 1988.

[Godlewski89] P. Godlewski and P. Camion, "Manipulations and errors, detection and localization," in *Lecture Notes in Computer Science 330; Advances in Cryptology: Proc. Eurocrypt'88,* C. G. Günther, Ed., Davos, Switzerland, May 25–27, 1988, pp. 97–106. Berlin: Springer-Verlag, 1988.

[Goldwasser84] S. Goldwasser, S. Micali, and R. L. Rivest, "A 'paradoxical' solution to the signature problem," in *Proc. 25th Ann. IEEE Symp. Foundations Computer Science,* Singer Island, FL, Oct. 24–26, 1984, pp. 441–448. Los Angeles, CA: IEEE Computer Society Press, 1984.

[Gordon85] J. Gordon, "How to forge RSA key certificates," *Electron. Lett.,* vol. 21, pp. 377–379, 1985.

[I'Anson90] C. I'Anson and C. J. Mitchell, "Security defects in CCITT Recommendation X.509—The directory authentication framework," *Computer Communication Review,* vol. 20, no. 2, pp. 30–34, April 1990.

[ISO8372] ISO 8372, *Information Processing—Modes of Operation for a 64-bit Block Cipher Algorithm*, Geneva: International Organization for Standardization, 1987.

[ISO8731-1] ISO 8731-1, *Banking—Approved Algorithms for Message Authentication—Part 1: DEA*, Geneva: International Organization for Standardization, 1987.

[DIS9797] ISO DIS 9797, *Data Integrity Mechanism Using a Cryptographic Check Function Employing a Block Cipher Algorithm*, Geneva: International Organization for Standardization, 1989.

[DP9798-3] ISO DP.2 9798-3, *Peer Entity Authentication Mechanisms, Part 3: Peer Entity Mutual Authentication Mechanisms Using a Public Key Algorithm*, Geneva: International Organization for Standardization, 1989.

[DP10118] ISO DP 10118, *Hash-Functions for Digital Signatures*, Geneva: International Organization for Standardization, 1989.

[Jefferies88] N. P. H. Jefferies, private communication, 1988.

[Jueneman83b] R. R. Jueneman, "Analysis of certain aspects of output feedback mode," in *Advances in Cryptology: Proc. Crypto'82*, D. Chaum, R. L. Rivest, and A. T. Sherman, Eds., Santa Barbara, CA, Aug. 23–25, 1982, pp. 99–127. New York: Plenum Press, 1983.

[Jueneman83] R. R. Jueneman, S. M. Matyas, and C. H. Meyer, "Message authentication with manipulation detection codes," in *Proc. 1983 IEEE Symp. Security and Privacy*, G. R. Blakley and D. Denning, Eds., Oakland, CA, April 1983, pp. 33–54. Los Angeles: IEEE Computer Society Press, 1983.

[Jueneman85] R. R. Jueneman, "Message authentication," *IEEE Communications Magazine*, vol. 23, no. 9, pp. 29–40, Sept. 1985.

[Jueneman87] R. R. Jueneman, "Electronic document authentication," *IEEE Network Magazine*, vol. 1, no. 2, pp. 17–23, April 1987.

[Jueneman87b] R. R. Jueneman, "A high-speed manipulation detection code," in *Lecture Notes in Computer Science 263, Advances in Cryptology*, Proc. Crypto '86, A. M. Odlyzko, Ed., Santa Barbara, CA, Aug. 11–15, 1986. Berlin: Springer-Verlag, pp. 327–346, 1987.

[RFC1114] S. Kent and J. Linn, "Privacy enhancement for Internet electronic mail: Part II—Certificate-based key management," *Internet RFC 1114*, Washington, D.C.: Internet Activities Board Privacy Task Force, Aug. 1989.

[Kohnfelder78] L. M. Kohnfelder, "A method for certification," MIT Laboratory for Computer Science, Cambridge, MA, May 1978.

[Kranakis86] E. Kranakis, *Primality and Cryptography*, Stuttgart: Teubner/Chichester, England: John Wiley, 1986.

[Lenstra90] A. K. Lenstra, H. W. Lenstra, Jr., M. S. Manasse, and J. M. Pollard, "The number field sieve," in *Proc. 22nd ACM Symposium on Theory of Computing*, pp. 564–572, 1990.

[Lenstra90a] A. K. Lenstra and M. S. Manasse, "Factoring by electronic mail," in *Lecture Notes in Computer Science 434; Advances in Cryptology; Proc. Eurocrypt'89*, J.-J. Quisquater and J. Vandewalle, Eds., Houthalen, Belgium, April 10–23, 1989, pp. 355–371. Berlin: Springer-Verlag, 1990.

[Lenstra91] A. K. Lenstra, H. W. Lenstra, M. S. Manasse, J. M. Pollard, "The factorization of the ninth Fermat number," preprint, dated February 27, 1991.

[RFC1040] J. Linn, "Privacy enhancement for Internet electronic mail, Part I: Message encipherment and authentication procedures," *Request for Comments (RFC) 1040* (Internet Activities Board), Jan. 1988.

[RFC1113] J. Linn, "Privacy enhancement for Internet electronic mail, Part I: Message encipherment and authentication procedures," *Request for Comments (RFC) 1113* (Internet Activities Board), Aug. 1989.

[RFC1115] J. Linn, "Privacy enhancement for Internet electronic mail, Part III: Algorithms, modes, and identifiers," *Request for Comments (RFC) 1115* (Internet Activities Board), Aug. 1989.

[Matyas79] S. M. Matyas, "Digital signatures—an overview", *Computer Networks,* vol. 3, pp. 87–94, 1979.

[Meijer83] H. Meijer and S. G. Akl, "Remarks on a digital signature scheme," *Cryptologia,* vol. 7, pp. 183–186, 1983.

[Merkle78] R. C. Merkle and M. E. Hellman, "Hiding information and signatures in trap-door knapsacks," *IEEE Trans. Inform. Theory,* vol. IT-24, pp. 525–530, 1978.

[Merkle79] R. C. Merkle, Secrecy, authentication and public key systems, Ph.D. Thesis, Stanford University, 1979.

[Merkle80] R. C. Merkle, "Protocols for public key cryptosystems," in *Proc. 1980 IEEE Symp. Security and Privacy,* Oakland, CA, April 14–16, 1980, pp. 122–134. Los Angeles, CA: IEEE Computer Society Press, 1980.

[Merkle82] R. C. Merkle, *Secrecy, Authentication and Public Key Systems,* Ann Arbor, MI: University of Michigan Press, 1982.

[Merkle89a] R. C. Merkle, "One way hash functions and DES," in *Lecture Notes in Computer Science 435; Advances in Cryptology: Proc. Crypto'89,* G. Brassard, Ed., Santa Barbara, CA, Aug. 20–24, 1989, pp. 428–446. Berlin; Springer-Verlag, 1990.

[Merkle89b] R. C. Merkle, "A software encryption function," Palo Alto Research Center (PARC) Xerox preprint, July 1989; revised (V2.0 of Snefru) Nov. 1989.

[Merkle90] R. C. Merkle, "A digital signature based on a conventional encryption function," preprint, 1987. An expanded and revised version, "A certified digital signature," appears in *Lecture Notes in Computer Science 435; Advances in Cryptology: Proc. Crypto'89,* G. Brassard, Ed., Santa Barbara, CA, Aug. 20–24, 1989, pp. 218–238. Berlin: Springer-Verlag, 1990.

[Meyer82] C. Meyer and S. M. Matyas, *Cryptography—A New Dimension in Computer Data Security,* New York: John Wiley & Sons, 1982.

[Mitchell88] C. J. Mitchell and M. Walker, "Solutions to the multi-destination secure electronic mail problem," *Computers and Security,* vol. 7, pp. 483–488, 1988.

[Mitchell89c] C. J. Mitchell, "Multi-destination secure electronic mail," *Computer Journal,* vol. 32, pp. 13–15, 1989.

[Mitchell89] C. J. Mitchell, P. D. C. Rush, and M. Walker, "A remark on hash functions for message authentication," *Computers and Security,* vol. 8, pp. 55–58, 1989.

[Mitchell89b] C. J. Mitchell, M. Walker, and P. D. C. Rush, "CCITT/ISO standards for secure message handling," *IEEE J. Selected Areas Commun.,* vol. 7, pp. 517–524, 1989.

[Mitchell90] C. J. Mitchell, "Authenticating multi-cast internet electronic mail messages using a bidirectional MAC is insecure" (submitted for publication).

[OSIS85] OSIS, *OSIS Security Aspects,* OSIS (Open Shops for Information Services) European Working Group, WG1, Final Report, Oct. 1985.

[Pohlig78] S. C. Pohlig and M. E. Hellman, "An improved algorithm for computing logarithms over GF(p) and its cryptographic significance," *IEEE Trans. Inform. Theory,* vol. IT-24, pp. 106–110, 1978.

[Quisquater89] J.-J. Quisquater and M. Girault, "2n-bit hash-functions using n-bit symmetric block cipher algorithms," in *Lecture Notes in Computer Science 434; Advances in Cryptology; Proc. Eurocrypt'89,* J.-J. Quisquater and J. Vandewalle, Eds., Houthalen, Belgium, April 10–13, 1989, pp. 102–109. Berlin: Springer-Verlag, 1990.

[Rabin78] M. O. Rabin, "Digitalized signatures," in: *Foundations of Secure Computation,* R. Lipton and R. DeMillo, eds., New York: Academic, pp. 155–168, 1978.

[Rabin79] M. O. Rabin, "Digitalized signatures and public-key functions as intractable as factorization," Massachusetts Institute of Technology Laboratory for Computer Science, Cambridge, MA, Technical Report, MIT/LCS/TR-212, Jan. 1979.

[Rivest78] R. L. Rivest, A. Shamir, and L. Adleman, "A method for obtaining digital signatures and public-key cryptosystems," *Commun. ACM,* vol. 21, pp. 120–126, 1978.

[Schöbi83] P. Schöbi and J. L. Massey, "Fast authentication in a trapdoor-knapsack public key cryptosystem," in *Lecture Notes in Computer Science 149; Cryptography: Proc. Workshop Cryptography,* T. Beth, Ed., Burg Feuerstein, Germany, March 29–April 2, 1982. pp. 289–306. Berlin: Springer-Verlag, 1983.

[Silverman87] R. D. Silverman, "The multiple polynomial quadratic sieve," vol. 48, pp. 329–399, 1987.

[Simmons88] G. J. Simmons, "How to ensure that data acquired to verify treaty compliance are trustworthy," *Proc. IEEE* vol. 76, pp. 621–627, 1988.

[Winternitz84a] R. S. Winternitz, "Producing a one-way hash function from DES," in *Advances in Cryptology: Proc. Crypto'83,* D. Chaum, Ed., Santa Barbara, CA, Aug. 22–24, 1983, pp. 203–207. New York: Plenum Press, 1984.

[Winternitz84b] R. S. Winternitz, "A secure one-way hash function built from DES," in *Proc. 1984 IEEE Symp. Security and Privacy,* Oakland, CA, April 29–May 2, 1984, pp. 88–90. Los Angeles, CA: IEEE Computer Society Press, 1984.

[Yuval79] G. Yuval, "How to swindle Rabin," *Cryptologia,* vol. 3, pp. 187–189, 1979.

CHAPTER 7

A Survey of
Information Authentication*

G. J. SIMMONS
Sandia National Laboratories
Albuquerque, New Mexico 87185

---

*This work was performed at Sandia National Laboratories and supported by the U.S. Department of Energy under contract no. DE-AC04-76DP00789.

Abstract—In both commercial and private transactions, authentication of information (messages) is of vital concern to all of the participants. For example, the party accepting a check usually insists on corroborating identification of the issuer—authentication of the originator, or as we shall say throughout this chapter, the transmitter—and the party issuing the check not only fills in the face amount in numerals, but also writes out the amount in script, and may even go so far as to emboss that part of the check to make it more difficult for anyone to subsequently alter the face amount appearing on an instrument bearing his valid signature, that is, a primitive means of providing for the later authentication of the communication or message. Although this example illustrates the two main concerns of the participants in the authentication of information, namely, the verification that the communication was originated by the purported transmitter and that it hasn't subsequently been substituted for or altered, it fails to illustrate perhaps the most important feature in the current use of authentication. The information conveyed on the check is inextricably linked to a physical instrument, the check itself, for which there exist legally accepted protocols to establish the authenticity of the signature and the integrity of what the issuer wrote in the event of a later dispute as to whether the check is valid or the signature genuine, independent of the information content (date, amount, etc.) recorded there. The contemporary concern in authentication, though, is with situations in which the exchange involves only information, that is, in which there is no physical instrument that can later be used to corroborate the authenticity of either the transmitter's identity or of the communication.

In deference to the origins of the problem of authentication in a communications context, we shall refer to the authenticated information as the message* and, as mentioned earlier, to the originator (of a message) as the transmitter. The message, devoid of any meaningful physical embodiment, is presented for authentication by a means that we shall call the authentication channel. This channel is by definition insecure, that is, all communications that pass through it are public and may even be intercepted and replaced or altered before being relayed on to the intended receiver. In the simplest possible authentication scheme the party receiving the message (the receiver) is also the one wishing to verify its authenticity; although, as we shall see, there are circumstances in which this is not the case. Authentication, however, is much broader than this communications-based terminology would suggest. The information to be authenticated may indeed be a message in a communications channel, but it can equally well be data in a computer file or resident software in a computer; it can be quite literally a fingerprint in the application of the authentication channel to the verification of the identity of an individual [14,22] or figuratively a "fingerprint" in the verification of the identity of a physical object such as a document or a tamper-sensing container [11]. In the broadest sense, authentication is concerned with establishing the integrity of information purely on the basis of the internal structure of the information itself, irrespective of the source of that information.

1 INTRODUCTION

Since information authentication often depends on complex cryptographic protocols and algorithms that cause the process of authentication itself to appear complex, it is useful to first discuss the general principles that underlie all authentication schemes before discussing the (unavoidably) complicated real authentication schemes. In the simplest terms possible, *authentication* is nothing more nor less than the determination by the authorized receiver(s), and perhaps the arbiter(s), that a particular message was most probably sent by the authorized transmitter under the existing authentication protocol and that it hasn't subsequently been altered or substituted for. Implicit in this statement is the fact that in any particular authentication protocol the receiver will accept as authentic only a fraction out of the total number of possible messages and that the transmitter will only use some subset (perhaps all) of this fraction to communicate with the authorized receiver(s). It should also be obvious that an opponent should not, in all probability, be able to select a message that the receiver will accept as authentic, otherwise he could impersonate the transmitter and/or substitute fraudulent messages of his choice for legitimate ones. The conditions determining the set of messages the receiver will accept and which of these the transmitter may use are what specifies a particular authentication scheme. As we will see, this commonly involves some form of encryption/decryption operation that the transmitter and receiver can do since they each know a secret cryptographic key, but that outsiders who do not know the key (probably) cannot. The following example, however, illustrates the essential features of message authentication without having to appeal to cryptography.

At the end of the last century and early in this one commercial codes were in widespread use to provide economy in telegraphic charges by encoding common (often

---

*This choice of terminology is not as straightforward as it might seem, and will be fully justified later.

very long) phrases into five-letter groups which were treated, and charged for, by the cable companies as single words. One of the best known of these was the Acme Commodity and Phrase Code [13] which consisted "of one-hundred thousand five-letter code [words] ciphers with at least two-letter difference between each and every [code] word. No transposition of any two adjoining letters will make another [code] word." Since the Acme Code is long since out of use and presumably forgotten, for our example we will assume that the transmitter and receivers have a copy and have agreed that the transmitter will only use and the receiver will only accept five-letter groups appearing in the Acme Code and that an opponent is ignorant of this code. Thus OGHAU would be accepted by the receiver as authentic and interpreted to mean "some are and some are not" while none of the 129 "nearby" five-letter groups AGHAU, . . . , OGHAS, GOHAU . . . OGAUA would be accepted since they do not appear in the code. Incidentally, it is interesting to note (from our vantage point of 1991) that the Acme Code was genuinely a precursor of modern error detecting and correcting codes, since a systematic procedure to correct "mutilations" was given [30]. If a codeword was received that did not appear in the code, it was assumed that this was the result of a mutilation (error) in transmission. The procedure constructed the five codewords that did appear in the code and differed in only one letter from the received codeword, and thus were the most likely codewords to have been sent. Context was then used to select one out of these five codewords, that is, a maximum likelihood detector in current terminology. For the purposes of our example, however, if an opponent knows nothing of the rule used by A. C. Meisenbach, the Acme Code designer, to construct acceptable (read authentic) codewords, but only knows that messages consist of five-letter groups, his probability of "choosing" an acceptable (to the receiver) codeword and hence his probability of deceiving the receiver would be

$$P_d = \frac{10^5}{26^5} \approx 0.0084$$

Although this example illustrates the essential notions involved in the authentication of information, that is, the definition of a restricted set of messages that the receiver will accept (as authentic) and of a subset (not necessarily a proper subset) of these messages that the transmitter will use, commercial codes (or similar fixed rules specifying the set of acceptable messages) have no useful authentication capability since the collection of acceptable messages would be either known to the opponent a priori or else quickly exposed by continued use.

It is at this point that cryptography commonly enters into authentication since it provides an easy to implement way for the transmitter and receiver to define the subset of messages that they will use and accept, respectively, dependent on the secret cryptographic key(s) known only to them that will consequently be opaque to an opponent who does not know the key(s).

For example, in a common U.S. military authentication protocol the transmitter and receiver each have matching sealed authenticators (note the use of the term "authenticator"), actually a short random sequence of symbols produced and distributed by the National Security Agency. They must also each have a common cryptographic keying variable (key) which must be protected to ensure its secrecy and integrity. The sealed authenticator packets are constructed so as to provide a positive indication (tattle-tale) if they are opened. Each communicant is responsible for the protection of his sealed authenticator and is administratively constrained from opening it until it is used to authenticate a message. Because of the sensitivity of these authenticators, that is, as will be apparent in a moment anyone having access to one could authenticate a fraudulent message, they must normally be handled under two-man control in the same way that the associated cryptographic keying variables must be handled which greatly complicates their generation, distribution, and control, and more importantly often limits the level of authentication that can be achieved in an exposed situation.

To authenticate a message, the transmitter opens his sealed packet, appends the enclosed authentication suffix to the message he wishes to authenticate, and then encrypts the resulting extended message with cipher or text feedback, using the cryptographic key he shares with the intended receiver, so that the effect of the appended authenticator is spread throughout the resulting cipher. In principle, the extended message could be block encrypted, but the size of the blocks involved rules this out in all practical applications. This encryption is done using a secret key (in a single-key cryptographic system) that the transmitter shares with the intended receiver(s). The resulting cipher is then transmitted as the authenticated message. The receiver, on receiving and decrypting the cipher using his copy of the secret key, opens his matching sealed authenticator and accepts the message as genuine if and only if the cipher decrypts to a string of symbols with the proper authenticating suffix, and otherwise rejects it as unauthentic. In this example, the subset of messages (ciphers) that the transmitter will use and that the receiver will accept are precisely those that decrypt to have the authenticating suffix. If the crypto algorithm is secure, an opponent who doesn't know the secret key(s) being used by the transmitter and receiver can do no better than to randomly choose a cipher in the hope that it will be accepted by the receiver. If there are r bits of information in the authenticator, an opponent (if he cannot break the "sealing" encryption algorithm) would have only a 2^{-r} probability of choosing (guessing) a cipher that would decrypt into a message ending with the unknown (to him) authentication suffix, and hence that would be accepted as authentic by the receiver.

This example illustrates an essential feature in all authentication schemes, namely, that authentication depends on the presence of redundant information, either introduced deliberately as in this example, or else inherently present in the structure of the message, that will be recognizable to the receiver. This results in the transmitter and receiver restricting their use to only a fraction of the total number of communications possible, that is, to those messages containing this redundant information; any others would be rejected by the receiver as unauthentic since the transmitter would not have sent them. As used in this example, and in widespread cryptographic use, the term "authenticator" denotes the redundant information appended to the message that is to be authenticated, which is functionally independent of the information content of the message itself, to an extended message that is then encrypted, etc. as described above. The object actually communicated by the transmitter to the receiver through the communications channel in this case is a "cipher." This use of the term "cipher" is in accordance with the accepted conventions of cryptography, since both the content of the original message and the authenticating redundant information must be concealed (kept secret) in the cipher, otherwise the appended authenticator, if it were revealed, could be used to authenticate an arbitrary fraudulent message.

On the other hand, an equally common use of the term authenticator, with a quite different meaning, occurs in connection with the authentication of electronic funds transfers in the Federal Reserve System. By directive of the Secretary of the Treasury

[3], all such transfers must be authenticated using a procedure that de facto depends on the Data Encryption Standard (DES) single-key cryptographic algorithm. The protocol includes precise format requirements, etc.; however, the essential feature for our purposes is that an authenticator is generated using a DES mode of operation known as cipher block chaining. The information to be authenticated is first broken into blocks of 64 bits each. The first block is added bitwise modulo two (exclusive-or) to a 64-bit initial vector, which is changed daily and kept secret by the member banks, and the sum encrypted using a secret DES key (known to both the transmitter and the receiver). The resulting 64-bit cipher is then exclusive-or'ed with the second block of text and the result encrypted to give a second 64-bit cipher, etc. This procedure is iterated until all blocks of the text have been processed. The final 64-bit cipher is clearly a function of the secret key, the initial vector, and of every bit of the text, irrespective of its length. This cipher, called a message authenticating code (MAC), is appended to the information being authenticated to form an extended message. The resulting extended message itself is normally sent in the clear, that is, unencrypted, although it may be super-encrypted if privacy is desired, but this operation is independent of the authentication function. The authenticator (MAC) can be easily verified by anyone in possession of the secret key and initial vector by simply repeating the procedure used by the transmitter to generate it in the first place. An outsider, however, cannot generate an acceptable authenticator to accompany a fraudulent message, nor can he separate an authenticator from a legitimate message to use with an altered or forged message since the probability of it being acceptable in either case is the same as his chance of "guessing" an acceptable authenticator, that is, 1 in 2^{64}. In this application, which is a classic example of an appended authenticator, the authenticator is a complex function of the information that it authenticates. The subset of acceptable extended messages in this case consists of those text-MAC pairs that pass the test of the MAC being related to the text by the secret DES key. Since this makes up only 2^{-64} of all possible extended messages, the probability of an opponent being able to "guess" an acceptable message is less than his chance of "guessing" the secret DES key: 1 in 2^{56}. This simplified discussion of the authentication security provided by a MAC is misleading (and in some instances untrue). The reader should refer to the chapter on Digital Signatures in this volume for a much more complete discussion of the security provided by appended authenticators.

In both of these uses of the term, the term "authenticator" denotes additional information communicated by the transmitter to enable the receiver to satisfy himself that the message should be accepted (as authentic). In the first case, the redundant information was appended to an unrelated message, and could therefore, if directly accessible to an opponent, be stripped off of one message and used to authenticate any other message. To prevent this, the resulting extended message was secured in a (block or feedback) cipher in which each bit of the cipher was a function not only of a secret encryption key but also of all of the bits of the extended message. In the second case, the redundant information was already by virtue of the generating procedure, a function of the secret key and initial vector as well as all of the bits in the information being authenticated, and hence, with high probability, inseparable from the original text in the sense that it has no better chance of being accepted as the authenticator for some other text than would any other randomly chosen 64-bit sequence.

This is a convenient point to comment on the terminology we will use in the balance of this chapter in discussing authentication. The preceding two examples indicate the confusing state of affairs as to the use of the term "authenticator." At least in both of these cases the term authenticator referred to the redundant information on which authentication depends, but in general even this isn't true.

It might at first seem a natural choice to call the actual sequence of symbols communicated by the transmitter to the receiver the authenticator paralleling the use of the term "cipher" in cryptography. Unfortunately, this leads to an immediate, and irresolvable, logical difficulty. In the Federal Reserve example, the transmitted sequence consisted of the original message concatenated with the 64-bit "authenticator" suffix. If the entire sequence consisting of both the information symbols and appended authenticator were to also be called an authenticator, there would be an unavoidable confusion between this authenticator and the appended authenticator it contains. A similar logical difficulty is clearly inherent in the other scheme as well. Consequently, the widespread use of the term authenticator, although not precise, has already preempted the most natural term to designate the sequence of symbols actually communicated by the transmitter to the receiver.

In the military authentication protocol, the sequence transmitted could be accurately described as a cipher. However, there is the same logical difficulty in using the term cipher to describe the total sequence of symbols communicated in the Federal Reserve protocol as there was in using the term authenticator in the military protocol. The appended authenticator in the Federal Reserve protocol can be properly described as a cipher, since it is produced by a block chaining encryption of the original message. However, if one described the entire sequence consisting of the concatenation of the original sequence of symbols and the appended authenticator (cipher) as a "cipher," there is both the problem of the confusion produced by the two uses of the term cipher, as well as the fact that the information being authenticated is not concealed in the inclusive cipher (extended message), contrary to the commonly accepted meaning of that term. Hence cipher is an even less satisfactory term than authenticator, for similar reasons.

The point of the preceding discussion was to make it clear that there is no completely satisfactory terminology for the primary object in the theory authentication, namely, the sequence of symbols actually transmitted through the authentication channel, and that each of the terms already in use for comparable objects in allied subjects is logically unacceptable. Nor does any constructed new word seem sufficiently natural to be persuasive. We have therefore settled on the cumbersome term "authenticated message," or where no confusion is likely, simply "message," as denoting the transmitted sequence of symbols. This conflicts with the equally natural use of the term "message" to denote the information that is to be authenticated. To avoid this, we tolerate the rather artificial device that the information conveyed by a message is the state of a hypothetical source. This convention has a precedent in error detecting and correcting codes where one speaks of a "message source" and/or "source encoding." The difference is that we are compelled to refer to the information that is being authenticated as a "state of the source," etc. In this convention, for example, the message in an appended authenticator scheme is of the form

$$m_{i,j} = s_i : f(\mathbf{e}_j, s_i)$$

where s_i is the ith state of the source, $\mathbf{e}_j$ is the jth encoding rule, and f is the authenticating function, so that $f(\mathbf{e}_j, s_i)$ is the authenticator that is to be appended to s_i when encoding rule $\mathbf{e}_j$ is being used.

The examples in the preceding paragraphs illustrate very clearly an essential dichotomy in all authentication schemes, namely, the division according to whether authentication is achieved with or without secrecy for the information being authenticated. In many applications secrecy isn't an important consideration, so it doesn't much matter whether the authentication is with or without secrecy. There are, however, applications in which authentication is essential but in which secrecy cannot be tolerated. One such application is the authentication of data taken to verify compliance with a comprehensive nuclear weapons test ban described in the chapter "How to Insure That Data Acquired to Verify Treaty Compliance Are Trustworthy" in this volume. We will return to this classification of authentication schemes according to whether authentication is achieved with or without secrecy later.

There is one final item of terminology that is important to understand for many contemporary applications of authentication. We pointed out above that the sealed authenticators used in the military authentication protocol had to be generated, distributed, and protected in precisely the same way that the cryptographic keying variables had to be. Furthermore, all of the schemes that depend on single-key cryptoalgorithms to define the acceptable sets of (authentic) messages, that is, that depend on the encryption operation to spread the collection of acceptable messages in what appears to an opponent to be a random and uniform manner in the set of all possible messages, requires that both the transmitter and receiver protect their copy of the cryptographic key to the same level of security demanded of the authentication scheme. One of the principal benefits of two-key or public key cryptography is that the decryption key does not have to be kept secret to have confidence in the authenticity of authenticated messages. This is true because it is by definition, or assumption, computationally infeasible to calculate the encryption key given the decryption key. It is, of course, essential that the integrity, against substitution or alteration, of the decryption key be insured during generation, distribution, and storage, at the same level of security demanded of the authentication scheme. If there were a lapse in the protection of the decryption key so that an opponent could substitute a fraudulent decryption key corresponding to an encryption key that he knows for the genuine key, he could then deceive the receiver by undetectably authenticating any fraudulent message he might wish. It might seem that if the integrity must be protected for the two-key cryptoalgorithm based schemes, then why not simply protect the secrecy also and use the much simpler and faster single-key cryptoalgorithms. One answer is that protecting secrecy often involves splitting up the information, that is, two-man control, or using trusted couriers to deliver sealed tattletale containers, etc. or repositories that require more than one combination lock be opened to gain access to the information. If only the integrity of the keying variable must be insured, then the key could be distributed by what is often called the "Merkle channel" (after Ralph Merkle who proposed it), that is, communicated over so many different public channels that it is improbable that any opponent could usurp them all: telephone, telex, radio, television, mail, etc. After the key is received, it only need be protected from substitution or alteration, for example, by keeping it in a regular single combination safe, or even more simply, by several responsible persons keeping a copy, and comparing these copies when needed. The descriptive term that has come to be accepted for this type of authentication in which the information that is used to verify the authenticity of a message doesn't have to be kept secret is "desensitized authentication," since the keying variables are no longer sensitive information.

2 THE THREAT(S)

In the discussion of authentication thus far, we have treated the problem as though it were strictly a matter of protecting the transmitter/receiver (the insiders) against deception by an opponent (the outsider). In other words, the communication takes place between mutually trusting and trustworthy parties over a channel that is assumed to be under the surveillance, and perhaps the control of a knowledgeable and sophisticated opponent who is capable of carrying out complex calculations and then either originating messages of his own devising or else of intercepting and modifying messages from the legitimate transmitter. In military and diplomatic schemes, this is almost always assumed to be the case, or at least the problem of insider cheating is considered to be an acceptable risk. In the commercial world almost the reverse is true, that is, the originator and the receiver (the insiders) to a communication are even more likely to be the ones trying to cheat each other than are outsiders.

It is perhaps useful in understanding insider deceits to think of the receiver as a stockbroker and the transmitter as one of the broker's customers. In this setting it is easy to believe that a customer might wish to disavow an order that he actually issued if it later turns out that the decision was a bad one that cost him money. Similarly, the broker, who is managing the customer's account, might very well wish to execute an order of his own devising when he received no such instructions from the customer, or even to execute orders contrary to the customer's instructions, to generate commissions for himself or in his judgment to make better investments. In either case, the function of authentication would be, in the event of a dispute between the broker and the customer as to whether the broker had faithfully carried out the customer's instructions, to make it possible for an impartial third party to decide who was, in all probability, liable. Although it won't be possible to discuss solutions to all of the possible authentication threats here, several will be illustrated in the applications that are discussed. Because the subject is so convoluted, it is desirable to describe each of the various deceptions as clearly as possible.

In the most general model of authentication there are four essential participants; of these four participants, the "insiders" are the transmitter (the authorized originator for messages), the legitimate receiver(s), and, depending on the particular authentication scheme being used, perhaps the arbiter(s). Whether the arbiter is an insider or an outsider depends on whether he is in possession of any privileged information, that is, information not available to one or more of the other participants. We mention in passing that for the unconditionally secure authentication codes that permit arbitration described here and in [25,26] the arbiter is necessarily an insider, while for the authentication channel that permits arbitration based on public key cryptographic techniques [21], that is, for computationally secure authentication with arbitration, the arbiter is an outsider. The chapter "Digital Signatures" by Mitchell, Piper, and Wild in this volume gives a comprehensive treatment of digital signature schemes in which the arbiter has no privileged information. The fourth participant, the opponent, is always an outsider who is assumed to have no privileged information, but who is assumed to be knowledgeable of the general authentication scheme being used by the transmitter and receiver (an extension of Kerckhoff's criteria in cryptology to authentication) and to be capable of sophisticated eavesdropping, computation, and message alterations. Given this general setting, there are (at least) eight types of cheating (attempted deceptions) that can occur.

The opponent can send a fraudulent message to the receiver in hopes of having it be accepted as an authentic communication from the transmitter before the legitimate transmitter sends any message at all. This type of attempted deception we will denote by I_0. He can also wait to observe ℓ authentic messages, $\ell \geq 1$, sent by the legitimate transmitter before he attempts to deceive the receiver; after which he can either substitute some other message in the stead of the last message intercepted or else forward it unchanged and then attempt to get some message of his choosing accepted as authentic by the receiver before the legitimate transmitter sends another authentic message. We denote this type of cheating by the notation I_ℓ and S_ℓ, $\ell \geq 1$. The cases I_0 and I_ℓ and S_ℓ, $\ell \geq 1$, are sufficiently different that we will describe them separately.

1. I_0. The opponent, based only on his knowledge of the general authentication scheme being used by the transmitter and receiver, can send a fraudulent message to the receiver when in fact no message has yet been sent by the transmitter. The probability of his succeeding in deceiving the receiver in this case is simply the value of the two-person game whose representation is the incidence matrix of the authenticating rules—mapping source information into messages—in the general authentication scheme [1,23,32]. The calculation of this probability is computationally easy even for large authentication schemes. I_0 is commonly referred to as (the opponent) impersonating the transmitter.

2. I_ℓ and S_ℓ. The opponent can wait to observe $\ell - 1$ legitimate messages from the transmitter which he allows to pass to the receiver without tampering with them. When he intercepts the ℓth message there are two courses of action available to him: He can either substitute some other message of his own devising in its stead or else he can forward it without modification to the receiver. He could then, based on what he has learned from the ℓ observations he has made of legitimate messages, send a message of his own choosing to the receiver before the legitimate transmitter sends the next authentic message, that is, he can attempt to impersonate the legitimate transmitter. The first type of deception is an ℓth order substitution attack, S_ℓ, where S_1 is commonly referred to as simply substitution, while the second type of deception is an ℓth order impersonation denoted by I_ℓ. The opponent's strategy in either of these cases, $\ell \geq 1$, is defined by conditional probabilities, that is, his decision as to which message to substitute or of the message he should use to maximize his chances of successfully impersonating the legitimate transmitter will be affected by the legitimate messages he has observed (and also by whether he knows or doesn't know the information being conveyed by the observed messages [1,20,23,32]). Already for the case of S_1 this problem is computationally difficult (even for modest-sized authentication schemes) unless very stringent conditions are imposed on the regularity and symmetry of the associated designs.

The present author has restricted attention in both his earlier work on authentication codes [20, 23–26] and in this chapter to deceits I_0 and S_1, that is, transmitter impersonation and/or message substitution. The reason for this is that for opponent deceits I_1 and I_ℓ and S_ℓ, $\ell > 1$, an ad hoc rule must be introduced to prevent the opponent from simply substituting a legitimate message already observed prior to the ℓth communication and hence known to be acceptable to the receiver, for the ℓth message—if they are different. Various other authors [4,12,18,23] have considered authentication codes for the cases $\ell > 1$. In all cases, the opponent wins if the receiver accepts the fraudulent message as being an authentic communication from the transmitter *and*, if $\ell > 0$, ends up being misinformed as to the transmitter's communicated information in consequence.

Insider cheating involves a participant who knows some piece of information about the authentication scheme not known to all of the other participants: the transmitter, receiver, or in some instances as noted above, the arbiter(s). Protection against transmitter or receiver cheating presupposes that there is an arbiter who will arbitrate disputes between them, that is, who will assign liability to the party most likely to be responsible. This arbiter, for the scheme described in this chapter to work, must be assumed to be unconditionally trustworthy. In other words, we assume that the arbiter will not misuse his privileged position to deceive either the transmitter or receiver. We will not consider the case of insider-arbiter cheating since the authentication codes described in this chapter provide no protection against this type of deceit. In the most general setting, though, arbiter cheating is a fourth type of deception that needs to be protected against in addition to the three that are considered here. Brickell and Stinson [2] have constructed authentication schemes capable of detecting (in probability) dishonest arbiters.

The receiver can cheat if he can successfully attribute a message of his own devising to the transmitter, that is, a message not sent by the transmitter. The adverb "successfully" means that when the transmitter later claims (correctly) that he didn't send the message in question, that the arbiter will rule against him. The receiver can wait to attribute a fraudulent message to the transmitter until after he has received ℓ, $\ell \geq 0$, legitimate messages. We denote this second form of cheating by the notation R_ℓ, $\ell \geq 0$.

3. R_ℓ. The receiver, using both the public knowledge of the general authentication scheme and his privileged information, claims to have received from the transmitter in an authentic message fraudulent information of his own devising. He is successful if and only if the arbiter later certifies the fraudulent message as being one that the transmitter could have sent under the existing protocol. If he attempts to cheat before any legitimate message has been sent, this is an R_0 deception. If he waits until after he has received ℓ legitimate messages, $\ell \geq 1$, from the transmitter, and then—using this additional information—attempts to cheat in the same way as before, this is said to be an R_ℓ deception.

The category R_ℓ, $\ell > 1$, will not be considered here for the same reason that the categories I_ℓ, $\ell \geq 1$, and S_ℓ, $\ell > 1$, were excluded from consideration for opponent deceptions.

4. T. The transmitter can attempt to cheat the receiver by sending a message that the receiver may accept and then claiming that he didn't send it, that is, by disavowing a legitimate message. He will be successful in this deceit if and only if when the receiver later claims to have been cheated the arbiter rules that the message is not one that the transmitter would have sent under the existing protocol and, in consequence, the receiver is held liable.

There is an exact parallel between the deceptions available to an opponent and to an arbiter since the object of either of them in attempting a deception is to cause the receiver to either accept a fraudulent or an altered message and to be misinformed thereby. If the arbiter is not privy to any privileged information, which, for example, is the case in the use of the Rivest–Shamir–Adleman (RSA) digital signature scheme described in the next section, then the arbiter's capabilities are the same as those of an opponent; that is, for I_ℓ- and S_ℓ-type deceptions, $\ell \geq 1$. If, however, the arbiter is an insider, with perhaps information not known to either the transmitter or receiver, then he will have a class of deceptions unique to him, which we denote by A_{I_ℓ} or A_{S_ℓ}.

5. A_{I_0}. The arbiter, using both the public knowledge of the general authentication scheme and his privileged information, chooses a fraudulent message that he sends to

the receiver when in fact no message has been sent by the transmitter. If this message is accepted the receiver will certainly be deceived (as to the transmitter's intent).

6. A_{I_ℓ} and A_{S_ℓ}. The arbiter can wait to intercept ℓ legitimate messages, $\ell \geq 1$, sent by the authorized transmitter and then using the additional information acquired from observing which messages have been used by the transmitter, either substitute in the stead of the ℓth message a fraudulent message that has a maximum chance of being accepted by the receiver or else forward the ℓth message unmodified and attempt to get a fraudulent message accepted as authentic before the legitimate transmitter sends the $(\ell + 1)$st authentic message. In this case, he (the arbiter) wins if and only if the receiver ends up being misinformed of the transmitter's intended communication.

This is an appropriate point to point out the second natural dichotomy of authentication schemes—depending on whether they permit arbitration of transmitter/receiver disputes as to the authenticity of messages or not.

In all of the authentication schemes discussed thus far, since the transmitter and receiver must both know the same secret (from the opponent) information (either the key in a simple key cryptographic algorithm, or a sealed authenticator and a cryptographic key or an initial vector and a cryptographic key, etc.) they can each do anything the other can do. In particular, because of this duality, the receiver cannot "prove" to a third party that a message he claims to have received from the transmitter was indeed sent by the transmitter, since he (the receiver) has the capability to utter an indetectable forgery, that is, the transmitter can disavow a message that he actually sent. Similarly, the receiver can claim to have received a message when none was sent, that is, to falsely attribute a message to the transmitter, who cannot prove that he didn't send the message since he could have. In the classic military authentication scheme this is an acceptable situation, since a superior commander doesn't worry that a subordinate will attribute an order to him that he didn't issue and the subordinate doesn't worry that his superior will disavow an order that he did send. There is, in fact, some rudimentary protection against this sort of cheating provided by the sealed authenticators the military uses since if either party can produce his unopened authenticator, it is prima facie evidence that he doesn't know its contents and hence could not have authenticated a message using its contents. This doesn't protect the transmitter from the receiver's stripping the authenticator from the legitimate message and attaching it to a fraudulent one—but again, in the military setting, this is not regarded as a serious problem. In many situations, and in almost all commercial and business applications, the primary concern is with insider cheating, for example, the person withdrawing cash from an ATM may not be the account holder or the amount shown on a properly signed and valid check may be altered to a larger figure, etc. If the authentication scheme permits arbitration, the arbiter's sole function is to certify on demand whether a particular message presented to him is authentic or not, that is, whether it is a message that the transmitter could have sent under the established protocol. He can never say that the transmitter did send the message—although the probability that it could have come from a source other than the authorized transmitter can be made as small as one likes—only that he could have under the established protocol.

3 A NATURAL CLASSIFICATION OF AUTHENTICATION SCHEMES

We have already pointed out two classifications for authentication schemes depending on whether authentication is achieved with or without secrecy for the information being authenticated, and whether it is possible to arbitrate a dispute between the transmitter and receiver as to the authenticity of a message. We now consider the most important classification of all arising from the source of the security of the authentication scheme itself.

Authentication schemes are classified as being computationally secure, provably secure, or unconditionally secure, terms that sound very much alike. In fact, they describe quite different bases for confidence in the integrity of the authentication.

A scheme is said to be *computationally secure* if the security depends on a would-be cheater carrying out some computation that in principle is possible but in which all of the known methods of attack require an infeasible amount of computation. A good example of a computationally secure authentication scheme is a widely used variation of the Federal Reserve electronic funds transfer (EFT) protocol described above, in which the only difference is that there is no daily initialization vector. The protocol for generating the appended authenticator in this case is described in the work by Meyer and Matyas, *Cryptography: A New Dimension in Computer Data Security* [15].

> The suggested technique for MAC generation is as follows: The first 64 bits of that portion of the transaction to be protected are block-encrypted using DES and the secret key. Then the next 64 transaction bits to be thus protected are exclusive-or'ed (modulo 2 added) with the just produced cipher. The result is then block encrypted using the same key, producing a new 64 bits of cipher. This procedure is continued until all critical transaction fields have been included. (The final data block will likely be less than 64 bits, so it is padded with zeros to make a full 64 bits prior to being exclusive-or'ed with the just-produced cipher.) Some subset of the final cipher, at least six decimal digits or five hexadecimal digits, serves as the MAC.

In this case an opponent, given the plaintext (state of the source) and matching MAC has all of the information needed to solve for the unknown DES key. DES keys are 8 bytes long, each byte consisting of 7 bits of actual key with the eighth bit being an unused parity check bit over the other 7 bits. With virtual certainty (≈ 0.996) for a 64-bit MAC there is only one key that will generate the observed MAC from the plaintext, therefore an opponent would only have to test at most the 2^{56} possible keys (with probability 0.996) to find the key used by the transmitter to generate the MAC and hence to be able to authenticate an arbitrary message of his choosing. This is the brute force cryptanalytic attack on the DES discussed by Diffie and Hellman [5] which remains at this time computationally infeasible to carry out. In other words, the security of the MAC generated by the protocol described in [15] is equivalent to the difficulty of carrying out a well-defined, but at present infeasible, computation. There is an important, but subtle, difference between the Federal Reserve EFT protocol and the one described here that has to do with the uncertainty introduced by the initialization vector. The net result is that the EFT protocol essentially requires cryptanalysis in depth requiring impractically many matched plaintext-MAC pairs to achieve the same degree of confidence in recovering the unknown DES key as for the simplified protocol. Both schemes are, however, only computationally secure.

An authentication scheme is said to be *provably secure* if it can be shown that breaking it implies that some other, presumed hard, problem, such as factoring suitable chosen large composite integers or extracting discrete logarithms in a finite field $GF(q)$ where q has been carefully chosen, etc. could be solved with comparable effort. We illustrate this concept using a provably secure authentication scheme equivalent in security to the difficulty of factoring.

The RSA cryptoalgorithm [17] is the most widely known and used of the two-key (originally public key) algorithms. This algorithm depends for its security on the difficulty of factoring suitably chosen large composite integers. A bare bones description of the algorithm* is the following. Two primes p and q, sufficiently large that their product (modulus) $n = pq$ is infeasible (impossible?) to factor, are randomly chosen, subject to some side conditions that are not relevant to the present discussion. An exponent, e, is randomly chosen such that

$$(e, (p - 1)(q - 1)) = 1$$

that is, so that e has no factor in common with $(p - 1)$ or $(q - 1)$ other than 1, and the multiplicative inverse exponent, d, calculated:

$$ed = 1 \pmod{(p - 1)(q - 1)}$$

Given e this is easy to do, requiring only $0(\log(n))$ work, using the Euclidean algorithm if one knows the factorization of n, and as hard as factoring n otherwise.

To establish an authentication channel using the RSA cryptoalgorithm, the pair $(d; n)$ is made public and the pair $(e; n)$ kept secret. It should be remarked that if one knows both e and d, n is easy to factor. Messages are integers, $m < n$, where the message is assumed to contain redundant authenticating information known to both the transmitter and receiver, and the opponent; say that all authentic messages must end with fifty 0's in the least significant bit positions. The RSA algorithm is a block encryption algorithm, defined by

$$c \equiv m^e \pmod{n}$$

where m is the extended message (including the appended authenticator 00 . . . 0) and c is the least positive residue of m^e modulo n. The receiver will accept as authentic any cipher, $\bar{c}$ that when decrypted using the key $(d; n)$

$$\bar{m} \equiv (\bar{c})^d \pmod{n}$$

yields a text, $\bar{m}$ with the suffix of fifty 0's. The probability that a randomly chosen cipher $\bar{c}$ will decrypt to a text $\bar{m}$ of this form is $2^{-50} \approx 8.9 \times 10^{-16}$ which is the confidence one would have in the authenticity of a message for this example. Clearly one could arrange for a higher or lower degree of confidence by varying the amount of prearranged redundant information in the authentic messages; limited above, of course, by the difficulty of factoring n, since if this is possible the secret e could be calculated from the public d and any message authenticated. This authentication scheme might at first appear to satisfy the definition of a provably secure system, however, this isn't the case since the implication is in the wrong direction; that is, the integrity of the authentication is no greater than factoring is difficult. What we would like to be able to say is that deception using this authentication scheme is at least as hard as factoring. In spite of an enormous amount of work on this and related questions of security dependent on

---

*The reader is referred to J. L. Massey's chapter "Contemporary Cryptology: An Introduction" or to the chapter on "Public Key Cryptography" by J. Nechvatal for a more complete discussion of the RSA cryptoalgorithm.

factoring, it is not known whether the decryption of almost all ciphers for arbitrary exponents, e, in the RSA encryption scheme is as hard as factoring. However, Rabin [16] has shown that if one uses the encryption function

$$c \equiv m(m + b) \pmod{n}$$

where b is an integer, $0 \leq b < n$, which is effectively the same as encrypting using the exponent $e = 2$ in the RSA algorithm, then decryption is not simply a consequence of being able to factor n but is actually equivalent. It should be pointed out that if the encryption exponent, e, is chosen to be 2 in an RSA scheme, that there is no corresponding decryption exponent, d, since there can be no solution to the congruence

$$2d \equiv 1 \pmod{(p - 1)(q - 1)}$$

and that consequently the Rabin scheme is slightly different from the RSA scheme. From the standpoint of a secrecy channel, Rabin's scheme, however, has the serious shortcoming that the authorized user is almost always left with an ambiguity among four potential messages. This is true since a quadratic residue, r, modulo $n = pq$, that is, a residue that is the square of some other residue, has four square roots if $(r, n) = 1$. For example, $2^2 \equiv 7^2 \equiv 8^2 \equiv 13^2 \equiv 4 \pmod{15}$. Williams has extended Rabin's work by showing that if the primes p and q are chosen so that $p \equiv 3 \bmod 8$ and $q \equiv 7 \bmod 8$, then there is an easy procedure for identifying one of the four roots to a quadratic equation as the distinguished root, that is, the intended message. The price that is exacted for this resolution of the ambiguity, besides the insignificant restriction on the primes p and q, is that the modulus, n, must be roughly four times larger than the largest message. The precise reasons for this are not important to the present discussion, the interested reader is referred to [34] for the full details. Since the Rabin–Williams scheme is a secrecy channel, not an authentication channel, it doesn't directly satisfy our needs. The reason why it is only a secrecy channel is easy to see; in the Rabin–Williams scheme the public users know the rules for transforming a message into a form suitable for encryption and that squaring (or in the most general form raising the transformed message to a publicly known even exponent) is the public encryption operation. The receiver, knowing the factorization of the modulus n, can recover the transformed message which he can then unambiguously interpret by reversing the initial transformation.

Williams does describe a digital signature scheme based on his and Rabin's technique that requires each user to have two of the private channels: one for signing and one for sealing messages. The basic idea is that the transmitter first carries out an operation that only he can do, that is, extracting a square root with respect to one of his moduli and then performs an operation that only the receiver can undo, that is, squaring with respect to the receiver's public encryption modulus. This cipher, along with a plaintext message identifying the transmitter, is sent to the receiver. The receiver, since he knows who the purported transmitter is, can carry out the inverse operations in reverse order to recover a message, which, if it contains the expected redundant (authenticating) information, could only have come from someone knowing the factorization of the identified transmitter's signing modulus—ipso facto the transmitter.

To see roughly how this scheme would work, assume that the source can take any one of a thousand equally likely states, labeled digitally 000,001, . . . , 999 and that the redundant (appended authenticator) information consists of three terminal 0's, that is, only ciphers that decrypt to texts of the form $xyz\,000$ will be accepted as authentic. In the Rabin–Williams scheme, not only must $p \equiv 3 \bmod 8$ and $q \equiv 7 \bmod 8$ but it must also be true that $2m + 1$ is a quadratic residue with respect to $n_t = p_t q_t$ and that $4(2m+1) < n_t$, where m is the message. In the present example $n_t > 7992004$. Therefore, let the transmitter's signing modulus be defined by the primes $p_t = 2027$ and $q_t = 3943$ and $n_t = 7992461$. The receiver's sealing (public encryption) modulus must be greater than $8n_t + 1$ in order to be certain that the cascaded encryption and decryption operations will not interfere with each other, that is, $n_r > 63939689$. Let $p_r = 7907$ and $q_r = 8087$ so that $n_r = 63943909$. Using these parameters, which were selected to make nearly optimal use of the Rabin–Williams channel bandwidth, any message of the form $xyz\,000$ can be authenticated with a probability of successful deception by an opponent of only 10^{-3}. The reason for describing this example in such detail was to highlight a common cost in authentication; ≈ 26 bits must be transmitted to communicate the superencrypted cipher (residue with respect to $n_r = 63943909$). This cipher when processed by the receiver conveys roughly 10 bits of information (one message out of 1000 equally likely triples) and provides roughly 10 bits of authentication (10^{-3} probability of deception) at the expense of an extra 6 bits having to be transmitted that cannot be used to either convey useful information or to confer additional security. The important point, though, is that authentication using the Rabin–Williams algorithm has been proven to be as secure as factoring is difficult.

The distinction between computationally secure schemes and provably secure schemes is a subtle one since in both cases the security is dependent on the computational difficulty of solving some hard problem. Perhaps the difference is only that in the first case one has reason to believe that the security is as great as a hard problem is difficult to solve, while in the other, one knows that it is at least that great. One important difference, though, is that in the case of a provably secure scheme, a proof must be given showing that subverting the security (for almost all cases) is equivalent to solving the hard problem. Because of this common dependence on computational infeasibility to achieve security, the present author combined computationally secure and provably secure authentication schemes into a single class of computationally secure schemes in his taxonomy for authentication schemes [27].

Figure 1 shows the natural taxonomy for authentication schemes based on the categories identified here; security, secrecy and arbitration. The divisions are:

- Unconditionally secure—computationally secure
- With secrecy—without secrecy
- With arbitration—without arbitration

In Fig. 1 for the four computationally secure categories, examples that have been discussed here and that are representative of each category are indicated. For the unconditionally secure categories which have yet to be discussed the references cited indicate where the first schemes illustrative of a category appeared in the literature. No one lays claim to the "brute force" solution for unconditionally secure authentication codes that provide for both secrecy and arbitration. Given the authentication codes for the other categories, it is obvious how to replicate a code sufficiently many times in a Cartesian product construction to achieve the desired result, however, the resulting constructions are so extravagantly wasteful of key (the information secretly

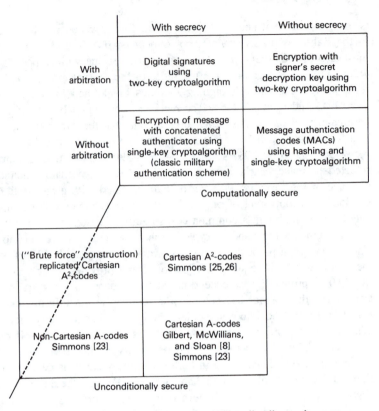

Figure 1 A natural taxonomy for authentication schemes.

exchanged between the insiders) that they are of no practical interest. The construction of efficient authentication codes for this category of authentication schemes is still an open problem.

In computationally and provably secure authentication schemes, the sets of acceptable messages are often determined (virtually constructed) by either appending a cryptographically related MAC to the information being authenticated, or else by first appending an unrelated authenticator and then cryptographically "sealing" the resulting extended message using either a single-key or a two-key cryptoalgorithm. Each choice of a key defines one subset of acceptable messages. In the case of authentication schemes based on single-key cryptography this is unavoidable since the only operation that the insiders (who know the key) can do, that outsiders cannot, is to encrypt or decrypt information using the secret key. In the case of two-key cryptographic techniques though, or especially in the case of a pure authentication channel, this need not be true. This is because an authentication channel can differ significantly from a secrecy channel, since in the one case it is only necessary that the receiver be able to verify that the authentication operation has been correctly carried out to establish that the communication is authentic, while in the other case he must be able to "invert" the operation to actually recover the information concealed in the cipher.

A well-known example of the latter type, that is, of using a public key encryption algorithm to define the set of acceptable messages by concealing them in ciphers is the digital signature scheme defined by Rivest, Shamir, and Adleman [17]. In this case, the information being authenticated is concealed by the authentication operation and revealed as an essential part of the process of verifying its authenticity. It is, of course, essential that only a predesignated fraction of the messages will be accepted as authentic: for example, those ending in a preagreed-on suffix. On the other hand, the digital signature scheme of El Gamal [6] is essentially an appended authenticator (MAC) to the message which need not be, and in fact can't be, decrypted by the receiver in the process of verifying that the message is authentic. Both of these schemes are computationally secure as are most of the other digital signature schemes based on two-key cryptographic techniques.

There are, however, a few provably secure authentication (digital signature) schemes [10,16,29,34] based on public key cryptographic algorithms. The digital signature scheme of Goldwasser, Micali, and Rivest [10] in addition passes a very strong security requirement; namely, it is secure against adaptive chosen message attacks. This is an appropriate point to remark that a digital signature is more than just a computationally secure or provably secure authentication with arbitration scheme. Anyone (having access to entirely public information) can verify the authenticity of a signature—not just the predesignated arbiter(s). It is an important point and should be clearly stated that the price paid to achieve unconditional security (in all presently known realizations) is to restrict the ability to authenticate messages to insiders, that is, to parties possessing some information not known to all of the other participants.

An authentication scheme is said to be unconditionally secure if the security is independent of the computing power or time an opponent can bring to bear, or equivalently if the opponent can do no better than to randomly choose a message on the chance that it may be acceptable to the receiver, irrespective of whether he attempts an I_0 or an I_1 or S_1 deception. Since the only known unconditionally secure authentication schemes are of the form of an explicit description of the various encoding (from states of the source into messages) rules, we will normally refer to such schemes as authentication codes. The notion of unconditionally secure authentication codes (similar in many respects to unconditionally secure encryption using one-time keys) is originally due to Simmons [23–24] although the subject now has a burgeoning literature. Because authentication codes are not nearly so well known as either computationally secure or provably secure schemes, we must develop the essential concepts for unconditionally secure schemes to illustrate one.

As the present author has pointed out elsewhere [23], unconditionally secure authentication codes are in a strict sense a mathematical dual to error detecting and correcting codes. In both cases, redundant information is introduced into the sequence of symbols that are transmitted, resulting in only a fraction out of the set of all possible sequences being available for use by the transmitter. In the one case, if the receiver receives a sequence that would not have been used by the transmitter, a fixed rule (usually a maximum likelihood detector) is invoked to decide which sequence was most likely transmitted, while in the other the receipt of a sequence that would not have been used by the transmitter under the agreed-on protocol is interpreted to mean that he didn't send it, or that if he did someone else has subsequently altered it, with the result that the message is to be rejected as unauthentic in either event.

Coding theory is concerned with schemes (codes) that introduce a redundancy in such a way that the most likely alterations to the encoded messages are in some sense close to the codeword they derive from. The receiver can then use a maximum likelihood detector to decide which (acceptable) message he should infer as having most likely been transmitted from the (possibly altered) codeword that was received. In other words, the object in coding theory is to cluster the most likely alterations of an acceptable codeword as closely as possible (in an appropriate metric) to the codeword itself, and disjoint from the corresponding clusters about other acceptable codewords to make unambiguous decoding possible.

Authentication codes are also concerned with schemes that introduce redundancy in the message, but in such a way that for any message the transmitter may send, the substitute or altered messages that the opponent may introduce using his optimal strategy are spread in what appears to be a random manner, that is, as uniformly as possible (again with respect to an appropriate metric) over the set of all possible messages. The theory of authentication codes is concerned with devising and analyzing schemes (codes) to achieve this "spreading." It is in this sense that coding theory and authentication theory are dual theories: one is concerned with clustering the most likely alterations of acceptable messages (codewords) as closely about the original codeword as possible and the other with spreading the optimal (to the opponent) alterations as uniformly as possible over the set of all messages.

Just as in error detecting and correcting codes, in either authentication codes or in constructive procedures for the authentication of digital information, for any fixed encoding rule, there is also a subset of acceptable, that is, authentic, sequences (those containing the specified redundant information) and a nonempty collection of sequences that do not contain the redundant information and hence that would be rejected as unauthentic. The difference is that whereas in coding theory there is but one encoding rule corresponding to the fixed code, in authentication codes there are many encoding rules from which the transmitter/receiver can choose the particular (secret) rule they will employ.

We illustrate these concepts with the smallest possible example. The source in our example is a fair coin toss observed by the transmitter, whose outcome, denoted by H or T, is the information (1 bit) the transmitter wishes to communicate to the receiver. The opponent wishes to deceive the receiver into either believing a legitimate coin toss has occurred when it hasn't, an I_0 deception, or else in misinforming him as to the outcome of the legitimate toss, an S_1 deception. An unconditionally secure authentication code to provide protection (1 bit) against both of these deceptions is the following:

$$
\mathbf{A} = \begin{bmatrix} H- \\ -H \end{bmatrix} \otimes \begin{bmatrix} T- \\ -T \end{bmatrix} = \begin{array}{c|cccc} & m_1 & m_2 & m_3 & m_4 \\ \hline e_1 & H & - & T & - \\ e_2 & H & - & - & T \\ e_3 & - & H & T & - \\ e_4 & - & H & - & T \end{array} \tag{1}
$$

There are four encoding rules, e_i, each of which encode the source states (H and T) into two out of the four possible messages, m_j; that is, 2 bits must be communicated through the channel to specify a message. To use this code to authenticate a communication of the outcome of a coin toss, the transmitter and receiver would choose (in secret from the opponent) one of the encoding rules with the uniform probability distribution on the e_i (their optimal authentication strategy) in advance of their need to authenticate a communication. The opponent knows the code (see Eq. (1)) and their strategy for choosing an e_i. If he chooses to impersonate the transmitter and send an unauthentic message when no message has yet been sent by the transmitter, it should be obvious that irrespective of which message, m_j, he chooses, the probability that it will correspond to an encoding of a source state under encoding rule e_i, and hence that it will be accepted by the receiver as an authentic message, is $\frac{1}{2}$. Similarly, if the opponent waits to observe a legitimate communication by the transmitter, his uncertainty about the encoding rule being used will drop from one of four equally likely possibilities to one of two. However, his probability of choosing an acceptable (to the receiver) substitute message will still be $\frac{1}{2}$. For example, if the opponent observes m_1 he knows that the transmitter and receiver are using either encoding rule e_1 or e_2.

In the first case m_3 would be an acceptable message to the receiver while m_4 would be rejected as unauthentic, while in the second case, exactly the opposite would be true. Hence, the opponent's probability of deceiving the receiver is $\frac{1}{2}$ irrespective of whether he impersonates the transmitter, an I_0 deception, or substitutes (modifies) legitimate messages, an S_1 deception. Another way of expressing this is that the opponent's a priori and a posteriori probabilities of success are both the same as the probability that a randomly chosen message will be acceptable to the receiver.

It should be pointed out that the code in Eq. (1) can be viewed as an appended authenticator scheme; functionally the same as the provably secure EFT scheme described above. Since messages m_1 and m_2 communicate only source state H, and messages m_3 and m_4 only source state T, the messages in this code are of the form shown above.

$$m_{i,j} = s_i : f(e_j, s_j) \tag{2}$$

where

$$
\begin{array}{c|cc}
 & \multicolumn{2}{c}{i} \\
j & H & T \\
\hline
e_1 & 0 & 0 \\
e_2 & 0 & 1 \\
e_3 & 1 & 0 \\
e_4 & 1 & 1 \\
\end{array}
\qquad (3)
$$

$$f(e_j, s_i)$$

Eq. (1) can be rewritten with the messages m_i in the form $s_i : f(e_j, s_i)$ that makes clear that this is an appended authenticator scheme

$$
A = \begin{array}{c|cccc}
 & H:0 & H:1 & T:0 & T:1 \\
\hline
e_1 & H & - & T & - \\
e_2 & H & - & - & T \\
e_3 & - & H & T & - \\
e_4 & - & H & - & T \\
\end{array}
$$

By the arguments already given, there is 1 bit of uncertainty as to authenticator that should accompany a source state in a message, even if the message conveying the other source state and matching acceptable authenticator has been observed. Irrespective of the computing power an opponent may have, he cannot reduce this uncertainty as to the acceptable appended authenticator for either an S_0 or I_1 deception.

This small example (1-bit source) can easily be extended to give an arbitrarily high degree of confidence against deception: $p_d = 1/n$, n an integer. Replace **A** in Eq. (1) by the Cartesian product of the $n \times n$ matrices

$$A = \begin{bmatrix} H & - & \cdots & - \\ - & H & & - \\ \vdots & & \ddots & \vdots \\ - & - & \cdots & H \end{bmatrix} \otimes \begin{bmatrix} T & - & \cdots & - \\ - & T & & - \\ \vdots & & \ddots & \vdots \\ - & - & \cdots & T \end{bmatrix} \tag{4}$$

This forms an authentication code with n^2 rows (encoding rules) and $2n$ columns (messages) in which the optimal strategy for the transmitter/receiver is still to choose an encoding rule e_i with the uniform probability distribution. The opponent's chances of success will be $1/n$ for either an I_0 or S_1 deception by the same arguments used before. The channel requires $\log_2(2n) = 1 + \log_2 n$ bits be communicated to convey the state of the source (1 bit) and to secure $1/n$ confidence in the authenticity of the received message (log n bits). The code is therefore said to be perfect in the sense that every bit of information communicated in a message is used to either convey the state of the source or else to confound the opponent. All of these codes are unconditionally secure.

Given the simple nature of the construction in the preceding paragraph, it might appear at first that there would be no difficulty in constructing unconditionally secure authentication codes for arbitrary sources and for any desired level of security. If the source can assume any one of k states, and if the desired level of security is $1/n$, n an integer, then the obvious generalization of Eq. (4) would be the k-fold repeated Cartesian product of $n \times n$ matrices

$$A = \begin{bmatrix} s_1 & - & \cdots & - \\ - & s_1 & & - \\ \vdots & & \ddots & \vdots \\ - & - & \cdots & s_1 \end{bmatrix} \otimes \begin{bmatrix} s_2 & - & \cdots & - \\ - & s_2 & & - \\ \vdots & & \ddots & \vdots \\ - & - & \cdots & s_2 \end{bmatrix} \otimes \cdots \otimes \begin{bmatrix} s_k & - & \cdots & - \\ - & s_k & & - \\ \vdots & & \ddots & \vdots \\ - & - & \cdots & s_k \end{bmatrix} \tag{5}$$

which yields a code with n^k rows (encoding rules) and nk columns (messages). This unconditionally secure authentication code is also perfect in the sense defined above since the number of bits required to specify a particular message out of nk equally likely messages is

$$\log_2(nk) = \log_2 k + \log_2 n$$

which is precisely the amount of information needed to convey the state of the source ($\log_2 k$ bits if all states are equally likely) and to present the opponent with an equivocation of log n bits as to the appended authenticator that the receiver will accept. This code is not very efficient, however, in another important respect which we will discuss below.

In the military sealed-authenticator authentication scheme described above, the information being authenticated (state of the source) was concealed in the cipher, that is, it was secret. On the other hand, in the Federal Reserve EFT authentication protocol, the information being authenticated was transmitted in the clear, that is, it was unencrypted or without secrecy. This distinction of whether the authentication is made with or without secrecy for the underlying information is an essential division of all authentication schemes [27], and can be of enormous importance for some applications such as treaty verification where secrecy is unacceptable. For the moment we merely remark that the Cartesian product constructions just described, since they lead to only appended authenticator schemes, are necessarily authentication without secrecy schemes. It is possible, however, to construct perfect authentication with secrecy codes corresponding to the authentication without secrecy schemes already exhibited. An authentication with secrecy code corresponding to the authentication without secrecy code shown in Eq. (1) is:

$$
\mathbf{A} = \begin{array}{c} \\ \mathbf{e}_1 \\ \mathbf{e}_2 \\ \mathbf{e}_3 \\ \mathbf{e}_4 \end{array}
\begin{array}{cccc}
m_1 & m_2 & m_3 & m_4 \\
\hline
H & - & T & - \\
- & H & - & T \\
T & - & - & H \\
- & T & H & -
\end{array}
\qquad (6)
$$

The source is the same as before; the channel still requires 2 bits to specify a particular message and the optimal strategies for the transmitter/receiver and for the opponent remain the same. The difference is that no matter which message the opponent may observe, he has 1 bit of uncertainty remaining as to the encoding rule the transmitter/receiver are using and thus 1 bit of uncertainty as to the source state. But there was only 1 bit of a priori uncertainty as to the source state before an observation was made; hence, this is a perfect authentication with secrecy scheme. It should be noted that if the opponent does somehow learn the state of the source and also observes the message that the transmitter uses to communicate this state to the receiver, he can then unambiguously identify the encoding rule they are using and hence substitute an acceptable message to misinform the receiver of the state of the source, an S_1 deception, with certainty of success.

There is a simple example that demonstrates in a striking way the difference between computationally secure authentication schemes and unconditionally secure schemes. To present it here, we need to anticipate a result that will be derived later in Section 4. This is a bound on the security of any authentication channel, whether computationally secure or unconditionally secure, first derived by Gilbert, McWilliams, and Sloan [8] (Eq. (28)) in a slightly different setting than used here, which says that the probability, P_d, of an opponent succeeding in either an I_0 or an S_1 deception is bounded below by

$$
P_d \geq \frac{1}{\sqrt{|\mathcal{E}|}}
$$

(see Eq. 28) where $|\mathcal{E}|$ is the total number of encoding rules in the authentication scheme, that is, the number of rows in the encoding matrix. In 1985, the present author reported [24] an apparently paradoxical situation in which this bound appeared to be violated in a computationally secure authentication scheme.

Let the source be a fair coin toss as before and let the sets of acceptable (authentic) messages be constructed by encrypting with a secret DES key the state of the source concatenated with a redundant authenticating suffix of sixty-three 1's. In other words, the only ciphers that the receiver will accept are those that decrypt to one of the two texts

$$
\overbrace{011 \ldots 1}^{63} \quad \text{or} \quad \overbrace{111 \ldots 1}^{63}
$$

communicating, say, tails or heads, respectively. As is well known, there are 2^{56} DES keys, and hence in the scheme just described 2^{56} encoding rules. Consequently, $|\mathscr{E}| = 2^{56}$, and the bound (Eq. (10)) says that even if the transmitter/receiver choose among the 2^{56} encoding rules optimally, they cannot limit the opponent's probability of successfully deceiving the receiver into accepting an unauthentic message to less than

$$P_d \geq \frac{1}{\sqrt{|\mathscr{E}|}} = \frac{1}{2^{28}} \approx 3.7 \times 10^{-9} \tag{7}$$

or roughly 4 parts in a billion.

In practice, the opponent's chances of success are dramatically less than Eq. (7) would suggest. There are 2^{64} possible ciphers (messages), only 2 of which are acceptable for any particular choice of a key (authentication encoding rule). Therefore, if the opponent merely selects a cipher at random and attempts to impersonate the transmitter, his probability of success is 2^{-63} or approximately 1 chance in 10^{19}. The question is, Can he do better? As far as an I_0 deception is concerned, the answer is essentially no, even if he has unlimited computing power. For each choice of an encoding rule, there are two (out of 2^{64}) ciphers that will be acceptable as authentic. Assuming that the mapping of the sequences $11 \ldots 1$ and $01 \ldots 1$ into 64-bit cipher sequences under DES keys is a random process, this says that the total expected number of acceptable ciphers (over all 2^{56} keys) is $\approx 2^{56.9888}$, that is, ε close to 2^{57}. Even if the opponent could feasibly carry out the enormous amount of computation that would be required to permit him to restrict himself to choosing a cipher from among this collection, his chances of having a fraudulent message be accepted by the receiver would still only be $\approx 2^{-56}$ or roughly 1 chance in 10^{17} which is what we meant when we said that the answer was essentially no since 10^{-17} isn't much different than 10^{-19} while both differ enormously from the bound (see Eq. (7)) of $\approx 10^{-9}$. The opponent could not do better, nor worse (in attempting to impersonate the transmitter) even if he possessed infinite computing power than choose a cipher randomly, with a probability distribution weighted to reflect the number of times each cipher occurs, from among the $\approx 2^{57}$ potentially acceptable ciphers.

However, the channel bound $p_d \geq |\mathscr{E}|^{-\frac{1}{2}}$ (31) applies to all authentication schemes, hence the apparent contradiction must arise in connection with the opponent's substitution, S_1, strategy. If the opponent waits to observe a legitimate message (cipher), can the information acquired by virtue of this observation be put to practical use to improve his chances of deceiving the receiver? Even if he doesn't know the state of the source, he knows that the cipher is the result of encrypting one of the two 64-bit sequences $111 \ldots 1$ or $011 \ldots 1$ with one of the 2^{56} DES keys. He also knows that with a probability of essentially 1 (≈ 0.996), there is only one key that maps the observed cipher into either of these two sequences, hence, he is faced with a classic "meet in the middle" cryptanalysis of DES as discussed by Diffie and Hellman [5].

Clearly if he succeeds in identifying the DES key, that is, the encoding rule being employed by the transmitter receiver, he can encrypt the other binary string and be certain of having it be accepted, and hence be certain of deceiving the receiver. The point, though, is that for him to make use of his observation of a message he must be able to determine the DES key the transmitter and receiver(s) are using, that is, he must be able to cryptanalyze DES. If he can do this, the expected probability of deceiving the receiver is e close to 1, the small deviation being attributable to the exceedingly small chance that two (or more) DES keys might have encoded source states into the same message (cipher). But even if this is the case (improbable though it may be), his probability of success in an S_1 deception is simply the probability that he selects the same DES key as the receiver and transmitter are using out of a set of two (or few) possible choices. Thus, we have the paradoxical result that the practical system in this example is some eight or nine orders of magnitude more secure than the theoretical limit simply because it is computationally infeasible for the opponent to carry out in practice what he should be able to do in principle. In this respect, practical message authentication is closely akin to practical cryptography where security is equated to the computational infeasibility of inverting from arbitrarily much known matching cipher text and plaintext pairs to solve for the unknown key, even though in principle there may be more than enough information available to insure a unique solution.

4 HOW INSECURE CAN UNCONDITIONALLY SECURE AUTHENTICATION BE?

We have already adopted a model for unconditionally secure authentication codes in which the encoding rules, e_i, are organized in an array, $\mathbf{A}$, specifying the available mappings of the source states into the set of all possible messages. The rows of $\mathbf{A}$ are indexed by encoding rules and the columns by messages. The entry in $a(e_i, m_j)$ is the element of $\mathcal{S}$ encoded by rule e_i into message m_j if such a source mapping exists under e_i and 0 otherwise. Every element of $\mathcal{E}$ appears in each row of $\mathbf{A}$ at least once and perhaps several times since the transmitter must be able to communicate an arbitrary state of the source to the receiver. Clearly, if an encoding rule uses all of the messages, that is, every message is acceptable to the receivers as an authentic communication, then an opponent is certain of success in an S_0 deception whenever that encoding rule is used. Therefore, in any strategy that holds the opponent to less than a certainty of success (in deceiving the receiver) any such encoding rule will not be used, and can therefore with no loss of generality be assumed to not occur in $\mathbf{A}$. Put another way, every encoding rule will have at least one unused message. Similarly, no message can occur in all encoding rules, otherwise the opponent could impersonate using that message with certainty of success. We now define another $|\mathcal{E}| \times |\mathcal{M}|$ matrix $\mathbf{X}$, in which

$$\chi(e_i, m_j) = \begin{cases} 1 & \text{if} \quad a(e_i, m_j) \in \mathcal{S} \\ 0 & \text{otherwise} \end{cases}$$

For example, for the authentication scheme in Eq. (1):

$$\mathbf{A} = \begin{array}{c} \\ s_1 \\ s_1 \\ \\ \\ \end{array} \begin{array}{cccc} m_1 & m_2 & m_3 & m_4 \\ - & s_2 & - \\ - & - & s_2 \\ s_1 & s_2 & - \\ s_1 & - & s_2 \end{array} \quad \text{and} \quad \mathbf{X} = \begin{vmatrix} 1 & 1 & 0 & 0 \\ 1 & 0 & 1 & 0 \\ 0 & 1 & 0 & 1 \\ 0 & 0 & 1 & 1 \end{vmatrix}$$

It is now easy to see the relationship of the impersonation "game," the I_0 deception, to the matrix $\mathbf{X}$. If m_j is an acceptable (authentic) message to the receiver when encoding rule e_i has been agreed to by the transmitter and receiver then $\chi(e_i, m_j) = 1$

and the cpponent has a probability of success of $p = 1$ if he communicates m_j to the receiver. Conversely, whenever $\chi(e_i, m_j) = 0$ he is certain the message will be rejected. It is certainly plausible, and in fact rigorously true, that the opponent's probability of success in impersonating the transmitter is the value, v_I, of the zero-sum game whose payoff matrix is **X**. It is possible to define a companion payoff matrix **Y** for the substitution game, although it is considerably more complex since the probabilities are now all conditioned based on the observed legitimate message, so that the size of the matrix **Y** grows as a power of $|M|$. The value of this game, v_s, is the probability that the opponent will be successful in deceiving the receiver through intercepting a message sent by the transmitter and substituting one of his own devising, that is, the probability of an S_1 deception being successful. Given an authentication code, the transmitter/receiver have the freedom to choose among the encoding rules and if some state(s) of the source can be encoded into more than one message under some of the encoding rules, a choice of which messages to use, that is, a splitting strategy. The opponent, on the other hand, can choose between impersonation and substitution with whatever probability distribution he wishes and then choose according to his optimal strategy which fraudulent message he will communicate to the receiver, either with no conditioning if he is impersonating the transmitter or else conditioned on the message he observed if he is substituting messages. Not surprisingly there exist authentication codes in which the optimal strategy for the opponent is either pure impersonation, pure substitution, immaterial mixes of the two, or most interestingly, essential mixing of both as well as examples in which splitting is essential in the transmitter/receiver's optimal strategies. The point of these remarks is that it has been shown [1,23,32] that an opponent's overall probability of success in deceiving the receiver, P_d, is simply the value of a game, G, whose payoff matrix is the concatenation of **X** and **Y**, and hence that

$$P_d = v_G \geq \max(v_I, v_s)$$

The notation we will use is:

Name	Set	Element	Variable	Entropy
Source	$\mathcal{S}$	s_i	S	$H(S)$
Message space	$\mathcal{M}$	m_j	M	$H(M)$
Encoding rules	$\mathcal{E}$	e_k	E	$H(E)$
Splitting strategies		$\pi(m_j \mid s_1 e_k)$	Π	$H(M \mid ES)$
Impersonation strategy	Q	q_j	Q	$H(Q)$

where

$P(X = x)$ is the probability that the random variable X takes the value x, as, for example, $P(M = m)$, $P(S = s)$, or $P(E = e)$

$|e_i| = \Sigma_{m \in \mathcal{M}} \chi(e_i, m)$ is the number of nonzero entries in the e_i row of either **A** or **X**

$|m_j| = \Sigma_{e \in \mathcal{E}} \chi(e, m_j)$ is the number of nonzero entries in the m_j column of either **A** or **X**

It is now an easy matter to state and prove several easy lower bounds for just how insecure an unconditionally secure authentication code can be. All of these results appear in [23], but are repeated here to make this chapter self-contained. The notation P_d was introduced earlier, and denotes the probability that an opponent will be successful in an I_0 or an S_1 deception.

Theorem 1.

$$P_d \geq \frac{\min_{\mathcal{E}} |e_i|}{|\mathcal{M}|} \tag{10}$$

Comment: As has already been noted, the opponent has available as part of his strategy the choice of whether to impersonate the transmitter or to substitute messages, that is, whether to attempt a type I_0 or S_1 deception. Therefore, the value of the concatenated game is at least as large as the value of either game alone (Eq (9)). What is actually proven in [23] is that for the impersonation game:

$$v_I \geq \frac{\min_{\mathcal{E}} |e_i|}{|\mathcal{M}|} \tag{11}$$

Corollary: Since $\min_{\mathcal{E}} |e| \geq |\mathcal{S}|$

$$P_d = v_G \geq \frac{|\mathcal{S}|}{|\mathcal{M}|} \tag{12}$$

Theorem 2.   Given an authentication scheme for which

$$v_G = \frac{\min_{\mathcal{E}} |e_i|}{|\mathcal{M}|} \tag{13}$$

in every optimal strategy, E, for the transmitter/receiver $P(E = e) = 0$ for any encoding rule for which $|e| > \min_{\mathcal{E}} |e_i|$. In other words, the transmitter/receiver will not use any encoding rule in which $|e| > \min_{\mathcal{E}} |e|$ if equality Eq. (13) holds.

Corollary: If for an authentication scheme

$$v_G = \frac{|\mathcal{S}|}{|\mathcal{M}|}$$

which by Theorem 1 can only happen if $\min|e| = |\mathcal{S}|$, then every optimal strategy for the transmitter/receiver, E, has $P(E = e_j) = 0$ for any encoding rule e_j for which $|e_j| > |\mathcal{S}|$.

Another way of stating the conclusion of the corollary is that if $v_G = |\mathcal{S}|/|\mathcal{M}|$ no splitting occurs in any encoding rule occuring in an optimal strategy! It is worth remarking that

$$v_G = \frac{\min_{\mathcal{E}} |e|}{|\mathcal{M}|}$$

does not imply that splitting does not occur in any of the encoding rules that occur in $\mathscr{E}$. What is true, by Theorem 2, is that in this case all of the encoding rules that occur (with positive probability) in an optimal strategy use the same number of messages.

Several other channel capacity theorems of similar flavor can be proven; however, we now turn to the primary object in this section of formulating lower bounds for the insecurity of unconditionally secure authentication schemes in terms of the entropies of the primary variables: S, E, M, Q, etc.

A trivial bound can be given in terms of $H(E)$. Since $H(E)$ is the total equivocation that the opponent has as to which encoding rule is being used by the transmitter/receiver, and since he could deceive the receiver with certainty if he only knew the rule they had chosen, we have

$$\log P_d = \log v_G \geq - H(E) \tag{14}$$

Eq. (14) isn't a particularly useful result since as we shall see later there is a much stronger bound in terms of $H(E)$. The bound of the following theorem is the main result in the theory of unconditionally secure authentication codes.

Theorem 3. (Authentication Channel Capacity)

$$\log P_d \geq H(MES) - H(E) - H(M) \tag{15}$$

Comment: The proof of Theorem 3 is similar in style to many channel capacity proofs in information theory in which multiple summations of the entropies for discrete events are manipulated, usually by interchanging the orders of summation to reexpress and regroup terms in a form where Jensen's inequality* can be invoked. Such proofs seem to be unavoidably tedious and long. The reader is referred to [23] where the (tedious and long) proof of Theorem 3 is given. Massey gives an elegant (and short) proof in his "Contemporary Cryptology: An Introduction" in this volume of a bound on the probability of an impersonation deception being successful. He uses a slightly different notation than is used here, but the reader should have no difficulty in converting results from either chapter into the form used in the other.

A variety of useful equivalent expressions can be derived from Eq. (15) using simple identities from information theory, for the cases of authentication either with or without secrecy. We illustrate the technique in Theorem 4 for the case of authentication with secrecy; that is, the opponent does not know the state of the source observed by the transmitter. This, of course, only matters when the opponent elects to substitute messages rather than to impersonate the transmitter.

---

*If $q(x)$ is a convex function on an interval (a, b) and $x_1, x_2, \ldots, x_n$ arbitrary real numbers $a < x_i < b$, and if $w_1, w_2, \ldots, w_n$ are positive numbers with $\Sigma w_i = 1$, then

$$g\left[\sum_{i=1}^{n} w_i x_i\right] \leq \sum_{i=1}^{n} w_i g(x_i)$$

Theorem 4. $H(MES) - H(E) - H(M)$ is equivalent to any of the following eight entropy expressions:

Equivalent Form	Equation
$H(M \mid ES) + H(S) - H(M)$	(16)
$\left\{\begin{array}{l} H(E \mid MS) - H(E) + H(MS) - H(M) \\ \text{or} \\ H(E \mid MS) - H(E) + H(S \mid M) \end{array}\right.$	(17) (18)
$\left\{\begin{array}{l} H(E \mid M) - H(E) \\ \text{or} \\ H(M \mid E) - H(M) \end{array}\right.$	(19) (20)
$H(ME \mid S) + H(S) - H(E) - H(M)$	(21)
$H(MS \mid E) - H(M)$	(22)
$H(ES \mid M) - H(E)$	(23)

Using the results of Theorem 4 it is possible to derive some (generally) weaker but enlightening lower bounds for the insecurity of unconditionally secure authentication codes. We first note that the total effective equivocation to the opponent playing the substitution game but without knowledge of the source state, that is, authentication with secrecy is no greater than $H(E \mid M)$ and as remarked above, the opponent's total effective equivocation if he knows the source state, that is, authentication without secrecy, is at most $H(E \mid MS)$.

Theorem 5.   For authentication with secrecy

$$\log v_G \geq -\frac{1}{2} H(E) \tag{24}$$

while for authentication without secrecy

$$\log v_G \geq -\frac{1}{2} \{H(E) - H(MS) + H(M)\} = -\frac{1}{2} \{H(E) - H(S \mid M)\} \tag{25}$$

The results contained in Theorem 5 are so important to the theory of authentication that it is perhaps worthwhile to indicate how they are obtained. For authentication with secrecy

$$\log v_G \geq \min\{\log v_I, -H(E \mid M)\} \tag{26}$$

while for authentication without secrecy

$$\log v_G \geq \min\{\log v_I, -H(E \mid MS)\} \tag{27}$$

In either Eq. (26) or Eq. (27) the bounds derived in Theorems 3 and 4 on the value of the impersonation game can be substituted, since the opponent's impersonation strategy is independent of whether he plays substitution with or without secrecy. Replacing the minimum on the right-hand side of the inequality by the average of the two bracketed terms either weakens the inequality if the terms are not identical or leaves it unaffected if they are. Therefore for authentication with secrecy, replacing v_I with the bound in Eq. (19) in Eq. (26) we get

$$\log v_G \geq \frac{1}{2}\{H(E \mid M) - H(E) - H(E \mid M)\} = -\frac{1}{2}H(E)$$

and similarly by replacing v_I with the bounds in Eq. (17) or Eq. (18) in Eq. (27) we get

$$\log v_G \geq \frac{1}{2}\{H(E \mid (MS) - H(E) + H(MS) - H(M) - H(E \mid (MS)\}$$

$$= -\frac{1}{2}\{H(E) - H(MS) + H(M)\}$$

or

$$\log v_G \geq \frac{1}{2}\{H(E \mid MS) - H(E) + H(S \mid M) - H(E \mid MS)\}$$

$$= \frac{1}{2}\{H(E) - H(S \mid M)\}$$

Corollary

$$P_d = v_G \geq \frac{1}{\sqrt{|\mathcal{E}|}} \tag{28}$$

As was mentioned above, the bound in Eq. (28) was first derived by Gilbert, McWilliams, and Sloan [8] under slightly more restrictive conditions and proven directly in the same generality used here by Brickell and Simmons in [1]. The more useful, and informative lower bounds on the insecurity of unconditionally secure authentication schemes are Eqs. (24) and (25) for authentication with and without secrecy, respectively.

The inescapable conclusion that must be drawn from the theoretical results just given is that a large number of encoding rules must be available in any unconditionally secure authentication code—on the order of $1/P_d^2$ at least—to realize a security of P_d and that these encoding rules must also have a well-defined structural interdependence to insure that the conditional entropy conditions be met to make this level of security available at all. As we have seen from the example at the end of the previous section, neither of these comments applies to computationally secure (and hence also to provably secure) authentication schemes.

5 THE PRACTICE OF AUTHENTICATION

We have already described two of the more straightforward applications of authentication to the problem of enabling a receiver on a communications channel to verify that in all probability a message received over the channel was originated by the legitimate (authorized) transmitter and that it hasn't been tampered with subsequently. In the one case an authenticator, known to both the transmitter and to the receiver, was appended to the information to be authenticated and then encrypted to produce a cipher which served to provide both secrecy and authentication. In the other case the information was partitioned into blocks which were then block-chain-encrypted to produce a MAC that was appended to the text. In this case only authentication was achieved, unless the extended message was then superencrypted to yield a cipher that would provide secrecy.

The most important point about these two schemes, though, is that if in either case a single-key cryptoalgorithm is used for the encryption and decryption functions, then the transmitter and receiver must be mutually trusting and trustworthy. This is unavoidable since for a single-key cryptoalgorithm the transmitter and receiver must both possess the same cryptographic keying variable and hence each be able to interchangeably do anything the other can. The more challenging case is when the transmitter and receiver cannot be assumed to trust each other, or even the more extreme case of when they must both be assumed to be untrustworthy and deceitful, that is, that if either can get away with cheating, they will cheat. The problem in either of these cases is further complicated by what action the transmitter or receiver will take if they conclude the other has cheated. If the action is unilateral, that is, if no third party need be convinced that cheating has occurred, the demands on the authentication scheme are entirely different, and much simpler, than if the authentication must be logically compelling to a third party or arbiter.

In a companion chapter to this one in this volume entitled "How to Insure That Data Acquired to Verify Treaty Compliance Are Trustworthy," a detailed account is given of an authentication scheme serving mutually distrusting and deceitful parties in which deceits must be logically demonstrable to third parties who themselves may wish to deceive either the transmitter or receiver, host and monitor, respectively, in the terminology of treaty verification schemes. This application of authentication is, to this author's knowledge, the most intricate and convoluted, and hence the most interesting, that has been made to date.

Rather than repeat, even in abbreviated form, the details of that application here, we describe instead a common business/commercial application of authentication with similar requirements, but with a quite different solution. Much more efficient, and hence applicable, techniques are described in the chapters "Digital Signatures" by Mitchell, Piper, and Wild and "Smart Card: A Standarized Security Device Dedicated to Public Cryptology" by Guillou, Quisquater, and Ugon, however the simplicity of the protocol described here recommends it as an illustrative example for an important area of application.

Consider the needs of the various participants in a credit card transaction at the point of sale. The merchant (or automated teller machine [ATM], etc.) will give up something of real value, that is, an item of merchandise, money, services, etc. if the transaction is completed in exchange for a record (information) evidencing the credit due from the customer. The merchant must be able to satisfy himself as to the customer's identity, the validity of his claimed account, and perhaps of the level of credit in that account as well as the integrity of the record he holds establishing the credit due to him. In other words, the merchant must suspect that the customer isn't who he claims to be, that the credit card (credential) presented is either a forgery or else that even if it is genuine that it doesn't belong to that customer, and that the customer will later disavow having made the purchase or if he does admit that a purchase was made, will aver that it was for a lesser amount or was made at a later date, etc.

The customer, on the other hand, needs to be able to satisfy himself that the record of the transaction is accurate, that is, it is for the proper amount and shows the correct date, etc.; it can only be presented for collection once; it can't be altered to increase the customer's liability; and the merchant can't later, as a result of any number of transactions with the customer, impersonate him to fraudulently make purchases, withdrawals, etc. on his (the customer's) account. In other words, the customer must suspect that the merchant will attempt to alter the record of the transaction or submit it for collection more than once, or to alter the dates of transactions, etc. He must also suspect that the merchant, or his agents, will accumulate records of his past transactions with the object of impersonating him to other merchants to collect goods, services, or monies in his (the customer's) name.

The bottom line is that a normal credit transaction provides the classic paradigm for an information exchange between mutually distrustful and untrustworthy participants. Since the resolution of a dispute over the validity of a claimed transaction will necessarily involve third parties, that is, a bank, a court, or an arbiter, etc.; a satisfactory solution should address all of the concerns of all of the participants and produce a record that can be logically evaluated to assign liability to the party that should, in all probability, be held liable.

We discuss how authentication of information can be used to satisfy these needs in several stages. First, we show how a customer can be identified in a way that can later be verified (as to the correctness of the identification) by third parties who were not parties to the transaction in real time. There are only two ways an individual can be identified, irrespective of whether the identification is made by a manned or an automated facility; these two ways depend on whether intrinsic or extrinsic identifying information is used.

Intrinsic information are the physical identifiers of the individual; fingerprints, voiceprint, retinal prints, signature and/or signature dynamics, hand geometry, physical appearance (usually as compared to a photograph), height, weight, distinguishing marks, etc. Intrinsic identifiers are inextricably linked to the person they identify, so that to successfully impersonate someone else, an imposter must succeed in mimicking their identifiers.

Extrinsic information is something that the right individual is known to know or be able to do, but which an imposter probably won't know or be able to do; for example, computer passwords, telephone credit card numbers, Swiss bank account numbers, personal identification numbers (PINs), signs or countersigns for sentries, etc. are all examples of extrinsic identifiers. Extrinisic identifiers are not linked to the person being identified, hence it is only necessary for a would-be imposter to learn what the identifiers are to undetectably impersonate the legitimate owner. Consequently, there is the logical problem that this type of identifying information, once exhibited by the owner to prove his identity, is potentially compromised and could be used by anyone, in particular the merchant or his agents in the present discussion, to undetectably impersonate the legitimate owner.

To have an acceptably secure identification scheme based on extrinsic identifiers, one needs to devise a protocol that will allow an individual to "prove" that he knows the secret piece(s) of information, whose possession is equated with his identity, without revealing anything about the information itself which could aid a would-be cheater to impersonate him. Several investigators have proposed identification schemes to accomplish this [7, 9, 19, 28] that depend on interactive-proof schemes, often referred to as zero-knowledge proofs, in which the individual responds to a series of queries in a way that the legitimate user could, but which an imposter (probably) could not.

Later we will describe a simple identity verification scheme that uses a public authentication channel to validate the public part of a private authentication channel belonging to the individual who wishes to prove his identity. He can then prove that he is who he claims to be by being able to generate authenticated messages in the private authentication channel which is possible (in probability) only if he knows the secret part of that private channel. The public and the private channels can be completely independent and can even be based on different authentication algorithms, or they can both be of the same type. This scheme also provides certified receipts for transactions whose legitimacy can later be verified by impartial arbiters who were not involved in the transaction itself.

First, however, we describe an identity verification scheme that uses an authentication channel to validate intrinsic identifiers for the customer to the merchant. Since we are considering point-of-sale protocols, we assume that the identification must be made on the basis of information supplied to the merchant by the customer. For practical reasons, we impose the additional restriction that very little communication to remote locations be required in advance of, and none at the time that the customer's identity is to be verified. Finally, there is the practical constraint, arising from enormous numbers of merchants with whom a customer may unexpectedly wish to do business as well as the number of persons and the turnover in the staff having access to the customer identification equipment in the store, that while the merchant's information must be protected from alteration or substitution, it is generally not possible to guarantee its secrecy. In other words, a pure authentication channel distinct and separate from a secrecy channel is what is involved.

In either case, that is, for either intrinsic or extrinsic identifying information, it is assumed that there exists some party or facility that is unconditionally trusted by both the customer(s) and the merchant(s); the issuer of validated (signed) identification credentials. This could be, depending on the application, a government agency, a credit card center or financial institution, a military command center, a centralized computer facility, etc. The trusted issuer first establishes a public, that is, desensitized, authentication channel to which he retains the (secret) authenticating function.

For simplicity, throughout the discussion in this section we will use the well-known authentication channel based on the RSA cryptoalgorithm for both the public (issuer) and the private (customer) channels, although authentication channels based on any other algorithm would serve equally well. As described in the body of the chapter, the issuer would choose a pair of primes p and q by the same standards used to compute a good RSA modulus, that is, so that it is computationally infeasible for anyone to factor n, and then calculate a matching pair of encryption/decryption exponents, e and d, and publish n and d as the public key. The issuer would keep e (and equivalently the factors p and q) secret; in fact, the security of the system against fraudulent claims of validated identity is no better than the quality of protection given to e by the issuer.

The solution to the problem of identifying the customer is now so simple as to be almost anticlimatic. The central facility, the issuer, is entrusted to first establish the identity and accuracy of the associated information for each potential customer to whatever degree of certainty is deemed necessary and then to generate an authenticated ID record that would be given to the customer as his identifying credential. This record will comprise intrinsic information, that is, personal attributes (photograph, fingerprints, hand geometry, voiceprint, retinal prints, signature, etc.) encrypted using the encrypt key of a two-key system, modulus n and exponent e in our example, along with extrinsic identifiers such as name, social security number, etc. User confidence is totally dependent on the issuer keeping the encrypt key secret, that is, maintaining the security of the authentication channel.

Depending on the application, this may require a two-man (or more) rule for access, or k of n shared key reconstructions, at the issuer's facility so that an improbably high level of collusion would have to occur for subversion to be possible. The decrypt key would be delivered as an authenticated, but not necessarily secret, message to all of the merchants, ATMs, etc. who would thereafter have to protect the integrity but not the privacy of the key. When, at some later time, a customer appears at a point of sale with a claimed identity, he would present the cipher record (credential) in his possession and permit his individual attributes to be reread by the merchant's point-of-sale equipment.

Using the decrypt key, the merchant would first decrypt the ID cipher and verify the authenticity of the cipher by the presence of the expected redundant information. He would next check for a suitable agreement between the individual attributes just measured and the decrypted information (intrinsic attributes) contained in the authenticated message. If an acceptable match is achieved, the identity of the customer will have been confirmed since the cipher could only have been generated using the secret encrypt key held by the enrollment station that was responsible for verifying the identity of the customer in the first place and the supposition is that an imposter could not adequately mimic the attributes of someone else to be accepted as them. The only advance communication required between the site and central facility is the authenticated (but not necessarily secret) exchange of the decrypt key to set up the public authentication channel.

The other channel of communication is the public one of the user bringing his own ID cipher to the site. Since authentication is possible in some two-key cryptosystems, that is, those in which there is a one-to-one mapping between plaintext and ciphertext spaces as in the RSA cryptoalgorithm, where the sender's key is known to be secure, the site can be certain (to the same level of confidence as the two-key cryptosystem is cryptosecure) that ID records it has received are authentic, that is, that they were issued and signed by the central issuing authority. Then, to the degree that the information in the ID records can identify an individual, the merchant can be confident of the identification. No communication with the central facility is required at the time that the individual is identified, and more importantly, no files of identifying information for possible customers need be transmitted to and stored at the merchant's place of business.

It is worth noting that the first reported application of public key cryptography techniques (fielded by the Sandia National Laboratories in 1978), made use of the authentication channel based on the RSA cryptoalgorithm, exactly as described here, to create trusted credentials that users could carry with them and present at the time they requested access, in this case to the very sensitive Zero Power Plutonium Reactor at Idaho Falls, Idaho [14,22]. The authenticated information in the credential included physical descriptors for the individual being identified as well as the details of the nature, type, duration, etc. of the access authorized. The object of this scheme was to make it possible for each user to carry with him what would have effectively been his entry in a trusted directory (a trusted credential in this case) at the remote site had a directory of potential users been feasible to compile, update, and protect at the remote site. The user-supplied information could be authenticated by the verifier at the site but would be of no assistance to anyone wishing to produce a fraudulent credential. In this particular application, the identification information was intrinsic to the user (hand geometry, body weight, etc.), however, in other applications [11] the same basic technique has been used with extrinsic information in a manner similar to the protocol to be described here.

The use of an authentication channel to validate to public receivers intrinsic identifying information "signed" by a trusted issuer has an enormous range of applications beyond the simple customer identification protocol just described. For example, every physical object at some level of fineness in detail is unique, that is, everything has a "fingerprint." Two pieces of paper of the same dimensions cut from the same roll may appear identical, but at a sufficient magnification, the cotton fibers lie in random and unreproducible three-dimensional patterns. The pattern at any selected location can be thought of as a fingerprint for that piece of paper. The important points are that every piece of paper has such a fingerprint by virtue of its existence, but that given a particular fingerprint it is impossible to produce another piece of paper with the same fingerprint. Unfortunately, it is not feasible to "read" the fiber layup of a piece of paper in any practical way to fingerprint a document. Any specific fine structure that can be placed deliberately in genuine objects, such as detailed engraving, etc., can also be cloned or placed in counterfeit objects. If, however, an object (a piece of paper, a physical container, a very large-scale integration [VLSI] chip, etc.) can be caused to have *a unique random fingerprint that is feasible to read but infeasible to clone*, that is, to reproduce, then we can use the protocol described above to insure that the object cannot be counterfeited or substituted for.

The issuing authority responsible for either producing the genuine articles, or for signing genuine articles for authenticity, reads and digitizes the unique intrinsic information (fingerprint) for each object, and then computes the matching authenticator using the secret encryption key for a public authentication channel. This authenticator is attached to the object, perhaps simply printed on it. Later, anyone can verify whether an article is genuine or not by rereading the fingerprint from the object and matching it against the authenticated fingerprint, either openly contained in the message if an authentication without secrecy channel was used, or else obtained by decrypting the authenticated cipher with the public key for the authentication channel. The physical fingerprint may have been degraded by wear and tear, improper location or alignment in reading, etc., however, the digital description of the fingerprint as it was originally read by the issuing authority will be recovered exactly.

The match between the two need not be perfect, and in all practical schemes will be far from perfect, however, if there is sufficient information content to the fingerprint large discrepancies can be tolerated with a very high confidence that counterfeit items will be detected and that genuine ones will not be rejected. The point is that a pure authentication channel makes it possible to transfer confidence in the identity or integrity of any piece of information from the issuing authority to publicly exposed remote locations. There are only two ways that undetectable counterfeits could be created in such schemes: Either the fingerprint of a genuine article would have to be cloned to match the genuine authenticator computed by the issuing authority or else the authentication channel would have to be cryptanalyzed to make it possible to "sign" the fingerprints read from bogus articles.

Returning to the point-of-sale problem, identifying the customer and verifying his account and credit rating, etc. satisfy some of the merchant's concerns, but not all of them. He also needs a record of the transaction that can be validated by impartial third parties in the event of a dispute; in other words, he needs the equivalent of a legal signature. If the transaction is occurring at a manned location, once the customer has been identified an actual signature will suffice. This is the way credit card transactions are handled presently. If, however, the transaction is either remote (conducted from a terminal or over a telephone, for example) or at an unmanned site such as an ATM or ticket vendor, etc. then the protocol must provide means for the customer to ''sign'' the charge slip, if significant liabilities are to be acceptable.

Conversely, the customer must end up with a copy of the record of the transaction that he can produce later to verify the amount, the date, the items purchased, the taxes paid, etc. for his own protection. All of these objectives can be satisfied by making a slight modification to the protocol with which credentials are created and employed and an addition to the information authenticated by the central issuing authority in the customer's credential. It should be pointed out that although the identity verification scheme just described has been successfully applied to a variety of very sensitive access control problems, such as the Zero Power Plutonium Reactor mentioned above or to the nuclear reactor fuel rod reprocessing plant at Savannah River, Georgia, etc., the extension of these techniques to provide verifiable and unforgeable transaction receipts has not yet been fielded. We include a description in this chapter, even though these techniques are not yet strictly ''practice,'' simply because they complement what is practice in a very logical way and illustrate the underlying principles clearly. The scheme described here is a simplified version of a general identification scheme proposed by Simmons [29] in 1989.

To make verifiable receipts possible, the central issuing authority in addition to setting up a public authentication channel as before, would also choose a polyrandom hashing function f that maps arbitrary strings of symbols to the range $[0, n)$. This function may well have unlimited life, and at the very worst, would be changed only after very long periods of use. By polyrandom we mean that f cannot be distinguished from a truly random function by any polynomially bounded computation. Many strong, single-key, cryptographic functions, such as the DES when used with a publicly known key in a block-chain encryption mode, appear to adequately approximate this condition. f is also made public by the issuer.

At the time the credential is created for customer i, in addition to checking his identity and verifying the correctness of all of the other information and intrinsic identifiers, etc., the issuer would require the customer to provide to him the public part of a private (i.e., belonging to the customer) authentication channel. By assumption (here) this would be a suitable RSA modulus, n_i, and a decryption exponent d_i; the matching encryption exponent e_i (and equivalently the factorization of n_i) the customer would, of course, keep secret. This string of information, I_i, must also include redundant information, such as message format, fixed fields of symbols common to all identifiers, etc. The issuer calculates

$$s_i \equiv m_i^d \pmod{n}$$

where

$$m_i = f(I_i)$$

and gives the credential (I_i, s_i) to customer i. No part of this credential need be kept secret. However, the customer must keep secret his private encryption exponent, e_i, corresponding to d_i. His security against impersonation is dependent on his protecting e_i, since his proof of identity in the scheme is equated to knowing e_i.

The public part of the issuer's authentication channel is the modulus n and decryption exponent d, the hashing function f, and a knowledge of the redundant information present in all of the I_i, which must be sufficient to prevent a forward search cryptanalytic attack [8] on the function f. This is a type of cryptanalytic attack on a two-key secrecy channel that has no counterpart in single-key cryptography that exploits the fact that the encryption key is publicly known to allow a would-be eavesdropper to precompute a table of likely plaintext/ciphertext pairs. Even though he cannot decrypt ciphers not appearing in the resulting table, he can replace any cipher that does appear with its matching plaintext. In the present case, this means that anyone wishing to fraudulently validate an identity could calculate $s_j^d = m_j$ for randomly chosen signatures s_j in the hopes of obtaining a hit with $f(I)$ for some usable I. By making I contain sufficient redundant information, the probability of success of this sort of attack can be made as small as desired.

When customer i wishes to prove his identity to vendor j, say to gain access to a restricted facility or to log onto a computer or to withdraw money from an ATM etc., he initiates the exchange by identifying himself to j by presenting his identification credential, (I_i, s_i) concatenated with a string of symbols, u_i, that describes or identifies the transaction i is requesting; u_i could be the date, the amount of the desired withdrawal, etc. j replies with his identification credential (I_j, s_j) concatenated with a string of symbols that describe the transaction from his standpoint; terminal ID, transaction number, confirmation of withdrawal amount, etc. Both i and j form the concatenation of u_i and u_j, $u = u_i, u_j$, and calculate the function $f(u)$ of the resulting string,

$$z = f(u)$$

In addition, j also calculates $f(I_i) = m_i$ and s_i^d. j accepts the credential (I_i, s_i) as valid if and only if

$$f(I_i) = s_i^d \pmod{n}$$

i can carry out a similar calculation to verify that the credential (I_j, s_j) was indeed issued by the issuing authority, however, this is unnecessary in the case being considered here, since i will get immediate delivery of goods or services from the merchant in exchange for a promise (on his part) to pay later. At this point in the protocol, j is confident that the customer identified in I_i can authenticate messages using the private authentication channel described therein, in other words, that customer i knows e_i matching the exponent d_i authenticated by the issuer. If the customer is who he claims to be, he can calculate

$$t_i = z^{e_i} \pmod{n}$$

using his private exponent e_i, which he communicates to j.

The merchant j calculates

$$t_j = z^{e_j} \pmod{n_i}$$

and sends t_j to the customer. Note that in both cases z is being used effectively as a one-time key, indeterminate to both i and j because of the polyrandom nature of f, to permit i to give to j an encrypted function of z in a form that will permit j to satisfy himself that i had to know e_i without providing any information whatsoever about e_i. j knows the identity claimed by i from I_i, which he accepts as valid if and only if the following identity is satisfied:

$$t_i^{d_i} = z \ (\text{mod } n_i) \tag{29}$$

If the person seeking to be recognized as customer i really is who he claims to be, that is, if he knows e_i, then in order for him to be able to impersonate i, that is, to cause Eq. (32) to be satisfied, he would have to be able to find a number x such that

$$x^{d_i} = z \ (\text{mod } n_i) \tag{30}$$

n_i or d_i are values signed by the issuer in I_i with only the authorized customer knowing e_i or equivalently the factorization of n_i. z is a pseudorandom number jointly determined by i and by j. Solving Eq. (29) without knowing e_i is equivalent to breaking the RSA cryptoalgorithm from ciphertext alone.

j keeps the 4-tuple $((I_i, s_i) : u, t_i)$ as his certified receipt for the transaction while the customer keeps the 4-tuple $((I_j, s_j) : u, t_j)$. Anyone, using only publicly available information, that is, n, d, and f, can later verify that the merchant's 4-tuple satisfies Eq. (29) which validates the transaction description and verifies that it was signed, that is, endorsed, by customer i. The customer's 4-tuple can be validated in a similar manner.

6 CONCLUSIONS

The problem of how to make it possible to decide whether a piece of information is what it purports to be or not is a much broader and more difficult question than this survey suggests. However, the method of solution remains the same, even in the most general case; namely, that one or more of the participants in an information-based system (which may include physically secure devices) perform operations on the information, the correctness of which can be verified by others, but that some of the other participants (probably) can't duplicate. What makes the problem so complex is the seemingly unlimited number of motives and methods for cheating in information-based systems that involve manipulating the information as a means.

Since this survey is a chapter in a volume on cryptology, we have emphasized those aspects of authentication that use cryptographic transformations (encryption and decryption) as the operation that persons in the know can do and others (probably) cannot. As was pointed out in the discussion of unconditionally secure authentication codes, though, there are alternative authentication operations available. The more difficult questions concerning the integrity (authenticity) of information in systems in which the participants may join forces and pool their privileged information to conspire to defraud others, lead into the area of provably secure protocols, which incidentally draws very heavily on cryptographic techniques for their realization; such questions were considered to be beyond the scope of this chapter.

Probably the most important development in authentication in recent years, growing out of the discovery of public key cryptography, is the recognition that the authentication channel can be separated from the secrecy channel. This made desensitized authentication possible, which removed the necessity that the originator (of the authenticated information) and the receiver trust each other unconditionally. This, in turn, was a first step in the evolution of authentication schemes and protocols in which no one need trust anyone (specifically), but only need trust that some number of the participants (but not specific ones) will perform in a trustworthy manner to be able to trust the integrity of the information itself. When viewed in this setting, cryptography is seen to be a much wider, and more vital, subject than in its classic setting of merely concealing information from outsiders (secrecy) or of preventing outsiders from spoofing the insiders (authentication). It is hoped that this survey has conveyed some notion of the breadth of this subject.

REFERENCES

[1] E. F. Brickell, "A few results in message authentication," presented at 15th Southeastern Conf. on Combinatorics, Graph Theory and Computing, (Baton Rouge, LA), March 5–8, 1984, in *Congressus Numerantium*; vol. 43, pp. 141–154, Dec. 1984.

[2] E. F. Brickell and D. R. Stinson, "Authentication codes with multiple arbiters," extended abstract in *Lecture Notes in Computer Science 330; Advances in Cryptology: Proc. Eurocrypt '88*, C. G. Günther, Ed., Davos, Switzerland, May 25–27, 1988, pp. 51–55. Berlin: Springer-Verlag, 1988.

[3] Department of the Treasury, "Electronic Funds and Securities Transfer Policy—Message Authentication," Directive signed by Donald T. Regan, Secretary of the Treasury, Aug. 16, 1984.

[4] M. De Soete, "Some constructions for authentication—secrecy codes," in *Lecture Notes in Computer Science 330; Advances in Cryptology: Proc. Eurocrypt '88*, C. G. Günther, Ed., Davos, Switzerland, May 25–27, 1988, pp. 57–75. Berlin: Springer-Verlag, 1988.

[5] W. Diffie and M. E. Hellman, "New directions in cryptography," *IEEE Trans. Inform. Theory*, vol. IT-22, no. 6, pp. 644–654, Nov. 1976.

[6] T. El Gamal, "A public key cryptosystem and a signature scheme based on discrete logarithms," *IEEE Trans. Inform. Theory*, vol. IT-31, no. 4, pp. 469–472, July 1985.

[7] A. Fiat and A. Shamir, "How to prove yourself: Practical solutions to identification and signature problems," in *Lecture Notes in Computer Science 263; Advances in Cryptology: Proc. Crypto '86*, A. M. Odlyzko, Ed., Santa Barbara, CA, Aug. 11–15, 1986, pp. 186–194. Berlin: Springer-Verlag, 1987.

[8] E. N. Gilbert, F. J. MacWilliams, and N. J. A. Sloane, "Codes which detect deception," *Bell Syst. Tech. J.*, vol. 53, no. 3, pp. 405–424, March 1974.

[9] O. Goldreich, S. Micali, and A. Wigderson, "Proofs that yield nothing but the validity of the assertion and the methodology of cryptographic protocol design," in *Proc. 27th Annu. Symp. Foundations Comput. Sci.*, Toronto, Canada, Oct. 27–29, 1986, pp. 174–187. Los Angeles, CA: IEEE Computer Society, 1987.

[10] S. Goldwasser, S. Micali, and R. L. Rivest, "A digital signature scheme secure against adaptive chosen-message attacks," *SIAM Jour. Computing*, vol. 17, no. 2, pp. 281–308, April 1988.

[11] C. L. Henderson and A. M. Fine, "Motion, intrusion and tamper detection for surveillance and containment," *Sandia National Laboratories Rept. SAND79-0792,* March 1980; also published by the International Safeguards Project Office for the International Atomic Energy Agency (IAEA) as *ISPO Report No. 91,* 1980.

[12] J. L. Massey, "Cryptography—a selective survey," presented at International Tirrenia Workshop on Digital Communications, Tirrenia, Italy, Sept. 1–6, 1985, in *Alta Frequenza,* vol. 55, no. 1, pp. 4–11, Jan.–Feb. 1986; also published in *Digital Communications,* E. Biglieri and G. Prati, Eds., pp. 3–25. Amsterdam: Elsevier, 1986.

[13] A. C. Meisenbach, *Acme Commodity and Phrase Code,* Acme Code Co., San Francisco, CA, 1923.

[14] P. D. Merillat, "Secure stand-alone positive personnel identity verification system (SSA-PPIV)," *Sandia National Laboratories Tech. Rept. SAND79-0070,* March 1979.

[15] C. H. Meyer and S. M. Matyas, *Cryptography: A New Dimension in Computer Data Security,* New York: Wiley, 1982.

[16] M. O. Rabin, "Digitized signatures and public-key functions as intractable as factorization," *Massachusetts Institute of Technology Laboratory for Computer Science, Tech. Rept. LCS/TR-212,* 1979.

[17] R. A. Rivest, A. Shamir, and L. Adleman, "A method for obtaining digital signatures and public-key cryptosystems," *Commun. Assn. Comput. Mach.,* vol. 21, no. 2, pp. 120–126, Feb. 1978.

[18] P. Schoebi, "Perfect authentication systems for data sources with arbitrary statistics," presented at Eurocrypt '86, Linköping, Sweden, May 20–22, 1986.

[19] A. Shamir, "Identity-based cryptosystems and signature schemes," in *Lecture Notes in Computer Science 196; Advances in Cryptology: Proc. Crypto '84,* G. R. Blakley and D. Chaum, Eds., Santa Barbara, CA, Aug. 19–22, 1984, pp. 47–53. Berlin: Springer-Verlag, 1985.

[20] G. J. Simmons, "Message authentication without secrecy," in *Secure Communications and Asymmetric Cryptosystems* (AAAS Selected Symposia Series), G. J. Simmons, Ed., pp. 105–139. Boulder, CO: Westview Press, 1982.

[21] G. J. Simmons, "Verification of treaty compliance—revisted," in *Proc. IEEE Computer Soc. 1983 Symp. on Security and Privacy,* G. R. Blakley and D. Denning, Eds., Oakland, CA, April 25–27, 1983, pp. 61–66. Los Angeles: IEEE Computer Society Press, 1983.

[22] G. J. Simmons, "A system for verifying user identity and authorization at the point-of-sale or access," *Cryptologia,* vol. 8, no. 1, pp. 1–21, Jan. 1984.

[23] G. J. Simmons, "Authentication theory/coding theory," in *Lecture Notes in Computer Science 196; Advances in Cryptology: Proc. Crypto '84,* G. R. Blakley and D. Chaum, Eds., Santa Barbara, CA, Aug. 19–22, 1984, pp. 411–431. Berlin: Springer-Verlag, 1985.

[24] G. J. Simmons, "The practice of authentication," in *Lecture Notes in Computer Science 219; Advances in Cryptology: Proc. Eurocrypt '85,* F. Pichler, Ed., Linz, Austria, April 1985, pp. 261–272. Berlin: Springer-Verlag, 1986.

[25] G. J. Simmons, "Authentication codes that permit arbitration," presented at 18th Southeastern Conf. on Combinatorics, Graph Theory and Computing, Boca Raton, FL, Feb. 23–27, 1987, in *Congressus Numerantium,* vol. 59, pp. 275–290, March 1988.

[26] G. J. Simmons, "Message authentication with arbitration of transmitter/receiver disputes," in *Lecture Notes in Computer Science 304; Advances in Cryptology: Proc. Eurocrypt '87*, D. Chaum and W. L. Price, Eds., Amsterdam, The Netherlands, April 13–15, 1987, pp. 151–165. Berlin: Springer-Verlag, 1988.

[27] G. J. Simmons, "A natural taxonomy for digital information authentication schemes," in *Lecture Notes in Computer Science 293; Advances in Cryptology: Proceedings of Crypto '87*, C. Pomerance, Ed., Santa Barbara, CA, Aug. 16–20, 1987, pp. 269–288. Berlin: Springer-Verlag, 1988.

[28] G. J. Simmons, "An impersonation-proof identity verification scheme," in *Lecture Notes in Computer Science 293; Advances in Cryptology: Proceedings of Crypto '87*, C. Pomerance, Ed., Santa Barbara, CA, Aug. 16–20, 1987, pp. 211–215. Berlin: Springer-Verlag, 1988.

[29] G. J. Simmons, "A protocol to provide verifiable proof of identity and unforgeable certified receipts," *IEEE Jour. Selected Areas Commun. Special Issue on Secure Communications*, vol. 7, no. 4, pp. 435–447, May 1989.

[30] G. J. Simmons, "How good is the Acme code?" *Supplementary volume to the Proceedings of the 4th Joint Swedish-Soviet International Workshop on Information Theory*, Visby, Sweden, Aug. 27–Sept. 1, 1989, pp. 24–30, 1989.

[31] G. J. Simmons, "A Cartesian product construction for authentication codes that permit arbitration," *Jour. Cryptology*, vol. 2, no. 2, pp. 77–104, 1990.

[32] D. R. Stinson, "Some constructions and bounds for authentication codes," in *Lecture Notes in Computer Science 263; Advances in Cryptology: Proc. Crypto '86*, A. M. Odlyzko, Ed., Santa Barbara, CA, Aug. 11–15, 1986, pp. 418–425. Berlin: Springer-Verlag, 1987.

[33] D. R. Stinson, "A construction for authentication secrecy codes from certain combinatorial designs," in *Lecture Notes in Computer Science 293; Advances in Cryptology: Proceedings of Crypto '87*, C. Pomerance, Ed., Santa Barbara, CA, Aug. 16–20, 1987, pp. 355–366, Berlin: Springer-Verlag, 1988. Also appears in *J. Cryptology*, vol. 1, no. 2, pp. 119–127, 1988.

[34] H. C. Williams, "A modification of the RSA public-key encryption procedure," *IEEE Trans. Inform. Theory*, vol. IT-26, no. 6, pp. 726–729, Nov. 1980.

SECTION 3

Protocols

CHAPTER 8

Overview of Interactive Proof Systems and Zero-Knowledge

J. Feigenbaum
A T & T Bell Laboratories
Murray Hill, New Jersey 07974

Abstract—In traditional computational complexity theory, the informal notion of *efficiently verifiable* sets of statements is formalized as *nondeterministic polynomial time* sets. Recently, an alternative formalization has emerged: sets with *interactive proof systems*. An interactive proof system is called *zero-knowledge* if it succeeds in proving the desired statements and nothing else. This chapter surveys definitions, examples, known results, and open problems in the area of interactive proof systems and zero-knowledge.

1 INTRODUCTION

Some basic problems in complexity theory and cryptography can be thought of as two-player games in which one player (the "prover") tries to convince the other player (the "verifier") of the truth of an assertion. Indeed, the complexity class NP can be formulated in these terms. To convince a verifier that a string x is in the language L, where L is in NP, the prover can provide a "short certificate" of membership; if x is not in L, then no "cheating prover" could produce a certificate because none exists. In cryptography, we have the example of authentication schemes. To gain access to a computer, a building, or some other secure facility, a user must convince an operating system, a guard, or some other type of player that he is who he claims to be and he is entitled to access. Thus the user can be thought of as a prover and the operating system as a verifier.

The work surveyed in this chapter addresses two questions about such games:

- How much *interaction* is needed for the prover to convince the verifier?
- How much *knowledge* does the verifier gain during the course of the interaction?

Let us look more closely at NP languages in terms of two-player games. For example, consider the language L of pairs (G, k), where G is a finite, undirected graph that contains a k-clique—that is, a k-vertex subgraph every two vertices of which are connected by an edge. The obvious way to play a game in which the prover's objective is to convince the verifier that a pair (G, k) is in L is to require the prover to produce a subset $\{v_1, \ldots, v_k\}$ of the vertex set of G and to require the verifier to check that every $\{v_i, v_j\}$, $1 \leq i < j \leq k$, is in the edge set of G. This game is very efficient in terms of interaction. The prover has to send only one message to the verifier in order to prove his claim. On the other hand, this game may be inefficient in terms of knowledge. After receiving this one message from the prover, the verifier has full knowledge of how to prove that (G, k) is in L; he could turn around and use this knowledge to convince a third party that (G, k) is in L, which is something he might not have been able to do before he received a message from the prover. Finally, note that this game involves no random choices and no probability of error: If (G, k) is in L, then, given enough computing power, the prover can always produce a k-clique, and the verifier will accept the prover's claim with probability 1; if (G, k) is not in L, then no k-clique exists, and the probability is zero that a dishonest prover can convince the verifier to accept.

It is fairly easy to formalize the statement that games with these three properties (one message from prover to verifier, full knowledge of the proof given to the verifier, and no probability of error) exactly characterize NP—that is, the assertions that can be proven with these games are those of the form "x is in L," where L is a language in NP.

What if we allow games in which the prover and verifier exchange more messages? What if we allow a small probability of error—that is, what if a true assertion is rejected once in while or a false assertion believed? Can we then devise games for languages that are not known to be in NP? Can we devise games that are more knowledge-efficient? For example, is it possible for the prover to convince the verifier that his assertion is true without giving him *any* knowledge of the proof?

These questions motivate the theory of *interactive proof systems* and *zero-knowledge*, which is surveyed in this chapter. In Section 2 we give formal definitions of these concepts; the definitions are taken from the seminal papers of Goldwasser, Micali, and Rackoff [24], Babai and Moran [6], and Ben-Or, Goldwasser, Kilian, and Widgerson [10]. Section 3 reviews two well-known examples of interactive proof systems, one drawn from cryptography and one from complexity theory. Major results of the theory are given in Section 4. Section 5 contains a brief discussion of related notions such as *arguments, program checkability,* and *instance-hiding;* these notions are related in two ways: Their study is motivated by intuitively similar concerns, and they are used in some of the proofs of the main results stated in Section 4. Finally, open problems are given in Section 6.

2 DEFINITIONS

The reader is assumed to be familiar with the basic notions of computational complexity theory. Refer to the Appendix for a brief review of these notions.

We are concerned with the following model of *interactive computation.* In a *one-prover interactive protocol* (P, V), the *prover P* and *verifier V* have a shared *input x,* two private sources of unbiased random bits (say r_P and r_V, respectively), and a way to send messages to each other. On input x, the *transcript* of the protocol is a sequence of *moves* $(\alpha_1, \beta_2, \alpha_3, \beta_4, \ldots, \alpha_{2t-1}, \beta_{2t})$, where $t = t(n)$ is a polynomially bounded function of $n = |x|$. There is also a polynomial l such that $|\alpha_{2i-1}|$ and $|\beta_{2i}|$ are bounded by $l(n)$, for $1 \le i \le t$. Each α_{2i-1} is a message from the verifier to the prover and is a polynomial-time computable function of x, r_V, and $(\alpha_1, \beta_2, \ldots, \alpha_{2i-3}, \beta_{2i-2})$. Each β_{2i} is a message from the prover to the verifier and is a function of x, r_P, and $(\alpha_1, \beta_2, \ldots, \beta_{2i-2}, \alpha_{2i-1})$; note that no restrictions are placed on this function except that its output be bounded in length by $l(n)$. After receiving β_{2t}, the verifier computes a polynomial-time function of x, r_V and the entire transcript; the output of this computation is either ACCEPT or REJECT.

Let $(P, V)(x)$ denote the verifier's final output. For every x, $(P, V)(x)$ is a random variable whose distribution is induced by the uniform distributions on the random variables r_P and r_V and the functions computed by P and V. Similarly, we denote by *View* (P, V, x) the random variable consisting of the transcript $(\alpha_1, \beta_2, \ldots, \alpha_{2t-1}, \beta_{2t})$ produced by an execution of the protocol, together with the prefix of r_V that the verifier consumes during the execution. More generally, if M is any probabilistic Turing machine, the output of M on input x is a random variable denoted by $M(x)$.

For any interactive protocol (P, V), we denote by (P^*, V) the interactive protocol in which the verifier behaves exactly as in the original protocol (P, V), but the prover computes the functions specified by P^*; this is a "cheating prover" if $P^* \ne P$. Similarly, we denote by (P, V^*) the interactive protocol in which the prover behaves exactly as in the original protocol, but the (potentially cheating) verifier computes the functions specified by V^*. The potentially cheating prover P^* is limited only in that the messages that it sends must be of length at most $l(n)$; the potentially cheating verifier V^* is limited to probabilistic polynomial-time computation.

Definition 2.1 (see also [24]): *The interactive protocol (P, V) recognizes the language L if, for all $x \in L$,*

$$\text{Prob}((P, V)(x) = \text{ACCEPT}) > 2/3$$

and, for all $x \notin L$, for all provers P^*,

$$\text{Prob}((P^*, V)(x) = \text{REJECT}) > 2/3$$

We denote by IP(k) the class of all languages recognized by interactive protocols with at most k moves. In the above discussion, $k = 2t$. IP(*poly*) is the union, over all polynomially bounded k, of IP(k); the shorthand IP is used for IP(*poly*). The statement that (P, V) is *a (one-prover) interactive proof system for the language L* is synonymous with the statement that (P, V) recognizes L.

Note that the definition of IP, like the definition of BPP, allows for error probability $1/3$. In both cases, the error probability can be made exponentially small by performing polynomially many repetitions of the computation and taking the majority answer. Note that these are *sequential* repetitions; so, in the case of IP, this procedure reduces the error probability at the expense of increasing the number of moves.

Definition 2.2 (see also [24]):   *The interactive proof system (P, V) for the language L is* **computational (resp. perfect) zero-knowledge** *if, for every verifier V^*, there exists a probabilistic, expected polynomial-time machine M_{V^*}, called the* **simulator,** *such that the ensembles $\{View\,(P, V^*, x)\}_{x \in L}$ and $\{M_{V^*}(x)\}_{x \in L}$ are* **computationally indistinguishable (resp. the same).**

Informally speaking, two ensembles are said to be *computationally indistinguishable* if no probabilistic polynomial-sized circuit family can tell them apart. The term is supposed to convey the idea that the ensembles are computationally "close" to being equal. Refer to Goldwasser et al. [24] for a formal definition. There is also an intermediate notion of *statistical zero-knowledge* and a corresponding notion of *statistical indistinguishability* of ensembles. Many of the results on perfect zero-knowledge that are stated in Section 4 below actually hold for statistical zero-knowledge. Refer to [2, 19] for details.

The statement that (P, V) is an interactive proof system for L can be viewed as a limitation on the prover: No matter what P^* does, it cannot force the verifier to accept a string x that is not in L, except with negligible probability. Similarly, the statement that (P, V) is zero-knowledge can be viewed as a limitation on the verifier: No matter what V^* does, the only distributions that it computes by interacting with P are distributions that it could have computed in expected polynomial time *without* interacting with P.

A more restricted form of interactive protocol, the *Arthur–Merlin protocol,* is considered in [6]. In an Arthur–Merlin protocol, two players, Arthur and Merlin, have a common input x of length n. Once again, there is a sequence of moves. In odd-numbered moves, Arthur simply sends to Merlin a string of $l(n)$ unbiased random bits. In even-numbered moves, Merlin sends to Arthur a string of length $l(n)$ that is *optimal* (in a sense that will be made precise shortly). After $k(n)$ moves, a deterministic polynomial-time Turing machine, the *referee,* decides the winner; so each polynomial-time referee determines an Arthur–Merlin protocol. Because Arthur's moves are random, Merlin's winning probability depends only on x—call this probability $W(x)$. It is required that $W(x)$ be greater that $\frac{2}{3}$ or less than $\frac{1}{3}$, for each x. The referee is known in advance to both Arthur and Merlin; because the referee is polynomial-time bounded and Merlin is not, Merlin can make *optimal* moves—that is, he can maximize $W(x)$.

Definition 2.3 (see also [6]):   *The* **language recognized by an Arthur–Merlin protocol** *is the set of all x for which $W(x) > \frac{2}{3}$.*

Think of Merlin's objective as "trying to convince" Arthur to accept the string x. Thus he is analogous to the prover in IP.

The class of languages recognized by Arthur–Merlin protocols with at most k moves is denoted AM(k), and AM(*poly*) is the union of AM(k) over all polynomially bounded k. Unlike the notation IP, the notation AM is used for AM(2). Merlin–Arthur protocols are also considered in [6]; they are the same as Arthur–Merlin protocols, except that Merlin makes the odd-numbered moves. So the sequence of movers for a language in MA(4) is MAMA; the sequence for a language in AM(3) is AMA.

Multiprover interactive protocols are defined analogously to one-prover interactive protocols. Instead of one prover P, there are m provers $P_1, \ldots, P_m$, where $m = m(n)$ is a polynomially bounded function of $n = |x|$. If the verifier V makes the tth move, he sends m messages $(\alpha_{1t}, \ldots, \alpha_{mt})$ simultaneously. For each i, $1 \leq i \leq m$, P_i receives

α_{it}, and P_i cannot overhear α_{jt}, for any $j \neq i$. Similarly, if the provers make the tth move, each P_i sends a message β_{it} to V, who receives all of $(\beta_{1t}, \ldots, \beta_{mt})$ simultaneously, and no P_i can overhear β_{jt}, for $j \neq i$. The notation IP(m, *poly*) is used for the class of languages recognizable by m-prover interactive protocols with polynomially many moves. Zero-knowledge for multiprover interactive protocols is defined in terms of simulation of transcripts, as it is in the one-prover case; there is one nonobvious part of the definition—the provers $P_1, \ldots, P_m$ use a shared random string that is unknown to the verifier. This is often interpreted as follows: The provers can get together in advance to "agree on a strategy" by picking a shared random string and deciding how to compute the β_{it}'s; during the execution of the protocol, however, the provers are "kept separate," which gives the verifier a chance to catch them in an inconsistency if they try to cheat.

The definitions given above are for (zero-knowledge) interactive proofs of *language membership*. The following notions also have rigorous definitions: proofs that the *functional value* $f(x)$ is what the prover claims it is (see, for example, [21]), and *proofs of knowledge* (see, e.g., [17]). The latter are particularly relevant to the original cryptographic motivation for zero-knowledge proof systems: The prover wants to convince the verifier that he "knows" a secret without revealing that secret. The example in Section 3.1 is a protocol for proofs of knowledge.

3 EXAMPLES

3.1 A Cryptographic Example: Proofs of Identity

The following example, a system for proofs of knowledge, is taken from Feige, Fiat, and Shamir [17]. It is designed to allow a community of users to *authenticate* themselves to each other. For each user, there is a pair (I, S). The public information I can be interpreted as the user's identity and the private information S as his secret key. The goal of the proof system is for a user with identity I to convince another user that he "knows" the corresponding key S without revealing anything about S beyond the fact that he knows it. The effectiveness of the proof system is based on the assumption that it is computationally infeasible to compute square roots modulo a large composite integer with unknown factorization; this is provably equivalent to the assumption that factoring large integers is difficult.

Modulus generation: A *trusted center* generates two large primes each congruent to 3 mod 4. The product m of these primes is published, but the primes are not.

Note that -1 is a quadratic non-residue modulo m—that is, there is no a such that $a^2 = -1 \mod m$. In what follows, $Z_m^*[+1]$ denotes the set of integers between 1 and m that are relatively prime to m and have Jacobi symbol $+1$ with respect to m.

Key generation: Each user chooses t_1 random numbers $S_1, \ldots, S_{t_1}$ in $Z_m^*[+1]$ and t_1 random bits $b_1, \ldots, b_{t_1}$. He sets I_j equal to $(-1)^{b_j}/S_j^2 \mod m$, for $1 \leq j \leq t_1$. This user's identity, which he publishes, is $I = (I_1, \ldots, I_{t_1})$, and his secret key, which he keeps private, is $S = (S_1, \ldots, S_{t_1})$.

Proofs of identity: User A authenticates himself to user B as follows. Let I be A's published identity and S his secret key. A and B repeat the following four steps t_2 times.

1. A picks a random R in $Z_m^*[+1]$ and a random bit c and sends $X = (-1)^c R^2 \bmod m$ to B.

2. B sends a random vector of bits $(E_1, \ldots, E_{t_1})$ to A.

3. A sends $Y = R \cdot \prod_{E_j=1} S_j \bmod m$ to B.

4. B verifies that $Y^2 \cdot \Pi E_j I_j \bmod m$ is equal to $\pm X$.

B believes that A is who he claims he is if the verification in Step 4 succeeds in each of the t_2 trials.

In [17], it is shown that this scheme provides *zero-knowledge proofs of knowledge* for the values $t_1 = O(\log \log m)$ and $t_2 = \Theta(\log m)$. A bit more care is needed when implementing the scheme in practice; for example, B should check in Step 2 that A actually sent an element of $Z_m^*[+1]$, lest a cheating prover just send $X = Y = 0$ in each of the t_2 executions (refer to [16] for more discussion of this issue). The most important feature of the scheme is that there is no need for the prover to have significant computational power. These identity proofs require only a few modular multiplications and can be implemented on smart cards.

3.2 A Complexity-Theoretic Example: Proofs for the PERM Function

The following interactive proof system is due to Lund, Fortnow, Karloff, and Nisan [30]. It provides a way for P to convince V that the permanent of an integer matrix is what P claims it is. Because the permanent function is complete for the complexity class #P (see also [36]), this proof system of Lund et al., combined with the seminal theorem of Toda [35] that $PH \subseteq P^{\#P}$, shows that every language in the PH has a one-prover interactive proof system and goes a long way toward a complete characterization of IP.

It is easy to see (and is shown in [30]) that it suffices to prove the values of permanents modulo a large prime p. Let $A = (a_{ij})$ be an $N \times N$ matrix over Z_p and $A_{1|i}$ denote the $(1, i)$-minor of A. Recall the formula

$$PERM(A) = \sum_{\sigma \in S_N} a_{1\sigma(1)} a_{2\sigma(2)} \cdots a_{N\sigma(N)}$$

for the permanent of A and the fact that $PERM(A)$ can be computed by cofactor expansion:

$$PERM(A) = \sum_{i=1}^{N} a_{1i} \cdot PERM(A_{1|i})$$

If $C = (c_{ij})$ and $D = (d_{ij})$ are $N \times N$ matrices over Z_p and x is an indeterminant, we denote by $(C + x(D - C))$ the $N \times N$ matrix whose ij^{th} entry is the degree-1 polynomial $c_{ij} + x(d_{ij} - c_{ij})$ in $Z_p[x]$. Notice that $PERM(C + x(D - C))$ is a polynomial $f(x)$ in $Z_p[x]$ of degree at most N. Furthermore, $f(0) = PERM(C)$ and $f(1) = PERM(D)$.

The following is a protocol for P to prove that $PERM(A)$ is equal to $a \bmod p$. Each stage of the protocol is either an *expand* stage or a *shrink* stage. The object that is expanded and shrunk is a list $\mathcal{L} = \langle (B_1, b_1), \ldots, (B_t, b_t) \rangle$ in which b_i is P's claimed value for $PERM(B_i) \bmod p$.

Initially, $\mathscr{L}$ contains only $\langle A, a \rangle$. P and V repeat the following steps until $\mathscr{L}$ consists of a single entry $\langle B, b \rangle$, where B is 1×1.

If $\mathscr{L} = \langle B, b \rangle$, where B is $r \times r$ and $r > 1$, then EXPAND:

V: Let $C_i = B_{1|i}$. Ask P for the permanents of C_i, $1 \leq i \leq r$.

P: Send claimed values $c_i = PERM(C_i)$ to V, $1 \leq i \leq r$.

V: If $b \neq \Sigma_{i=1}^{r} b_{1i} c_i \mod p$, then REJECT and terminate the protocol. Otherwise, set $\mathscr{L}$ to $\langle (C_1, c_1), \ldots, (C_r, c_r) \rangle$.

If $\mathscr{L}$ contains two or more pairs, then SHRINK:

V: Take the first two pairs (C, c) and (D, d) from $\mathscr{L}$ and ask P for the permanent of $C + x(D - C)$.

P: Send a claimed value $g(x)$ for $PERM(C + x(D - C))$.

V: If $g(0) \neq c$ or $g(1) \neq d$, then REJECT and terminate the protocol. Otherwise, choose a random $s \in Z_p$, send it to P, and replace the first two pairs of $\mathscr{L}$ with $(C + s(D - C), g(s))$.

It is clear that an honest P will always convince V that $PERM(A)$ is what he claims it is. The essential reason that a cheating P^* cannot convince V to accept a wrong value with high enough probability is that $f(x) = PERM(C + x(D - C))$ has low degree. Specifically, it has degree $r \leq N$, where N is the dimension of the original input. If P^* sends a degree-r polynomial $g \neq f$ in some shrink stage, then g and f can agree on at most r points; if p is sufficiently large with respect to r, then f and g are unlikely to agree on a random $s \in Z_p$. Thus, it is likely that $g(s) \neq PERM(C + s(D - C))$ and that P^* will have to lie in subsequent stages. The matrices whose permanents he lies about get smaller and smaller, and eventually V can catch him.

4 KNOWN RESULTS

One of the most exciting recent developments in complexity theory is the complete characterization of the language-recognition power of interactive proof systems.

Theorem 4.1: AM($poly$) = IP = PSPACE.

The fact that AM($poly$) is contained in IP is clear from the definitions, because Arthur and Merlin of Definition 2.3 are just special cases of the verifier and prover of Definition 2.1. The equality of IP and AM($poly$) was proven by Goldwasser and Sipser [25]; actually, they proved the sharper result IP($k(n)$) $\subseteq$ AM($k(n) + 2$), for all polynomially bounded functions k. The upper bound IP $\subseteq$ PSPACE is also clear from the definitions: IP is contained in the class of languages accepted by *games against nature*, which was defined earlier by Papadimitriou [33] and shown to be equal to PSPACE. The fact that the PSPACE-complete language quantified Boolean formulas (QBF) and hence every PSPACE language, has an interactive proof system was obtained by Shamir [34]; his work relies heavily on the work of Lund et al. discussed in Section 3.2; in particular, it uses a similar low-degree polynomial trick.

Given the widespread interest in interactive proof systems and zero-knowledge, it is clear why a definitive theorem like IP = PSPACE would be exciting. But why was it surprising? One reason is that it is a rare example of a *nonrelativizing* result in complexity theory; see [3] for a discussion of relativization and its relationship to the characterization of IP. Another reason for the surprise is the technical simplicity of Shamir's result and the results leading to it; some of the key ingredients are discussed briefly in Section 5.

Now that we know exactly which languages have interactive proof systems, it is natural to ask how much interaction is required for languages of interest. It is clear from the definitions that $AM(k(n)) \subseteq AM(k(n) + 1)$ for every polynomially bounded function k. Babai and Moran have shown that these inclusions are not always proper.

Theorem 4.2 (see also [6]): *For any polynomially bounded function $k(n) \geq 2$, $AM(2k(n)) = AM(k(n))$. In particular, $AM(c) = AM(2)$, for any constant $c \geq 2$.*

The known proof systems for PERM and QBF involve a polynomial number of moves. It is unknown whether QBF is provable in a constant number of moves; the following results address this question.

Theorem 4.3 (see also [6]): $AM(2) \subseteq \Pi_2^p$.

Thus, if QBF does have a constant-move interactive proof system, all of PSPACE is contained in the second level of the polynomial-time hierarchy. So such a proof system for QBF may be impossible to obtain and, in any case, would require a revolutionary new idea. A weaker, but still revolutionary, consequence would follow from the weaker assumption that every language in coNP has a constant-move interactive proof system.

Theorem 4.4 (see also [13]): *If* coNP $\subseteq AM(2)$, *then the* PH *collapses at the second level.*

How serious a restriction is the requirement of zero-knowledge? The answer depends on which definition of zero-knowledge is used.

Theorem 4.5: *If one-way functions exist, then every language in* IP *has a computational zero-knowledge proof system.*

This general result follows from Impagliazzo and Yung's theorem [26] that every language in IP has a computational zero-knowledge proof system if *bit-commitment schemes* exist and Naor's theorem [31] that one-way functions yield bit-commitment schemes. Impagliazzo and Yung's work relies heavily on earlier work by Goldreich, Micali, and Wigderson [23] and Brassard, Chaum, and Crépeau [14].

Theorem 4.6: PZK, *the class of languages with perfect zero-knowledge proof systems, is contained in* AM $\cap$ coAM.

The inclusion of PZK in coAM is due to Fortnow [19], and its inclusion in AM is due to Aiello and Håstad [2]. Theorems 4.3 and 4.6 together imply that PZK is contained in $\Sigma_2^p \cap \Sigma_2^p$; so there is no reason to believe that PZK contains all of IP.

What about multiple provers? The first basic result, due to Ben-Or, Goldwasser, Kilian, and Wigderson [10], is that two provers suffice.

Theorem 4.7: *For any polynomially bounded function m,* $IP(m(n), poly) = IP(2, poly)$.

From now on, we denote by MIP the language class IP(2, *poly*). This class has been fully characterized in terms of traditional complexity classes by Babai, Fortnow, and Lund [5]:

Theorem 4.8:   MIP = NEXP.

Unlike the one-prover case, two-prover interactive proof systems can always be made perfect zero-knowledge.

Theorem 4.9 *Every language in* MIP *has a perfect zero-knowledge, two-prover interactive proof system.*

Theorem 4.9 is due to Ben-Or, Goldwasser, Kilian, and Wigderson. For a complete proof, see [27].

So two-prover systems seem to be different from one-prover systems with respect to zero-knowledge. They also seem to be different with respect to move-complexity. Kilian [28] has shown recently that two provers and c moves suffice for any language in MIP, provided that error probability ε is tolerated—here ε is an arbitrarily small constant and c is a constant that depends on ε. The best possible result would be that any language in MIP can be recognized in two moves with exponentially small error probability.*

5 RELATED NOTIONS

In most practical cryptographic situations, it does not make much sense to allow the prover to use unlimited computing power; for this reason, results such as "all of PSPACE has computational zero-knowledge proofs if one-way functions exist" are essentially impractical, beautiful as they are. In the model of Brassard, Chaum, and Crépeau [14], all parties are limited to reasonable computing power. The only advantage that the prover has over the verifier is a short piece of advice, such as the factorization of a large integer. Of course, limiting the prover's power reduces the class of languages that can be handled from PSPACE down to MA. On the other hand, under suitable cryptographic assumptions, these limits *increase* the class of problems that can be handled in *perfect* zero-knowledge—that is, the problems in which the prover's secret is protected unconditionally. It is shown in [14] that all languages in MA (and, *a fortiori,* all languages in NP) admit perfect zero-knowledge protocols of this type, assuming that *unconditionally concealing bit-commitment schemes* exist. (This is in sharp contrast with the result of Fortnow [19] mentioned in Section 4.) Brassard, Crépeau, and Yung [15] have shown that all languages in NP admit constant-move, perfect zero-knowledge protocols of this type, under the same assumption.

It is important to understand that the protocols in [14, 15] are *not* IP-protocols—it would be easy for an arbitrarily powerful prover to cheat. In other words, these protocols are *computationally convincing,* as opposed to proof-systems, which are *statistically* convincing. To distinguish computationally convincing protocols from proof

---

*Note added in proof: This result has been proven by Uriel Feige.

systems, the former are referred to as *arguments*. There is a strong duality between computational zero-knowledge IP-protocols and perfect zero-knowledge arguments for NP-complete problems. However, one important aspect of this duality breaks down. After a computational zero-knowledge IP-protocol is completed, the verifier has complete information about the prover's secret, albeit in enciphered form. Working *offline* for long enough, the verifier may be able to decipher that secret. In sharp contrast, a dishonest prover must break the cryptographic assumption *online,* while the protocol is taking place, if he wishes to cheat in an argument. As a consequence, an algorithm capable of factoring large integers in a month of Cray time would be a severe threat to the security of computational zero-knowledge IP-protocols but of little practical consequences for perfect zero-knowledge arguments.

M. Blum defined the related notion of *program checkability*. The goal of the work on program checkability (see, for example, [9, 11, 12]) is to be able to check whether a program is correct on a specific input. One way to do this is to regard the program as a prover in an interactive proof system and design a *checker* that plays the role of the verifier. Note that this changes the nature of both the honest prover and the cheating prover; both are constrained to be non-self-modifying functional programs, and the honest prover must be a program that computes the function being checked. Because of the restriction on the honest prover, languages in IP are not necessarily checkable; in fact, Beigel and Feigenbaum [9] give a "natural" complexity-theoretic hypothesis that implies the existence of a language in IP that is not checkable. Because of the restriction on the cheating prover, checkable languages are not necessarily in IP; in fact, Babai, Fortnow, and Lund [5] show that EXP-complete languages are checkable, whereas they are not in IP unless EXP = PSPACE. The fact that checkable languages are all in NEXP follows from the results in [20].

Abadi, Feigenbaum, and Kilian [1] defined the related notion of *instance-hiding schemes.** In such a scheme, a polynomial-time *querier* interacts with an *oracle* in order to learn the value $f(x)$, for some function f that the querier cannot compute on its own, *without revealing x to the oracle*—specifically, revealing nothing (in the information-theoretic sense) except the length of x. There is an approximate analogy with zero-knowledge proof systems, in which the prover convinces the verifier of the truth of a statement, without actually telling him the proof. In [1], it is shown that no NP-hard function has an instance-hiding scheme, unless the PH collapses at the third level. Beaver and Feigenbaum [7] considered a *multioracle* version of instance-hiding schemes; in this version, the querier asks questions of multiple oracles, and the oracles "cannot talk to each other," by analogy with the multiple provers in MIP. Here there are positive results: If the querier can talk to $n + 1$ oracles, where n is the length of the secret query, then *any* function has an instance-hiding scheme (see [7]; this construction was subsequently improved to work with $n/\log n$ oracles; see [8]).

The instance-hiding schemes in [7] have a special form that is of complexity-theoretic significance; they are *locally random reductions*. Informally, f is k-locally random reducible (k-lrr) to g if there is a probabilistic polynomial-time algorithm that maps an arbitrary instance x in the domain of f to a set of random instances $y_1, \ldots, y_k$ in the domain of g in such a way that $f(x)$ is computable in polynomial time from

---

*"Instance-hiding" is current terminology; in [1], these are called *encryption schemes for functions.*

$g(y_1), \ldots, g(y_k)$ and the distribution of each of the random instances y_i depends only on $|x|$; a formal definition can be found in [8]. If $g = f$, the reduction is called a *k-random-self-reduction* (*k*-rsr). It is shown in [7] that every Boolean function is $(n + 1)$-lrr to a sequence $\{g_n\}_{n \geq 1}$ of polynomials, where *degree* $(g_n) = n$. This was the first use of the fateful *low-degree polynomial trick* that was to be used by Lund et al. and by Shamir in the proof systems discussed above.

Building on Beaver and Feigenbaum [7], Lipton [29] observed that any f that is itself a low-degree polynomial (including the #*P*-complete function PERM) is random-self-reducible. PERM is also *length-decreasing self-reducible*—that is, computing the permanent of A can be reduced in polynomial time to computing permanents of matrices of smaller size; this is done by co-factor expansion, as discussed in Section 3.2. Blum, Luby, and Rubinfeld [12] showed that any function that is both random-self-reducible and length-decreasing self-reducible has a program checker; by the earlier observation of Fortnow, Rompel, and Sipser [20], such a function also has a multiprover interactive protocol. Nisan [32] put all these observations together into the statement that PERM (and, from Toda [35], the entire PH) is in MIP. This dramatic sequence of results culminated in the complete characterization of the language-recognition power of IP (see also [30, 34]) and MIP (see also [5]) that was discussed in Section 4.

6 OPEN PROBLEMS

One of the most interesting class of problems remaining concerns the necessary power of the prover. For example, in an interactive proof system for a PSPACE-complete language, the prover can be taken to be a PSPACE machine (see [34] for details). The analogous statements can be made about $\oplus$P-complete languages (see also [4]) and #*P*-complete functions (see also [30]). Furthermore, in a two-prover interactive proof system for an EXP-complete language, both provers can be taken to be EXP machines (see also [5]).

> **Question 1:** Is there a one-prover interactive proof system for coSAT in which the prover is given only enough power to answer SAT queries?
>
> **Question 2:** Is there a two-prover interactive proof system for coSAT in which both provers are given only enough power to answer SAT queries? This is equivalent to the question of whether there is a program checker for SAT.
>
> **Question 3:** An easier question than 1 or 2: Is there a one- or two-prover interactive proof system for coSAT in which the power of the prover(s) is somewhere in the PH?
>
> **Question 4:** Does every NEXP-complete language S have a two-prover interactive proof system in which both provers are given only enough power to answer queries about membership in S?
>
> The answer to Question 2 would be "yes" if SAT were random-self-reducible. It is known that, unless the PH collapses, any random-self-reduction for SAT would have to be *adaptive*—that is, if instance x is mapped to a set $y_1, \ldots, y_k$ of random instances, then instance y_{i+1} has to depend on the answer to instance y_i (see [18] for details).

Question 5: Is there an adaptive random-self-reduction for SAT?

Question 6: Find other sufficient conditions for interesting classes of provers, besides random-self-reducibility and length-decreasing self-reducibility.

There are also two obvious questions remaining about zero-knowledge.

Question 7: Theorem 4.5 says that one-way functions are sufficient to ensure computational zero-knowledge proofs for all of PSPACE. Is there a weaker assumption that also suffices? For example, does the assumption that P $\neq$ NP suffice to ensure computational zero-knowledge proofs for every language in NP (perhaps with a PSPACE-complete prover)?

Question 8: Do one-way functions imply the existence of perfect zero-knowledge arguments for NP-complete languages?

So far, all that is known about the question of whether more moves allow proof systems to recognize more languages is Theorems 4.1, 4.2, and 4.4. Further results on the move hierarchy would be interesting.

Finally, one can ask which other notions in theoretical computer science can be fruitfully related to interactive proof systems, program checking, random-self-reducibility, and so on. One good candidate is machine-learning.

ACKNOWLEDGMENT

I am extremely grateful to Gilles Brassard for his prompt and thoughtful comments on an earlier draft.

REFERENCES

[1] M. Abadi, J. Feigenbaum, and J. Kilian, "On hiding information from an oracle," *J. Comput. Syst. Sci.*, vol. 39, pp. 21–50, 1989.

[2] W. Aiello and J. Håstad, "Perfect zero-knowledge languages can be recognized in two rounds," *Proc. 28th IEEE Symposium on Foundations of Computer Science*, pp. 439–448, 1987. Expanded version to appear in *J. Comput. Syst. Sci.*

[3] E. Allender, "Oracles vs. proof techniques that do not relativize." *Proc. Special Interest Group on Algorithms '90*, (T. Asano, T. Ibaraki, H. Imai, T. Nishizeki, eds.), Berlin: Springer-Verlag, *Lecture Notes In Computer Science* 450, pp. 39–52, 1990.

[4] L. Babai and L. Fortnow, "A characterization of #P by arithmetic straight line programs," *Proc. 31st IEEE Symposium on Foundations of Computer Science*, pp. 26–34, 1990.

[5] L. Babai, L. Fortnow, and C. Lund, "Nondeterministic exponential time has two-prover interactive protocols," *Proc. 31st, IEEE Symposium on Foundations of Computer Science*, pp. 16–25, 1990.

[6] L. Babai and S. Moran, "Arthur–Merlin games: A randomized proof system and a hierarchy of complexity classes," *J. Comput. Syst. Sci.*, vol. 36, pp. 254–276, 1988.

[7] D. Beaver and J. Feigenbaum, "Hiding instances in multioracle queries," *Proc. 7th Symposium on Theoretical Aspects of Computer Science,* (C. Choffrut, T. Lengauer, eds.), Berlin: Springer-Verlag *Lecture Notes In Computer Science* 415, pp. 37–48, 1990.

[8] D. Beaver, J. Feigenbaum, J. Kilian, and P. Rogaway. "Security with low communication overhead," *Proc. CRYPTO '90* (S. Vanstone, ed.) Berlin: Springer-Verlag, *Lecture Notes In Computer Science,* (in press).

[9] R. Beigel and J. Feigenbaum, "Improved bounds on coherence and checkability," Yale University Tech. Rept. YALEU/DCS/TR-819, Computer Science Department, Sept. 11, 1990. To appear in *Computational Complexity.*

[10] M. Ben-Or, S. Goldwasser, J. Kilian, and A. Wigderson, "Multiprover interactive proof systems: How to remove intractability assumptions," *Proc. 20th Symposium on Theory of Computing,* Association of Computing Machinery, pp. 113–131, 1988.

[11] M. Blum and S. Kannan, "Designing programs that check their work," *Proc. 21st Symposium on Theory of Computing,* Association of Computing Machinery, pp. 86–97, 1989.

[12] M. Blum, M. Luby, and R. Rubinfeld, "Self-testing/correcting with applications to numerical problems," *Proc. 22nd Symposium on Theory of Computing,* Association of Computing Machinery, pp. 73–83, 1990.

[13] R. Boppana, J. Håstad, and S. Zachos, "Does coNP have short interactive proofs?," *Inform. Proc. Lett.,* vol. 25, pp. 127–132, 1987.

[14] G. Brassard, D. Chaum, and C. Crépeau, "Minimum disclosure proofs of knowledge," *J. Comput. Syst. Sci.,* vol. 37, pp. 156–189, 1988.

[15] G. Brassard, C. Crépeau, and M. Yung, "Everything in NP can be argued in perfect zero knowledge in a bounded number of rounds," in *Proc. 16th International Colloquim on Automata, Languages, and Programming,* Berlin: Springer-Verlag, *Lecture Notes In Computer Science, 372,* Ed. G. Ausiello, M. Dezani-Ciancaglini, S. R. D. Rocca, pp. 123–136, 1989. Expanded version to appear in *Theor. Comput. Sci.*

[16] M. Burmester and Y. Desmedt, "Remarks on the soundness of proofs, *Electron. Lett.,*vol. 25 pp. 1509–1511, 1989.

[17] U. Feige, A. Fiat, and A. Shamir, "Zero knowledge proofs of identity," *J. Cryptology,* vol. 1, pp. 77–94, 1988.

[18] J. Feigenbaum and L. Fortnow, "On the random-self-reducibility of complete sets," University of Chicago Tech. Rept. 90-22, Computer Science Department, Aug. 20, 1990. Also in *Proc. 6th IEEE Structures,* pp. 124–132, 1991. Expanded version to appear in *SIAM J. Comput.*

[19] L. Fortnow, "On the complexity of perfect zero-knowledge," in *Advances in Computing Research—Vol. 5: Randomness and Computation* (S. Micali, ed.), Greenwich CT: Johnson Associates Inc. Press, pp. 327–343, 1989.

[20] L. Fortnow, J. Rompel, and M. Sipser, "On the power of multiprover interactive protocols," *Proc. 3rd IEEE Structures,* pp. 156–161, 1988.

[21] Z. Galil, S. Haber, and M. Yung, "Minimum-knowledge interactive proofs for decision problems," *SIAM J. Comput.,* vol. 18, pp. 711–739, 1989.

[22] M. Garey and D. Johnson, *Computers and Intractability: A Guide to the Theory of NP-Completeness,* San Francisco: Freeman, 1979.

[23] O. Goldreich, S. Micali, and A. Wigderson, "Proofs that yield nothing but their validity and a method of cryptographic protocol design," *Proc. 27th IEEE Sympo-*

sium on Foundations of Computer Science, pp. 174–187, 1986. Expanded version to appear in *J. Assoc. Comput. Mach.*

[24] S. Goldwasser, S. Micali, and C. Rackoff, "The knowledge complexity of interactive proof systems," *SIAM J. Comput,* vol. 18, pp. 186–208, 1989.

[25] S. Goldwasser and M. Sipser, "Public coins vs. private coins in interactive proof systems," in *Advances in Computing Research—Vol. 5: Randomness and Computation* (S. Micali, ed.), Greenwich CT: Johnson Associates Inc. Press, pp. 73–90, 1989.

[26] R. Impagliazzo and M. Yung, "Direct minimum-knowledge computations," *Proc. Crypto '87,* Berlin: Springer-Verlag, *Lecture Notes In Computer Science* 293, C. Pomerance, Ed., pp. 40–51, 1988.

[27] J. Kilian, *Use of Randomness in Algorithms and Protocols,* Ph.D. thesis, Massachusetts Institute of Technology: MIT Press, 1989.

[28] J. Kilian, Interactive Proofs Based on the Speed of Light, unpublished manuscript, Nov. 1990.

[29] R. Lipton, "New directions in testing," *Distributed Computing and Cryptography,* DIMACS Series on Discrete Mathematics and Theoretical Computer Science, AMS/ACM, 2 (1991) pp. 191–202.

[30] C. Lund, L. Fortnow, H. Karloff, and N. Nisan, "Algebraic methods for interactive proof systems," *Proc. 31st IEEE Symposium on Foundations of Computer Science,* pp. 2–10, 1990.

[31] M. Naor, "Bit commitment using pseudo-randomness," *Proc. CRYPTO* 89, (G. Brassard, ed.), Berlin: Springer-Verlag *Lecture Notes In Computer Science* 435, pp. 128–136, 1989. Expanded version to appear in *J. Cryptology.*

[32] N. Nisan, "CoSAT has multiprover interactive proofs," e-mail announcement, Nov. 22, 1989.

[33] C. Papadimitriou, "Games against nature," *J. Comput. Syst. Sci.,* vol. 31, pp. 288–301, 1985.

[34] A. Shamir, "IP = PSPACE," *Proc. 31st IEEE Symposium on Foundations of Computer Science,* pp. 11–15, 1990.

[35] S. Toda, "On the computational power of PP and $\oplus$P," *Proc. 30th IEEE Symposium on Foundations of Computer Science,* pp. 514–519, 1989.

[36] L. Valiant, "The complexity of computing the permanent," *Theor. Comput. Sci.,* vol. 8, pp. 189–201, 1979.

APPENDIX
COMPLEXITY THEORETIC BACKGROUND

The complexity classes that are most central to the ideas discussed in this chapter are as follows:

P: deterministic polynomial time; those membership problems that can be computed efficiently with no errors,

NP: nondeterministic polynomial time; those membership problems that can be verified efficiently,

#P: counting functions associated with NP sets; for example, the set of all satisfiable propositional formulas is called SAT and is in NP, and the function that maps a formula to the number of truth assignments that satisfy it is in #P,

PSPACE: deterministic polynomial space,

EXP: deterministic exponential time,

NEXP: nondeterministic exponential time, and

BPP: probabilistic polynomial time; those membership problems that can be computed efficiently with error probability at most $1/3$.

Other classes of interest include PH (the polynomial-time hierarchy Σ_i^p, Π_i^p, $i \geq 0$) and $\oplus P$ ("parity-P"). The class Σ_i^p, is referred to as the "i^{th} level" of the PH.

The following inclusions are known.

$$P \subseteq NP \subseteq \ldots \subseteq \Sigma_i^p \subseteq \ldots \subseteq PH \subseteq BPP^{\oplus P} \subseteq P^{\#P} \subseteq PSPACE \subseteq EXP \subseteq NEXP$$

The only inclusion known to be proper is $P \subsetneq EXP$. One often hears that all these inclusions are "believed" to be proper, but "belief" (like "knowledge") is very hard to define mathematically. Suffice it to say that any resolution of the question of whether these inclusions are proper (e.g., a theorem of the form "the PH collapses at the second level" or "PSPACE is contained in the second level of the PH") would require a revolutionary new idea. See Garey and Johnson [22] for a thorough introduction to these issues.

We use the notation coC for the class of languages whose complements are in C and the notation coS for the complement of language S. Thus coSAT is the set of all unsatisfiable propositional formulas. The symbol Π_i^p is used for $co\Sigma_i^p$.

Finally, we will need the notion of *one-way function*. Intuitively, a function is one-way if it is easy to compute but hard to invert. For cryptography theory, we need that it is *often* hard to invert. One standard way to formalize this is as follows. Let f be computable in deterministic polynomial time. Then f is one-way if, for all polynomials q, for all probabilistic polynomial-time algorithms I (for "inverter"), for all sufficiently large n, the probability that $f(I(f(x))) = f(x)$ is less than $1/q(n)$. This probability is computed over uniformly chosen x of length n and uniform coin-toss sequences for I.

CHAPTER 9

An Introduction to Shared Secret and/or Shared Control Schemes and Their Application*

G. J. SIMMONS
Sandia National Laboratories
Albuquerque, New Mexico 87185

*This work was performed at Sandia National Laboratories and supported by the U.S. Department of Energy under contract number DE-AC04-76DP00789.

It may seem odd to have a chapter on secret sharing appear in a volume devoted to cryptology, but as we shall see, secret sharing is intimately related to two problems common to all cryptosystems, namely, the problems of key management and key distribution. The justification for this chapter, however, is even more direct since the larger subject of this volume is how to insure that information-based systems either cannot be caused to function improperly through tampering with the information they act on, or else to detect (in probability) if they have been caused to function improperly by such tampering. The integrity of information (from unauthorized scrutiny or disclosure, manipulation or alteration, forgery, transmission, false dating, etc.) is commonly provided for by requiring operation(s) on it that one or more of the participants or elements in the system, who have access to some private piece(s) of information not known to all of the other participants, can carry out, but which (probably) can't be carried out by anyone who doesn't have access to the private information.

Encryption/decryption in a single cryptoalgorithm is a paradigm of such an operation, with the common key being the private (secret) piece of information. If the transmitter can rely on the receiver protecting and keeping secret his copy of the key, then he—the transmitter—can be confident that he can communicate information to the receiver in private, that is, with secrecy. Conversely, if the receiver can rely on the transmitter keeping his copy of the key secret, then he—the receiver—can be confident that the ciphers he receives and decrypts into acceptable messages must have been sent by the (legitimate) transmitter. Although it is implicit in single-key cryptosystems, it is almost never stated explicitly that the transmitter and the receiver must unconditionally trust each other since either can do anything that the other can. What this means is that the receiver could falsely claim to have received an encrypted message from the transmitter that the transmitter would be unable to prove (to an impartial third party or arbiter) that he hadn't sent. Conversely, the transmitter could send an encrypted message to the receiver and then disavow having sent it and the receiver would likewise be unable to prove that he hadn't generated the encrypted message himself. In this case the transmitter must unconditionally trust the receiver to not falsely attribute messages to him that he did not send, and the receiver must unconditionally trust the transmitter to not disavow messages that he did send. There are situations in which the participants are all willing to unconditionally trust each other, but there are many more in which they are not.

Even if it can't be assumed that all of the elements in a system are trustworthy, so long as there exists at least one identified unconditionally trustworthy element (individual or device), it is generally possible to devise protocols to transfer trust from this element to other elements of unknown trustworthiness to make it possible for users to trust the integrity of the information in the system, and hence to trust the system function, even though they may not trust all of the elements. A paradigm for such a protocol is the cryptographic key distribution system described in American National Standards Institute (ANSI) X9.17 [R1] which makes it possible for users who have had no previous contact, nor any reason to trust each other, to trust a common cryptographic session key because they each unconditionally trust the key distribution centers (KDC). In reality, however, there is generally no unconditionally trustworthy element—be it a person or a device—only elements of tolerable trustworthiness, or circumstances that compel them to be accepted as trustworthy as in the case of customers' acceptance of the security of automated teller machines (ATMs), and of user identification via personal identification numbers (PINs).

The more common (and hence the more realistic) situation is that there are no identified unconditionally trustworthy elements in a system. Instead, the most that can be assumed is that while any specific element may be suspect, that is, possibly subject to either deliberate or inadvertent compromise, and hence untrustworthy insofar as the faithful execution of the part of the protocol entrusted to it, that at any given time there are some (unidentified) elements in the system that are trustworthy. Under these circumstances there is apparently only one way to improve the confidence one can have in the integrity of the system over the confidence one has in the integrity of the individual elements, and that is by introducing some form of redundancy—on the assumption that the more elements that must collude or be compromised in order for a deception to be possible, or to go undetected, the less likely the deception is to occur or to escape detection if it does occur. To protect against random failures of devices, say in fault-tolerant networks, this task is commonly achieved by parallel or by series-parallel operation of redundant elements or by even more complex logical interconnections.

In the case of individuals, though, since the failure may be both deliberate and clandestine, redundancy typically takes the form of requiring the concurrence of two or more knowledgeable persons to carry out an action. A paradigm for this would be the well-known two-man control rule that the United States enforces for critical military actions. This is the simplest possible example of secret sharing or shared control since in this case there are only two parties, each of whom knows a private piece of information that when combined with the piece known only to the other party suffices to allow access to (or control of) a weapons system, but each of which individually provides its holder no greater chance of access or ability to use the weapon than what an outsider who knows nothing at all about the secret controlling information would have.

In this chapter, we will develop the general principles of secret sharing and/or shared control as a means for increasing the confidence in the proper functioning of information-based systems and examine a number of the characteristics such systems need to have for various real-world applications over and above the simple property of enforcing the concurrence of a predesignated number of the participants before a secret piece of information can be recovered or a jointly controlled event initiated. These schemes are all unconditionally secure in the sense that the security they provide is independent of the computing time or power that an opponent may bring to bear on subverting the system, or, put in another way, even with infinite computing power would-be cheaters can do no better than guess (with a uniform probability distribution on the choices available to them) at the controlling secret information.

1 INTRODUCTION

In 1979, Blakley [10] and Shamir [53] independently devised shared secret schemes for the same application: robust key management for cryptosystems. They were concerned both with the problem of legitimate users being locked out of the system, that is, of the key being lost for some reason, and of unauthorized users getting in. Although their approaches to solving these problems were quite different, as we shall see in a moment, the essential notion is the same in both cases. Given a cryptographic key, that is, the secret piece of information, one wishes to construct ℓ related pieces of information with the property that any set of k of these pieces will suffice to recover the original secret, but such that no subset of $k - 1$ or fewer of them will reveal it. Given such a construction, the pieces of information can then be distributed privately and securely to the ℓ participants in the secret sharing scheme. If there are ℓ of the private pieces of information (generally called ''shares'' by those authors whose work derives from Shamir's or ''shadows'' by those who use Blakley's model) and k are required to reconstruct the secret, the scheme is called either a k-out-of-ℓ shared secret scheme or a (k, ℓ)-threshold scheme, respectively. Thereafter, as many as $\ell - k$ of the private pieces of information could be lost or unavailable when needed and the key could still be reconstituted, while no security breach of fewer than k of the pieces could reveal the key to an unauthorized user.

Since we will generalize the notion of shared control (of information and hence of actions dependent on that information) in ways that cannot be easily viewed as threshold schemes, we will use the term shared secret scheme since this emphasizes the essential feature of shared capability as opposed to the more restrictive notion of simply requiring the concurrence of a specified number of the participants. Both the Blakley and the Shamir constructions realize k-out-of-ℓ shared secret schemes, however, as we have already mentioned their constructions are fundamentally different. We will first describe their schemes in detail and then make use of these differences to illustrate several concepts important to shared secret and/or shared control schemes in general.

Shamir's construction is the easier to explain and is algebraic in nature—based on interpolation of a polynomial defined over a finite field, $GF(q)$. Figure 1 illustrates the basic idea. Given k points in the two-dimensional plane, (x_i, y_i) $i = 1, 2, \ldots, k$, there is a unique $(k - 1)$st degree polynomial, $\mathbf{P}^{k-1}(x)$, for which $\mathbf{P}^{k-1}(x_i) = y_i$ for all i. If the secret (key) is taken to be an element $p \in GF(q)$, it can be partitioned into ℓ shares

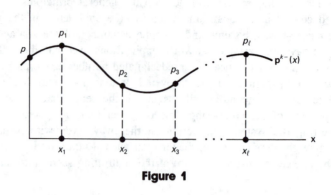

Figure 1

as follows. A $(k - 1)$st degree polynomial $\mathbf{P}^{k-1}(x)$ is chosen randomly. $\mathbf{P}^{k-1}(x)$ has the representation

$$\mathbf{P}^{k-1}(x) = \sum_{i=0}^{k-1} a_i x^i \tag{1}$$

where the constant term a_0 is the secret p; that is, $\mathbf{P}^{k-1}(0) = a_0 = p$. The ℓ shares, $p_i = (x_i, y_i)$, $\ell \geq k$, are calculated by evaluating Eq. (1) at ℓ distinct points x_i, $x_i \neq 0$:

$$p_i = y_i = \mathbf{P}^{k-1}(x_i) \tag{2}$$

Given any subset of k of these points, it is computationally easy using Lagrange interpolation to solve for $a_0 = p$ in Eq. (1), that is, to recover the secret. Given only $k' \leq k - 1$ of the private pieces of information (points on $\mathbf{P}^{k-1}(x)$) a_0 could equally likely be any element in $GF(q)$ since for each choice of a point p' on the y-axis there will be equally many $(k - 1)$st degree polynomials through the subset of k' points and p'. Consequently, any collusion of fewer than k of the participants (the insiders) will have no better chance of determining the secret than will an outsider who has no privileged information at all. To all unauthorized groupings (fewer than k participants) including outsiders, the secret is equally likely to be any point on the line $(0, y)$, that is, the probability of guessing p will be $1/q$. A shared secret scheme in which insiders (and all unauthorized groupings of insiders) have no advantage over outsiders in guessing the secret are said to be *perfect*. Shamir's polynomial interpolation shared secret scheme is perfect in this sense.

The private information content of each share in a perfect shared secret must be at least as great as the information content (equivocation) of the secret itself. This is easy to see. Assume that some share has less private information in it than does the secret, and consider an arbitrary collusion of $k - 1$ participants not including the deficient share. Since the system is perfect, the $k - 1$ of them together are, by definition, equally uncertain of the secret as an outsider. On the other hand, if they knew the missing (deficient) share they could recover the secret, since they are all participants in a k-out-of-ℓ shared secret scheme, but their uncertainty about the missing share value is less than their uncertainty about the secret: a contradiction. Therefore, the private information content in each share in a perfect secret sharing scheme is necessarily at least as great as the uncertainty (information content) of the secret itself. If equality holds, the shared secret scheme is said to be *ideal*. As described, Shamir's scheme is not ideal since each private point consists of a pair of coordinate values, that is, two elements in $GF(q)$, whereas the secret is a single element in $GF(q)$. This is not an essential difference, however, and we will return to this point below. For the moment we only wish to introduce and define the concept of ideal shared secret schemes.

Shamir recognized that it might be desirable in some circumstances for participants to have differing capabilities to recover the secret. He proposed to accomplish this by giving more than one point on the polynomial $\mathbf{P}^{k-1}(x)$ to the more capable participants. If $k = 4$, so that four points are required to recover $\mathbf{P}^3(x)$ then by giving pairs of points to participants in Class I and single points to participants in Class II a scheme could be set up in which any subset that included two participants from Class I or four participants from Class II or one participant from Class I and any two from Class II would be able to recover the secret, while any collection of participants and outsiders that did not satisfy one of these conditions would be unable to do so. Schemes of this sort in which the differing capability of the participants private pieces of information to assist in the recovery of the secret are a function of their information content we define as *intrinsic*. An intrinsic scheme with more than a single class of capability cannot be perfect.

The converse to an intrinsic scheme is an *extrinsic* one wherein the differing capability of the private pieces of information to assist in the recovery of the secret are a consequence of their functional relationship to the other private pieces of information, not on their differing information contents. An ideal scheme is necessarily an extrinsic one, but an extrinsic scheme may not be ideal since the private pieces of information, while they all contain the same amount of information, may all contain more bits of information than are in the concealed secret. Brickell and Davenport [19] have introduced the terminology of *strongly ideal* shared secret schemes for those in which the secret point and the private pieces of information are each drawn from the same space with the same probability distribution, that is, in which all of the points in some geometric object are equally likely to be the secret as they are to be a private piece of information. We will have to wait until later to exhibit convincing examples of extrinsic schemes since we aren't yet ready to deal with differing capabilities for the participants. We will also show how to realize arbitrarily complex but perfect shared secret schemes. For the time being, though, these are the only properties of shared secret schemes we wish to point out using the Shamir constructions.

Blakley's construction for shared secret schemes is geometric. To motivate his terminology of "shadows" for the private pieces of information, consider the construction shown in Fig. 2. The secret (information) in this case is the point p in the three-dimensional space E^3. p has been projected (along the x-axis) from infinity to generate the "shadow," p_{yz}, of p in the plane yz. Similarly, p is projected parallel to the y- and to the z-axes to generate the shadows p_{xz} and p_{xy}, respectively. The private pieces of information in this case are the three shadows p_{xy}, p_{xz}, and p_{yz}. The resulting scheme is a (2, 3)-threshold scheme. It should be mentioned that this is not precisely the way Blakley uses the term "shadow," although it is equivalent. We will give his general definition shortly, after the concept is made intuitively clear. Participant i, $i = x$, y, or z, knows that his shadow is the result of projecting the point p parallel to the i-axis, that

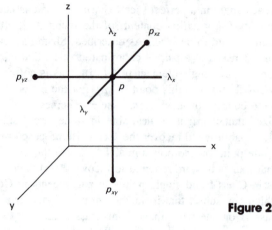

Figure 2

is, he knows that p is a point on the line λ_i through his shadow (private point) and parallel to the i-axis. Since the three lines, λ_i, all intersect in the point p, any two of them suffice to determine it. Any one of the participants alone, however, knows only that p is a point on a line that he (alone) knows.

This small example already suffices to illustrate an important characteristic of Blakley's construction for shared secret schemes—subsets of $k - 1$ or fewer participants (the insiders), including individual participants, have an advantage over outsiders in guessing at the secret information because of their knowledge of the private information. The setting for Blakley's shared secret schemes is not in a Euclidean space E^n, but rather in a finite projective space $PG(n, q)$, but more importantly, his construction of the shadows is slightly less intuitive than what we have done in the small 2-out-of-3 scheme. Continuing with this example for the moment, in Blakley's setting the space would be $PG(3, q)$, in which there are $q + 1$ points on a line and $q^3 + q^2 + q + 1$ points in all. An outsider's chances of guessing p (assuming, of course, that p was chosen originally with a uniform probability distribution on the points in $PG(3, q)$) would be $(q - 1)/(q^4 - 1) \approx 0(q^{-3})$, while the probability of any one of the participants guessing p would be $(q - 1)/(q^2 - 1) \approx 0(q^{-1})$. Consequently, this example—and the Blakley shared secret scheme in general—is not perfect.

There isn't anything special about the choice of the three lines λ_x, λ_y, and λ_z in Fig. 2 to be parallel to the x-, y-, and z-axes, respectively. This choice was made to simplify the example. In the projective space, $PG(3, q)$, there are no parallel lines, but the projectors of p could equally well have been any other three distinct lines lying on the point p. When looked at in this generality, if p is a finite point of $PG(3, q)$, then the $q^2 + q + 1$ lines lying on p are simply the projectors of p from each of the $q^2 + q + 1$ points in the plane at infinity, that is, the lines defined by p and the points in the plane at infinity. Any pair of these lines intersect in the point, p, and hence the lines can be used as the private pieces of information in a 2-of-ℓ shared secret scheme; $\ell \leq q^2 + q + 1$.

In complete generality, the private pieces of information can be considered to be the projection of the point p from a (private) point in one plane, π_1, onto another plane, π_2, that is, the shadow of p in π_2 under such a projection. p cannot be a point in the plane π_1 which is why in the example we choose p to be any finite point and π_1 to be the plane at infinity. We have described this construction in such tedious detail to motivate Blakley's terminology for the private shares as "shadows." If one takes not points in the plane at infinity, but lines in the plane, then the geometric objects defined by these lines and p would be all of the planes on p in $PG(3, q)$. In this case any three of the planes not on a common line through p would define p by their intersection. In general, k randomly chosen hyperplanes—$(k - 1)$-dimensional subspaces in PG(k, q)— will probably define, that is, intersect in, a single point.

Blakley's construction for a k-out-of-ℓ shared secret scheme makes use of this simple geometric result. The secret, p, is a randomly chosen point in $PG(k, q)$. The ℓ shadows are chosen from among the $(q^k - 1)/(q - 1)$ hyperplanes (subspaces of projective dimension $k - 1$) lying on p. If q is sufficiently large and ℓ is not too large, the probability of any subset of k of these shadows having more than one point in common is very close to zero. If this statistical confidence isn't adequate, as Blakley points out, it is an easy matter to test to insure that this independence condition is satisfied by the chosen set of shadows. Blakley's construction in $PG(3, q)$ would therefore not be a set of lines, λ_i, on p, but rather a set of hyperplanes, π_i on p defining a 3-out-of-ℓ threshold scheme as shown in Fig. 3.

There are $q^2 + q + 1$ such planes (on the point p) so that any ℓ, $3 \le \ell \le q^2 + q + 1$, is possible for a 3-out-of-ℓ scheme in such a construction. As already remarked, this is not a perfect scheme since outsiders can only guess at p being a point in $PG(3, q)$ while any insider has only to guess at p being a point on the plane (shadow) that he knows. The reader can now appreciate why we developed the Blakley implementation of shared secret schemes in the way we did, since the planes in Fig. 3 are not suggestive of "shadows" of p, while the points p_{ij}, $i \ne j$, in Fig. 2 are.

Blakley's shared secret schemes are, in general, far from ideal since each participant must have enough private information to identify his shadow (a $k - 1$)-dimensional hyperplane). This requires k coordinate values, that is, k elements from

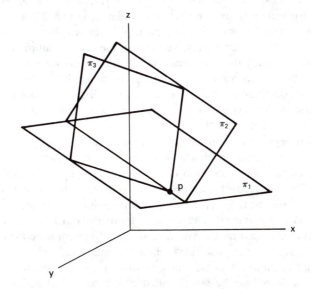

Figure 3

$GF(q)$. On the other hand, a collusion of $(k-1)$ participants can only have linear uncertainty about the secret since their $(k-1)$ hyperplanes intersect in a line that must include the secret point p. This means that there is precisely one coordinate value of uncertainty, that is, 1 out of $q+1$ equally likely values, for the secret—in the worst case of an improper collusion of $k-1$ of the participants—while each participant is responsible for k, $k \geq 2$, elements from $GF(q)$ as his private information.

The reason for describing the shared secret schemes of Shamir and of Blakley at such great length is not purely historical. As we shall see in the next section, their constructions typify two fundamentally different, but general, approaches to shared secret schemes, both of which we will need to generalize and understand for later application. We should point out that a modest literature (at least the over 60 papers cited in the Bibliography on shared secret schemes at the end of this chapter) has been generated since the notion was first discovered by Shamir and Blakley. However, since all of the schemes proposed thus far are subsumed in the general models given in the next section, and since the purpose of this chapter is to provide a tutorial introduction to shared secret and/or shared control schemes, we will not attempt a survey of the literature. As a partial compensation for this, the bibliography includes every paper on the subject known to the author—not just the ones cited as references here.

2 THE GENERAL MODEL(S)

There are two essentially different ways in which pieces of information related to another, secret, piece of information can be constructed and distributed among a group of participants so that designated subsets of the participants will be able to recover the secret piece of information, while no collection of participants that doesn't include one of these subsets can. As it happens, the schemes discovered by Shamir and Blakley provide ready examples for each of these classes of shared secret schemes. In some schemes, the set of possible values for the secret, consistent with all of the private pieces of information that have been exposed, remains unchanged until the last required piece of private information becomes available, at which point the unique value for the secret suddenly becomes the only possibility, as was the case for the Shamir construction. In others, as each successive piece of the private information is exposed, the range of possible values that the secret could assume narrows, until finally when the last required piece of private information becomes available, the secret will have been isolated and identified as was the case for the Blakley construction. There are numerous examples of each type of system and applications in which the most natural solution involves an intermixing of the two.

We begin our development of the general models with a discussion designed to clarify the essential difference between the two classes of schemes. Without loss of generality, the secret can always be considered to be a point, p, in some suitably chosen space. The function of a shared secret scheme is to make it possible for authorized subsets of the participants to identify this point, but to make it improbable that any collection of participants and/or outsiders that doesn't include an authorized subset of the participants can do so. We have seen that the constructions of both Shamir and of Blakley satisfy these conditions when the authorized subsets are defined by a simple threshold condition; that is, that k or more (out of ℓ) of the participants must concur before the secret can be reconstituted or the controlled action initiated. This includes the special case of unanimous consent, $k = \ell$.

In real-world applications, however, the concurrences that one needs to be able to insure exist before the shared control can be initiated may be much more complex: For example, a bank might want to require the concurrence of two vice-presidents or of three senior tellers for it to be possible to authenticate an electronic funds transfer (EFT)—with the natural requirement that a vice-president should also be able to act in the stead of a senior teller, that is, that any vice-president and any two senior tellers should also be able to authenticate an EFT. To realize these more general concurrence schemes (or access structures in the terminology of Benaloh and Leichter [3], Brickell and Davenport [19], or Brickell and Stinson [20]), the model for shared secret schemes must be much extended. Since the discussion of these extensions will (unavoidably) be both lengthy and detailed, the reader may find it helpful to know in advance approximately what we are going to do.

In Shamir's threshold scheme, each participant holds as his private piece of information a point on a polynomial $\mathbf{P}^{k-1}(x)$ which can be thought of as a geometric object—in this case the function $y = \mathbf{P}^{k-1}(x)$—that "points" to the secret point, p, on the y-axis. In a strict sense, $\mathbf{P}^{k-1}(x)$ is "spanned" by any k of the private pieces of information. We will generalize this notion so that a geometric object (usually an m-dimensional subspace of the containing space $\mathbf{S}$) spanned by the set of private points held by each of the authorized concurrences of participants will point to the secret point, p, in some other geometric object, also usually a subspace of S. Most of our constructions will be of this simple form, although the geometry of the two objects involved may be exceedingly complex.

In Blakley's scheme, each participant holds, as his private piece of information, a geometric object—actually a hyperplane in his construction, but we want to deliberately suggest a more general construction—that he knows contains the secret point p. In a simple threshold scheme, the objects are chosen so that any subset of k of them will intersect in, that is, identify, the unique point p. If this approach to shared secret schemes were generalized, one would require that each collection of the geometric objects (private pieces of information) held by one of the authorized concurrences of participants should also unambiguously identify p by their intersection. We will not pursue the generalization of Blakley-type shared secret schemes, however, since these schemes can never be perfect, because as the number of participants in a collusion increases, the uncertainty about the secret, p, must decrease since p is in each of the privately held geometric objects and hence in their intersection as well.

Consider the simplest possible example of a shared secret scheme, a 2-out-of-ℓ scheme. A generalized version of the Shamir construction for this case is shown in Fig. 4. We say generalized because the secret is no longer the constant term of the shared polynomial, that is, the intersection of the curve $y = \mathbf{P}^{k-1}(x)$ with the y-axis, but rather the point of intersection of two subspaces—lines in this case.

The private pieces of information are points on a line V_i, whose intersection with a line V_d is the index point, p, at which the shared secret information is defined: $p = (x_p, y_p)$. We call the reader's attention to an important change in terminology that has been introduced here. In Shamir's paper and in our discussion of his construction, the secret was defined to be a (randomly chosen) point in the ground field, $p \in GF(q)$. Similarly, the constant term, a_0, of the polynomial $\mathbf{P}^{k-1}(x)$ was equally likely to be any element of $GF(q)$ since $\mathbf{P}^{k-1}(x)$ was randomly chosen from all of the $(k-1)$-st degree polynomials over $GF(q)$. This made it natural to equate p with a_0. In the construction of Fig. 4, p is a point on the (publicly known) line V_i. However, since p satisfies a known

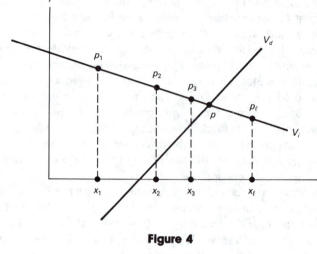

Figure 4

linear constraint there is only one degree of freedom, that is, of uncertainty, to it. In other words, the uncertainty about p is only $H(p) = \log_2(q)$ bits—in the affine plane $AG(2, q)$—just as it was for the Shamir construction. We assume that some entropy-preserving function is evaluated at p to recover the secret. One suitable function would be the projection of p onto the y-axis, which, of course, is identical with the Shamir construction when V_d is the y-axis. There are many such entropy-preserving transformations, which as we will see later is an essential characteristic of this construction for shared secret schemes.

As we have pointed out elsewhere [55] and will discuss in detail below, the information needed to specify the points, p_i, is not all of the same type insofar as its security requirements are concerned. If we use the obvious specification of the point p_i in the affine plane $AG(2, q)$ by its coordinates (x_i, y_i), then it is sufficient (for the security of the shared secret scheme in this example) that each participant keep secret one of the coordinate values (say y_i), and that he merely insure the integrity of the other coordinate value against substitution, modifications, etc. With this convention for partitioning the private pieces of information into secret and nonsecret parts, V_i cannot be parallel in the affine plane to the y-axis since in that case the nonsecret parts would all be the same, $x_i = x_p$ for all i, and V_i itself, and hence p, could be deduced from the exposed (nonsecret) parts of any pair of private pieces of information.

The point we wish to make using this simple scheme has to do with the probability of some improper (i.e., unauthorized) collection of persons recovering the secret. An outsider who knows only the geometric nature of the scheme, that is, the line V_d and that there is a line V_i whose intersection with V_d determines the unknown point p and the public parts of the private pieces of information, cannot restrict the possible values for p beyond the fact that it is a point on V_d. Since each of the q points of V_d has the same number of lines on it that are not parallel to the y-axis and hence could be the unknown line V_i, it should be obvious that the opponent can be held to an uncertainty about the secret of

$$H(p) = \log(q) \tag{3}$$

that is, his "guessing probability" of choosing p in a random drawing using a uniform probability distribution on the points of V_d. This is the best (security) that can be achieved by any scheme for concealing p.

Now consider the uncertainty faced by one of the participants: an insider. He knows his private piece of information, the point p_i on V_i, the public abscissas x_j, $j \neq i$, for the other participant's private pieces of information and the line V_d. Each point, p', on V_d determines a unique line lying on both p' and p_i that could be the unknown (to the participant) line V_i. Clearly, his uncertainty about the secret is the same as that of an outsider who has no access to any privileged information

$$H(p) = \log(q) \tag{4}$$

These are the only two meaningful improper groupings of persons in this example since no combination of outsiders with an insider is more capable (in improperly recovering the secret) than is the insider alone. Consequently, this is a perfect 2-out-of-ℓ scheme.

We next consider the Blakley construction for a 2-out-of-ℓ shared secret scheme shown in Fig. 5. In this case, the private pieces of information are lines all of which are concurrent on the secret point p. To an outsider, every point in the plane is equally likely to be the point p, hence his uncertainty about p is

$$H(p) = \log(q^2) = 2\log(q) \tag{5}$$

An insider, on the other hand, knows that p must be a point on the line that is his private piece of information. Hence, his uncertainty is only

$$H(p) = \log(q) \tag{6}$$

Consequently, this scheme is not perfect, since the insiders have an advantage (in cheating) over an outsider. Both schemes, however, provide the same minimum level of security against unauthorized recovery of the secret information. From that standpoint alone, they would appear to be equally good. There are other factors that need to be considered, such as the amount of secret information each participant must be responsible for—or even the information content of the part whose integrity must be

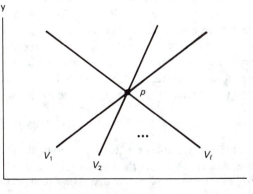

Figure 5

insured—and the information content of the secret itself. Since the plane is two-dimensional in both points and lines, it is easy to see that in either example the participant need only keep the equivalent of one coordinate value, that is, $\log(q)$ bits, secret about his private piece of information and to insure the integrity of a like amount of information. Thus, the two schemes are equivalent with respect to these parameters. It is at least plausible to measure the security of a shared secret scheme by the least uncertainty faced by any unauthorized person or grouping of persons about the secret; in other words, $\log(q)$ bits in both examples as well.

The bottom line to this discussion of the two small examples is that while they are certainly different (not just superficially in the geometric implementations) they are also alike in important respects. It is the differences that we wish to understand to better understand shared secret schemes.

In the first example, where the secret was defined at the point p on the line V_d, the security of the scheme was measured by the uncertainty about p, which, as we saw, was the same for an outsider or for any one insider. We now examine this example from a different standpoint. Although it may seem strange at first to do so, the line V_i can be thought of as being a cryptographic key that encrypts the plaintext (ordinate) of a private piece of information into the ciphertext (abscissa). This is consistent with our convention that the ordinate is the information being protected (kept secret) and that the abscissa can be exposed. The secret information is the decryption of a known ciphertext (the ordinate, x_p, of the point p) using the key V_i. In this simple geometric construction of a 2-out-of-ℓ shared secret scheme if V_i is to define a cryptotransformation, it must not be parallel to either the y-axis or the x-axis in order for it to define a one-to-one mapping (i.e., a nonsingular linear transformation) of the y-axis onto the x-axis.

While it is essential for the simple shared secret scheme described here, that V_i be restricted to not be parallel to the y-axis—since as we explained earlier the exposed parts of the private pieces of information would reveal V_i in that case—it isn't necessary that V_i be restricted to not be parallel to the x-axis as well. If V_i is parallel to the x-axis, then the secret coordinate of the point p would satisfy $y_p = y_i$ for all i, but this could only be discovered if two or more of the participants compared (exposed) their secret pieces of information. But in this example, any two participants have been assumed to have the capability to reveal the secret, not just to recover V_i. Therefore, there is no necessity to exclude lines parallel to the y-axis in this example since we are only using the one-way nature of the encryption operation without any requirement that it also be invertible—which will be satisfied so long as to each choice of a plaintext y_p, every ciphertext is an equally likely pre-image.

Looked at in this way, we can calculate the uncertainty about the (secret) key to the various combinations of individuals. An outsider knows only that V_i is a line in the plane not parallel to the y-axis, that is, one of the $q^2 + q$ lines in the plane less the q lines parallel to the y axis, or q^2 lines in all. Hence his uncertainty about the key is

$$H(V_i) = \log(q^2) = 2\log(q) \tag{7}$$

which is twice his uncertainty about y_p, the encrypted value of the ordinate of p. Note that in this interpretation V_d has effectively been restricted to be the line parallel to the y-axis lying on x_p. It should be noted that the outsiders' uncertainty about p, $H(p)$, in the first example is the same as his uncertainty about V_i, $H(V_i)$ in this example.

There are $q + 1$ lines through each point on the line V_d, one of which is V_d itself, and hence not a candidate to be the key V_i. Therefore the q^2 potential keys (lines in $AG(2, q)$ not parallel to the y-axis) are uniformly distributed q at a time on each of the q points of V_d; hence

$$H(p) = \log(q) \tag{8}$$

In other words, since the set of q^2 lines that could be the unknown key, V_i, are uniformly distributed on the points on V_d (q on each point) and are all equally likely to be the key, an opponent's chance of determining the secret by "guessing" at the value of the key is exactly the same as his chance of "guessing" the value of the secret in the first place: $\log(q)$ in either case.

Next consider the situation of an insider. He knows a point on the unknown key, V_i. There are $q + 1$ lines through this point, q of which are potential keys. Consequently, there is a one-to-one association between the potential keys (given his insider information) and the possible values for the secret cipher. Thus for the insider

$$H(V_i) = H(p) = \log(q) \tag{9}$$

The point is that in the first example it was the uncertainty about the key that was eroded with the exposure of successive pieces of the private information, that is, of plaintext/ciphertext pairs in the present setting (only one such pair is possible in this small example; we are anticipating the general case in this remark), however, the uncertainty about the secret index point, $H(p)$ or more precisely $H(y_p)$, remains the same for any grouping other than one able to uniquely identify the key. So long as the surviving candidate keys uniformly map each cipher into all possible plaintexts, the uncertainty about the secret plaintext remains the same, even though the uncertainty about the key decreases with each successive piece of private information that becomes available. The second example has no intermediate key, so it is the uncertainty about the secret point, p, that is directly eroded by the exposure of successive pieces of private information. When viewed in this way, a very close relationship exists between cryptanalysis in depth (with the key as the depth component) and shared secret schemes.

The entire purpose of this discussion was to support the following observation: The information contained in each of the private pieces of information constrains the values that some other variable can take. If this variable is itself the secret, then the shared secret system cannot be perfect, since in that case, unauthorized groupings of insiders would necessarily have an advantage over outsiders in guessing at the value of the secret. If, however, the variable is an intermediate function, out of a family of functions, satisfying suitable constraints such as being entropy preserving over the space in which the secret is located, then the scheme can be perfect. Although we won't make direct use of the principle here, we are in fact faced with the problem of devising cryptosystems with the unusual property that they are immune to cryptanalysis in depth (against the key as the depth component) for all "improper" groupings of plaintext/ciphertext pairs, but cryptanalyzable with certainty of success in recovering the key given any set of plaintext/ciphertext pairs that includes at least one of the prescribed concurrences.

The examples shown in Figs. 4 and 5 contain all of the essential features for the general models we will use for the two types of shared control schemes, irrespective of how complex the required concurrence may be or of what other properties the shared

secret scheme may be required to have. In the first type of scheme, there is one geometric object (an algebraic variety, generally a linear subspace in some higher dimensional space), which can be determined given any subset of the points in it that includes
at least one of the specified concurrence groupings, which intersects another object in a
single point, p, at which the secret is defined. While p is a point in both of the sets, the
first set is always secret (until it is reconstructed by an authorized concurrence among
the participants) while in many applications the other is publicly known, a priori. We
therefore refer to the geometric object (set of points) whose determination isn't shared
among the participants as the domain variety, V_d, since the secret (argument) can be
thought of as being a point concealed in its domain. The object determined by the
shared information can be thought of as indicating (in the sense of pointing to) the
secret point p in V_d. We therefore call this object the indicator (variety), V_i. Stripped of
the inessential (to the present discussion) details of how sets of points can be chosen in
a variety V_i so that any of the designated subsets of them will suffice to reconstruct all
of V_i, the configurations of interest are of the form shown in Fig. 6.:

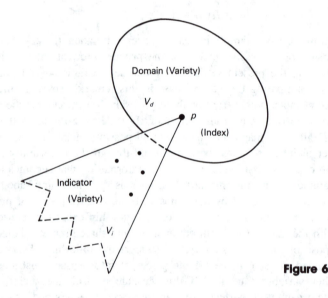

Figure 6

where $\dim(V_i) + \dim(V_d) = \dim(S)$ and $\dim(V_i \cap V_d) = 0$.

A single shared secret scheme may have either several indicators, or several domains, depending on the nature of the concurrence that is being realized. In the example
of Fig. 4, the indicator was the line V_i, determined by any pair of the points on it. V_i
could have been equally well replaced by a quadratic curve (determined by any three of
its points) or a cubic (determined by any four of its points not lying on a quadratic
curve, i.e., of rank four), etc. This, in fact, is the implementation of shared secret
schemes originally proposed by Shamir [53]. In this paper V_i and V_d will usually be
restricted to be linear subspaces of some higher-dimensional containing space, **S**. In this
case if $\dim(V_i) = k - 1$ and $\ell = k$, so that all of the private points are needed to span
V_i, a k-out-of-k unanimous consent scheme results.

We will only consider a special case of the second type of scheme, namely, unanimous consent or k-out-of-k shared control schemes, which we will find useful later in a
protocol to make it possible for a group of mutually distrustful participants—who don't

trust anyone else either—to establish a shared secret or shared control scheme that they must (logically) trust. Given k, the scheme is implemented in the k-dimensional space $S = PG(k, q)$. The private pieces of information are k randomly chosen hyperplanes, that is, $(k - 1)$-dimensional subspaces of S, which almost certainly (with increasing q) intersect in only a single point p. If necessary, a particular choice for the k hyperplanes could be tested to verify that their intersection does not contain a line, but this is probably unnecessary. This is the natural generalization of the construction in Fig. 4 with only a pair of lines in the plane, that is, $\ell = 2$, and of Fig. 5 with a triple of planes in 3-space, that is, $\ell = 3$, etc. Although shared secret schemes of this type can be made for any ℓ, $k \le \ell \le (q^{k-1})/(q - 1)$, we will not make use of them for values of $\ell > k$.

For much of this chapter we will be concerned only with shared secret schemes of the first type where S is a projective space $PG(n, q)$ over some finite field $GF(q)$. However, our examples will generally be constructed in affine spaces $AG(n, q)$, because of the closer analogy to the more familiar Euclidean spaces. Occasionally this will require a note of explanation to deal with special cases, but this should cause no confusion.

Our constructions make essential use of a simple result in projective geometry known as the rank formula:

$$r(U) + r(V) = r(U \cap V) + r(U \cup V) \tag{10}$$

true for all subspaces U and V of the containing space $S = PG(n, q)$. $r(x)$ denotes the rank of the subspace x. Note that $r(x) = \dim(x) + 1$, and that the empty subspace has rank 0, and consequently dimension -1. It is easy to see that Eq. (10) does not hold in affine spaces. In $AG(3, q)$ there are pairs of parallel lines, that is, pairs of lines that do not intersect, but whose union is only a plane: $r(U \cap V) = 0$ and $r(U \cup V) = 3$ so that

$$r(U) + r(V) = 2 + 2 \ne 0 + 3 = r(U \cap V) + r(U \cup V)$$

From the standpoint of geometric intuition, Eq. (10) is more accessible if rank is replaced by dimension:

$$\dim(U) + \dim(V) = \dim(U \cap V) + \dim(U \cup V) \tag{11}$$

Constructions where $V_i \cup V_d = S$ and $V_i \cap V_d = p$, p a point, make it possible to construct perfect (k, ℓ)-threshold scheme for any choice of k and ℓ. For a given q the dimension and hence the cardinality of the domain space V_d is chosen so that the probability of any collusion being able to (randomly) guess the secret point p, even though V_d is publicly known, is small enough to satisfy the security requirements. Let $\dim(V_d) = m$. The containing space, S, is then chosen to be n-dimensional, where $n = m + k - 1$ and V_i is a $k - 1$ dimensional subspace chosen so that $\dim(V_i \cap V_d) = 0$, etc. The private pieces of information in such a construction are points in V_i distinct from p, such that every subset of k of them span V_i and no subspace spanned by fewer than k of them lies on p. This guarantees that every subset of k (or more) participants can recover V_i and hence p, but that every point in V_d will remain an equally likely candidate to any collusion involving fewer than k participants. We will discuss this construction again in Section 5, which is devoted to implementing shared control schemes: It is described here so that we will have a ready example of (k, ℓ)-threshold schemes to use in Sections 3 and 4. The following four examples illustrate this result.

A pair of distinct lines, λ_1 and λ_2, in a plane define (span) the plane by their union and a point, p, by their intersection (Fig. 7):

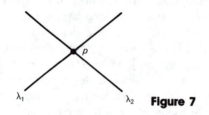

λ_1 λ_2 **Figure 7**

Similarly a plane, π, and a line, λ, not in the plane in a 3-space define (span) the 3-space by their union and a point, p, by their intersection (Fig. 8):

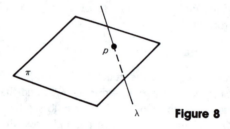

λ **Figure 8**

In the plane or in a three-dimensional space, these are the only possible configurations of this type. The following pair of examples in a four-dimensional space, **S**, indicates the usefulness of Eq. (11). In a four-dimensional space any pair of planes π_1, and π_2, that do not lie in a common three-dimensional subspace, intersect in a single point. Since they do not lie in a common three-dimensional subspace the dim $(U \cup V) = 4$, so that we have

$$\dim(U \cap V) = \dim(U \cup V) - \dim(U) - \dim(U) = 4 - 2 - 2 = 0$$

hence, $U \cap V = p$, p a point. We will represent this four-dimensional construction with Fig. 9:

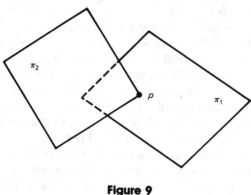

Figure 9

Similarly, a three-dimensional subspace, ω, and a line, λ, which span **S** must have a point, p, in common as well (Fig. 10):

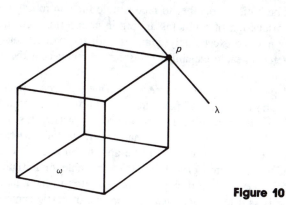

Figure 10

In Section 4 we will continue this discussion of geometric results essential to shared secret schemes, however, for the moment the four configurations in Figs. 7–10 suffice to illustrate everything we wish to show about threshold schemes. We have already seen how a 2-out-of-ℓ scheme can be implemented using the configuration in Fig. 7 by choosing ℓ points on one of the lines (none of which are the point p) as the private pieces of information, etc. This is the only type of shared control scheme that can be implemented using this configuration. The situation is different, however, for the configuration shown in Fig. 8. A 2-out-of-ℓ scheme is the only possibility if the private pieces of information are taken to be points on the line λ, however, several quite different concurrence schemes are possible if the private pieces of information are points in the plane π. For example, if the points are chosen so that they are distinct from p and no three of them are collinear and no pair of them are collinear with p, then any subset of three of the points will define π and hence p, but no subset of one or two of them will provide any information about p at all. This is one implementation for a perfect 3-out-of-ℓ shared secret scheme. p in this case is an unknown point on the (assumed publicly known) line λ. An identical set of remarks holds for the configuration in Fig. 9 except that in this case the secret can be any point in a (assumed publicly known) plane instead of on a line. The index point, p, since it must satisfy the linear constraints to lie in the plane V_d has an entropy of $H(p) = 2\log_2(q)$, that is, the same as that of a randomly chosen point in $GF(q) \times GF(q)$. The entropy-preserving transformations on p would therefore effectively map $p \in V_d$ onto $GF(q) \times GF(q)$. We will return to these configurations in much greater detail, however, these brief remarks should make clear the general nature of the constructions we will use for shared secret schemes.

3 CONSTRUCTING CONCURRENCE SCHEMES

In the application of shared control schemes, normally the first thing to be specified is the identification or description of the set of participants in the scheme and of those subsets (concurrences) of them who are authorized to initiate the controlled action. This could be as simple as requiring the concurrence of both participants where only two are involved as is the case for opening and relocking of safety deposit boxes where both the box holder's key and the banking institution's key are required or in the U.S. policy of two-man control of the information used to control nuclear weapons where two pieces of information must be brought together to reconstitute the controlling information. Or, as we will see in a discussion of problems arising in connection with treaty-controlled actions, the schemes may be exceedingly complex with the participants having greatly different and noninterchangeable capabilities insofar as recovering the secret information is concerned.

Given a set of participants, a *concurrence scheme* (or *access structure* [3,19,20]) is a specification of those subsets of participants who are to be able to initiate the controlled action. A *collusion* is defined to be any set of participants or of participants and outsiders (to the shared control scheme) that doesn't include at least one set from the concurrence scheme. No collusion should be able to initiate the controlled action. We remind the reader that if the probability of being able to initiate the controlled action is the same for every collusion of participants as it is for an outsider, the scheme is perfect. A scheme is said to be *monotone* if every set of participants that includes at least one subset from the concurrence scheme is also able to initiate the controlled action. While this is certainly a natural condition to impose on shared control schemes, that is, if A and B together can initiate an action, then A, B, and C together should also be able to do so, nonmonotone schemes have been considered [8]. We will only consider monotone schemes here, however.

In the most extreme case, a concurrence scheme could take the form of a tabulation of all of the subsets of participants who are supposed to be able to recover the secret or initiate the controlled action. For a monotone scheme there is no reason to list sets that properly contain one or more sets also in the concurrence scheme; in other words, *ABC* would not need to be listed if any one of the sets *AB* or *AC* or *BC* is in the scheme. In many cases (covering most applications encountered thus far), there are much more succinct descriptions of the concurrence scheme than a listing of the subsets in it. For example, unanimous consent schemes where all of the participants must concur for the secret to be reconstituted or the controlled action initiated are described simply as unanimous consent or k-out-of-k shared control schemes; the second terminology has the advantage of identifying the number of participants involved. The identification of a concurrence scheme as a k-out-of-ℓ shared control or (k, ℓ) threshold scheme is much more compact than the tabulation of the $\binom{\ell}{k}$ k-sets that make up a (set containment) basis for the scheme. All of the concurrence schemes we will be concerned with here have such succinct descriptions as an artifact of their origin in applications where the specification of the desired control is similarly succinct.

The simplest class of schemes (to describe) are k-out-of-k unanimous consent schemes for which two implementations have already been given: in one of which k linearly independent points are chosen to span a $(k-1)$-dimensional subspace, V_i, of **S** that points to the secret point p in another subspace V_d, $\dim(\mathbf{S}) \geq k$, and in the other, k independent hyperplanes are chosen in **S**, $\dim(\mathbf{S}) = k$, whose intersection is the point p. There is another way to construct a unanimous consent scheme which is already widely used, and which is a useful building block in the construction of other, more complex, concurrence schemes.

To be consistent with the discussion of the other two examples of implementations of unanimous consent schemes, we will also examine a 2-out-of-2 concurrence scheme in this case as well. If it is desired that a specific two persons (controllers) must concur for a vault to be opened (or a weapon enabled or a missile fired) then each of these controllers could—during the initialization of the locking mechanism in the vault door—enter a randomly chosen k-digit number whose value is kept secret by the controller who chose it. The mod 10 sum of these two private and secret k-digit numbers would be the secret k-digit combination needed to open the vault. The subsequent entry of any pair of k-digit numbers whose mod 10 sum is equal to the secret combination determined by the two controllers would open the vault door. Clearly the probability that an outsider or either of the two insiders (controllers) alone being able to open the vault on the first try would be 10^{-k}. In this control scheme, two controllers are involved, and both must (in probability) concur in order for the controlled event to be initiated. This approach has in fact been used by the United States to protect critical shared command and control information.

In general, in a space whose cardinality is suitable for the concealment of the secret, that is, in which the probability of selecting a randomly chosen secret (point) in a subsequent random drawing provides an acceptable level of security for the controlled action, each participant chooses a random point as his contribution to the unanimous consent control scheme. If q is large enough to satisfy the security requirements there is no reason to not choose **S** to be one-dimensional (Fig. 11):

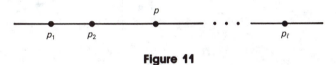

Figure 11

During the initialization of the mechanism that implements the shared control, each of the participants secretly enters the point he has selected and the sum (field, modular, exclusive-or, etc.) of all of the points becomes the jointly defined secret value. Since this procedure is an obvious generalization of Vernam encryption, the secret is unconditionally secure from discovery (or recovery) by any concurrence of fewer than all of the participants so long as the sum operation is an entropy-preserving mapping. To see that this is true, assume the worst-case scenario in which all of the participants but one conspire in an attempt to initiate the controlled action without the cooperation of the single missing participant. They can calculate the point that is the sum of all of their contributions, however, every point in the space is equally likely to be the secret point depending on the point chosen by the missing participant, since the sum operation is entropy-preserving. In other words, even in this worst-case scenario, the best that the would-be cheaters can do is to "guess" at the value of the secret using a uniform probability distribution on the points in the space. Clearly, this is the best that can be achieved with any shared control scheme.

There has been considerable interest in characterizing various classes of concurrence schemes. Brickell [18] and Brickell and Davenport [19] have discussed the characterization of ideal and strongly ideal secret sharing schemes while Benaloh and Leichter [3] have shown that there exist monotone secret sharing schemes that are not ideal. We shall restrict attention here, however, to only two general types of concurrence schemes—multilevel and multipart (or compartmented) schemes—and to arbitrary combinations of these types, since these are adequate to satisfy all applications known to the author.

A multilevel scheme is characterized by the participants having differing capabilities to initiate the controlled action. We described one such scheme earlier wherein
either two vice-presidents or three senior tellers at a bank could authenticate an EFT. If
there is no requirement that a vice-president also be able to act in the stead of a senior
teller, it would be trivial to set up a shared control scheme with this concurrence. Separate and independent 2-out-of-ℓ_2 and 3-out-of-ℓ_3 shared control schemes could be set
up for the two classes of participants, such that both schemes revealed the same point p.
Assuming that q is sufficiently large that the guessing probability of $1/(q + 1)$ for
choosing p in a random drawing of the points on a line is adequately secure, V_d can be
taken to be a line. The dimension of the two indicator varieties must be $k - 1$, that is,
1 for Class I and 2 for Class II. Schematically, this may be shown as in Fig. 12:

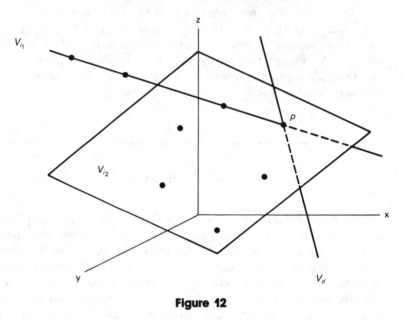

Figure 12

where the line V_{i_1} does not lie in the plane V_{i_2}, but both V_{i_1} and V_{i_2} intersect the line V_d
in the point p. The private pieces of information for the Class I participants consist of
points on V_{i_1} distinct from p, while the private pieces of information for the Class II
participants consist of points in the plane V_{i_2}, also distinct from p, but chosen so that no
three of them are collinear and no pair of them are collinear with p.

In Section 5, which is devoted to the implementation of shared control schemes,
we will discuss the more difficult questions concerning how many participants there can
be in each of the classes; that is, the bounds on ℓ_2 and ℓ_3, but for the moment we are
only concerned with the essential notions underlying multilevel shared control schemes.
In the implementation of Fig. 12, the classes are clearly independent, that is, no combination of a point on V_{i_1} and a pair of points in V_{i_2} can define a plane that intersects V_d
at p since this would require that the pair of points in V_{i_2} be collinear with p, contrary
to the way the private points were chosen.

If, however, we want it to be possible for any member of Class I in cooperation
with any pair of members from Class II to be able to recover p, which is the concur-

rence scheme described earlier, then V_{i_1} must be a line in the plane V_{i_2} and the choice of the private points for the Class II participants must satisfy two additional constraints. No Class II point can be on the line V_{i_1} and no pair of them can be collinear with a point assigned as a private piece of information to any Class I participant. This latter requirement is essential; otherwise the private pieces of information held by this set of three participants (one from Class I and two from Class II) would only span a line skew to V_d and hence would be unable to recover p. We shall see below that all of these constraints are easy to satisfy. Schematically, the construction for this two-level shared control scheme in which the members of the more capable class can also act as members of the less capable class is of the form (Fig. 13):

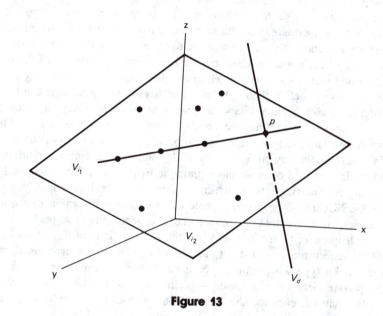

Figure 13

where the private points in the indicators V_{i_1} and V_{i_2} have been chosen subject to the constraints just described. Although we won't discuss the details of how to construct other multilevel schemes at this time, these two examples should at least make plausible the assertion that the general construction for k-out-of-ℓ shared control schemes can be extended in an obvious—and feasible—way to realize general multilevel schemes.

In all of the shared control schemes described thus far, the participants, even though they may have differing capabilities to recover the secret, are "on the same team." It is easy to conceive of situations, however, in which it is desirable that some action require a preselected level of concurrence by two or more parties for the action to be executed where the interests of the different parties may be quite divergent. For example, a treaty might require that two individuals out of a Russian control team and two individuals out of a U.S. team agree that a treaty-controlled action is to be taken before it could be initiated. What is different about such a compartmented scheme from either the simple k-out-of-ℓ schemes or the multilevel schemes, is that no matter how many of the participants of one nationality (compartment or part) concur, the action is to be inhibited unless the prescribed number of the other nationality also concur.

Clearly, there is nothing special about partitioning the private information into only two parts (compartments). The specific application will determine how many parts are needed to effect the type of concurrence desired.

Again we will use the smallest example that illustrates the essential notions to a multipart or compartmented scheme: Two parties (nations) are involved and they must both concur in order for a controlled action to be initiated. If each nation has only a single controller who is empowered to act as their representative, in other words, to act solely on his own to contribute his nation's input to the shared control scheme, a simple 2-out-of-2 unanimous consent scheme will suffice. We have already shown three ways to realize such unanimous consent schemes in the configurations in Figs. 4, 5, and 11: using only two private points on the line V_i in the first case, two private lines lying on the point p in the second, and two private points in S whose "sum" is p in the third. The fact that there are two nations or two parties involved is immaterial. There are two participants and both must concur for the controlled action to be initiated. If, however, each nation has not a controller but a control team, and in view of the importance of the controlled action each nation requires some degree of concurrence among its control team members before their national input to the control scheme can be made, the situation is quite different. Assume that each control team, of, say, the United States and the USSR, consists of four members and the concurrence of at least two of them is required before their national commitment can be made. In this case a collusion of as many as five participants, say all four of the U.S. controllers in collusion with one of the USSR controllers, should be unable to initiate the treaty-controlled action while a concurrence of only four participants consisting of any two of the U.S. controllers with any two of the USSR controllers should be able to do so. We will exhibit four quite different implementations of shared control schemes that realize this two-part concurrence scheme.

It is easier to see how to adapt the constructions in Figs. 5 and 11 to realize the desired concurrence than it is to see how to adapt the construction shown in Fig. 4. In the Blakley-type scheme in Fig. 5, instead of each national controller being given—as his private piece of information—the line on p that is his nation's input to the shared control scheme, each member of a national control team would be given a point on the national line as his private piece of information (Fig. 14).

If one of the participants (in either team) were given the point p as his private piece of information, he would have no way of knowing that this was the shared (se-

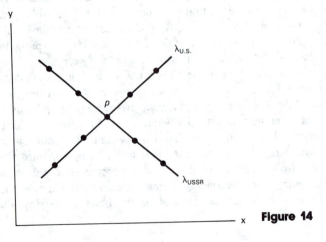

Figure 14

cret) point. This would be equally true for all collusions that did not include a pair of members from the other team. However, a collusion involving the participant who knows p and a pair of members of the other team could infer that p is the secret, since p lies on the line their points determine and hence must be the intersection of the two national lines. Therefore, only q of the $q + 1$ points on each line are available for use as private pieces of information: $L = q$ for each control team in this case. The resulting scheme is ideal, but not perfect, since any collusion including a pair of members from one of the national teams can restrict p to be a point on a line that they can unilaterally determine. It does have the desired property, though, that no collusion can recover the secret p, with a security equal to the probability of guessing an element in $GF(q)$, that is, $1/q$. We remind the reader that in this implementation, there are no indicator and domain varieties even though the secret point is defined by the intersection of a pair of subspaces.

The 2-out-of-2 unanimous consent scheme based on the construction shown in Fig. 11 is easy to adapt to an implementation of the desired two-party concurrence scheme. The line, λ, on which the 2-out-of-2 scheme is implemented, is embedded in the plane and the national lines, $\lambda_{U.S.}$ and λ_{USSR}, on which the 2-out-of-4 shared national control schemes are implemented, are used to indicate the two input points $p_{U.S.}$ and p_{USSR} on λ and hence to define p (Fig. 15). It should be pointed out that although p is an arbitrary point on the line λ, that λ is not a domain variety, V_d, in the sense we have defined V_d above, since the private pieces of information are not used to span an indicator variety that points to p.

Neither of the points $p_{U.S.}$ and p_{USSR} can be used as one of the private pieces of information. If both were used, then the two individuals who were assigned the points $p_{U.S.}$ and p_{USSR} could recover p without needing the concurrence of any of the other participants. If only one of the points, say $p_{U.S.}$ was assigned, then this member of the U.S. team in cooperation with any two members of the USSR control team could recover p, in violation of the concurrence requirements. Therefore, only q points on each of the national lines are available for use as the private piece of information (in $PG(2, q)$). This scheme is perfect—unlike the scheme in Fig. 14—since every point on the line λ is an equally likely candidate to be p to either an outsider or to any collusion of insiders, and realizable in a two-dimensional space: $S = PG(2, q)$ or $S = AG(2, q)$.

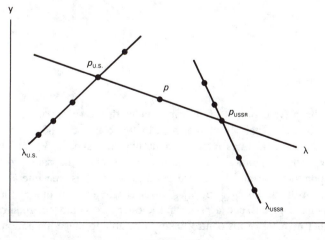

Figure 15

If the Shamir type of shared secret scheme is modified to realize the concurrence desired in this case, there are (at least) two possible constructions. With no loss of generality, we restrict the domain, V_d, to be a line in both constructions so as to keep the security the same in all of the constructions. A 2-out-of-ℓ scheme, $\ell > 2$, can only define a line as the shared subspace, hence the only thing that the national control teams can do (which others cannot) is to reconstruct the shared national lines $\lambda_{U.S.}$ and λ_{USSR}. One can either use these two lines (linear subspaces) as indicators to point to a pair of points in some other linear space which then determines a line in that space that can be used to point to the secret point p, or else take the linear subspace spanned by the two lines directly as an indicator, V_i, to point to p.

The first approach is intuitively the easier and also the one that lends itself most readily to generalization. The construction in Fig. 4 shows one way to implement a Shamir-type 2-out-of-2 unanimous consent scheme in a plane. We must somehow cause the two lines determined by the U.S. and the USSR control teams to in turn define a line, V_i, that indicates the point p in V_d. This must be done in such a way that all of the concurrence and security requirements are satisfied at the same time. One way to do this would be to make S be three-dimensional and to embed the plane, π, in which V_i and V_d occur in S (Fig. 16):

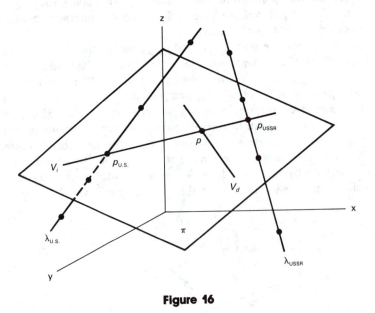

Figure 16

The private pieces of information in this case are points on the national lines, $\lambda_{U.S.}$ and λ_{USSR}, neither of which is in the plane π. The points, $p_{U.S.}$ and p_{USSR}, at which the national lines intersect π define the indicator V_i, and hence its intersection with V_d, in the point p. Any pair of points from the same control team define the corresponding line and hence the point at which this line intersects π; $p_{U.S.}$ or p_{USSR}. Neither of the points $p_{U.S.}$ or p_{USSR} in π can be used as one of the private points since a participant given such a point and knowing π would not require the concurrence of any other member of his control team to be able to make his nation's contribution to the overall concurrence scheme. Therefore, as in the previous construction, q out of the $q + 1$ points on each of the national lines can be used as private pieces of information. Clearly, this scheme realizes the desired concurrence and security objectives with a security equal to the probability of guessing an element in $GF(q)$, that is, $1/q$. This scheme is perfect, since no collusion has any better chance of determining p than does an outsider—but not ideal.

Another way to implement the desired concurrence in this case would be to take **S** to be a four-dimensional space, and to let V_i and V_d be three-dimensional and one-dimensional subspaces of **S**, respectively, chosen so that; $\dim(V_i \cup V_d) = 4$ and $\dim(V_i \cap V_d) = 0$, that is, so that V_i and V_d intersect only in a point, p. Now if $\lambda_{\text{U.S.}}$ and λ_{USSR} are any pair of skew lines in V_i neither of which lies on p, we have the geometric basis for constructing a concurrence scheme with the desired properties (Fig. 17):

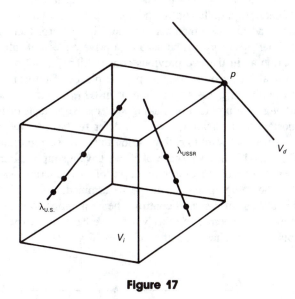

Figure 17

The private pieces of information are points on the two national lines $\lambda_{\text{U.S.}}$ and λ_{USSR}, chosen to satisfy a condition which we will describe momentarily. Clearly by the rank or dimension formula, the pair of skew lines span the three-dimensional subspace V_i, so that any subset of participants that includes at least a pair of members from each national team will be able to reconstruct V_i from the points they jointly know, and hence to recover p. The more difficult problem is to choose the private points in such a way as to insure that no subset of participants not satisfying this concurrence condition can do so.

It is an easy to see result in enumerative geometry that given two skew lines, λ_1 and λ_2, in a three-dimensional space and any point, p, not on either of the lines, that there is a single line lying on p and on both of the skew lines. p and λ_1 define a plane, π_1. Similarly p and λ_2 define a plane π_2. These planes are distinct since λ_1 and λ_2 are skew, hence they intersect in a line, λ^* which—by construction—lies on p and intersects both λ_1 and λ_2, say in points p_1 and p_2, respectively. If both of these points were

used as private pieces of information, then since they determine the line λ^* that intersects V_d at p, the two participants who were assigned p_1 and p_2 (one from the U.S.
team and one from the USSR team) would be able to recover the secret without anyone
else's assistance. Therefore it is not possible to use both of these points as private
pieces of information. We will now show that it is not possible to use either of them.

Assume that p_1 on $\lambda_{U.S.}$ has been issued as a private piece of information to a
member of the U.S. control team. Consider what can be inferred by this participant in
collusion with at least two members of the USSR control team. They can determine the
line λ_{USSR} that is skew to the line V_d, which is publicly known. λ_{USSR} and V_d together
span a three-dimensional subspace which must contain the line λ^* since it contains two
points on it; p_2 and p. By the result cited above from enumerative geometry, p_1, which
is a point in the 3-space defined by the pair of skew lines λ_{USSR} and V_d and on neither
of the lines, lies on a unique line intersecting both of the lines λ_{USSR} and V_d. Where it
intersects the line, λ_{USSR} is of no interest, however, it is known a priori that it intersects
V_d at p. Therefore a collusion of the participant who knows p_1 as his private piece of
information and any two members of the other control team could recover p. Consequently neither p_1 nor p_2 can be used as a private piece of information and we have the
same situation as in the two previous constructions; q out of the $q + 1$ points on each
of the national lines can be used as private pieces of information with a security equal
to the probability of guessing an element in $GF(q)$, that is, $1/q$.

Although the problem of making systems be perfect, or ideal, or extrinsic, etc. is
very important to applications, we will defer further discussion of these topics for the
moment. Instead we want to call attention to the most important feature of the construction in Figs. 15–17. In a subspace of $\mathbf{S}$, we have a simple 2-out-of-2 unanimous consent
scheme. Such a scheme requires the input of two functionally related points, $p_{U.S.}$ and
p_{USSR}, for the secret point, p, to be determined. Each of the national concurrence
schemes is a 2-out-of-4 shared control scheme pointing to the corresponding input point
for the 2-out-of-2 scheme. In other words, we have constructed a lattice, organized as a
tree in this case, of shared control schemes (Fig. 18):

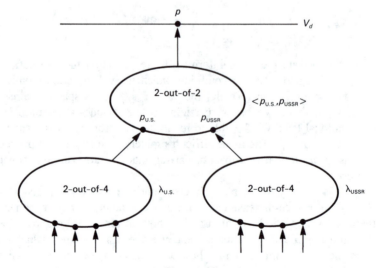

Figure 18

The notation $<p_{U.S.}, p_{USSR}>$ indicates the subspace of **S**—in this case a line—spanned by the objects (points) inside the brackets. Since the inputs to all of the shared control schemes we are considering are points in some space **S** or in subspaces of **S**, and the outputs are also points in **S**, we can construct shared control schemes of almost arbitrary complexity in this way. Let T be an arbitrary inverted tree, that is, with the edges all directed toward the root of the tree and let each node be a shared control scheme (k-out-of-k unanimous consent, k-out-of-ℓ threshold, k_i-out-of-ℓ_i multilevel—with or without interlevel capabilities—or multipart) with the input points for each node being defined by the shared control schemes on the nodes adjacent to it whose edges are directed into it and its output being an input point in the control scheme adjacent to it and above it in the inverted tree. The root control scheme defines the final indicator that points to the secret point, p, in V_d.

There are undoubtedly monotone shared control schemes not realizable in this way, although we know of no such example, however, all applications we have encountered can be satisfied with concurrence schemes that can be implemented as a tree of simple shared control schemes.

4 THE GEOMETRY OF SHARED SECRET SCHEMES

There are nongeometric constructions for shared secret schemes, most of which also have natural geometric interpretations. Asmuth and Bloom [2] generalized Vernam encryption to a modular cryptographic key sharing scheme. Several researchers have based secret sharing schemes on various properties of combinatorial designs: block designs [5,52], tactical configurations [33], perpendicular arrays [58,60], etc. Others have noted that the error-correcting property of error detecting and correcting codes can be used to recover the secret (codeword) from a perturbed version of it that can be formed by some subset of the participants, each of whom has partial information about the codeword [16, 47, 67]. Almost all of these constructions are examples of matroids, so that a few researchers have generalized shared secret schemes to their most abstract setting; matroids [18,19,61] or to general algebraic or logical systems [4,39,44,48,62,63]. In keeping with the development thus far in this chapter, we will only treat geometric schemes here, and, except for one application to make it possible for a group of participants who don't trust each other, nor any outside party, to set up a shared control scheme that they must (logically) trust, we will consider only the even more restricted class of extrinsic (ideal) systems in which the private pieces of information and the shared secret pieces of information are all points in a single space **S**.

The reader is already familiar with our preferred model for shared secret schemes: Any concurrence of the participants should possess a set of points that span a subspace (an algebraic variety in the more general setting) that intersects another (publicly known) subspace (algebraic variety) at the secret point p. What we plan to do in this section is to show how several plausible and/or real concurrence schemes can be realized. First, however, we state explicitly an important result that was only implicit in an earlier discussion. In the configuration shown in Fig. 7, **S** is two-dimensional and the two varieties, λ_1 and λ_2, are both lines, so that it is immaterial which of them is taken to be the subspace V_i and which to be V_d. In the configuration shown in Fig. 8 which we repeat here for the reader's convenience, **S** is three-dimensional and the subspaces are a plane, π, and a line, λ.

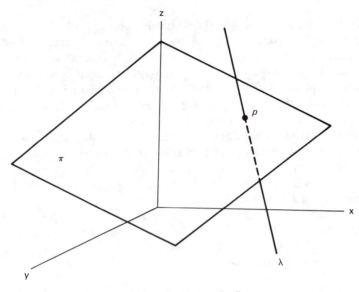

Figure 8 (repeated)

In this case, it makes a great deal of difference—from the standpoint of the concurrence schemes that can be realized—whether the indicator V_i is taken to be the line or the plane. Either choice is possible. This example illustrates an important duality principle: that for any geometric configuration consisting of two subspaces U and V of a projective space $\mathbf{S}$ in which $\dim(U \cup V) = \dim(\mathbf{S})$, $\dim(U \cap V) = 0$ and $\dim(U) \neq \dim(V)$, there exist families of dual shared secret schemes [55] depending on whether U or V is chosen to be the indicator V_i. If $V_i = \lambda$ in the present example, then the only concurrence possible would be a 2-out-of-ℓ shared control, $\ell \geq 2$. If $V_i = \pi$, however, at least four distinct concurrence schemes are possible.

If the private points (in π) are chosen to be different from p and such that no three of them are collinear and no pair lie on a line through p, then we have a perfect 3-out-of-ℓ scheme (Fig. 19). Although an explanation is probably unnecessary, the reason the points in π must satisfy these conditions is, first, if any pair of them is collinear with p, then, since the line they define must be in π, its intersection with V_d would have to be p—and the pair of participants knowing the scheme, but not π, would know this. Second, if any three of the points are collinear, then since the line they define does not intersect V_d, that is, is skew to V_d, these three participants would jointly be unable to define π and hence to recover p.

We remarked above that since the number of points on a line (in $PG(n, q)$) is $q + 1$, the order, q, of the ground field $GF(q)$ must be chosen sufficiently large that the probability of a random choice of points on V_d identifying p, namely, $1/(q + 1)$, will be small enough to satisfy the security requirements for the shared control scheme. There is another constraint, which is almost always trivially satisfied if the security constraint has been met: this is the number of participants that must be accommodated in the scheme. In the present case (the scheme shown in Fig. 19), the maximum number of points that can be chosen in a plane in $PG(3, q)$ such that no three of them are collinear is $q + 1$ if the characteristic of $GF(q)$ is odd, that is, if q is a power of an odd prime, and $q + 2$ if $q = 2^\alpha$ for some positive integer α.

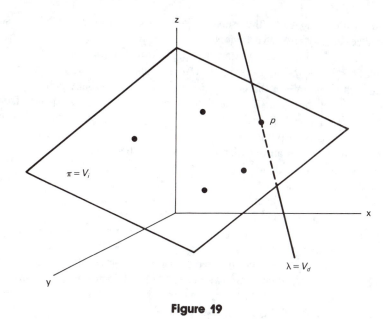

Figure 19

In the first case (q odd), such a maximal set of $q + 1$ points define a conic, Ω in the plane and the second case (q even), a conic, Ω, and its nucleus, η; that is, the point of common intersection of the $q + 1$ tangents to the conic. We will only be concerned with fields of odd characteristic. In this case, it is easy to determine a (sharp) upper bound, L, for the maximum number of participants such a 3-out-of-ℓ shared control scheme can accommodate. If p is an interior point of Ω, that is, if p does not lie on any tangent to Ω, then $L = (q + 1)/2$. This is obvious since any line on p and one point of Ω must also lie on another point of Ω; otherwise it would be a tangent to Ω, contradicting the assumption that p is an interior point. One and only one of this pair of points can be used as a private point. If p is an exterior point of Ω, then two tangents to Ω lie on p. The remaining $q - 1$ points can be paired by the same argument as before and one member of each pair used as a private point as well as the two points of tangency, or $(q + 3)/2$ points in all: $L = (q + 3)/2$. Because in applications, one normally wants high security which means large values of q and relatively few participants, it is unlikely that the order of the ground field will ever be dictated by the number of participants.

Continuing with the discussion of the concurrence schemes that can be realized using the configuration shown in Fig. 8, $\pi = V_i$ is also spanned by a line, λ^*, in π and any point, p', in π not on λ^*. Using this fact, we can realize a (plausible) two-part concurrence scheme. Assume that the United States wished to set up a shared control scheme with its 15 North American Treaty Organization (NATO) allies to control the first use of nuclear weapons in NATO such that at least two of the allies (other than the United States) had to concur before nuclear weapons could be used, but so that the United States retained an absolute veto power over this decision. This would mean that even if all fifteen of the European parties wanted to initiate the use of the nuclear weapons committed to NATO they would be unable to do so without U.S. concurrence, but it also means that the United States or the United States and any single ally would be unable to do so by themselves. The two-part concurrence scheme shown in Fig. 20 realizes this type of shared control.

λ^* is a line in π that does not pass through the point p, that is, λ^* is skew to the line $\lambda = V_d$ in S. $p_{U.S.}$ is a point in the plane distinct from p and not on any line through p and one of the private points on λ^*. Clearly, any pair of the points on λ^* will determine it, but since it is skew to V_d every point on V_d is equally likely to be p to any collusion of non-U.S. participants. Since $p_{U.S.} \neq p$, and the ℓ lines defined by $p_{U.S.}$ with the ℓ private points on λ^* are all skew to $\lambda = V_d$, all collusions are equally unable to recover p as is an outsider; therefore the scheme is perfect. The point $p_{U.S.}$ and the line λ^* determine π and hence p, so that this configuration (Fig. 20) does realize the desired two-part shared control scheme. In this case, the line determined by the pair of points p and $p_{U.S.}$ intersects λ^* in a point that cannot be used as one of the private pieces of information. Therefore $L = q$ in this case, with $q + 1$ points in all in π being usable as private pieces of information. We mention this because we will encounter this same bound in other cases also.

In the previous section, we discussed a two-level shared control scheme, shown in Fig. 13, in which there were two classes of participants and the authorized concurrences were subsets that included either two members of Class I or three members of Class II or any member of Class I and two members of Class II. This scheme is also one of those that can be realized using the configuration shown in Fig. 8. We will not repeat the analysis of the security of the scheme here, but will briefly derive the bounds on the number(s) of participants such a scheme can support.

The first condition on the points chosen for the private pieces of information for the members of Class II is that no three of them can be collinear. As we have already remarked, for q odd this means that at most the $q + 1$ points on some conic, Ω, out of the $q^2 + q + 1$ points in π can be used (Fig. 21).

Since the line $\lambda = V_{i_1}$ must lie on p, and p cannot be used as a private piece of information, otherwise its holder, knowing V_d, would know that he had the secret, at

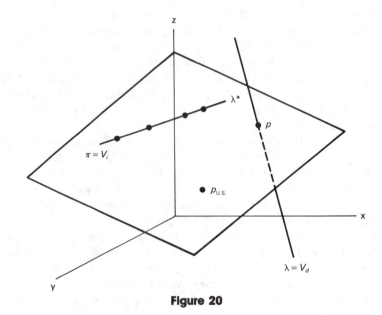

Figure 20

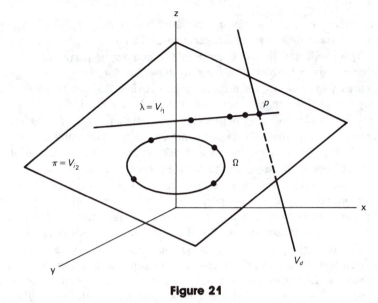

Figure 21

most q of the $q + 1$ points on λ can be used. The ℓ_3 Class III points on Ω determine $\binom{\ell_3}{2} = 0(\ell_3^2)$ distinct lines (secants of Ω) each of which intersects λ in a point that cannot be used as a private piece of information for the Class I participants—otherwise these three participants who form an authorized concurrence would only be able to determine a line skew to V_d which would leave all points on V_d equally likely candidates to be p, contrary to the concurrence requirements for the scheme. It is certainly conceivable that for $\ell_3 > \sqrt{q}$, all of the points on λ could be denied for use and Class I would be vacuous. In fact, if either λ or the set of ℓ_3 points on Ω are chosen randomly and ℓ_3 is only slightly larger than $\sqrt{q}$, this will almost certainly be the case. The surprising result is that by choosing the ℓ_3 points properly, it is possible to accommodate as many as q participants even when $\ell_3 = (q + 1)/2$. It is inappropriate to a survey report such as this to go into the mathematical detail needed to fully appreciate this result. The interested reader is referred to the mathematical literature [53,R2] for these details. We will describe only one of the several results of this sort.

For q odd, given any conic, Ω, in the projective plane $\pi = PG(2, q)$, and an exterior line, λ, to Ω, that is, a line that has no point in common with Ω, then λ induces a partitioning of the $q + 1$ points on Ω into sets of k points each, for every k that divides $q + 1$, such that the $\binom{k}{2}$ secants to Ω that they determine intersect λ in only k points! For example, if $q = 23$, then an exterior line λ partitions the 24 points on Ω into six sets of 4 points, four sets of 6 points, three sets of 8 points, and two sets of 12 points so that for any one of these sets, the 6, 15, 28, or 66 secants they define will intersect λ in only, 4, 6, 8, or 12 points, respectively. If the ℓ_3 points for the Class II participants are chosen in this way, then the remaining 20, 18, 16, or 12 points on λ can all be used either as the secret or private pieces of information for the Class I participants. The bottom line for the result just cited is that q out of the $q^2 + q + 1$ points in the plane can be used as private pieces of information in those cases where ℓ_3 is chosen to be a divisor of $q + 1$. A similar result holds, with different geometric constraints, for the divisors of $q - 1$ and of q. The latter is only meaningful when $q = p^\alpha$, where p is a prime and α is a positive integer greater than 1.

We will not discuss bounds on the number of participants for any of the other schemes, since it should be clear by now that for the values of q needed to satisfy security requirements, the number of participants that can be accommodated will probably be much larger than any application will ever need.

It is worth pointing out that there is another choice for the private pieces of information for a 2-out-of-ℓ shared control scheme which can be based on the configuration in Fig. 22. Any pair of distinct lines in π span the plane, so we could use any ℓ distinct lines—not on p—as the private pieces of information.

Since a plane is two-dimensional in either points or lines, that is, the same amount of information is needed to specify an arbitrary one of either type, the scheme in Fig. 22 is equivalent to the 2-out-of-ℓ scheme based on points on a line (in π) as was done in the scheme shown in Fig. 13. The only possible advantage is that the maximum number of participants, L, can be greater in this case. One of the points on the line defined by the pairs of private points in the scheme shown in Fig. 13 must be the secret point p, so $L = q$ in that case. On the other hand, the number of lines in a projective plane is $q^2 + q + 1$, while the number that lies on the point p is $q + 1$. Therefore there are q^2 lines in the plane that do not lie on p and $L = q^2$ in this case. As we have already remarked, though, q is dictated by security constraints that cause it to be so large that it is inconceivable that an application would ever require $L > q$. This case has been included only for completeness. We will not consider any other schemes in this section in which the private pieces of information are not points in $\mathbf{S}$.

The preceding discussion should give some indication of the rich diversity of shared control schemes that can be implemented from even a small geometric configuration of two subspaces that span $\mathbf{S}$ and have only a single point in common.

The balance of this section will be devoted to a discussion of a bare minimum of geometric results needed to analyze or to appreciate the analysis of the security of

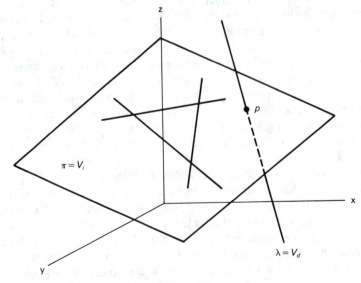

Figure 22

shared secret schemes. It can safely be skipped over if the reader is either already familiar with the geometry of finite spaces (affine and projective) or if he is only concerned with the essential notions of shared secret schemes and not with the details of their implementation.

In the discussion of the simple 2-out-of-ℓ shared secret scheme shown in Fig. 4 (which is repeated here), there were two (interrelated) points to the discussion of the scheme's security. The first had to do with showing that the scheme was perfect, that is, equally secure to all collusions—an outsider or a single insider in this case. The other had to do with the fact that the information content (in the information theoretic sense) of the private pieces of information did not all have to be communicated and protected as secret information. These two points must be considered for even the most general schemes as well.

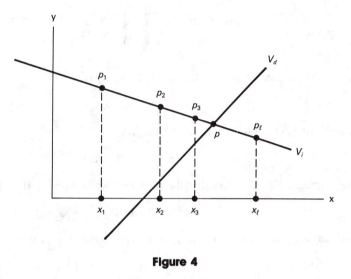

Figure 4

In the scheme in Fig. 4, it is public knowledge that the secret is a point p on V_d that has been chosen with a uniform distribution over all of the points in the subspace. Hence, the probability of an outsider successfully identifying p in a similar manner, when $\mathbf{S} = AG(2, q)$, is

$$P_d = \frac{1}{|V_d|} = \frac{1}{q} \tag{12}$$

Although the answer is obvious in this case, consider whether an outsider could improve his chances (of identifying p) by using his knowledge of the scheme itself and of the publicly exposed abscissas, x_j, of the participant's private pieces of information. He knows only that the indicator, V_i, is a line that is not parallel to the y-axis, but there are q such lines through every point on V_d and on each point on the vertical lines (x_j, y) for each of the x_j. Since these are all equally likely to be V_i, his odds of being able to guess V_i are only $1/q^2$, hence he would be better off to guess at p. The important point, however, is that the number of possible choices (for the line V_i) is the same for each possible choice of p. This would also have been true if the space had been $\mathbf{S} = PG(2, q)$, and in general for any space $\mathbf{S}$.

We have used the term "subspace" imprecisely—and will continue to do so—to indicate a geometric object in an affine space that is either a subspace or is isomorphic to a subspace. These objects are properly called *flats*, and only those flats that include the origin are actually subspaces. In $AG(2, q)$, for example, only the $q + 1$ lines that contain the point $(0, 0)$ are subspaces. The other lines are 1-flats, the 1 indicating that they are one-dimensional objects, isomorphic to $AG(1, q)$, etc. We will use this precision of language only in the stating of the next few results.

The result we need is a well-known [R3,R4] formula that enumerates the number of k-dimensional subspaces or k-flats of $PG(n, q)$ or $AG(n, q)$ that contain a given t-dimensional subspace (flat), respectively. To this end, define the function $\varphi(n, k; q)$:

$$\varphi(n, k; q) = \begin{cases} 1 & k = 0 \\ \displaystyle\prod_{i=0}^{k-1} \left[\frac{q^n - q^i}{q^k - q^i}\right] & 1 \leq k \leq n \end{cases} \tag{13}$$

The number of distinct k-flats in $AG(n, q)$ is

$$q^{n-k} \varphi(n, k; q) \tag{14}$$

and the number containing a given t-flat is

$$\varphi(n - t, k - t; q) \tag{15}$$

For example, Eq. (14) says that the total number of lines, 1-flats, in $AG(2, q)$ is

$$q^{2-1} \varphi(2, 1; q) = q^2 + q$$

and that the number of lines lying on a given point, Eq. (15), is

$$\varphi(2, 1; q) = q + 1$$

results we have already made use of.

Similarly, the number of k-dimensional subspaces in $PG(n, q)$ is

$$\varphi(n + 1, k + 1; q) \tag{16}$$

and the number containing a given t-dimensional subspace is

$$\varphi(n - t, k - t; q) \tag{17}$$

For example, Eq. (16) says that the number of one-dimensional subspaces in $PG(2, q)$ is

$$\varphi(3, 2; q) = q^2 + q + 1$$

and the number lying on a given point is

$$\varphi(2, 1; q) = q + 1$$

To see how these formulas can be used to calculate the security of a shared control scheme, let $\dim(V_i) = m$, $\dim(V_d) = s$ where as usual, $\dim(V_i \cup V_d) = \dim(S) = n$ and $\dim(Vi \cap Vd) = 0$. Assume that a collusion of k' of the participants pool their private pieces of information in an effort to improperly recover p. Together their points span a subspace T, where $\dim(T) = t \leq k' - 1$. $T \cap V_d = \phi$, since $T \subset V_i$ and $V_i \cap V_d = p$; otherwise the collusion would be able to identify p. Hence the subspace spanned by the k' points and an arbitrary point, p', in V_d, $<T \cup p'>$, will have dimension $t + 1$;

$$\dim <T \cup p'> \, = t + 1 \leq k'$$

Irrespective of whether we are in $AG(n, q)$ or $PG(n, q)$ Eqs. (15) and (17) state that the number of $(t + 1)$-dimensional subspaces (flats) lying on all $k' + 1$ of these points is uniformly $\varphi(n - t - 1, k - t - 1; q)$ for every choice of a point in V_d. In other words, every point in V_d is equally likely to be p given the private information available to any collusion, just as it is in the absence of any insider information. Neither can the collusion improve their chances by guessing at V_i among all of the m-dimensional subspaces that contain the k' known points and an assumed point, p', in V_d, since there are uniformly many of these for each choice of p' as well. Hence such schemes are perfect.

It is a little more difficult to show that it is only necessary for each participant to keep secret an amount of information equal to the information content of the secret itself. We have already shown that in a perfect shared secret scheme, which is the only type being considered here, at least this much secret information must be contained in each private piece of information. Given a shared secret scheme of this type, it may be the case that the subspaces V_i and V_d are not full rank with respect to the natural coordinate system. This possibility was why, in the scheme shown in Fig. 4, we excluded the $q + 1$ lines parallel to the y-axis, in $AG(2, q)$, from being used as the indicator V_i, since if V_i were chosen to be one of these lines and the abscissas, x_j, publicly exposed, then even an outsider would be able to determine p.

This geometric argument is easy to see, but the essential difficulty is that the functional expression for V_i in this case is singular with respect to the x parameter; that is, given x_j one cannot solve for y_j. Since $V_i \cup V_d = \mathbf{S}$—the two subspaces span $\mathbf{S}$—and $\dim(V_i) + \dim(V_d) = m + s = n = \dim(\mathbf{S})$ there is a rigid transformation, τ (consisting of rotations and translations) that maps p into the origin and has rank n, that is, any m of the coordinates of the points in V_i can be used to compute the other s coordinates, or any s of the coordinates of a point in V_d can be used to compute the other m.

If the initial subspaces V_i and V_d do not satisfy this condition, we will choose a transformation τ so that τV_i and τV_d do, after which an arbitrary (but fixed) set of s of the coordinate values in the private pieces of information must be kept secret while the other m can be publicly exposed. The integrity of the exposed coordinates must be insured since if an opponent could modify these (without being detected) he could deny an authorized concurrence the ability to initiate the controlled action by causing them to reconstruct an incorrect value for p. The geometric construction for shared secret schemes described here is both perfect and ideal, the latter implying the first as was pointed out above.

The significance of minimizing the amount of information that must be kept secret is that in real-world applications, this information must often either be committed to memory, or else be recoverable from a private cipher (by the participant) using a mnemonically reconstructed one-time key. In either case, there is a great premium on minimizing the amount of information that the participant must memorize.

5 SETTING UP SHARED SECRET SCHEMES

The discussion thus far has been solely concerned with how, and whether, a shared secret scheme can be designed to realize the desired concurrence and security specifications. The tacit assumption has been that when it came to setting up the scheme that there would be some third party, which could be either an individual or a device, that is unconditionally trusted by all the participants in the scheme. This trusted party would first choose the secret (piece of information) and then construct and distribute in secret to each of the participants the private pieces of information that are to be their shares in the shared secret or control scheme. These private pieces of information would be constructed as described at such length in this chapter to realize the desired concurrence scheme. Given the existence and assistance of such an unconditionally trusted third party, there is no conceptual difficulty in realizing any of the shared control schemes described here.

For many applications, however, such an autocratic scheme is not possible since there is no one who is trusted by all of the participants, and in the extreme case, no one who is trusted by anyone else. It is worth noting that in commercial and/or internation(al) applications, this situation is more nearly the norm than the exception. In the absence of a trusted party or authority, no one can be trusted to know the secret and hence, until now, it has appeared to be impossible to construct and distribute the private pieces of information needed to realize a shared control scheme.

The problem of setting up shared secret schemes in the absence of a trusted third party has been ignored by researchers in this area with the single exception of a paper by Meadows [48]. Meadows's paper discusses this problem and examines at even greater length the twin questions of how new participants can be enrolled in an already existing shared control scheme and of how previously enrolled participants can be cut out. To accomplish this, Meadows uses a construction that she calls a rigid linear threshold scheme which makes it possible for a predetermined number of the existing participants to delegate their capability to a new member—essentially to vote him into membership. Meadows's constructions do not appear to be related to the approach to be presented here, especially so since her primary proposal depends on a secure (unconditionally trustworthy) black box to replace the services of an unconditionally trustworthy key distribution center. Meadows attributes the question of whether a shared secret scheme can be set up without the assistance of a trusted key distribution center to Chaum. Chaum posed the question in a talk given in the Rump Session of Crypto'84 that was not published in the Proceedings for that meeting. The important point, however, is that Meadows's work appears to be the only prior published reference to the problem of how a shared control scheme can be set up without the assistance of a trusted key (share) distribution center.

The single exception to what has just been stated is that a way has been known (and used) for several years to insure unanimous consent before a controlled action can be initiated. These are schemes of the sort discussed in connection with the construction shown in Fig. 11 in which each participant secretly chooses a point—randomly and with a uniform probability distribution—from the same space, S, in which the secret point p occurs, and the "sum" of these private points is defined to be p. This implementation of shared control is in fact used by the United States for the control of sensitive command and control information. For anything other than a unanimous consent scheme, however, there must be some sort of functional dependence enforced between the participants' private pieces of information reflecting the structure of the authorized concurrences—even though the participants don't trust each other so that cooperation can't be assumed. Until now, the only way a functional dependence could be enforced between uncooperative and distrusting parties was with the assistance of a third party who had to be unconditionally trusted by all of the participants to set up the shared control scheme.

The essential notion to a protocol devised by Ingemarsson and Simmons [38] to make it possible for parties that don't trust each other—or anyone else either—to set up a shared control scheme that they must (logically) trust is that the secret (piece of information) will be jointly determined in a unanimous concurrence scheme from inputs chosen and made privately by each of the participants. Each of these inputs is to be equally influential in determining the value of the secret and is itself to be kept secret by its contributor. Shared control schemes of this sort in which each participant has an equal influence on the determination of the secret (i.e., the information equivalent of the democratic principal of "one man, one vote") will be referred to as *democratic schemes*, as contrasted with *autocratic schemes* which presuppose the availability of an unconditionally trusted third party.

Once the secret has been determined, each participant may, if he wishes, devise private shared control schemes by means of which he can distribute among the other participants private pieces of information that would make it possible for some groupings of them to reconstruct his contribution to the determination of the secret. Since he is acting to protect his own interests, he can insure by means of the shared control schemes that he devises that only concurrences of participants whom he would be willing to trust to act in his stead will be able to reconstruct his contribution to its determination. This is sufficient to guarantee to each participant that his private contribution will be accessible only to concurrences that either include him as a member or else which are acceptable to him. Since each participant acts similarly, the net result is that by using this protocol, parties that do not trust each other can jointly set up a shared control scheme that they do trust.

The general protocol for a group of mutually distrustful participants to use to set up a shared control scheme that they must logically trust, without the assistance of any outside party, is therefore:

1. The participants first set up a democratic unanimous consent scheme; that is, one in which they each contribute equally to the determination of the secret point, p, and hence in which all of their private inputs (contributions) must be made available for the secret to be reconstructed. This step doesn't require that anyone trust anyone else.

2. After the unanimous consent scheme is in place, any of the participants who trust some concurrence(s), that is, subsets of the other participants, to faithfully represent their interests can then create private shared secret schemes to distribute information about the private (and secret) contribution they made to the determination of p. Thereafter those concurrences can act in their stead—and more importantly, no other collections of participants can do so. In setting up these private shared secret schemes each participant is acting as his own "trusted authority" to protect his own interests, so that he need trust no one else insofar as the delegation of the capability to act in his stead is concerned. In this way, each participant can guarantee that only concurrences that either include him as a member or else that include a subset of the other participants whom he trusts to represent his interests will be able to initiate the controlled action or to recover the shared secret. But this is precisely the risk a participant would have had to accept if there had been an unconditionally trusted central authority to set up the shared control scheme in the first place.

The net result is that democratic shared control schemes of arbitrary complexity (of control) can be established that accurately reflect the placement of trust (or lack of it) by the participants in each other. In other words, every shared control scheme that would be acceptable to the participants and which could be set up by a mutually trusted authority (in Meadows's terminology, a trusted KDC), can also be set up as a democratic scheme by the participants themselves without anyone having to accept a greater risk of their interests being abused than they would have had to accept for a trusted authority to set up the scheme instead.

Given this insight into the nature of democratic shared controls schemes, a question of primary importance is how the initial unanimous consent scheme can be set up. Ingemarsson and Simmons [38] found that of all of the ways to implement unanimous consent schemes, there were apparently only two—inequivalent—ways to set up a democratic unanimous consent scheme.

Either way, of course, can serve as the starting point for setting up more complex schemes using the protocol described above. The first is simply a generalization of the example given earlier.

1. In a space whose cardinality is adequate for the concealment of the secret, that is, in which the probability of selecting a randomly chosen secret (point) in a subsequent random drawing provides an acceptable level of security for the controlled action, each participant chooses a random point as his contribution to the unanimous consent control scheme. During the initialization of the mechanism that implements the shared control, each of the participants secretly enters the point he has selected and the sum (field, modular, exclusive-or, etc.) of all of the points becomes the jointly defined secret point, p.

2. In a k-dimensional finite space where k is the total number of participants in the scheme and the cardinality of the space is chosen such that a hyperplane provides an adequate concealment for the secret point, each participant randomly chooses a hyperplane as his private contribution to the determination of the secret (point). The intersection of these k hyperplanes is then used to define the secret point, p. The important point is that k hyperplanes almost certainly (with q) intersect in only a single point, so that the protocol described here will almost certainly define a unique value for the secret in a democratic shared control scheme. In the event that they do not, this would be detected during the initialization phase of setting up the shared control scheme and the participants would have to make another (random) choice of their inputs (hyperplanes).

It might at first appear that these two protocols for setting up unanimous consent schemes are in some sense simply two versions of a single scheme; especially so in view of the geometric duality between points and hyperplanes when these objects are the private choices of inputs in the two protocols. It is easy, however, to show that this cannot be the case.

In the first scheme, the uncertainty about the secret is the same for an outsider as it is for every combination of fewer than all of the participants: namely, it is equally likely to be any point in the containing space. Furthermore, there is no relationship between the dimension of the space in which the secret is concealed and the number of participants in the shared control scheme. The only requirement is that the number of points in the space be large enough that the probability of choosing the secret (one) at random will be sufficiently small. In other words, a k-out-of-k scheme could be implemented in a one-dimensional space as well as any other, even if the dimension of the containing space is greater than k.

On the other hand, in the second unanimous consent scheme the dimension of the space must equal the total number of participants, k. Otherwise the intersection of the k randomly and independently chosen hyperplanes will almost certainly over- or under-determine a point; that is, the hyperplanes will either not have a common point of intersection or else they will intersect in a subspace of higher dimension. More importantly, however, outsiders and all proper subsets of the insiders will be faced with substantial differences in uncertainty about the secret. An outsider knows only that p is some point in $PG(k, q)$, where all points are equally likely, that is, an uncertainty of $O(q^{-k})$. Any one of the participants, however, knows that p must be a point in the hyperplane he chose, that is, a point in an $(k - 1)$-dimensional subspace that is an uncertainty about p of only $O(q^{-(k-1)})$. Similarly, any pair of participants together could reduce the uncertainty about p to being a point in the $(k - 2)$-dimensional subspace that is the intersection of the two hyperplanes they chose, etc.

There are other, geometric, arguments to show the inequivalence of these two protocols for setting up unanimous consent schemes in the absence of trust, but none so easy to see as this information-based argument.

The smallest example that fully illustrates the protocol would be a 2-out-of-3 threshold scheme: For instance, a scheme in which any two out of three vice-presidents at a bank can open the vault door but in which no one of them alone can do so. The constructions for 2-out-of-ℓ shared control schemes shown in Figs. 4 and 5 require the assistance of a trusted third party to set them up initially, and as we have already mentioned, don't seem to be realizable without such outside assistance. We will now show how the three vice-presidents can set up such a scheme using either of the two unanimous consent democratic schemes described earlier. In either case, the combination (secret) can be thought of as a point in some suitable space.

For the first unanimous consent scheme the secret can be taken to be any point, p, on a line. Each of the three participants secretly and randomly chooses a point, p_i, on λ. p is defined to be the field sum of the three points:

$$p = \sum_{i=1}^{3} p_i \tag{18}$$

Clearly Σ satisfies the definition of an entropy-preserving sum, since as any single summand, p_i, ranges over all $q + 1$ possible values, with the other two points remaining fixed, so does the sum p.

The inescapable conclusion that follows from the acceptability of a 2-out-of-3 threshold scheme is that each participant is willing to trust the other two participants jointly to only initiate the controlled action (i.e., to open the vault door in the present example) when they should. By the same token, the need for a 2-out-of-3 concurrence presupposes a lack of confidence in what a single individual might do. Consequently each participant (in this example) must logically be willing to share his private input to the secret between the other two participants in such a way that they can jointly reconstruct his contribution, but are individually not only unable to do so but are totally uncertain of it. To do this, each participant constructs a private 2-out-of-2 scheme of the sort described earlier, that is, he randomly chooses a pair of points whose sum is his contribution to the democratic shared secret scheme, and gives (in secret) each of the other participants a different one of these points. A convenient way to represent this implementation of the protocol is:

$$
\begin{array}{c c c c}
 & 1 & 2 & 3 \\
\hline
1 & p_1 & p_{21} & p_{31} \\
2 & p_{12} & p_2 & p_{32} \\
3 & p_{13} & p_{23} & p_3
\end{array}
$$

where the three points in column i are all chosen by participant i—subject to the condition that $p_i = \sum_{j \neq i} p_{ij}$. The three entries in row j are known to participant j: the entry on the diagonal because he chose it and the off-diagonal entries because they are the private pieces of information (points) given to him by the other participants. Clearly, any two participants have between them all the information needed to compute $p = \sum p_i$, while any one of them is totally uncertain as to the value of p. Although we have described the protocol starting with the establishment of the democratic unanimous consent scheme, the scheme would probably be implemented in reverse order. Participant i would choose at random the two points p_{ij}, $j \neq i$, and then calculate his input, p_i, to the unanimous consent scheme $p_i = \sum_{j \neq i} p_{ij}$, etc.

To set up the other type of unanimous consent scheme each participant chooses at random a plane in a projective 3-space $PG(3, q)$. As was noted above, since there are three participants, the second type of scheme is only possible in a three-dimensional space. With virtual certainty (with increasing size of q), the three randomly and independently chosen planes intersect in only a single point. This point, p, is the jointly determined secret (combination) $p = \cap_{i=1}^{3} \pi_i$. This protocol defines a 3-out-of-3 unanimous concurrence scheme (such as was shown in Fig. 3) in the case $\ell = 3$, because the three vice-presidents acting together can cause the secret to be reconstructed within the vault door mechanism at any time by reentering their private pieces of information (planes).

As before, each participant also sets up a private 2-out-of-2 shared secret scheme to distribute information about the plane he chose to the other participants, constructed so that they can jointly reconstruct his plane, but individually cannot do so. One way he could do this is by choosing any pair of distinct lines lying in his plane and giving a different one of these lines to each of the other participants. Since the lines are distinct, taken together they span the plane. Hence any two vice-presidents have between them the capability to reconstruct all three planes and thus redefine p.

The private pieces of information for each participant will be the plane he chose and the two lines given to him by the other vice-presidents. Since the pair of lines are shares in a perfect 2-out-of-2 scheme defining his secret plane, each vice-president is assured that a successful concurrence must either include him as a participant or else include both of the other vice-presidents. A convenient way to represent this implementation of the protocol is

$$
\begin{array}{c c c c}
 & 1 & 2 & 3 \\
\hline
1 & \pi_1 & \lambda_{21} & \lambda_{31} \\
2 & \lambda_{12} & \pi_2 & \lambda_{32} \\
3 & \lambda_{13} & \lambda_{23} & \pi_3
\end{array}
$$

where the lines (off-diagonal) entries in column i are chosen by participant i—subject to the condition that they span the plane π_i, $\pi_i = \cap_{j \neq i} \lambda_{ij}$. The three entries in row j are known to participant j: The entry on the diagonal because it is the plane he chose and the off-diagonal entries because they are the private pieces of information (lines) given to him by the other participants.

This small example illustrates the bare bones of the protocol from the standpoint of making sure that no participant can increase his capability beyond what is acceptable to the other participants.

The bottom line is that the protocol described in this section permits democratic shared secret schemes, which must logically be trusted, to be set up by mutually distrustful parties without outside assistance. In addition, no participant is required to accept a greater risk of the secret information being misused than what he would have had to be willing to accept if there had existed a trusted authority to set up the scheme instead. Clearly, this is the most that can be hoped for from any protocol. It should also be pointed out that if other protocols for setting up democratic unanimous consent schemes (to the two described here) are discovered, the protocol can be adapted to use them as well.

6 KEY DISTRIBUTION VIA SHARED SECRET SCHEMES

Shared secret schemes were originally devised by Shamir and Blakley as a means of key management in cryptosystems—to insure against authorized users being denied use of the system due to a loss of the key and to prevent unauthorized persons from being able to use the system as a result of the key being compromised. In this section we will explore shared secret schemes as a means of handling key distribution for some applications.

The primary motivation for developing the multilevel schemes discussed above was to provide protection from what is known as a decapitation attack. Although this terminology is suggestive, it may not be self-explanatory. If there is information that is held by a higher level of command that must be communicated to lower levels in order for some action to be initiated—such as arming warheads, launching missiles, etc.—an adversary may attempt to prevent the action from being initiated by destroying the higher level in a surprise attack before the information can be disseminated to the lower levels of command. This is called a *decapitation attack*. There is another aspect to decapitation attacks in that if the higher levels of command are destroyed, the lower levels either may not know what to do or else be ineffectual in their response. We are not concerned with these consequences to a decapitation attack, however, but only with the situation in which the lower levels of command are rendered incapable of carrying out an action, that they are otherwise able to do, because one or more pieces of information needed to initiate the action were prevented from reaching them by the attack on the superior command.

Multilevel shared control schemes were devised specifically to solve this problem in situations in which the lower level of authority (and hence responsibility) at a lower level of command could be satisfactorily compensated for by requiring an increased level of concurrence for the action to be initiated. There are certainly many situations in which this is the case and for which the multilevel schemes discussed earlier provide a means of solution. There are actions, however, in which either the law doesn't permit a delegation of authority (and hence capability), irrespective of the concurrence that might be required for it to be exercised, or else in which the party(ies) holding the capability are not willing to delegate the capability under normal circumstances. The crucial words in this description are "under normal circumstances," implying that there are circumstances that would either permit such a delegation to be made or in which such a delegation would be acceptable. The problem, from the standpoint of the shared secret or shared control scheme, is the same in either case.

Consider a missile battery at which there are a dozen officers. The consequences of a missile being launched without proper authority would be so great that in normal times (peacetime or in lower levels of alert) the capability to initiate such an action is to be held at a higher level of command, perhaps by the president; in other words, the policy is that even if all of the officers at the battery believe that a missile should be launched, they should not be able to do so without requesting authorization from the superior commander (and more importantly, could not do so without being given the launch enable codes).

In the absence of a shared control scheme, the only way that the superior commander could protect against a decapitation attack on his headquarters (and him) would be to preemptively enable the missiles as a part of going to an advanced state of alert. But these are precisely the circumstances in which there is the greatest concern that something might go wrong and a missile could be launched when it shouldn't have been. Requiring the concurrence of k of the battery officers, say two of them, for a launch is a way of increasing confidence in the proper execution of the plan of battle. What is needed is a scheme in which, under normal circumstances, only the superior commander has the capability to enable a missile launch, but that would allow him, when intelligence inputs or other early warnings indicate, to delegate a k-out-of-12 shared control of the launch to the battery to prevent a decapitation attack from succeeding. The problem is: How does he establish the k-out-of-12 shared control scheme at the time the battery goes to an advanced state of alert? If no advance arrangements have been made, the 12 pieces of private information would have to be communicated to the battery officers in a secure and authenticated manner, at a time (advanced stage of alert) when communications are apt to be both congested and disrupted. Even if the information could be communicated to the battery at that time, the risk of human error in dealing with unfamiliar codes of a size at the limits of mnemonic aids to memorization would be high.

Ideally, it should be possible to distribute the private pieces of information in advance of a need to use the shared control scheme, but with the constraint that in normal circumstances even if all of the participants were to violate their trust and pool their private pieces of information they would still have no better chance of recovering the secret than an outsider would have of simply guessing it. In such a scheme, at the time the battery is put in an advanced state of alert, a single piece of information (one share in the terminology introduced earlier) would need to be communicated by the superior commander to activate the prepositioned k-out-of-12 shared control scheme so that any k of the officers would thereafter be able to launch the missiles.

The important point to this discussion is that almost all of the information needed to implement the shared control (the private pieces of information) could be communicated in advance of the need during a time of low tension and reliable communications. Since there is no special urgency during this setup phase, the communication could even be handled by courier or by having the officers in the subordinate command come to headquarters to be given their private pieces of information. At a time when the battery is going to a state of advanced alert, that is, a period of high tension when communications are at a premium, only a single share (the minimum amount) of information needs to be communicated to activate the scheme.

This same example (a missile battery) can also be used to illustrate the other extended capability to shared control schemes that is also the subject of this chapter. There are two ways this need can arise. First, consider the case in which not all of the missiles have the same launch enable code. The problem in this case is to devise a scheme that can be prepositioned which will allow any one, or any selected subset, of the launch enable codes to be activated in a shared control scheme without affecting the quality of control of the unreleased missiles. Clearly, this result could be accomplished by prepositioning a shared control scheme for each launch enable code, but almost equally clearly this would be a completely unacceptable solution because each participant would be required to remember several private pieces of information, each of which is near the limit of even mnemonically aided recall. What is needed is a way that the same pieces of private information can be used to recover different pieces of secret information.

Another equally important problem has to do with how the battery can stand down from an advanced state of alert, where standing down means reverting to the kind of control that existed prior to the alert. For this to be possible, the scheme must provide both a capability for the superior commander to activate the shared control scheme (delegate authority) and to deactivate it, that is, to rescind his delegation of authority, if the circumstances change so that an advanced state of alert is no longer warranted. If the system is to truly revert to the same type and quality of control after a recall that it had prior to the alert without changing the private pieces of information involved in the shared control scheme, then both the activating information that is to be sent by the superior commander and the enabling code that the missile will respond to must change with each delegation of authority, irrespective of whether the delegated capability was exercised or not.

Although it is inappropriate to the purpose of this discussion to say much about the practical problems of implementing shared control schemes that change with time or use, it is perhaps worthwhile remarking that there are two ways (at least) to achieve this. The simplest scheme would be for the enable codes to change automatically as a function of time, say once each day. Another approach would be for the mechanism in the missile controller that carries out the calculation of the secret information from the private pieces of information to have a stored list of enabling values, only one of which would be operational at a time. In a scheme of this type, the activating piece of information that is sent by the superior commander would have to correspond to the current value of the secret if the shared control is to be operable. If a recall is received (from the superior command), its entry would advance the store to the next stored value for the secret and output a piece of information that could only be obtained by executing this protocol. This would return the missile to a condition wherein only the superior commander could enable it for a launch or delegate its release if the battery was later put in an advanced state of alert again. The old value of the secret information would become invalid so that stale values of the activating information would not be operable. The unique output that could only be obtained by properly carrying out the recall protocol could be returned to the superior command to verify that the control system had been returned to its prealert status.

The point of this lengthy discussion of the simple, but plausible, example of a missile battery was to illustrate as clearly as possible the two essential features to the schemes that are the subject of this selection:

1. It should be possible to preposition all of the private information needed for the shared control scheme subject to the condition that even if all of the participants were to violate the trust of their position and collaborate with each other, they would have no better chance of recovering the secret information than an outsider has of guessing it; that is, the scheme should be perfect.

2. It should be possible to activate the shared control scheme once it is in place by communicating a single share of information, and for many applications, it should also be possible to reveal different secrets (using the same prepositioned private pieces of information) by communicating different activating shares of information.

To achieve multilevel and multipart shared control schemes, we modified the indicator, V_i, in some cases to multiple independent indicators (Fig. 12) or to multiple geometrically nested indicators (Fig. 13) or to multiple functionally related indicators (Figs. 16 and 17), etc. To achieve the shared control capabilities just described, we will find it necessary to modify the domain variety V_d; in some cases by withholding it, that is, keeping it secret from the participants and the outsiders alike, until the controlled action is to be authorized, and in others by having multiple domains—either independent or functionally related, etc. Roughly speaking, extended concurrence is achieved by refinements of V_i while extended control is achieved by refinements of V_d.

An obvious way to implement a shared control scheme such that even a collusion of all of the participants will be powerless to recover the secret until they are later enabled to do so, is to field the private pieces of information, but to withhold the identification of the domain V_d until such time as the scheme is to be activated. That way even if all of the insiders should conspire to pool their private pieces of information in an attempt to recover the secret before the domain is revealed, the most they will be able to do is reconstruct the indicator V_i and hence to learn that p is a point in the subspace V_i instead of being an arbitrary point in S, which is all that an outsider knows about p. There is a problem, however, with this simple approach which is best illustrated using two small examples of shared secret schemes analyzed above.

In the example of a 2-out-of-ℓ shared secret scheme shown in Fig. 4, both V_i and V_d were lines in the plane S. Since a plane is two-dimensional in both points and lines, two shares of information, that is, the identification of two elements from $GF(q)$, are required to specify either one in the affine plane $AG(2, q)$. In other words, the same amount of information would have to be communicated to identify p as a general point in S as would have to be communicated to identify V_d. This might seem to indicate that two shares of information would need to be communicated to activate the scheme, instead of the information theoretic minimum of a single share. However, given V_i (or else V_d) p is no longer an arbitrary point in the plane but rather an unknown point on a line, whose specification on that line requires only one share of information. Similarly, given that p is constrained to be a point on V_i, V_d no longer need be free to be an arbitrary one of the q^2 lines in the plane not parallel to the y-axis, but can instead be restricted to be one out of a set of q lines in which one line lies on each point of V_d.

One easy way to do this would be to preposition an x-coordinate, x_d, different from the point at which V_i intersects the x-axis at the time the scheme is set up. Later, when the scheme is to be activated, the y-intercept of the line V_d through the points x_d and p is all that would need to be communicated to permit V_d to be determined. Thereafter, any two of the participants could recover V_i using their private points, and hence recover the secret p.

If we construct a 3-out-of-ℓ scheme based on the configuration shown in Fig. 9 in which **S** is four-dimensional and V_d is a plane, the problem is even more difficult to deal with since 4-space is six-dimensional in planes while the secret is only two-dimensional in information content. While a similar, but more complex, resolution is possible in which four out of the six needed shares of information would be prepositioned along with the private pieces of information required to set up the shared secret scheme, and the remaining two shares communicated at the time the scheme is to be activated, there is a more efficient (and general) way to implement such schemes. Table 1, tabulating the dimension of the space of m-flats in an n-dimensional space, suggests how difficult this problem can become. If V_i and V_d were both three-dimensional, which only permits a 4-out-of-ℓ shared control, the space of 3-flats is already twelve-dimensional, meaning that 12 shares of information are required to identify V_d.

TABLE 1.   DIMENSION OF THE SPACE OF m-FLATS IN AN n-DIMENSIONAL SPACE

n	0	1	2	3	4	5	6	7
1	1							
2	2	2						
3	3	4	3					
4	4	6	6	4				
5	5	8	9	8	5			
6	6	10	12	12	10	6		
7	7	12	15	16	15	12	7	
8	8	14	18	20	20	18	14	8

In Section 2, which was devoted to the general model(s) for shared secret schemes, we discussed at length the differences between the two approaches to realizing such schemes. In one of these approaches a concurrence of points (the private pieces of information) spanned an indicator, V_i, that pointed out the secret point p, while in the other approach, p was defined by the intersection of a concurrence of geometric objects, each of which contained p. Our principal observation was that while the first scheme was inherently perfect, the other could never be if p was regarded as the secret itself, but that both schemes provided the same level of security in a very natural sense for the shared secret information. In particular, if the information revealed by the second type of scheme was viewed as a cryptographic key, the equivalence in security between the two types of schemes was easy to see.

The problem, arising in our earlier identification of V_i with a cryptographic key, is that the revelation of the secret was equated with the identification of the point p at which the secret is determined. p is not itself the secret, but rather some entropy-preserving function which when evaluated with p as an argument reveals the actual secret. In several examples this function was taken to be either the projection of p onto one or more of the natural coordinate axes or else the value of the variables parameterizing a surface at p. Instead, for the present application consider p to be a normal cryptographic key, say a 56-bit key for the Data Encryption Standard (DES), and let the information that is to be communicated to enable the system be a cipher, which when decrypted with the key will reveal the secret plaintext. Clearly, this implementation solves both of the objectives of a prepositioned shared secret scheme. If the participants cheat and misuse their private pieces of information, all that they can do is recover the shared cryptographic key. Since the cipher hasn't yet been communicated, they have no information whatsoever about the secret plaintext. On the other hand, any plaintext whatsoever can be revealed without having to change the private pieces of information, simply by communicating the cipher that will decrypt with the fixed (shared) key into the desired text.

To illustrate this implementation, consider again the simple 2-out-of-ℓ scheme shown in Fig. 4 when used with the DES encryption algorithm. The plane in this case would be $AG(2, 2^{56})$. Each private piece of information would consist of 112 bits, 56 of which would have to kept secret by the participant, and 56 of which need only be protected against substitution, alteration, or destruction or loss. V_d, or rather two shares of information (112 bits) adequate to determine V_d, would be prepositioned at the time the scheme was set up. After this has been done, any two of the participants, using their private pieces of information, could determine V_i and hence recover p which in this case would be a 2-tuple in $AG(2, 2^{56})$.

There is no reason to not use the simplest entropy-preserving function available; namely, let the secret DES key be the y-coordinate of p, since by the constraints on the construction in this example all 2^{56} possible values are equally likely. Thus an authorized concurrence could recover the DES key at any time after the scheme was set up, however, they could not recover the secret(s) until such time as a cipher was communicated.

On the other hand, the secret (plaintext) would be secure even if the cipher had been communicated unless an authorized concurrence of the participants cooperated to recover the key and decrypt the cipher. In those applications where it was either tolerable or acceptable that a proper concurrence be able to recover the secret at any time after the scheme was fielded, the cipher(s) could be prepositioned along with the private pieces of information. In situations such as those considered in this section, the cipher(s) could be withheld until it is desired that the scheme be enabled, at which time the minimum of only one share of information would have to be communicated for each secret that is to be revealed.

7 CONCLUSIONS

When it comes to implementing a shared control scheme for a specific application, there is an important point that needs to be considered, but which hasn't been; namely, when there is more than one implementation for a shared secret scheme satisfying the concurrence and security requirements, which is the best. To illustrate: Assume that the management of a bank that has a dozen vice-presidents wishes to set up a shared control scheme so that a (arbitrary) pair of the vice-presidents must concur for the vault to be opened, and that the security of this scheme is to be such that the likelihood of any collusion (an outsider or a vice-president or an outsider and a vice-president) being able to open the vault will be 1 in a million or less on their first attempt. Since $2^{20} = 1,048,576 > 10^6$, such a scheme could be implemented using the 2-out-of-ℓ scheme shown in Fig. 4 with $q = 2^{20}$. p is an arbitrary point on the line V_d, so the security is just the probability, P_s, of guessing p

$$P_s = \frac{1}{|V_d|} = \frac{1}{q} < 10^{-6}$$

as desired. This also means that the vault lock would have to accept 2^{20} different combinations for this result to be true. We will assume therefore that the input mechanism for the vault accepts binary strings; that is, the vault door is opened by a 20-bit combination. The private pieces of information (points on V_i), on the other hand, require two shares or 40 bits for their specification. As we have made clear in earlier discussions, these 40 bits do not all need to be kept secret: 20 bits must be kept secret while it suffices for the participant to merely insure the integrity of the other 20.

In the scheme of Fig. 4, implemented in $AG(2, q)$, there are q points on each line, $q - 1$ of which could be used as private pieces of information, that is, the scheme could accommodate over a million vice-presidents when only a dozen are needed. This is a wasted capability, bought at the expense of increasing the amount of information (in the information-theoretic sense) in the private pieces of information. If instead, a 2-out-of-ℓ scheme were implemented using the configuration shown in Fig. 8 with V_i being the line λ and V_d being the plane π, then $q = 2^{10}$ would suffice, since $|V_d| = 2^{20}$, etc. as before. In this case, only 30 bits would be needed in the private pieces of information; of these 30 bits, 20 would still have to be kept secret, while only 10 would need to have their integrity assured. This is still an extravagant scheme though since over a thousand vice-presidents could still be accommodated. If, however, S is taken to be a six-dimensional space over $q = 2^4$ and V_d taken to be a five-dimensional subspace of S, then $|V_d| = 2^{20}$ as before, and V_i would be a line with only $q = 16$ points on it. In this case, the dozen vice-presidents could be accommodated (with three slots to spare) and the private pieces of information would consist of only 24 bits.

A more informative example is provided by the three two-part schemes shown in Figs. 15–17. The concurrence implemented in all three of these schemes requires that both of the national inputs be available in order for the controlled action to be initiated, where each national input is itself controlled by a 2-out-of-4 threshold scheme. The subspace, V_d, was chosen to be a one-dimensional subspace in each case. Hence, if the security requirement is that the likelihood of a collusion being able to identify p on the first try, is to be no better than 1 in a million, then $|V_d| = q > 10^6$. As in the previous example, this requirement could be satisfied by taking $q = 2^{20}$, etc. Since the space S in which the constructions are made is two-dimensional (Fig. 15), three-dimensional (Fig. 16), and four-dimensional (Fig. 17), respectively, this would mean that the private pieces of information would have to be 40-, 60-, and 80-bit binary numbers, only 20 bits of each of which would have to be kept secret and the integrity of the remaining bits insured. It is possible, however, to do much better than this. A 2-out-of-ℓ, $\ell > 2$, threshold scheme can only define a line in S irrespective of the dim(S). Since in each construction we showed that only $q - 1$ out of the q points on a national line (in $AG(n, q)$) could be used as private pieces of information, to accommodate four control team members $q \geq 4$. $q = 2^2$ would therefore be one possibility. If we set $q = 2^2$, then $\dim(V_d)$ must be at least 10 to meet the security requirements.

The equivalent generalization of the configuration in Fig. 15 requires an eleven-dimensional space, S, in which V_d is a ten-dimensional subspace. The lines $\lambda_{U.S.}$ and λ_{USSR} are lines in S that do not lie in V_d but are not skew to it either and hence intersect V_d in single points, $p_{U.S.}$ and p_{USSR}. Just as in the two-dimensional construction these can be any pair of points in V_d, so that their sum p is also (equally likely to be) any point in V_d. The private pieces of information in this case are 22-bit binary numbers, all numbers being equally likely.

The generalization of the configuration shown in Fig. 16 is a little more difficult to visualize. As before, V_d is a ten-dimensional subspace in S and p is any point in V_d. The line V_i is an arbitrary line in S lying on p, but not in V_d. Let T be the subspace $<V_i, V_d>$ spanned by V_i and V_d. In other words, $\dim(T) = 11$. $p_{U.S.}$ and p_{USSR} are any pair of distinct points on V_i, not equal to p. $\lambda_{U.S.}$ and λ_{USSR} are a pair of lines lying on $p_{U.S.}$ and p_{USSR}, respectively, but not in the subspace T. $\lambda_{U.S.}$ and λ_{USSR} satisfy the condition that $\lambda_{U.S.} \subset <\lambda_{USSR}, T>$ and $\lambda_{USSR} \subset <\lambda_{US}, T>$. In other words, S, which is the space spanned by $\lambda_{U.S.}$, λ_{USSR}, and T is twelve-dimensional. This construction is both easy to do and uniform independent of the choices of lines, points, etc. The private pieces of information in this case would be 24-bit binary numbers—all numbers being equally likely, etc., as in the previous case.

Finally, the generalization of Fig. 17 requires that $\dim(S) = 13$ with V_d being a ten-dimensional subspace and V_i a three-dimensional subspace, where $\dim(V_i \cup V_d) = 13$ and $\dim(V_i \cap V_d) = 0$. In this case, the private pieces of information would be 26-bit binary numbers—with the same conditions regarding secrecy and integrity as before.

The whole point in developing this second example has been to illustrate another criteria, other than the information content of the private pieces of information, for deciding which among several implementations of equivalent shared secret schemes is best. Assume that the application needed to accommodate more than two parties, but that the concurrence of (any) two was sufficient for the controlled event to be initiated. The first implementation (the generalization of the configuration in Fig. 15) cannot add even one more party, that is, it cannot even be extended to a 2-out-of-3 party concurrence, since the sum operation (used to define p) is restricted to a unanimous consent scheme.

On the other hand, in the second implementation (the generalization of the configuration in Fig. 16) $q - 1$ out of the q points on the line V_i could be used as national inputs, hence a total of three national parties could be accommodated in this case in a 2-out-of-3 party concurrence scheme. Obviously, each of the national points on V_i must be distinct, otherwise any two parties whose national lines defined the same point would be unable (together) to define V_i and hence to recover p, in contradiction of the concurrence requirements.

The analysis of the third implementation is a bit more difficult. Given that the ten-dimensional subspace, V_d, has been identified and a point p randomly chosen in it, the two national lines are chosen to be both skew to V_d and to each other, and hence they define a three-dimensional subspace V_i that intersects V_d only in the point p. To add another party to the concurrence scheme so that any two of the parties could identify p, the private pieces of information for the members of this party's control team would have to be points on a third line in V_i skew to V_d and to each of the lines $\lambda_{U.S.}$

and λ_{USSR}. The question of how many parties could be accommodated by such a scheme is thus equivalent to the geometric question: Given a point p in a three-dimensional affine space, T, what is the maximum number of pairwise skew lines that can be chosen in T such that no line lies on p? We first answer this question in $PG(3, q)$ and then use this result to answer the question in $AG(3, q)$. It is easy to see that there are at most q^2 such lines in $PG(3, q)$. The number of points in the space is

$$\varphi(3, 0; q) = \frac{q^4 - 1}{q - 1} = q^3 + q^2 + q + 1$$

Since the lines are pairwise skew, no point is on two of the lines. There are $q + 1$ points on each line, hence an upper bound for the number of pairwise skew lines is

$$\# \leq \frac{\varphi(3, 0; q)}{q + 1} = q^2 + 1 \tag{19}$$

If equality holds in Eq. (19), the lines partition the points of T, hence p must be on one of these lines and the other q^2 could be used as the national lines for different national parties. It is well known that such partitions (called spreads) of $PG(3, q)$ by lines do exist [R3, R4], hence the generalization of the configuration shown in Fig. 17 could accommodate as many as q^2 parties in $PG(3, q)$. Given a spread of $q^2 + 1$ lines in $PG(3, q)$, let π be an arbitrary plane. π contains $q^2 + q + 1$ points, each of which must be on one line of the spread, but there are only $q^2 + 1$ lines in the spread so that one line of the spread must be in π. π could not contain two lines of the spread since they could not then be skew lines. Therefore every plane in the space contains one line of the spread and intersects every other line in a distinct point. To answer the question for $AG(3, q)$, take an arbitrary spread in $PG(3, q)$ and delete the plane at infinity. This leaves q^2 skew lines, each of which has q points, that partition the points of the space. p lies on one of these lines. The remaining $q^2 - 1$ pairwise skew lines could all be used as national lines in a 2-out-of-ℓ multiparty shared control scheme; hence as many as 15 parties could be accommodated.

The point of this example has been that while the information content of the private pieces of information differed very little in this case, the number of parties that could be accommodated varied significantly; 2, $q - 1$, and $q^2 - 1$, respectively. If the capability to enroll new parties (even one additional party) is a significant consideration, this would rule out using the scheme that minimizes the information content of the private pieces of information.

The bottom line to this is that the concurrence scheme determines the dimension of V_i while the security requirement dictates the cardinality of V_d. There is in general a range of geometric implementations satisfying both of these conditions, with the final choice being determined by which scheme is optimal by some other measure; in the first example, by minimizing the information content of the private pieces of information and in the second by whether the desired number of parties could be accommodated by the scheme. Different applications will have differing criteria as to which implementation is "best." The point of these concluding remarks is that for a given set of concurrence and security requirements, there are in general many different implementations for shared control schemes that satisfy the requirements—if an implementation exists at all—and that the decision as to which of these is the best must come from other considerations.

References

[R1] *American National Standard, X9.17-1985*, "Financial institution key management (Wholesale)," American Bankers Association, Washington D.C. April 4, 1985.

[R2] W. Jackson, "On designs which admit specific automorphisms," Ph.D. Thesis, Mathematics Department, University of London, Royal Holloway and Bedford New College, 1989.

[R3] J. W. P. Hirschfeld, *Projective Geometries Over Finite Fields*, Oxford Mathematical Monographs, Oxford: Clarendon Press, 1979.

[R4] J. W. P. Hirschfeld, *Finite Projective Spaces of Three Dimensions*, Oxford Mathematical Monographs, Oxford: Clarendon Press, 1985.

Bibliography (Shared Secret Schemes)*

[1] C. A. Asmuth and G. R. Blakley, "Pooling, splitting and reconstituting information to overcome total failure of some channels of communication," *Proc. IEEE Computer Soc. 1982 Symp. Security and Privacy*, Oakland, CA, April 26–28, 1982, pp. 156–169. Los Angeles: IEEE Computer Society Press, 1982.

[2] C. Asmuth and J. Bloom, "A modular approach to key safeguarding," *IEEE Trans. Inform. Theory*, vol. IT-29, no. 2, pp. 208–210, March 1983.

[3] J. Benaloh and J. Leichter, "Generalized secret sharing and monotone functions," in *Lecture Notes in Computer Science 403; Advances in Cryptology: Proc. Crypto '88*, S. Goldwasser, Ed., Santa Barbara, CA, Aug. 21–25, 1987, pp. 27–35. Berlin: Springer-Verlag, 1990.

[4] J. C. Benaloh, "Secret sharing homomorphisms: Keeping shares of a secret secret," in *Lecture Notes in Computer Science 263; Advances in Cryptology: Proc. Crypto '86*, A. M. Odlyzko, Ed., Santa Barbara, CA, Aug. 11–15, 1986, pp. 251–260. Berlin: Springer-Verlag, 1987.

[5] L. Berardi, M. DeFonso, and F. Eugeni, "Threshold schemes based on criss-cross block designs," private communication available from G. J. Simmons.

[6] L. Berardi and F. Eugeni, "Geometric structures, cryptography and security systems requiring a quorum," *Proc. 1987 ATTI del Primo Simposio Nazionale su Stato e Prospettive della Ricerca Crittografica in Italia*, Rome, Italy, Oct. 30–31, 1987, pp. 127–133, 1987; in Italian, English translation available from G. J. Simmons.

[7] A. Beutelspacher, "Enciphered geometry: Some applications of geometry to cryptography," Proc. Combinatorics '86, in *Annals of Discrete Mathematics*, vol. 37, A. Barlotti, M. Marchi, and G. Tallini, Eds., pp. 59–68. Amsterdam: North-Holland, 1988.

[8] A. Beutelspacher, "How to say 'no'," in *Lecture Notes in Computer Science 434; Advances in Cryptology; Proc. Eurocrypt '89*, J.-J. Quisquater and J. Vandewalle,

---

Note: This bibliography includes all of the papers on shared secret or threshold schemes that the author is aware of. Although only a few of the references appearing here are cited in this chapter, it has been included for its own value to other researchers.

Eds., Houthalen, Belgium, April 10–13, 1989, pp. 491–496. Berlin: Springer-Verlag, 1990.

[9] A. Beutelspacher and K. Vedder, "Geometric structures as threshold schemes," 1986 IMA Conference on Cryptography and Coding, Cirencester, England, in *Cryptography and Coding*, H. J. Beker and F. C. Piper, Eds., Oxford: Clarendon Press, pp. 255–268, 1989.

[10] G. R. Blakley, "Safeguarding cryptographic keys," *Proc. AFIPS 1979 Natl. Computer Conf.*, New York, vol. 48, pp. 313–317, June 1979.

[11] G. R. Blakley, "One-time pads are key safeguarding schemes, not cryptosystems: Fast key safeguarding schemes (threshold schemes) exist," *Proc. IEEE Computer Soc. 1980 Symp. on Security and Privacy*, Oakland, CA, April 14–16, 1980, pp. 108–113. Los Angeles: IEEE Computer Society Press, 1980.

[12] G. R. Blakley and R. D. Dixon, "Smallest possible message expansion in threshold schemes," in *Lecture Notes in Computer Science 263; Advances in Cryptology: Proc. Crypto '86*, A. M. Odlyzko, Ed., Santa Barbara, CA, Aug. 11–15, 1986, pp. 266–274. Berlin: Springer-Verlag, 1987.

[13] G. R. Blakley and C. Meadows, "Security of ramp schemes," in *Lecture Notes in Computer Science 196; Advances in Cryptology: Proc. Crypto '84*, G. R. Blakley and D. Chaum, Eds., Santa Barbara, CA, Aug. 19–22, 1984, pp. 411–431. Berlin: Springer-Verlag, 1985.

[14] G. R. Blakley and L. Swanson, "Security proofs for information protection systems," *Proc. IEEE Computer Soc. 1981 Symp. on Security and Privacy*, Oakland, CA, April 27–29, 1981, pp. 75–88. Los Angeles: IEEE Computer Society Press, 1981.

[15] J. R. Bloom, "A note on superfast threshold schemes," preprint, Texas A & M University, Department of Mathematics, 1981.

[16] J. R. Bloom, "Threshold schemes and error correcting codes," in *Abstracts of Papers Presented to the American Mathematical Society*, vol. 2, 1981, p. 230.

[17] J. Bos, D. Chaum, and G. Purdy, "A voting scheme," presented at Crypto '88, Santa Barbara, CA, August 21–25, 1988. Presented at Rump Session but not published in the Proceedings of the conference (copies available from the authors).

[18] E. F. Brickell, "Some ideal secret sharing schemes," presented at 3rd Carbondale Combinatorics Conference, Oct. 31, 1988, Carbondale, IL; in *J. Combinatorial Math. and Combinatorial Computing*, vol. 6, pp. 105–113, Oct. 1989. Also in *Lecture Notes in Computer Science 434; Advances in Cryptology; Proc. Eurocrypt '89*, J.-J. Quisquater and J. Vandewalle, Eds., Houthalen, Belgium, April 10–23, 1989, pp. 468–475. Berlin: Springer-Verlag, 1990.

[19] E. F. Brickell and D. M. Davenport, "On the classification of ideal secret sharing schemes," in *Lecture Notes in Computer Science 435; Advances in Cryptology: Proc. Crypto '89*, G. Brassard, Ed., Santa Barbara, CA, Aug. 20–24, 1989, pp. 278–285. Berlin: Springer-Verlag, 1990.

[20] E. F. Brickell and D. R. Stinson, "The detection of cheaters in threshold schemes," in *Lecture Notes in Computer Science 403; Advances in Cryptology: Proc. Crypto '88*, S. Goldwasser, Ed., Santa Barbara, CA, Aug. 21–25, 1987, pp. 564–577. Berlin: Springer-Verlag, 1990.

[21] E. F. Brickell and D. R. Stinson, "Some improved bounds on the information rate of perfect secret sharing schemes," University of Nebraska, Department of Computer Science, Report Series no. 106, May 1990; *J. Cryptology* (in press).

[22] D. Chaum, "Computer systems established, maintained, and trusted by mutually suspicious groups," Memo. No. UCB/ERL/M79/10, University of California, Berkeley, Electronics Research Laboratory, 1979; also D. Chaum, Ph.D. dissertation in Computer Science, University of California, Berkeley, 1982.

[23] D. Chaum, "How to keep a secret alive: Extensible partial key, key safeguarding, and threshold systems," in *Lecture Notes in Computer Science 196; Advances in Cryptology: Proc. Crypto '84,* G. R. Blakley and D. Chaum, Eds., Santa Barbara, CA, Aug. 19–22, 1984, pp. 481–485. Berlin: Springer-Verlag, 1985.

[24] D. Chaum, C. Crepeau, and I. Damgård, "Multiparty unconditionally secure protocols," 4th SIAM Conf. Discrete Math., San Francisco, CA, June 13–16, 1988, abstract appearing in *SIAM Final Program Abstracts: Minisymposia,* no. M-28, p. A8, 1988.

[25] B. Chor, S. Goldwasser, S. Micali, and B. Awerbuch, "Verifiable secret sharing and achieving simultaneity in the presence of faults," *Proc. 26th IEEE Symp. Found. Comp. Sci.,* Portland, OR, pp. 383–395, Oct. 1985.

[26] B. Chor and E. Kushilevitz, "Secret sharing over infinite domains," in *Lecture Notes in Computer Science 435; Advances in Cryptology: Proc. Crypto '89,* G. Brassard, Ed., Santa Barbara, CA, Aug. 20–24, 1989, pp. 299–306. Berlin: Springer-Verlag, 1990.

[27] R. A. Croft and S. P. Harris, "Public-key cryptography and re-usable shared secrets," 1986 IMA Conference on Cryptography and Coding, Cirencester, England, in *Cryptography and Coding,* H. J. Beker and F. C. Piper, Eds., pp. 255–268, Oxford: Clarendon Press, 1989.

[28] G. I. Davida, R. A. DeMillo, and R. J. Lipton, "Protecting shared cryptographic keys," *Proc. IEEE Computer Soc. 1980 Symp. on Security and Privacy,* Oakland, CA, April 14–16, 1980, pp. 100–102. Los Angeles: IEEE Computer Society Press, 1980.

[29] Y. G. Desmedt and Y. Frankel, "Threshold cryptosystems," in *Lecture Notes in Computer Science 435; Advances in Cryptology: Proc. Crypto '89,* G. Brassard, Ed., Santa Barbara, CA, Aug. 20–24, 1989, pp. 307–315. Berlin: Springer-Verlag, 1990.

[30] M. De Soete, "Geometric threshold schemes," presented at "Course on Geometries, Codes and Cryptography, organized by the *Centre International des Sciences Mécaniques,* June 19–23, 1989, Udine, Italy, to appear in *Lecture Notes* of the course.

[31] M. De Soete, J. Quisquater, and K. Vedder, "A signature with shared verification scheme," in *Lecture Notes in Computer Science 435; Advances in Cryptology: Proc. Crypto '89,* G. Brassard, Ed., Santa Barbara, CA, Aug. 20–24, 1989, pp. 253–262. Berlin: Springer-Verlag, 1990.

[32] M. De Soete and K. Vedder, "Some new classes of geometric threshold schemes," in *Lecture Notes in Computer Science 330; Advances in Cryptology: Proc. Eurocrypt '88,* C. G. Günther, Ed., Davos, Switzerland, May 25–27, 1988, pp. 57–76. Berlin: Springer-Verlag, 1988.

[33] A. Ecker, "Tactical configurations and threshold schemes," preprint (available from author).

[34] P. Feldman, "A practical scheme for non-iterative verifiable secret sharing," *Proc. 28th Annu. Symp. Foundations of Computer Science,* Los Angeles, Oct. 12–14, 1987, pp. 427–437. Los Angeles: IEEE Computer Society Press, 1987.

[35] D. K. Gifford, "Cryptographic sealing for information secrecy and authentication," *Commun. ACM*, vol. 25, no. 4, pp. 274–286, April 1982.

[36] S. Harari, "Secret sharing systems," in *Secure Digital Communications*, G. Longo, Ed., Vienna: Springer-Verlag, pp. 105–110, 1983.

[37] I. Ingemarsson and G. J. Simmons, "How mutually distrustful parties can set up a mutually trusted shared secret scheme," *Intl. Assoc. Cryptologic Research (IACR) Newsletter*, vol. 7, no. 1, pp. 4–7, Jan. 1990.

[38] I. Ingemarsson and G. J. Simmons, "A protocol to set up shared secret schemes without the assistance of a mutually trusted party," in *Lecture Notes in Computer Science* 473; *Advances in Cryptology, Proc. Eurocrypt '90*, Aarhus, Denmark, May 21–24, 1990; I. Damgård, Ed., pp. 266–282. Berlin: Springer-Verlag, 1991.

[39] M. Ito, A. Saito, and T Nishizeki, "Secret sharing scheme realizing general access structure" (in English) *Proc. IEEE Global Telecommun. Conf., Globecom '87*, Tokyo, 1987, pp. 99–102. Washington, DC: IEEE Communications Soc. Press, 1987. Also appeared in *Trans. IECE Japan*, vol. J71-A, no. 8, 1988 (in Japanese).

[40] M. Ito, A. Saito, and T. Nishizeki, "Multiple assignment scheme for sharing secret," preprint (available from T. Nishizeki).

[41] E. D. Karnin, J. W. Greene, and M. E. Hellman, "On secret sharing systems," IEEE Intl. Symp. Inform. Theory, Session B3 (Cryptography), Santa Monica, CA, February 9–12, 1981, *IEEE Trans. Inform. Theory*, vol. IT-29, no. 1, pp. 35–41, Jan. 1983.

[42] S. C. Kothari, "Generalized linear threshold scheme," in *Lecture Notes in Computer Science 196; Advances in Cryptology: Proc. Crypto '84*, G. R. Blakley and D. Chaum, Eds., Santa Barbara, CA, Aug. 19–22, 1984, pp. 231–241. Berlin: Springer-Verlag, 1985.

[43] K. Koyama, "Cryptographic key sharing methods for multi-groups and security analysis," *Trans. IECE Japan*, vol. E66, no. 1, pp. 13–20, 1983.

[44] C. S. Laih, L. Harn, and J. Y. Lee, "Dynamic threshold scheme based on the definition of cross-product in an N-dimensional linear space," in *Lecture Notes in Computer Science 435; Advances in Cryptology: Proc. Crypto '89*, G. Brassard, Ed., Santa Barbara, CA, Aug. 20–24, 1989, pp. 286–298. Berlin: Springer-Verlag, 1990.

[45] C. S. Laih, J. Y. Lee, and L. Harn, "A new threshold scheme and its applications in designing the conference key distribution cryptosystem," *Infor. Processing Lett.*, vol. 32, pp. 95–99, 1989.

[46] C. Matsui, K. Tokowa, M. Kasahara, and T. Namekawa, "Notes on (K,N) threshold scheme," *Proc. Joho Riron To Sondo Ooyo Kenkyukai, VII-th Symposium*, Kinugawa, Japan, November 5–7, 1984, pp. 158–163 (in Japanese); in *The VII-th Symposium on Information Theory and Its Applications* (English translation available from G. J. Simmons).

[47] R. J. McEliece and D. V. Sarwate, "On sharing secrets and Reed-Solomon codes," *Commun. ACM*, vol. 24, no. 9, pp. 583–584, Sept. 1981.

[48] C. Meadows, "Some threshold schemes without central key distributors," *Congressus Numerantium*, vol. 46, pp. 187–199, 1985.

[49] M. Merritt, "Key reconstruction," in *Advances in Cryptology: Proc. Crypto '82*, D. Chaum, R. L. Rivest, and A. T. Sherman, Eds., Santa Barbara, CA, Aug. 23–25, 1982, pp. 321–322. New York: Plenum Press, 1983.

[50] M. Mignotte, "How to share a secret," Workshop on Cryptography, Burg Feuerstein, Germany, March 29–April 2, 1982, in *Cryptography*, vol. 149, T. Beth, Ed., pp. 371–375. Berlin: Springer-Verlag, 1983.

[51] R. von Randow, "The bank safe problem," *Discrete Appl. Math.*, vol. 4, pp. 335–337, 1982.

[52] P. J. Schellenberg and D. R. Stinson, "Threshold schemes from combinatorial designs," *Journal of Combinatorial Mathematics and Combinatorial Computing*, vol. 5, pp. 143–160, 1989.

[53] A. Shamir, "How to share a secret," *Massachusetts Institute of Technology Technical Report MIT/LCS/TM-134*, May 1979. (See also *Commun. ACM*, vol. 22, no. 11, pp. 612–613, Nov. 1979.)

[54] G. J. Simmons, "Robust shared secret schemes or 'how to be sure you have the right answer even though you don't know the question'," 18th Annu. Conf. Numerical Mathematics and Computing, Sept. 29–Oct. 1, 1988, Winnipeg, Manitoba, Canada; appeared in *Congressus Numerantium*, vol. 68, pp. 215–248, May 1989.

[55] G. J. Simmons, "How to (really) share a secret," in *Lecture Notes in Computer Science 403; Advances in Cryptology: Proc. Crypto '88*, S. Goldwasser, Ed., Santa Barbara, CA, Aug. 21–25, 1987, pp. 390–448. Berlin: Springer-Verlag, 1990.

[56] G. J. Simmons, "Sharply focused sets of lines on a conic in PG(2,q)," presented at 20th Southeastern International Conference on Combinatorics, Graph Theory and Computing, Boca Raton, FL, Feb. 20–24, 1989; in *Congressus Numerantium*, vol. 73, pp. 181–204, Jan. 1990.

[57] G. J. Simmons, "Prepositioned shared secret and/or shared control schemes," in *Lecture Notes in Computer Science 434; Advances in Cryptology; Proc. Eurocrypt '89*, J.-J. Quisquater and J. Vandewalle, Eds., Houthalen, Belgium, April 10–23, 1989, pp. 436–467. Berlin: Springer-Verlag, 1990.

[58] D. R. Stinson and S. A. Vanstone, "A combinatorial approach to threshold schemes," in *Lecture Notes in Computer Science 293; Advances in Cryptology: Proceedings of Crypto '87*, C. Pomerance, Ed., Santa Barbara, CA, Aug. 16–20, 1987, pp. 330–339. Berlin: Springer-Verlag, 1988.

[59] D. R. Stinson and S. A. Vanstone, "A combinatorial approach to threshold schemes," *SIAM J. Disc. Math*, vol. 1, no. 2, pp. 230–236, May 1988. (This is an expanded version of the paper that appeared in *Lecture Notes in Computer Science 293; Advances in Cryptology: Proceedings of Crypto '87*, C. Pomerance, Ed., Santa Barbara, CA, Aug. 16–20, 1987, pp. 330–339. Berlin: Springer-Verlag, 1988.)

[60] M. Tompa and H. Woll, "How to share a secret with cheaters," in *Lecture Notes in Computer Science 263; Advances in Cryptology: Proc. Crypto '86*, A. M. Odlyzko, Ed., Santa Barbara, CA, Aug. 11–15, 1986, pp. 133–138. Berlin: Springer-Verlag, 1987.

[61] T. Uehara, T. Nishizeki, E. Okamoto, and K. Nakamura, "Secret sharing systems with matroidal schemes," *Trans. IECE Japan*, vol. J69-A, no. 9, pp. 1124–1132, 1986 (in Japanese; English translation available from G. J. Simmons) presented at the 1st China-USA International Conference on Graph Theory and its Applications, Jinan, China, June 1986. English summary by Takao Nishizeki available as *Technical Report TRECIS8601*, Department of Electronic Communications, Tohoku University, 1986.

[62] H. Unterwalcher, "Threshold schemes based on systems of equations," *Öster-reichische Akademie der Wissenschaften, Math.-Natur. Klasse, Sitzungsber. Abt. II*, vol. 196 (4–7), Vienna, pp. 171–180, 1987.

[63] H. Unterwalcher, "A department threshold scheme based on algebraic equations," in *Contributions to General Algebra 6*, Dedicated to the memory of Wilfried Nöbauer, pp. 287–298. Stuttgart, Federal Republic of Germany: Vienna, Verlag B. G. Teubner, 1988.

[64] W. D. Wallis, "Not all perfect extrinsic secret sharing schemes are ideal," *Australasian J. Combinatorics*, vol. 2, pp. 237–238, Sept. 1990.

[65] H. Yamamoto, "On secret sharing schemes using (k, L, n) threshold scheme," *Trans. IECE Japan*, vol. J68-A, no. 9, pp. 945–952, 1985 (in Japanese); also published as "Secret sharing system using (k, L, n) threshold scheme," *Electronics and Communications in Japan*, part 1, vol. 69, no. 9, pp. 46–54, 1986.

[66] H. Yamamoto, "On secret sharing communication systems with two or three channels," *IEEE Trans. Inform. Theory*, vol. IT-32, no. 3, pp. 387–393, May 1986.

[67] H. Yamamoto, "Coding theorem for secret sharing communication systems with two noisy channels," *IEEE Trans. Inform. Theory*, vol. IT-35, no. 3, pp. 572–578, May 1989.

[68] A. C. Yao, "How to generate and exchange secrets," in *27th Annual Symp. Foundations of Computer Sci.*, Toronto, Canada, Oct. 27–29, 1986, pp. 162–167. Los Angeles: IEEE Computer Society Press, 1986.

SECTION 4

Cryptanalysis

CHAPTER 10

Cryptanalysis
A Survey of Recent Results

E. F. BRICKELL
Sandia National Laboratories
Albuquerque, New Mexico 87185

A. M. ODLYZKO
A T & T Bell Laboratories
Murray Hill, New Jersey 07974

Abstract—In spite of the progress in computational complexity, it is still true that cryptosystems are tested by subjecting them to cryptanalytic attacks by experts. Most of the cryptosystems that have been publicly proposed in the last decade have been broken. This chapter outlines a selection of the attacks that have been used and explains some of the basic tools available to the cryptanalyst. Attacks on knapsack cryptosystems, congruential generators, and a variety of two-key secrecy and signature schemes are discussed. There is also a brief discussion of the status of the security of cryptosystems for which there are no known feasible attacks, such as the Rivest–Shamir–Adleman (RSA), discrete exponentiation, and Data Encryption Standard (DES) cryptosystems.

1 INTRODUCTION

The last decade has seen explosive growth in unclassified research in all aspects of cryptology, and cryptanalysis has been one of the most active areas. Many cryptosystems that had been thought to be secure have been broken, and a large collection of mathematical tools useful in cryptanalysis has been developed. The purpose of this survey is to present some of the recent attacks in a way that explains and systemizes the cryptanalytic techniques that are used, with the hope that they will be useful in assessing the security of other cryptosystems.

Most of the discussion in this chapter is devoted to public key systems. This reflects the general developments in cryptography over the last decade. At the beginning of the 1970s only classic (single-key) cryptography was known, but very little unclassified research was being done on it. The reasons for this lack of interest were manifold.

There did not seem to be much need for commercial encryption. The vast body of classified work in cryptography discouraged researchers who naturally like to discover new results. Finally, perhaps the most important factor was that despite the development of the beautiful Shannon theory of secrecy systems, and the use of some tools from abstract algebra, generally speaking cryptography appeared to consist of a large bag of tricks, without a coherent mathematical framework.

The situation changed drastically in the 1970s. First, with growth in communications and the proliferation of computers, the need for cryptographic protection became widely recognized. Second, the invention of public key cryptography by Diffie and Hellman appeared to provide an answer to the commercial need for security that avoided some of the disadvantages of classical cryptography, such as the difficulty of key management. Furthermore, this development galvanized the research community because it appeared to open up a brand new field, and it presented the exciting promise of using new tools from the rapidly developing field of computational complexity to develop systems with simple mathematical descriptions. The security of these systems would depend on the intractability of well-known problems, and hopefully would eventually lead to proofs of unbreakability of such systems.

Ironically, the promise of provable security through reduction to well-known mathematical problems has not only not been fulfilled, but instead, the fact that attacks on the new cryptosystems could be formulated as mathematically attractive problems, and that various tools from computational complexity, number theory, and algebra could be brought to bear on them, has resulted in the breaking of many systems. The old one-time pad remains the only system that is known to be unconditionally secure.

The ideal proof of security for a public key cryptosystem would be to show that any attack that has a nonnegligible probability of breaking the system requires an infeasible amount of computation. While no public key system has been shown to satisfy this strong definition of security, the situation is not completely bleak. Many systems have been developed whose security has been proved to be equivalent to the intractability of a few important problems, such as factoring integers, that are almost universally regarded as very hard. (Many of the systems that have been broken were derived from these presumably secure ones by weakening them to obtain greater speed.) Furthermore, the extensive work of the last decade, both in cryptography itself and in general computational complexity, has given cryptologists a much better understanding of what makes a system insecure. The aim of this chapter is to distill some of the lessons of this research.

Before outlining the contents of this chapter, we have to explain what we mean by saying that a cryptographic system is insecure. One can define a fairly precise notion in terms of a polynomial fraction of instances of the system being decipherable in polynomial time. Such an approach is unsatisfactory for two reasons, however. One is that in practice one has to build systems of fairly limited size, and so one cannot assume that asymptotic properties apply. A more serious reason is that for many cryptanalytic attacks, no rigorous proofs of effectiveness exist. Instead one relies on heuristics and experimental evidence; for example, one shows that a reduced-size version of a proposed cryptosystem can be broken relatively fast on a small general-purpose computer, and then one argues that since the effort involved in the attack does not increase too fast with the size of the problem, even the full-size cryptosystem is insecure from a determined attacker. This approach is occasionally used also in other areas of computational complexity (e.g., factoring polynomials or integers), where the best practical algorithms rely on unproved assumptions. Use of such approaches in cryptography is very easy to justify. Because cryptosystems often protect very sensitive information and once adapted, are difficult to change, it is important that they be above suspicion. We will see below, for example, that attacks on some of knapsack cryptosystems depend on being able to find very short nonzero vectors in lattices. In general, it is not known just how difficult a task it is to find such vectors, and the known polynomial time algorithms are not guaranteed to find such vectors. On the other hand, these algorithms usually work much better than they are guaranteed to, and moreover, there has been a lot of progress recently on obtaining improved algorithms. Therefore it seems prudent in assessing the security of the knapsack cryptosystems to assume that one can find even the shortest nonzero vector in a lattice relatively fast.

The remainder of this chapter is organized as follows: Sections 2 and 3 discuss the cryptanalysis of knapsack cryptosystems. Section 4 contains the cryptanalysis of the Ong, Schnorr, and Shamir (OSS) signature schemes, and Section 5 that of the Okamoto–Shiraishi scheme. Section 6 briefly mentions several two-key cryptosystems that have been broken. Sections 7–9 describe what is known about the security of the Rivest–Shamir–Adleman (RSA) algorithms, discrete exponentiation, and the McEliece cryptosystem. The next two sections deal with the cryptanalysis of some single key systems, with Section 10 covering the remarkable success in breaking congruential generators, and Section 11 discussing the remarkable lack of success in breaking DES. Section 12 briefly discusses the successful cryptanalysis of the fast data encipherment algorithm (FEAL). Finally, Section 13 contains some miscellaneous comments.

2 KNAPSACK CRYPTOSYSTEMS

Knapsack cryptosystems are based on the knapsack (or more precisely the subset sum) problem, that is, given a set of integers (or weights) $a_1, \ldots, a_n$ and a specified sum s, find a subset of $\{a_1, \ldots, a_n\}$ that sums to exactly s, or equivalently find a 0–1 vector $(x_1, \ldots, x_n)$ such that $\sum_{i=1}^{n} x_i a_i = s$. We will sometimes refer to the set of weights as a knapsack. Merkle and Hellman [105] discovered a way to use the knapsack problem as the basis for a two-key cryptosystem. Although the knapsack problem is NP-hard [58], there are knapsacks for which the problem is easy. An example of an easy knapsack which was used in [105] is a superincreasing sequence, that is, a sequence of positive integers $b_i, \ldots, b_n$ such that $b_j > \sum_{i<j} b_i$, for $1 < j \le n$.

The basic technique for using the knapsack problem as a two-key cryptosystem is straightforward.

- Public key: Positive integers $a_1, \ldots, a_n$.
- Private key: A method for transforming $a_1, \ldots, a_n$ into an easy knapsack.
- Message space: n-dimensional 0–1 vectors $(x_1, \ldots, x_n)$.
- Encryption: $s = \sum_{i=1}^{n} x_i a_i$.
- Decryption: Solve the knapsack problem with weights $a_1, \ldots, a_n$ and sum s.

Merkle and Hellman used a superincreasing sequence as an easy knapsack and disguised it with one or more modular multiplications. Specifically, an easy knapsack

$b_1, \ldots, b_n$ can be disguised with a modular multiplication by selecting $M > \Sigma_{i=1}^n b_i$ and W with $(W, M) = 1$, and computing

$$a_i \equiv b_i W \pmod{M} \tag{2.1}$$

Any solution $(x_1, \ldots, x_n)$ to the knapsack problem $\Sigma_{i=1}^n x_i a_i = s$ is also a solution to the knapsack problem $\Sigma_{i=1}^n x_i b_i = s'$ where $s' \equiv sW^{-1} \pmod{M}$ and $0 \leq s' < M$. Merkle and Hellman [105] further observed that the disguising operation could be *iterated* many times. For instance, given the above knapsack $a_1, \ldots, a_n$, a new knapsack $c_1, \ldots, c_n$ could be formed by choosing a new modulus $M_2 > \Sigma_{i=1}^n a_i$ and multiplier W_2 with $(W_2, M_2) = 1$ and defining $c_i \equiv a_i W_2 \bmod M_2$. The knapsack $c_1, \ldots, c_n$ is called a *double-iterated knapsack*.

The designer of the system can further complicate matters by permuting the weights before publishing them. For clarity of exposition, we will assume that the weights are not permuted, but we will discuss the permutation when it is relevant.

One reason to be suspicious about the security of knapsack cryptosystems is that they are basically linear. Specifically $\Sigma_{i=1}^n x_i a_i + \Sigma_{i=1}^n y_i a_i = \Sigma_{i=1}^n (x_i + y_i) a_i$. In fact, if (as we may assume) not all the a_i are even, then by looking at the least significant bit of the ciphertext s we obtain a bit of information about the plaintext, although this usually does not yield even a single bit of the plaintext. Although there is no attack on the cryptosystem based just on this linearity, it should raise questions about its security because linearity in cryptosystems is known to be dangerous. Another cause for suspicion is due to a result of Brassard [18]. Essentially it says that if the problem of breaking a cryptosystem is **NP**-hard, then **NP = CoNP**. When the Merkle–Hellman knapsack cryptosystem was proposed, the only attack known was to use an algorithm which would solve any knapsack problem. If one believes that **NP ≠ CoNP**, then it seems likely that there is an attack on the Merkle–Hellman knapsack cryptosystem that runs faster than algorithms that solve the general knapsack problem. This suspicion does not apply to RSA, since factoring integers is not believed to be **NP**-hard.

These suspicions were extended by various authors. Herlestam [71] observed by using simulations that often a single bit of the message could be easily recovered. Shamir [150] showed that Merkle–Hellman knapsacks in which the modulus M has close to n bits can be broken easily, and [149] that compact knapsacks (i.e., general knapsacks with few weights c_i and with coefficients x_i that are allowed to vary over a wider range than just the set $\{0, 1\}$) ought to be avoided. Amirazizi, Karnin, and Reyneri [6] also showed that compact knapsacks are insecure, but with an even more powerful argument than that of [149], since they were able to use the theorem of H. W. Lenstra [97] that integer programming in a fixed number of variables is solvable in polynomial time. (This key result [97] was proved at the end of 1980 and became widely known right away, although it was not published until much later.) Shamir and Zippel [151] showed using continued fractions that if the modulus M was known, a cryptanalyst could break the single iterated system. Ingemarsson [73] developed a method of successive reduction modulo suitably chosen integers which seemed to apply to a wide class of knapsacks. However, none of these attacks could convincingly be shown to apply to the Merkle–Hellman system.

Eier and Lagger [51] and independently Desmedt, Vandewalle, and Govaerts [46] made a key observation that led eventually to the complete demise of these knapsack systems. From Eq. (2.1), there exist integers $k_1, \ldots, k_n$ such that

$$a_i U - k_i M = b_i \tag{2.2}$$

where $U \equiv W^{-1} \bmod M$. Therefore

$$\frac{U}{M} - \frac{k_i}{a_i} = \frac{b_i}{a_i M} \tag{2.3}$$

and so all of the k_i/a_i are close to U/M. Furthermore, as was apparently realized by the authors of [51] and [46], the actual values of U and M are not important, since if one finds any pair of integers u and m with $u/m - U/M$ small, one can use u and m to decrypt the knapsack. For an arbitrary collection of integers a_i it is highly unlikely that there would exist k_i such that all of the k_i/a_i would be close together. This seemed to provide a way to attack the Merkle–Hellman system.

Shamir [146] completed the cryptanalysis of the single iterated Merkle–Hellman system by making two more observations. Since the a_i's are superincreasing, $a_i < M2^{i-n}$. Hence, for small i, the k_i/a_i are extremely close together; from Eq. (2.2) we see that

$$|b_i k_1 - b_1 k_i| \le M2^{i-n} \tag{2.4}$$

Only a few (three to four) of these inequalities uniquely determine the k_i's, and once the k_i's are found, it is easy to break the system. The system (Eq. (2.4)) is an instance of integer programming with a small number of variables. Therefore the Lenstra [97] integer linear programming algorithm can find the k_i's fast. For this attack, it is necessary for the cryptanalyst to know which of the public weights correspond to the smallest elements in the superincreasing sequence. If the knapsack was permuted before it was published, he would not know this. Since he only needs to know the three or four smallest elements, however, he can find them in polynomial time ($O(n^3)$ or $O(n^4)$) by trying all possibilities.

The Shamir attack sketched above was universally accepted as valid when it was announced, although nobody up to that time had implemented the Lenstra integer programming algorithm. (In fact, as will be explained below, for the standard version of the Merkle–Hellman system, in which M has about $2n$ bits, one can use continued fractions to find the k_i.) Furthermore, the Shamir attack did not seem to generalize to other knapsack systems. These problems were soon overcome, though, because Adleman [3] found that the Lovasz lattice basis reduction algorithm [93] could be used instead of the Lenstra integer programming algorithm, and this enabled him to break the Graham–Shamir knapsack cryptosystem (see [151] for a definition). The introduction of this new tool, the Lovasz algorithm, was the main key to most of the major breakthroughs that were achieved in analyzing knapsacks. The Lovasz algorithm and more efficient ones that were derived later by Radziszowski and Kreher [134] and by Schnorr ([140,141]) are now among the basic tools of constructive diophantine approximation, and will be discussed after we introduce some definitions.

2.1 Diophantine Approximation

Simultaneous diophantine approximation is the study of approximating a vector of reals $(\theta_1, \ldots, \theta_n)$ by a vector of rationals $\left(\frac{p_1}{p}, \ldots, \frac{p_n}{p}\right)$ all having the same denominator. An approximation $\left(\frac{p_1}{p}, \ldots, \frac{p_n}{p}\right)$ to a vector of rationals $\left(\frac{q_1}{q}, \ldots, \frac{q_n}{q}\right)$ is said to be an unusually good simultaneous diophantine approximation (UGSDA), if $\left|p\frac{q_i}{q} - p_i\right| \le q^{-\delta}$ for some

$\delta > \frac{1}{n}$. Lagarias [86] has justified this definition by showing that unusually good simultaneous diophantine approximations are indeed unusual.

For breaking knapsack-type cryptosystems, we are interested in the algorithmic question of finding unusually good simultaneous diophantine approximations that are known to exist. For $n = 1$, continued fractions can be used to find UGSDA. The set of convergents to the continued fraction expansion of $\frac{q_1}{q}$ contains every rational $\frac{r_1}{r}$ such that $r < q$ and $\left| r \frac{q_1}{q} - r_1 \right| \leq \frac{1}{r}$. Thus if $\frac{p_1}{p}$ is an UGSDA to $\frac{q_1}{q}$ then $\frac{p_1}{p}$ will be a convergent.

More surprisingly, continued fractions can also be used to find UGSDA for $n = 2$. To see this let $\left(\frac{q_1}{q}, \frac{q_2}{q} \right)$ be a pair of rationals that have an UGSDA $\left(\frac{p_1}{p}, \frac{p_2}{p} \right)$. Let

$$c_i = q_i p - q p_i$$

Then

$$\left| c_i \right| < q^{\frac{1}{2}}$$

Taking these equations mod q, we obtain

$$c_i \equiv q_i p \bmod q$$

Let us assume for now that the greatest common divisor $\mathrm{GDC}(q_2, q) = 1$. Then

$$\frac{c_1}{c_2} \equiv \frac{q_1}{q_2} \bmod q$$

Let $x \equiv \frac{q_1}{q_2} \bmod q$. Then $c_2 x \equiv c_1 \bmod q$, and there exists a y such that

$$c_2 \frac{x}{q} - y = \frac{c_1}{q}$$

and

$$\left| \frac{c_1}{q} \right| < \left| c_2 \right|^{-1}$$

Therefore we can find $\frac{y}{c_2}$ as a convergent in the continued fraction expansion of $\frac{x}{q}$. Using c_2, we can find $p \equiv q_2^{-1} c_2 \bmod q$ and then p_1 and p_2 are easily determined. If the GCD $(q_2, q) = d \neq 1$, then one can replace q and q_2 by $\frac{q}{d}$ and $\frac{q_2}{d}$ and proceed as above. This provides an attack on the single iterated Merkle–Hellman cryptosystem for certain parameters. In particular, if $M < 2^{2n-8}$ and if $b_1 > M/2$, then using Eq. (2.4) we see that $\left(\frac{k_2}{k_1}, \frac{k_3}{k_1} \right)$ is an UGSDA to $\left(\frac{b_2}{b_1}, \frac{b_3}{b_1} \right)$. The knapsack cryptosystem of Henry [70] can be broken by using continued fractions.

Finding UGSDA is related to finding short vectors in a lattice. Given a set of n independent vectors in R^n, $b_1, \ldots, b_n$, a lattice L is the set of points

$$L = \left\{ \sum_{i=1}^{n} z_i \, \mathbf{b_i} : z_i \in \mathbb{Z} \right\}$$

The vectors $\mathbf{b}_1, \ldots, \mathbf{b}_n$ are said to be a basis for L. Consider the lattice L generated by the row vectors $\mathbf{b}_0, \ldots, \mathbf{b}_n$ of the matrix

$$\begin{pmatrix} \lambda & p_1 & p_2 & \cdots & p_{n-1} & p_n \\ 0 & -p & 0 & \cdots & 0 & 0 \\ \vdots & & & & & \\ 0 & 0 & 0 & \cdots & -p & 0 \\ 0 & 0 & 0 & \cdots & 0 & -p \end{pmatrix}$$

where λ is a real number between 0 and 1. There is an obvious relationship between short vectors in this lattice and UGSDA to $\left(\frac{p_1}{p}, \ldots, \frac{p_n}{p}\right)$ since a vector $\mathbf{v} = \Sigma_{i=0}^n q_i \mathbf{b}_i$ in L has length $\|\mathbf{v}\| = \sqrt{\Sigma_{i=1}^n (q_0 p_i - q_i p)^2 + \lambda^2 q_0^2}$. ($\lambda$ should be chosen small enough so that λ_{q_0} is not the largest term in this sum.)

Although the problem of finding the shortest vector in a lattice is not known to be NP-hard, there is no known polynomial time algorithm for solving it. There are, however, polynomial time algorithms for finding relatively short vectors in a lattice. The first such algorithm was due to Lovasz [93]. For a lattice with a basis in which all coefficients in the basis are integers with absolute value $< B$, the algorithm is guaranteed to terminate in $O(n^6 (\log B)^3)$ bit operations and produce a vector $\mathbf{v}$ such that $\|\mathbf{v}\|^2 \leq 2^n \|\mathbf{u}\|^2$ where $\mathbf{u}$ is the shortest nonzero vector in the lattice. In practice, modifications of the Lovasz algorithm run much faster than this, usually about $O(n(\log B)^3)$ steps, and produce vectors that are much closer to the length of the shortest vector in the lattice. In particular, on all of a small set of test cases which came from knapsack cryptosystems, an implementation of the Lovasz algorithm by Brickell [21] found vectors that could be used to break the cryptosystem even though the vectors needed were only about $1/n$ times the length of the original basis vectors in the lattice. There are also other lattice basis reduction algorithms due to Schnorr [141] that produce vectors that are guaranteed to be closer in length to the shortest vector in the lattice, but these algorithms are slower.

2.2 Multiple-Iterated Knapsacks

An I-iterated knapsack cryptosystem is one in which I modular multiplications are used to disguise an easy knapsack. For an I-iterated knapsack, there are I independent UGSDA. These UGSDA were studied by Brickell, Lagarias, and Odlyzko [25] and more extensively by Lagarias [86,87] and were later used to break the multiple iterated knapsack by Brickell [21].

These UGSDA can be used to break all of the knapsack cryptosystems [8,20,47,50,115,123,143,144,145,161] that have been proposed that rely on modular multiplications as a disguising technique. See the surveys by Brickell [23] and Desmedt [44] for more details.

2.3 Low-Density Attacks

The density of a knapsack $a_1, \ldots, a_n$ in which $A = \max \{a_1, \ldots, a_n\}$ is defined to be $\frac{n}{\log_2 (A)}$. There are algorithms due to Brickell [19] and to Lagarias and Odlyzko [88] for solving knapsacks of low density. The Lagarias-Odlyzko [88] algorithm consists of looking for short vectors in the lattice L generated by the row vectors in the matrix

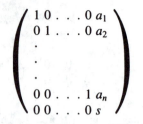

where s is the sum for the knapsack problem. In [88], the algorithm is analyzed with the Lovasz basis reduction algorithm [93] being used to find the short vectors in L. The polynomial time algorithm will solve almost all knapsack problems of density $< \frac{2}{n}$. (Frieze [54] has obtained a simpler proof of this result.) In practice, the algorithm is successful on knapsacks of much higher density, but the densities for which the algorithm succeeds does appear to go to 0 as n increases. Using more efficient lattice basis reduction algorithms [134,140,141] would increase the critical density below which this attack succeeds.

3 GENERALIZED KNAPSACK CRYPTOSYSTEMS

In this section, we will examine several cryptosystems that have been proposed which use ideas similar to those used in the knapsack cryptosystems.

3.1 Lu–Lee Systems

The Lu–Lee Cryptosystem [98]:

- Public key: c_1, c_2, r, M_1, M_2 all positive integers.
- Messages: Integers m_1, m_2 such that $0 < m_1 < M_1$, $0 < m_2 < M_2$.
- Encryption:

$$E(m_1, m_2) = c_1 m_1 + c_2 m_2 \bmod r \qquad (3.1.1)$$

- Decryption: The parameters were chosen so that the encryption function is one-to-one on the message space. The knowledge of the private key allows easy decryption.

This cryptosystem was broken by Adleman and Rivest [4], Goethals and Couvreur [60], and Kochanski [81]. Adiga and Shankar [2] suggested a modification of this scheme.

The modified Lu–Lee cryptosystem [2]:

- Public key: c_1, c_2, r, M all positive integers.
- Messages: Positive integers $m < M$.
- Encryption: Pick $m_1 < M_1$, $m_2 < M_2$, and compute

$$E(m) = m + c_1 m_1 + c_2 m_2 \bmod r \qquad (3.1.2)$$

- Decryption: Same remarks as above apply.

For both of these systems, cryptanalysis by solving Eqs. (3.1.1) or (3.1.2) by integer linear programming is immediate. Kannan's integer linear programming algorithm [78] runs in $O(n^{9n} \log r)$ in the worst case on problems with n variables and integer coefficients bounded by r. Since $n \leq 4$ for Eqs. (3.1.1) and (3.1.2), Kannan's algorithm is a viable threat to these systems.

3.2 Niederreiter Cryptosystem

Niederreiter [114] proposed a knapsack-type cryptosystem using algebraic coding theory.

- Private key: H, an $(n - k)$ by n parity check matrix of a t-error correcting linear (n, k) code, C, over GF(q) with an efficient decoding algorithm. P, an $n \times n$ permutation matrix. M, a nonsingular $(n - k) \times (n - k)$ matrix.
- Public key: $K = MHP$ and t.
- Messages: n dimensional vectors $\mathbf{y}$ over GF(q) with weight $\leq t$.
- Encryption: $\mathbf{z} = K\mathbf{y}^T$.
- Decryption: Since $\mathbf{z} = K\mathbf{y}^T = MHP\mathbf{y}^T$, $M^{-1}\mathbf{z} = HP\mathbf{y}^T = H(\mathbf{y}P^T)^T$. Use the decoding algorithm for C to find $\mathbf{y}P^T$ and thus $\mathbf{y}$.

This cryptosystem is said to be of knapsack type because the encryption can be viewed as picking t columns from the matrix K and forming a weighted sum of these t column vectors.

We will mention three cryptanalytic attacks on this system. In the first attack, for a ciphertest, $\mathbf{z}$, we pick a submatrix J of K consisting of $(n - k)$ columns of K. We then compute $\mathbf{y}' = J^{-1}\mathbf{z}$. If all of the t columns that were added to form $\mathbf{z}$ are in J, then $\mathbf{y}'$ will be the encrypted message, that is, $\mathbf{y}'$ will satisfy $K\mathbf{y}' = \mathbf{z}$ and have at most t nonzero entries. The probability of this occurrence is $\rho = \binom{n-k}{t} / \binom{n}{t}$. Thus the expected number of times we must repeat this procedure before we are successful is $\frac{1}{\rho}$. There are two examples mentioned in [114]. For the first $n = 104$, $k = 24$, $t = 15$, and so $\frac{1}{\rho} = 72$. For the second example, $n = 30$, $k = 12$, $t = 9$, and $\frac{1}{\rho} = 295$.

Another attack on this cryptosystem is based on a deterministic linear algebra procedure. It is easy to find some vector $\mathbf{w}$ such that $K\mathbf{w} = \mathbf{z}$. Once $\mathbf{w}$ is found, we must have $\mathbf{w} = \mathbf{y} + \mathbf{c}$, for some codeword $\mathbf{c}$ in C. We can write C as the direct sum of two subspaces C_1 and C_2, with C_1 of dimension $[k/2]$, and list all the codewords of C_1 and C_2 (approximately $q^{k/2}$ in each case). Then, for each $\mathbf{c}_1$ in C_1, we only need to check whether $\mathbf{w} - \mathbf{y} - \mathbf{c}$ is in C_2. In both of the examples presented in [114], this procedure would be very fast.

There is another attack based on the low-density algorithm of [88] that can be used if GF(q) is a prime field, that is, q is a prime. Let $\mathbf{v}_1^T, \ldots, \mathbf{v}_n^T$ be the n column vectors of K. Let $v_i = (v_{i1}, v_{i2}, \ldots, v_{i,n-k})$. Let r be an integer. Let L be the lattice generated by the row vectors in the matrix

$$
Q = \begin{pmatrix}
1 & 0 & \ldots & 0 & 0 & rv_{11} & rv_{12} & \ldots & rv_{1,\,n-k} \\
0 & 1 & \ldots & 0 & 0 & rv_{21} & rv_{22} & \ldots & rv_{2,\,n-k} \\
 & \cdot & & & & & & & \\
0 & 0 & \ldots & 1 & 0 & rv_{n1} & rv_{n2} & \ldots & rv_{n,\,n-k} \\
0 & 0 & \ldots & 0 & 1 & rz_1 & rz_2 & \ldots & rz_{n-k} \\
0 & 0 & \ldots & 0 & 0 & rq & 0 & \ldots & 0 \\
0 & 0 & \ldots & 0 & 0 & 0 & rq & \ldots & 0 \\
 & \cdot & & & & & & & \\
0 & 0 & \ldots & 0 & 0 & 0 & 0 & \ldots & rq
\end{pmatrix}
$$

The vector $\mathbf{y}^* = (y_1, \ldots, y_n, 0, 0, \ldots, 0)$ is a vector in the lattice and has at most t nonzero entries. If $r \geq t$, then $\mathbf{y}^*$ will be the shortest vector in the lattice. (Since K generates a t-error correcting linear code, there cannot be two vectors, $\mathbf{y}_1^*$ and $\mathbf{y}_2^*$, such that $K\mathbf{y}_1^* = K\mathbf{y}_2^*$ and such that Hamming weight of each $\mathbf{y}_1^*$ and $\mathbf{y}_2^*$ is $\leq t$.) Although the lattice basis reduction algorithm is not guaranteed to find $\mathbf{y}^*$, this attack does cast suspicion on the security of the cryptosystem.

3.3 Goodman–McAuley Knapsack Cryptosystem

The Goodman–McAuley [61] knapsack cryptosystem uses modular multiplication to disguise an easy knapsack that is substantially different from those discussed above.
The Goodman–McAuley knapsack cryptosystem:

- Public key: Integers $b_1, \ldots, b_n$, q and p.
- Private key: Integers h, r satisfying $h \geq r + q$; primes $p_1, \ldots, p_n$ such that $p_i \geq 2^h$ for $1 \leq i \leq n$; $p = \prod_{i=1}^n p_i$; non-negative integers a_{ij} for $1 \leq i, j \leq n$, such that $\sum_{j=1}^n a_{ij} < 2^r$ for $1 \leq i \leq n$, and the matrix $A = (a_{ij})$ is nonsingular; non-negative integers a_i such that $a_i \equiv a_{ij} \bmod p_j$ for $1 \leq i \leq n$, $1 \leq j \leq n$; W relatively prime to p such that $b_i \equiv W a_i \bmod p$, for $1 \leq i \leq n$.
- Messages: $\mathbf{m} = (m_1, \ldots, m_n)$ such that $0 \leq m_i \leq 2^q$
- Encryption: $c = \sum_{i=1}^n m_i b_i \bmod p$.
- Decryption: Let $\mathbf{d} = (d_1, \ldots, d_n)$ where $d_i \equiv cW^{-1} \bmod p_i$. Then $\mathbf{m} = \mathbf{d}A^{-1}$.

If n is small, this cryptosystem can be broken by the Lenstra [97] or Kannan [78] linear programming algorithms. There is also a GCD attack that works for small r, and enables the cryptanalyst to recover all the secret information. Since $p_j \mid a_i - a_{ij}$, we have $p_j \mid b_i - a_{ij}W$, and so $p_j \mid a_{ij}b_k - a_{kj}b_i$. Since a_{kj}, $a_{ij} < 2^r$, the cryptanalyst can find (y, z) such that $p_j \mid \text{GCD}(p, yb_k - zb_i)$ by checking all pairs y, z with $0 \leq y, z \leq 2^r$. If for each pair j, l there exist $0 \leq y, z \leq 2^r$ such that $p_j \mid yb_k - zb_i$ and $p_l \mid yb_k - zb_i$, then the cryptanalyst will find all of the p_j's by taking GCDs. If this is not the case, he can pick a different k and continue. (Note that if $p_j p_l \mid y_1 b_k - z_1 b_i$ and $p_j p_l \mid y_2 b_k - z_2 b_i$, then $y_1 z_2 \equiv y_2 z_1$, mod $p_j p_l$. Since $0 < y_1, z_2 < p_j p_l$, this implies $y_1 z_2 = y_2 z_1$. Hence if the GCD attack fails to separate p_j and p_l, then it must be the case that $\frac{a_{ij}}{a_{kj}} = \frac{a_{il}}{a_{kl}}$. But this cannot be true for all k since A is nonsingular.)
Even if n and r are chosen large enough so that the above attacks will not work, the cryptosystem can still be broken using lattice basis reduction. Consider the lattice spanned by the rows $\mathbf{v}_0, \ldots, \mathbf{v}_n$ of the matrix

$$\begin{pmatrix} b_1 & b_2 & \ldots & b_n & \varepsilon \\ p & 0 & \ldots & 0 & 0 \\ 0 & p & \ldots & 0 & 0 \\ & \vdots & & & \\ 0 & 0 & \ldots & p & 0 \end{pmatrix}$$

Let k_{ij} be the integers satisfying $b_i W^{-1} - k_{ij}p_j = a_{ij}$. Then $b_i W^{-1} \frac{p}{p_j} - k_{ij}p = a_{ij} \frac{p}{p_j}$. Thus, there are n vectors in the lattice of the form $W^{-1} \frac{p}{p_j} \mathbf{v}_0 - \sum_{j=1}^n k_{ij}\mathbf{v}_j =$

$(x_1, \ldots, x_n)$ where $\sum_{i=1}^{n} x_i < 2^{-q}p$. Even if the Lovasz basis reduction algorithm does not produce these vectors, but instead finds vectors $\mathbf{u}_i = (x_{i1}, \ldots, x_{in})$ that satisfy $\sum_{j=1}^{n} |x_{ij}| < 2^{-q}p$, the cryptanalyst can still use these vectors to break the system.

3.4 Pieprzyk Knapsack Cryptosystem

Pieprzyk [124] designed a knapsack-type cryptosystem based on polynomials over GF(2). In the following description, all polynomials are over GF(2).
The Pieprzyk Knapsack Cryptosystem:

- Public key: Polynomials $k_1(x), \ldots, k_n(x)$ and integer d.
- Private key: Polynomials $\Psi(x), \phi_1(x), \ldots, \phi_n(x), p_1(x), \ldots, p_n(x), a(x)$ such that for $1 \leq i \leq n$, deg $\phi_i(x) = d + 1$, $\phi_i(x)$ is irreducible, $p_i(x) \equiv 1 \mod \phi_i(x)$, $p_j(x) \equiv 0 \mod \phi_i(x)$ if $i \neq j$, deg $\Psi(x) \geq \sum_{i=1}^{n} \deg\phi_i(x) + d$ and $\Psi(x$ is irreducible, and $k_i(x) \equiv p_i(x)a(x) \mod \Psi(x)$.
- Messages: $M = (m_1(x), \ldots, m_n(x))$ where deg $m_i(x) \leq d$.
- Encryption: $c(x) = \sum_{i=1}^{n} m_i(x)k_i(x)$.
- Decryption: Let $c'(x) = c(x)a^{-1}(x) \mod \Psi(x)$. $c'(x) \equiv \sum_{i=1}^{n} m_i(x)p_i(x) \mod \Psi(x)$. Since deg $(\sum_{i=1}^{n} m_i(x)p_i(x))$ deg $< \Psi(x)$, then $c'(x) = \sum_{i=1}^{n} m_i(x)p_i(x)$. So $m_j(x) = c'(x) \mod \phi_j(x)$.

The Pieprzyk knapsack cryptosystem is similar to the Goodman–McAuley knapsack cryptosystem except that integers have been replaced by polynomials. It can be broken by a similar GCD attack as well. However, in this case a much simpler solution is available. As is the case with the Luccio–Mazzone system described in Section 6.5, this cryptosystem can be broken by simple linear algebra. Note that the encryption does not involve any modular reductions. In fact, encryption is a linear transformation of the plaintext, and the matrix giving this transformation can be constructed easily from the coefficients of the polynomials $k_i(x)$. Since decryption is guaranteed to work, this matrix must have full rank.

3.5 Chor–Rivest Knapsack Cryptosystem

The Chor–Rivest knapsack cryptosystem [32] is the only knapsack cryptosystem that has been published that does not use some form of modular multiplication to disguise an easy knapsack. There is no feasible method known for breaking this system.
The Chor–Rivest cryptosystem:

- Public key: Integers $c_0, c_1, \ldots, c_{p-1}, p, h$, where p is a prime power, $h \leq p$, and finding discrete logarithms in $GF(p^h)$ is feasible.
- Private key: $f(x)$ is a monic irreducible polynomial over $GF(p)$ of degree h, $GF(p^h)$ will be implemented as $GF(p)[x]/f(x)$, t is a root of $f(x)$, g is a generator of the multiplicative group of $GF(p^h)$; for $\alpha \in GF(p)$, a_α is an integer such that $g^{a_\alpha} = t + \alpha$, π is a one-to-one map from $\{0, 1, \ldots, p - 1\}$ into $GF(p), b_i = a_{\pi(i)}$, d is an integer, $0 \leq d \leq p^h - 2$, and $c_i = b_i + d$.

- Messages: Vectors $M = (m_0, \ldots, m_{p-1})$ of non-negative integers such that $\sum_{i=0}^{p-1} m_i = h$.
- Encryption: $E(M) \equiv \sum_{i=0}^{p-1} m_i c_i \bmod p^h - 1$.
- Decryption: Compute $r \equiv E(M) - hd \equiv \sum_{i=0}^{p-1} m_i b_i \bmod p^h - 1$. Then $g^r = \prod_{i=0}^{p-1} g^{m_i b_i}$. Since we are implementing $GF(p^h)$ as $GF(p)[x]/f(x)$, g^r is represented as a polynomial in x of degree $< h$. Now $\theta(t) = \prod_{j=0}^{p-1} (t + \pi(j))^{m_j}$ is represented by a polynomial of degree h, and $\theta(x) = g^r$ in $GF(p)[x]/f(x)$. So $\theta(x) = u(x) + f(x)$ in $GF(p)[x]$, where $u(x)$ represents g^r. Thus by factoring $u(x) + f(x)$, the values of $m_0, \ldots, m_{p-1}$ can be obtained.

Chor and Rivest show that if some of the secret information is revealed, then the system is insecure. In particular, the cryptosystem is insecure if g and d are known in some model of $GF(p^h)$, or if t is known, or if π and d are known. They also mention an attack with nothing known that runs in $O(p^2 \sqrt{h}\, h^2 \log p)$. However, this attack is infeasible for the parameters they suggest (e.g., $p = 197$ and $h = 24$).

If d is known, then it is possible to reduce the cryptanalysis problem to the problem of finding the root of a very high-degree polynomial. Since d is known, the b_i's are also known. It can be shown that one can assume that the cryptanalyst knows $\pi^{-1}(0)$ and $\pi^{-1}(1)$. Without loss of generality, suppose $\pi(0) = 0$ and $\pi(1) = 1$. Then $g^{b_0} = t$. Let $e_0 = b_0^{-1} \bmod p^h - 1$. Then $t^{e_0} = g$, and $t^{e_0 b_1} = g^{b_1} = t + 1$. A somewhat similar argument can be used if d is not known. In both cases, though, one has to find the root of a very high-degree trinomial. Rabin [133] and Ben-Or [11], for example, have shown that a root of a polynomial of degree w over $GF(p)$ can be found in $O(w \log w \log \log w \log p)$ operations, but these algorithms are infeasible here since w is of the order of p^h. No faster method for finding a root of a trinomial is known.

4 THE ONG–SCHNORR–SHAMIR (OSS) SIGNATURE SCHEMES

Ong, Schnorr, and Shamir [121] proposed a signature scheme based on polynomial equations modulo n. Their motivation was to develop a scheme that requires little computation for generation and verification of signatures, an area where the RSA scheme is deficient.

- Public key: Polynomial $P(x_1, \ldots, x_d)$ and modulus n.
- Private key: A method of solving $P(x_1, \ldots, x_d) \equiv m \bmod n$ for $x_1, \ldots, x_d$ using only a small number of multiplications, additions, and divisions mod n.
- Messages: $m \in \mathbb{Z}_n$.
- Signature: $x_1, \ldots, x_d$ such that $P(x_1, \ldots, x_d) \equiv m \bmod n$.
- Verification: Check that $P(x_1, \ldots, x_d) \equiv m \bmod n$.

This scheme generated a great deal of interest when it was first announced [120] using a polynomial P of degree 2. In fact the authors offered $100 reward for its cryptanalysis. This reward was won by Pollard [127], but this did not deter the authors, who within a few months described a cubic version. This was also broken by Pollard [127], which caused the authors to publish a quartic version [122]. This version was broken by

Estes, Adleman, Kompella, McCurley, and Miller [53] and independently by Schnorr [127], and there have been no more schemes of this type proposed. We will give a brief exposition of the Pollard attack on the quadratic version.

4.1 Cryptanalysis of the Quadratic OSS Signature Scheme

The quadratic version proposed in [120] uses the polynomial $P(x_1, x_2) = x_1^2 + kx_2^2$ where the private key is an integer u such that $k = u^2$. To forge a signature to m, it is necessary to find x, y such that $x^2 + ky^2 = m$.

Note that the signature scheme is multiplicative, that is, if $x_1^2 + ky_1^2 = m_1$ and $x_2^2 + ky_2^2 = m_2$, then $x = x_1x_2 - ky_1y_2$ and $y = x_1y_2 + x_2y_1$ is a solution to $x^2 + ky^2 = m_1m_2$.

The Pollard algorithm:

1. Do steps (2) and (3), below, until m and k are small enough so that $x^2 + ky^2 = m$ can be solved with $x, y \in \{0, 1\}$, or until m is a square.

2. Replace m by a number $< 2\sqrt{k}$.

3. Interchange m and k by using $x \leftarrow \frac{x}{y}$ and $y \leftarrow \frac{1}{y}$.

4. Solve with $x, y \in \{0, 1\}$ and use the transformations of (2) and (3) to work back to the original equation.

To explain step (2), first find m_0 such that $m_0 = m \bmod n$, $m_0 \equiv 3 \bmod 4$, m_0 is prime, and $-k$ is a quadratic residue mod m_0. The integer m_0 is found by examining the integers in the sequence $m, m + n, m + 2n$, until an integer is found that satisfies all the conditions. Assuming appropriate randomness conditions on this set of integers, a success is expected with $O(\log n)$ trials. Solve $x_0^2 \equiv -k \bmod m_0$ and thus $x_0^2 + k = m_0m_1$.

Next we want to find $m_2 < 2\sqrt{k}$ and x_1, y_1 such that $x_1^2 + y_1^2 k = m_1m_2$. Let $Q = m_1^{1/2} k^{-1/4}$. Use continued fractions to find a, b with $|a| < Q$ such that $\left|\frac{x_0}{m_1} + \frac{b}{a}\right| \leq \frac{1}{aQ}$. Setting $x_1 = x_0a + m_1b$ and $y_1 = a$ satisfies the requirements.

Using the multiplicative property, the problem of solving $x^2 + ky^2 \equiv m \bmod n$ reduces to solving $x^2 + ky^2 \equiv m_2 \bmod n$ and this satisfies step (2).

$O(\log \log k)$ iterations of steps (2) and (3) will result in a small enough k and m so that a solution is easily found.

4.2 Other OSS Schemes

After Pollard broke the quadratic and cubic OSS schemes, Ong, Schnorr, and Shamir developed a scheme using fourth-degree polynomials which was essentially a quadratic scheme over a quadratic number field. This scheme was broken [53] by reducing its cryptanalysis to the cryptanalysis of the quadratic scheme over the integers.

The success that the cryptanalysts have had with the OSS schemes does not imply that there are no secure signature schemes of this type. However, it is enough evidence to create strong suspicions about the security of any such schemes. Also, as their complexity increases, their speed improvement over the RSA and Rabin schemes decreases. For these reasons, there has been no further search for new OSS-type signature schemes.

5 THE OKAMOTO–SHIRAISHI SIGNATURE SCHEME

The Okamoto–Shiraishi signature scheme [119] is based on the difficulty of finding approximate kth roots mod n. This signature scheme is interesting because (as is case with the OSS schemes) it is possible to generate these signatures much faster than RSA signatures. It has also been used as an example of a subliminal channel [154].

- Private key: Factorization of $n = p^2q$.
- Public key: n, a small integer k, and a one-way function h.
- Messages: m in the domain of h.
- Signature: s such that $s^k - h(m) \equiv \delta \mod n$ where $|\delta| \leq n^{2/3}$.

This scheme was originally proposed for $k = 2$. This version was quickly broken by Brickell and DeLaurentis [24]. The techniques of this attack also extend to $k = 3$. Shamir [148] found a different method to break the $k = 2$ case. We will present his method and also the Brickell and DeLaurentis [24] method for $k = 3$.

For $k = 2$, the forger is given $M = h(m)$ and wants to find s such that $s^2 - M \equiv \delta \mod n$ for some δ satisfying $|\delta| \leq n^{2/3}$. The forger picks r and computes x such that $1 \leq x < n^{1/3}$ and $2rx - M + r^2 \equiv \gamma \mod n$ for $\gamma < O(n^{2/3})$. Such an x does not exist for all choices of r (e.g., $r = (n + 1)/2$). However if the $n^{2/3}$ different valid signatures to M mod n are randomly distributed over the interval $[0, n]$, then we expect that an x will exist for most choices of r. If an x exists, it can be found through a variation of the extended Euclidean algorithm ([119] middle bits methods). Given x, $s = r + x$ is then a valid signature.

For the cubic scheme, again let $M = h(m)$. Pick $r = [\sqrt[3]{n}]$ (i.e., $r = (n/3) + \theta$ for $|\theta| \leq 1/2$.) Compute $z = M - r^3 \mod n$, and let $x = $ nearest integer to $z^{1/3}$ that is divisible by 3 (i.e., $x = z^{1/3} + \varepsilon$ for $|\varepsilon| \leq 3/2$). Then $s = r + x$ is a valid signature to m, since

$$s^3 \equiv r^3 + 3r^2x + 3rx^2 + x^3 \mod n$$
$$\equiv r^3 + 3\left(\frac{n}{3} + \theta\right)^2 x + 3\left(\frac{n}{3} + \theta\right) x^2 + z + 3z^{2/3} \varepsilon + 3z^{1/3} \varepsilon^2 + \varepsilon^3 \mod n$$
$$\equiv M + 3\theta^2x + 3\theta x^2 + 3z^{2/3} \varepsilon + 3z^{1/3} \varepsilon^2 + \varepsilon^3 \mod n$$
$$\equiv M + \delta \mod n \quad \text{for } |\delta| \leq O(n^{2/3})$$

Although this specific attack is easily guarded against by disallowing signatures that are close to $\frac{n}{3}$, the basic attack can be generalized. For example, let $r = [un/v]$ for an arbitrary rational $\frac{u}{v}$. Pick x as above except divisible by v^2. If ε is small enough, then $r + x$ will be a valid signature, otherwise pick a different u, v and try again.

Okamoto [117] also proposed an encryption scheme based on similar ideas. Again n is an integer of the form $n = p^2q$. The public key also contains an integer $u = a + bpq$ where $0 < a < \frac{1}{2}\sqrt{pq}$. This system can be broken by using u^2 mod n to solve for a. We have

$$u^2 \equiv a^2 + 2abpq \equiv 2au - a^2 \mod n$$

Solve $0 < a < n^{1/3}$ and $|u^2 - 2au \mod n| < n^{2/3}$ by the methods mentioned earlier. After Shamir discovered how to break this scheme (his attack is discussed in [118]),

Okamoto [118] modified the cryptosystem. In the new system, u is chosen in a different manner. A message (m_1, m_2) for $0 < m_i < n^{1/9}$, $i = 1$, 2 is encrypted as $c \equiv (m_1 u + m_2)^l \bmod n$. Vallee, Girault, and Toffin [158,159] cryptanalyzed this modified scheme for any l by using lattice basis reduction.

6 ADDITIONAL BROKEN TWO-KEY SYSTEMS

In this section, we will discuss several two-key cryptosystems that were broken soon after their publication. Several of them relied on composition of polynomials over finite fields. In addition to the specific attacks on such schemes that are mentioned below, there are now some fairly general techniques for decomposing polynomials developed by von zur Gathen, Kozen, and Landau [59] that cast suspicion on all similar schemes.

6.1 Matsumoto–Imai Cryptosystem

The Matsumoto–Imai cryptosystem [101] uses polynomials over GF(2^m). The private key consists of secret information about the public encryption polynomial.

- Private key: $E(X) = a(b + X^\alpha)^\beta$.
- Public key: $E(X) = \sum_{i=0}^{2^m-2} e_i X^i$.
- Messages: M in GF(2^m).
- Encryption: $C = E(M)$.
- Decryption: Use the private key to solve for M.

Matsumoto and Imai suggested that the Hamming weight of β should be small so the public key is not too long. Delsarte, Desmedt, Odlyzko, and Piret [45] showed that the public polynomial $E(X)$ would have a special form and this form would actually reveal the private key (or at least something that was functionally equivalent to the private key).

6.2 Cade Cryptosystem

The Cade [28] cryptosystem also uses polynomials over GF(2^m), for $m = 3r$. Let $M(x) = x^{q+1}$, where $q = 2^m$.

- Private key: $T(x) = a_0 x + a_1 x^q + a_2 x^{q^2}$, $S(x) = b_0 x + b_1 x^q + b_2 x^{q^2}$ where S and T are chosen to be invertible. $P(x) \equiv SMT(x) \bmod (x^q - x)$.
- Public key: $P(x) = p_{00} x^2 + p_{10} x^{q+1} + p_{11} x^{2q} + p_{20} x^{q^2+1} + p_{21} x^{q^2+q} + p_{22} x^{2q^2}$.
- Messages: M in GF(2^m).
- Encryption: $C = P(M)$.
- Decryption: Use the private key to solve for M.

James, Lidl, and Niederreiter [74] have shown that the private variables $a_0, \ldots, b_2$ can be found from the public key. Cade [29] has since used similar ideas to develop a much more complicated cryptosystem.

6.3 Yagisawa

Yagisawa [164] described a cryptosystem that combined exponentiation mod p with arithmetic mod $p - 1$. Brickell [22] showed that it could be broken without finding the private key.

To construct a public key in Yagisawa's cryptosystem, a designer picks a prime p and integers k_1, k_2, A, and B such that $2 \le A$, $B \le p - 2$, $0 \le k_1 \le p - 2$, $0 \le k_2 \le p - 2$, and $\text{GCD}(k_1 - k_2, p - 1) = 1$. He then picks integers β_1 and β_2 such that $\beta_1 + \beta_2 k_1 \equiv 1 \bmod p - 1$ and computes $C \equiv B^{\beta_1 + \beta_2 k_2} \bmod p$ and $D \equiv B^{\beta_1 + \beta_2} \bmod p$.

The public key will consist of $(A, B, C, D, k_1, k_2, p)$. To encrypt a message (X_1, X_2, X_3) where $0 \le X_i \le p - 2$, one computes (Y_1, Y_2, Y_3) where $Y_1 \equiv X_1 + X_2 + X_3 \bmod p - 1$, $Y_2 \equiv k_1 X_1 + k_2 X_2 + X_3 \bmod p - 1$, $F \equiv B^{X_1} C^{X_2} D^{X_3} \bmod p$, and $Y_3 \equiv A^F X_3 \bmod p$.

The designer, given (Y_1, Y_2, Y_3), can compute $F \equiv \beta_1 Y_1 + \beta_2 Y_2 \bmod p$ and hence can compute X_3, and then X_1 and X_2.

Even though the cryptanalyst does not know β_1 and β_2, he can decrypt in much the same manner because he can actually find B^{β_1} and B^{β_2}. To do this he first computes $r \equiv (k_1 - k_2)^{-1} \bmod p - 1$. Since p is prime, for all X, $X^{r(k_1 - k_2)} \equiv X \bmod p$. $(C^{k_1} B^{-k_2})^r \equiv B^{\beta_1(k_1 - k_2)r} \equiv B^{\beta_1} \bmod p$ and $(C^{-1} B)^r \equiv B^{\beta_2(k_1 - k_2)r} \equiv B^{\beta_2} \bmod p$. Hence the cryptanalyst can also compute F.

6.4 Tsujii–Matsumoto–Kurosama–Itoh–Fujioka Cryptosystem

Tsujii, Matsumoto, Kurosama, Itoh, and Fujioka [157] have devised a public key cryptosystem in which encryption is the evaluation of some rational functions. They remark that if a certain polynomial in a small (e.g., 4) number of variables could be factored, then their system is insecure. Unfortunately, multivariate polynomials can be factored in polynomial time, as was shown by Lenstra [94] and others.

6.5 Luccio–Mazzone Cryptosystem

In our early discussions of knapsack cryptosystems we noted that in general, their linearity was a reason to be suspicious of them. A surprisingly large number of cryptosystems that have been proposed either formally or informally have succumbed to attacks based on this weakness, for example, the Pieprzyk cryptosystem (see Section 3.4). As another example, Luccio and Mazzone [99] have proposed a system (which is not really a two-key system, though) for sending information simultaneously to several receivers. Each receiver i, $1 \le i \le n$, has a secret key (k_i, c_i) known only to himself and the sender, and a large prime p is public. To send message m_i to receiver i, for $1 \le i \le n$, the sender finds an $(n - 1)$-degree polynomial $f(z)$ in $\text{GF}(p)$ [2] such that $f(k_i) = c_i m_i \bmod p$ and broadcasts the coefficients of $f(z)$. Receiver i then obtains

$$m_i \equiv c_i^{-1}(f(k_i)) \bmod p$$

As was noted by Hellman [68], this system is very insecure, as the coefficients of the polynomial $f(z)$ are a linear transformation of the messages $(m_1, \ldots, m_n)$, and so a knowledge of n or slightly more ciphertext-plaintext pairs suffices to break the system.

7 THE RSA CRYPTOSYSTEM

The cryptosystem found by Rivest, Shamir, and Adleman [138] is the best known two-key cryptosystem. A message is encrypted as $f(m) = m^e \bmod n$ where n is a composite integer that is usually chosen as the product of only two primes, p and q, and e is relatively prime to $(p - 1)(q - 1)$. Both n and e are public, while p and q have to be kept secret. If the cryptanalyst can factor n, he can decrypt messages just as easily as the intended user. With the exception of some special situations discussed below, it is not known how to break the RSA system without factoring n. However, this has not been proved, although there are some interesting results of Alexi, Chor, Goldreich, and Schnorr [5] that say that recovering even a single bit of information from an RSA ciphertext is as hard as deciphering the full message.

Since there are several very good surveys of integer factoring algorithms (e.g., Lenstra and Lenstra [92] and Pomerance [128]), we will not go into details, but will only sketch briefly how effective those algorithms are and what precautions need to be taken in choosing the parameters of an RSA cryptosystem. We will also briefly mention some recent developments that could have dramatic impact on this area.

It has long been recognized that the primes p and q which give the public modulus $n = pq$ have to be carefully chosen, so that, for example, $p - 1$, $p + 1$, $q - 1$, and $q + 1$ all have relatively large prime factors. However, it is easy to find primes that satisfy these conditions, as was shown by Williams and Schmid [163] and Gordon [63]. It has also been shown that using very small public encryption exponents is insecure. It has recently been shown that precaution must also be taken in choosing the secret exponent, d. Wiener [160] has proven that if $e < n$, and $d < n^{1/4}$, then d can be easily determined, and thus n can be factored.

Integer factorization has advanced significantly in the last decade. When RSA was invented, the largest "hard" integer (i.e., an integer that did not have many prime factors that were either small or of special form that allows them to be split off easily) that had been factored up to then was under 40 (decimal) digits in length. Right now, hard integers of over 110 digits are being factored. This progress is due to advances in both amounts of computing power that are available and theory. As far as hardware is concerned, the most striking development has been the successful implementation of factoring algorithms on networks of workstations. This work was pioneered by Caron and Silverman [30], and extended by A. Lenstra and M. Manasse [96]. In their recent factorization of a 111-digit integer, Lenstra and Manasse used roughly the computing power of a 300 mips (million instructions per second) machine running for a year. What was remarkable about this was that this computation was accomplished in several weeks, employed machines from around the whole world, and used only spare time on them. This is in contrast to the situation a few years ago, when it seemed that one needed to have access either to supercomputers or to special-purpose machines like that proposed in [129] to factor large integers. Since every factor of 10 increase in computing power allows one to factor integers slightly over 10 decimal digits longer, and the Lenstra-Manasse implementation is relatively portable and extendible to networks with many more machines, one can expect that in the very near future, networks of workstations around some universities or industrial laboratories could be used in their idle time to factor 130-digit integers in a few weeks or months of elapsed time. In particular, it seems very likely that the RSA challenge cipher will be broken in the next year or so, since it involves factoring an integer of 129 digits. Since workstations are becoming more powerful very rapidly (much more rapidly than supercomputers, say) and computer networks are proliferating very fast, and are going to be much more easily accessible than special-purpose machines like that of Pomerance, Smith, and Tuler [129], one should not regard even 140-digit moduli as safe from present day algorithms.

While one can make fairly good projections about the development of technology and how that will affect the security of the RSA cryptosystem, it is much harder to be certain about theoretical developments. Most of the advances in factoring in the last decade have been due to new ideas, not faster machines. Then, for a while, theoretical advances slowed down. Most of the fast factoring algorithms that have been considered until recently have been shown (under various assumptions) to run in time

$$\exp\left((1 + o(1))\left((\log n)(\log \log n)\right)^{1/2}\right)$$

as $n \to \infty$ for the "hard" integers n that are of interest in cryptography. This was explained on technical grounds as being due to all these algorithms relying in one way or another on the density of so-called "smooth" integers (integers with only small prime factors). Recently, however, a new method was suggested by J. Pollard, developed further by H. Lenstra, and implemented by A. Lenstra and M. Manasse [95]. It is referred to as the number field sieve. It is very practical when it is applied to factoring so-called Cunningham integers, that is, integers n of the form

$$n = a^k \pm 1$$

where a is small and k is large. If we let

$$M(n, r) = \exp\left((r + o(1))(\log n)^{1/3}(\log \log n)^{2/3}\right)$$

then the number field sieve factors Cunningham integers n in time

$$M(n, 1.526. . .)$$

This algorithm is fast not only asymptotically, but also in practice, although it is quite complicated to implement, and A. Lenstra and Manasse have used it to factor Cunningham integers of about 150 decimal digits.

The number field sieve can also be extended to factor general integers. The best currently known method of doing this yields a running time estimate of

$$M(n, 2.080. . .)$$

Although asymptotically this is still far better than other algorithms, the point at which this method would be faster than algorithms such as the quadratic sieve appears to be in the vicinity of 150 decimal digits. On the other hand, the number field sieve is a very recent invention, and so it is likely that substantial improvements might occur which would make it practical.

One of the fascinating questions about RSA is whether it is as secure as factoring. There are several modifications and restrictions of RSA for which this has been proven (Rabin [132], Williams [162]), but it has never been shown for RSA itself. There are however, no known attacks on RSA that are faster than factoring the modulus.

Some of the protocols for using RSA have been broken. They are described in "Protocol Failures in Cryptosystems," by J. H. Moore [107] in this book.

7.1 Variations on the RSA Cryptosystem

While the basic RSA cryptosystem has resisted all attacks, that is not true for all variants of it. Kravitz and Reed [84] have proposed using irreducible binary polynomials in place of the primes p and q. That is, $p(z)$ and $q(z)$ are two secret irreducible polynomials over GF(2) of degrees r and s, respectively, the public modulus is the polynomial $n(z) = p(z)\, q(z)$, and the public encryption exponent e is chosen to be relatively prime to $(2^r - 1)(2^s - 1)$. This system can be broken by factoring $n(z)$, which is usually quite easy to do. However, a further weakness exists in this system, and was already noted in [84], and more extensively by Delsarte and Piret [41] and by Gait [57], namely, that the decryption exponent is the multiplicative inverse of e modulo 1 of $(2^u - 1)(2^{t-u} - 1)$, $1 < u < t/2$, where $t = r + s$ is the degree of $n(z)$. Thus the number of possible decryption exponents grows only linearly with the number of bits in the public key.

8 DISCRETE EXPONENTIATION

In the seminal paper of Diffie and Hellman [48] which started two-key cryptography, they suggested using exponentiation modulo a prime as a public key exchange algorithm. Let p be a prime and α a primitive element mod p. Alice chooses a random integer a and Bob a random integer b. Alice sends α^a mod p to Bob. Bob sends α^b mod p to Alice. Then both can compute α^{ab} mod p. There have been numerous extensions of this basic scheme. The scheme clearly extends to finite fields [126]. Shmuely [153] and McCurley [102] have studied this idea mod n when n is composite. Miller [106] and Koblitz [80] have extended this idea of elliptic curves. El Gamal [52] developed techniques for using discrete exponentiation directly for encryption and signatures.

The security of the discrete exponentiation cryptosystems is based on the difficulty of the discrete logarithm problem, that is, given α, β, find x such that $\alpha^x = \beta$. There have been significant advances in algorithms for finding discrete logarithms in finite fields, particularly in GF(2^n), where a striking advance was made by Coppersmith [34]. These results are surveyed in [116] and [103]. With current algorithms, the complexity of finding discrete logarithms in a prime field GF(p) for a general prime p is essentially the same as the complexity of factoring an integer n of about the same size where n is the product of two approximately equal primes ([36,90]). In particular, the number field sieve can also be extended to compute discrete logs in prime fields, but so far it is only practical when the prime is a factor of a Cunningham integer [62]. However finding discrete logarithms in GF(2^k) is considerably easier.

When utilizing finite fields GF(q), whether q is prime or $q = 2^k$, it is necessary to ensure that $q - 1$ has a large prime factor, as otherwise it is easy to find discrete logarithms in GF(q). This restriction is similar to the need to choose the secret primes in the RSA system carefully [63, 163].

To date, there are no subexponential algorithms for finding discrete logarithms in elliptic curves.

9 THE McELIECE CRYPTOSYSTEM

In 1978, McEliece [104] introduced a two-key cryptosystem based on error correcting codes. An implementation of this scheme would be two to three orders of magnitude faster than RSA. It has two major drawbacks. The key is quite large and it increases the bandwidth. For the parameters suggested by McEliece [104], the key would have 2^{19} bits, and a ciphertext would be twice as long as a message.

Let d_H denote the Hamming distance. The following is a description of the Mc-Eliece cryptosystem for parameters n, k, t.

- Private key: G'—a $k \times n$ generator matrix for a Goppa code that can correct t errors; P—an $n \times n$ permutation matrix; S—a $k \times k$ nonsingular matrix.
- Public key: $G = SG'P$, a $k \times n$ matrix.
- Messages: k-dimensional vectors over GF(2).
- Encryption: $c = mG + z$ for z a randomly chosen n-dimensional vector over GF(2) with Hamming weight at most t.
- Decryption: Let $\mathbf{c'} = \mathbf{c}P^{-1}$. Using a decoding algorithm for the Goppa code, find $\mathbf{m'}$ such that $d_H(\mathbf{m'}G, \mathbf{c'}) \le t$. Then $\mathbf{m} = \mathbf{m'}S^{-1}$.

McEliece suggested that for $n = 1024$, t should be 50. For $n = 2^r$, the maximum $k = 2^r - rt$.

The security of this scheme is based on the NP-completeness of the general decoding problem for linear codes [12]. The only attacks on the system so far have come from improvements in algorithms that would decode any error correcting code.

An obvious attack on this system is to pick k columns of the matrix G. Let G_k, $\mathbf{c}_k$, z_k be restrictions onto these k columns. If $z_k = 0$, then $mG_k = \mathbf{c}_k$, and m can be found by linear algebra. A given choice of k columns can be checked in k^3 operations (assuming that fast matrix multiplication is not used) to see if it gives an appropriate m. For $n = 1024$ and $t = 50$, the expected number of operations before a success is about $2^{80.7}$. However, Adams and Meijer [1] showed that for $n = 1024$, $t = 37$ is the optimum value based on this attack, and for this value of t, the expected number of operations is about $2^{84.1}$.

Lee and Brickell [91] modified this attack. They found that after picking k columns, it was more efficient to check if z_k had at most two 1's. Against this attack, for $n = 1024$ $t = 38$ is optimal. For $t = 37$ or 38, the expected number of operations is about $2^{73.4}$.

To the best of our knowledge, there have been no successful attempts to cryptanalyze this system which examined possible leakage of the structure of Goppa codes into the public key.

Rao and Nam [135] have proposed using a variant of the McEliece scheme as a single-key cryptosystem. The key consists of a matrix G' generated in exactly the same manner as in the McEliece scheme, and a set F of possible error vectors. A message m is encrypted by picking a $z \in F$ at random and forming $c = mG' + z$. Rao and Nam give two methods of selecting the set F. Hin [72] showed how to break the Rao–Nam system for one of these methods and Struik and Tilburg [156] for the other. Both of these attacks used a chosen plaintext attack in which the cryptanalyst needs $|F|$ different encryptions of a fixed message m.

The Rao-Nam system could be modified slightly by using a pseudo-random function f, and letting $z = f(m)$ so that there is only one encryption for each message m. It is not known if the above attacks could be modified so that they would also break this system.

10 CONGRUENTIAL GENERATORS

A *congruential generator* is a method of generating a sequence $s_0, s_1, \ldots$ where s_i is computed by the recurrence

$$s_i \equiv \sum_{j=1}^{k} \alpha_j \phi_j(s_0, \ldots, s_{i-1}) \bmod m$$

Research in the last few years has uncovered serious weaknesses in using congruential generators as secure pseudorandom number generators. Methods have been found for cryptanalyzing congruential generators in which the cryptanalyst knows the functions ϕ_j but not the coefficients α_j and the modulus m. We will examine these results in Section 10.1.

The simplest congruential generator, the *linear congruential generator*, has the form

$$s_i \equiv as_{i-1} + b \bmod m$$

A *truncated congruential generator* generates a sequence $x_0, x_1, \ldots$ where x_i is the leading t bits of s_i for some sequence s_i produced by a congruential generator. Alternately, we could determine the x_i by some window of t of the bits of the s_i. Truncated linear congruential generators have recently been shown to be insecure even if the parameters a, b, and m are secret. We will examine these results in Sections 10.2 and 10.3.

There have been no attacks proposed for truncated nonlinear congruential generators.

10.1 Congruential Generators (Nontruncated)

We will evaluate the security of congruential generators relative to a variation of a known plaintext attack. We will assume that the cryptanalyst knows the functions ϕ_j, but does not know the coefficients α_j or the modulus m. The cryptanalyst is given $s_1, \ldots, s_{i-1}$. He tries to guess s_i. After he guesses, he is told the correct value. We will say that such an attack breaks the cryptosystem if there is a bound that is polynomial in $\log m$ and k on the running time of the attack and on the number of errors that are made by the cryptanalyst.

The cryptanalysis of congruential generators was started by Boyar [125] when she found how to break linear congruential generators. (Knuth [79] had an earlier result, but his algorithm was exponential in $\log m$.) Boyar also showed how to break quadratic and cubic congruential generators. Lagarias and Reeds [89] then extended Boyar's result by showing that the same algorithm would break any congruential generator, where $k = 1$ and ϕ is a polynomial depending only on s_{i-1}. Recently, Krawczyk [85] has proven how to break any congruential generator, in which the functions ϕ_j are computable over the integers in time polynomial in $\log m$.

Krawczyk's algorithm is only a slight modification of Boyar's and we will present it here because of its simplicity. The basic idea that Krawczyk introduces is that he does not try to find the α_j's.

Let

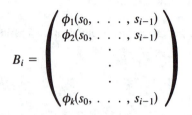

$$B_i = \begin{pmatrix} \phi_1(s_0, \ldots, s_{i-1}) \\ \phi_2(s_0, \ldots, s_{i-1}) \\ \cdot \\ \cdot \\ \cdot \\ \phi_k(s_0, \ldots, s_{i-1}) \end{pmatrix}$$

The first idea used by both Boyar and Krawczyk is that for all but possibly k values of i, there exist integers γ_j, $j = 1, \ldots, i$ such that $\gamma_i \neq 0$ and $\gamma_i B_i = \sum_{j=0}^{i-1} \gamma_j B_j$. Then $\gamma_i s_i \equiv \sum_{j=0}^{i-1} \gamma_j s_j \bmod m$. Thus, either s_i can be predicted (in the case that $\gamma_i s_i \equiv \sum_{j=0}^{i-1} \gamma_j s_j$) or a multiple of m can be computed after the correct value of s_i is given. The size of such a multiple of m will be polynomial in $\log m$ and k.

Once we know a multiple of m, we do the following for each i. Let $\hat{m}$ be the current multiple of m that is known.

1. Given s_{i-1}, try to express B_i as $B_i \equiv \sum_{j=0}^{i-1} \gamma_j B_j \bmod \hat{m}$.
2. If (1) is successful, compute p as $p \equiv \sum_{j=0}^{i-1} \gamma_j s_j \bmod \hat{m}$ and if $p \neq s_i$, then replace $\hat{m}$ by GCD($\hat{m}, p - s_i$).

Krawczyk has shown that if $p \neq s_i$, then $\hat{m} \neq$ GCD($\hat{m}, p - s_i$). He also showed that for a fixed $\hat{m}$, step (1) fails at most $k \log \hat{m} + 1$ times. From these results, it follows that this algorithm breaks these congruential generators in polynomial time.

10.2 Linear Truncated Congruential Generators with Known Parameters

In this section, we will consider the security of truncated linear congruential generators in which the cryptanalyst knows the parameters a, b, and m. These generators were shown to be insecure by Frieze, Hastad, Kannan, Lagarias, and Shamir [55,56,67]. All of the known attacks are attacks on linear congruential generators in which some constant fraction of the bits of each s_i are used as the pseudorandom sequence. The attacks are all based on lattice basis reduction. Each of the attacks that we will describe has been proven to break certain truncated linear congruential generators. However, it has not been determined whether these attacks would also be effective against most truncated linear congruential generators.

Let s_i be a sequence generated by

$$s_i \equiv as_{i-1} + b \bmod m \tag{10.1}$$

Let $n = \log_2 m$. For $0 < \beta < 1$ such that βn is an integer, we can write

$$s_i = x_i 2^{\beta n} + y_i \tag{10.2}$$

so that y_i is the lower βn bits of s_i and x_i is the high-order $(1 - \beta)n$ bits of s_i.

To evaluate the security of these sequences, we will assume that the cryptanalyst knows $x_1, \ldots, x_{i-1}$, and he wants to predict x_i. For the remainder of this section, we will assume that $b = 0$, for if $b \neq 0$, we could examine the sequence $\hat{x}_i = x_i - x_{i-1}$. This sequence is essentially the truncation of the sequence $\hat{s}_i = s_i - s_{i-1}$ which is generated by $\hat{s}_i \equiv a\hat{s}_{i-1}$ mod m. If we could predict the sequence $\hat{x}_i$, then we could also predict the sequence x_i.

Let L be the lattice spanned by the vector $(m, 0, \ldots, 0)$ and by the $k - 1$ vectors

$$(a^{i-1}, 0, \ldots 0, -1, 0, \ldots 0), \text{ for } i = 2, \ldots k$$

where the -1 is in the i'th coordinate. All vectors $\mathbf{w} = (w_1, \ldots, w_k)$ in L satisfy

$$\sum_{i=1}^{k} w_i s_i \equiv 0 \text{ mod } m \tag{10.3}$$

The attack consists of two steps. First, find a reduced basis for L, of vectors $\mathbf{w}^j, j = 1, \ldots, k$. We have

$$\sum_{i=1}^{k} w_i^j s_i = \sum_{i=1}^{k} w_i^j x_i 2^{\beta n} + \sum_{i=1}^{k} w_i^j y_i \tag{10.4}$$

If

$$\left| \sum_{i=1}^{k} w_i^j y_i \right| < \frac{m}{2} \tag{10.5}$$

for $j = 1, \ldots, k$, then since we know that each equation in Eq. (10.4) is 0 mod m, and we know the x_i, we get k independent equations over the integers for the s_i, $i = 1, \ldots, k$.

If the vectors in the reduced basis satisfy Eq. (10.5), then this attack will be successful.

Theorem 10.1[55]:   Let m be squarefree, $\varepsilon > 0$, and k be a given integer. There exist constants c_k and $C(\varepsilon, k)$ such that if $m > C(\varepsilon, k)$ and if $(1 - \beta)n > n(\frac{1}{k} + \varepsilon) + c_k$, then the reduced basis found by the Lovasz algorithm will satisfy Eq. (10.5) for at least $1 - O(m^{-\frac{\varepsilon}{2}})$ of the possible coefficients a.

The constant $c_k = O(k^2)$ and $C(\varepsilon, k) = e^{2k^{d_0 \varepsilon^{-1}}}$ for some constant d_0. Frieze, et al. also have a similar result for m which are almost squarefree and they have proved Theorem 10.1 for $k = 3$ and any m. It is an interesting question to determine if this attack will work for $k > 3$ and m an integer that is not almost squarefree, for example, $m = 2^n$. To prove that the attack will work in this case appears to need different proof techniques than those used in [55]. The attack that has been described will also be effective against truncated linear congruential generators in which some block of bits other than the most significant bits are used for the pseudorandom sequence. However, in this case, the algorithm is not quite as efficient, and asymptotically twice as many bits are needed to break the system.

It would also be interesting to determine whether this attack would be successful when the modulus m is not so large compared with k. This could probably be established by experimental evidence, but to our knowledge, there has been no computational experience with this algorithm.

10.3 Truncated Linear Congruential Generators with Unknown Parameters

In this section, we assume that the cryptanalyst does not know the parameters a, b, and m. Boyar [16] showed that if only a few bits ($O(\log \log m)$) were truncated, then her attack would still work. Stern [155] has recently discovered an extension of the FHKLS method that will break the truncated linear congruential generators (LCGs) when a constant fraction of the bits have been truncated.

Let us first consider his algorithm when m is known. Let $\mathbf{v}_i$ be the vector $(x_{i+1} - x_i, x_{i+2} - x_{i+1}, x_{i+3} - x_{i+2})$. In part 1 of the algorithm, use the algorithm of Hastad, Just, Lagarias, and Schnorr [66] to find a short integer relation

$$\sum_{i=1}^{k} \lambda_i v_i = 0$$

Let $\mathbf{w}_i$ be the vector $(s_{i+1} - s_i, s_{i+2} - s_{i+1}, s_{i+3} - s_{i+2})$. Then, let

$$\mathbf{u} = \sum_{i=1}^{k} \lambda_i \mathbf{w}_i$$

Stern has shown that if k is at least $\sqrt{6(1 - \beta) \log m}$, then for most a, $\mathbf{u}$ will be the zero vector. If $\mathbf{u}$ is the zero vector, then part 1 is successful.

If part 1 is successful, then the cryptosystem is insecure under a type of known plaintext attack. Assume that the cryptanalyst knows the values of $x_0, \ldots, x_{i-1}$, and that he is also given the h least significant bits of x_i. From this information, he is asked to predict the next bit of x_i. Stern has shown that if part 1 was successful, then out of the $(1 - \beta)n$ bits of x_i, the expected number of mistakes is only $\sqrt{6(1 - \beta) \log \frac{m}{2}}$.

Now we will consider the case when the cryptanalyst does not know a or m. Stern has shown that if part 1 is successful, then the polynomial $P(z) = \sum_{i=0}^{k-1} \lambda_i z^i$ satisfies $P(a) \equiv 0 \bmod m$. Stern suggests that by repeatedly using part 1, we could obtain many such polynomials and use them to determine m and a. Stern could prove that this method would work based on an assumption that involves the randomness of the polynomials P. Lacking a proof of this assumption, it would be interesting to also test this algorithm.

11 DES

The most remarkable news about the cryptanalysis of DES [112] is that there are no substantial attacks to mention. See Konheim's book [83] for a complete description of the algorithm. Although DES has been the U.S. standard for almost 10 years, and been the focus of many attempts at cryptanalysis, [17,49] it remains unbroken. The fastest attacks known at this time require $|K|/2$ encryptions where $|K| = 2^{56}$ is the total number of possible keys.

11.1 Cryptanalytic Attacks on Weakened DES

There has been some success in breaking weakened DES-like cryptosystems. Grossman and Tuckerman [64] showed DES could be made weak by modifying the method in which the S-boxes were used.

Another way to weaken DES is to shorten the number of rounds from the 16 that were proposed. Andelman and Reeds [7] developed a general technique for cryptanalyzing substitution-permutation cryptosystems which worked extremely well on networks with only three or four rounds. Chaum and Evertse [31] found a known plaintext attack on a six-round DES that is faster than exhaustive key search. Davies [37] exploited some nonrandom structures that he found in the S-boxes of DES that enabled him to break an eight-round DES using 2^{40} known plaintext messages. Biham and Shamir [14] have recently announced a chosen plaintext attack that can break an eight-round DES with 2^{18} chosen plaintext-ciphertext pairs. Their method extends to a 15-round DES, which can be broken with 2^{52} chosen plaintext-ciphertext pairs. However, for the full 16-round DES, their method requires more plaintext-ciphertext pairs than the 2^{55} encryptions needed for an exhaustive key search.

Although there has been no success against the full DES algorithm, there has been cryptanalytic success in breaking one of the proposed modes of operation of DES [113]. In output feedback mode (OFB), DES is used to generate a pseudorandom sequence, which is then used as a one-time pad to encrypt the message. It makes use of a function, $f_{k, m}$, where k is any valid DES key and $1 \le m \le 64$. $f_{k, m}(x) = x$ shifted left m bits and concatenated with the leftmost m bits of $E_k(x)$. ($E_k(x)$ is the DES encryption of x using key k). To generate a sequence $s_1, \ldots$ using OFB, a key, k, and an initial 64-bit vector x_0 are chosen. Then for $i \ge 1$, $s_i = E_k(x_{i-1})$ and $x_i = f_{k, m}(x_{i-1})$. Davies and Parkin [38] observed that for a fixed key k and $m = 64$, the function, f, is a permutation. The expected cycle size of a random permutation on N elements is $N/2$. However, if $m \le 63$, then f is not a permutation. The expected cycle size for a random function on N elements is only about $N^{1/2}$. Therefore only $m = 64$ should be considered secure for OFB.

11.2 Cycles in DES

Kaliski, Rivest, and Sherman [77] examined DES to see if any of several properties held. As an example, they wanted to determine whether the 2^{56} permutations E_k for $k \in K$ formed a subgroup. That is, for any two keys k_1 and k_2, is there another key k_3 such that $E_{k_1}(E_{k_2}(x)) = E_{k_3}(x)$ for all messages x. It was quite important to determine if DES had these properties, because if any one of them held, there would be an attack on DES that would require only $\sqrt{|K|}$ operations. By examining the results of some cleverly designed experiments on DES, they concluded that it was extremely unlikely that DES had any of these properties.

Additional cycling experiments have been performed by Moore and Simmons [108,109]. Soon after DES was released, four keys were labeled as weak keys. (These keys had the first 28 bits identical and also the last 28 bits identical.) In addition, several other keys were labeled as semiweak keys [75]. Coppersmith [35] and Moore and Simmons found some remarkable properties of these keys. In particular, they were able to find fixed points or antifixed points, that is, messages such that the encryption of the message is either the message itself or the complement of the message. Unfortunately, it is not apparent how to apply these results to give any information about other keys.

Quisquater and Delescaille [130] constructed an algorithm for finding collisions in DES. A collision is a message, m, and a pair of keys, k_1, k_2, such that both keys encrypt the message to the same ciphertext. Using their algorithm, they discovered many collisions in DES. It is not known how the existence of collisions can be used to aid in the cryptanalysis of DES.

11.3 Structural Properties of DES

The S-boxes introduce nonlinearity into the DES. There are eight S-boxes in the DES, each of which is a set of four permutations on sixteen elements. In the first public analysis of DES by Hellman, Merkle, Schroeppel, Washington, Diffie, and Schweitzer [69], there were several properties noted that were satisfied by all of the S-boxes. It was obvious that the S-boxes were not chosen at random, but there is no known cryptographic weakness resulting from these properties. Shamir [147] discovered an additional property of the S-boxes that at first looked very suspicious. However, Brickell, Moore, and Purtill [26] showed that this additional property was the result of the design properties noted by Hellman et al. [69] and Brickell et al. [26].

11.4 Birthday Attacks

There have been some cryptanalytic attacks based on the so called "birthday paradox." If $\alpha\sqrt{n}$ are drawn with replacement from a set of size n, the probability that two of them will be a match is about $1 - e^{-\alpha^2/2}$. This means that in a random group of 24 people, the probability that two will have the same birthday is about $1/2$. This is an old and well-understood concept and it has been the essential point of some recent cryptanalytic attacks.

Rabin [131] described a scheme for authenticating data using any block cipher as a hash function and RSA for a signature of the hashed value. Yuval [165] showed that this system could be broken with a birthday attack. Let n be the size of the image space of the hash function. (For DES, $n = 2^{64}$.) Suppose that Alice has two messages, x, a message that Bob wants to sign, and y, a message that Alice wants signed but Bob is not willing to sign. Alice prepares about $\sqrt{n}$ different slight variations of x and of y and computes the hash functions of each of them. With high probability, she will find variations $\hat{x}$ and $\hat{y}$ that hash to the same point. She gives $\hat{x}$ to Bob to be signed, but she can now use the signature of $\hat{x}$ as a signature for $\hat{y}$.

Another version of the birthday attack can be used to break this system even if Alice only has access to one valid signature and cannot obtain any additional ones. In the Rabin scheme, a text, $M_1, \ldots, M_r$, is signed by picking an H_0 at random. Then, $H_i = E_{M_i}(H_{i-1})$ for $i = 1, \ldots, r$. Finally, the pair (H_0, H_r) is signed using RSA.

Suppose that Alice is given the signature for a pair (H_0, G). She then picks $M_1, \ldots, M_{r-2}$ to be anything she likes, and computes $H_i = E_{M_i}(H_{i-1})$ for $i = 1, \ldots, r - 2$. Alice picks $\sqrt{n}$ X's and computes $E_X(H_{r-2})$ for each X. She picks $\sqrt{n}$ Y's and computes $D_Y(G)$ for each Y. With high probability, she will find a pair (X, Y) such that $E_X(H_{r-2} = D_Y(G)$. Then the signature for (H_0, G) will be a valid signature for $M_1, \ldots, M_{r-2}, X, Y$.

Davies and Price [39] proposed an iterated form of the Rabin scheme to avoid this latter birthday attack. They proposed going through all of the messages twice. Coppersmith [33] showed that this scheme is still susceptible to a birthday attack. See Jueneman [76] for a survey of these results.

12 FAST DATA ENCIPHERMENT ALGORITHM

The fast data encipherment algorithm (FEAL) was proposed by Shimizu and Miyaguchi [152] at Eurocrypt '87 as an alternative to DES for use in software. FEAL is a four-round substitution-permutation cryptosystem with a 64-bit key. Den Boer [42] soon found an attack on FEAL that requires only 10,000 chosen plaintexts. This has since been improved by Murphy [110] to an attack which needs only 20 chosen plaintexts. FEAL has since been modified to become FEAL-N [111], where N is the number of rounds. The methods that Biham and Shamir [14] developed can be used to break FEAL-8 with less than 2000 chosen plaintexts, and to break FEAL-N for $N \le 31$ with fewer chosen plaintexts than the number of encryptions needed in an exhaustive key search.

13 ADDITIONAL COMMENTS

In this section we will mention a few additional results, but without any details. The need to protect computer files has created a need for very efficient secure cryptosystems. However, many of the cryptosystems designed and sold to fill this need have been shown to be insecure. Reeds and Weinberger [136] have shown how to break the UNIX crypt command using a ciphertext-only attack. Kochanski [82] studied five security products designed for the IBM personal computer. He found them to be extremely insecure. He broke all of them using only a PC and without any knowledge about the encryption algorithms that was not provided by the manufacturer with the purchase of the product. He broke four of them with a ciphertext-only attack. A purchaser of these products should be very skeptical about their claims of security.

Rivest, Adleman, and Dertouzos [137] introduced several privacy homomorphisms. Essentially, a *privacy homomorphism* is an encryption function in which desired operations on plaintext messages can be achieved by performing corresponding operations on ciphertext messages. For example, $E(m) \equiv m^e$ mod n, the RSA encryption function, is a privacy homomorphism since $E(m_1) * E(m_2) \equiv E(m_1 m_2)$ mod n. There are four other privacy homomorphisms mentioned in [137]. Brickell and Yacobi [27] showed that two of these can be broken with ciphertext-only attacks and the other two can be broken with known plaintext attacks.

Although many of the encryption machines used during World War II were broken during the war, new techniques for breaking them are still being discovered. The techniques of Andelman and Reeds [7] for cryptanalyzing rotor machines and the comprehensive book covering cryptanalysis of World War II–era encryption machines by Deavours and Kruh [40] are excellent examples.

Schnorr [142] proposed an algorithm for constructing a string, $G_n(x)$ of length $2n2^{2n}$ bits from a random seed, x, of length $n2^n$ bits. He claimed that no statistical test that depended on fewer than $2^{o(n)}$ bits could distinguish $G_n(x)$ from a random bit string. However, Rueppel [139] has demonstrated a statistical test that depends on only $4n$ bits that does distinguish (with very high probability) $G_n(x)$ from a random string. Furthermore, Rueppel has shown that the seed, x, can be computed in time $O(n2^n)$ using only $n2^n + O(1)$ bits of $G_n(x)$. Thus, Schnorr's random number generator expands the randomness of the seed by at most a constant number of bits.

Matyas and Shamir [100] developed a novel idea for encrypting video signals. A randomly generated curve that passes through all pixels of a video signal is used to transmit the video picture. The light values at the pixels are then sent in the clear. Hastad [65] showed that this method was insecure if the same curve is used to transmit many pictures. Bertilsson, Brickell, and Ingemarsson [13] showed that it was insecure if many different curves were used to transmit similar pictures. Together, these results indicate that this scheme is unlikely to be secure without some major modifications.

A radically different concept for a cryptosystem has been proposed by Bennett, Brassard, Breidbart, and Wiesner [10]. They call it quantum cryptography and its security is based on the uncertainty principle of quantum physics. (A very complete list of references on this subject can be found in the report of Bennett and Brassard [9].) If such systems become feasible, the cryptanalytic tools discussed here will be of no use.

ACKNOWLEDGMENTS

The authors would like to thank Joan Boyar, D. Coppersmith, J. Hastad, H. W. Lenstra, Jr., and C. Schnorr for useful comments.

REFERENCES

[1] C. M. Adams and H. Meijer, "Security-related comments regarding McEliece's public-key cryptosystem," in *Lecture Notes in Computer Science 293; Advances in Cryptology: Proceedings of Crypto '87,* C. Pomerance, Ed., Santa Barbara, CA, Aug. 16–20, 1987, pp. 224–230. Berlin: Springer-Verlag, 1988.

[2] B. S. Adiga and P. Shankar, "Modified Lu-Lee cryptosystem," *Electron. Lett.,* vol. 21, no. 18, pp. 794–795, Aug. 29, 1985.

[3] L. M. Adleman, "On Breaking Generalized Knapsack Public Key Cryptosystems," *Proc. 15th ACM Symposium on Theory of Computing,* pp. 402–412, 1983.

[4] L. M. Adleman and R. L. Rivest, "How to break the Lu-Lee (COMSAT) public-key cryptosystem," MIT Laboratory for Computer Science, July 1979.

[5] W. Alexi, B. Chor, O. Goldreich, and C. P. Schnorr, "RSA and Rabin functions: Certain parts are as hard as the whole," *SIAM J. Comput.,* vol. 17, pp. 194–209, 1988.

[6] H. R. Amirazizi, E. D. Karnin, and J. M. Reyneri, "Compact knapsacks are polynomially solvable," extended abstract in *Crypto '81 Abstracts,* IEEE Workshop on Communications Security, Santa Barbara, CA, Aug. 24–26, 1981. Reprinted in *ACM SIGACT NEWS,* vol. 15, pp. 20–22, 1983.

[7] D. Andelman and J. Reeds, "On the cryptanalysis of rotor machines and substitution-permutation networks," *IEEE Trans. Inform. Theory,* vol. IT-28, no. 4, pp. 578–584, July 1982.

[8] B. Arazi, "A trapdoor multiple mapping," *IEEE Trans. Inform. Theory,* vol. 26, no. 1, pp. 100–102, Jan. 1980.

[9] C. H. Bennett and G. Brassard, "Quantum public key distribution reinvented," *SIGACT News,* vol. 18, no. 4, pp. 51–53, Summer 1987.

[10] C. H. Bennett, G. Brassard, S. Breidhart, and S. Wiesner, "Quantum cryptography, or unforgeable subway tokens," in *Advances in Cryptology: Proc. Crypto '82*, D. Chaum, R. L. Rivest, and A. T. Sherman, Eds., Santa Barbara, CA, Aug. 23–25, 1982, pp. 267–275. New York: Plenum Press, 1983.

[11] M. Ben-Or, "Probabilistic algorithms in finite fields," in *Proc. 22nd IEEE Found. Computer Sci. Symp.*, pp. 394–398, 1981.

[12] E. R. Berlekamp, R. J. McEliece, and H. C. A. van Tilborg, "On the inherent intractability of certain coding problems," *IEEE Trans. Inform. Theory*, vol. IT-24, pp. 384–386, 1978.

[13] M. Bertilsson, E. F. Brickell, and I. Ingemarsson, "Cryptanalaysis of video encryption based on space-filling curves," in *Lecture Notes in Computer Science 434; Advances in Cryptology; Proc. Eurocrypt '89*, J.-J. Quisquater and J. Vandewalle, Eds., Houthalen, Belgium, April 10–23, 1989, pp. 403–411. Berlin: Springer-Verlag, 1990.

[14] E. Biham and A. Shamir, "Differential cryptanalysis of DES-like cryptosystems," extended abstract in *Advances in Cryptology: Proc. Crypto '90*, S. Vanstone, Ed., Santa Barbara, CA, Aug. 11–15, 1990. Berlin: Springer-Verlag (in press) to appear in *J. Cryptology*, vol. 4, no. 1, 1991.

[15] J. Boyar, "Inferring sequences produced by pseudo-random number generators," *J. ACM*, vol. 36, pp. 129–141, 1989.

[16] J. Boyar, "Inferring sequences produced by a linear congruential generator missing low-order bits," *J. Cryptology*, vol. 1, no. 3, pp. 177–184, 1989.

[17] D. K. Branstead, J. Gait, and S. Katzke, "Report of the workshop on cryptography in support of computer security," *NBSIR 77-1291*, National Bureau of Standards, September 21–22, 1976, Sept. 1977.

[18] G. Brassard, "A note on the complexity of cryptography," *IEEE Tran. Inform. Theory*, vol. IT-25, pp. 232–233, 1979.

[19] E. F. Brickell, "Solving low density knapsacks," in *Advances in Cryptology: Proc. Crypto '83*, D. Chaum, Ed., Santa Barbara, CA, Aug. 22–24, 1983, pp. 25–37. New York: Plenum Press, 1984.

[20] E. F. Brickell, "A new knapsack based cryptosystem," paper presented at Crypto '83, Santa Barbara, CA, Aug. 22–24, 1983.

[21] E. F. Brickell, "Breaking iterated knapsacks," in *Lecture Notes in Computer Science 196; Advances in Cryptology: Proc. Crypto '84*, G. R. Blakley and D. Chaum, Eds., Santa Barbara, CA, Aug. 19–22, 1984, pp. 342–358. Berlin: Springer-Verlag, 1985.

[22] E. F. Brickell, "Cryptanalysis of the Yagisawa public key cryptosystem," *Abstracts of Papers*, Eurocrypt '86, May 20–22, 1986.

[23] E. F. Brickell, "The cryptanalysis of knapsack cryptosystems," in *Applications of Discrete Mathematics*, R. D. Ringeisen and F. S. Roberts, Eds., pp. 3–23. Philadelphia: Society for Industrial and Applied Mathematics, 1988.

[24] E. F. Brickell and J. M. DeLaurentis, "An attack on a signature scheme proposed by Okamoto and Shiraishi," in *Lecture Notes in Computer Science 218; Advances in Cryptology: Proc. Crypto '85*, H. C. Williams, Ed., Santa Barbara, CA, Aug. 18–22, 1985, pp. 28–32. Berlin: Springer-Verlag, 1986.

[25] E. F. Brickell, J. C. Lagarias, and A. M. Odlyzko, "Evaluation of the Adelman attack on multiple iterated knapsack cryptosystems," in *Advances in Cryptology:*

Proc. Crypto '83, D. Chaum, Ed., Santa Barbara, CA, Aug. 22–24, 1983, pp. 39–42. New York: Plenum Press, 1984.

[26] E. F. Brickell, J. H. Moore, and M. R. Purtill, "Structure in the S-boxes of the DES," extended abstract in *Lecture Notes in Computer Science 263; Advances in Cryptology: Proc. Crypto '86*, A. M. Odlyzko, Ed., Santa Barbara, CA, Aug. 11–15, 1986, pp. 3–8. Berlin: Springer-Verlag, 1987.

[27] E. F. Brickell and Y. Yacobi, "On privacy homomorphisms," in *Lecture Notes in Computer Science 304; Advances in Cryptology: Proc. Eurocrypt '87*, D. Chaum and W. L. Price, Eds., Amsterdam, The Netherlands, April 13–15, 1987, pp. 117–126. Berlin: Springer-Verlag, 1988.

[28] J. J. Cade, "A public key cipher which allows signatures," paper presented at 2nd SIAM Conference on Applied Linear Algebra, Raleigh, NC, 1985.

[29] J. J. Cade, "A modification of a broken public-key cipher," in *Lecture Notes in Computer Science 263; Advances in Cryptology: Proc. Crypto '86*, A. M. Odlyzko, Ed., Santa Barbara, CA, Aug. 11–15, 1986, pp. 64–83. Berlin: Springer-Verlag, 1987.

[30] T. R. Caron and R. D. Silverman, "Parallel implementation of the quadratic scheme," *J. Supercomputing*, vol. 1, no. 3, pp. 273–290, 1987.

[31] D. Chaum and J. Evertse, "Cryptanalysis of DES with a reduced number of rounds," in *Lecture Notes in Computer Science 218; Advances in Cryptology: Proc. Crypto '85*, H. C. Williams, Ed., Santa Barbara, CA, Aug. 18–22, 1985, pp. 192–211. Berlin: Springer-Verlag, 1986.

[32] B. Chor and R. Rivest, "A knapsack type public key cryptosystem based on arithmetic in finite fields," *IEEE Trans. Inform. Theory*, vol. 34, pp. 901–909, 1988.

[33] D. Coppersmith, "Another birthday attack," in *Lecture Notes in Computer Science 218; Advances in Cryptology: Proc. Crypto '85*, H. C. Williams, Ed., Santa Barbara, CA, Aug. 18–22, 1985, pp. 14–17. Berlin: Springer-Verlag, 1986.

[34] D. Coppersmith, "Fast evaluation of logarithms in fields of characteristic two," *IEEE Trans. Inform. Theory*, vol. IT-30, pp. 587–594, 1984.

[35] D. Coppersmith, "The real reason for Rivest's phenomenon," in *Lecture Notes in Computer Science 218; Advances in Cryptology: Proc. Crypto '85*, H. C. Williams, Ed., Santa Barbara, CA, Aug. 18–22, 1985, pp. 535–536. Berlin: Springer-Verlag, 1986.

[36] D. Coppersmith, A. M. Odlyzko, and R. Schroeppel, "Discrete Logarithms in $GF(p)$," *Algorithmica*, vol. 1, no. 1, pp. 1–16, 1986.

[37] D. W. Davies, "Investigation of a potential weakness in the DES algorithm," unpublished manuscript circulated in July 1988.

[38] D. W. Davies and G. I. P. Parkin, "The average cycle size of the key stream in output feedback encipherment," in *Advances in Cryptology: Proc. Crypto '82*, D. Chaum, R. L. Rivest, and A. T. Sherman, Eds., Santa Barbara, CA, Aug. 23–25, 1982, pp. 97–98. New York: Plenum Press, 1983.

[39] D. W. Davies and W. L. Price, "The application of digital signatures based on public key cryptosystems," *NPL Report DNACS 39/80*, National Physical Laboratory, Teddington, Middlesex, England, Dec. 1980.

[40] C. A. Deavours and L. Kruh, *Machine Cryptography and Modern Cryptanalysis*, Dedham, MA: Artech House, 1985.

[41] P. Delsarte and P. Piret, "Comment on 'Extension of RSA cryptostructure: A Galois approach'," *Electron. Lett.*, vol. 18, pp. 582–583, 1982.

[42] B. Den Boer, "Cryptanalysis of F.E.A.L.," in *Lecture Notes in Computer Science 330; Advances in Cryptology: Proc. Eurocrypt '88*, C. G. Günther, Ed., Davos, Switzerland, May 25–27, 1988, pp. 293–299. Berlin: Springer-Verlag, 1988.

[43] D. E. R. Denning, *Cryptography and Data Security*, Menlo Park, CA: Addison-Wesley, 1983.

[44] Y. Desmedt, "What happened with knapsack cryptographic schemes?" in *Performance Limits in Communication, Theory and Practice*, J. K. Skwirzynski, Ed., Boston/Dordrecht/London: Kluwer, pp. 113–134, 1988.

[45] P. Delsarte, Y. Desmedt, A. Odlyzko, and P. Piret, "Fast cryptanalysis of the Matsumoto–Imai public key scheme," in *Lecture Notes in Computer Science 209; Advances in Cryptology: Proc. Eurocrypt '84*, T. Beth, N. Cot, and I. Ingemarsson, Eds., Paris, France, April 9–11, 1984, pp. 142–149. Berlin: Springer-Verlag, 1985.

[46] Y. G. Desmedt, J. P. Vandewalle, and R. J. M. Govaerts, "A critical analysis of the security of knapsack public key algorithms," *IEEE Trans. Inform. Theory*, vol. IT-30, no. 4, pp. 601–611, July 1984. Also in *Abstract of Papers, IEEE Intern. Symp. Inform. Theory* (Les Arcs, France), pp. 115–116, June 1982.

[47] Y. Desmedt, J. Vandewalle, and R. Govaerts, "A general public key cryptographic knapsack algorithm based on linear algebra," in *Abstract of Papers, IEEE Intern. Symp. Inform. Theory*, St. Jovite, Quebec, Canada, Sept. 26–30, 1983, pp. 129–130, 1983.

[48] W. Diffie and M. E. Hellman, "New directions in cryptography," *IEEE Trans. Inform. Theory*, vol. IT-22, pp. 644–654, 1976.

[49] W. Diffie and M. E. Hellman, "Exhaustive cryptanalysis of the NBS data encryption standard," *Computer*, vol. 10, pp. 74–84, 1977.

[50] A. DiPorto, "A public key cryptosystem based on a generalization of the knapsack problem," presented at Eurocrypt '85, Linz, Austria, April 9–11, 1985.

[51] R. Eier and H. Lagger, "Trapdoors in knapsack cryptosystems," in *Lecture Notes in Computer Science 149; Cryptography: Proc. Workshop Cryptography*, T. Beth, Ed., Burg Feuerstein, Germany, March 29–April 2, 1982, pp. 316–322. Berlin: Springer-Verlag, 1983.

[52] T. El Gamal, "A public key cryptosystem and a signature scheme based on discrete logarithms," *IEEE Trans. Inform. Theory*, vol. IT-31, no. 4, pp. 469–472, July 1985.

[53] D. Estes, L. M. Adleman, K. Kompella, K. S. McCurley, and G. L. Miller, "Breaking the Ong–Schnorr–Shamir signature schemes for quadratic number fields," in *Lecture Notes in Computer Science 218; Advances in Cryptology: Proc. Crypto '85*, H. C. Williams, Ed., Santa Barbara, CA, Aug. 18–22, 1985, pp. 3–13. Berlin: Springer-Verlag, 1986.

[54] A. M. Frieze, "On the Lagarias-Odlyzko algorithm for the subset sum problem," *SIAM J. Comput.*, vol. 15, no. 2, pp. 536–539, May 1986.

[55] A. M. Frieze, J. Hastad, R. Kannan, J. C. Lagarias, and A. Shamir, "Reconstructing truncated integer variables satisfying linear congruences," *SIAM J. Comput.*, vol. 17, pp. 262–280, 1988.

[56] A. M. Frieze, R. Kannan, and J. C. Lagarias, "Linear congruential generators do not produce random sequences," in *Proc. 25th IEEE Symp. Foundations of Computer Science,* Singer Island, FL, Oct. 24–26, 1984, pp. 480–484. Los Angeles: IEEE Computer Society Press, 1984.

[57] J. Gait, "Short cycling in the Kravitz-Reed public key encryption system," *Electron. Lett.,* vol. 18, pp. 706–707, 1982.

[58] M. R. Garey and D. S. Johnson, *Computers and Intractability: A Guide to the Theory of NP—Completeness,* San Francisco: W. H. Freeman, 1979.

[59] J. von zur Gathen, D. Kozen, and S. Landau, "Functional decomposition of polynomials," in *Proc. 28th IEEE Symp. Foundations of Computer Science,* Los Angeles, CA, Oct. 12–14, 1987, pp. 127–131. Los Angeles: IEEE Computer Society Press, 1987.

[60] J. M. Goethals and C. Couvreur, "A cryptanalytic attack on the Lu–Lee public-key cryptosystem," *Philips J. Res.,* vol. 35, pp. 301–306, 1980.

[61] R. M. Goodman and A. J. McAuley, "A new trapdoor knapsack public key cryptosystem," in *Lecture Notes in Computer Science 209; Advances in Cryptology: Proc. Eurocrypt '84,* T. Beth, N. Cot, and I. Ingemarsson, Eds., Paris, France, April 9–11, 1984, pp. 150–158, Berlin: Springer-Verlag, 1985. Also in *IEE Proc.,* vol. 132, pt. E, no. 6, pp. 289–292, Nov. 1985.

[62] D. M. Gordon, "Discrete logarithms in $GF(p)$ using the number field sieve" (in press).

[63] J. A. Gordon, "Strong primes are easy to find," in *Lecture Notes in Computer Science 209; Advances in Cryptology: Proc. Eurocrypt '84,* T. Beth, N. Cot, and I. Ingemarsson, Eds., Paris, France, April 9–11, 1984, pp. 216–233. Berlin: Springer-Verlag, 1985.

[64] E. Grossman and B. Tuckerman, "Analysis of a Feistel-like cipher weakened by having no rotating key," IBM Research Rept. RC 6375, Jan. 31, 1977. Also in *Proc. ICC 78.*

[65] J. Hastad, personal communication.

[66] J. Hastad, B. Just, J. Lagarias, and C. P. Schnorr, "Polynomial time algorithms for finding integer relations among real numbers," *SIAM J. Comput.,* vol. 18, pp. 859–881, 1989.

[67] J. Hastad and A. Shamir, "The cryptographic security of truncated linearly related variables," in *Proc. 17th ACM Symp. Theory of Computing,* Providence, RI, May 6–8, 1985, pp. 356–362. New York: Association for Computing Machinery, 1985.

[68] M. E. Hellman, "Another cryptanalytic attack on 'a cryptosystem for multiple communication'," *Inform. Processing Lett.,* vol. 12, pp. 182–183, 1981.

[69] M. E. Hellman, R. C. Merkle, R. Schroeppel, L. Washington, W. Diffie, S. Pohlig, and P. Schweitzer, "Results on an initial attempt to cryptanalyze the NBS data encryption standard," *Tech. Rept. SEL 76-042,* Stanford University, 1976.

[70] P. S. Henry, "Fast implementation of the knapsack cipher," *Bell Labs. Tech. J.,* vol. 60, pp. 767–773, May/June 1981.

[71] T. Herlestam, "Critical remarks on some public key cryptosystems," *BIT,* vol. 18, pp. 493–496, 1978.

[72] P. J. M. Hin, "Channel-error-correcting privacy cryptosystems," Ph.D. Thesis, Delft Univ. of Technology (1986, in Dutch).

[73] I. Ingemarsson, "A new algorithm for the solution of the knapsack problem," in *Lecture Notes in Computer Science 149; Cryptography: Proc. Workshop Cryptography*, T. Beth, Ed., Burg Feuerstein, Germany, March 29–April 2, 1982, pp. 309–315. Berlin: Springer-Verlag, 1983.

[74] N. S. James, R. Lidl, and H. Niederreiter, "Breaking the Cade cipher," in *Lecture Notes in Computer Science 263; Advances in Cryptology: Proc. Crypto '86*, A. M. Odlyzko, Ed., Santa Barbara, CA, Aug. 11–15, 1986, pp. 60–63. Berlin: Springer-Verlag, 1987.

[75] R. R. Jueneman, "Analysis of certain aspects of output feedback mode," in *Advances in Cryptology: Proc. Crypto '82*, D. Chaum, R. L. Rivest, and A. T. Sherman, Eds., Santa Barbara, CA, Aug. 23–25, 1982, pp. 99–127. New York: Plenum Press, 1983.

[76] R. R. Jueneman, "Electronic document authentication," *IEEE Networks*, vol. 1, no. 2, pp. 17–23, April 1987.

[77] B. S. Kaliski, R. L. Rivest, and A. T. Sherman, "Is the data encryption standard a group? (results of cycling experiments on DES)," *J. Cryptology*, vol. 1, no. 1, pp. 3–36, 1988.

[78] R. Kannan, "Improved algorithms for integer programming and related lattice problems," in *Proc. 15th ACM Symp. Theory of Computing*, pp. 193–206, 1983.

[79] D. E. Knuth, *Deciphering a Linear Congruential Encryption, Tech. Rept. 024800*, Stanford University, 1980.

[80] N. Koblitz, "Elliptic curve cryptosystems," *Mathematics of Computation*, vol. 48, pp. 203–209, 1987.

[81] M. J. Kochanski, "Remarks on Lu and Lee's proposals," *Cryptologia*, vol. 4, no. 4, pp. 204–207, 1980.

[82] M. J. Kochanski, "A survey of data insecurity packages," *Cryptologia*, vol. 11, no. 1, pp. 1–15, Jan. 1987.

[83] A. G. Konheim, *Cryptography, A Primer*, New York: John Wiley, 1981.

[84] D. Kravitz and I. Reed, "Extension of RSA cryptostructure: A Galois approach," *Electron. Lett.*, vol. 18, pp. 255–256, 1982.

[85] H. Krawczyk, "How to predict congruential generators," in *Lecture Notes in Computer Science 435; Advances in Cryptology: Proc. Crypto '89*, G. Brassard, Ed., Santa Barbara, CA, Aug. 20–24, 1989, pp. 138–153. Berlin: Springer-Verlag, 1990.

[86] J. C. Lagarias, "Knapsack public key cryptosystems and diophantine approximation," in *Advances in Cryptology: Proc. Crypto '83*, D. Chaum, Ed., Santa Barbara, CA, Aug. 22–24, 1983, pp. 3–23. New York: Plenum Press, 1984.

[87] J. C. Lagarias, "Performance analysis of Shamir's attack on the basic Merkle–Hellman knapsack cryptosystem," in *Lecture Notes in Computer Science 172; Proc. 11th Internat. Colloquium on Automata, Languages and Programming (ICALP)*, J. Paredaens, Ed., pp. 312–323. Berlin: Springer-Verlag, 1984.

[88] J. C. Lagarias and A. M. Odlyzko, "Solving low density subset sum problems," *J. Assoc. Comp. Mach.*, vol. 32, pp. 229–246, 1985.

[89] J. C. Lagarias and J. Reeds, "Unique extrapolation of polynomial recurrences," *SIAM J. Comput.*, vol. 17, pp. 342–362, 1988.

[90] B. A. LaMacchia and A. M. Odlyzko, "Computation of discrete logarithms in prime fields," *Designs, Codes, and Cryptography*, vol. 1, pp. 46–62, 1991.

[91] P. J. Lee and E. F. Brickell, "An observation on the security of McEliece's public-key cryptosystem," in *Lecture Notes in Computer Science 330; Advances in Cryptology: Proc. Eurocrypt '88,* C. G. Günther, Ed., Davos, Switzerland, May 25–27, 1988, pp. 275–280. Berlin: Springer-Verlag, 1988.

[92] A. K. Lenstra and H. W. Lenstra, Jr., "Algorithms in number theory," in *Handbook of Theoretical Computer Science,* J. Van Leeuwen, Ed., pp. 673–716. Cambridge, MA: MIT Press, 1990.

[93] A. K. Lenstra, H. W. Lenstra, Jr., and L. Lovasz, "Factoring polynomials with rational coefficients," in *Math. Annalen 261,* pp. 515–534, 1982.

[94] A. K. Lenstra, "Factoring multivariate polynomials over finite fields," *J. Computer System Sci.,* vol. 30, no. 2, pp. 235–248, Apr. 1985.

[95] A. K. Lenstra, H. W. Lenstra, Jr., M. S. Manasse, and J. M. Pollard, "The number field sieve," in *Proc. 22nd ACM Symposium on Theory of Computing,* pp. 564–572, 1990.

[96] A. K. Lenstra and M. S. Manasse, "Factoring by electronic mail," in *Lecture Notes in Computer Science 434; Advances in Cryptology; Proc. Eurocrypt '89,* J.-J. Quisquater and J. Vandewalle, Eds., Houthalen, Belgium, April 10–23, 1989, pp. 355–371. Berlin: Springer-Verlag, 1990.

[97] H. W. Lenstra, Jr., "Integer programming with a fixed number of variables," *Math. Operations Res.,* vol. 8, no. 4, pp. 538–548, Nov. 1983.

[98] S. C. Lu and L. N. Lee, "A simple and effective public-key cryptosystem," *COMSAT Tech. Rev.,* pp. 15–24, 1979.

[99] F. Luccio and S. Mazzone, "A cryptosystem for multiple communication," *Inform. Proc. Lett.,* vol. 10, pp. 180–183, 1980.

[100] Y. Matyas and A. Shamir, "A video scrambling technique based on space filling curves," in *Lecture Notes in Computer Science 293; Advances in Cryptology: Proceedings of Crypto '87,* C. Pomerance, Ed., Santa Barbara, CA, Aug. 16–20, 1987, pp. 392–397. Berlin: Springer-Verlag, 1988.

[101] T. Matsumoto and H. Imai, "A class of asymmetric crypto-systems based on polynomials over finite rings," in *Abstracts of Papers, IEEE Intern. Symp. Inform. Theory,* St. Jovite, Quebec, Canada, Sept. 26–30, 1983, pp. 131–132, 1983.

[102] K. S. McCurley, "A key distribution system equivalent to factoring," *J. Cryptology,* vol. 1, pp. 95–106, 1988.

[103] K. S. McCurley, "The discrete logarithm problem," in *Cryptography and Computational Number Theory (Proc. Symp. Appl. Math.),* C. Pomerance, Ed., pp. 49–74. Providence, RI: American Mathematics Society, 1990.

[104] R. J. McEliece, "A public-key cryptosystem based on algebraic coding theory," *DSN Progress Rept. 42-44,* Jet Propulsion Laboratory, pp. 114–116, 1978.

[105] R. C. Merkle and M. E. Hellman, "Hiding information and signatures in trapdoor knapsacks," *IEEE Trans. Inform. Theory,* vol. 24, no. 5, pp. 525–530, Sept. 1978.

[106] V. S. Miller, "Use of elliptic curves in cryptography," in *Lecture Notes in Computer Science 218; Advances in Cryptology: Proc. Crypto '85,* H. C. Williams, Ed., Santa Barbara, CA, Aug. 18–22, 1985, pp. 417–426. Berlin: Springer-Verlag, 1986.

[107] J. H. Moore, "Protocol failures in cryptosystems," this volume.

[108] J. H. Moore and G. J. Simmons, "Cycle structure of the DES for keys having palindromic (or antipalindromic) sequences of round keys," *IEEE Trans. Software Eng.*, vol. SE-13, no. 2, pp. 262–273, Feb. 1987.

[109] J. H. Moore and G. J. Simmons, "Cycle structure of the DES with weak and semiweak keys," in *Lecture Notes in Computer Science 263; Advances in Cryptology: Proc. Crypto '86*, A. M. Odlyzko, Ed., Santa Barbara, CA, Aug. 11–15, 1986, pp. 9–32. Berlin: Springer-Verlag, 1987.

[110] S. Murphy, "The cryptanalysis of FEAL-4 with twenty chosen plaintexts," *J. Cryptology*, vol. 2, no. 3, pp. 145–154, 1990.

[111] S. Miyaguchi, A. Shiraishi, and A. Shimizu, "Fast data encipherment algorithm FEAL-8," *Rev. Electrical Commun. Lab.*, vol. 36, no. 4, pp. 433–437, July 1988.

[112] National Bureau of Standards, "Encryption algorithm for computer data protection," *Federal Register*, vol. 40, pp. 12134–12139, March 17, 1975.

[113] National Bureau of Standards, "DES Modes of Operation," *Federal Information Processing Standard*, U.S. Department of Commerce, FIPS Pub. 81, Washington, DC, 1980.

[114] H. Niederreiter, "Knapsack-type cryptosystems and algebraic coding theory," in *Problems of Control and Information Theory*, vol. 15, no. 2, pp. 159–166, 1986.

[115] A. M. Odlyzko, "Cryptanalytic attacks on the multiplicative knapsack cryptosystem and on Shamir's fast signature system," *IEEE Trans. Inform. Theory*, vol. IT-30, no. 4, pp. 594–601, July 1984.

[116] A. M. Odlyzko, "Discrete logarithms in finite fields and their cryptographic significance," in *Lecture Notes in Computer Science 209; Advances in Cryptology: Proc. Eurocrypt '84*, T. Beth, N. Cot, and I. Ingemarsson, Eds., Paris, France, April 9–11, 1984, pp. 224–314. Berlin: Springer-Verlag, 1985.

[117] T. Okamoto, "Fast public-key cryptosystems using congruent polynomial equations," *Electron. Lett.*, vol. 22, no. 11, pp. 581–582, 1986.

[118] T. Okamoto, "Modification of a public-key cryptosystem," *Electron. Lett.*, vol. 23, no. 16, pp. 814–815, 1987.

[119] T. Okamoto and A. Shiraishi, "A fast signature scheme based on quadratic inequalities," in *Proc. IEEE Symp. Security and Privacy*, pp. 123–132, 1985.

[120] H. Ong and C. P. Schnorr, "Signatures through approximate representations by quadratic forms," in *Advances in Cryptology: Proc. Crypto '83*, D. Chaum, Ed., Santa Barbara, CA, Aug. 22–24, 1983, pp. 117–132. New York: Plenum Press, 1984.

[121] H. Ong, C. P. Schnorr, and A. Shamir, "An efficient signature scheme based on quadratic equations," in *Proc. 16th ACM Symp. Theory of Computing*, pp. 208–216, 1984.

[122] H. Ong, C. P. Schnorr, and A. Shamir, "Efficient signature schemes based on polynomial equations," in *Lecture Notes in Computer Science 196; Advances in Cryptology: Proc. Crypto '84*, G. R. Blakley and D. Chaum, Eds., Santa Barbara, CA, Aug. 19–22, 1984, pp. 37–46. Berlin: Springer-Verlag, 1985.

[123] M. Petit, *Etude mathématique de certains systèmes de chiffrement: Les sacs à dos*, (Mathematical study of some enciphering systems: The knapsack, in French), Ph.D. Thesis, Université de Rennes, France, 1982.

[124] J. P. Pieprzyk, "On public-key cryptosystems built using polynomial rings," in *Lecture Notes in Computer Science 219; Advances in Cryptology: Proc. Eurocrypt*

'85, F. Pilcher, Ed., Linz, Austria, April 1985, pp. 73–80. Berlin: Springer-Verlag, 1986.

[125] J. Boyar Plumstead, "Inferring a sequence generated by a linear congruence," in *Proc. 23rd IEEE Symp. Foundations Computer Science*, pp. 153–159, 1982.

[126] S. Pohlig and M. E. Hellman, "An improved algorithm for computing logarithms over $GF(p)$ and its cryptographic significance," in *IEEE Trans. Inform. Theory*, vol. IT-24, p. 106–110, 1978.

[127] J. M. Pollard and C. P. Schnorr, "An efficient solution of the congruence $x^2 + ky^2 = m(\text{mod } n)$," *IEEE Trans. Inform. Theory*, vol IT-33, no. 5, pp. 702–709, Sept. 1987.

[128] C. Pomerance, "Fast, rigorous factorization and discrete logarithm algorithms," in *Discrete Algorithms and Complexity*, D. S. Johnson, T. Nishizeki, A. Nozaki, and H. S. Wilf, Eds., New York: Academic Press, pp. 119–143, 1987.

[129] C. Pomerance, J. W. Smith, and R. Tuler, "A pipe-line architecture for factoring large integers with the quadratic sieve algorithm," *SIAM J. Comput.*, vol. 17, 387–403, 1988.

[130] J.-J. Quisquater and J. P. Delescaille, "How easy is collision search? Application to DES," in *Lecture Notes in Computer Science 434; Advances in Cryptology; Proc. Eurocrypt '89*, J.-J. Quisquater and J. Vandewalle, Eds., Houthalen, Belgium, April 10–23, 1989, pp. 429–435. Berlin: Springer-Verlag, 1990.

[131] M. Rabin, "Digitalized signatures," in *Foundations of Secure Computation*, R. A. DeMillo, D. P. Dobkin, A. K. Jones, and R. J. Lipton, Eds., pp. 155–168. New York: Academic Press, 1978.

[132] M. Rabin, "Digitalized signatures and public key functions as intractable as factorization," Massachusetts Institute of Technology, Laboratory for Computer Science, *Report MIT/LCS/TR-212*, Cambridge, MA, Jan. 1979.

[133] M. Rabin, "Probabilistic algorithms in finite fields," *SIAM J. Comput.*, vol. 9, pp. 273–280, 1980.

[134] S. P. Radziszowski and D. L. Kreher, "Solving subset sum problems with the L^3 algorithm," *J. Combin. Math. Combin. Comput.*, vol. 3, pp. 49–63, 1988.

[135] T. R. N. Rao and K. H. Nam, "Private-key algebraic-coded cryptosystem," in *Lecture Notes in Computer Science 263; Advances in Cryptology: Proc. Crypto '86*, A. M. Odlyzko, Ed., Santa Barbara, CA, Aug. 11–15, 1986, pp. 35–48. Berlin: Springer-Verlag, 1987.

[136] J. A. Reeds and P. J. Weinberger, "File security and the UNIX system crypt command," *AT&T Bell Lab. Tech. J.*, vol. 63, no. 8, pp. 1673–1683, Oct. 1984.

[137] R. L. Rivest, L. Adleman, and M. L. Dertouzos, "On data banks and privacy homomorphisms," in *Foundations of Secure Computation*, R. A. DeMillo, D. P. Dobkin, A. K. Jones, and R. J. Lipton, eds., New York: Academic Press, pp. 169–179, 1978.

[138] R. L. Rivest, A. Shamir, and L. Adleman, "A method for obtaining digital signatures and public key cryptosystems," *Commun. ACM*, vol. 21, pp. 120–126, April 1978.

[139] R. A. Rueppel, "On the security of Schnorr's pseudo random generator," in *Lecture Notes in Computer Science 434; Advances in Cryptology; Proc. Eurocrypt '89*, J.-J. Quisquater and J. Vandewalle, Eds., Houthalen, Belgium, April 10–23, 1989, pp. 423–428. Berlin: Springer-Verlag, 1990.

[140] C. P. Schnorr, "A more efficient algorithm for a lattice basis reduction," *J. Algorithms,* vol. 9, pp. 47–62, 1988.

[141] C. P. Schnorr, "A hierarchy of polynomial time lattice basis reduction algorithms," *Theoretical Computer Sci.,* vol. 53, pp. 201–224, 1987.

[142] C. P. Schnorr, "On the construction of random number generators and random function generators," in *Lecture Notes in Computer Science 330; Advances in Cryptology: Proc. Eurocrypt '88,* C. G. Günther, Ed., Davos, Switzerland, May 25–27, 1988, pp. 225–232. Berlin: Springer-Verlag, 1988.

[143] P. Schöbi and J. L. Massey, "Fast authentication in a trapdoor-knapsack public key cryptosystem," in *Lecture Notes in Computer Science 149; Cryptography: Proc. Workshop Cryptography,* T. Beth, Ed., Burg Feuerstein, Germany, March 29–April 2, 1982, pp. 289–306. Berlin: Springer-Verlag, 1983.

[144] A. Shamir, "A fast signature scheme," *MIT, Laboratory for Computer Science Report RM—107,* Cambridge, MA, July 1978.

[145] A. Shamir, "The strongest knapsack-based cryptosystem," presented at Crypto '82.

[146] A. Shamir, "A polynomial time algorithm for breaking the basic Merkle-Hellman cryptosystem," in *IEEE Trans. Inform. Theory,* vol. IT-30, no. 5, pp. 699–704, Sept. 1984.

[147] A. Shamir, "On the security of DES," in *Lecture Notes in Computer Science 218; Advances in Cryptology: Proc. Crypto '85,* H. C. Williams, Ed., Santa Barbara, CA, Aug. 18–22, 1985, pp. 280–281. Berlin: Springer-Verlag, 1986.

[148] A. Shamir, personal communication, Oct. 1985.

[149] A. Shamir, "The cryptographic security of compact knapsacks," *MIT/LCS/TM-164,* Cambridge MA: Massachusetts Institute of Technology, 1980.

[150] A. Shamir, "On the cryptocomplexity of knapsack systems," in *Proc. 11th ACM Symp. Theory Comput.,* pp. 118–129, 1979.

[151] A. Shamir and R. Zippel, "On the security of the Merkle-Hellman cryptographic scheme," *IEEE Trans. Inform. Theory,* vol. 26, no. 3, pp. 339–340, May 1980.

[152] A. Shimizu and S. Miyaguchi, "Fast data enciphement algorithm FEAL," in *Lecture Notes in Computer Science 304; Advances in Cryptology: Proc. Eurocrypt '87,* D. Chaum and W. L. Price, Eds., Amsterdam, The Netherlands, April 13–15, 1987, pp. 267–271. Berlin: Springer-Verlag, 1988.

[153] Z. Shmuely, "Composite Diffie–Hellman public-key generating systems are hard to break," *Tech. Rept. no. 356.* Technion—Israel Institute of Technology Computer Science Department, Feb. 1985.

[154] G. J. Simmons, "A secure subliminal channel (?)" in *Lecture Notes in Computer Science 218; Advances in Cryptology: Proc. Crypto '85,* H. C. Williams, Ed., Santa Barbara, CA, Aug. 18–22, 1985, pp. 33–41. Berlin: Springer-Verlag, 1986.

[155] J. Stern, "Secret linear congruential generators are not cryptographically secure," in *Proc. 28th Symp. Foundations of Computer Science,* Los Angeles, CA, Oct. 12–14, 1987, pp. 421–426. Los Angeles: IEEE Computer Society Press, 1987.

[156] R. Struik and J. van Tilburg, "The Rao–Nam scheme is insecure against a chosen-plaintext attack," in *Lecture Notes in Computer Science 293; Advances in Cryptology: Proceedings of Crypto '87,* C. Pomerance, Ed., Santa Barbara, CA, Aug. 16–20, 1987, pp. 445–457. Berlin: Springer-Verlag, 1988.

[157] S. Tsujii, K. Kurosawa, T. Itoh, A. Fujioka, and T. Matsumoto, "A public-key cryptosystem based on the difficulty of solving a system of non-linear equations," *TSUJII Lab. Tech. Memorandum, no. 1*, 1986.

[158] B. Valee, M. Girault, and P. Toffin, "How to guess *l*-th roots modulo *n* when reducing lattice bases," in *Applied Algebra, Algebraic Algorithms and Error-Correcting Codes (Proc. 6th Internat. Conf, AAECC-6)*, T. Mora, Ed., Rome, Italy, July 4–8, 1988, pp. 427–442. Berlin: Springer-Verlag, 1989.

[159] B. Vallee, M. Girault, and P. Toffin, "How to break Okamoto's cryptosystem by reducing lattice bases," in *Lecture Notes in Computer Science 330; Advances in Cryptology: Proc. Eurocrypt '88*, C. G. Günther, Ed., Davos, Switzerland, May 25–27, 1988, pp. 281–291. Berlin: Springer-Verlag, 1988.

[160] M. J. Wiener, "Cryptanalysis of short RSA secret exponents," *IEEE Trans. Inform. Theory*, vol. IT-36, pp. 553–558, 1990.

[161] M. Willett, "Trapdoor knapsacks without superincreasing structure," *Inform. Process. Lett.*, vol. 17, pp. 7–11, July 1983.

[162] H. C. Williams, "A modification of the RSA public-key encryption," in *IEEE Trans. Inform. Theory*, vol. IT-26, no. 6, pp. 726–729, 1980.

[163] H. C. Williams and B. Schmid, "Some remarks concerning the MIT public-key cryptosystem," *BIT*, vol. 19, pp. 525–538, 1979.

[164] M. Yagisawa, "A new method for realizing public-key cryptosystem," *Cryptologia*, vol. 9, no. 4, pp. 360–380, Oct. 1985.

[165] G. Yuval, "How to swindle Rabin," *Cryptologia*, vol. 3,, no. 3, pp. 187–190, July 1979.

CHAPTER 11

Protocol Failures
in Cryptosystems*

J. H. MOORE
Sandia National Laboratories
Albuquerque, New Mexico 87185

*This chapter first appeared in the *Proceedings of the IEEE*, vol. 76, no. 5, May 1988.

Abstract—When a cryptoalgorithm is used to solve data security or authentication problems, it is implemented within the context of a protocol that specifies the appropriate procedures for data handling. The purpose of the protocol is to ensure that when the cryptosystem is applied, the level of security or authentication required by the system is actually attained. In this chapter, we survey a collection of protocols in which this goal has not been met, not because of a failure of the cryptoalgorithm used, but rather because of shortcomings in the design of the protocol. Guidelines for the development of sound protocols will also be extracted from the analysis of these failures.

1 INTRODUCTION

At one time, cryptography was an area of interest only in military and diplomatic circles, with the possible exception of a few eccentric souls with a curiosity for the bizarre. However, in this age of electronic mail, electronic transfer of funds, and huge databases of sensitive medical and personal histories stored in computers with dial-up capabilities, the use of cryptography is widespread enough to touch everyone in our society. Keeping step with these technological developments which require secrecy and authentication, cryptographers have developed new algorithms using complex mathematical systems. These algorithms often require quite sophisticated computing capabilities for their implementation and are designed to withstand attack by equally sophisticated opponents with nearly unlimited resources available to them. However, the mere existence of strong cryptoalgorithms is not enough to solve the problems for which they were developed. A cryptoalgorithm must be used within a set of rules or procedures, known as a *protocol*, which insures that the algorithm will actually provide the security and/or authentication required by the system. Consequently, the development of a system to protect data secrecy and/or integrity actually involves two areas of analysis: the design of strong cryptoalgorithms, and the design of sound protocols.

From the point of view of this chapter, the design of a protocol includes the specification of the characteristics of the cryptoalgorithms that may be used in the protocol without degradation of the security of the system. This specification might be so detailed as to actually define the algorithm that should be used, in which case the protocol should include guidelines for the choice of parameters necessary for that algorithm to protect the data to which it is applied. However, the specification could simply be a list of properties that the cryptoalgorithm must satisfy in order for the protocol to provide the security or authentication for which it was designed.

In this chapter, we will consider examples of protocol failures. By this we mean instances in which the protocol fails to provide the advertised level of security or authentication. Since the examples considered do not use cryptoalgorithms that are inherently weak, the failure involves the protocol design. In all cases, the rules specified in the protocol are sufficient to establish the desired secure data communications if there were no cheating. Unfortunately, those who wish to attack the system are not restricted to only the operations specified by the protocol. As long as they abide by the rules of the protocol while using the system, they can perform any side calculations or manipulations with the data using the information available to them. Although these examples of protocol failures are of interest in and by themselves, more than intellectual curiosity motivates this paper. Careful analysis of these examples should provide insight into principles of protocol design. The extraction of such principles could lead to guidelines for development of future protocols that are resistant to the type of attacks demonstrated in this chapter. Therefore, as we proceed through these examples, we will try to extract some general guidelines that will be summarized in the last section of this chapter.

2 THE NOTARY PROTOCOL

We begin our survey of protocol failures with an example that seems to have first made the cryptocommunity aware that some attacks on cryptosystems were not really revealing weaknesses of the algorithms but rather weaknesses of the protocols calling for their use. This protocol was designed to allow a message to be signed by an entity A, in a way that allowed others to verify at a later date that that message was in fact signed by A. Because in such a concept, A seems to behave as a notary public, we call this the notary protocol.

To set up a notary protocol, A must choose Rivest-Shamir-Adleman (RSA) parameters: primes p and q, and encryption and decryption exponents e and d satisfying $e \cdot d = 1 \mod \phi(n)$, where $n = pq$. The value of n as well as the public exponent e are published, while d and the factorization of n are kept secret by A.

To sign a document M, the notary uses the private exponent to compute a signature $S = M^d \mod n$. The signature is then appended to the document much as a notary seal is placed on a paper document. Anyone can use the public information to verify that $S^e = M \mod n$. Since only A has access to the value d used to create S, the protocol claims to have provided proof that only A could have calculated the signature S.

However, as we pointed out by Davida [3] and Denning [4], it is possible to use the protocol to obtain a forged signature on a document. There are several methods for doing this, but we will include only a simple version here to illustrate the nature of the failure.

For this attack, a forger can use the notary protocol to obtain a signature on a document M, which can be manipulated to obtain a forgery for a signature on another document P. To do this, the forger arbitrarily chooses a value X and computes $Y = X^e$ mod n. He can then use this value to modify the document P on which he wants a signature by calculating $M = YP$. By having the notary sign M, the forger obtains $S = M^d$ mod n. Using this signature, a forged signature S' for P can be obtained by calculating $S' = SX^{-1}$. Notice that the signature for P would be P^d mod n, and that

$$M^d = (YP)^d = Y^dP^d \text{ mod } n$$

By the definition of Y, we know that $Y^d = X$ mod n, so that the above equation becomes $S = XP^d$ mod n, or equivalently, $P^d = SX^{-1} = S'$ mod n. Therefore, the claimed value of this protocol, that is the ability to produce a signature that could only have been obtained if A in fact signed the document, is not attained.

All of the attacks on this protocol rely on the fact that RSA uses a mathematical function, namely, exponentiation, which preserves the multiplicative structure of the input. In fact, the crux of these attacks is that, for any choice of X, M, d, and n,

$$(XM)^d = X^dM^d \text{ mod } n$$

This means that any attempt to modify the protocol to defeat these attacks will have to destroy the ability of a forger to make use of this multiplicative structure. A more general version of Davida's attack has been developed by Desmedt and Odlyzko [9] which can be applied to protocols calling for RSA encryption. In their attack, by obtaining the decryption of a particular collection S of ciphertexts, the cryptanalyst can decrypt further messages by representing them as products of the members of S. In a sense, this makes the ultimate use of the multiplicative structure to defeat a protocol using RSA. It is curious that there are other protocols [2], designed to solve different problems, which in fact exploit this multiplicative property to the benefit of the legitimate users of the system. Apparently then, this structure does not represent an intrinsic weakness of the cryptosystem, although the application in a particular protocol should consider the effect of this structure on the system.

The essence of the flaw in this protocol design, from which a useful design principle can be deduced, is that the signature in an authentication scheme should be the encryption of a message chosen from a selected subset of all possible messages to which the encryption function applies. That is, in this example, the only documents that a notary will sign must possess certain predetermined structure. When the signature is verified, that structure must also be present in the document if the signature is to be accepted as authentic. For this particular example, the set of possible messages is Z_m, the ring of integers modulo m, for some modulus m. To eliminate the ability of the forger to make use of the multiplicative structure, the set of acceptable messages must not be an ideal of Z_m. By this we mean that multiplying any acceptable message by any other message should not necessarily produce an acceptable message. The purpose of such a requirement is to make the creation of forgeries extremely unlikely since the multiplicative structure of the entire system will not carry over to the set of acceptable messages. Protocols for other signature schemes should consider the type of structure which the set of acceptable messages must possess in order to protect against forgeries.

A mathematical model for the system involved in the protocol would make the description of the required properties much easier to state. Some theoretical protocol analysis has begun this type of modeling in [5] and [10], and the results should be valuable in the prevention of similar protocol failures. However, that work is beyond the scope of this chapter, so the reader is referred to the literature for more details.

3 THE COMMON MODULUS PROTOCOL FAILURE

Next we consider a protocol using RSA, which has been proposed on several occasions. In this system, a central keying authority (CKA), would generate two good primes, p and q, calculate the modulus $M = p \cdot q$, and generate encryption/decryption pairs $\{e_i, d_i\}$. Each subscriber in the system would be issued a secret key d_i, along with the public information which consists of the common modulus M and the complete list of public keys $\{e_i\}$. Anyone possessing this public information can send a message X to the nth subscriber by using the RSA encryption algorithm with the public key e_n as the encryption exponent. That is, the ciphertext becomes

$$Y = X^{e_n} \bmod M$$

The protocol is designed to protect the secrecy of the message X sent to subscriber n, since only the nth subscriber knows the secret key, d_n, which allows the decryption of Y. A signature channel is also available in this system, since the nth subscriber can sign a message X by encrypting it using the secret exponent d_n. Both X and the signature, $S = X^{d_n} \bmod M$, is then made public. Anyone can then verify that X was signed by the nth user by using the public exponent e_n to check that $S^{e_n} = X$. As long as only the nth subscriber can uniquely produce the correct signature, that is, a message that when encrypted using the public exponent e_n yields X, the verifier can be confident that the message was authentically signed. By using a common modulus M for all subscribers, the key management is simplified, since only the decryption exponents $\{d_i\}$ must be protected.

However, the use of a common modulus poses several problems for this protocol. As was pointed out by Simmons [21], if a message is ever sent to two subscribers whose public encryption exponents e_i and e_j happen to be relatively prime, then the message can be recovered without breaking the cryptosystem. To demonstrate this, consider the effect of encrypting the message X using e_i and e_j

$$Y_i = X^{e_i} \bmod M$$

$$Y_j = X^{e_j} \bmod M$$

Since e_i and e_j are relatively prime, integers r and s can be found using the Euclidean algorithm, so that

$$re_i + se_j = 1$$

Obviously, either r or s must be negative and for this discussion we will assume that $r < 0$ and write $r = -1 \cdot |r|$. We can also assume that Y_i and Y_j are relatively prime to M, since if this were not true, the Euclidean algorithm could be used to factor the modulus, thereby breaking the cryptosystem. Since Y_i is relatively prime to M, we can once again use the Euclidean algorithm to calculate the multiplicative inverse of Y_i mod M. The following calculation shows how the message is then recovered

$$[Y_i^{-1}]^{|r|} \cdot [Y_j]^s = [X^{e_i}]^{-|r|} \cdot [X^{e_j}]^s = X^{re_i + se_j} = X \bmod M$$

Consequently, the protocol fails to protect the secrecy of the message X sent to two subscribers whose public keys are relatively prime. Notice that this does not break the cryptosystem in the traditional sense, since the ability to read the message X does not transfer to an ability to read arbitrary messages encrypted with the same system.

The use of a common modulus also makes this protocol vulnerable to two other attacks in which a subscriber can break the cryptosystem. Once the cryptosystem has been broken, of course, the privacy channel fails since such a subscriber can therefore decrypt messages intended for other users and the signature channel fails since he can also forge the signature of a user without detection. The first attack of this type involves a probabilistic method for factoring the modulus, while the second uses a deterministic algorithm for calculating the encryption/decryption exponent without factoring the modulus. Both of these attacks are described in detail by DeLaurentis [6], and we will only outline the concepts in this chapter.

The basic idea used to factor the modulus, is the identification of a square root of $1 \bmod M$. By this we mean a number b, satisfying

1. $b^2 = 1 \bmod M$,
2. $b \neq \pm 1 \bmod M$,
3. $1 < b < M - 1$.

If such a number can be found, then the modulus M can be factored in the following way. Since $b^2 = 1 \bmod M$

$$b^2 - 1 = 0 \bmod M, \text{ or}$$
$$(b + 1)(b - 1) = 0 \bmod M, \text{ or}$$
$$(b + 1)(b - 1) = sM = spq, \text{ for some integer } s.$$

However, $1 < b < M$, so that $0 < b - 1 < b + 1 < M = pq$ must hold. These inequalities make clear that p and q cannot both divide either $b - 1$ or $b + 1$. Hence, the greatest common divisor of $b + 1$ and M must be p or q. Applying the Euclidean algorithm will thus yield a factorization of M. Consequently the attack on this system now centers on a method for finding a nontrivial square root of $1 \bmod M$.

Let e_1 and d_1 be the encryption and decryption exponents for a user of the system. By the definition of these exponents, $e_1 d_1 = 1 \bmod \phi(M)$. Thus, $e_1 d_1 - 1$ must be some integer multiple of $\phi(M)$, and the Euclidean algorithm will allow us to find non-negative integers c and k so that $e_1 d_1 - 1 = c \cdot \phi(M) = 2^k \varphi$ where φ is odd. Consider the following procedure for finding a nontrivial square root of $1 \bmod M$:

1. Choose an integer a such that $(a, M) = 1$ and $1 < a < M - 1$.
2. Find the smallest positive integer, j, which satisfies $a^{2^j \varphi} = 1 \bmod M$. (Since $2^k \varphi$ is a multiple of $\phi(M)$, we know that such an integer does exist.)
3. Let $b = a^{2^{j-1} \varphi}$. If $b \neq -1 \bmod M$, then it is a nontrivial square root of 1.

4. If $b = -1 \bmod M$, return to step 1.

DeLaurentis has shown that this procedure will fail at most half of the time, so that the expected number of trials before a nontrivial square root is found is no greater than 2. Therefore, an insider can break the cryptosystem within this protocol with unacceptably high probability, by using information that each subscriber in the system must have. The attack that we have just described is quite significant since it points out that the knowledge of one encryption/decryption exponent pair for a given modulus M is sufficient to allow the factorization of M.

The second type of failure breaks the cryptosystem by demonstrating that a subscriber can use his own public and private keys to generate the private key of another user. That is, given a public encryption exponent e_1, the holder of an encryption/decryption pair e_2, d_2 can find an integer n such that $e_1 n = 1 \bmod \phi(M)$, without actually knowing $\phi(M)$.

To find such an n, it is enough to find an integer which is relatively prime to e_1 and is a multiple of $\phi(M)$. This is verified by noting that if n and e_1 are relatively prime, then there are integers r and s satisfying $rn + se_1 = 1$. If n is also a multiple of $\phi(M)$, then $se_1 = 1 \bmod \phi(M)$.

Consider the following procedure for finding such an integer, in which the only values needed are e_1, e_2, and d_2:

1. Using the Euclidean algorithm, find the greatest common divisor f of e_1 and $e_2 d_2 - 1$.
2. Let $n = (e_2 d_2 - 1)/f$.

It is obvious that n is relatively prime to e_1. By definition, we know that e_1 is relatively prime to $\phi(M)$. Since f is a divisor of e_1, it must also be relatively prime to $\phi(M)$. However, $nf = e_2 d_2 - 1$ is a multiple of $\phi(M)$, which means n must also be a multiple of $\phi(M)$. The above procedure then yields a decryption exponent for e_1. Since the computational difficulty of this procedure has been shown to be at worst $O[(\log M)^2]$, this does pose a viable threat to the system. Once again, the information available to a legitimate user of the system is actually sufficient to break the cryptosystem. Of course, such a user is not operating strictly within the protocol designer's concept of a user, but the information required is obtainable by the user without stepping out of the bounds of the protocol.

The net conclusion of the above three attacks must be that the common modulus protocol fails miserably to protect the secrecy of messages or to provide a means for authenticating the signatures of users of the system. Therefore, in designing new protocols with the RSA algorithm, the use of a common modulus should be avoided. However, these attacks provide several more detailed guidelines for the development of further secure protocols. From the first attack, we note that the protocol designer must consider what an opponent can do with a collection of ciphertext whose plaintexts are related (in the example they were equal) or whose keys are related (in the example they were relatively prime). With respect to the RSA algorithm, two guidelines for applications of the algorithm were established.

1. Knowledge of one encryption/decryption pair of exponents for a given modulus gives rise to a probabilistic algorithm for factoring the modulus whose expected number of trials before success is no greater than 2.

2. Knowledge of one encryption/decryption pair of exponents for a given modulus M gives rise to a deterministic algorithm for calculating other encryption/decryption pairs without having to first determine $\phi(M)$.

4 THE SMALL EXPONENT PROTOCOL FAILURE

Another commonly suggested protocol using RSA involves the use of a small exponent for the public key in order to make the calculations for encryption fast and inexpensive to perform. The general setting for this type of protocol involves a large communication network in which the messages sent between two users should not be readable to other users in the system or by outsiders. The protocol specifies that the ith user should choose two large primes, p_i and q_i, and publish their product n_i as the modulus for an RSA algorithm used in communicating with him. An encryption/decryption pair, $\{e_i, d_i\}$ is chosen and one of these, say d_i, is published. We are interested in the case when the encryption exponent is chosen to be a small integer. For some applications, this is a very appealing choice since the implementation with a small exponent can be simpler and quicker to operate. However, such a specification causes the protocol to fail if the exponent is d and the same message is sent to at least d users. This observation has been made by several people, at least Blum, Lieberherr, and Williams, and seems to have become part of the cryptanalytic folklore. To illustrate the problem, consider the case when $d = 3$. Suppose that user 1, whose public exponent is 3, decides to send a message M to users 2, 3, and 4. The ciphertexts then are

$$C_2 = M^3 \bmod n_2$$
$$C_3 = M^3 \bmod n_3$$
$$C_4 = M^3 \bmod n_4$$

If n_2, n_3, and n_4 are relatively prime, the Chinese remainder theorem will enable the calculation of $M^3 \bmod (n_2 n_3 n_4)$ from the knowledge of C_2, C_3, C_4. But $M^3 < n_2 n_3 n_4$, so M can be recovered. If n_2, n_3, and n_4 are not relatively prime, then the attacks from the common modulus protocol apply. Thus, even an eavesdropper has enough information to recover the message.

A common technique for salvaging this protocol is to never send exactly the same message by using something like a time stamp concatenated to the message before the encryption takes place. In our example above, the new ciphertexts in this protocol would be

$$C_2 = (2^{|t_2|}M + t_2)^3 \bmod n_2$$
$$C_3 = (2^{|t_3|}M + t_3)^3 \bmod n_3$$
$$C_4 = (2^{|t_4|}M + t_4)^3 \bmod n_4$$

where t_2, t_3, and t_4 are times associated with each message. This foils the previously described attack.

However, Hastad [11] has shown that this may not vary the plaintext enough to overcome the weakness inherent in the low exponent protocol. In his paper, Hastad showed that a system of modular polynomial equations

$$P_i(x) = 0 \bmod n_i, \ 1 \le i \le k$$

of degrees no greater than d can be solved in polynomial time if the number of equations is larger than $d(d + 1)/2$. Thus in the case of an exponent of 3, if the message, adjusted by a time stamp, is sent to at least seven members of the network, the message may no longer be secret to an eavesdropper. Of course, for this attack to be viable, the time stamps must be known. However, the time stamps involve a small number of bits as compared with the total size of the message, so that these could be estimated before applying Hastad's algorithm. Obviously, the security of the system cannot reasonably depend on the secrecy of these time stamps alone.

Hastad's results require some lattice theory which is beyond the scope of this chapter. Therefore, we will not delve deeper into the actual techniques used to attack the secrecy of messages in such a network. The interested reader can investigate the details by referring to the original paper. However, it is important to note that the proof of the result requires the use of an algorithm for lattice basis reduction [17] which has only recently been developed and has had a rather extensive effect on cryptosystem analysis. The reader is referred to Brickell and Odlyzko's chapter, "Cryptanalysis: A Survey of Recent Results," which also appears in this volume, for further discussion of the impact of this algorithm on cryptanalysis.

The protocol failure considered in this section emphasizes the point made in the last section that the protocol designer must consider the information that can be gained from a collection of ciphertexts whose plaintexts are related (here equal or differing by time stamps) or whose keys are related (here the same exponent with relatively prime moduli).

5 THE LOW ENTROPY PROTOCOL FAILURE

In both the low exponent and the common modulus protocol failures, some mathematical properties of the cryptoalgorithm enabled the attacker to cause the protocol to fail. It might be surmised at this point, that all protocol failures are brought about by using some mathematical idiosyncrasies of the cryptoalgorithm. However, this is not always the case and in this section, we will examine a failure that is not really dependent on the particular algorithm used.

The purpose of a secrecy channel is to hide the meaning of a message from all but the authorized receivers. A strong cryptoalgorithm is employed to make decryption of the message infeasible for an outsider. However, if only a small number of messages can possibly be sent, the meaning of the message might be discernible without decryption. This is particularly a problem when a public key system is used, as was pointed out by Holdridge and Simmons [22].

When a public key system is used in a protocol providing for a secrecy channel, the encryption key is publicly known so that anyone can send a message through the channel that can be understood only by the intended receiver, who possesses the private decryption key. But, if only a small number of messages are meaningful, an opponent could precompute the encryption of those messages. Then when an encrypted message is sent through the channel, only a search through these precomputed values must be made to realize the meaning of the message. The simplest example of such a failure would occur if there were only two possible messages to be sent, such as "yes" and "no." Anyone possessing the public encryption key could encrypt these two messages and then when a "secret" message was sent through the system, easily distinguish between the two possible plaintext messages. Obviously, this is an extreme example, but the potential problem is well illustrated by it.

The above discussion makes clear that the message space to which a secrecy channel using public key encryption is applied, must be large enough to preclude an attack in which the set of all possible messages is pre-encrypted by an adversary who can then intercept ciphertext and recover the plaintext by exhaustive search. However, the encryption of all messages in the space is not required for this attack to work. If the ciphertext for a significant part of the message space is precomputed, the meaning of a given encrypted message may be discernible by simply matching, whenever possible, ciphertext in the message with the precomputed table, without decryption of the rest of the message. The definition of "significant" in the previous statement depends on the nature of the particular message space used in the system.

The title for this section refers to low entropy as the cause of this protocol failure. Entropy is a measure of uncertainty in the message space in the sense that it measures the amount of information that must be determined about a given message in order to make sense of it. If the message space has low entropy, only a small amount of information about a given message will reveal the **total** information contained in the message. Using this now-defined concept, we can state that the protocol failure examined in this section is the use of a public key system with a message space of low entropy.

In the paper referenced above, Holdridge and Simmons pointed out and demonstrated this protocol failure in an application proposed by Bell Telephone Laboratories for use in secure telephony [12,13]. In this system, each subscriber would enter his encryption key in a public key directory, keeping his decryption key secret. The messages to be encrypted in this system consist of digitized voice transmissions in a mobile radio telephone net. To speak with a given subscriber in the net, the sampled and digitized speech signal of the sender would be encrypted with the subscriber's encryption key, available from the public directory. On receipt, the subscriber could use his privately held decryption key to recover the plaintext message.

However, voice signals have a rather narrow bandwidth thus enabling an attack on the system since most of the information being transmitted occurs in a rather small collection of data. This combined with the ability of the human mind to use the redundancy, caused by the intersymbol and interword dependencies, of our language to ascertain the meaning of a sample of corrupted speech signals, makes the protocol for this system fail to protect the secrecy of messages.

To be more specific, in their demonstration, Holdridge and Simmons used digitized speech data that was formatted into 32-bit blocks in preparation for encryption. Thus the actual message space for the encryption algorithm consisted of all 32-bit words. However, they were able to determine that the entropy of such speech-derived messages was more like 16 to 18 bits, rather than 32. Using a precomputed table of about 100,000 ciphers, they reconstructed parts of the original data from various enciphered messages. Then using interpolation to fill in missing data, they created approximations to the original plaintext data. Finally these approximations were made into audio tapes and played to several listeners who were able to recover at least 90% of the original messages.

The particular public key algorithm used has no real bearing on the protocol failure. The cryptosystems used were not broken by such an attack, while the secrecy channel failed. Once again, it is merely a failure of the protocol—namely, the use of a public key encryption system to protect messages drawn from a message space with low entropy—to provide the advertised privacy.

6 A SINGLE KEY PROTOCOL FAILURE

Thus far, in our survey of protocol failures, all of the examples have used public key algorithms. To avoid the erroneous conclusion that single-key encryption systems are impervious to protocol failures, we will consider in this section, a protocol using DES that fails to provide the authentication for which it was designed.

Whenever encrypted data are transmitted over nonsecure lines, the vulnerability of the data to manipulation by an intruder must be considered. Even though such an intruder may not be able to decrypt the ciphertext, he may be able to use the characteristics of the protocol to manipulate the data to his benefit.

A solution to this problem is to append to the ciphertext a code that is a function of the actual plaintext message and that can be checked by the receiver against the result of applying that function to the received message. Agreement of these two values would validate that the message received had not been altered in transmission. If this function is chosen appropriately, an opponent should find it difficult to modify the ciphertext and/or the code without failing this validity check.

Two quite distinctive techniques for computing such a code have been considered. The first calculates a Message Authentication Code (MAC) by using a modified version of an encryption function. This function requires a secret key, different from the one used for preparation of the ciphertext, but chooses only some of the bits from the ciphertext as the code in an effort to keep the total message size as small as possible. The second method computes a Manipulation Detection Code (MDC), using a function that requires no secret information. The security advantages of the MAC over the MDC are apparent, but for some applications, a further encryption step or a complication of the key management system may place an unbearable burden on the communication channel.

A protocol for protecting information encrypted with DES with an MDC was proposed for inclusion in a federal standard for data communications. The proposal was quite appealing, probably because of its simplicity, and was mentioned in several publications [20] and books [19] before it was discovered that it failed to detect several types of manipulations that appear to be viable areas of concern.

The setting in which this protocol was designed to operate is as follows. The data to be transmitted are divided into n blocks each consisting of k bits, where k varies between 1 and 64. Denote these plaintext blocks by $X_1, X_2, \ldots, X_n$. After the blocks have been encrypted using the appropriate mode of DES to form the ciphertext blocks, $Y_1, Y_2, \ldots, Y_n$, the MDC is formed, Y_{n+1}, which is the exclusive-or sum of the n plaintext blocks. This block can be used by the authorized receiver to verify that the data received were not tampered with during transmission. The receiver simply decrypts the data and calculates the exclusive-or sum of these blocks, comparing the result to the MDC. If they agree, the receiver will conclude that the ciphertext had not been manipulated.

To illustrate the failure of this protocol, consider an application using Cipher Block Chaining as the mode of DES. In this mode, a 64-bit initializaton vector Y_0 and a 56-bit key K are exchanged secretly between those wishing to communicate. The ciphertext then consists of n blocks, $Y_1, Y_2, \ldots, Y_n$, each consisting of 64 bits, calculated by

$$Y_i = Y_{i-1} \oplus E(K, X_i)$$

where $E(K, X_i)$ represents encryption with DES under the key K of the message X_i. The MDC is computed as

$$Y_{n+1} = \bigoplus_{i=1}^{n} X_i$$

When the message is received, the first n blocks, $Z_1, Z_2, \ldots, Z_n$, are decrypted using the reverse operations. That is, the plaintext received is calculated as

$$W_{i+1} = W_i \oplus D(K, Z_i)$$

where $D(K, Z_i)$ represents decryption with DES under the key K of the message Z_i. The exclusive-or sum of these n blocks is then formed and compared with Y_{n+1}. If these are identical, the receiver will conclude that the message had not been manipulated.

Of course, this check merely confirms that $\oplus X_i = \oplus W_i$ so that an intruder can modify the message without detection as long as these sums remain equal. An obvious technique for spoofing this system is to rearrange the order of the blocks. Since the sum of the decrypted text will not change, this manipulation will not be detected. The seriousness of this problem depends on the nature of the plaintext. If the message was a transfer of $50,000 from your account to mine, I would certainly like to show that the account numbers were not rearranged, resulting in the transfer of $50,000 from my account to yours. If the account identification numbers happened to fall on 64-bit boundaries, the above protocol would not detect such a change. Secondly, blocks can be inserted without detection as long as they occur in pairs, since the exclusive-or sum of a block with itself yields a zero block and therefore will not affect the MDC. This could be advantageous to the intruder, particularly if he knew an encrypted block of data that could change the value of a deposit to his account from $1000 to $1,000,000. The severity of these failures would have to be analyzed for any particular application, but regardless, the protocol does not provide the advertised protection against undetected manipulation.

In Output Feedback Mode, the DES is used to generate a cryptographic bit stream which is then exclusive-or summed with the plaintext. To be specific, the plaintext is divided into n blocks of 64 bits each. Letting R_0 denote the initialization vector, n blocks of pseudorandom bits are found by

$$R_i = E(K, R_{i-1})$$

The ciphertext blocks are then obtained by

$$Y_i = X_i \oplus R_i, \qquad 1 \le i \le n$$

The MDC, Y_{n+1}, is once again calculated as the exclusive-or sum of the plaintext blocks. The receiver of the ciphertext calculates the values of R_i for as many blocks of data as were received. These are then exclusive-or summed with the ciphertext blocks. As long as the sum of these decrypted blocks is the same as Y_{n+1}, the message is accepted without question. Once again, the intruder can manipulate the data without detection, provided the sum of the decrypted blocks remains the same as Y_{n+1}.

It is easy to see that rearrangement of the order of the ciphertext blocks will not be detected by using the MDC, although such a rearrangement may produce, on decryption, blocks of random numbers, which may or may not be interpreted as valid data. By the same token, any blocks can be substituted as long as the exclusive-or sum of all the ciphertext blocks remains fixed. To see this, let $Z_1, Z_2, \ldots, Z_n$ be the received ciphertext blocks and assume that the exclusive or sum $Z_1 \oplus Z_2 \oplus \ldots \oplus Z_n$ is the same as $Y_1 \oplus Y_2 \oplus \ldots \oplus Y_n$. In this case when the ciphertext is decrypted, the recovered plaintext will be $W_1, W_2, \ldots, W_n$, which satisfies

$$
\begin{aligned}
W_1 &\oplus W_2 \oplus \ldots \oplus W_n \\
&= R_1 \oplus R_2 \oplus \ldots \oplus R_n \oplus Z_1 \oplus Z_2 \oplus \ldots \oplus Z_n \\
&= R_1 \oplus R_2 \oplus \ldots R_n \oplus Y_1 \oplus Y_2 \oplus \ldots \oplus Y_n \\
&= X_1 \oplus X_2 \oplus \ldots \oplus X_n = Y_{n+1}
\end{aligned}
$$

The intruder thus has considerable flexibility in choosing messages to substitute in the ciphertext, although the decrypted values may be jibberish. Once again, the type and format of the plaintext will determine whether or not the received message would be accepted as genuine, but the protocol, as proposed, does not provide adequate protection against data manipulation. Obviously, further constraints will be required by this protocol to insure that such manipulations will not allow the acceptance of tampered ciphertext by the receiver.

Similar techniques also show that the MDC proposed does not detect insertions by pairs with Cipher Feedback Mode. The interested reader is preferred to a paper by Jueneman, Meyer, and Matyas [15] for details of all attacks on this protocol for authentication of messages using DES.

7 SUMMARY AND ANALYSIS

We have surveyed a number of protocol failures which have hopefully covered a broad enough spectrum to convince the reader that the problems are widespread and that much can be learned from considering these past failures. In fact, many reports of the ''breaking'' of cryptosystems, may be better described as the ''revealing'' of protocol failures. The distinction seems to be that when a weakness is reported in a cryptosystem, the effect of which is to merely limit the scope of application or more clearly define the range of parameters that should be used for the algorithm, then the flaw discovered probably represents a protocol failure. However, if the effect is to leave the cryptosystem useless in any setting or to so severely restrict the possible range of parameters that the definition of a strong cryptofunction is infeasible, then the cryptosystem is actually ''broken.'' The point which we appear to be belaboring is not merely a question of semantics, since the reaction by the cryptocommunity to these two results will be drastically different. A protocol failure can lead to the definition of new guidelines for the use of a particular algorithm or class of algorithms, whereas a broken cryptosystem simply removes a given algorithm from consideration by protocol designers.

The examples of protocol failures considered in this chapter seem to naturally fall into three distinct classifications based on the type of flaw which was revealed. The first classification is characterized by the identification of a weakness in the cryptoalgorithm, of the type discussed in the last paragraph, as applied in the protocol. The low exponent and common modulus protocols using RSA are examples of this class of failures. We were able to identify several restrictions on the use of RSA by analyzing these failures. For emphasis and for completeness, we will restate those here.

1. Knowledge of one encryption/decryption pair of exponents for a given modulus gives rise to a probabilistic algorithm for factoring the modulus whose expected number of trials before success is no greater than 2.

2. Knowledge of one encryption/decryption pair of exponents of a given modulus M gives rise to a deterministic algorithm for calculating other encryption/decryption pairs without having to first determine $\phi(M)$.

3. A common modulus should not be used in a protocol using RSA in a communication network. (This follows from the first two points above as well as from the other attack discussed in this section on the common modulus protocol failure.)

4. Given a collection of k modular equations of the form $(X + t_i)^d \bmod n_i$, where $k > d(d + 1)/2$ and the $\{t_i\}$ are known, then X can be found in time polynomial in both k and $\log n_i$. The implication for RSA of this is that the exponents chosen in a protocol should not be small.

The second class of protocol failure is caused by the oversight of some principle applicable to a broad class of cryptoalgorithms, as seen in the low entropy or the notary protocol failure examples. In the case of the low entropy protocol failure, a general rule—that public key encryption should not be applied to protect messages that are drawn from a message space with low entropy—was not heeded by the protocol. With the notary example, a broad principle is exposed, namely, that applying encryption to plaintext to form signatures requires that the plaintext possess some verifiable structure to make forgeries difficult to obtain. When analyzing a particular implementation of a signature scheme, the nature of this structure may be obvious and even easy to describe. Unfortunately, for a generic signature scheme, the properties that the set of acceptable messages used to form signatures must possess, are difficult to state without reference to the specific algorithm being used. This makes a precise statement of the principle for signature schemes, which should be extractable from the notary protocol failure, difficult to express. However, the current trends in theoretical protocol analysis have been moving in the direction needed to formally state the principle required for a secure signature protocol.

The last classification, and probably the one from which the least amount of information can be elicited, is the protocol failure in which the designer simply overstated the amount of security that the protocol can provide. Such is the case in the single-key example in which a code to enable a receiver to detect ciphertext manipulation was calculated with no secret input by the sender. The designer of the protocol should have clearly stated the level of manipulation detection that such a code could provide, and as pointed out in our discussion of that scheme, the level of protection against undetected tampering was questionably low. Jueneman [14] has continued to work on modifications to this code, with some level of success, but not without considerably more complications to the calculations for the code. These complications are probably unavoidable simply from the application of the colloquial adage "you can't buy something for nothing." Protocol designers, however, should try to be as precise as possible in their claims about the level of security or authentication which can be expected from a given protocol.

Other papers surveying protocols have included some analysis of the problems which arise in verifying security. A two-step approach for protocol designers was suggested in [7] and [8] which bears repeating here.

1. Identify explicit cryptographic assumptions.

2. Determine that any successful attack on the protocol requires the violation of at least one of those assumptions.

This is a reasonable and straightforward set of guidelines, which really should be regarded as essential analysis by protocol designers. Of course, in practice, the proof required in step (2) may be quite difficult, if not impossible. The following modified version of these two steps might provide even more information to potential users of the protocol, while still providing the benefits of the analysis as originally stated.

1. Identify **all** assumptions made in the protocol.
2. For each assumption in step (1), determine the effect on the security of the protocol if that assumption were violated.

It is not clear that the first steps in the two versions are really different. However, the discussion that followed the identification of the original steps in the paper from which they were taken, leaves open the possibility that only the mathematical assumptions would be considered in the first version. The intention in step (1) of the modified version is to clarify even the setting that the protocol designer may have assumed would exist for the application of the protocol. The second step in the modified version could be thought of as an expansed version of the original step (2). This expansion provides real benefits, in that it helps to clarify the purpose of each assumption, helps to identify those elements of the protocol that are most critical, and makes modification of the protocol simpler to accomplish, if in practice some assumption cannot be met. The basic level of analysis for a protocol designer should include these steps. By themselves they will not prevent protocol failures, but should help in the early detection of flaws.

The restrictions for applications of RSA enumerated earlier in this section, may be simply specific applications of a set of more general principles for protocol design. In describing these general principles, the notation $E_i(M_j)$ will be used to denote the encryption with a fixed algorithm E, using the ith key, of a message M_j. The possible relationships among a collection of these ciphertexts should always be considered, especially when the encryption functions may be related because the keys satisfy some known relationship, or when the plaintexts satisfy some known relationship. That is, protocol designers should consider the following collections of ciphertext, which may be available to an opponent through the protocol, analyzing their effects on security.

1. A collection $\{E_i(M)\}$, where various keys are used to encrypt the same message, particularly if the keys used in this collection are related.
2. A collection $\{E(M_j)\}$, where the same key is used to encrypt messages M_j which satisfy some known relationship.
3. A collection $\{E_i(M_j)\}$, where various keys are used to encrypt known variations M_j of the same message. This is particularly interesting if the keys used also satisfy some known relationship.

The first principle was the basis of an attack on the common modulus protocol, while the third principle was involved in the low exponent protocol failure. These should actually be used in the analysis of all cryptoalgorithms, with any results serving as guidelines for the application of the algorithm by protocol designers.

Research in the area of cryptographic protocols has also been increasing with some interesting results. The basic approach used in this research has been to make a few simplifying assumptions, then mathematically model the setting and operations of the protocol. The mathematical model gives the researcher the tools to prove statements about the security of protocols which fit the model. One line of research has produced a computer program which searches for security vulnerabilities in protocols for key distribution [16]. Although still under development, it has demonstrated some level of success by its ability to rediscover some known failures in key distribution protocols. Some important theoretical work has been done in [1], [5], and [10] in which the results are more abstract. In each of these papers, restricted settings have been considered in order to obtain results, but progress is being made to give some foundation to the area of security analysis of protocols. Meanwhile, protocol designers may have to rely upon analysis of past errors as a major source of guidelines for strong cryptographic protocol development.

REFERENCES

[1] R. Berger, S. Kannan, and R. Peralta, "A framework for the study of cryptographic protocols," in *Lecture Notes in Computer Science 218; Advances in Cryptology: Proc. Crypto '85*, H. C. Williams, Ed., Santa Barbara, CA, Aug. 18–22, 1985, pp. 87–103. Berlin: Springer-Verlag, 1986.

[2] D. Chaum, "Untraceable electronic mail, return addresses and digital pseudonyms," *Commun. ACM*, vol. 24, no. 2, pp. 84–88, Feb. 1981.

[3] G. I. Davida, "Chosen signature cryptanalysis of the RSA(MIT) public key cryptosystem," *Tech. Rept. TR-82-2*. Milwaukee: University of Wisconsin Department of Electrical Engineering and Computer Science, Oct. 1982.

[4] D. E. Denning, "Digital signatures with RSA and other public key cryptosystems," *Commun. ACM*, vol. 27, pp. 388–392, April 1984.

[5] D. Dolev and A. C. Yao, "On the security of public key protocols," *IEEE Trans. Inform. Theory*, vol. IT-29, no. 2, pp. 198–208, March 1983.

[6] J. M. DeLaurentis, "A further weakness in the common modulus protocol for the RSA cryptoalgorithm," *Cryptologia*, vol. 8, no. 3, pp. 253–259, July 1984.

[7] R. A. DeMillo, G. L. Davida, D. P. Dobkin, M. A. Harrison, and R. J. Lipton, *Applied Cryptology, Cryptographic Protocols, and Computer Security Models, Proc. Symposia Appl. Math.*, Providence: American Mathematical Society, vol. 29, 1983.

[8] R. A. DeMillo and M. J. Merritt, "Protocols for data security," *Computer*, vol. 16, no. 2, pp. 39–50, Feb. 1983.

[9] Y. Desmedt and A. M. Odlyzko, "A chosen text attack on the RSA cryptosystem and some discrete logarithm schemes," in *Lecture Notes in Computer Science 218; Advances in Cryptology: Proc. Crypto '85*, H. C. Williams, Ed., Santa Barbara, CA, Aug. 18–22, 1985, pp. 516–522. Berlin: Springer-Verlag, 1986.

[10] S. Even, O. Goldreich, and A. Shamir, "On the security of ping-pong protocols using the RSA," in *Lecture Notes in Computer Science 218; Advances in Cryptology: Proc. Crypto '85*, H. C. Williams, Ed., Santa Barbara, CA, Aug. 18–22, 1985, pp. 58–72. Berlin: Springer-Verlag, 1986.

[11] J. Hastad, "On using RSA with low exponent in a public key network," in *Lecture Notes in Computer Science 218; Advances in Cryptology: Proc. Crypto '85,* H. C. Williams, Ed., Santa Barbara, CA, Aug. 18–22, 1985, pp. 403–408. Berlin: Springer-Verlag, 1986.

[12] P. S. Henry, "Fast implementation of the knapsack cipher," *Bell Syst. Tech. J.,* vol. 60, pp. 767–773, 1981.

[13] P. S. Henry and R. D. Nash, "High speed hardware implementation of the knapsack cipher," paper presented at Crypto '81, IEEE Workshop on Communications Security, Santa Barbara, CA, Aug. 24–26, 1981.

[14] R. R. Jueneman, "A high speed manipulation detection code," in *Lecture Notes in Computer Science 263; Advances in Cryptology: Proc. Crypto '86,* A. M. Odlyzko, Ed., Santa Barbara, CA, Aug. 11–15, 1986, pp. 327–346. Berlin: Springer-Verlag, 1987.

[15] R. R. Jueneman, S. M. Matyas, and C. H. Meyer, "Message authentication with manipulation detection codes," in *Proc. 1983 IEEE Symp. Security and Privacy,* R. Blakley and D. Denning, Eds., Oakland, CA, April 1983, pp. 33–54. Los Angeles: IEEE Computer Society Press, 1983.

[16] J. K. Millem, S. C. Clark, and S. B. Freedman, "The interrogator: Protocol security analysis," *IEEE Trans. Software Eng.,* vol. SE-13, no. 2, pp. 274–278, Feb. 1987

[17] A. K. Lenstra, H. W. Lenstra, and L. Lovasz, "Factoring polynomials with integer coefficients," *Mathematische Annalen,* vol. 261, pp. 513–534, 1982.

[18] M. J. Merritt, "Cryptographic protocols," Ph.D. Thesis GIT-ICS-83/06, Georgia Institute of Technology, 1983.

[19] C. H. Meyer and S. M. Matyas, *Cryptography: A New Dimension in Computer Data Security.* New York: Wiley, 1982, pp. 457–458.

[20] Proposed Federal Standard 1026, "Telecommunications: Interoperability and security replacements for the use of the data encryption standard in the physical and data link layers of data communications," National Communications System, Washington, DC, draft of June 1, 1981.

[21] G. J. Simmons, "A 'weak' privacy protocol using the RSA cryptoalgorithm," *Cryptologia,* vol. 7, pp. 180–182, 1983.

[22] G. J. Simmons and D. B. Holdridge, "Forward search as a cryptanalytic tool against a public key privacy channel," in *Proc. 1982 Symp. on Security and Privacy,* R. Schell, Ed., pp. 117–128. Los Angeles: IEEE Computer Society Press, 1982.

SECTION 5

Applications

CHAPTER 12

The Smart Card
A Standardized Security Device Dedicated to Public Cryptology

Louis Claude Guillou, Michel Ugon,
and Jean-Jacques Quisquater

The smart card will be an important tool in the hand of
mankind. It will be a major usage for chip technology.

J. Svigals

At first glance, a smart card appears to be simply an improved traditional credit card. But a smart card is in reality a multipurpose, tamper-resistant security device. Some consider it to be either the ultimate incorruptible cell resisting virus attacks or a fourth level in the hierarchy after the host computer, the departmental computer, and the personal computer. As a matter of fact, these two concepts are not exclusive.

Smart cards are already in widespread public use. Through this user-friendly technology, cryptology is invading our everyday life. This invasion has a large influence on security in various fields of applications, not only in banking, but also in the areas of health, pay television, telephone, home computers, data processing, communication network, and more generally, information technology.

1 INTRODUCTION

Traditional financial cards rely on embossed characters and magnetic stripes for information storage. The relevant existing standards (International Standards Organization [ISO] 7811) specify the characters and stripes in so much detail that there is no additional degree of freedom for any further evolution.

Smart cards rely on VLSI chip technology not only for information storage, but for information processing as well. A microcircuit is embedded in the plastic base of existing smart cards. As illustrated in Fig. 1, the microcircuit consists of an electronic chip bonded to a circuit board and connected to electrical contacts on the board. The relevant existing standards (ISO/International Electrotechnical Commission [IEC] 7816) do not specify the size or the performance of the chip, but rather deal with the

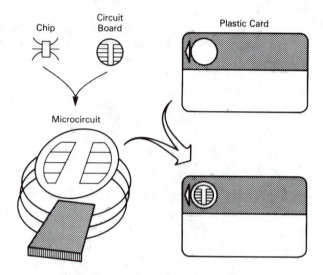

Figure 1　Integrated circuit card with contacts.

specification of the interface through which secure transactions are negotiated between the outside world and the embedded electronic circuits.

In the course of a transaction involving a smart card, the card delivers information (stored data, computation results) and/or modifies its contents (data storage, event memorization): The built-in electronic circuits both process data and store information in internal memory. Trade-offs between cost and performance of existing chips dedicated to smart cards are related to the state of the art in VLSI technology and to the current needs of the applications.

Advances in semiconductor technology modify the trade-offs between cost and performance. Smart cards improve in both memory size and processing power at the same rate as any other microprocessor while terminals remain unchanged owing to the standardized interface.

These technological trends definitely enhance both the physical and logical security of smart cards.

- Better integration (about a factor of ten every 5 years) enhances physical security by making it more difficult to physically probe and recover information from the VLSI chips dedicated to smart cards,
- Additions in processing power (central processing unit [CPU], random access memory [RAM]) and in operating systems (read only memory [ROM]) enhance logical security by allowing the implementation of more and more elaborate cryptographic algorithms and protocols in smart cards.

The more our society becomes computerized, the greater are the risks from banking fraud, economic sabotage, industrial spying, etc. The inescapable conclusion is that our computerized open systems require additional security.

Cryptography is a powerful security tool in the field of information technology. However, the expansion of public cryptologic knowledge is moderated by governmental and political concerns aiming at controlling the spread of cryptologic technology and devices; expressed most often in the form of embargos or export controls.

The smart card, which stores, processes, and controls internal cryptographic algorithms [1,2], as we will see, suggests solutions that may satisfy both national regulations and commercial needs.

2 COMPREHENSIVE APPROACH

The smart card is a portable (or detachable) file system that plays an active role in a transaction, has large possibilities for giving or proving its identity, and incorporates many features ensuring physical and logical security [2].

In many applications [3,4,5,6], the smart card plays an active role in a security system, storing secrets, and providing an easy opportunity to change algorithms without changing the entire system.

2.1 What a Smart Card Is

The chip embedded in a smart card is a single-chip microcomputer (MCU). A MCU is a computer system in miniature integrated onto a single piece of silicon. The only computerlike resources it lacks are the external human or machine interface (I/O) devices such as keyboards, displays, disk drives, etc. The chip embedded in smart cards is in fact a ''secure'' MCU. But there are major differences between a secure MCU and a general-purpose MCU [7].

In a general-purpose MCU, different operating modes can be selected by the user. For example, during an operation in ''expanded mode,'' the internal data and address buses are connected to the input/output (I/O) pins for accessing memories and resources outside the chip; and during an operation in ''write mode,'' the control of the internal buses is taken over by the outside world for the purpose of modifying the contents of an internal nonvolatile memory; other special modes may be used for testability.

In a secure MCU, after the device has been tested and passed as fully functional by the semiconductor manufacturer, the only possible mode must be the ''use-mode,'' under exclusive control of the user software in the on-board ROM. The internal buses must never be accessible through the I/O pins.

This is the main difference between secure MCUs and general-purpose MCUs: A secure MCU has the built-in capability to prevent, by various means, unauthorized access to the CPU, the memories, the buses, and any data being stored or processed within the device at any time.

Thus, the original title of this section ''What a Smart Card Is'' might better have been ''What a Secure MCU Is.''

2.2 What a Smart Card Does

The five basic operations of the smart cards are

1. Input data
2. Output data

3. Read data from nonvolatile memory (NVM)

4. Write or erase data in NVM

5. Compute a cryptographic function

Each of these five operations is related to the logical security of the card, but operations (2) and (4) are particularly sensitive. Operation (2) delivers data and results to the outside world, and operation (4) modifies the content of the NVM.

For example:

• Cryptographic secret keys are to be used by the microcomputer but are not to be output.

• Some data in the nonvolatile memory may give the right to access some resources, therefore precautions must be taken before writing or erasing.

• The result of a cryptographic computation may be a control word delivered to the outside (to descramble television signals, for example); therefore precautions must be taken before computing and outputting it.

The card must be sure that the right card holder is present during some operations, or that the received command has been formulated by the right card issuer. Various security mechanisms and techniques are used by the card for checking these facts, ranging from personal identification numbers (PINs) and message authentication codes (MACs) to sophisticated digital signatures and authentication schemes. These mechanisms are based on both cryptographic and noncryptographic techniques. The card may react upon detecting some types of fraud attempts. For example, the program may be such that three unsuccessful PIN presentations block the card, i.e., inhibit its further use. The increase of memory sizes and computation speeds together with the sophistication of the physical security features make possible the use of a set of more and more elaborate mechanisms for ensuring logical security of smart cards.

As illustrated by the specifications of the French bank cards [8], a well-structured use of these mechanisms and techniques during the basic operations makes it possible to organize several physical zones in the nonvolatile memory.

• An open zone, accessible without any control

• A protected zone, where a password is needed for writing, but where reading is free

• A confidential zone, where a password is needed for reading

• A secret zone, containing PINs, passwords, and cryptographic keys

In more recent masks [9], the nonvolatile memory is organized at the logical level rather than at the physical level. Therefore the zones are far less visible in these masks. The exact physical location of a given file in a NVM is immaterial and hence has no precise meaning in these masks.

3 STANDARDIZATION

Any discussion of standardization requires a good understanding of the International Organization for Standardization. For easy reference a comprehensive overview of the ISO organization and procedures is given in Appendix A.

In 1980 the French standards institution (Association Française de Normalisation [Afnor]) proposed a new work item (NWI), Interface of Integrated Circuit Cards with Contacts. In October 1981, this NWI was included by technical committee Information Systems (TC97) in the program of work of the subcommittee Identification Cards (SC17) which then created working group Integrated Circuit Cards with Contacts (WG4). The following participant members are very active: France, the United States, Japan, Germany, the United Kingdom, Canada, Italy, Denmark, the Netherlands, as well as the following liaison members: International Association for Microcircuit Cards (Intamic), Mastercard International, VISA International, Eurocheque, and International Air Transport Association (IATA). Other participant members of SC17 are Australia, Belgium, Czechoslovakia, Norway, South Africa, Sweden, Switzerland, Turkey, and the USSR.

Since 1987, the work in the field of information technology has been carried out through a joint technical committee established by ISO together with the International Electrotechnical Commission (IEC): ISO/IEC JTC1 Information Technology.

The smart card interface is now being standardized as a multipart standard [ISO/ IEC 7816] prepared by ISO/IEC JTC1/SC17/WG4. As a result of this work, several parts of the standard are now available and future parts are in progress.

3.1 Standardization of Physical Characteristics

A draft proposal (DP) was registered in 1983. A draft international standard (DIS) was registered in 1985 and approved in 1986. The final international standard (IS) (ISO 7816/1) was published in 1987. The result will probably be merged in a more general standard (ISO 7810).

Smart cards must be pliable; the contacts must be sufficiently conductive; smart cards must also resist mechanical stresses like falls, torsion, and bending and be resistant to static electricity and to exposure to various types of radiation such as x-rays, ultraviolet (UV) light, and electromagnetic fields. These physical characteristics are very precisely specified in the existing standards.

3.2 Standardization of Contact Location

In 1981, Afnor proposed a location and an assignment of six operational contacts plus two contacts reserved for future use. These are located on the front of the card, near the upper left corner, as shown in Fig. 2. This location corresponds to the minimum mechanical constraints for the microcircuit when the card is under torsion and bending stresses.

A DP was registered in 1984. A DIS was registered in 1985 and approved in 1986, but five votes were negative: Japan, Germany, United States, United Kingdom,

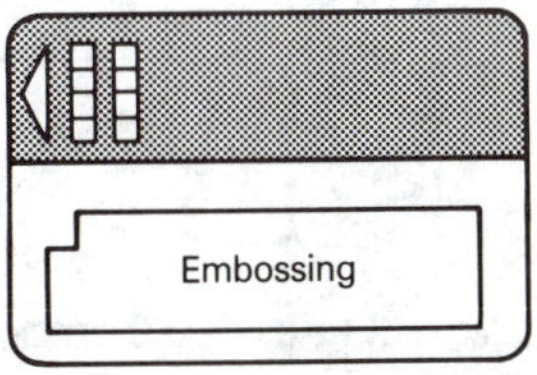

Figure 2   Upper location in front.

and Canada. Finally, an agreement was reached in 1987 and the final (ISO 7816/2) was published in 1988.

The unanimous agreement quickly reached on major points must be stressed:

- Type (surface contacts, and not edge contacts)
- Shape (minimum rectangular surface)
- Pattern (relative positions)
- Electrical functions and contact assignment

The exact location of the contacts was debated at length. While being the one mainly in use, the upper left-hand corner location proposed by Afnor and shown in Fig. 2 is now *transitional* in the ISO. The changeover to the new standard will occur sometime in the early 1990s. After that time a lower location has been adopted as shown in Figs. 3–5.

The standard refers to a corner, on any side of the card. Upper and lower locations form a regular pattern shown in Fig. 3. The two locations are deduced from each other by a rotation of the card in the plane. The same microcircuit, which consists of a chip connected to a contact board, may be used in any location. The standard preserves the existing chips. Hence, as long as all the contacts are in front, dual connectors are useful.

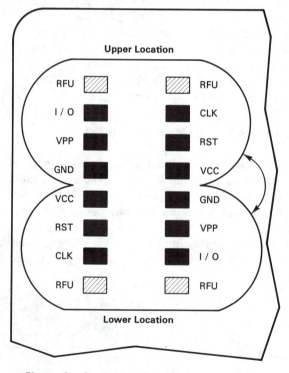

Figure 3 Contact assignment compatibility.

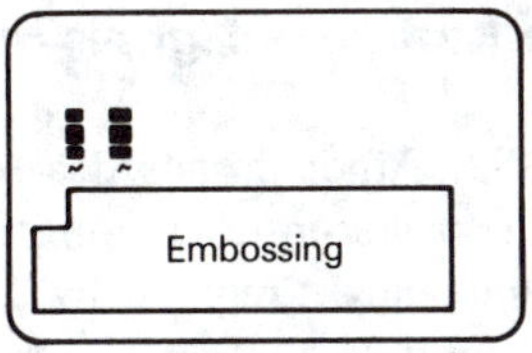

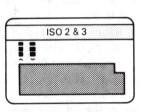

Figure 4 Lower location on front.

But there are in fact two final lower locations: in front and on rear, as described in Figs. 4 and 5. The most probable ultimate location is now the lower one on rear, as shown in Fig. 5.

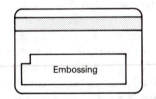

Figure 5 Lower location on back.

From the beginning, Japan disagreed with the Afnor proposal because the proposed contact placement conflicted with their placement of magnetic stripes. Their magnetic stripes are currently on the front of the card, as shown in Fig. 6.

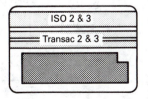

Figure 6 Front of a Japanese card.

Mastercard and Visa also objected to the Afnor proposal, pointing out that U.S. banks reserve the front of the card for identifying features: names, logos, and holograms. The technical aspects should be on the rear of the card: magnetic stripes, signature panel, as well as the electrical contacts of smart cards.

These marketing considerations raised a difficulty in France. Before ISO2 and ISO3 were standardized, Transac invented and designed the magnetic stripes T2 and T3 shown in Fig. 7. The French banks will surrender these stripes as soon as possible. The delay agreed on in 1990 for the transitional location means that these stripes will not exist on cards issued after that transition. The magnetic stripes T2 and T3 will have completely disappeared by the end of 1992.

Figure 7 Back of a French card.

3.3 Standardization of Signals and Protocols

In October 1982, Afnor proposed a set of electrical characteristics, a reset procedure followed by an answer-to-reset from the card, and an exchange protocol for processing subsequent commands. Prepared by Task Force WG4/TF1 created in 1984, a first DP was registered in 1985. A first DIS was registered in 1986 and approved in 1988. The final IS (ISO/IEC 7816/3) has been unanimously agreed on and was published in 1989. The basic Afnor proposal has been considerably amended in its presentation, but its technical content is the basis of the ISO. Electrical characteristics of the contacts now include NMOS, CMOS, and HCMOS technologies.

Transactions between the outside world and the embedded microcomputer are conducted through six electrical contacts detailed in Fig. 3. With respect to contact GND (ground) used as reference voltage, signals must be correctly provided to four contacts: RST (reset), VCC (power supply), VPP (programming voltage), and CLK (clock), in order to exchange data in a half-duplex mode on contact I/O (input/output).

Each transaction with a card consists of the following successive steps:

1. Activation of the contacts by the device
2. Resetting of the card by the device
3. Answer-to-reset by the card
4. Optional selection of a protocol type
5. Processing of successive commands according to the scenario of the transaction
6. Deactivation of the contacts by the device

The notion of command has to be carefully explained: Through a command, the outside world instructs the card to carry out some elementary action. Security plays a major part during any transaction with a card. At each command, the card decides either to continue or to stop according to the results of internal computations and according to the internal context of the transaction.

From the physical point of view, the card is a slave and the outside world is the master with control effected through the electrical contacts. But from the logical point of view, the card has autonomy of decision based on its processing power and its operating system.

During answer-to-reset, during subsequent option selection, as well as during processing of commands, data on I/O are organized in asynchronous characters transmitted in half-duplex mode. Each character consists of 10 consecutive bits: a start bit followed by 8 data bits completed by an even parity bit, as shown by Fig. 8. A minimum guardtime must be ensured before the next character to make it possible to resynchronize the receiver between subsequent characters.

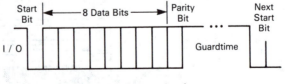

Figure 8 Character frame.

In those industrial fields in which smart cards promise to have the widest application, many struggles occur during standardization. Adopting some suggested modifications could eliminate existing cheap chips and/or create difficulties for existing cards. With strategic rather than technical underlying motivations, the following two struggles are characteristic of such competitions. The reference frequency has been strongly debated: This frequency is provided on CLK by the interface device to exchange data on I/O at a rate of 9600 bps. Some major silicon chip manufacturers, like Motorola, consider 4 MHz to be a lower threshold while Afnor proposed 3.579545 MHz. Basic in NTSC television sets, the frequency is the most used in the world at 10% under 4 MHz. The standard specifies that the bit duration is to be 372 clock cycles during answer-to-reset. This duration corresponds to the frequency proposed by Afnor for 9600 bps on I/O.

Japan and Germany proposed 4.9152 MHz (512 times 9600 Hz, the bit duration being then 512 clock cycles) as being a frequency more consistent with other telecommunications standards. The only argument with some technical basis is the use of standard universal asynchronous receiver transmitters (UART) on the I/O. This argument is not very compelling since, on the one hand, many existing UARTs use other frequencies such as 3.68, 4.02, and even 3.57 MHz. On the other hand, the timers suggested by Japan for accommodating UARTs show that these devices are not well suited for managing data in a half-duplex mode on one unique short line like I/O.

The *exchange protocol* has also been strongly debated. The standard specifies the asynchronous character protocol originally resulting from an agreement reached in France in 1981 between Honeywell Bull, Philips, and Schlumberger under the authority of the French PTT administration and with the technical expertise of CCETT. In this character protocol, an error signal is inserted by the receiver in the guardtime of any erroneous character. This error signal asks for an immediate repetition of the disputed character. Excluding the practical use of classic UARTs but well adapted to a local connection, this protocol is simple, efficient, and inexpensive: a few bytes of buffer and less than 200 bytes of handler. Japan argued, however, that the same end-to-end protocol should manage the exchanges between a host computer and a multiplicity of cards.

Some experts suggested that error detection and error recovery should be more sophisticated. In their opinion, vibrations in a car might disturb the exchanges between a card and a radiotelephone set. This argument is easily refuted by pointing out that other contacts like RST may also be influenced by such vibrations and that the consequences of spurious resets cannot be handled by an exchange protocol. This is a problem of connector design.

Japan and Germany have suggested that successive asynchronous characters in the same direction should be organized in blocks of characters with a redundancy checksum of one or two characters. Since 1987, all the experts have been working on a totally new block protocol inserted in 1991 in the standard. In a block protocol, error diagnoses are more complex and so the handler is more expensive: from 600 to 700 bytes depending on the services provided by the protocol. The card must store blocks in both directions: in reception for redundancy checking before processing and in transmission for a possible repetition. The block length in a card is limited by the RAM size. Also a block protocol is not efficient when the blocks are short. The recent publication of the standard has seriously reduced the interest in a block protocol: General use of the character protocol suggests that it is easily adapted to the large spectrum of present applications.

3.4 Additional Standardizations in WG4

A future fourth part of the standard (WD 7816/4) is in preparation. It will ensure inter-industrial interchange. Several points are under consideration.

- File architecture and related security
- Global access method to information
- Consistency in command coding
- Provisions in answer-to-reset for naming chip manufacturers, types, and masks, as well as card manufacturers and card issuers

The first point is essential. And the problems are now considered in the right order. Objects and entities inside the card must be clearly defined and characterized by their security attributes before naming and coding commands addressed to these objects and entities.

The second point is also important. Cards may initialize automatic processes in general-purpose devices. Examples of such processes are:

- Automatic dialing and automatic connection to a remote database
- Security instructions delivered to a terminal for confidentiality and integrity purposes
- Software loading from a card into a terminal

These specifications should accommodate the future evolution of devices, like pay television decoders using smart cards and card acceptor devices connected to electronic directory terminals (Lécam and Minitel in France, [10]).

A further step beyond part 4 should be the standardization of a smart card interpretive language (SCIL) for which interpreters should be implemented in the resident firmware of the terminals. Such a language should facilitate the writing of applications, for example in point-of-sale terminals using different processors.

3.5 Standardizations in ISO Outside WG4

In 1985, the ISO Banking technical committee adopted two new work items (NWI): Data Contents of Messages Exchanged with Integrated Circuit Cards, and Security Architecture of Banking Systems Using Integrated Circuit Cards.

In May 1986, this technical committee (TC68) felt the subject important enough for restructuring itself and creating the subcommittee Financial Transaction Cards, Related Media and Operations (SC6) with two working groups (WG5 and WG7) dealing with the two NWIs.

The work (ISO 9992) on Messages Exchanged with Integrated Circuit Cards (TC68/SC6/WG5) includes five parts:

- Concepts and structures
- Functions
- Messages (commands and responses)

- Common data for interchange
- Data elements

The work (ISO 10202) on Security Architecture of Banking Systems Using Integrated Circuit Cards (TC68/SC6/WG7) includes seven parts:

- Card life cycle
- Transaction process
- Cryptographic key relationships
- Secure application modules
- Use of algorithms
- Cardholder verification
- Key management

The first ISO documents (DIS 9992/1 and DIS 10202/1) are now two interim standards awaiting publication.

In October 1988, the joint technical committee (JTC1) adopted a NWI, Interface of Contactless Integrated Circuit Cards. This NWI was assigned to a new working group. Contactless integrated circuit cards (SC17/WG8). The work is just beginning. American Telephone and Telegraph (AT&T) (United States), GEC (United Kingdom), Valvo (Germany), and Dai Nippon (Japan) have all made proposals for contactless integrated circuit card standards. Several other proposals are presently under consideration as well.

3.6 Other Standardizations at the European Level

In the European Broadcasting Union (EBU), a working group (V5) is looking for an agreement on a satellite pay television system. Such a system includes a security device, currently named conditional access subsystem (CASS). Two approaches are described: CASSs are either detachable or buried in receivers. Smart cards have been carefully considered as potential detachable CASSs, while secure MCUs are a basis for buried CASSs.

In addition, still at the European level, currently in Commission Européenne des Postes et Telecommunications (CEPT) and in European Telecommunications Standards Institute (ETSI), a working group is drafting the specifications of a subscriber identification module (SIM) for the cellular digital radiotelephone system. These SIMs have turned out to be either smart cards, or plug-in security devices. The same secure MCUs may be used in both cases.

3.7 Standardization of Security Techniques in ISO

The ISO structures for standardizing cryptographic tools have been highly variable and very sensitive to the political context concerning cryptology and security [11] particularly to the U.S. context. The data encryption standard (DES) was published in 1977 [12] by the National Bureau of Standards, now the National Institute of Standards and Technology (NIST). In 1981, the American National Standards Institute (ANSI) adopted the DES as a U.S. commercial standard and published it as X3-92. This is a unique situation where a cryptographic algorithm has been submitted to public scrutiny and standardized. In 1980, ISO technical committee Information Systems (TC97) created a special working group Data Encryption (WG1). The goal was the international standardization of the DES. In 1984, the technical committee replaced this special working group with the subcommittee, Data Cryptographic Techniques (SC20). This subcommittee in turn created three working groups:

- Secret Key Algorithms and Applications (WG1)
- Public Key Systems and Modes of Use (WG2)
- Use of Encipherment Techniques in Communication Architectures (WG3)

By January 1986, the DES algorithm had reached the status of accepted draft international standard, under reference DIS 8227, First Data Encipherment Algorithm DEA1, but in May 1986, the ISO council decided to not publish the standard although publication had been imminent. The chief argument against publication was that the adoption of the DEA1 as a standard might encourage overdependence on the DES which was already an attractive enough target for potential criminal cryptanalysts. The council decided, in fact, to stop standardizing cryptographic algorithms. Several experts tried to define what a cryptographic algorithm is with only partial success. Finally, the subcommittee adopted the principle of a register where both secret and published encipherment algorithms could be introduced and recorded. The algorithm register is now being itself standardized (DIS 9979).

At the same time, in SC20/WG2, Public Key Cryptosystems, the embargo on data encipherment algorithms was extended to include a draft proposal on the Rivest–Shamir–Adleman (RSA) algorithm and to a draft technical report surveying the state of the art in public key cryptography. The contents of this technical report have been disclosed [13]. In WG 2, the work is concentrating on solutions to integrity problems, while no more work is being done on solutions to confidentiality problems.

These decisions were in line with the political situation in the United States. In 1985, commercial as well as governmental cryptography was centralized under responsibility of the National Security Agency. However, in 1987 the situation changed in the United States and the responsibilities for these two areas were separated [14]. The governmental applications were assigned to NSA and the commercial applications to NBS.

In June 1989, the joint technical committee (JTC1) decided to disband SC20 and to create a subcommittee entitled Security Techniques (SC27). The first meeting of SC27 occurred in Stockholm in April 1990. This subcommittee in turn created three working groups:

- Security Services and Guidelines (WG1)
- Security Mechanisms (WG2)
- Evaluation Criteria (WG3)

A fundamental problem remains, however; the border between experts standardizing systems and protocols and experts standardizing security techniques, has still to be clarified in the new organization of JTC1. Therefore, to coordinate its work on security, the joint technical committee also created in June 1989 a special working group on security (SWG-S) which met in Rennes, France, in October 1989. This special working group is still in existence.

It appears that a majority of commercial needs can be met simply by protecting the integrity of information. Consequently in ISO work, priority is given to integrity techniques: identification, authentication, and signature. In open systems, public key cryptosystems make possible standardizable solutions for ensuring integrity. The practical implementations of these schemes more often than not require the use of personal portable security devices, like smart cards.

4 TECHNOLOGY

During the past decade, enormous advancements have occurred in the semiconductor industry: greatly increased performance and memory sizes, and correspondingly great reductions of cost and power consumption. The number of transistors per chip was multiplied by 400 between 1970 and 1985 while the dimensions of a transistor were divided by 2 every four years as shown in Fig. 9. The evolution of random access memories and microprocessors results from advances in a number of areas, such as computer-aided design, photolithography, etching, ion implantation, process mastering, and scanning electron beam microscope. It can be confidently predicted that the next decade will follow the same trend.

In the 1980s, a breakthrough occurred with the development of CMOS technology which consumes much less power, and which also provides this capability at an acceptable cost. No doubt the next step in this development, high-speed CMOS (HCMOS) technology, will be an important part of the integrated circuit market before the end of this century, as indicated in Fig. 10.

Many factors influence the rapidity of the evolution of semiconductor technology; more complex circuits involve more sophisticated constraints and know-how in manufacturing, as well as huge investments and learning curves for mass production. As an

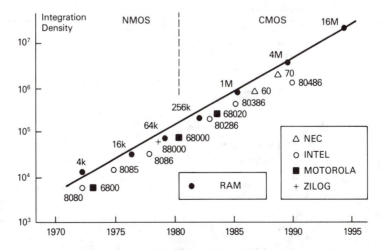

Figure 9 Memories and microprocessors.

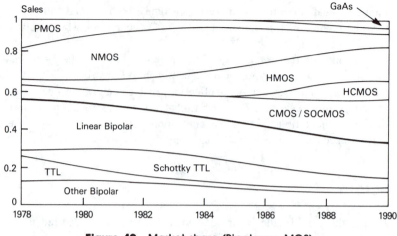

Figure 10 Market share (Bipolar vs. MOS).

example, research and development costs are multiplied by a factor of roughly 3 for each new technology. This explains the large size and the small number of the semiconductor companies that lead the market. As a matter of fact, it is agreed upon that it takes seven to ten years from research laboratory to large-scale production of integrated circuits. This phenomenon is the source of much confusion between future possibilities and the present reality. Figure 11 shows the delay from research phase to mass production of RAMs in MOS technology.

In the development of semiconductor technology, the simple structures were naturally developed before complex very large-scale integration (VLSI) circuits. Consequently, the best integration has been achieved in dynamic RAMs because of small size and low complexity of the elementary cell in the logic circuitry. However, because of cost and reliability considerations, complex chips involve compromises between different technologies for processors and for memories.

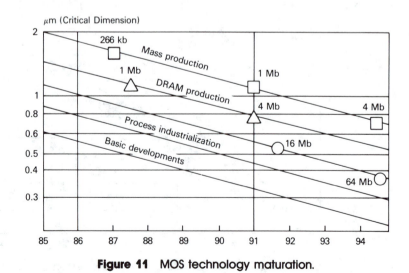

Figure 11 MOS technology maturation.

Because a secure MCU is a complete system, including CPU, RAM, ROM, and nonvolatile programmable memory, one can understand why such VLSI circuits are a relatively recent development. Several functional blocks are gathered on the same substrate, including chip security features as well as all the resources needed by the application. Cost and reliability considerations induce trade-offs limiting ambitious designs and memory sizes. A generally agreed-upon limitation is that the chip size should not exceed 20 mm^2 if one is to obtain a reliable card at a reasonable price. While this is achievable with today's VLSI technology, smart card chips will undoubtably follow the same general trends as the rest of the semiconductor industry to achieve greater density and functions in the future.

4.1 Nonvolatile Programmable Memories

In smart cards, the built-in electronics include a nonvolatile programmable memory (NVM). Each cell of NVM is originally in logical state ONE; it may be turned to logical state ZERO by an electrical process under control of the built-in electronics itself. Data stored in the NVM vary from one card to another and changes during card life. Any NVM area may always be selectively erased by turning to ZERO all the bits in the area, but the possibility of returning back to original state ONE depends on the NVM technology used. At the beginning of the smart card story, 10 years ago, two technologies were considered for implementing the NVM: bipolar and MOS. Although it is quicker, bipolar technology is much more power consuming. It also uses more silicon to realize the same function. An elementary bipolar transistor is illustrated in Fig. 12. More importantly, the bipolar writing process physically destroys a metallic fuse in each NVM cell so that a cell once written into cannot be erased. Broken and intact fuses are visible in Fig. 13. Because the status of a fusable link can be read in a

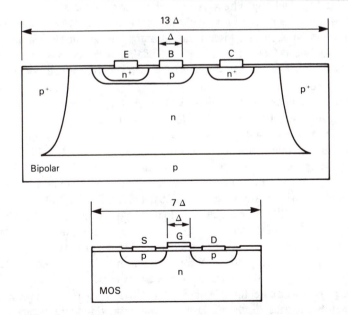

Figure 12 Comparison of transistor layouts.

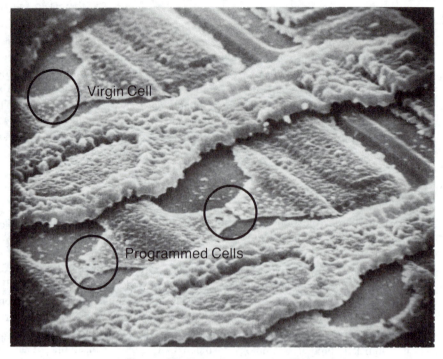

Figure 13   Fuses in a bipolar NVM.

magnified image, a bipolar NVM cannot hide secrets from microscopic examination. This technology is therefore not appropriate for a secure smart card storing cryptographic keys and algorithms. Nevertheless, at the onset of smart cards, bipolar technology was considered: the reasons in favor of this technology were writing irreversibility and ability to support a logic array of gates on the same substrate as the memory.

As illustrated in Fig. 14, a MOS NVM does not have the same security problem as did the bipolar NVM: Since writing is reversible, a cell content cannot be read optically, but can only be determined by electrically accessing internal buses. Moreover, for large volumes and low costs, MOS technology is better suited than bipolar technology to the integration of a NVM together with a microprocessor on the same substrate.

In MOS technology, the NVM may be built either in electrically programmable read-only memory (EPROM) or in electrically erasable PROM (EEPROM). If in EPROM technology, the return from logical state ZERO to logical state ONE is not selective: Erasing radiations affects the whole NVM contents. But in smart cards, for obvious security reasons, such a global process would kill the card. As illustrated in Fig. 15, an EPROM cell uses the floating gate avalanche MOS (FAMOS) technology. To write a bit, high voltage is applied to the control gate and to the drain in order to cause high-energy electrons to avalanche through the drain junction. Quantum theory says there is a small probability for an electron to jump across a potential barrier higher than the energy of the electron, a process referred to as tunneling since the electron's energy is inadequate for it to cross the barrier. Therefore some of the electrons may tunnel through the insulating layer of silicon dioxide and be trapped inside the floating gate. This causes an increase in the threshold voltage yielding an off-state transistor.

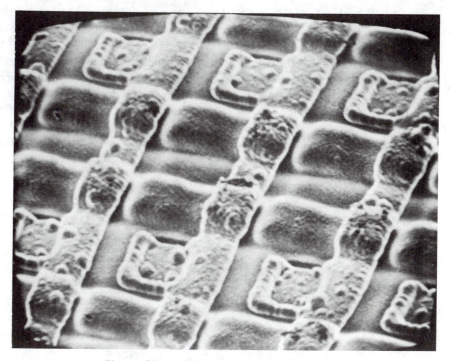

Figure 14   Invisible contents of a MOS NVM.

Erasing is done by ionizing the insulated layer with UV light during 20 min. to recover the virgin state by discharging the floating gate.

The programming voltage and the writing time are consequences of the thickness of the oxide which reflects the accuracy of the technology. At the early stage of EPROM technology, 25 V were needed. Subsequently this voltage was decreased to 21 V, and currently only 15–12 V are needed. During the same period, the writing time decreased from 50 ms to less than 10 ms. The writing time in EPROM is now of the same order of magnitude as the writing time in EEPROM. Despite this trend toward lower programming voltages, the energy needed for writing in EPROMs at the moment precludes the incorporation of a high-voltage generator on the chip as is common practice in EEPROM technology. In EEPROM Technology, the binary information stored in

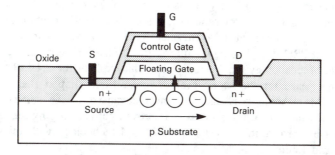

Figure 15   EPROM technology.

each cell (or block of cells) of NVM may be selectively inversed by an electrical process. The relevance of this technology to smart cards is that erasable reusable cards are very desirable.

An EEPROM cell uses the reversible Fowler-Hordheim effect to extract electrons by tunneling through a very thin oxide layer. This effect requires a high electric field, greater than 10^7 V/m, able to extract electrons from a doped semiconductor. As shown in Fig. 16, the structure of the most widespread EEPROM in use today is similar to the FAMOS structure. A cell includes a very thin layer of silicon dioxide under the poly-silicon floating gate. The thickness of this insulating layer is difficult to master in mass production. This is the reason why this technology, which was born ten years ago, is only emerging now.

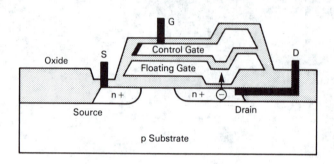

Figure 16   EEPROM technology.

To erase a cell means to force electrons from the floating gate and selectively discharge the gate. This is achieved by applying the programming voltage on the drain and keeping the gate at 0 V, thus allowing electrons to tunnel back to the drain. The write-erase cycles progressively destroy the thin layer of dioxide through which the electrons must tunnel: Every time, a few electrons are trapped within defects of the silicon dioxide. The number of cycles is limited to around 10^4. This phenomenon has to be carefully considered according to the smart card application requirements.

The programming pulse is 18–20 V with a duration of 1–10 ms. The small amount of energy required allows use of a voltage converter on the chip itself. EEPROM cards do not use contact VPP, consequently smart card terminals accepting exclusively such cards are much simpler.

If a smart card generates the programming voltage internally, then stored information may be modified without any control by the outside world, especially during power-on and power-off sequences. In such a card the chip designer must take this possibility into account to avoid undesirable perturbations in the NVM.

EEPROM technology requires two transistors for each cell of NVM whereas EPROM technology requires only one. This explains the 2:4 EPROM/EEPROM ratio in memory size for the same level of integration. Simpler than EEPROM technology, EPROM technology is also the most advanced in the semiconductor industry; the EPROM manufacturing process is improved permanently by the feedback of an important mass production. No doubt, EEPROM technology will follow the same trends in the near future.

4.2 Smart Cards in the Integrated Circuit Card Family

The NVM contents evolve during card life under control of the built-in electronics. Depending on an increasing complexity, three types of integrated circuit cards are illustrated in Fig. 17.

Figure 17   Integrated circuit card family.

The same system may support different types of cards. This is illustrated by European public telephones which accept all three types of cards. The number of cards in the following paragraphs were valid in 1990, but are increasing rapidly. For example, in France more than 5 million télécartes are presently manufactured each month; mid 1991.

In France, sixty million anonymous memory cards, named *télécartes,* have been produced for the seventy thousand French public phones in use. These public phones also accept the five million banking smart cards in circulation as well as the one million personal smart cards, named *Cartes Pastel,* delivered to phone subscribers by France-Telecom.

In Germany, more than three million anonymous memory cards, named *telekards,* have been produced for use with German public phones. The Bundespost is now manufacturing smart cards for its public telephone system.

The simplest cards are those specific to a single application. Economic considerations, though, dictate against making each card be totally unique. Instead, the manufacturing process capitalizes on the fact that smart cards are inherently multipurpose devices. The microcomputers are programmed by masks during the manufacturing process. Sharing chip production among several masks to satisfy different applications is easy; designing a new mask is not too complicated, and the same line produces smart card chips, irrespective of which mask is used. However sharing a very simple memory chip for several different applications can cause severe security problems.

The smart card operating system deals with different commands and with the general security of the whole system. Since chip design is not recurrent, software development cost may well exceed hardware cost. Similar to personal computers, the most important part of a smart card system lies in the software, as illustrated in Fig. 18. A poor software design can induce weak security, inefficient functions, erroneous data, deadlocks, and many other potential problems. On the other hand, a good software design provides the user with qualified operations and additional functions.

As a matter of fact, the user is not buying a hardware chip, but rather a complete smart card product providing functional solutions to his problems. The efficiency of a smart card operating system is not only related to ROM size, but also to the virtuosity of the software designer who finally specifies the technical configuration of the card.

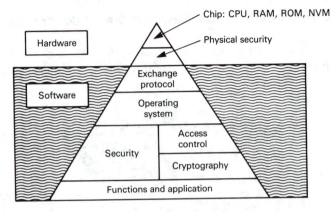

Figure 18 Smart card environment application.

4.3 Self-Programmable One-Chip Microcomputer

The first smart cards were produced in March 1979 as the result of a successful collaboration between CII Honeywell Bull and Motorola. These smart cards included two chips: a 2716 EPROM memory and a 3870 microprocessor originally designed by Fairchild. This dual-chip stage was essential to prove the feasibility of the concept and to convince potential users to start experiments. These dual-chip cards also played an important role in the initialization of applications and in the development of various other elements in systems using smart cards.

Despite the fact that it is always possible to assemble several standard components in a plastic card, the natural solution is a one-chip one because of cost, security, and reliability of the final product. Indeed, the microcircuit manufacturing is simplified; the risk of failure is seriously reduced; and there are no wires from one chip to another that might allow access to internal buses for an attacker to exploit. Thus, the security is taken into account in the design of the chip.

A chip dedicated to smart cards must be able to execute an internal routine for writing to itself in its NVM. Such a chip is termed a self-programmable one-chip microcomputer (SPOM).

Figure 19 describes the architecture invented by Honeywell Bull for managing registers on internal buses in such a way that the processor remains in control while holding the right address and the right content on the ports directed to the NVM.

The cooperation between Motorola and Honeywell Bull continued with the development of a SPOM. The first silicon SPOM was operational in 1981. Since 1982, Motorola has produced more than twenty million SPOMs in East Kilbride, Scotland. Since 1985, Thomson has also been producing SPOMs in Le Rousset, France. In all these SPOMs, successful trade-offs between cost and performance are a result of the know-how gathered from the first dual-chip cards. The Bull CP8 and Philips are currently manufacturing cards with these SPOMs which are about 18 mm^2 in size.

Recently, Honeywell Bull and Philips, the two major companies involved in smart card development from the beginning, decided to join their efforts and know-how. They created a common development team with a goal of designing and implementing a high-security, multiapplication card operating system called TB100 [9].

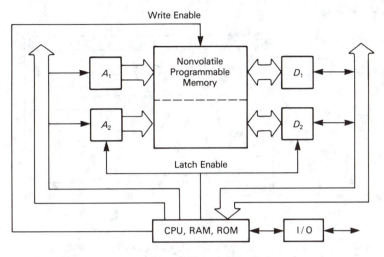

Figure 19 SPOM architecture.

The choice of the central processing unit (CPU) is an important decision in the design of such a chip. Currently Motorola is proposing a family of SPOMs based on a classic 8-bit CPU: the 6805. From the beginning, Motorola has used the same CPU. Now with a new SPOM family named ST16XYZ, Thomson is also moving toward the same CPU: the 6805. The consequences of these industrial decisions are important.

One advantage of choosing a classic CPU is the availability of very complete development tools which simplify software production. Another advantage is hardware evolution inside a large family of microprocessors. Mass production makes these step-ups easier for the transition from NMOS to HCMOS. In addition to NVM, a SPOM also includes two other types of memory: RAM and ROM. The RAM stores contexts and intermediate results during computations. The ROM stores the smart card operating system written by mask during chip manufacturing process at the factory. Memory cells differ not only in their function, but also in the amount of silicon "real estate" required for their realization. On the SPOM illustrated in Fig. 20, a cell of RAM is roughly 20 times larger than a cell of EPROM which in turn is roughly three times larger than a cell of ROM.

The four spare contacts on the left of the SPOM in Fig. 20 are additional I/O contacts available for connecting other devices inside the card and for using the chip in different environments.

Today, high-speed CMOS (HCMOS) technology is in production. In either HCMOS or CMOS technology, a switch consists of a pair of complementary NMOS and PMOS transistors. Such a switch is shown in Fig. 21. The gates of the two transistors are wired together and the input voltage is applied to both of them. The responses are complementary: A signal activating one transistor deactivates the other one and vice versa. The drain of the NMOS transistor is wired to the source of the PMOS transistor: Together they deliver the output signal. The source of the NMOS transistor is connected to a low-voltage line; and the drain of the PMOS transistor is connected to a high-voltage line.

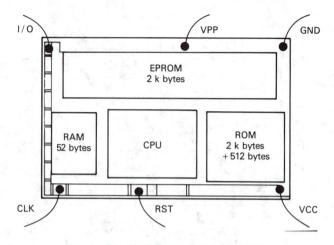

Figure 20   SPOM die (MC6805SC03).

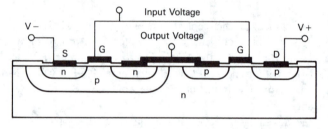

Figure 21   CMOS switch.

The two most important features of CMOS technology are the following ones [15]:

(i) CMOS devices consume less power than previous NMOS devices: No current passes between the two lines except during the short periods when the input signal is switched.

(ii) CMOS devices are also less susceptible to ambient electrical noise than NMOS devices: A spurious signal would have to be twice as great to force a CMOS device into an error setting as would be required to cause an error setting in an NMOS device.

With these new HCMOS designs, the range of SPOMs becomes broader. Memory sizes of the existing SPOMs are summarized in Table 1.

In the near future, some new SPOMs will include arithmetic operators running in parallel with the main processors. These operators are designed for multiplying and exponentiating large integers modulo large integers. Two such projects were publicly described in 1989 [16,17]. New SPOMs of this type are presently under development by Philips, Honeywell Bull, and Siemens. Those types of SPOMs will be well adapted to processing public key algorithms and zero-knowledge schemes.

TABLE 1 SUMMARY OF EXISTING SPOMS (MEMORY SIZES IN BYTES)

SPOM Type	NVM		ROM	RAM
	EPROM	EEPROM		
Motorola				
68HC05SC01	1K		1.6K	36
68HC05SC03	2K		2K	52
68HC05SC11	8K		6K	128
68HC05SC21		3K	6K	128
68HC05SC23		512	3K	96
68HC05SC24		1K	3K	128
SGS-Thomson				
ST1002	1K		2K	44
ST1834	4K		3K	76
ST16402		2K	4K	256
ST16612		2K	6K	160
S9	8K		4K	256
Hitachi				
65901		2K	3K	128
6483108		8K	10K	256
Oki				
62720		2K	3K	128
62780		8K	6K	192

5 SECURITY

Absolute security does not exist, no more for the smart card than for any other computing device. However, security may be enhanced by a coherent set of physical and logical features. Several secret key algorithms are currently used in the numerous masks of existing smart cards. A reasonable question is: Why are there so many masks and so many algorithms? The answer is in part due to the widely differing card capabilities, and in part due to the variety of crypto algorithms available. There are two main families of masks: key-carrier cards KC0, KC1, KC2, and multipurpose cards M4, M9, MP, M64, B1, B2, D1, D2, and TB100. There are also evolutionary stages in the algorithms: from the noninversible one-way function Telepass1 and the semireversible function TDF, up to the fully reversible functions Telepass2, Videopass, and DES.

We shall not describe in detail either the masks or the algorithms (only DES [12] has been made public). Our description is restricted to the functional evolution of masks and algorithms so as to give an overview of secret key cryptology in smart cards.

As an introduction to security, we will first describe some physical features of the chip itself and next some aspects of cardholder identification.

5.1 Chip Security Features and Card Life Cycle

There is a hidden problem inherent to smart cards. The problem arises because a card must be considered in the course of its existence to have three phases: a birth, a life, and a death. This is unique in the data processing world, and one must resist the temptation to consider the NVM as simply a conventional database without at the same time considering the creation process of smart cards and the associated rights it represents. This confusion comes from the broad use of mass memories such as disks, which have to be formatted by the user and then loaded with information. It seems relatively easy to securely handle data if only one application is able to be run in a card. If several applications may run in the same card, there is indeed a risk that other people may access or tamper with data belonging to a user.

In this section, we consider some basic security features in chip production and chip life. The file architecture of the NVM in a smart card is developed in Section 5.10.

Since the design of the very first chips for smart cards, two approaches have been considered to the problem of chip testing. These are:

(1) Each chip supports about twenty additional test contacts, and tests are conducted under control of the outside world; or

(2) each chip supports one or two test contacts, and tests are conducted by an internal self-testing program written in a small extra ROM (about a half-kilobyte).

During the development of the manufacturing process, both efficiency and flexibility of testing require the 20 additional test contacts. More information can be gained through measuring internal electrical signals in this way. Self-testing, though, is more economical. Therefore the number of test contacts is reduced when the manufacturing process is mature enough.

Before cutting wafers on SPOM03 production lines, a 512-byte internal routine is activated through two specific test contacts, visible in Fig. 20 near RST, under the CPU. The NVM of each validated component receives various information: locks, codes, erasure indicators, chip serial number, while nothing is written in rejected components. The two test contacts are then systematically destroyed by breaking fuse links buried in the silicon. An equivalent operation exists for any secure SPOM.

This operation, which eliminates non-user modes on valid chips, also positively disables invalid chips where nothing has been written.

As a matter of fact, only the self-testing routine may write these erasure indicators to be tested by the card before executing any command in user mode during any transaction. If such an erasure indicator is erased, either by accident or by violation, then the chip is definitely disabled. Such NVM cells are constructed so as to be the most sensible ones to erasing radiations. This is an example of the current reliability philosophy of using weak-link/strong-link designs to enhance reliability, since the weak-link is designed to disable the device before the operational strong-links can be subverted.

Valid chips are then inserted into cards during the process of card manufacture. A manufacturing code or key is used for protecting chips from the time of chip manufacturing to card issuing. Throughout the operational card life, several testers in the chip determine readiness: voltages, clock frequency, light, temperature are all measured. These indications may also be used by the operating system to increase security. The mapping of memory addresses should be controlled by the internal program itself, and not be accessible to outside control.

Whatever the physical security systems, system designers must carefully consider the potential consequences of chip violations. Secret keys must be as diversified as possible, tied to user identification number and/or chip serial number. A successful violation then compromises only one user and does not endanger the whole system; thus reducing the risk of widespread fraud. These aspects of logical security are strongly related to cryptology.

5.2 Cardholder Identification

The identification of the cardholder has several aspects depending on the scenario of the transaction to be performed with the card.

In this section we will focus on the identification of the cardholder by the card, considering that the card itself is authenticated by other means. The problem of card authentication is developed in a subsequent chapter. We restrict our consideration to the simplest case where a cardholder is paying a retailer with a smart credit card, and where the smart card has to identify the person attempting to use it to be the authorized cardholder according to the security policy of the payment.

There are several ways for a card to identify the cardholder. The simplest one is to carry out a direct personal identification number (PIN) check inside the card. When a PIN is required, no operation can take place in the card without the presentation of the correct information. The card internally compares the PIN presented by the user with the reference PIN written in a secret area of its NVM. The card keeps the result of this comparison secret until after the results can be entered into its memory. If the result is incorrect, then this fact must be recorded in the NVM to total the number of successive erroneous attempts to use the card. When this number reaches a predetermined value (1–7, depending on the security policy), the card is blocked and cannot be used thereafter. If the result is correct, then the external behavior must be the same as above in order to not reveal the test result before it has been recorded.

By systematically recording the result, the card prevents a fraudulent (unauthorized) user from deriving any benefit from observing a difference in the card's actions, no matter how slight.

At least 1 bit must be written in the NVM whenever an access protected by a PIN is made.

In some masks such as MP and TB100, the PIN may be enciphered to foil attempts to eavesdrop on this confidential information. A security module is located in the pinpad of the point-of-sale terminal. The card produces a random number. The security module then computes a message from the random number and the identity claimed by the card using an internal master key. The card tests the message and reacts as above for memorizing the result. This method also provides an authentication of the terminal by the card.

In the same manner, an MP card is able to cooperate with an external biometric identification device which increases the authentication abilities in a system. Identification may be performed by fingerprint, retina pattern, dynamic signature, or any physical characteristic of the individual. For this purpose, the card must deliver a reference pattern to the external checking device and the dialogue between the card and the checking device has to be randomized and encrypted.

To close these considerations of cardholder identification, we give an example that costs only NVM memory and requires no processing power in the card. The identification of the cardholder by the retailer may also involve a device displaying digital pictures. For this purpose, the card must store a compressed digital photograph which has been signed by the authority and tied to the chip serial number of the card. After having authenticated the card by other means, the retailer gets the signed reference picture, checks the associated signing appendix, and checks the picture visually to authenticate the cardholder.

5.3 Secret Key One-Way Function

Algorithm Telepass1 was designed by Honeywell Bull in 1979. This unpublished algorithm is a one-way function coded by about 200 bytes in masks M4 and M9.

Mask M4 is a general-purpose mask, not dedicated to a particular application. The corresponding cards may be turned into banking cards under personalization B0, and into Pastel cards for public phone subscribers under personalization B03. Each M4 card holds only one cryptographic key. This unique secret key is generally computed by diversifying a secret master key by the chip serial number; such a computation is performed in security devices protecting the secret master key.

Telepass1 computes a result R (64 bits) from four variables: an external argument E (48 bits), an address @ (16 bits) of any nonsecret word in the card, the content of this word (32 bits), and the unique internal secret key S (three words of 32 bits in B0) (Fig. 22). The access to a nonsecret word is either free or conditioned by the previous presentation of the personal identification number. Before involving a confidential word in a computation, the PIN must have been correctly presented.

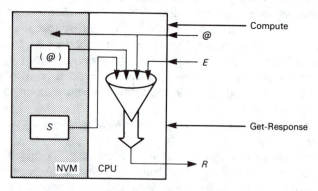

Figure 22 Telepass1 algorithm.

Two types of security devices are currently involved in systems using M4 and M9 cards.

Mother cards are used for personalizing individual cards by computing one diversified key for each issued card. Mother cards transfer the diversified keys to the outside world.

Security modules internally recompute diversified keys for controlling results computed by cards. Security modules do not output such diversified keys which are systematically used in subsequent internal computations.

5.4 Dynamic Authentication by Security Modules

The Telepass1 algorithm allows a dynamic authentication of cards by security modules, as illustrated in Fig. 23. At each authentication, the module picks at random a string of

48 bits and transmits it to the card. The security module checks that the response from the card corresponds to the internal result obtained by reconstructing the diversified key depending on the chip serial number of the card and then computing the response depending on the random challenge sent to the card. This result is internally compared with the response received from the card. Only 1 bit (yes or no) is transmitted to the outside world.

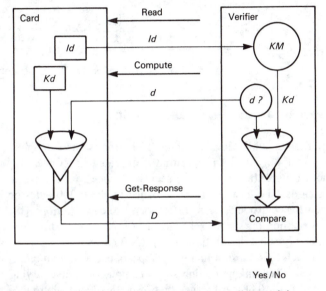

Figure 23 Authentication by a security module.

When the content of a confidential word in a card is involved, an authentication becomes an identification: The security module verifies that the right card has been activated with the right PIN. A similar authentication scheme may be implemented with any secret key algorithm.

Despite using a one-way function and only a single key per card, M4 cards provide several functions dealing with both confidentiality and integrity in both on-line and off-line operations. For confidentiality, Telepass1 allows the management of secret keys between a central mother card and a set of distributed remote M4 cards. These secret keys may be used to protect both data and programs. For integrity, Telepass1 allows one to verify the content of a word in a card. This function is used in access control systems: The relevant word represents an access right or an authorization. This function is also used in management and payment systems for controlling the result of a write command. For example, certificates are stored by retailers during payment operations: Such certificates may be used later for resolving disputes.

A pseudosignature is obtained by appending to a message a certificate involving the content of a confidential word in the card and the hashing of a message. The reduced length of the variable (48 bits) may be compensated for by using the algorithm twice on a twice-repeated hashing to give twice as many hashed bits.

The violation of a card does not endanger the whole system. But the violation of a security module has major consequences: Any other card issued by the same authority may then be subverted.

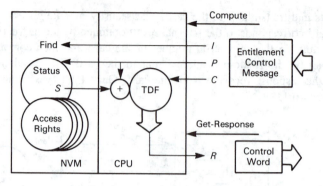

Figure 24   Control operation by TDF algorithm.

5.5 Semi-Inversible Secret Key Algorithm

The pair of unpublished algorithms TDF (for twisted double field) was designed by CCETT in 1980. The user algorithm is executed in KC0 cards. It is the left-inverse of the mother algorithm executed in security devices. KC0 cards are currently called key-carrier cards. The user algorithm is coded in about 300 bytes on mask KC0 which was designed for controlled access to broadcast teletext Antiope [18,19]. At present, broadcast information on stock exchange rates is sold in France, where access is controlled on a monthly subscription basis using KC0 cards.

In each KC0 card, a hierarchy appears between a unique issuing key and up to 32 service keys. The unique issuing key is computed by diversifying a secret master key, as in the M4 card. The card issuer then uses this key to introduce new service keys into the cards it issues and for managing the status of existing service keys in the cards. The service keys are not diversified in a controlled access broadcast application. Each service key in a card is associated with a status limiting its use. Such a status is a set of conditions, such as periods for authorizations based on subscription, or credit amounts for authorizations based on a pay-per-view scheme. A service key and its status represent an authorization or an access right. In a broadcast environment, a service key and its status is generally referred to as an entitlement.

The user algorithm of TDF in KC0 cards computes a result R (61 bits) from three variables: an external cryptogram C (127 bits), an external parameter P (23 bits), and an internal secret key S (127 bits) selected by its name (3 bytes). A control operation is shown in Fig. 24. In the pay television technology, the three external variables (name, parameter, cryptogram) are generally referred to as a message (14 bytes), and more specifically, as an entitlement control message (ECM). The result is referred to as a control word (CW).

5.6 Invertible Secret Key Algorithms

The know-how gained from Telepass1 and TDF is gathered in two unpublished algorithms named Telepass2 and Videopass, designed in 1984 by Bull CP8. Telepass2 is used in masks B1 and B2, which are the property of the French banks. Videopass, as well as Telepass2, computes a result R (64 bits) from three variables: an external cryptogram C (64 bits), a parameter P (32 bits), and an internal secret key S (128 bits) selected by its name (3 bytes). The parameter is either an external variable provided to the card or an internal nonsecret word in the card. These two algorithms are invertible: We speak of a "user algorithm" in the one direction and of a "mother algorithm" in the opposite direction.

Mask B1 provides only user cards because it performs only the user direction of the algorithm. However, both directions are programmed in masks B2 and KC1 on about 250 bytes. In B2 and KC1 user cards, a lock restricts the algorithm to the user direction. This user lock is written during card personalization. A mother card may still execute the algorithm in both directions because the user lock is not written in its NVM.

Each user card holds a unique issuing key and several service keys. The hierarchy that appeared in mask KC0 has been significantly improved in KC1. These masks, B1, B2, and KC1, include all the functions previously developed for M4, M9, and KC0. In addition, two new techniques were introduced: cryptowriting and dissymetrization. Described below, these two techniques are based on a practical use of redundancy.

Algorithm DES was programmed in 1986 by Philips in mask D1 on less than 700 bytes. Before that practical proof by realization, it was thought that algorithm DES was too expensive in RAM and ROM for realistic implementations in smart cards. But the limited resources in the SPOMs have put pressure on cryptologist programmers to use this limited memory space very efficiently.

Algorithm DES is present in several multipurpose masks which separate keys for confidentiality and keys for integrity: D1, D2 by Philips, M64 by Schlumberger, MP by Bull CP8, and TB100 by a cooperation between Bull CP8 and Philips. In these masks, DES may be replaced very easily by any algorithm requiring resources equal to or less than those required by the DES. These five masks also use cryptowriting and dissymmetrization, discussed below.

5.7 Cryptowriting and Dissymmetrization

The mechanism called *cryptowriting,* or *secure writing,* is a generalization of the subscription management mechanism invented in KC0 cards. After a computation involving its secret issuing key, the card checks the redundancy of the resulting 64 bits: if it is correct, then a secret word of 32 bits is recovered and written in the NVM.

The mechanism called *dissymmetrization* (literally—enforcing dissymmetry or asymmetry) was introduced in the masks B1, B2, and KC1. A mother card must store two different copies of the same key to compute the algorithm in both directions with this key: In one storage, the key bytes are written in the opposite direction of the other storage. A key may thus be stored either in the user direction or in the mother direction. This mechanism is very efficient for controlling the mother cards and the user cards in the system.

Redundancy is used in both directions, either inserted by mother cards toward user cards for managing rights, or inserted by user cards toward mother cards for certification purposes. Even if the results are transmitted to the outside world, the insertion of redundancy prevents a mother card which has the key stored in the mother direction from simulating a user card in which the key is stored in the user direction.

This asymetric property is reminiscent of public keys. This is not surprising since complexity of computations are the basis for security in both cases: factoring large integers on the one hand and investigating NVM contents on the other hand.

5.8 Conditional Access to Audiovisual Services

The principles developed during the KC0 study [18,19], are the basis for standardizing a pay television system by the European Broadcasting Union (EBU) [20]. The conditional access system is only one element in a strategy leading to new European television standards which are the basis for generalizing direct broadcasting satellites and later, high-definition television (HDTV) pictures.

In the conditional access system illustrated in Fig. 25, all the receivers receive the same signal consisting of scrambled components and access control parameters.

The service components are scrambled before broadcast. The scrambling method depends on the component and its coding. Each scrambling operation is controlled by a control word (CW) typically randomly modified every 10 seconds. The cryptograms of the control words are multiplexed in the broadcast signal. Control word updating is anticipated by sending these cryptograms slightly in advance of when they become effective. The signal also transmits a clear synchronization for the descrambling process. The conditional access to service components is thus reduced to the conditional access to transient control words.

Access cards implement access rights, also called entitlements. The European Broadcasting Union (UER/EBU) has adopted a vocabulary [20] describing the various entities illustrated in Fig. 25.

The *entitlement management messages* (EMMs) are produced by management centers under the authority of a card issuer. Each EMM consists of a card number, a management parameter, a service name, and a management cryptogram.

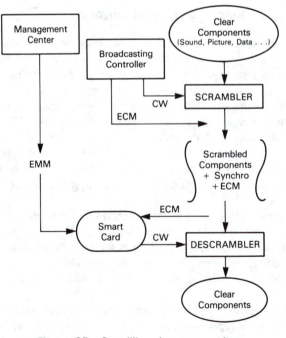

Figure 25 Conditional access system.

The *entitlement control messages* (ECM) are produced by broadcast controllers under the authority of a service broadcaster. Each ECM consists of a service name, a control parameter, and the cryptogram of a control word.

The key hierarchy in KC0 cards results in two types of security devices for producing cryptograms using the mother algorithm of TDF.

- The management security devices are used by the card issuer for managing keys and rights.
- The control security devices are used by the service broadcaster for controlling access rights.

As used by a management center, management security devices first reconstruct diversified issuing keys from the secret master key and, second, compute the mother algorithm of TDF under control of a reconstructed diversified issuing key. With such a management security device, the card issuer computes personalized cryptograms where redundancy plays the main part. In a KC0 card, the computation involves the issuing key and the result is accepted as correct if the same bit string is generated twice. Such a result is not transmitted to the outside world. If the result is correct, then the card executes the command according to information given partly in the management parameter and partly in the resulting bit string. If the result is incorrect, then the card stops and waits for a reset by the interface. Owing to such secure mechanisms, the EMMs securely initiate special actions in KC0 cards, such as the canceling of a subscription.

Each card may thus securely authenticate its issuer, without any assumption regarding network and terminal security. Such personalized cryptograms may be transmitted on any network: mail, data network, telephone, television, etc. One speaks of over-the-air addressing when these management data are multiplexed in the television channel itself.

Owing to TDF semireversibility, access rights are remotely and securely managed in KC0 cards. As used by broadcast controllers, control security devices do not diversify keys; they compute only the mother algorithm of TDF. With such a control security device, a service broadcaster computes cryptograms of the control word in use for each service key in use and for the limiting conditions coded by the parameter. Each such cryptogram is associated with the name of its service key and its control parameter (a date or a cost), which constitutes an ECM. These ECMs are mandatorily multiplexed in the television channel with the scrambled service components.

If a card receives such a control message (key name, parameter, cryptogram), it first searches for the service key. It then verifies that the conditions indicated by the parameter are compatible with the conditions indicated by the stored status of the service key: for example, the broadcast date lies in a stored subscription period; or in another access mode, a new session is automatically opened by the card, thus reducing the internal amount of credit. If these conditions are satisfied, then the card reconstructs the control word from the cryptogram. Finally, the card delivers the control word upon a get-response command. Lasting less than 1 second, such a transaction with the card is performed once every 10 seconds. Owing to TDF semireversibility, for a given program, the same control word is enciphered in as many cryptograms as there are service keys in use at the same time.

At the present in Europe, in conditional access to audiovisual services on direct broadcasting satellites, several complete systems are in competition. The system in use in France is named Eurocrypt; the corresponding specifications have been published under authority of the French government [21]. In conjunction with these developments, a new key-carrier card named KC2 has been designed. This card uses a family of unpublished secret key algorithms different from algorithm TDF.

5.9 Control of Algorithm Execution in Mask KC2

Mask KC2 is the latest mask in the key-carrier family. It has been designed for conditional access to audiovisual services on direct broadcasting satellites [21]. There is no reason for publishing the cryptographic algorithms used in KC2 cards. Consequently, these algorithms are kept secret.

The execution of the algorithm in KC2 user cards is controlled systematically. A message (either ECM or EMM) consists of three successive fields. The first field of a message contains various fields of data indicating either parameters to be checked for entitlement control or actions to be taken for entitlement management. The second field of a message consists of a variable number (0, 1, or 2) of cryptograms enciphered in electronic codebook (ECB) mode [22] under a confidentiality key. The last field of a message is a redundancy block, also named message authentication code (MAC), and computed in cipher block chaining (CBC) mode [22] for authenticating the complete message under an integrity key.

If a user card receives a message, then it uses an integrity key for checking the message authentication code (MAC) before doing anything else. This mechanism generalizes the previous *cryptowriting*. If the MAC is incorrect, the card stops and waits for a reset from the outside world. If the MAC is correct, the card continues. Neither intermediate nor final results of a computation involving an integrity key are transmitted to the outside world.

The subsequent operation may be:

1. The computation of a pair of control words (the cryptograms of the current and next control words are present in the message),
2. The secure writing of a new secret key, such as storing a new service key (the corresponding cryptogram is in the message),
3. An update of rights associated with an existing service key, such as storing a new subscription period (in this case, no cryptogram is present in the message).

If a message picked at random is submitted to such a user card for decipherment, then the probability of getting a result is about 2^{-64}. The cryptographic decipherment by user cards is thus a function which is null almost everywhere.

User cards with such a property are similar to artillery shells without their fuses. User cards are unable to work together. User cards can only work under control of a security device (also called a mother card) for either managing or controlling rights.

Just as in the military, artillery rounds are secured by locking up their fuses, only the security devices have to be controlled in order to secure the whole card system.

5.10 Logical Architecture of Card Operating Systems

The consecutive masks, from M4 to MP and TB100, on the one hand, and from KC0 to KC2, on the other hand, are more and more elaborate. Each M4 card holds only one key. In KC0 and KC1, in essence, the issuing key is different from the multiple service keys. In MP, TB100, and KC2, keys for confidentiality are distinct from keys for integrity. In masks D1 and D2, the hierarchy of keys is more sophisticated than in B1 or KC1 cards, but the management of rights is less elaborate than in KC0 or KC1 cards. Masks MP and TB100 aim at replacing M4 and M9 with substantial improvements to file management providing a total independence between data files.

In masks MP and TB100, the possibility of extending file structure by creating new files at any time during the life of the card provides a high degree of flexibility, allowing not only the implementation of applications not envisioned at the time the system was fielded, but also the introduction of new applications in already issued cards.

The independence between data files associated with flexible file management is the basis of any high-security multiapplicaton card operating system. This evolution clarifies the security architecture which is presently mature enough to be standardized. Security is fundamental in the logical architecture of smart cards. The security cannot be granted on an existing data file organization as in the existing operating system, disk operating system (DOS), UNIX, or FINDER. The difference takes place mainly in the security management which has to be taken into account in the model from the beginning of the design.

Because a file is fathered by another file, the essential creation process has to be protected. In other words, the right to create or to access a file has to be transmitted by heredity to enforce the independence between applications. This does not compel a son to have the same rights as its father, because it has the freedom to choose its way except for the creation procedure. The transmission of hereditary rights is managed by specific attributes that are transmitted to the son by the father, like chromosomes of a living creature.

The commands affect the objects and the entities specified by the security architecture. The set of commands should be defined afterward to permit compatibility and interchange between cards supporting different applications. TC68/SC6/WG5 (DIS 9992/1) introduced the notions of files, with a common data file (CDF) and application data files (ADF), according to the following set of definitions:

- File—organized set of data elements
- Common data file—unique mandatory file containing the common data elements stored in the card and used to describe the card, the card issuer and the cardholder
- Application data file—optional file supporting one or more services.

In this application-oriented structure, the CDF may clearly be interpreted either as a directory indicating the partition of NVM for applications or as the master in a security architecture. The role of the ADF is not clear and is in fact rather ambiguous. This structure is illustrated in Fig. 26.

JTC1/SC17/WG4 (WD 7816/4) introduced the notions of master file (MF), dedicated file (DF), subdedicated file (SF), and elementary file (EF), with the following set of definitions.

- File—set of elements, having logical attributes related to security and access methods, and created under common rules
- Master file—unique and mandatory file containing control information and all the other files

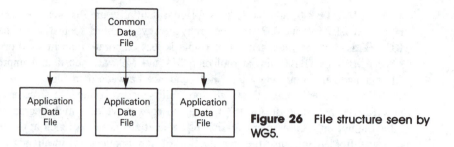

Figure 26 File structure seen by WG5.

- Dedicated file, subdedicated file—file containing control information and other files
- Elementary file—file having a security policy under which no other file may be created

Each file, except the MF, is the son of only one other file: either the MF or a (sub)dedicated file.

These definitions proposed in JTC1/SC17/WG4 are not in contradiction with the previous ones proposed in TC68/SC6/WG5; but are more general and less related to the banking point of view. This revised structure can only be interpreted in terms of security. In fact, WG4 is standardizing the security architecture in the operating system of a smart card. This structure is illustrated in Fig. 27.

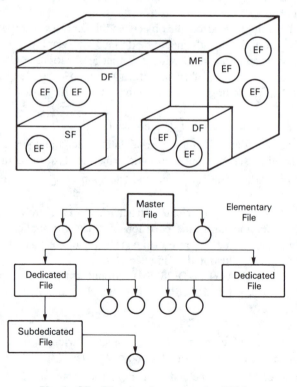

Figure 27 File structure seen by WG4.

These notions are illustrated by existing masks: KC2 (key-carrier), MP, and TB100 (multipurpose). In a KC2 card, the card issuer, the service managers, and the service providers are clearly identified and associated with levels of files. In a broadcast environment, the card issuer controls the MF and the creation of DFs; each service manager controls the creation and the evolution of EFs in its own (sub)dedicated files; each program broadcaster either controls access rights or consumes credits in EFs. Therefore each KC2 card can support several independent ''application'' files. A new DF (alias ADF) can be created at any time under control of the MF (alias CDF). The MF and each DF contain a bunch of keys: up to eight management keys (for managing access rights and updating keys) and up to eight control keys (for deciphering control words). The first management key in a DF is mandatorily written under control of a management key of the MF. The entitlements, along with various names and addresses, are stored in EF. In addition, in the MF, EFs may hold parameters for a general-purpose device. Such parameters are security information, software to be downloaded, or con- nection information to access a remote management center. For example, in the French videotex system called *Télétel,* a card reader called *Lécam* connected to a terminal called *Minitel* may thus automatically dial, connect, and access a remote application as long as the elementary data file created for this purpose respects the presentation of the data and one of the access methods recognized by the *Lécams* [6,10].

The problem of providing a means for global access to information in the cards has to be solved in the context of a security architecture. For example, how is it to be possible to access in any card an elementary file containing a phone number for auto- matic dialing? Each MP or TB100 card supports several independent application files. A new DF (alias ADF) can be created at any time under control of the MF (alias CDF). The security policy of the MF and of each DF is based on a set of nine independent diversified cryptographic keys: one issuer key, four secondary keys used for authenti- cating the service provider and for securing operations, and four secondary keys for signature and authentication of the card. The MF, as well as each DF, also contains a set of other elementary files (EF) storing various data.

Mask TB100 is a superset of MP mask. In each file, four erase keys are added for the erasement of the EEPROM memory, and three dedicated keys are added for digital signatures.

In both cards, the creation of the MF which is the birth of the card makes use of the test contacts before their destruction.

6 EVOLUTION OF CARD AUTHENTICATION

Two methods are currently used for authenticating existing banking cards and Pastel cards. The discussion of secret key one-way functions introduced one such method as was illustrated in Fig. 23. The corresponding dialogue is dynamic, but a secret master key is used in the security module. This method is mainly used for online authentication when a small number of security modules are easily protected in secure areas near central computers. Banking smart cards that use this approach are presently being mass issued in France and Norway. Each such card holds an authentication value which has been written during personalization. Similar certified identities are also used on Pastel cards issued by France Telecom for public phone subscribers.

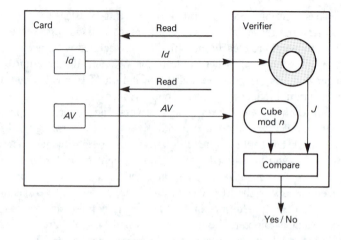

Figure 28   Static card authentication by RSA.

A second method for authenticating cards is illustrated in Fig. 28. Each card holds either its authentication value or its certified identity. A public key is used in the verifying devices. The corresponding dialogue is static. This method does not avoid the cloning of existing cards, and it is used only for assessing a local visual authentication.

Any new method should have the advantages (while eliminating the inconveniences) of both of these two existing methods. The verifying entities should use public keys, and the dialogue should be dynamic.

As shown later (Fig. 29), zero-knowledge techniques are one such method where each proving entity—implemented as a user card—privately uses a personal secret accreditation (analogous to certified identities and authentication values).

6.1 Present Use of Public Key Algorithms

During any financial transaction on point-of-sale terminals or on handheld certificators, a numerical value is provided by the card to the card-accepting-device; the verifying entity raises it to the cube modulo an integer (the public modulus) published by the card issuer. The value is accepted when the result repeats twice the same bit string consisting of various information such as chip serial number, bank account number, service code, and validity period. Figure 28 illustrates this process. If the result is inconsistent, then the point-of-sale terminal rejects the card. The result must be consistent for the transaction to be continued. The scheme is secure against forgery since to create a number whose cube is of the required form is equally difficult with the factoring of the modulus.

According to the RSA algorithm [23], the prime factors of the composite integer are involved in the computation of such authentication values. The prime factors are stored, protected, and used during card personalization by security devices named Camelias, while the composite integer published by the issuer is known and used by any verifying device. The composite integers presently in use are 320 bits long ($\approx$97 decimal digits).

As a result of advances in factorization techniques over the past few years, a 320-bit composite integer is no longer secure against a network of workstations (DEC or SUN) with a few day's computation. The fact still remains that the scheme is part of the specifications written in 1983 [8]. At the very first revision of these banking specifications, larger composite integers should be introduced.

ISO/IEC JTC1/SC27/WG2, Security Techniques, Security Mechanisms, is preparing a standard on a Digital Signature Scheme Giving Message Recovery (DIS 9796). This standard may be used for specifying an accreditation that generalizes the authentication value of a banking card and the certified identity of a *Pastel* card. Redundancy rules are more elaborate; odd and even exponents are specified, thus extending the RSA algorithm and known generic attacks against the RSA algorithm are eliminated.

6.2 Zero-Knowledge Techniques

The first practical zero-knowledge scheme was proposed by Fiat and Shamir [24] in 1986. In their scheme, computations were reduced to a near minimum. The security level grows exponentially with the product of the number of interactions (challenge/response pairs) by the number of accreditations (accreditation is a better name for authentication value and certified identity). Any desired level of security may thus be achieved as a compromise between the number of accreditations stored in the card and the number of successive successful interactions required for an acceptance. However, in the design of smart cards one must be concerned with both exchanges and storage: The exchanges with the outside world are time-consuming, while the NVM is an expensive resource. Therefore, minimization of computations alone does not seem to be the best optimization.

A second solution suited to smart cards was published in 1988 by Guillou and Quisquater [25,26]. In this protocol, storage and exchange are reduced to an absolute minimum: only one accreditation and only one interaction with the outside world. The computations required in the Guillou–Quisquater scheme are greater than is required by the Fiat–Shamir scheme for the same level of security, but only by a factor of approximately three. We give a sketch of the Guillou–Quisquater scheme. Each card is characterized by its own set of credentials (a better name for what we have been calling the card's identity). A set of credentials consists of data specified at the application level, such as bank account number, chip serial number, validity period, and service code. More generally, the set of credentials of a proven entity includes at least a validity period and a distinguished name. The set of credentials Cr is transformed into a longer integer, termed representative J of the same size as modulus n. The transformation from Cr into J is specified by publicly known redundancy rules. Such public rules are being standardized in ISO/IEC 9796 (DIS 9796). The public key of the accrediting entity consists of a public exponent v and a composite integer n. Therefore, any accreditation, denoted by B, is the secret solution to a public equation. DIS 9796 specifies such an accreditation.

$$JB^v \equiv 1 \pmod{n}$$

Today, a good size for the composite integer is 512 bits, and then the set of credentials of a card may be as long as 256 bits.

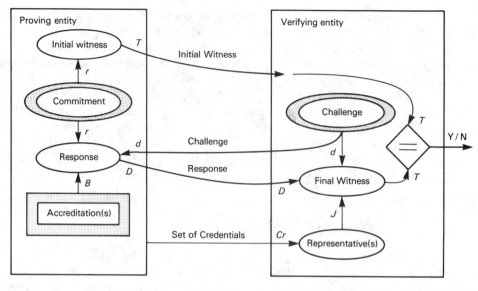

Figure 29 General zero-knowledge interactive authentication mechanism.

Figures 29 and 30 show an authentication in three moves with a card claiming the set of credentials *Cr*.

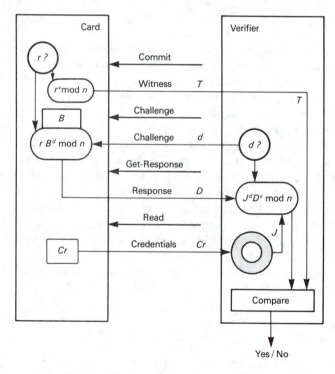

Figure 30 Card authentication by the Guillou–Quisquater scheme.

The sequence of actions by the card and by a verifying entity in proving the identity of the card are:

1. For each transaction, the card privately and randomly selects a new integer, r, in the ring of integers mod n. This integer is termed the *commitment r*. The card then privately computes the v-th power modulo n of commitment r. The result is called the *initial witness T. As a first move, the card transmits this initial witness to the verifier.*

$$T \equiv r^v \pmod{n}$$

2. Then, and not before, the verifier randomly selects an integer, d, between zero and $v - 1$. This integer is called the *challenge d. As a second move, the verifier transmits this challenge to the prover.*

3. The card then computes the product mod n of commitment r by the *d-th power of accreditation B. The result is the response D. As a last move, the card transmits this response to the verifier.*

$$D \equiv rB^d \pmod{n}$$

4. Finally, the verifying entity computes the product mod n of the v-th power of response D by the d-th power of representative J. The result is called the *final witness T'. The authentication succeeds if and only if initial and final witnesses are equal modulo n.*

$$T' = D^v J^d \pmod{n}$$

If all steps have been properly executed,

$$T' = D^v J^d = (rB^d)^v J^d = r^v B^{dv} J^d = r^v (JB^v)^d \equiv r^v = T \pmod{n} \qquad (1)$$

since B was constructed (secretly) to satisfy the equation

$$JB^v \equiv 1 \pmod{n}$$

The following three statements are crucial to understanding the conditions imposed in the protocol and why the verifying entity learns nothing about the underlying secret accreditation in the process.

1. If a cheater could guess the challenge, d, then he would have a winning strategy. If he knew what the challenge from the verifier would be before he had to commit himself to the initial witness he could construct a T to satisfy the test in (4) without having to know B.

2. A judge cannot distinguish enrolled data corresponding to successful verifications from enrolled masquerades where the challenges have been asked before fixing the witnesses.

3. The knowledge of two responses D_1 and D_2 to two distinct challenges d_1 and d_2 for the same witness T is equivalent to the knowledge of the k-th power of accreditation B where k is the greatest common divisor of v and $d_2 - d_1$.

(1) and (2) require that guessing the challenges should be impossible (or at least extremely improbable) because a successful guess would allow cheating. Similarly, guessing the commitments should also be impossible because that would compromise the secrecy of the accreditation. It is a little more difficult to explain why the proving entity must use a new and randomly chosen initial witness for each transaction and respond only to a single challenge to each witness.

Let the public key of the accrediting entity be a modulus, n, and a public verification exponent, v. The modulus is the product of two distinct primes p and q large enough to insure that n will be infeasible to factor. The public verification exponent v is chosen to be a prime number that does not divide $p - 1$ or $q - 1$; that is, such that

$$(v, (n)) = 1$$

We will show how the verifying entity could determine the proving entity's secret accreditation B if he were to respond to two challenges using the same witness, T. Let d_1 and d_2 be the two challenges (integers) such that

$$0 \le d_1 < d_2 < v$$

and let D_1 and D_2 be two responses to the challenges d_1 and d_2 respectively for the same witness T. Then

$$D_1^y J^{d_1} \equiv D_2^y J^{d_2} \equiv T \pmod{n}$$

or

$$\left(\frac{D_2}{D_1}\right)^v J^{d_2-d_1} \equiv 1 \pmod{n} \tag{2}$$

Given any pair of positive integers x and y, the congruence

$$ax - by = \pm (x, y) \tag{3}$$

always has a solution, where a is a reduced residue modulo y; that is, $0 < a < y$, and b is a reduced residue modulo x. (x, y) denotes the greatest common divisor of x and y. Equation (3) is a specialized form of Bezout's identity* (after Etienne Bezout) and the unknowns a and b are known as the Bezout coefficients. It is easy to calculate a and b using the extended Euclidean algorithm.

Replacing x by $d_2 - d_1$ and y by v in Eq. (3), we get

$$a(d_2 - d_1) - bv = \pm 1 \tag{4}$$

since $(v, d_2 - d_1) = 1$. Equation (4) can be solved to find the reduced residues a and b.

If Eq. (2) is raised to the exponent a, we get

$$\left(\frac{D_2}{D_1}\right)^{av} J^{a(d_2-d_1)} = \left(\frac{D_2}{D_1}\right)^{av} J^{\pm 1+bv} \equiv 1 \pmod{n}$$

---

*According to Gauss, Lagrange, and Legendre, the Bezout identity was discovered earlier by Bachet de Méziriac but bears Bezout's name. Such is the history of science.

or

$$J^{\pm 1} \left[\left(\frac{D_2}{D_1} \right)^a J^b \right]^v \equiv 1 \pmod{n}$$

But since the secret accreditation B is known to be the (supposedly secret) solution to the public equation

$$JB^v \equiv 1 \pmod{n}$$

we have

$$B^{\pm 1} \equiv \left(\frac{D_2}{D_1} \right)^a J^b \pmod{n}$$

Therefore, if the proving entity were to respond to two challenges, d_1 and d_2, chosen as described above, an opponent could then solve for the proving entity's secret accreditation B. Consequently the proving entity must construct a new and random initial witness T for each transaction if B is to be kept secret.

The size of the public exponent is determined by the security requirements of the particular application: A cheater has at most one chance out of v to deceive a verifier, and a verifier has at least $v - 1$ chances out of v to detect a cheater.

In a local verification where the verifying entity retains the card if it fails an authentication exchange, the public exponent may (need to) be as large as $2^{17} + 1$. But if the cardholder itself is retained by the verifying entity, then the public verification exponent may even be reduced to be as small as $2^8 + 1$ or $2^4 + 1$.

Cards and verifiers perform similar operations, with the same complexity of computation. If a card can be authenticated, it can also authenticate other cards. With such schemes, banking security modules should be personalized as retailer cards. All these cards, user cards as well as retailer cards, should authenticate each other in a very symmetrical way.

6.3 New Signatures

By using general principles first suggested by Fiat and Shamir [2,3] the previous method can be adapted in a natural way to a digital signature scheme by using a hash function. A good hash function must be one-way and collision-resistant, in the sense that finding a collision is very difficult; that is, practically impossible. A collision of the function "hash" is a pair of distinct arguments x' and x'' such that

$$\text{hash}(x') = \text{hash}(x'')$$

ISO/IEC JTC1 SC27/WG2 is now standardizing hash functions for digital signatures (DP10118).

In several signature schemes, the hashing of the message is the input to the inverse of a trapdoor permutation such as the RSA algorithm. Such schemes are susceptible to the *birthday attack* where a cheater looks in advance for a collision of two messages: one favorable to the signer and the other one favorable to himself. When the cheater obtains the signature of the message favorable to the signer, it can reveal the second message favorable to himself. In these schemes, the hashing of the message must be long enough to avoid a search for collisions using a birthday attack. Hash results of 64 bits are too short. In such signature schemes, the hashing of the message must be longer: The suggested length is 128 bits.

But in signature schemes that are derived from zero-knowledge proof techniques, the random challenge selected by the verifying entity is replaced by a pseudorandom challenge computed by hashing together the message and the initial witness. When message and witness are hashed together into a 64-bit challenge, the same pseudorandom variable (challenge d) is involved at both the beginning and the end of the verification process. In this case the birthday attack seems to be irrelevant. Therefore a 64-bit hashing is secure in this case.

The size of the exponent v is now around 64 bits (instead of 17 or fewer as before), which corresponds to the entropy of the challenge d. This is the price one must pay for giving up an interactive protocol.

Figure 31 illustrates the G–Q signature process: Message M is signed by an appendix consisting of initial challenge d, response D, and credentials Cr. In such a scheme, challenge d represents redundancy while response D represents randomness.

Figure 32 illustrates the G–Q verification process which begins by recomputing representative J from the set of credentials Cr according to public redundancy rules. Then final witness T' is obtained by computing mod n the product of the v-th power of response D by the d-th power of representative J. Then final challenge d′ is obtained by hashing message M and final witness T'. A signature is valid if and only if initial and final challenges are equal.

Let the public exponent be $2^{64} + 1$ (a product of two primes). Let the Hamming weight of challenge d be limited to 32. For example, by choosing as challenge d the complementary value of hash(T, M) when the result contains more than 32 1's. Then the final witness T′ may be computed by squaring response D 64 times, interleaved with at most 32 multiplications by representative J, plus a last multiplication for adjusting with response D. The computational complexity of a verification is therefore less

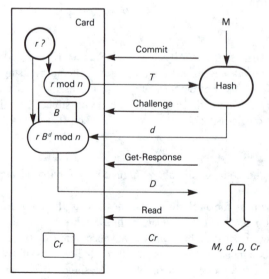

Figure 31 Guillou–Quisquater signature scheme.

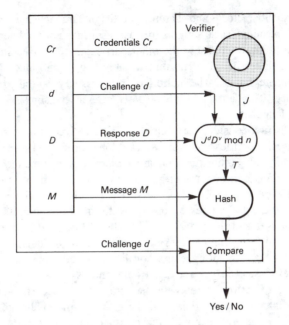

Figure 32 Verification of a Guillou–Quisquater signature.

than 100 multiplications mod n. Signature and verification processes have the same level of complexity. Therefore, if a card can sign, it can also verify. These authentication and signature schemes are based on identity. It is an evolution of methods published separately in 1984 by Shamir [23] and by a French banking card organization [8].

The cards in the Guillou–Quisquater scheme hold secrets that are not cryptographic keys. An accreditation does not provide immediate confidentiality services. The method cannot be directly derived from its basic goal—integrity. The RSA scheme does not have this property. In the RSA scheme, the secret prime factors can be used for both integrity and confidentiality purposes. This property that the secret is exclusively usable for integrity purposes is in tune with political and governmental requirements. Indeed, a fair integrity (signature and authentication) scheme should not make any assumption as to the integrity (morality and good citizenship) of its potential users.

6.4 Multisignatures

In multisignature schemes, several signing entities collaborate in signing the same message. In a first—trivial—solution, each signing entity signs separately; but that is not cooperation. In a second solution, the signing entities sign successively, one after another, on a progressive basis: The present signing entity signs the result produced by the previous ones. But this method introduces between entities an order that is unnatural in some applications. Intermediate significant results are a potential cause of dispute. Later on, an additional signing entity may always sign as the new last one. A progressive process intrinsically excludes the notion of simultaneity which looks very attractive for avoiding several problems resulting from signature repudiation.

In a satisfactory and natural multisignature scheme, the intermediate results should be meaningless so long as the last result has not yet been obtained. Each signing entity should know precisely the other participants involved in the multisignature process. Also, there should be no undetectable way to introduce at a later stage an additional ultimate signer. The notion of progressivity is thus replaced by a notion of simultaneity. This attractive property is offered by signature schemes derived from zero-knowledge.

Let us first consider the general solution derived from zero-knowledge techniques. Several signing entities collaborate in a global process. These signing entities may depend on different authorities that are members of a directory system as described in CCITT in X509 and now under going standardization by ISO (ISO9594/8). Each signing entity proposes an initial witness. Then a global initial challenge is produced by hashing all the initial witnesses together with the common message to be signed. All the initial witnesses and the global initial challenge are then sent to each signing entity for verification of the initial challenge and construction of individual responses. Finally, the message is signed by an appendix consisting of the global initial challenge, all the individual responses, and all the sets of credentials.

The verification begins with the computation of each final witness from each response, each set of credentials, and the global initial challenge. Then all the final witnesses are hashed together with the message so as to obtain the final challenge. The signature is accepted if and only if the initial and final challenges are equal.

Let us now consider two smart cards issued by the same accrediting entity: Each card stores its unique accreditation related to its own set of credentials. The accrediting entity has published a public key consisting of two integers n and v. Both cards, with sets of credentials Cr_1 and Cr_2, cooperate on the same personal computer to produce a global signature of message M.

The signature protocol consists of the following steps:

1. Global initial witness T is the product mod n of both individual initial witnesses, T_1 and T_2.
2. Global initial challenge d is the hashing of global initial witness T and message M.
3. Global response D is the product mod n of both individual responses D_1 and D_2.

Let us write the corresponding equations:

$$T = T_1 T_2 \bmod n; \quad d = \text{hash}(T, M); \quad D = D_1 D_2 \bmod n$$

Message M is signed by the appendix, consisting of global initial challenge d, global response D, and two sets of credentials Cr_1 and Cr_2.

This method may be extended to any number of participants. The signing appendix then consists of one global initial challenge, one global response, and all the sets of credentials.

The verification is performed in the usual way. The global representative is the product mod n of all the representatives. Global final witness T' is computed as the product mod n of the v-th power of global response D by the d-th power of the global representative. Finally, global final challenge d is computed as the hashing of global final witness T' and message M. The signature is accepted if and only if initial and final challenges are equal.

As a matter of fact,

$$T' = D^v(J_1J_2)^d = (r_1B_1^d r_2 B_2^d)^v J_1^d J_2^d \mod n$$
$$= (J_1B_1^v)^d (J_2B_2^v)^d (r_1r_2)^v = T_1T_2 = T \mod n$$

A cosignature is attractive for retail banking applications. The financial message may consist of a date and an amount, plus a serial number for the buyer and a serial number for the seller. This message is signed simultaneously by the retailer and by the customer. The result is an electronic check which may be verified by any other card, including both the retailer card and the customer card.

6.5 Probable Future Evolution of Smart Cards

Examining the development of smart card systems, we see a strong clue to the future evolution of smart cards: *Cards will interact without sharing secrets!*

Cryptology in the cards will soon include new authentication and signature methods derived from zero-knowledge techniques. New access methods which are standardizable are now being developed. In addition, the ISO working group on public keys is preparing a first working draft on zero knowledge techniques. In open systems, the standardization of authentication and signature tools is an important issue that does not seem to conflict with political and governmental considerations. This evolution has a good influence on banking card systems by making more nearly symmetric the personalization of user cards and security devices; anonymous secrets can be replaced by personalized secrets. The security devices may thus be considered as retailer cards.

Let us give an illustrative practical example of the practical use of this cryptologic development in another field of application. Health cards record a lot of confidential data about cardholders. These data should always be easily accessible when the PIN is presented correctly. However, the cardholder may select fields of data to be accessible by other methods, even though he cannot present his PIN. A health card should authenticate accredited physician cards and the cards of other rescue personnel. To provide access to information that has been previously selected by the cardholder in a proper fashion, the card should only recognize the public keys made public by the relevant emergency medical personnel.

Continuing this line of reasoning, the general public will probably buy smart cards in the future as it does calculators today. Having purchased such a card, the consumer would visit his service provider and after filling out a form and signing a contract, the service provider would write an accreditation in the personal card of the user. The same card would hold several accreditations from different authorities for different purposes. Using the appropriate accreditation, the consumer would later be able to access the corresponding service either by interactively authenticating himself or by signing. This scenario describes what we believe to be both a possible and probable evolution of smart card systems toward open systems where multiapplication cards will be the property of their users. We don't see why bankers should have to deal with all the side effects due to a card expiring on a certain date, such as refunding taxable units of parking or telephone messages. If a card is the personal property of its holder, the problem of refunding small amounts is to be solved directly between the cardholder and the service provider. Such situations modify considerably the economy of the systems. The economy of smart card systems is related in a very sensitive way to the cost of the cards and to the security architecture of the global systems. Up to the present, side effects due to shared security such as rigid expiration dates have slowed down the synergy between applications of smart cards.

7 CONCLUSIONS

The development of smart card systems raises a major question related to cryptography. Smart cards to be used by the general public are very efficient security tools because their logical security is based on cryptographic techniques. The manufacture, use, and export of cryptographic materials are subject to national regulations and export controls because cryptography has national security ramifications.

A crucial question therefore (to the development and application of smart cards) is how are the commercial needs to be accommodated while at the same time satisfying the governmental concerns?

A card is an element of solution to a problem, and the whole solution has to be considered. The following schematic example shows three generic levels; authentication, keying, and (de)scrambling. If confidentiality is required on a communication network, then a session key may well be created by exponentiating in finite fields, as suggested by Diffie and Hellman [28]. An intruder may well be active on the communication path. But such an intruder is immediately detected if, before doing anything else, all the keying data elements are authenticated (e.g., by a zero-knowledge protocol). The session key is subsequently used for scrambling/descrambling the transmitted signal.

(De)scrambling computations are to be performed by dedicated devices such as radiotelephones and decoders. For example, the picture and sound signals of a pay television program are scrambled according to their respective natures and the codings fixed by the television standards. Solutions to confidentiality problems appear to be essentially related to the network and to the service provided on the network. Scrambling mechanisms are standardized according to the application. National regulations may influence the solutions.

Authenticating computations are performed by detachable security devices such as smart cards. Authentication is always performed in reference to an authority such as a card issuer. Essentially related to the individuals, solutions to integrity problems must be international in scope. Even if the technology evolves apace, a European payment card must continue to be usable in the United States and vice-versa. Therefore the standardization of authentication mechanisms is required.

Keying computations may be performed either by the dedicated devices or by the personal cards. Keying and authentication should not be confused because the commercial cryptographic techniques separate as much as possible integrity (certification, authentication, identification, signature) and confidentiality (secrecy, discretion).

In this context, what is the role of smart cards? There are two main approaches depending on whether secret key or public key techniques, respectively, are used. In making such a decision, system designers must be guided by practical trade-offs between cost and performance. The smart card controls the use of any internal cryptographic algorithm (i.e., one able to provide confidentiality). Cards KC2 provide a good illustration: Any cryptogram submitted for decipherment to a card KC2 must be included in a message terminated by a message authentication code (MAC). Deciphering occurs if and only if the associated MAC is correct. Hence, KC2 user cards cannot interact for keying a communication. Cards KC2 very well fit access control problems such as pay television. Such smart cards are the domain of customized secret algorithms that ensure independence between applications. But the corresponding mother cards still have to be controlled. The management of security in this approach cannot be generalized to an open context.

There is a strong need for standards based on public key and zero-knowledge techniques. If these standards are restricted to integrity, and cannot be derived for confidentiality purposes, then they should not conflict with governmental and political policy. For example, some authentication schemes are by nature restricted to integrity: An accreditation is a secret but not a key. However, the corresponding implementations should also avoid any potential misuse, even though the basic operations (multiplying and exponentiating large integers modulo large integers) are also the basic operations of various confidentiality schemes. If such an authentication scheme is efficiently performed in an ''open'' operating system, then the corresponding efficient arithmetic operator (a dedicated chip, for example) may be misused for keying any pair of users on an open communication network by exponentiating in finite fields. Smart cards, however, must be tamper-resistant. If the mask is designed correctly, then the card's computational power, including the built-in arithmetic operators, cannot be misused for other purposes. Therefore the integrity standards (authentication, signature, and key management) based on public key and zero-knowledge techniques should be completed by national agreements authorizing the corresponding implementations. Some masks for SPOMs should obtain such an agreement.

The approach just described makes it possible to predict the concept of ''basic common cards'' to be issued by a public authority (like a bank of issue) and to be freely sold to the general public for general purposes. On such an ''agreed-upon'' card, the user will ask for the introduction of several various applications, such as payment and credit by a bank, public transportation by an authority, access by a service provider, telephone privileges granted by an operator, personal files, etc. The study, specification, experimental trials, and standardization of these basic common cards will almost certainly become a major international development over the next few years.

APPENDIX A: ISO PRESENTATION

ISO has three official languages (don't try to spell ISO in any of them!):

- English: International Organization for Standardization
- French: Organisation Internationale de Normalisation
- Russian: Международная Организация по Стандартизации

A.1 ISO STRUCTURE

The ISO is the specialized international agency for standardization, comprising the national member bodies of about ninety countries.

A national member body is the most representative organization for standardization in its country. The United States is represented by ANSI (American National

Standards Institute), France by Afnor (Association Française de Normalisation), the United Kingdom by BSI (British Standards Institution), and the USSR by GOST.

ISO is administered by a council consisting of a president, a vice-president, a treasurer, and eighteen member bodies. The council creates technical committees (TC) in main fields of interest. Each TC then creates subcommittees (SC). Each SC creates working groups (WG). And finally, each WG may create task forces (TF). A central secretariat in Geneva prepares the ballots and edits and prints the standards. This pyramidal structure is illustrated in Fig. A.1.

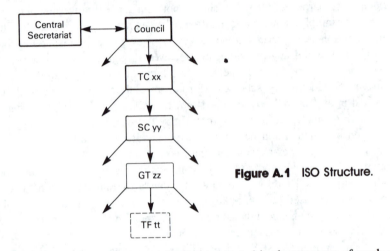

Figure A.1 ISO Structure.

At TC and SC levels, member bodies interested in the program of work are either participants (P-members) or observers (O-members).

Decisions are taken in association with liaisons (L-members) like IEC (the International Electrotechnical Commission), CCITT (le Comité Consultatif International Télégraphique et Téléphonique), and more generally, any international organization interested in the program of work.

The votes are organized at SC and TC levels, with one ballot per country. L-members do not vote, but their technical comments are considered.

At WG and TF levels, there is no ballot, only experts who are nominated by P-, O-, and L-members and who draft all the documents.

A.2 ISO PROCEDURE (Figure A.2)

0	**WI**	New Work Item
1	**WD**	Working Draft
2	**DP**	Draft Proposal
3 to 5	**DIS**	Draft International Standard
6 to 8	**IS**	International Standard

Figure A.2.   ISO procedure (stage numbers).

Any P-, O-, or L-member may propose a new work item (WI) together with estimated resources, dates, and targets. After approval by a ballot at TC level, the WI is included in the program of work of the relevant SC which assigns it to a WG.

The WG produces a document as soon as possible. After reaching a consensus, the preliminary document is transmitted to the SC as a working draft (WD). Through a resolution at a plenary meeting, the SC approves the document for registration as a draft proposal (DP).

The DP number is that of the future International Standard. A DP is circulated at SC level for a three-month ballot. Results and comments are sent to experts to consider the positive comments and to resolve the negative ballots. The goal is to reach unanimity or at the very least a substantial consensus.

After approval, a DP becomes a draft international standard (DIS): It is translated, edited, and printed under central secretariat control, and then balloted for 6 months at TC level. A DIS must be approved by at least 75% of the members voting, including a majority of P-members.

After approval, a DIS is transmitted to council with a comprehensive report for final decision. After council acceptance, the International Standard (IS) is published.

Every 5 years after publication, each IS is subject to revisions to confirm, amend, or withdraw the IS.

Several successive DP or DIS versions may be balloted; amendments due to negative comments of a member may offset a positive vote of another member; industrial and commercial ventures may modify the position of a member; compromises obtained during a meeting may be denied at the next meeting. Thus technical experts often become involved in strategic games without being prepared.

The elaboration of an IS is a very long story (5–10 years): Any efficient standard is largely supported by the members.

GLOSSARY

CLK. Clock line, one of the six contacts standardized by ISO.

CPU. Central process in 1 unit.

Die, pl. dice. Individual device (microcomputer or other) on silicon.

EEPROM. NVM—may be erased by applying special voltage.

EPROM. NVM—may be erased by exposure to UV light.

GND. Ground, reference voltage, one of the six contacts standardized by ISO.

I/O. Input/output communication line, one of the six contacts standardized by ISO.

MAC. Message authentication code; artificial redundancy used to check message authenticity.

Mask. Medium used to convert customers' application software (ROM code) to a pattern on silicon and, by extension, the application software itself.

MCU. Single-chip microcomputer unit.

Microcomputer. A system containing a microprocessor, various memories, and other peripheral devices.

NVM. Nonvolatile memory.

PIN. Personal identification number.

RAM. Random access memory.

ROM. Read-only memory.

RST. Reset line, one of the six contacts standardized by ISO.

Smart card. A card, which looks like a credit card, but which contains a microcomputer.

SPOM. Self-programmable one-chip microcomputer, a type of secure MCU.

Test mode. Special operating mode for an MCU for testing by the manufacturer prior to shipping to customer.

User mode. Normal operating mode for MCUs, the only mode used in a smart card.

VCC. Power supply line, one of the six contacts standardized by ISO.

VPP. Programming voltage line, one of the six contacts standardized by ISO.

Wafer. Slice of silicon which, after processing, contains typically hundreds of individual dice.

REFERENCES

[1] L. C. Guillou and M. Ugon, "Smart card: A highly reliable and portable security device," in *Lecture Notes in Computer Science 263; Advances in Cryptology: Proc. Crypto'86*, A. M. Odlyzko, Ed., Santa Barbara, CA, Aug. 11–15, 1986, pp. 464–479. Berlin: Springer-Verlag, 1987.

[2] M. E. Haykin and R. B. Warnar, *Smart Card Technology, New Methods for Computer Access Control*, NIST 500–175. Gaithersburg, MD: National Institute of Standards and Technology, Sept. 1988.

[3] D. W. Davies, "Smart cards, digital signatures, and negotiable documents," in *Proceedings of the International Conference on Secure Communications Systems*, London, UK, February 22–23, 1984, pp. 1–4. London: Institution of Electrical Engineers, 1984.

[4] A. G. Mason, "Conditional access for broadcasting," in *Proceedings of IBC'88, International Broadcasting Convention (Conf. Publ. No. 293)*, Brighton, UK, Sept. 23–27, 1988, pp. 328–332. London: Institution of Electrical Engineers, 1988.

[5] J. Svigals, "Improved security with integrated circuit cards," *J. Inform. Syst. Management*, vol. 5, no. 2, pp. 32–38, 1988.

[6] A. Turbat, "The smart card, an ace in France's telecom system," *Telephony*, vol. 204, no. 26, pp. 78, 80, 82, 86, June 27, 1983.

[7] M. Paterson, *Secure Single Chip Microcomputer Manufacture*, Eng. Bull. EB400/D. Phoenix, AZ: Motorola Semiconductor, 1990.

[8] *Spécifications et normes de la carte à mémoire bancaire*. Paris: Groupement des Cartes Bancaires, Jan. 1984.

[9] M. Ugon and P. Schnabel, *TB100, The Highly Secure Multipurpose Smart Card Family*, Bull. CP8. Trappes, France, Jan. 1990.

[10] J. F. Briend and J. J. Plancke, "French PTT Minitel and Lécam programme," *Philips Telecommun. Data Systems Rev.*, vol. 45, no. 2, pp. 10–26, June 1987.

[11] W. L. Price, "Standards for data security: A change of direction," in *Lecture Notes in Computer Science 293; Advances in Cryptology: Proc. Crypto'87*, C. Pomerance, Ed., Santa Barbara, CA, Aug. 16–20, 1987, pp. 3–8. Berlin: Springer-Verlag, 1988.

[12] *Data Encryption Standard*, FIPS PUB 46. Gaithersburg, MD: National Institute of Standards and Technology, April 1987.

[13] L. C. Guillou, M. Davio, and J.-J. Quisquater, "Public-key techniques: Randomness and redundancy," *Cryptologia*, vol. 23, no. 2, pp. 167–189, April 1989.

[14] *Defending Secrets, Sharing Data; New Locks and Keys for Electronic Information*, OTA-CIT-310. Washington, D.C.: U.S. Congress, Office of Technology Assessment, Oct. 1987.

[15] R. McIvor, "Smart cards," *Scientific American,* vol. 253, no. 5, pp. 130–137, Nov. 1985.

[16] J. K. Omura, "A smart card to create electronic signatures," in *Proceedings of BOSTONICC/89, IEEE International Conference on Communications,* vol. 3, Boston, MA, June 11–14, 1989, pp. 1160–1164. New York: IEEE, 1989.

[17] J.-J. Quisquater, D. de Waleffe, and J.-P. Bournas, "Corsair, a chip card with fast RSA capability," in *SMART CARD 2000: The Future of IC Cards; Proc. IFIP WG 11.6 Internat. Conf.,* D. Chaum and I. Schaumuller-Bichl, Eds., Laxenburg, Austria, Oct. 19–20, 1987. Amsterdam: North Holland, 1989.

[18] L. C. Guillou, "Radiodiffusion à péage pour application au télétexte ANTIOPE," in *Actes du congrès de Liège,* Belgique, Nov. 24, 1980.

[19] L. C. Guillou, "Smart cards and conditional access," in *Lecture Notes in Computer Science 209; Advances in Cryptology: Proc. Eurocrypt'84.* T. Beth, N. Cot, and I. Ingemarsson, Eds., Paris, France, April 9–11, 1984, pp. 480–489. Berlin: Springer-Verlag, 1985.

[20] *Spécification des systèmes de la famille MAC/paquets,* Document Technique 3258. Bruxelles: Centre technique de l'UER/EBU, Oct. 1986.

[21] *Systèmes d'accès conditionnel pour la famille MAC/paquet,* EUROCRYPT. République Française: Ministère des PTT, Ministère de l'Industrie, Ministère de la Culture, March 1989. (Available on request at CCITT, Rennes.)

[22] *DES Modes of Operation,* FIPS PUB 81. Gaithersburg, MD: National Institute of Standards and Technology, Dec. 1980.

[23] R. L. Rivest, A. Shamir, and L. Adleman, "A method for obtaining digital signatures and public-key cryptosystems," *Commun. ACM,* vol. 21, no. 2, pp. 120–126, Feb. 1978.

[24] A. Fiat and A. Shamir, "Unforgeable proofs of identity," in *Proceedings, SECURICOM'87: 5th Worldwide Congress on Computer and Communications Security and Protection,* Paris, France, March 4–6, 1987, pp. 147–153. Paris: Societe d'edition et d'organisation d'expositions professionnelles, 1987.

[25] L. C. Guillou and J.-J. Quisquater, "A practical zero-knowledge protocol fitted to security microprocessor minimizing both transmission and memory," in *Lecture Notes in Computer Science 330; Advances in Cryptology: Proc. Eurocrypt'88,* C. G. Günther, Ed., Davos, Switzerland, May 25–27, 1988, pp. 123–128. Berlin: Springer-Verlag, 1988.

[26] J.-J. Quisquater and L. C. Guillou, "Des procédés d'authentification basés sur une publication de problèmes complexes et personnalisés dont les solutions maintenues secrètes constituent autant d'accréditations," in *Proceedings of SECURICOM'89: 7th Worldwide Congress on Computer and Communications Security and Protection,* Paris, France, March 1–3, 1989, pp. 149–158. Paris: Société d'édition et d'organisation d'expositions professionnelles, 1989.

[27] A. Shamir, "Identity-based cryptosystems and signature schemes," in *Lecture Notes in Computer Science 196; Advances in Cryptology: Proc. Crypto'84,* G. R. Blakley and D. Chaum, Eds., Santa Barbara, CA, Aug. 19–22, 1984, pp. 47–53. Berlin: Springer-Verlag, 1985.

[28] W. Diffie and M. Hellman, "New directions in cryptography," *IEEE Trans. Inform. Theory,* vol. 22, pp. 644–654, 1976.

How to Insure That Data Acquired to Verify Treaty Compliance Are Trustworthy*

G. J. Simmons
Sandia National Laboratories
Albuquerque, New Mexico 87185

*This chapter first appeared in the *Proceedings of the IEEE*, vol. 76, no. 5, May 1988.

Abstract—In a series of papers [6–8] this author has documented the evolution at the Sandia National Laboratories of a solution to the problem of how to make it possible for two mutually distrusting (and presumed deceitful) parties, the host and the monitor, to both trust a data acquisition system whose function it is to inform the monitor, and perhaps third parties, whether the host has or has not violated the terms of a treaty. The even more important question of what data will adequately show compliance (or noncompliance) and of how this data can be gathered in a way that adequately insures against deception will not be discussed here. We start by assuming that such a data acquisition system exists, and that the opportunities for deception that are the subject of this chapter lie only in the manipulation of the data itself, that is, forgery, modification, retransmission, etc. The national interests of the various participants, host, monitor and third parties, at first appear to be mutually exclusive and irreconcilable, however we will arrive at the conclusion that it is possible to simultaneously satisfy the interests of all parties. The technical device on which this resolution depends is the concatenation of two or more private authentication channels to create a system in which each participant need only trust that part of the whole that he contributed. In the resulting scheme, no part of the data need be kept secret from any participant at any time; no party, nor collusion of fewer than all of the parties can utter an undetectable forgery; no unilateral action on the part of any party can lessen the confidence of the others as to the authenticity of the data and finally third parties, that is, arbiters, can be logically persuaded of the authenticity of data. Thus, finally after nearly two decades of development a complete technical solution is in hand for the problem of trustworthy verification of treaty compliance.

1 INTRODUCTION

The best known example of a treaty verification system is the series of systems developed at the Sandia National Laboratories to monitor compliance by the Russians with a proposed comprehensive nuclear test-ban (CTB) treaty [6–8]. Although the data acquisition system (RECOVER) developed to enable the International Atomic Energy Agency (IAEA) in Vienna to remotely monitor the compliance of a worldwide network of power reactors with the terms of their licensing agreements is less well known, it must satisfy precisely the same objectives for the participants as the CTB verification system. There have also been similar systems designed for arms control purposes for the Arms Control and Disarmament Agency (ACDA) and for continuous inventory by the Nuclear Regulatory Commission (NRC) of plutonium during fuel rod reprocessing by commercial facilities, which share many of the same system objectives. In this chapter though, we shall use as the paradigm for such monitors the system for verifying compliance with a CTB treaty, i.e., a treaty banning all underground nuclear weapons testing. Although the problem has been described elsewhere, we repeat the essential points here, primarily to make clear the conflicting interests of the various participants.

2 VERIFICATION OF A COMPREHENSIVE TEST BAN TREATY

For over two decades, the United States and the Soviet Union have explored, and on occasion negotiated, the details of a comprehensive nuclear test-ban treaty as a means to slowing the arms race. The immediate object of a comprehensive test-ban treaty would be to stop *all* testing of nuclear weapons, thereby essentially freezing the weapons technology at its state of development at the time the treaty takes effect, and hence eventually reducing the chance that another round of the arms race might occur based on yet another major improvement in nuclear weapons technology. Test-ban treaties prohibiting surface, ocean, and space testing of nuclear weapons are in effect and have been abided with by both sides in precisely those areas where verification of compliance by what has been euphemistically called "national means" is possible. In other words, the nation doing the monitoring, for our purposes the U.S., is limited to those observations that are possible from its sovereign territory or from the territories of its allies or from space using satellites. The most reliable technique for detecting underground tests, and essentially the only direct measurement method that can be used at a distance, is to measure the ground motions resulting from the underground detonation using seismic sensors. Unfortunately the threshold of yield for seismic detection at teleseismic distances, from the U.S. or from Scandinavia, is high enough that meaningful weapon development could be carried out below the teleseismic detection threshold. Just how small an underground test can be detected is a function of many things, some that are under the control of the tester, such as the geology of the test site, decoupling chambers for the detonation, time of the test, etc., and some that can be jointly agreed to in the terms of the treaty such as how close the monitoring stations can be to known test ranges and of the physical emplacement of the seismic sensors. Since the purpose of a comprehensive test-ban treaty is to slow the arms race by stopping the development (proof testing) of new nuclear weapons technology, such a treaty is logically feasible only if each party can be confident that the other cannot continue clandestine testing, and hence development of new weapons, to gain an advantage over the other. It is generally accepted by nuclear weapons designers that there is a lower limit to the size of the detonations needed to conduct meaningful weapon development programs, for the purpose of argument say one kiloton, and that tests involving yields below this limit are unlikely to have a significant impact for new weapon systems. The bottom line, which has been recognized by both sides in the negotiations, is that unlike previous treaties in which national means of verification were available, and adequate, that verification of compliance with a comprehensive test-ban treaty would necessitate emplacing seismic monitoring stations within the sovereign borders of the country being monitored (the host) and/or his allies. This would require a radical departure from previous treaty protocols since it would be necessary for the host to cooperate to make it possible for the monitor to verify compliance with the terms of the treaty. Protocols of this sort, anticipated in the verification means for the comprehensive test-ban treaty discussed here and in the SALT II (Strategic Arms Limitation Treaty) where each side would have had to cooperate in order for the other to verify the number of launch vehicles fielded, are apparently about to be realized for the first time in the intermediate-range nuclear forces (INF) treaty between the United States and Russia that has just been signed. With suitable placement of the sensors, seismic techniques that the U.S. has proven by monitoring underground tests at the Nevada Test Site to be capable of detecting subsurface tests and of discriminating the signals from naturally occurring seismic background, are available so that either nation could be extremely confident that no meaningful violations of the treaty could go undetected. Consequently, if it were possible to have appropriately sited seismic monitoring stations within the host's territory manned by the monitoring country's personnel a proven means of verifying compliance with an underground test-ban treaty exists. The difficulty, however, is that continuously manned installations are unacceptable.

A problem that the Sandia National Laboratories has worked on for over two decades has been to develop an unmanned seismic monitoring system, Figs. 1 and 2, that could satisfy the national interests of all parties. It is not difficult to physically secure the seismic sensor package in subsurface emplacements as shown in Fig. 2 since the seismic sensors themselves would detect any attempt to gain physical access to them long before they were in jeopardy. Hence only the data stream sent through an open communications channel would be subject to possible manipulation. From the viewpoint of the monitor, an opponent, usually assumed to be the host for the sensor emplacement, but possibly a third party desiring to undermine the treaty, may either introduce fraudulent or altered messages. For example, the host, if he can do so without being detected, may wish to substitute innocuous seismic records in the stead of incriminating ones that would reveal that tests had been conducted in violation of the terms of the treaty, thus lulling the monitor into erroneously believing that the treaty was being abided with. Conversely, he may wish instead to introduce spurious incriminating messages indicating that tests have occurred when in fact none have been carried out thereby misleading the monitor into erroneously reporting nonexistent violations. This latter stratagem is especially significant when only a limited number of on-site verification inspections are permitted the monitor.*

---

*Note added in proof: To provide information required for more reliable monitoring of underground nuclear detonations, the U.S. on August 17, 1988, conducted an underground nuclear weapons test at the Nevada test site monitored and instrumented on site by Soviet scientists. The USSR reciprocated on September 14, 1988, with a test shot at their Semipalatinsk test site monitored on site by U.S. scientists.

Figure 1 Prototype seismic monitoring station for verification of CTB (Alaskan installation). Reprinted with permission from Sandia National Laboratories.

Therefore, in order for a system to be acceptable to the U.S. it must be very improbable that anyone, either the Russians or a third party, could utter an undetectable forgery, that is, the messages must be capable of being authenticated (by the U.S.) as having originated with the seismic sensors that the U.S. had emplaced, and also that the data have not subsequently been tampered with. If only the U.S. objective had to be met, this would be an easy problem to solve. A conventional, that is, single-key, cryptosystem could be emplaced by the U.S. along with the seismic sensors in the borehole, and as in military communications systems a known authenticator appended prior to either block or cipher feedback stream encryption.* The resulting data stream (cipher)

---

*The reader is referred to the chapter ''A Survey of Information Authentication'' by G. J. Simmons in this volume for a more complete discussion of the standard military authentication protocol.

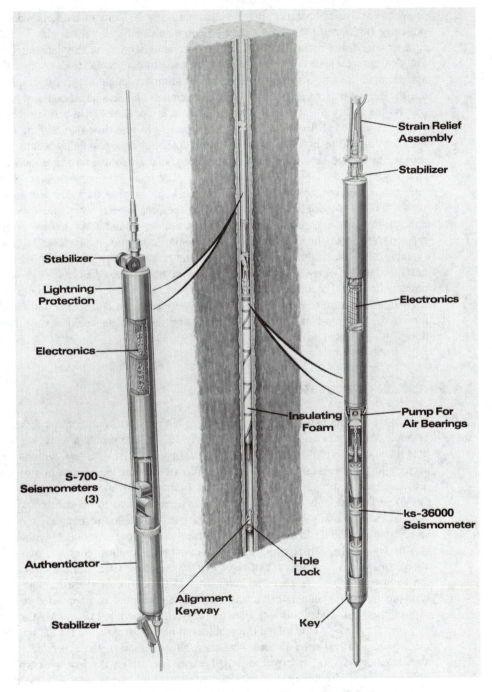

Figure 2. Downhole seismometer package. Reprinted with permission from Sandia National Laboratories.

would, of course, be inscrutable to the Russians, but easily authenticated by the U.S.
Such a system, however, would be totally unacceptable to the Russians for sound logical

reasons. Each seismic station gathers approximately 10^8 bits of data each day, and as presently envisioned these data would be communicated by satellite relay in either real time or near real time to receivers in the continental U.S. The Russians might suspect that this communications channel was being used to communicate data other than that agreed to in the test-ban treaty. Since in conventional single-key cryptography, if one has the decrypt key to enable him to decrypt ciphers, he also has the ability to encrypt, that is, the ability to utter fraudulent ciphers, it is not possible in a single-key crypto-system to give the Russians the capability to verify in real time that nothing other than what was agreed to by treaty is being communicated. One possibility would be for the monitor to change keys with each transmission, that is, to use what are known as session keys, and to give the Russians the key used in a session immediately after the cipher was received and authenticated by the U.S., so that they (the Russians) could verify that the previous cipher decrypted to the proper text, which they would know either from their own corroborating seismic sensors and/or from receipt of an unencrypted version of the message from the monitor's sensors. Unfortunately (and unacceptably), in view of the high data rate, for any practical keying period this requires the host to trust the monitor with too many bits of information before he can verify that nothing has been concealed in the transmission. Of course the host could refuse to cooperate for future transmissions if he detected deception by the monitor, but depending on the time and nature of the transmission the damage could already have been done.

3 VERIFICATION WITHOUT SECRECY

It was at this point that studies of "message authentication without secrecy" were begun at the Sandia Laboratories in the early 1970s. The problem as it was viewed at that time was to find a means for authenticating digital messages* without requiring secrecy for the message itself [7,10]. Recall that the first discussion of two-key (read also public key) cryptography in the open literature [3] appeared several years later (1976) so that the only tool available for a system that was to be shared with the Russians at that time was conventional single-key cryptographic techniques, applied so as to approximate the desired end result of authentication (to the monitor) without secrecy (to the host). The compromise solution, found by Simmons, Stewart, and Stokes in 1974 [10], was to form an authenticator that was much shorter than the message, where the authenticator was made to be a function of the entire message through a hashing type function. This authenticator was then block encrypted and appended to the unencrypted message. Today this appended authenticator would be called a MAC or message authenticating code. This solved the problem of making it possible for the host to monitor messages in real time as they were transmitted; however, the appended (encrypted) authenticator was still inscrutable until he was later given the key with which it had been encrypted. Ironically, the host and monitor could each trust this system to the same level of confidence for the same reason. The monitor trusts the authentication

_____

*The terms message and data are used interchangeably in this chapter since there is no chance of confusion; however, the reader should be aware that the term "message" normally means the authenticated information or data, not the raw data itself.

since to create a forgery the host would have to invert from a known plaintext/cipher pair, that is, break the cryptosystem by cryptanalysis, to find the key used by the monitor. On the other hand, the host is satisfied that the monitor did not conceal information in the preceding transmission if the key he is given generates the authenticator that was transmitted since in order to conceal information in the authenticator the monitor would have had to solve for the (unique?) key relating the plaintext and the desired bogus authenticator; that is, to have solved precisely the same problem on which the host bases his confidence in the authenticator.

To shorten the periods of implicit trust required of the host, smaller blocks of information can be authenticated, at the expense of having to have a unique session key for the encryption of each block. However, keys can be generated sequentially by the same cryptoalgorithm used to encrypt the authenticator, so that for all intents there are an unlimited number of session keys available. This makes it feasible to process shorter blocks of data using a unique session key for each block, with a flow of session keys being made available to the Russians after essentially only the delay of a two-way satellite relay link. In the limit, with block size and the two-way delay, such a scheme approximates very well a true message authentication without secrecy system.

The second iteration in the evolution of treaty verification systems was made shortly after Diffie and Hellman proposed public key cryptosystems in 1976 [3]. Two-key cryptography provided a ready-made solution to the problem of message authentication without secrecy, since the fundamental attribute of two-key cryptography is the separation of the secrecy channel from the authentication channel; both of which are inextricably linked in single-key cryptosystems. In two-key cryptography, the encrypt and decrypt keys are not only different, but it is also computationally infeasible to determine at least one of the keys from a knowledge of the other key, even with arbitrarily many matched plaintext message/cipher pairs. If the receiver (decrypt) key cannot be deduced from a knowledge of the transmitter (encrypt) key, then the transmitter key may be publicly exposed, so long as the receiver key is kept secret, without jeopardizing the transmitter's ability to communicate in secret to the receiver, although the receiver cannot authenticate the source of the communication, that is, cannot be sure of the origin of the ciphers he receives. This is the secrecy channel. Conversely, if the transmitter's encrypt key cannot be recovered from a knowledge of the receiver's decrypt key, etc., then, although secrecy is impossible, the receiver can be confident that the communication originated with the purported transmitter and that the message has not been altered in transit to the same level of confidence that the transmitter can be relied on to keep the encrypt key secret. This is the authentication channel.

Given the availability of an authentication channel an obvious solution to the authentication without secrecy problem would be for the U.S. to install the (secret) authentication function along with the seismic sensor package in the borehole. The downhole package would also be equipped with a variety of sensors designed to detect any attempt to tamper with the package or with the information processing subsystem and to volatilize the secret keying variable if tampering is detected. The decrypt key would be shared with the Russians and perhaps with third parties or arbiters such as the United Nations, etc. The messages would consist of the seismic data along with agreed-on identifiers, station ID number, date, clock, message number, etc., that are required, not only for their obvious utility, but also to provide the redundant information needed by the U.S. to authenticate the messages. This redundant information would of course be known in advance by the Russians so that there would be no possibility of hiding covert communications in what was claimed to be simply an overt authenticator. The

Russians could decrypt the transmission in real time, perhaps even delaying the transmission in a data buffer for the time required to decrypt it, to satisfy themselves that nothing other than the agreed upon siesmic data and prearranged formatting information were present. Thus no part of the transmission would need to be kept secret from the Russians at any time. Similarly, the U.S. would decrypt the cipher on receipt and accept the transmission as authentic if and only if the expected redundant formatting information or the deliberately introduced (but publicly known) authenticating information was present. This scheme depends only on the availability of an authentication channel, separate from the secrecy channel, and hence is not dependent on any particular two-key cryptoalgorithm. At the Sandia National Laboratories, however, we have chosen to use the Rivest-Shamir-Adleman (RSA) cryptoalgorithm [4]. The interested reader is referred to either the chapter ''Contemporary Cryptology: An Introduction'' by J. L. Massey appearing in this volume or to any of several references [2, 9] for a detailed discussion of the application of the RSA cryptoalgorithm to message authentication. We describe only the bare essentials here, since it will be necessary to refer to some of the associated parameters in subsequent sections.

In the RSA system, the user chooses a pair of primes p and q so large that factoring $n = pq$ is beyond all projected computational capabilities. p and q are kept secret. He also chooses a pair of numbers e and d, where $(e, \phi(n)) \equiv 1$ and $ed \equiv 1$ mod $\phi(n)$; $\phi(n) = (p - 1)(q - 1)$.* In other words, e and d are multiplicative inverses in the group of residue classes modulo $\phi(n)$. As already mentioned, for an authentication channel, the encrypt key e is kept secret, while the decrypt key d and the modulus, n, may be publicly exposed.

A message $m < n$ is encrypted in this system to the cipher c by the transmitter, using the encrypt key (e, n), by the rule

$$m^e \equiv (\text{mod } n)$$

and c is decrypted by the authorized receiver, using the decrypt key (d, n), by the rule

$$c^d \equiv (\text{mod } n)$$

Authentication, as we have already pointed out, is based on the receiver finding information already known to him in the decrypted cipher. For example,

if $p = 36756001033$
and $q = 110411555503$
so that the modulus $n = pq = 4058287248123404834599$
and $\phi(n) = 4058287247976237278064$,
then for the encrypt key $e = 1897225149044257283231$
the matching decrypt key $d = 15551$.

Using these cryptovariables, the message 1234567890 with the authenticator SANDIA,

$$m = 1234567890\text{SANDIA} = 1234567890291124141911$$

---

*$\phi(x)$ is the Euler phi function of x (x, a positive integer) and is simply the number of positive integers less than x that have no factor other than 1 in common with x.

would encrypt to the cipher

$$c \equiv m^e \equiv 1768576565013192607710 \pmod{4058287248123404834599}$$

while c would decrypt to recover the message

$$m \equiv c^d \equiv c^{15551} \equiv 1234567890SANDIA \pmod{4058287248123404834599}$$

The appended authenticator SANDIA has been encoded and decoded by the simple numeric substitution: $A = 11$, $B = 12$, $\cdots$, etc. In this example, only ciphers that decrypt to numbers ending in $\cdots$ 291124141911, that is, to the encoding of SANDIA, would be accepted as authentic transmissions. The probability that a randomly chosen cipher would be accepted as authentic in this case is $\approx 3 \times 10^{-9}$ or one chance out of 26^6.

The cryptosecurity of the RSA system is based on the difficulty (infeasibility?) of factoring suitably constructed and sufficiently large composite moduli, n. Obviously, if an opponent can factor n to recover p and q, he can then calculate the multiplicative inverse e of d using the Euclidean algorithm just as the user did to set up the system and hence be able to encrypt, that is, to authenticate, messages. Since computing the multiplicative inverse e of d from a knowledge of only d and n is essentially the same as factoring n or determining $\phi(n)$, e is as secure as factoring n is difficult. Therefore, so long as the factors p and q, and the encrypt key e are kept secret, the authentication channel based on the RSA system is thought to be as secure as factoring, which with reasonable conditions imposed on the choices for p and q is now generally accepted to be a computationally infeasible problem.

In the most direct application of the RSA-based authentication channel to insuring the trustworthiness of the seismic data acquired to verify compliance with a comprehensive test-ban treaty, the U.S. would choose the primes p and q and one of the exponents e or d and then calculate the inverse exponent (d or e, respectively) using the Euclidean algorithm. As part of the initialization procedure by the U.S., $n = pq$ and e would be securely entered into the downhole seismic package. The decryption key d and n would be given to the Russians and perhaps to third parties, and of course retained by the U.S. In operation, the seismic data as well as the redundant identifying information would be block-chain encrypted by the downhole package using the secret encrypt key e and the publicly known modulus n. The host can now satisfy himself that there is no covert communication by decrypting the cipher and verifying that only the previously agreed upon redundant information and the seismic data are present. Recall that he is assumed to know the actual seismic data (message) either from his own sensors or from data links to the monitor's sensors placed ahead of the authentication operation. The monitor, on the other hand, can be certain of the authenticity of a message (containing message numbers, clock readout, etc.) since by hypothesis neither the host nor any third party can compute e from the exposed n and d. Thus the host need not trust the monitor at all, while the monitor is free to introduce as much redundant (but prearranged with the host) information as required to provide authentication confidence.

4 VERIFICATION WITH ARBITRATION

Unfortunately, although the system just described allows the monitor to authenticate messages to whatever level of confidence he desires while at the same time permitting the host to reassure himself that no unauthorized information is concealed, it leaves unanswered another problem that could defeat the purpose of a treaty verification system. If unilateral response by the monitor, such as abrogation of a treaty or resumption of atmospheric testing of nuclear weapons as the U.S. did in 1962 in response to the Soviet's 1961 violation of the Joint Understanding of a moratorium on such tests, is the only action to result from a detection by the monitor of a violation of the agreement, the system just described suffices. If, however, the action to be taken by the monitor in the event that a violation is detected involves convincing third parties or arbiters, such as the United Nations, NATO, etc., then it must be impossible for the monitor to forge messages. Otherwise, the host could disavow an incriminating message as being a forgery fabricated by the monitor, an assertion that the monitor could not disprove if he has the known ability to encrypt messages and hence to create undetectable forgeries.

In 1980, research at Sandia was redirected to solving the authentication with arbitration problem and the related problem of preventing unilateral actions by the host from making it impossible for the monitor to prove the authenticity of a message. For arbitration to be possible, it clearly must be the case that neither party (host or monitor) is in possession of, nor capable of calculating by any feasible amount of computation, the encryption exponent e, since they could then utter undetectable forgeries. As long as this possibility exists it is impossible for the monitor to logically compel a third party to accept the authenticity of a message. There are a class of unconditionally secure authentication schemes that permit arbitration of transmitter/receiver (host/monitor) disputes* that depend on the availability of an arbiter that both parties unconditionally trust. Unfortunately, there is no arbiter who is unconditionally trusted by both the U.S. and Russia, so we must settle for only computationally secure authentication with arbitration schemes. Various schemes were considered in which the host and the monitor each contributed to the key in such a way that the result was unknown to both. Since there are no scenarios in which the objectives of the monitor and of the host are both furthered by their collaborating to create forgeries that would be accepted as authentic by third parties, this joint generation of key at first appears plausible. For example, they might each enter (in secret from all other participants) in the downhole data processing package a binary crytographic key and the exclusive-OR of the two keys could be used as the secret key e. In effect, the actual key e would be the Vernam encryption of each parties' key with the unknown (one-time) key of the other party which, as is well known, insures that the result is mathematically demonstrably cryptosecure to each party. In other words, for randomly chosen input keys neither the host nor the monitor could infer anything about e from their knowledge of the random component they had selected, hence neither is capable of uttering an undetectable forgery. The host, however, could still cheat in the following way. He could test with impunity, and when incriminating records were exhibited by the monitor, claim that his contribution to the key had been compromised; i.e., that one of his people had defected, his files had been rifled, etc. In fact, if he is brazen enough, he could simply publish some number that he claims to have been his contribution and thereby destroy the ability of the monitor to

---

*The reader is referred to the chapter "A Survey of Information Authentication" by G. J. Simmons in this volume for a description of unconditionally secure arbitration codes that permit arbitration.

prove the authenticity of any messages. If the published number is not the correct one, the monitor, using his contribution to the key, could verify that the released number was bogus, but would not be able to prove this to anyone, since they could not be convinced that the monitor was telling the truth about his number. The point is that in such a scheme the host can unilaterally make it possible for the monitor to generate undetectable forgeries, and hence make it impossible for the monitor to prove to an unbiased third party that he did not do so.

Therefore, a solution to the authentication with arbitration problem must both make it possible for the monitor to logically compel third parties to accept the authenticity of messages and make it extremely improbable that the host can by any unilateral action lessen the monitor's ability to convince third parties. The solution to these problems which constituted the third iteration of treaty verification systems [7] was to have the downhole equipment nondeterministically generate p and q in secret from all parties and then select an e (again nondeterministically and in secret from all parties). Only n and d, which is calculated using the secret values of p, q, and e, are revealed. We have mentioned the need for selecting ''good'' primes, the most obvious condition being the magnitude of the numbers, but also such that $p - 1$ and $q - 1$ have large prime factors, etc. All of these criteria can be programmed in, along with a nondeterminate random number generator that provides an unknown seed to start the prime generation process. For example, if a 100-bit seed is needed, a random process such as radioactive decay, could be observed for 100 intervals of sufficient length that many decays would occur in each interval. At the end of each interval a 0 or 1 is entered in the corresponding bit position according to whether an even or odd number of particles had been counted. Using the resulting random seed, the next larger ''good'' prime would be found and used as p or q. e could be generated in a similar manner and d calculated using the Euclidean algorithm. The decryption key n and d would be output at the end of the initialization process to the monitor, the host and to any arbiters needed. In such a system only the downhole equipment could generate authentic messages, and unlike the earlier systems, all of the objectives described thus far for each of the parties are realized.

1. No party can forge messages that would be accepted as authentic.
2. No part of the message is concealed from the host, or from specified third parties.
3. The host, the monitor, and third parties are all able to independently verify the authenticity of messages.
4. No unilateral action by any of the participants can lessen the confidence of any other party as to the authenticity of messages.

5 VERIFICATION IN THE PRESENCE OF DECEIT

For a time it was thought that the system just described had solved the treaty verification problem [7]. In principle, that is, so far as information security was concerned, this was true. Unfortunately, it is not true in practice, not because of any logical flaw in the system, but rather because of the practical impossibility of realizing the required properties in a mutually convincing way. We have discussed at length how either the host or the monitor, if they can learn the encryption key, can create forgeries to their benefit and to the detriment of the other party. It is not even necessary that the key be directly compromised, but only that it be a computationally feasible task to recover it from the information that is exposed. Consequently, the party that builds the downhole equipment potentially has an enormous, probably insurmountable, advantage over the other. Recognizing this, several protocols of the "take any card" sort were devised in which the party making the equipment would provide several sets of equipment, one of which would be selected by the other party and installed in the borehole under joint control and the others of which could be operated or dissected by the other party to convince themselves that they all operated exactly as they should. One of the problems with all such schemes is whether there exists any random number generator that can be convincingly shown to be nondeterministic by observing the output. A good case in point are maximal period n stage linear feedback shift registers whose output sequences satisfy most tests for randomness, but whose future output can be completely predicted with polynomial (in n) difficulty after only $2n$ bits of the output are observed, using Berlekamp's algorithm [1]. The point is that it is a much easier task to exploit a known bias than to detect an unknown one. Because of this, neither party is apt to trust a key generator built by the other. Intricate schemes were considered to get around this problem by having the parties share in the key generation process, the simplest of which is merely the downhole equivalent of the exclusive-OR technique discussed earlier. In this proposal the Russian sequence generator, representing their interests, would present n bits of equivocation to the U.S. and *vice versa*. In fact, there is nothing logically wrong with this approach to jointly generating a key that presents n bits of uncertainty to each party. The problem that is not solved by this scheme is that this key, once generated, must be used in a piece of cryptoequipment built by one of the participants. As anyone who has ever struggled with the Tempest certification of electronic equipment knows, it is difficult to the point of impossibility to be certain that all of the "sneak" channels for leaking information have been plugged and for the present application the problem is much worse. The natural suspicion of the party who did not build the equipment is that the one who did will have deliberately introduced time jitter, crosstalk, amplitude modulation or some other form of leakage for keying information, which can be made to be arbitrarily difficult to detect unless one knows the nature of the leak. The conclusion was that this problem could only be solved if each of the parties had an opportunity to process the signal in equipment that they supplied before the cipher was sent up the borehole. Furthermore, jointly generated keys were ruled out, since whoever's equipment operated last on the data could conceivably telegraph the key by subtle modulation. Thus each party's equipment operates only on and with information known to him. In the resulting system, the U.S. cryptosystem would first encrypt the data stream and forward the cipher stream to the Russian cryptographic equipment. Presumably, they would buffer, reclock, and gate the cipher stream so as to insure that only the overt channel is available. This is actually a workable scheme—logically. The problem remains that if the host reveals his key, or claims that it has been revealed, the monitor would be unable to persuade an arbiter that a message is authentic since he could know the other parties' key and hence might have the capability to utter a forgery and consequently cannot prove that he did not. A solution, and indeed perhaps the only logically complete one, is to require at least three participants; the host, the monitor, and the arbiter(s). Each supplies a two-key authentication channel, with the secret encryption key stored securely downhole and the decryption key shared with all parties. While it true that each party can perform any operations that his, supposedly secret, downhole equipment can, this does not make it possible for him to utter an acceptable forgery since the other cryptosystems are inscrutable to him. Furthermore, any party by publicizing the secret information he was supposed to protect can only make it possible for the other parties to duplicate the actions of two out of the three or more encryption systems. This concatenated encryption system renders it impossible for the host to disavow incriminating messages by unilaterally compromising his key. From the monitor's standpoint, even if the host and the arbiter(s) collude to deceive him, he will still be able to establish, to his satisfaction, the authenticity of messages. In the improbable event that all of the other parties gang up on the monitor, the monitor will still know whether a message is authentic or not but will be unable to persuade impartial (and uninvolved) observers that he is telling the truth. In other words, the worst that can happen, from the monitor's standpoint, for this fourth-generation system is that he can with low probability find himself in the same situation that he was faced with with certainty in the third-generation system. It should be noted that the three participants need not use the same cryptoalgorithm, the same key sizes, etc. All that is required is that each provide a two-key authentication channel and share their decode key with all other participants. The properties that the resulting system have are the following:

1. No party nor cabal of parties can forge messages that would be accepted as authentic by others.
2. No part of the message is concealed, in particular from the host.
3. The host, the monitor, the arbiters, and other third parties are all able to independently verify the authenticity of messages and to logically prove their authenticity to others.
4. No unilateral action by any of the participants can lessen the confidence of any other party as to the authenticity of messages.
5. No benefit or advantage accrues to the supplier of the hardware.

6 CONCLUDING REMARKS

No further subtleties to the technical problem of how to make the data acquired in various treaty verification systems be trustworthy have been found in the last several years so that there is reason to believe that the problem has finally been solved. In the application addressed in this chapter there is no information to be gained from covertly communicating the identity of the seismic site from which a particular piece of data came, since that is known *a priori* to all participants. Therefore, the order in which the various parties have their encryption operations concatenated does not matter. There are treaty verification systems in which these conditions do not hold, that is, in which the identity of the site from which the data came must be concealed from one or more of the participants, and in which, consequently, the order of the concatenation is vital to the system functioning. It is the author's intention to investigate concatenated cryptosystems, that is, either authentication channels or secrecy channels or mixes of the two, as a generic means by which mutually distrustful and deceitful parties can realize a data communications system they both can trust under a wide variety of circumstances. However, the classic problem of message authentication without secrecy in which disputes can always be logically arbitrated, as typified by a system to verify compliance with a comprehensive nuclear weapons test-ban treaty, appears to have been fully solved by concatenated authentication channels as described here.

References

1. E. R. Berlekamp, *Algebraic Coding Theory.* New York: McGraw-Hill, 1968.
2. D. E. R. Denning, *Cryptography and Data Security.* Reading, MA: Addison-Wesley, 1982.
3. W. Diffie and M. E. Hellman, "New directions in cryptography," *IEEE Trans. Informat. Theory,* vol. IT-22, no. 6, pp. 644–654, Nov. 1976.
4. R. A. Rivest, A. Shamir, and L. Adleman, "A method for obtaining digital signatures and public-key cryptosystems," *Commun. Ass. Comput. Mach.,* vol. 21, no. 2, pp. 120–126, 1978.
5. G. J. Simmons, "Symmetric and asymmetric encryption," *Computing Surveys,* vol. 11, no. 4, pp. 305–330, Dec. 1979.
6. ———, "Secure communications in the presence of pervasive deceit," *Proc. IEEE Computer Society 1980 Symp. on Security and Privacy* (G. Davida, ed.) (Oakland, CA), Apr. 14–16, 1980, pp. 84–92; 1980.
7. ———, "Message authentication without secrecy," *Secure Communications and Asymmetric Cryptosystems,* G. J. Simmons, Ed. Boulder, CO: Westview Press, 1982, pp. 105–139.
8. ———, "Verification of treaty compliance—Revisited," *Proc. IEEE Computer Society 1983 Symp. on Security and Privacy* (R. Blakley and D. Denning, eds.) (Oakland, CA), Apr. 25–27, 1983, pp. 61–66, 1983.
9. ———, "Cryptology," *Encyclopaedia Britannica 16th Edition.* Chicago, IL: Encyclopaedia Britannica, Inc., 1986, pp. 913–924B.
10. G. J. Simmons, R. E. D. Stewart, and P. A. Stokes, "Digital data authenticator," Patent Application SD2654, S42640, June 30, 1972.

Index

Editor's Biography

Gustavus J. Simmons received the Ph.D. degree in mathematics from the University of New Mexico, Albuquerque. He is Senior Fellow for National Security Studies at the Sandia National Laboratories, Albuquerque, NM. Earlier he was Manager of the Applied Mathematics Department and Supervisor of one of two divisions at Sandia devoted to the command and control of nuclear weapons. In all these positions he has been primarily concerned with questions of information integrity arising in national security: command and control of nuclear weapons, verification of compliance with various arms control treaties, individual identity verification at sensitive facilities, etc. His research has been primarily in combinatorics and graph theory and in the applied topics of information theory and cryptography, especially as applied to message authentication and systems design to achieve this function. His current research is aimed at devising information dependent protocols whose function can be trusted even though no specific inputs or participants can be. The need for such protocols arises frequently in questions of national security ranging from simple two-man control schemes for nuclear weapons to arbitrarily complex concurrence schemes for the initiation of various treaty controlled actions. Within the defense community he has pioneered in applying these techniques to the command and control of nuclear weapons.

Dr. Simmons was the recipient of the U.S. Government's E. O. Lawrence Award in 1986. The accompanying citation reads in part: "In the political climate that has emerged in the nuclear era, increasing importance in the design of nuclear weapons must be placed on control features including verification, authentication, and positive use control. This is the first time that achievements in this field, of vital importance to national security, have been recognized by a Lawrence Award. . . . " In that same year, he also received the Department of Energy Weapons Recognition of Excellence Award for "Contributions to the Command and Control of Nuclear Weapons."

Dr. Simmons was awarded an honorary Doctorate of Technology in May 1991 by the University of Lund (Sweden) in recognition of his contributions to communications science and to the field of information integrity. The diploma cites him as "The Father of Authentication Theory."

Dr. Simmons has published more than 120 papers and books, many of which are devoted to the analysis and application of asymmetric encryption techniques or to message authentication, and has been granted several patents for inventions in this area. At the invitation of the editors, he wrote the section on cryptology that appears in the 16th edition of the Encyclopaedia Britannica. He is an editor for *Journal of Cryptology, Ars Combinatoria,* and *Codes, Designs and Cryptography.*